MICRO- AND NANOENGINEERED GUM-BASED BIOMATERIALS FOR DRUG DELIVERY AND BIOMEDICAL APPLICATIONS

MICRO- AND NANOENGINEERED GUM-BASED BIOMATERIALS FOR DRUG DELIVERY AND BIOMEDICAL APPLICATIONS

Edited by

SOUGATA JANA

Department of Pharmaceutics, Gupta College of Technological Sciences, Asansol, West Bengal, India; Department of Health and Family Welfare, Directorate of Health Services, Kolkata, West Bengal, India

SUBRATA JANA

Department of Chemistry, Indira Gandhi National Tribal University, Amarkantak, Madhya Pradesh, India

ELSEVIER

Elsevier
Radarweg 29, PO Box 211, 1000 AE Amsterdam, Netherlands
The Boulevard, Langford Lane, Kidlington, Oxford OX5 1GB, United Kingdom
50 Hampshire Street, 5th Floor, Cambridge, MA 02139, United States

Library of Congress Cataloging-in-Publication Data
A catalog record for this book is available from the Library of Congress

British Library Cataloguing-in-Publication Data
A catalogue record for this book is available from the British Library

ISBN: 978-0-323-90986-0

For information on all Elsevier publications
visit our website at https://www.elsevier.com/books-and-journals

Publisher: Matthew Deans
Acquisitions Editor: Sabrina Webber
Editorial Project Manager: Chiara Giglio
Production Project Manager: Prem Kumar Kaliamoorthi
Cover Designer: Greg Harris

Typeset by STRAIVE, India

Contents

6. Gum kondagogu as a potential material for micro- and nanoparticulate drug delivery

Rimpy and Munish Ahuja

7. Gum-based nanoparticles in cancer therapy

Maria John Newton Amaldoss and Reeta

8. Gum-based micro- and nanobiomaterials in gene delivery

M.R. Rekha

9. Locust bean gum-based micro- and nanomaterials for biomedical applications

R.S. Soumya, K.G. Raghu, and Annie Abraham

10. Alginate microspheres: Synthesis and their biomedical applications

Nguyen Thi Thanh Uyen, Syazana Ahmad Zubir, Tuti Katrina Abdullah, and Nurazreena Ahmad

11. Biomedical applications of cashew gum-based micro- and nanostructures

Gouranga Nandi and Subhankar Mukhopadhyay

12. Dextran-based micro- and nanobiomaterials for drug delivery and biomedical applications

Yeliz Basaran Elalmis, Ecem Tiryaki, Burcu Karakuzu Ikizler, and Sevil Yucel

13. Gum arabic-based nanocarriers for drug and bioactive compounds delivery

Neda Aliabbasi, Morteza Fathi, and Zahra Emam-Djomeh

14. Tamarind gum as a wall material in the microencapsulation of drugs and natural products

Erik Alpizar-Reyes, Stefani Cortés-Camargo, Angélica Román-Guerrero, and César Pérez-Alonso

15. Tree gum-based nanostructures and their biomedical applications

K.P. Akshay Kumar, Rohith K. Ramakrishnan, Miroslav Černík, and Vinod V.T. Padil

16. Application of micro- and nanoengineering tragacanth and its water-soluble derivative in drug delivery and tissue engineering

Azam Chahardoli, Nasim Jamshidi, Aliasghar Varvani, Yalda Shokoohinia, and Ali Fattahi

17. Development of Persian gum-based micro- and nanocarriers for nutraceutical and drug delivery applications

Rassoul Kadkhodaee and Nassim Raoufi

Contributors

Tuti Katrina Abdullah School of Materials and Mineral Resources Engineering; Biomaterials Research Niche Group, School of Materials and Mineral Resources Engineering, Engineering Campus, Universiti Sains Malaysia, Nibong Tebal, Pulau Pinang, Malaysia

Annie Abraham Department of Biochemistry, University of Kerala, Kariavattom, Thiruvananthapuram, Kerala, India

Nurazreena Ahmad School of Materials and Mineral Resources Engineering; Biomaterials Research Niche Group, School of Materials and Mineral Resources Engineering, Engineering Campus, Universiti Sains Malaysia, Nibong Tebal, Pulau Pinang, Malaysia

Munish Ahuja Drug Delivery Research Laboratory, Department of Pharmaceutical Sciences, Guru Jambheshwar University of Science and Technology, Hisar, Haryana, India

K.P. Akshay Kumar Future Energy and Innovation Laboratory, Central European Institute of Technology, Brno University of Technology, Brno, Czech Republic; Department of Applied Chemistry, Cochin University of Science & Technology, Kochi, Kerala, India

Neda Aliabbasi Functional Food Research Core (FFRC), Transfer Phenomena Laboratory (TPL), College of Agriculture & Natural Resources, University of Tehran, Karaj, Iran

Erik Alpizar-Reyes Chemical Engineering Department, Faculty of Chemistry, Autonomous Mexico State University, Toluca, Estado de México, Mexico; LabMAT, Department of Civil and Environmental Engineering, University of Bío-Bío, Concepción, Chile

Maria John Newton Amaldoss Australian Centre for Nanomedicine; Adult Cancer Program, Lowy Cancer Research Centre; Prince of Wales Clinical School, UNSW Sydney, Sydney, NSW, Australia

Nikhil R. Bali University Department of Pharmaceutical Sciences, Rashtrasant Tukadoji Maharaj Nagpur University, Nagpur, Maharashtra, India

Aishik Banerjee Department of Pharmaceutical Technology, Brainware University, Kolkata, West Bengal, India

Chinmoy Baruah Department of Chemistry, School of Basic Sciences, The Assam Kaziranga University, Jorhat, Assam, India

Manas Bhowmik Department of Pharmaceutical Technology, Jadavpur University, Kolkata, West Bengal, India

Ruchika M. Bondre University Department of Pharmaceutical Sciences, Rashtrasant Tukadoji Maharaj Nagpur University, Nagpur, Maharashtra, India

Miroslav Černík Institute for Nanomaterials, Advanced Technologies and Innovation (CXI), Technical University of Liberec (TUL), Liberec, Czech Republic

Azam Chahardoli Department of Biology, Faculty of Science, Razi University, Kermanshah, Iran

Ana Ćirić University of Belgrade, Faculty of Pharmacy, Belgrade, Serbia

Stefani Cortés-Camargo Nanotechnology Department, Technological University of Zinacantepec, Zinacantepec, Estado de México, Mexico

Ana Letícia Rodrigues Costa Department of Materials and Bioprocess Engineering, School of Chemical Engineering, University of Campinas, Campinas, Brazil

Lucimara Gaziolla de la Torre Department of Materials and Bioprocess Engineering, School of Chemical Engineering, University of Campinas, Campinas, Brazil

Ljiljana Djekic University of Belgrade, Faculty of Pharmacy, Belgrade, Serbia

Yeliz Basaran Elalmis Department of Bioengineering, Yildiz Technical University, Istanbul, Turkey

Zahra Emam-Djomeh Functional Food Research Core (FFRC), Transfer Phenomena Laboratory (TPL), College of Agriculture & Natural Resources, University of Tehran, Karaj, Iran

Morteza Fathi Function Health Research Center, Lifestyle Institute, Baqiyatallah University of Medical Sciences, Tehran, Iran

Ali Fattahi Pharmaceutical Sciences Research Center, Health Institute, Kermanshah University of Medical Sciences, Kermanshah, Iran

Burcu Karakuzu Ikizler Department of Bioengineering, Yildiz Technical University, Istanbul, Turkey

Siddhesh S. Jadhav University Department of Pharmaceutical Sciences, Rashtrasant Tukadoji Maharaj Nagpur University, Nagpur, Maharashtra, India

Nasim Jamshidi Students Research Committee, Kermanshah University of Medical Sciences, Kermanshah, Iran

Sougata Jana Department of Pharmaceutics, Gupta College of Technological Sciences, Asansol; Department of Health and Family Welfare, Directorate of Health Services, Kolkata, West Bengal, India

Rassoul Kadkhodaee Department of Food Nanotechnology, Research Institute of Food Science and Technology, Mashhad, Iran

Megha N. Karemore University Department of Pharmaceutical Sciences, Rashtrasant Tukadoji Maharaj Nagpur University, Nagpur, Maharashtra; Department of Pharmaceutics, Dadasaheb Balpande College of Pharmacy, Besa, Nagpur, India

De-Qiang Li College of Chemistry and Chemical Engineering, Xinjiang Agricultural University, Urumchi, Xinjiang; Beijing Key Laboratory of Lignocellulosic Chemistry, Beijing Forestry University, Beijing, P.R. China

Jun Li College of Chemistry and Chemical Engineering, Xinjiang Agricultural University, Urumchi, Xinjiang, P.R. China

Sreejan Manna Department of Pharmaceutical Technology, Brainware University, Kolkata, West Bengal, India

Subhankar Mukhopadhyay Key Laboratory of Modern Chinese Medicines, China Pharmaceutical University, Nanjing, P.R. China

Gouranga Nandi Division of Pharmaceutics, Department of Pharmaceutical Technology, University of North Bengal, Raja Rammohunpur, West Bengal, India

Vinod V.T. Padil Institute for Nanomaterials, Advanced Technologies and Innovation (CXI), Technical University of Liberec (TUL), Liberec, Czech Republic

César Pérez-Alonso Chemical Engineering Department, Faculty of Chemistry, Autonomous Mexico State University, Toluca, Estado de México, Mexico

K.G. Raghu Biochemistry and Molecular Mechanism Laboratory, Agroprocessing and Technology Division, CSIR-National Institute for Interdisciplinary Science and Technology (NIIST), Thiruvananthapuram, Kerala, India

Rohith K. Ramakrishnan Institute for Nanomaterials, Advanced Technologies and Innovation (CXI), Technical University of Liberec (TUL), Liberec, Czech Republic

Nassim Raoufi Department of Food Science and Biotechnology, Zhejiang Gongshang University, Hangzhou, P.R. China

Reeta St. Soldier Institute of Pharmacy, Jalandhar, Punjab, India

M.R. Rekha Division of Biosurface Technology, Biomedical Technology Wing Sree Chitra Tirunal Institute for Medical Sciences & Technology, Thiruvananthapuram, Kerala, India

Rimpy Drug Delivery Research Laboratory, Department of Pharmaceutical Sciences, Guru Jambheshwar University of Science and Technology, Hisar, Haryana, India

Angélica Román-Guerrero Biotechnology Department, Autonomous Metropolitan University-Iztapalapa, México City, Mexico

Jayanta K. Sarmah Department of Chemistry, School of Basic Sciences, The Assam Kaziranga University, Jorhat, Assam, India

Yalda Shokoohinia Ric Scalzo Institute for Botanical Research, Southwest College of Naturopathic Medicine, Tempe, AZ, United States; Pharmaceutical Sciences Research Center, Health Institute, Kermanshah University of Medical Sciences, Kermanshah, Iran

R.S. Soumya Biochemistry and Molecular Mechanism Laboratory, Agroprocessing and Technology Division, CSIR-National Institute for Interdisciplinary Science and Technology (NIIST), Thiruvananthapuram, Kerala, India

Ecem Tiryaki Department of Bioengineering, Yildiz Technical University, Istanbul, Turkey; Department of Physical Chemistry, Biomedical Research Center, Southern Galicia Institute of Health Research and Biomedical Research Networking Center for Mental Health, Universidade de Vigo, Vigo, Spain

Nguyen Thi Thanh Uyen School of Materials and Mineral Resources Engineering; Biomaterials Research Niche Group, School of Materials and Mineral Resources Engineering, Engineering Campus, Universiti Sains Malaysia, Nibong Tebal, Pulau Pinang, Malaysia

Aliasghar Varvani Students Research Committee, Faculty of Pharmacy, Kermanshah University of Medical Sciences, Kermanshah, Iran

Feng Xu Beijing Key Laboratory of Lignocellulosic Chemistry, Beijing Forestry University, Beijing, P.R. China

Nikhil Y. Yenorkar University Department of Pharmaceutical Sciences, Rashtrasant Tukadoji Maharaj Nagpur University, Nagpur, Maharashtra, India

Sevil Yucel Department of Bioengineering, Yildiz Technical University, Istanbul, Turkey

Syazana Ahmad Zubir School of Materials and Mineral Resources Engineering; Biomaterials Research Niche Group, School of Materials and Mineral Resources Engineering, Engineering Campus, Universiti Sains Malaysia, Nibong Tebal, Pulau Pinang, Malaysia

Editors' biography

Dr. Sougata Jana has a B. Pharm degree (Gold Medalist) from West Bengal University of Technology (WBUT), Kolkata, an M. Pharm degree (Pharmaceutics) from Biju Patnaik University of Technology (BPUT), and a PhD degree from Maulana Abul Kalam Azad University of Technology (MAKAUT), West Bengal, India. He received the "M.N. Dev Memorial Award" for securing the highest marks in the state of West Bengal in 2005 from the IPA Bengal branch, Kolkata, India. He has worked in the pharmacy field for 14 years, including teaching, research, and health services, among other roles. He received the "Best Poster Presentation Award" at the 21st West Bengal State Science and Technology Congress (2014), Burdwan, and "Outstanding Paper Award" at the 1st Regional Science and Technology Congress (2016), Bardhaman Division, organized by DST, Govt. of West Bengal and Bankura Christan College, West Bengal, India. He has published 30 research and review articles in different national and international peer-reviewed journals. He has edited 10 books and published more than 45 book chapters in different edited books for Elsevier, Springer, Wiley VCH, CRC Press, and the Taylor & Francis group. He is a reviewer for various peer-reviewed international journals (Elsevier, Wiley, Springer, Taylor & Francis, Dove Press, etc.). He is a life member of the Association of Pharmaceutical Teachers of India (APTI) and has an Associateship from the Institution of Chemists (AIC), India. He successfully guided 17 postgraduate students on their research projects. He is working in the field of drug delivery science and technology including modification of synthetic and natural biopolymers, microparticles, nanoparticles, semisolids, and interpenetrating network (IPN) systems for controlled drug delivery.

Dr. Subrata Jana is presently working as an Associate Professor at the Department of Chemistry, Indira Gandhi National Tribal University (Central University), Amarkantak, Madhya Pradesh, India, and his current research focuses on the design and synthesis of artificial receptors for the recognition of anions, cations, and *N*-methylated protein residue. He is also working on biodegradable polymeric-based carrier systems for the delivery of drug molecules in collaboration with pharmaceutical scientists. He has published about 40 research papers in peer-reviewed international journals and contributed more than 20 book chapters in different edited books published by internationally renowned publishers. He is currently serving as Executive Editor of *Mekal Insights*, the official research journal of Indira Gandhi National Tribal University. He has also served as a reviewer for the *International Journal of Biological Macromolecules* (Elsevier) and the *Journal of PharmaSciTech and Current Pharmaceutical Design* (Bentham).

He obtained his PhD in organic chemistry from the Indian Institute of Engineering Science and Technology (IIEST), Shibpur, India. He then moved to the University of Victoria, Canada, to work with Professor Fraser Hof on supramolecular and medicinal chemistry as a postdoctoral fellow. He then also worked with Dr. Kenneth J. Woycechowsky at the University of Utah, USA, on protein engineering and enzyme catalysis as a postdoctoral research associate. Overall, he has extensively studied the supramolecular behavior of the host-guest interaction and synthesis of heterocyclics like pyrimidines, naphthyridines, quinoline, and diazepines, by exploiting microwave protocol for green chemical synthesis.

Preface

The emphasis of this book is on drug/biomolecule carriers using micro- and nanoengineered biomaterials based on gums and biopolymers. Currently natural gums and polymers are widely used as bio-carriers for the delivery of drugs/biomolecules to the targeted site for prolonged release and desired therapeutic effect. Natural gums and polymers are important because they are readily available from natural sources that make them biodegradable, biocompatible, and nontoxic in nature. Natural gums and polymers can also be synthetically modified with other polymers in the presence of cross-linking agents to develop scaffolds/matrix/composites/interpenetrating polymer networks using microtechnology and nanotechnology. This book also focuses on the biological importance of such materials in gene delivery, cancer therapy, and tissue engineering.

In summary, we believe that the selected topics cover the key aspects of micro- and nanoengineering of gum-based biomaterials, with special emphasis on drug delivery and biomedical applications. The book is an important resource for academics as well as pharmaceutical, materials science, chemical science, life science, and biotechnology scientists working in the field of biopolymer-related materials for drug delivery, in addition to medical and other health care professionals in these fields. This book is composed of 18 chapters, which are as follows:

Chapter 1 focuses on nanomedicine approaches and strategies for gum-based stealth nanocarriers and is written by Dr. Nikhil R. Bali and coauthors.

Chapter 2 considers micro- and nanoscale drug delivery systems based on xanthan gum hydrogels and is authored by Dr. Ljiljana Djekic.

Chapter 3 discusses chitosan-based nanoengineered drug delivery systems, and is written by Sreejan Manna.

Chapter 4 reports on pectin-based micro- and nanomaterials in drug delivery and is by Dr. De-Qiang Li.

Chapter 5 highlights gellan gum nanoparticles in drug delivery and is authored by Dr. Lucimara Gaziolla de la Torre.

Chapter 6 is a description by Dr. Munish Ahuja of gum kondagogu as a potential material for micro- and nanoparticulate drug delivery.

Chapter 7 looks at gum-based nano- and microparticles in cancer therapy and is by Dr. Maria John Newton Amaldoss and coauthor.

Chapter 8 is presented by Dr. M.R. Rekha on gum-based micro- and nanobiomaterials in gene delivery.

Chapter 9, on locust bean gum-based micro- and nanomaterials for biomedical applications, is reported by Dr. Annie Abraham.

Chapter 10 discusses alginate microspheres; their synthesis and biomedical applications are described by Dr. Nurazreena Ahmad.

Chapter 11 is written by Dr. Gouranga Nandi and coauthor on biomedical applications of cashew gum-based micro- and nanostructures.

Chapter 12 discusses dextran-based micro- and nanobiomaterials for drug delivery and biomedical applications and is authored by Dr. Yeliz Basaran Elalmis.

Chapter 13 is written on gum arabic-based nanocarriers for drug and bioactive compounds delivery by Dr. Zahra Emam-Djomeh.

Chapter 14 is composed by Dr. César Pérez-Alonso on tamarind gum as a wall material in the microencapsulation of drugs and natural products.

Chapter 15 focuses on tree gum-based nanostructures and their biomedical applications, as described by Dr. Vinod V.T. Padil.

Chapter 16 highlights micro- and nanoengineering of tragacanth gum and its derivative in drug delivery and tissue engineering, written by Dr. Ali Fattahi.

Chapter 17, authored by Dr. Rassoul Kadkhodaee, investigates the development of Persian gum-based micro- and nanocarriers for nutraceutical and drug delivery applications.

Chapter 18, on guar gum-based hydrogel and hydrogel nanocomposites for biomedical applications, is written by Dr. Jayanta K. Sarmah.

This book would not have been possible without the support and contribution of all the authors mentioned above and their respective teams.

Finally, the editors would also like to acknowledge the very efficient and friendly staff at Elsevier, especially Dr. Simon Holt, Chiara Giglio, and the staff who provided support throughout the entire production process.

Sougata Jana
Subrata Jana
December 2021

CHAPTER

1

Nanomedicine approaches and strategies for gum-based stealth nanocarriers

Nikhil R. Bali[a], *Megha N. Karemore*[a,b], *Siddhesh S. Jadhav*[a], *Ruchika M. Bondre*[a], *and Nikhil Y. Yenorkar*[a]

[a]University Department of Pharmaceutical Sciences, Rashtrasant Tukadoji Maharaj Nagpur University, Nagpur, Maharashtra, India [b]Department of Pharmaceutics, Dadasaheb Balpande College of Pharmacy, Besa, Nagpur, India

1.1 Introduction

Nanomedicine, in simple terms, refers to the application of nanotechnology in the treatment of diseases (Prasad, 2012); however, with the staggering progress of nanotechnology in various aspects of life, nanomedicine is having crucial revolutionary impacts on the health care system as well (Lee and Ventola, 2012). Often, nanomedicine involves the use of nanomaterials and/or nanoanalytical techniques for diagnosis, imaging, treatment, disease therapy (nanotheranostic). and integrated medical nanosystem which may be of use in future for monitoring and complex repairs in the body at molecular and cellular levels (Riehemann et al., 2009). Nevertheless, currently nanomedicine is being employed in pharmaceutical and biomedical research as nanoparticulate drug delivery systems or nanocarriers with approximate sizes ranging from 1 to 1000 nm (Bali and Salve, 2020; Khan et al., 2019; Wong et al., 2012), not only to deliver bioactive agents in the body, but also for targeted delivery of drugs, genes, and peptides (Bali et al., 2021; Patra et al., 2018; Shinde et al., 2020). Among various nanocarriers, the use of nanoparticles as navigators has received considerable interest over the past few decades due to their exceptionally small size and unique features responsible for improved pharmacokinetics and biodistribution (Bali and Salve, 2019, 2020), decreased toxicities, improved drug solubility and stability, along with controlled and site-specific delivery of therapeutic agents to active sites (Mishra et al., 2010).

Micro- and Nanoengineered Gum-Based Biomaterials for Drug Delivery and Biomedical Applications
https://doi.org/10.1016/B978-0-323-90986-0.00018-2

Despite the immense advantages of nanoparticulate-based drug delivery systems, their effectiveness to reach targeted area at a desired concentration depends upon their survival inside the biological environment (Wani et al., 2020). Any particle entering the body is being encountered by a mononuclear phagocytic system (MPS) or reticuloendothelial system (RES), which renders the particle inactive by destroying it; nanoparticles are not an exception (Thau et al., 2020; Wani et al., 2020). This defense mechanism of the body, i.e., the immune system, considers such nanocarriers as foreign invaders and rapidly eliminates them from blood circulation via phagocytosis carried out by macrophages (Gustafson et al., 2015). To overcome this, different approaches have emerged which protect nanoparticles from such invasion by the body's defense system and enauew their prolonged *in-vivo* circulation. One such approach is surface modulation of nanoparticles (Butcher et al., 2016; Khanna, 2012). Surface modification by means of hydrophobic and charged particles is known to undergo higher opsonization (the process of recognizing and targeting invading particles for phagocytosis) compared to hydrophilic and neutrally charged particles (Carrstensen et al., 1992; Müller et al., 1992; Norman et al., 1992; Roser et al., 1998), thus surface coating or modification of nanoparticles by hydrophilic polymers or by any other means shields these nanoparticles from interaction with biological proteins called opsonins that prevent the uptake of these nanoparticles by macrophages. This results in their prolonged blood circulation time, which in turn facilitates their effective site-specific drug delivery to targeted organs, tissues, or cells (Salmaso and Caliceti, 2013). Among the various polymers employed for surface modulation of nanoparticles, polyethylene glycol (PEG) (also called polyethylene oxide, PEO) remains the gold standard due to its hydrophilic and flexible nature (Owens and Peppas, 2006). Apart from PEG, alternative polymers include polypropylene glycol (PPG) (also called polypropylene oxide, PPO), polylactic acid (PLA), polyhydroxyethyl-L-asparagine (PHEA), N-(2-hydroxypropyl) methacrylamide (HPMA), polyvinyl pyrrolidone (PVP), poly(2-oxazoline) (POx), poly(zwitterions), etc., which make the nanoparticles smart enough to remain undetectable by RES; that is, they remain stealthy and thus bypass the immune evasion and thereby instant elimination from the blood circulation (Amoozgar and Yeo, 2012; Hadjesfandiari and Parambath, 2018; Wani et al., 2020).

Though these hydrophilic polymers form a so-called "stealth layer/coating" on the surface of nanocarriers, these nanocarriers still suffer from several intrinsic drawbacks due to their problems related to biocompatibility and biodegradability (Kang et al., 2015). In this regard, the use of natural polymers to modify the surface coating of nanocarriers has opened up new avenues for nanoparticulate drug delivery systems. Due to their hydrophilic nature, they are able to form dense conformational repulsive clouds over the surface of nanoparticles which prevents their interaction with other macromolecules even at low polymer concentrations (Torchilin, 1998), resulting in their improved physicochemical properties and stability in a biological milieu, thus preventing their rapid clearance from the body. Moreover, their natural origin renders them biocompatible, biodegradable, and nontoxic in nature (Bhardwaj et al., 2000; Iravani, 2020). In recent years, the research focus has shifted toward the use of natural gums and their derivatives for development of various pharmaceutical dosage forms and drug delivery systems (Bali and Salve, 2020; Deka et al., 2020; Karemore and Avari, 2019; Karemore and Bali, 2021; Kora et al., 2010; Novac et al., 2014; Rahimi et al., 2018; Rao et al., 2018) owing to their inherent compatibility with the biological system, together with their high potential for biological and biomimetic effects (Kang et al., 2015). Moreover, the basic polysaccharide moiety of these natural gums (Deshmukh and Aminabhavi, 2014) has been known to play a pivotal role in biological communication or signaling between cells, since most of the cellular surfaces are composed of complex carbohydrates or polysaccharide molecules (Sharon and Lis, 1993). In addition to this, the hydrophilic nature of these natural gums (polysaccharide molecules) has been reported to reduce unspecific protein adsorption on nanocarriers due to steric repulsion, thus prolonging their circulation time in the bloodstream, and hence they were regarded as a potential biodegradable substitute for PEG (Knop et al., 2010). This protein-repellent property, together with their ability to shield the nanoparticles from opsonization, enables natural gums to be promising elements for development of future therapeutics. With this concept in mind, the present chapter focuses on various approaches to deliver drugs using nanocarriers and the strategies to deliver these agents effectively to target site employing various natural gums. Furthermore, a brief understanding of commonly used gums for pharmaceutical application in nanocarriers is also discussed. Before investigating gums and their approaches in stealth coating, an understanding of the concept of opsonization and the need for stealth coating of nanocarriers is first required.

1.2 Need for stealth nanocarriers

Although nanocarriers have been recognized as potential drug cargo for targeted or site-specific delivery, the biggest obstacle to their application is their rapid clearance from the circulation via the immune system (Gustafson et al., 2015). The body's immune system acts as a protective barrier which safeguards the *in-vivo* environment from the attack of extraneous particles through various defense mechanisms regulated by a mononuclear phagocytic system (MPS) or reticuloendothelial system (RES) (Kao and Juliano, 1981; Riabov et al., 2018; Senior, 1987). When any foreign invader or extraneous particle gains access to the blood circulation, it is directed to different organs such as the liver, spleen, lungs, and other parts of the RES where it is taken up (engulfed) by the phagocytic cells (Nie, 2010). However, phagocytic cells as such do not have the ability to bind or recognize foreign entities, so surface-bound proteins, also known as opsonins, facilitate detection of extraneous particles entering the body to the immune system (Frank and Fries, 1991). These opsonins are nothing but serum protein molecules which adsorb on the surface of foreign particles and enable their identification by phagocytic cells via a process termed as opsonization (Wani et al., 2020). Following opsonization and interaction of foreign particles with the immune cells, phagocytosis takes place which leads to engulfment, eventual destruction, and ultimate elimination of foreign materials from the bloodstream (Nie, 2010).

Similarly, when nanoparticles get access into the bloodstream, they are also considered as extraneous particles by the immune system, resulting in their phagocytosis and ultimate clearance from the body. The moment nanoparticles get exposed to the bloodstream, serum proteins with maximum mobility predominantly approach and get adsorbed on the surface of nanoparticles, leading to the formation of a primary protein-complex termed a soft corona; this binds even the less frequent and low-affinity proteins onto the surface of nanoparticles (Rampado et al., 2020). After some time, these low-affinity proteins are substituted by relatively high-affinity proteins, resulting in the formation of a new protein-nanoparticle complex known as a hard corona (Casals et al., 2010; Fleischer and Payne, 2014; Wani et al., 2020). This

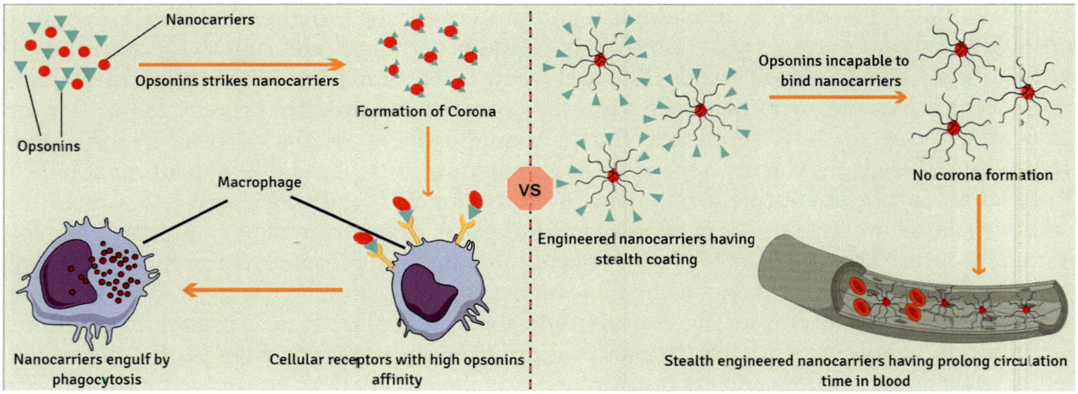

FIG. 1.1 The mechanism of opsonization and macrophage capture of nanocarrier vs. the role of stealth coating in prolonged circulation of engineered nanocarriers.

protein corona bestows a new identity on the nanoparticles, which facilitates their biorecognition and interaction with the receptors on the surfaces of phagocytic cells (Lee et al., 2014; Mahmoudi et al., 2011). This leads to macrophage internalization, lysosome formation, digestion, and eventual elimination of nanoparticles by the immune cells (particularly macrophages) via receptor-mediated phagocytosis (Moghimi and Hamad, 2008; Moghimi and Patel, 1998). For prolonged and targeted delivery and even for enhanced permeability and retention (EPR) effects in cancer therapeutics, it is essential that nanoparticles should escape such protein-corona formation thereby rapid clearance by macrophages. Since the overall stability, fate, and thereby organ uptake and clearance of nanoparticles depend upon opsonin binding, it is important to develop strategies that avoid opsonin-nanoparticulate interaction and subsequent corona formation. One such approach is surface coating of nanoparticles with protective polymer coats, which prevent opsonin binding and hence their recognition by the immune system, thereby imparting a "stealth characteristic" to nanoparticles and increasing their long-term circulation in the bloodstream (Fig. 1.1). Of the various available polymers, stealth coating with electrically neutral hydrophilic polymers is known to shield nanoparticles from uptake via RES compared to hydrophobic and/or charged nanoparticles which are prone to opsonization. Thus, for prolonged blood circulation, grafting of nanocarriers with hydrophilic polymers remains an effective strategy to combat macrophage capture and immune clearance.

1.3 Hydrophilic polymers as shielding agent for nanocarriers

When administered intravenously, nanocarriers have been known to interact massively with endothelium vessels as well as cells and blood proteins in the biological milieu, resulting in their rapid removal from the circulation mostly by the mononuclear phagocyte system (Brannon-Peppas and Blanchette, 2004; Cao et al., 2020; Nie, 2010). In order to prevent their capture by macrophages and endow nanocarriers with long circulation properties, new technologies aimed at the surface modulation of their physicochemical features have been

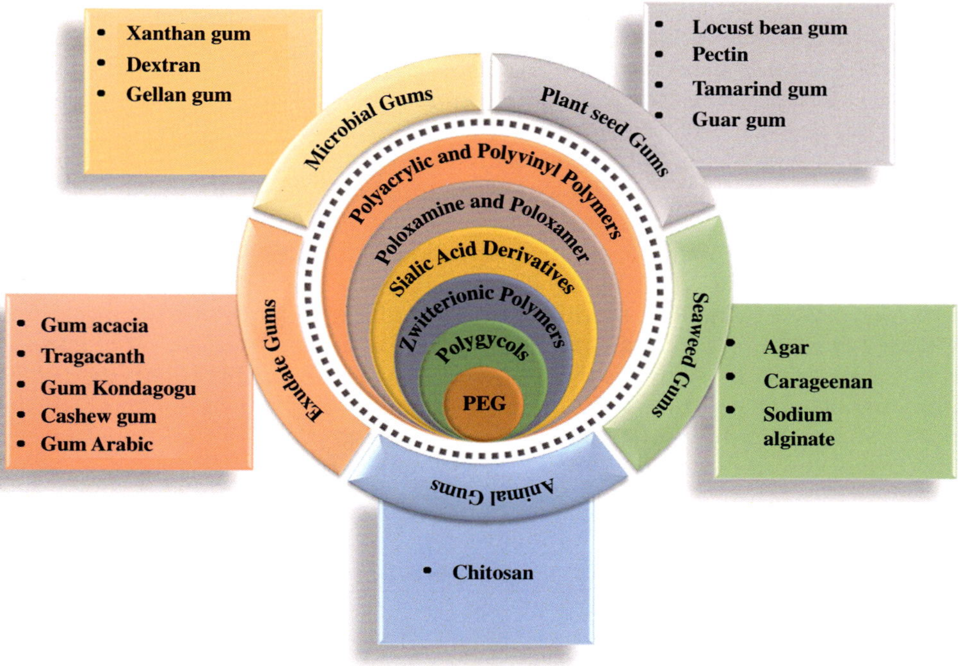

FIG. 1.2 Schematic representation of various gums (based on their natural origin) and polymers (or conjugated moiety) employed for surface modulation of nanoparticles.

developed. In particular, stealth nanocarriers can be obtained by polymeric coating, which prevents their recognition by the immune system, thus improving the blood residence time (Tang et al., 2019). So far, many efforts have been made to yield stealth products by modification of the surface properties of nanocarriers with polymers that prevent opsonin interactions and subsequent phagocyte clearance (Fam et al., 2020). Either natural and semisynthetic polysaccharides or synthetic polymers with high flexibility and high hydrophilicity have been used to confer stealth properties to nanoparticles. Of these, dextran (Dex), polysialic acid (PSA), hyaluronic acid (HA), chitosan (CH), and heparin remain the most-used natural polysaccharides whereas polyvinyl pyrrolidone (PVP), polyvinyl alcohol (PVA), polyacrylamide (Pam), poly(ethylene glycol) (PEG), and PEG-based copolymers such as poloxamers, poloxamines, and polysorbates are the synthetic polymers used to impart stealthiness to nanocarriers (Amoozgar and Yeo, 2012; Fam et al., 2020; Salmaso and Caliceti, 2013). As evidenced in the literature, polymers with a neutral or zwitterionic nature together with high hydrophilicity and high flexibility have been known to create a physical barrier which curtails the biorecognition of adsorbed opsonins by macrophages, thus conferring stealthiness to nanocarriers (Schlenoff, 2014). Among numerous natural polymers, the use of natural gums for various biomedical applications has been recognized due to their biodegradability, abundant availability, nontoxicity, and low cost (Deshmukh and Aminabhavi, 2014). Their natural origin from various sources such as seaweeds, animals, microbes, higher plants, and plant extracts (Fig. 1.2) together with the polysaccharide nature of these macromolecules

(Aravamudhan et al., 2014; BeMiller, 2012) and the ease of conjugation or complexation with other biological macromolecules enable wide applicability of these carbohydrate molecules in the design and development of novel drug delivery systems (Alvarez-Lorenzo et al., 2013; Buschmann et al., 2013; Goodarzi et al., 2013; Lemarchand et al., 2004; Liu et al., 2008; Mizrahy and Peer, 2012; Posocco et al., 2015; Wen and Oh, 2014). In addition, the ability of these gums to adhere via electrostatic interaction with glycocalyx on the cell membrane thereby imparting cell-cell interaction and biological communication together with ligand activities of some gums (leading to efficient cellular uptake of nanoparticles coated with such gums/polysaccharides owing to specific interactions with various receptors on target cells) (Jiang et al., 2008; Li et al., 2012; Rivkin et al., 2010; Yuan et al., 2009), their protein-repellent property and ease of chemical modification with different molecules all these properties make these natural gums an attractive biomaterial for surface engineering of nanoparticles (Passirani et al., 1998a, b). With this thought, the following section highlights the applicability and potential of some natural gums and polysaccharides in the field of nanomedicine as an alternative surface modification strategy to confer stealth properties to nanocarriers.

1.4 Gum-based stealth nanocarriers: An alternative approach for transportation of payloads to targeted sites

Although PEG remains the most representative polymeric material for stealth coating of nanocarriers, its interference with cellular uptake and numerous other disadvantages (Knop et al., 2010) demands alternative approaches and strategies for surface modification of nanocarriers. In this respect, use of natural polymers and gums provides new insights for steric hindrance and/or masking of charges via formation of a "conformational cloud" on nanoparticulate surface. Being hydrophilic in nature, these natural gums not only enhance the water solubility of molecules but also prevent aggregation of nanoparticles by steric stabilization owing to their long polymer chain, which aids their stability (Martin, 2006). With this in mind, the following section presents the most promising hydrophilic polysaccharides and gums that can be investigated as natural alternatives to PEG for different biomedical applications, and their properties and potentials as stealth coating materials (Table 1.1).

1.4.1 Xanthan gum

Xanthan gum (GX) is a natural polysaccharide and fermented product of gram-negative bacterium *Xanthomonas campestris*, which is chemically composed of β-1,4-D-glucopyranose glucan along with pendant trisaccharide side chain (Dumitriu, 2005; García-Ochoa et al., 2000). Due to its exceptional properties like relatively good viscous rheology, pH, as well as salt resistance properties (Patel et al., 2020), it is used for many purposes including as a suspending agent (Bumphrey, 1986), and a stabilizer for food, cosmetics, and pharmaceuticals (Katzbauer, 1998; Patel et al., 2020). A few of its useful characteristics like biocompatibility and softness make it a suitable candidate for delivery of drug and proteins (Patel et al., 2020). One of the key roles played by xanthan gum is as a stabilizer of nanoparticles like iron and gold nanoparticles (Ebrahiminezhad et al., 2018; Muddineti et al., 2016). There are two

TABLE 1.1 Various gum-based nanoformulation for targeted delivery of therapeutic moiety.

Sr. no.	Gum	Surface charge	Types of nanoparticles	Conjugated moiety	Targeted site/disease	References
1	Xanthan gum	Anionic	Gold nanoparticles	PEG	Enhanced cellular uptake cytotoxicity in murine melanoma (B16F10) cells	Muddineti et al. (2016)
			Gold nanoparticles		Showing three times more cytotoxicity in A549 cells	Pooja et al. (2014)
			Silver nanoparticles		Improves rheumatoid arthritis, significantly reduces reactive oxygen species	Rao et al. (2018)
2	Carrageenan	Anionic	Magnetic NPs		Improves stability along with antihemolytic activity	Shanmuga et al. (2015)
			Gold nanoparticles		Great improvement of cytotoxic activity and induction of apoptosis on cancer cells rather than on normal cells	Chen et al. (2018)
3	Chitosan	Cationic	Multibranched nano vehicle consisting of magnetic nanoparticles (Fe_3O_4)	mPEG	Shows enhanced cellular uptake of both drugs, MCF7 breast cancer cells with reduced toxicity	Rahimi et al. (2018)
			Chitosan nanoparticles	PEG	Increased insulin uptake in Caco-2 cells	Zhang et al. (2008)
			Polymeric nanoparticles using poly(lactic acid) (PLA)	PEG	Decreased macrophage uptake and improved prolongation of blood circulation	Sheng et al. (2009)
			PLGA NPs		Reduction in opsonization and phagocytic uptake	Amoozgar et al. (2012)
4	Pectin	Anionic	Silver nanoparticles		Act as reducing as well as stabilizing agent that prevents aggregation of nanoparticle by slowing its deposition	Tummalapalli et al. (2015)
			Silver nanoparticles		Avoid agglomeration of nanoparticles due to high viscosity and lead to formation of well-dispersed stabilized silver nanoparticles	Kong et al. (2008)

Continued

TABLE 1.1 Various gum-based nanoformulation for targeted delivery of therapeutic moiety—cont'd

Sr. no.	Gum	Surface charge	Types of nanoparticles	Conjugated moiety	Targeted site/disease	References
			Self-assembled pectin-conjugated eight-arm polyethylene glycol-dihydroartemisinin nanoparticles	Polyethylene glycol-dihydroartemisinin	Hydrophilic carriers of pectin and eight-arm PEG improved stability, water solubility, and controlled drug release	Liu et al. (2017)
			Ceftizoxime-loaded nanocarrier		Pectin nanocarrier exhibits better zone of inhibition than free ceftizoxime with no signs of toxicity	Kumar et al. (2020)
			Gold nanoparticles		Intracellular accumulation of doxorubicin coated gold nanoparticles enhances through receptor-mediated endocytosis	Borker and Pokharkar (2018)
			Silver nanoparticles		Showed better antibacterial as well as antibiofilm activity against gram-negative *E. coli* and gram-positive *S. epidermis*	Pallavicini et al. (2017)
5	Gellan gum	Anionic	Gold nanoparticles		More uptake of nanoparticles was observed in cancer cells with no characteristic changes in hematological, biochemical, and histopathological study	Dhar et al. (2008)
			Silver nanoparticles		Stability problem associated with silver nanoparticles was overcome by using gellan gum that acts as a capping agent and prevents aggregation of nanoparticles	Dhar et al. (2012)
			Gold nanorods		Stabilize nanomaterial by avoiding aggregation	Vieira et al. (2015)
			Gold nanorods		Gellan gum-coated gold nanorods possesses better stability and it can combine hyperthermia with diagnostic features for cancer treatment	Varvarà et al. (2020)
6	Gum kondagogum	Anionic	Silver nanoparticles		Antibacterial activity	Rastogi et al. (2015)
			Silver nanoparticles	Sulfide moiety	Antibacterial activity	Dasari and Guttena (2016)

7	Locust bean gum	Both anionic and cationic	Locust bean sulfated nanoparticles	Sulfated moiety	Antigen delivery in mucosal level	Braz et al. (2017)
			LBG microspheres	Carboxymethylated derivative	Targeted colon in colitis	Karakus et al. (2019)
8	Alginate gum	Anionic	Alginate nanoparticles	Octaarginine	Intestine	Li et al. (2020)
			Alginate conjugate nanoparticles	Amid moiety	Glioma cell line	Joshy et al. (2018)
			Phenylalanine ethyl ester self-assemble nanoparticle	Phenylalanine ethyl ester moiety	Vitamin B2 delivery	Zhang et al. (2016)
			Alginate nanoparticles		Brain	Haque et al. (2014)
			Polyplex nanoparticles		Lymph node	Brezaniova et al. (2017)
9	Cashew gum	Anionic	Self-assembled nanoparticles	Acetyl moieties	Self-assembled indomethacin characterization	Pitombeira et al. (2015)
			Acetylated CG nanoparticles	Acetylated moiety	Antiinflammatory action	do Amaral Rodrigues et al. (2019)
			Silver nanoparticles	Phthalated moiety with phthalate anhydride	Antibacterial activity	Oliveira et al. (2019)
			Acetylated cashew gum nanoparticles	Acetylated moiety	Antibacterial activity	Amorim et al. (2019)
10	Dextran gum	Anionic	Self-assembled nanoparticles	Ester moiety with N,N-carbonyldiimidazole	Antiinflammatory	Hornig et al. (2009)
			Aldehyde dextran nanocarriers	Aldehyde functionalization with Schiff base linkage	Tumor spheroid anticancer activity	Sagnella et al. (2014)
			Nonpolymer-bound paclitaxel or paclitaxel–CM dextran conjugates	Carboxymethyl dextran conjugate by Gly-Gly-Phe-Gly linker	Colon targeting	Sugahara et al. (2008)
			Dextran nanoparticles	Dodecylamine moiety	Anticancer	Wasiak et al. (2016)

Continued

TABLE 1.1 Various gum-based nanoformulation for targeted delivery of therapeutic moiety—cont'd

Sr. no.	Gum	Surface charge	Types of nanoparticles	Conjugated moiety	Targeted site/disease	References
			Silver nanoparticles	Dextran polyacrylamide moiety	Anticancer activity in leukemia	Telegeev et al. (2017)
11	Gum arabic	Anionic	Curcumin-loaded gum arabic-sodium alginate nanoparticles	1,1-Diphenyl-2-picrylhydrazyl (DPPH)	Anticancer activity against human liver cancer cells (HepG2) in colon cancer (HT29), lung cancer (A549) and breast cancer (MCF7) cells	Hassani et al. (2020)
			Multifunctional pH-sensitive magnetic nanoparticles	Adipic dihydrazide	Synchronous cancer therapy and sensing	Banerjee and Chen (2008)
			Insulin nanoparticles		Oral protein delivery	Avadi et al. (2010)
			Magnetic iron oxide nanoparticles (MNP) coated with arabic gum	Ferric chloride and ferrous chloride	Tumor imaging and targeted intratumoral drug delivery	Zhang et al. (2009)
			Superparamagnetic iron oxide nanoparticles	Meso 2,3-dimercaptosuccinic acid (DMSA)	MRI contrast agent	Palma et al. (2015)
			Curcumin-loaded liposomes	PEGylated lipid (polyethylene glycol moiety attached to a phospholipid)	Hydrocolloid enriched liposomal systems for formulation of functional foods	Li et al. (2020)
			Arabic gum-capped gold nanoparticles	Folic acid	Cancer cell targeting	Devi et al. (2015)
			Gum arabic-capped copper nanoparticles	L-ascorbic acid	Antimicrobial and photocatalytic dye reduction efficacy	Chawla et al. (2020)
12	Tamarind gum	Nonionic	Tamarind seed nanoaggregates		Ophthalmic delivery	Dilbaghi et al. (2013)
			Carboxymethyl tamarind kernel polysaccharide nanoparticles	Calcium chloride	Ocular drug delivery	Kaur et al. (2012)

13	Gum tragacanth	Anionic	Thymol-loaded bipolymeric (gum tragacanth/chitosan) nanocarriers		Antiinflammatory and antioxidant activity	Sheorain et al. (2018)
			Muco-adhesive hybrid nanocarriers	Lecithin	Enhancing amphotericin B (AmpB) oral bioavailability	Jabri et al. (2018)
			Silver nanoparticles		Determination of capecitabine in human plasma samples	Hajizadeh et al. (2020)
			Silver nanoparticles		Antibacterial activity	Kora and Arunachalam (2012)
			Gold nanoparticles		Bactericidal activity	Rao et al. (2017)
			TiO_2 nanoparticles	Glutaraldehyde	Removal of methylene blue from simulated colored solution	Ranjbar-Mohammadi et al. (2019)
14	Guar gum	Nonionic	Vitamin E d-A-tocopheryl poly (ethylene glycol) succinate (*E*-TPGS) nanoparticles	TPGS	Central nervous system illness and cerebrovascular disease	Liu et al. (2018)
			Guar gum-coated chitosan nanoparticles		Tuberculosis	Goyal et al. (2016)
			Carboxymethyl guar gum nanoparticles	Trisodium trimetaphosphate	Preliminary in vitro investigations	Dodi et al. (2016)
			Carboxymethyl guar gum nanoparticles		Used in pharmaceutical and drug delivery	Gupta and Verma (2014)
			Selenium-incorporated guar gum nanoparticles		H9c2 cardiomyoblast	Soumya et al. (2013)
			Palladium nanoparticles		Catalyst in reduction reactions and degradation of azo dyes	Anjum et al. (2020)
			Copper oxide nanoparticles		Copper nanoparticulates	Gajalakshmi et al. (2018)
			Guar gum-stabilized gold nanoparticles		Colon cancer	Ganeshkumar et al. (2018)
			Platinum nanoparticles		Catalytic reduction of *p*-nitrophenol	Pandey and Mishra (2014)

possible ways to stabilize nanoparticles: forming stearic repulsion among the formed nanoparticles, or slowing deposition of nanoparticles by enhancing the viscous nature of a solvent to prevent aggregation. Stearic repulsion is caused by adsorbing xanthan gum on the surface of formed nanoparticles, which helps to prevent aggregation of nanoparticles (Muddineti et al., 2016). High viscous behavior of xanthan gum at low concentration plays a key role in prevention of aggregation by increasing solvent viscosity; this is due to its high molecular weight and hydrogen bonding interactions (Comba et al., 2011; Patel et al., 2020; Xue and Sethi, 2012). Certain properties of gold nanoparticles like their inert and biocompatible nature makes them suitable candidates for drug delivery in cancer treatment, but development of stable gold nanoparticles is a major challenge. This drawback of gold nanoparticles can be solved by coating of xanthan gum. Muddineti et al. (2016) developed gold nanoparticles of curcumin by using xanthan gum as a stabilizer and ascorbic acid as a reducing agent along with PEGylation by polyethylene glycol-thiol (mPEG800-SH). Capping of xanthan gum helps to enhance cellular uptake due to the presence of mannose moiety in it, which facilitates receptor-mediated endocytosis. Xanthan gum and PEG coating not only solve the problem of low stability but also provide certain advantages like biocompatibility, longer circulation time, reducing immunogenic responses, i.e., stealth effect of nanoparticles, along with nontoxicity at all tested concentrations (Muddineti et al., 2016). Pooja et al. (2014) developed gold nanoparticles of doxorubicin hydrochloride (DOX) which are stabilized by xanthan gum. In this study, properties of xanthan gum as a reducing and capping agent were utilized to obtain stable nanoparticles without using harsh chemicals. DOX-loaded gold nanoparticles showed three times more cytotoxicity in A549 cells than free DOX (Pooja et al., 2014). Rao et al. (2018) prepared silver nanoparticles (AgNPs) of Bergenin stabilized by xanthan gum (GX), which significantly decreases reactive oxygen species compared to control groups. AgNPs, due to their attractive chemical and biological properties, become the prime choice. In this case, GX used as a stabilizing and reducing agent provides numerous advantages like lack of environmental hazards, easy availability, cost-effectiveness, and biocompatibility as compared to synthetic agents. This study confirms GX stabilized AgNPs presented a multitargeted nano drug delivery system for treatment of rheumatoid arthritis (Rao et al., 2018). Kumari et al. (2015), instead of using synthetic reducing and stabilizing agents, used GX to stabilize palladium nanoparticles employing green synthesis.

1.4.2 Carrageenan

Carrageenan (CG) is a sulfated polysaccharide, mainly anionic in nature, obtained by extraction of various genera of red algae like *Agardhiella*, *Chondrus crispus*, *Eucheuma*, *Furcellaria*, *Gigartina*, *Hypnea*, *Iridaea*, *Sarconema*, and *Solieria* (Campo et al., 2009; McHugh, 2003; Zia et al., 2017), which all belong to the Florideophyceae family. CG is known to have numerous applications including as a gelling, emulsifying, and stabilizing agent, and its properties such as biodegradability, nontoxicity, high viscosity, and gelling capacity make it a suitable candidate for use in the pharmaceutical and medical fields (Quito et al., 2020). Depending upon the source of extraction and solubility, CG is categorized into six basic forms: kappa (κ-), iota (ι-), lambda (λ-), mu (μ-), nu (ν-), and theta (θ-) carrageenan. Out of these, kappa (κ-), iota (ι-), and lambda (λ-) remain the pharmaceutically and commercially important forms

(Campo et al., 2009; Necas and Bartosikova, 2013). Over the past few decades, carrageenan has attracted the attention of researchers due to its various bioactive properties such as antiviral, antioxidant, antibacterial, antitumor, and immunomodulatory activity, and antithrombotic, anticoagulant, and antihyperlipidemic action (Quito et al., 2020). Shanmuga et al. (2015) synthesized and characterized magnetic nanoparticles for drug delivery application. One of the major drawbacks of iron oxide nanoparticles was their low stability due to oxidation resulting in aggregation of nanoparticles; this problem was overcome by coating κ-CG, which results in the formation of stable magnetic nanoparticles (NPs), as confirmed by hemolytic assay (Shanmuga et al., 2015). Chen et al. (2018) fabricated gold nanoparticles of epirubicin (EPI-CAO-AuNPs) for pH-triggered delivery to cancer cells. CG, which was used as a capping agent, also acts as a biocompatible reductant for green synthesis of EPI-CAO-AuNPs. A study showed that EPI-CAO-AuNPs greatly improved cytotoxic activity and induction of apoptosis on cancer cells rather than on normal cells (Chen et al., 2018).

1.4.3 Chitosan

Chitosan is a natural, deacetylated product of chitin obtained from exoskeletons of certain insects like shrimps, crabs, as well as cell walls of some fungi (Cheung et al., 2015). It is a linear polysaccharide, mainly polycationic in nature, that is composed of β-(1–4)-linked D-glucosamine and *N*-acetyl-D-glucosamine (Kumar et al., 2004). Due to its unique properties as well as its biocompatible and nontoxic nature and ease of degradation inside the body, it is a suitable candidate for various pharmaceutical and medical applications (Cheung et al., 2015). Moreover, its antimicrobial (Martins et al., 2014), antioxidant (Ngo and Kim, 2014), and antitumor activity (Karagozlu and Kim, 2014), together with its capacity of incorporating various macromolecules like proteins with the help of number of interactions (ionic cross-linking, desolvation, or ionic complexation), make it a suitable choice to be used as a drug carrier. Furthermore, its applicability to transport across cell epithelia by opening tight junctions makes it a suitable candidate for selecting a carrier for making nanoparticles (Janes et al., 2001). Rahimi et al. (2018) utilized properties of chitosan to develop a multibranched ionic liquid chitosan nano-vehicle incorporating doxorubicin (DOX) and methotrexate (MTX) for treatment of cancer showing a stealth mechanism. These multibranched nano-vehicles consist of magnetic nanoparticles (Fe_3O_4) coated with chitosan along with mPEG grafting, which have the advantage of targeted delivery of magnetic nanoparticles using a magnetic field to the targeted area while the coating of biodegradable polymer reduces the problem of toxicity. This nano-vehicle not only provides stealth properties but also shows enhanced cellular uptake of both drugs; this was confirmed by using SDS-PAGE and hemolysis assay (Rahimi et al., 2018). Zhang et al. (2008) fabricated injectable chitosan nanoparticles of insulin grafted by PEG. These PEGylated chitosan nanoparticles, which were prepared by electrostatic interaction, showed increased insulin uptake in Caco-2 cells (Zhang et al., 2008). Sheng et al. (2009) fabricated polymeric nanoparticles using poly (lactic acid) (PLA) with a chitosan coating along with PEG attachment in order to improve their stability as well as circulation time. It was observed that chitosan alone as well as in combination with PEG improves surface hydrophilicity and stability of formed nanoparticles. Two of the major problems associated with PLA NPs was rapid opsonization and clearance from body upon intravenous administration; these were overcome by the synergistic action of chitosan coating along with PEG (PEG/PDC NPs). An *in-vitro* study on phagocytosis by mouse peritoneal macrophage (MPM), *in-vivo* blood clearance, and biodistribution showed enhancement of blood circulation time ($t_{1/2}$ 63.5 h) along with a small amount of NPs metabolized by the liver upon IV administration (Sheng et al., 2009). Amoozgar et al. (2012) utilized properties of chitosan as a pH-sensitive biopolymer to form a stealth coating on paclitaxel (PTX/PLGA-LMWC2 − 4 k NP). It was observed that PLGA NPs with negative charge do not have any interaction with cancer upon incubation with them; on the other hand, a coating of low molecular chitosan (LMWC) on PLGA NPs remained with cancer cells at pH 6.2 and was shown to have stealth properties at the same pH, indicating that LMWC showed a pH-sensitive stealth property (Amoozgar et al., 2012).

1.4.4 Pectin

Pectin is an anionic heteropolysaccharide also called poly (1,4-galactouronic acid) that is formed in the cell walls of terrestrial plants, consisting of OH, COOH, and $COOCH_3$ functional groups as the backbone of the pectin chain, which is prone to functionalization (Tummalapalli et al., 2015). Structurally, it consists of four substructures: homogalacturonan (HG), rhamnogalacturonan I (RG-I), xylogalacturonan (XG), and rhamnogalacturonan II (RG-II) (Zhao and Zhou, 2016). Due to the presence of a large number of polar hydroxyl groups and charged carboxyl groups, pectin is hydrophilic in nature with widespread use as a stabilizing and reducing agent in nanoformulations. It is used in biomedical applications for drug delivery and gene delivery, and in tissue engineering due to its biodegradability, structural malleability, and cytocompatible gelling ability; it is also used as soluble dietary fiber in the human diet as a constituent in fruits and vegetables (Zhao and Zhou, 2016). Due to its viscous rheology, pectin alone or in combination with alkenyl succinic anhydride prevents nanoparticle aggregation by slow and controlled deposition. Tummalapalli et al. (2014) developed silver nanoparticles employing oxidized pectin by using facile and green synthesis. In this, oxidized pectin acts as a reducing as well as a stabilizing agent that prevents aggregation of nanoparticles by slowing their deposition. Results indicate that oxidized pectin stabilized nanoparticles with formation of a face centered cubic crystal structure (Tummalapalli et al., 2015). Kong et al. (2008) developed silver nanoparticles using pectin as a stabilizing as well as a reducing agent by a microwave-hydrothermal technique, and nanoparticles were studied by transmission electron microscopy. A study showed that pectin avoids agglomeration of nanoparticles due to high viscosity and leads to formation of well-dispersed stabilized silver nanoparticles upon uniform heating (Kong et al., 2008). Most anticancer drugs have certain problems like poor water solubility and difficulties in controlled drug release, which restrict their use in clinical application. To overcome these problems, Liu et al. (2017) developed self-assembled pectin-conjugated eight-arm polyethylene glycol-dihydroartemisinin nanoparticles (PPDH) for anticancer combination therapy. It was observed that the presence of hydrophilic carriers of pectin and eight-arm PEG improved stability, water solubility, and controlled drug release. An increase in cellular uptake, strong cell apoptosis induction, and cell viability inhibition capacity in 4 T1 cells and in MCF-1 cells were observed in an *in-vitro* assay of nanoparticles, leading researchers to conclude that PPDH nanoparticles surprisingly suppress tumor growth by removing bulk tumor cells in a 4 T1 orthotopic tumor murine model and give satisfying effects with no hypersensitivity reactions (Alvarez-Lorenzo et al., 2013; Buschmann et al., 2013; Goodarzi et al., 2013; Lemarchand et al., 2004; Liu et al., 2008; Mizrahy and Peer, 2012; Posocco et al., 2015; Wen and Oh, 2014). Kumar et al (2020) developed ceftizoxime, a third-generation cephalosporine antibiotic-loaded nanocarrier, for treatment of bacterial infection using pectin as a biocompatible, nontoxic, and high stable polymer. Results revealed that this pectin nanocarrier exhibited a better zone of inhibition than free ceftizoxime due to high efficacy for longer time, while a toxicity study indicated no signs of toxicity against the fibroblast cell line, suggesting controlled delivery of ceftizoxime from the pectin nanocarrier system (Kumar et al., 2004). Borker and Pokharkar (2018) fabricated pectin-capped gold nanoparticles for delivering doxorubicin for cancer treatment, where pectin was used as a stabilizing as well as a reducing agent. It was observed that intracellular accumulation of doxorubicin-coated gold nanoparticle enhances through receptor-mediated endocytosis while an increase in delivery of doxorubicin in the perinuclear and nuclear regions of cells occurred by clathrin-mediated endocytosis. It was concluded that pectin-capped gold nanoparticles enhance the efficacy of DOX and lower its systemic toxicity (Borker and Pokharkar, 2018). Pallavicini et al. (2017) developed pectin-capped silver nanoparticles for determination of antibacterial, antibiofilm, and wound-healing properties. It was found that pectin, which was used as a stabilizing and coating material in formulations of silver nanoparticles, weakly attached to bacterial membranes at minimum silver concentration, and thus showed better antibacterial as well as antibiofilm activity against gram-negative *E. coli* and gram-positive *S. epidermis*. Moreover, wound-healing activity tested against fibroblast revealed that pectin-capped silver nanoparticles improved fibroblast proliferation and were found to be very effective medication for wound healing and for treatment of vulnerable surgical site tissues (Pallavicini et al., 2017).

1.4.5 Gellan gum

Gellan gum is an anionic polysaccharide produced by the micro-organism *Sphingomonas elodea* with a linear tetrasaccharide repeating unit of glucuronic acid, rhamnose, and glucose residues. It is soluble in water and insoluble in ethanol, and shows high viscosity in fresh water which decreases with increases in temperature. It possesses certain properties such as being easily converted into heat-resistant gel, is noncytotoxic, and can be efficiently inoculated into tissue (Zhang et al., 2014). In the metal nano formulation, it is used as a reducing as well as a stabilizing agent as it reduces the toxicity of gold nanoparticles. It finds application as a gelling agent, suspending agent, vehicle in ocular drug delivery, and also in production of confectionary jellies, sugar confectionary, and milk beverages (Sworn, 2009). Dhar et al. (2008) developed gellan gum-reduced gold nanoparticles for assessment of cellular uptake and toxicity. Gellan gum was used as a capping agent and to overcome the toxicity problem associated with gold nanoparticles. In this study, gellan gum nanoparticles were labeled by Texas red and administered in rats via the oral route and evaluated for any significant effect in embryonic fibroblast cells and in human glioma cells. Result revealed that greater uptake of nanoparticle was observed in cancer cells with no characteristic changes in hematological, biochemical, and histopathological study. It was concluded that gellan gum-reduced nanoparticles did not show any toxicity, so the nanoparticle formulation can be potentially used as a safe carrier for delivery of biologicals (Dhar et al., 2008). Dhar et al. (2012) developed gellan gum-reduced silver nanoparticles to determine their cytotoxicity against mouse fibroblast cells and evaluated for *in-vitro* diffusion and stability in varying pH and electrolyte concentrations. The stability problem associated with silver nanoparticles was overcome by using gellan gum, which acts as a capping agent and prevents aggregation of nanoparticles. Results indicate that gellan gum-reduced nanoparticles show stability in varying pH and electrolyte concentrations, and an in vitro diffusion study showed that at minimum amounts, nanoparticles were localized in the skin and penetrated the receptor compartment. As per toxicity concerns, nanoparticles did not show cytotoxicity against embryonic fibroblast cells, hence they are potentially used in topical application (Dhar et al., 2012). Vieira et al. (2015) developed gellan gum-coated gold nanorods and assessed their stability in different pH and in salt concentrations. Gold nanoparticles were capped with low acyl gellan gum by using a layer-by-layer-based technique; this imparted subsequent deposition of poly acrylic acid, polyallylamine HCl, and gellan gum, which permitted formation of a gellan gum hydrogel-like shell. Low acyl gellan gum possesses unique properties; for example, it stabilizes nanomaterial by avoiding aggregation, and its biocompatibility and gelling property are of benefit in capping of silver nanorods. Stability studies in different pH and salt concentrations revealed that stability of nanorods increases at varying pH. Maximum stability was obtained at pH 6; also at higher salt concentrations, nanorods show better stability and ionic strength compared to electrolyte-coated nanorods. *In-vitro* studies concluded that gellan gum-capped nanorods did not show any cytotoxic effect and they were localized inside lysosomes. From the stability and biocompatibility of gellan gum-capped nanorods it was concluded that such a drug delivery system can be useful in biomedical and tissue engineering (Vieira et al., 2015). Varvarà et al. (2020) developed gellan gum and lipoic acid-coated gold nanorods for treatment of cancer. In this, gellan gum and lipoic acid were used as stabilizing agents. Gellan gum-coated gold nanorods possess better stability and can combine hyperthermia with diagnostic features so can be used for cancer treatment. In addition, optical properties of gellan gum-coated nanorods may profitably be combined with advantageous material features to attain selective photothermal therapy (PTT). In HCT116 human colon cancer cells, a photothermal effect was determined on two types of gold nanorods: gellan gum-coated nanorods and in ester derivative of gellan gum and lipoic acid. From the results, it was observed that nanorods coated with gellan gum derivative were prone to aggregation, whereas gellan gum and lipoic acid-coated gold nanorods were able to reach a convenient temperature for hyperthermia and revealed lower efficacy in *in-vitro* photothermal therapy. Due to better stability of gellan gum and lipoic acid-coated gold nanorods, it can be used in the *in-vivo* photothermal therapy of cancer since gellan gum shell offer to load bioactive molecules for drug delivery purposes. Hence, gellan gum and lipoic acid-coated gold nanorods serve as an auspicious choice for further use in *in-vivo* photothermal therapy of cancer (Varvarà et al., 2020).

1.4.6 Gum kondagogu

Gum kondagogu is a naturally occurring polysaccharide which is obtained from the bark of *Cochlospermum Gossypium*, belonging to the Bixaceae family. It consists of uronic acid and certain functional groups like hydroxyl, acetyl, carbonyl, and carboxylic groups. Basically, it consists of rhamnose, galacturonic acid, glucuronic acid, β-D-galactopyranose, α-D-glucosopyranose, galactose, arabinose, mannose, and fructose (Sashidhar et al., 2015). It finds application as a reducing agent and acts as a stabilizer in the preparation of silver nanoparticles. Dasari and Guttena (2016) formulated silver sulfide nanoparticles for determination of antibacterial activity. In this study, they synthesized silver sulfide nanoparticles using gum kondagogu as a capping agent and characterized the formulation. From characterization, they reveal that the synthesized silver sulfide nanoparticles are spherical, stable, and exhibit significant antibacterial action on both the gram classes of bacteria (*E. coli*, *P. aeruginosa*, *S. aureus*, and *Bacillus thuringiensis*) (Dasari and Guttena, 2016). Rastogi et al. (2015) developed silver nanoparticles of gum kondagogu and assessed the antibacterial activity of both aminoglycosidic and ciprofloxacin antibiotics. From their results, they concluded that silver nanoparticles were stabilized by gum kondagogu by preventing it from aggregation. Thus, silver nanoparticles of ciprofloxacin and aminoglycosidic show a synergistic effect against gram classes of bacteria with better antibacterial activity (Rastogi et al., 2015).

1.4.7 Locust bean gum

Locust bean gum is a white, water-soluble powder obtained from the carob tree and belongs to the legume family. It is a high molecular weight polysaccharide that consists of galactomannan having two units: galactose and mannose. Locust bean galactomannan is composed of a linear chain of (1→4)-linked D-mannopyranosyl units with (1→6)-linked-D-galactopyranosyl residues as side chains. It is less viscous and its viscosity decreases below pH 4 and above pH 9. It is used as a stabilizer, thickener, and additive in food, pharmaceuticals, cosmetics, and in dietary fiber-enriched food products. In nanoformulations, it is used as a nanoparticle stabilizer which prevents formation of aggregates and sedimentation of nanoparticles (Barak and Mudgil, 2014). Braz et al. (2017) developed sulfated locust bean gum nanoparticles for oral immunization of both HE and Ovalbumin. Locust bean gum possesses a neutral charge, thus shows a hindering effect in the polyelectrolyte complexation method, so surface modification was carried out using sulfated moiety which imparts a negative charge, which eases nanoparticle formation. These sulfated nanoparticles were then complexed with chitosan by means of electrostatic interaction for antigen delivery. The results of a cytotoxicity study carried out in Caco-2 cells revealed that there is no alteration in metabolic activity on exposure, but a moderate negative effect was observed on cell membrane integrity (Braz et al., 2017). Karakus et al. (2019) developed polymeric nanoparticles using PEGylated ester locust bean gum, using the ultrasonic irradiation method. PEGylated-LBG was used for coating nanosized and spherical amphiphilic rosin ester nanoparticles. The results indicate that the size of nanoparticles depends on the miscibility of PEG-LBG blends based on the intrinsic viscosity values in different conditions, and these PEG-LBG/RE polymeric nanoparticles find application in pharmaceutical and biomedical study for increasing the therapeutic potency and biocompatibility of nanodrugs (Karakus et al., 2019).

1.4.8 Alginate gum

Alginate is a linear polysaccharide obtained from brown seaweed consisting of alternating blocks of 1,4-linked α-L-guluronic and β-D-mannuronic acid residues (Brezaniova et al., 2017) which can be easily transformed into gel; hence it is a very useful material for production of hydrogels (Paques et al., 2014). It is a biocompatible, nonimmunogenic, and nontoxic material, thus it finds application as a stabilizer, viscosity increasing agent, release modifying agent, and taste masking agent in the pharmaceutical industry. It is also used in the food industry as an emulsifier, texturizer, adjuvant, humectant, and surfactant. In nanoformulation, it act as a stabilizer as it stabilizes nanoparticles by slowing deposition of nanoparticles by enhancing viscosity of solvent to prevent aggregation of nanoparticles and the other way by means of stearic interaction (Szekalska et al., 2016). Joshy et al. (2018) developed amide functionalized alginate nanoparticles for efficient delivery of zidovudine in HIV treatment. Zidovudine has certain drawbacks like low bioavailability, short half-life, and hydrophobicity; to overcome these disadvtanges, zidovudine was encapsulated in modified alginate nanoparticles. Amide derivative of alginate was used for preparation of nanoparticle by a chemical cross-linking method using Pluronic F-68 as a stabilizer which showed stability, high entrapment, and prolonged drug release (Joshy et al., 2018). In another study performed by Zhang et al. (2016), self-assembled nanoparticles of modified alginate were formulated to deliver riboflavin (vitamin B2). Alginate was modified by forming phenylalanine ethyl ester conjugate and used for fabrication of nanoparticles. The results indicated that conjugated nanoparticles do not show any cytotoxic effect, with the advantages of high drug encapsulation and drug release in bursts as well as in a controlled manner; thus it was concluded that nanoparticles based on PEA conjugates can potentially be used as a nanosized novel delivery system for the poorly water-soluble nutrient vitamin B2 (Zhang et al., 2016). Haque et al. (2014) formulated and evaluated alginate nanoparticles of venlafaxine for depression. Nanoparticles were prepared by an ionic gelation technique and administered via an intranasal route. It was observed that at pH 4.5–5, the alginate solution interacted efficiently with chitosan to form compact smaller nanoparticles, and it was concluded that alginate nanoparticles show better stability with efficacious delivery of venlafaxine in the brain in significant quantities compared to free drug via intranasal drug delivery (Haque et al., 2014).

1.4.9 Cashew gum

Cashew gum is a polysaccharide obtained from the tree *Anacardium occidentale* and is comprised of D-galactose (72%–73%), D-glucose (11%–14%), D-arabinose (4.6%–5.0%), D-rhamnose (3.2%–4.0%), and D-glucuronic acid (4.7%–6.3%) (Pitombeira et al., 2015). These polysaccharides contain amine chain of D-galactose (1–3) linked to galactose and glucose side chains. Cashew gum is known to show antimicrobial, antidiarrheal, antiinflammatory, and gastroprotective effects (do Amaral Rodrigues et al., 2019). Because of the presence of a hydroxyl group in the structure of cashew gum, it shows hydrophilic behavior, which is responsible for chemical modification and introduction of new functional groups (Oliveira et al., 2019). It is used as a stabilizer and as a capping agent for nanoparticles. Stabilization of nanoparticles occurs by adsorbing gum on the surface of nanoparticles which prevents their aggregation due to stearic or electrostatic effects (Amorim et al., 2019). Pitombeira et al. (2015)

developed indomethacin-loaded self-assembled nanoparticles using acylated cashew gum. In this study, acetyl moiety was chemically led into cashew gum polysaccharide for the activation of amphiphilic properties, which were then used for the preparation of self-assembled nanoparticles by a dialysis method. It was observed that the formulation presented good colloidal stability compared to free drug and the nanoparticles showed an initial burst release in the first 2 h and controlled release of indomethacin for up to 72 h (Pitombeira et al., 2015). In another study, performed by Oliveira et al. (2019), silver nanoparticles were prepared for assessment of antimicrobial activity using phthalate cashew gum by green synthesis. In this, cashew gum was used as a reducing and stabilizing agent. Silver nanoparticles prepared by chemical modification (phthalate cashew gum) showed surface charge electrostatic interaction with the bacterial cell wall causing changes in permeability, disturbed respiration, and leakage of intracellular content; it was concluded that pthalated polymers are an innovative approach for formulation of stable silver nanoparticles with potential antimicrobial activity (Oliveira et al., 2019). Amorim et al. (2019) developed copper nanoparticle using cashew gum for the determination of antibacterial activity and cytotoxicity against a 4 T1 mouse mammary tumor cell line. In this study, cashew gum was used for stabilizing nanoparticles by adsorbing on nanoparticles to prevent their aggregation. Results indicated that nanoparticles show optimum particle size; in addition, the cell viability assay demonstrated that the nanoparticles inhibited the growth of 4 T1 LUC and NIH 3 T3 cells. Hence, copper nanoparticles stabilized by cashew gum can be used as an antimicrobial agent with lower cytotoxic effects than their $CuSO_4 \cdot 5H_2O$ precursor (Amorim et al., 2019). do Amaral Rodrigues et al. (2019) developed acetylated cashew gum nanoparticles for assessment of antibacterial activity of epiisopiloturine alkaloid. These alkaloids had low solubility in water, which causes a problem in bioavailability; therefore, to overcome this problem, chemical modification was carried out. The results revealed that nanoparticles prepared with acetylated cashew gum showed good physical as well as chemical stability and no macroscopic changes (creaming, sedimentation, or flocculation) were observed; it was thus concluded that nanoparticles prepared with acetylated cashew gum showed remarkably efficient incorporation of epiisopiloturine with enhanced therapeutic benefit and showed better antibacterial activity (do Amaral Rodrigues et al., 2019).

1.4.10 Dextran gum

Dextran is a naturally occurring polysaccharide having appealing characteristics to be incorporated into nanoparticle formulation (Sagnella et al., 2014). It mainly consists of a 1,6-glucose unit with differing branches depending on the bacterial strain from which dextran is produced (Hornig et al., 2009). Dextran is considered to be a physiologically harmless polymer due to its worthwhile properties like being biocompatible, biodegradable, nonimmunogenic, and nonantigenic (Hornig et al., 2009). Clinically it is used as a plasma expander in the treatment of shock; pharmaceutically it is used in the preparation of gels in bead form for size exclusion, ion exchange, and hydrophobic chromatography. Due to the presence of hydroxyl and ether groups, dextran gum can be used in metallic nanoformulations for stabilization of nanoparticles by tightly binding to nanoparticles via electrostatic and stearic interactions, which prevents their aggregation (Diem et al., 2017). Sagnella et al. (2014)

developed a dextran-based doxorubicin nanocarrier for the management of tumors. In this study, dextran was functionalized with aldehyde and combined with doxorubicin to form an aldehyde-dextran-doxorubicin conjugate with Schiff base linkage. The results revealed that nanoparticles prepared with an aldehyde-dextran-doxorubicin conjugate showed better stability and improved tumor penetration compared to free drug (Sagnella et al., 2014). Hornig et al. (2009) developed self-assembled nanoparticles using a hydrophobic dextran derivative by a dialysis technique. They broadly established the use of macromolecular prodrugs with nanoparticulate drug delivery devices, where they simply modified dextran by performing esterification with carboxylic acid by *in situ* activation with N,N-carbonyldiimidazol (CDI). Ibuprofen and naproxen were functionalized with a dextran derivative. Stability of nanoparticles was achieved by means of dextran as a stabilizer. The study concluded that self-assembled nanoparticles of ibuprofen and naproxen prepared using modified dextran showed better stability against hydrolysis and can be easily incorporated into cells. In physiological condition particle shows stability so that their liability of enzymatic degradation via dextranase is a main requisite for its macromolecular prodrug use (Hornig et al., 2009). Diem et al. (2017) developed spherical-shaped gold nanoparticles (AuNPs) that are stabilized by dextran. Nanoparticles were synthesized using an irradiation method and analyzed for the influence of pH, Au^{3+}, and dextran concentrations on the size of AuNPs. Dextran was incorporated to overcome the stability problem associated with metal nanoparticles. It was observed that gold nanoparticles form aggregates after long-term storage, so dextran was used to prevent aggregation, creating stearic and electrostatic hindrance over nanoparticles. From the results, it was identified that an increase in pH as well as in the concentration of dextran decreases the size of nanoparticles while an increase in the concentration of Au^{3+} increases the size of nanoparticles; it was thus concluded that dextran imparts biocompatibility to gold nanoparticles and hence dextran-stabilized gold nanoparticles can be applied in biomedicine and pharmaceutics, specifically in X-ray contrast imaging and in photothermal cancer therapy (Diem et al., 2017).

1.4.11 Gum arabic

Gum arabic (GA) is a popular natural polysaccharide containing galactose and arabinose, derived from the stems and branches of the *Acacia Senegal* tree, also known as acacia gum (Hassani et al., 2020). It is a highly water-soluble, biocompatible, biodegradable phytochemical glycoprotein polymer commonly used as a stabilizer that has many uses in the food, pharmaceutical, and cosmetic industries (Avadi et al., 2010; Banerjee and Chen, 2008; Zhang et al., 2009). Basically it is a heterogeneous polymer composed of three main components: low protein content of arabinogalactan (90%); high protein content of arabinogalactan (10%); and high protein content of glycoproteins (<1%) (Palma et al., 2015; Zhang et al., 2009). Owing to its strongly branched structure containing 1,6-galactopyranose residues, galactose, and arabinofuranose, its solution viscosity is lower than that of other exudate gums; it only becomes viscous at high concentrations (30%–50%, w/v). GA is also used as an emulsifier because it is extremely surface-active due to the presence of an amphiphilic protein backbone, although it is typically hydrophilic; therefore, GA can provide steric stabilization without making the bulk too viscous (Li et al., 2020). Currently, to functionalize and stabilize the nanoparticles, GA has been used. In particular, the molecules of GA contain charged groups (amine and carboxyl) that can physically adsorb to the surface of a nanoparticle. In order to enhance colloidal stability, GA induces steric repulsion between the nanoparticles. Indeed, GA-coated gold nanoparticles have been shown to be stable in human serum albumin solution and strong ionic conditions, as shown by Kattumuri et al. (2007) (Kattumuri et al., 2007). Also as demonstrated by Kannan et al. (2006), surface modification of gum arabic nanoparticles offers a new avenue for the production of diagnostic or therapeutic hepatocyte-specific agents for the treatment of liver cancer and various other hepatocytic disorders (Kannan et al., 2006). Chawla et al. (2020) prepared copper nanoparticles capped with arabic gum and L-ascorbic acid used as a reducing agent. The study showed that the nanoparticles were found to be suitable for enhanced antimicrobial and photocatalytic dye reduction efficiency. In addition, gum arabic-capped copper nanoparticles were found to be extremely oxidatively stable (Chawla et al., 2020). Hassani et al. (2020) reported curcumin-loaded gum arabic-sodium alginate nanoparticles prepared by an ionotropic gelation technique. The study revealed that gum arabic has a potent antioxidant property. Based on the findings obtained from an MTT cytotoxicity assay, it can be concluded that nanoparticles of gum arabic-sodium alginate-loaded curcumin have therapeutic potential for prevention and treatment of solid malignancies such as cancer of the liver, breast, cervix, and skin (Hassani et al., 2020). Devi et al. (2015) prepared gold nanoparticles capped with arabic gum and leaves extract of *Vitex negundo*, which are used as capping and reducing agents, respectively. In the present research, gold nanoparticles were produced from green synthesis using a sunlight-mediated method. For tumor-targeted drug delivery, epirubicin-loaded arabic gum-capped gold nanoparticles (*E*-GNPs) have been synthesized and then functionalized with folic acid. Fa-E-GNPs have shown enhanced cytotoxic effects on the lung adenocarcinoma (A549) cell line, indicating that folic acid functionalization has contributed to cancer cell targeting and, in turn, increased epirubicin uptake by cancer cells (Devi et al., 2015).

1.4.12 Tamarind gum

Tamarind gum is a natural plant polysaccharide which is obtained from the endosperm seed of *Tamarindus indica* Linn. (Fabaceae family). Rao et al. (1946) first documented the extraction method for tamarind gum, and they extracted it from tamarind seed kernel powder (Nayak and Pal, 2017). Tamarind kernel powder (TKP) is formed by removing the seed's outer cover and grinding the white cream kernels (Sahoo et al., 2017). Chemically it consists of a (1,4)-b-D-glucan backbone substituted by a-D-xylopyranose and b-D-galactopyranosyl (1,2)-a-D-xylopyranose side chains connected (1,6) to glucose residues, where 55.4% of glucose, 28.4% of xylose, and 16.2% of galactose units are present, leading to a molar ratio of 2.8:2.25:1.0 (glucose:xylose:galactose) (Nayak and Pal, 2011). It is a branched polysaccharide that is neutral and nonionic, has water solubility, is hydrophilic, capable of forming a gel, and has mucoadhesive properties. Furthermore, tamarind gum is biodegradable, biocompatible, nonirritant, and noncarcinogenic (Nayak and Pal, 2017; Sangnim et al., 2018). Tamarind gum is used in medicinal, cosmetic, and food applications as a potential biopolymer. It has been tested and used extensively in various drug delivery applications as powerful pharmaceutical excipients in recent years. Tamarind gum is used in formulations of oral, colon, ocular, and nasal drug delivery systems. Different functionally derivatized tamarind gums have recently been highly promising as potential pharmaceutical excipients in various types of enhanced drug delivery systems, primarily due to their improved stability (lower degradability). While tamarind gum is commonly used in various drug delivery formulations, it has some potential disadvantages, such as unpleasant odor, dull color, low water solubility, and aqueous environment pattern of rapid degradability. Tamarind gum has been functionally derivatized by chemical treatment with a number of functional groups to resolve these limitations, such as carboxymethyl, acetal, hydroxyl alkyl, thiol, polymer grafting, etc. (Nayak and Pal, 2017). Dilbaghi et al. (2013) prepared tamarind seed xyloglucan nanoaggregates loaded with tropicamide for ophthalmic delivery. The present study was planned to optimize tragacanth seed xyloglucan (TSX) and poloxamer concentrations for the preparation of TSX nanoaggregates with minimum particle size and maximum efficiency of encapsulation. As it has a backbone chain with branching of xylose and galactoxylose substituents, tragacanth seed xyloglucan has excellent mucoadhesive properties that give it a mucin-like configuration. Because of these properties, it has been used to provide sustained relief from dry eye symptoms. Moreover, it increases the rate of healing of corneal wounds, increases corneal accumulation due to its mucomimetic effect, and plays a major role in ophthalmic applications (Dilbaghi et al., 2013). Kaur et al. (2012) developed tropicamide-loaded carboxymethyl tamarind kernel polysaccharide nanoparticles prepared by an ionotropic gelation technique. The results showed that carboxymethyl tamarind kernel powder concentration and calcium chloride had a major synergistic effect on particle size and percent efficiency of encapsulation. In addition, studies for ex vivo bioadhesion and eye tolerance showed the mucoadhesive nature of nanoparticles (Kaur et al., 2012).

1.4.13 Gum tragacanth

Gum tragacanth (TG) is a natural anionic polysaccharide which is obtained from replenishable sources. It is a wound exudate obtained from certain trees. The gum is extracted from stems and branches of different species of *Astragalus gummifer* and other Asian species of Astragalus, such as *Astragalus gossypinus* and *Astragalus microcephalus*. The exudate hardens and can be collected as flakes or curled ribbons after a couple of weeks. In particular, TG is a complex blend of branched acidic hetero-polysaccharides with molecular weight up to 850 kDa (Nazarzadeh Zare et al., 2019). It has been commonly used as an emulsifier, stabilizer, gelling agent, and thickener in various sectors such as the food, pharmaceutical, and cosmetics industries. It is recognized as safe for consumption and usage and has received the "generally recognized as safe" (GRAS) notation (Nejatian et al., 2020; Sheorain et al., 2018). It is a water-soluble mixture that is extremely acid-resistant, tasteless, and odorless. It comprises of two fractions named "Tragacanthic acid" and "Tragacanthin," which are biodegradable and nontoxic. The easy isolation of the two polysaccharides means that they are physically linked and not bonded chemically (Jabri et al., 2018). Some of the tragacanth fractions are insoluble in water; however, one of the compounds, called bassorin, has the capacity to swell and form a gel when exposed to water (Hajizadeh et al., 2020). Gum tragacanth produces D-galactose, D-xylose, L-fucose, L-arabinose, D-galacturonic acid, and L-rhamnose sugars upon hydrolysis (Nazarzadeh Zare et al., 2019). For the preparation of aqueous suspension of insoluble substances, the high viscosity of TG in water is useful. Kora and Arunachalam (2012) prepared silver nanoparticles using the aqueous extract of gum tragacanth by green synthesis. In this study, it was proposed that the water-soluble components in gum serve as reducing and stabilizing agents. Gum tragacanth, due to its attractive properties, such as nontoxic existence, being GRAS, natural availability, and higher resistance to microbial attacks and long shelf life, has become the choice for synthesis and stabilization of silver nanoparticles. Thus, as the study demonstrates, the gum plays a dual role of reducing and stabilizing silver ions (Kora and Arunachalam, 2012). Rao et al. (2017) developed green gold nanoparticles stabilized by gum tragacanth to increase the solubility of naringin. The nanoparticles loaded with drug were analyzed for possible enhanced bactericidal potential of naringin. The study indicates that AuNPs cargos stabilized by gum tragacanth increase the therapeutic efficacy of naringin by boosting its bactericidal activity through its destabilizing effect on the morphology of the bacterial surface (Rao et al., 2017). Ranjbar-Mohammadi et al. (2019) produced gum tragacanth-based TiO_2 nanoparticles using glutaraldehyde as a cross-linking agent. This study assessed the performance of gum tragacanth, an available, eco-friendly, low-cost, abundant biopolymer, as a suitable material for immobilization of TiO_2 nanoparticles for removal of cationic dye from wastewater. In order to immobilize TiO_2 nanoparticles for dye removal processes, the promising findings suggest GT biopolymer as a proper substrate (Ranjbar-Mohammadi et al., 2019).

1.4.14 Guar gum

Guar gum is a natural polysaccharide extracted by grinding guar beans of *Cyamopsis tetragonolobus* (Liu et al., 2018). It is a nonionic, water-soluble polysaccharide, combination of two sugars: galactose and mannose. Chemically it is comprised of a linear chain of β-1,4-linked mannose residues to which galactose residues are 1,6-linked at every second mannose, forming short side branches (Goyal et al., 2016). Guar gum is one of the extensively researched potential carrier materials for site-specific drug delivery due to its nontoxic, biodegradable, biocompatible, and mucoadhesive nature (Dodi et al., 2016). It has been used widely in the fields of biomedical, pharmaceutical, food processing, and cosmetic applications due to its natural abundance, low cost, and other prudent functionalities (Gupta and Verma, 2014). In pharmaceutical manufacturing, guar gum acts as a binder, disintegrant, suspending agent, thickening agent, and stabilizing agent (Soumya et al., 2013). It is highly appreciable in colon-specific drug delivery due its property of retarding the drug release as well as being susceptible to degradation in the colonic environment (Chourasia and Jain, 2004). Guar gum has the ability to stabilize metal nanoparticles, e.g., gold, platinum, palladium. The structure of guar gum was found to be highly complex because it contains hydroxyl groups which have the ability to bind metal clusters. In aqueous solution, guar gum has a very strong propensity to form a gel. This gel is extremely viscous and can therefore entrap, protect, and stabilize the synthesized nanoparticles for a longer period of time in its efficient mesh of gel and serve as an excellent surface capping agent in the process (Kumar et al., 2012). Anjum et al. (2020) prepared palladium nanoparticles using a green method of synthesis with guar gum as a capping and reducing agent. In the present study, the FT-IR study showed various reduction agent functional groups and demonstrated that guar gum is capable of stabilizing as well as reducing Pd(II) to Pd(0) (Anjum et al., 2020). Gajalakshmi et al. (2018) developed guar gum-stabilized copper oxide nanoparticles by a one-step chemical method which demonstrates the *in situ* method of synthesizing stabilized copper oxide nanofluids using guar gum polymer, resulting in improved thermal conductivity and antibacterial properties (Gajalakshmi et al., 2018). Ganeshkumar et al. (2018) reported that gold nanoparticles were successfully prepared with guar gum, a biopolymer using a green chemistry microwave-based method for the first time, and were first used for targeted drug delivery for colon cancer. In this study, guar gum-stabilized gold nanoparticles were prepared and observed to possess hemocompatibility (Ganeshkumar et al., 2018). Pandey and Mishra (2014) reported a simple and green process for synthesizing highly stable platinum nanoparticle dispersions. Guar gum as a reducing and capping agent precursor in an aqueous medium which is a natural, nontoxic, and eco-friendly biopolymer was used. Moreover, the prepared guar gum platinum NPs in aqueous solution had excellent colloidal stability and exhibited superior catalytic activity toward the reduction of *p*-nitrophenol (p-NP) (Pandey and Mishra, 2014). Soumya et al. (2013) demonstrated selenium-incorporated guar gum nanoparticles (SGG) prepared by a nanoprecipitation technique and characterized its effect on H9c2 cardiomyoblast to confirm the safety of nanoparticles on a biological system. In the present research, GG was selected as a carrier due to its widespread use as fiber for reducing blood cholesterol levels, obesity, and hyperglycemia (Soumya et al., 2013).

1.5 Concluding remarks

Nanomedicine has an extensive history that can be traced back through several years. There is a dire need for a nanoparticulate system to ensure accurate delivery of therapeutic agents to targeted cells. It is really a groundbreaking interdisciplinary approach of nanobiology, material science, and biotechnology. Nanocarriers filled with drug moiety have several benefits over traditional treatments. Plenty of drugs have problems due to involvement of intracellular entry and enzymatic degradation, moreover to it, body has its own endogenous defense mechanism, i.e., RES that helps human beings to protect from all types of external intruders (foreign particles), and nanocarriers is not an exception to this defense system. To bypass RES, stealth nanocarriers are essential, to maneuver through the body according to their targeted sites. For this, engineered surface modification of nanocarriers with hydrophilic and biodegradable polymer can be a strategy to improve therapeutic effects of drugs. Biodegradable polymers such as natural gums and their derivatives not only play a vital role in engineering of stealth nanocarriers but also reduce toxicity. In addition, they create repulsive forces to prevent interaction with macromolecules and increase their stability in biological systems, which prevents them from speedy clearance from the human body.

Many scientists are working on dual-targeting nanocarriers as an alternative to single-targeting nanocarriers because they bind two separate target sites. Additionally, this approach bolsters nanocarriers' affinity for their target site, which subsequently helps to minimize off-target binding to cells. Researchers are presently also engaged in finding out whether nanoparticles can cause long-term toxicity, and whether they undergo metabolic and degradative processes. For this, after delivering drug to the nucleus and other sensitive organelles, knowing the fate of the drug is important.

Ultimately, picobiology is the intended goal of researchers who embark on a journey into nanobiology. Even, in nature, things like atoms and molecules can be found all the way down to the nanometer level. Many single molecules are found in the picometer (pm) range. Volume-wise, vitamins also exist in this range. Science revolves around objects that fall in the range of 1–100 pm, which is the next step after nanoscience. However, the key question is: when will we reach picotechnology?

References

Alvarez-Lorenzo, C., Blanco-Fernandez, B., Puga, A.M., Concheiro, A., 2013. Crosslinked ionic polysaccharides for stimuli-sensitive drug delivery. Adv. Drug Deliv. Rev. 65 (9), 1148–1171. https://doi.org/10.1016/j.addr.2013.04.016.

Amoozgar, Z., Yeo, Y., 2012. Recent advances in stealth coating of nanoparticle drug delivery systems. Wiley Interdiscip. Rev. Nanomed. Nanobiotechnol. 4 (2), 219–233. https://doi.org/10.1002/wnan.1157.

Amoozgar, Z., Park, J., Lin, Q., Yeo, Y., 2012. Low molecular-weight chitosan as a pH-sensitive stealth coating for tumor-specific drug delivery. Mol. Pharm. 9 (5), 1262–1270. https://doi.org/10.1021/mp2005615.

Amorim, A., Mafud, A.C., Nogueira, S., Jesus, J.R., Araújo, A.R.d., Plácido, A., Brito Neta, M., Alves, M.M.M., Carvalho, F.A.A., Rufino Arcanjo, D.D., Braun, S., López, M.S.P., López-Ruiz, B., Delerue-Matos, C., Mascarenhas, Y., Silva, D., Eaton, P., Almeida Leite, J.R.S., 2019. Copper nanoparticles stabilized with cashew gum: antimicrobial activity and cytotoxicity against 4T1 mouse mammary tumor cell line. J. Biomater. Appl. 34 (2), 188–197. https://doi.org/10.1177/0885328219845964.

Anjum, F., Gul, S., Khan, M.I., Khan, M.A., 2020. Efficient synthesis of palladium nanoparticles using guar gum as stabilizer and their applications as catalyst in reduction reactions and degradation of azo dyes. Green Process. Synth. 9 (1), 63–76. https://doi.org/10.1515/gps-2020-0008.

Aravamudhan, A., Ramos, D.M., Nada, A.A., Kumbar, S.G., 2014. Natural polymers: polysaccharides and their derivatives for biomedical applications. In: Natural and Synthetic Biomedical Polymers. Elsevier Inc, pp. 67–89, https://doi.org/10.1016/B978-0-12-396983-5.00004-1.

Avadi, M.R., Sadeghi, A.M.M., Mohammadpour, N., Abedin, S., Atyabi, F., Dinarvand, R., Rafiee-Tehrani, M., 2010. Preparation and characterization of insulin nanoparticles using chitosan and Arabic gum with ionic gelation method. Nanomedicine 6 (1), 58–63. https://doi.org/10.1016/j.nano.2009.04.007.

Bali, N.R., Salve, P.S., 2019. Selegiline nanoparticle embedded transdermal film: an alternative approach for brain targeting in Parkinson's disease. J. Drug Deliv. Sci. Technol., 54. https://doi.org/10.1016/j.jddst.2019.101299.

Bali, N.R., Salve, P.S., 2020. Impact of rasagiline nanoparticles on brain targeting efficiency via gellan gum based transdermal patch: a nanotheranostic perspective for Parkinsonism. Int. J. Biol. Macromol. 164, 1006–1024. https://doi.org/10.1016/j.ijbiomac.2020.06.261.

Bali, N.R., Shinde, M.P., Rathod, S.B., Salve, P.S., 2021. Enhanced transdermal permeation of rasagiline mesylate nanoparticles: design, optimization, and effect of binary combinations of solvent systems across biological membrane. Int. J. Polym. Mater. Polym. Biomater. 70 (3), 158–173. https://doi.org/10.1080/00914037.2019.1706507.

Banerjee, S.S., Chen, D.H., 2008. Multifunctional pH-sensitive magnetic nanoparticles for simultaneous imaging, sensing and targeted intracellular anticancer drug delivery. Nanotechnology 19 (50). https://doi.org/10.1088/0957-4484/19/50/505104.

Barak, S., Mudgil, D., 2014. Locust bean gum: processing, properties and food applications-a review. Int. J. Biol. Macromol. 66, 74–80. https://doi.org/10.1016/j.ijbiomac.2014.02.017.

BeMiller, J.N., 2012. Glycoscience chemistry and chemical biology. In: Fraser-Reid, B., Tatsuta, K., Thiem, J. (Eds.), Gums and Related Polysaccharides. Springer, pp. 1–16. http://www.springerreference.com/index/chapterdbid/135002.

Bhardwaj, T.R., Kanwar, M., Lal, R., Gupta, A., 2000. Natural gums and modified natural gums as sustained-release carriers. Drug Dev. Ind. Pharm. 26 (10), 1025–1038. https://doi.org/10.1081/DDC-100100266.

Borker, S., Pokharkar, V., 2018. Engineering of pectin-capped gold nanoparticles for delivery of doxorubicin to hepatocarcinoma cells: an insight into mechanism of cellular uptake. Artif. Cells Nanomed. Biotechnol. 46 (2), 826–835. https://doi.org/10.1080/21691401.2018.1470525.

Brannon-Peppas, L., Blanchette, J.O., 2004. Nanoparticle and targeted systems for cancer therapy. Adv. Drug Deliv. Rev. 56 (11), 1649–1659. https://doi.org/10.1016/j.addr.2004.02.014.

Braz, L., Grenha, A., Ferreira, D., Rosa da Costa, A.M., Gamazo, C., Sarmento, B., 2017. Chitosan/sulfated locust bean gum nanoparticles: in vitro and in vivo evaluation towards an application in oral immunization. Int. J. Biol. Macromol. 96, 786–797. https://doi.org/10.1016/j.ijbiomac.2016.12.076.

Brezaniova, I., Trousil, J., Cernochova, Z., Kral, V., Hruby, M., Stepanek, P., Slouf, M., 2017. Self-assembled chitosan-alginate polyplex nanoparticles containing temoporfin. Colloid Polym. Sci. 295 (8), 1259–1270. https://doi.org/10.1007/s00396-016-3992-6.

Bumphrey, G., 1986. "Extremely useful" new suspending agent. Pharm. J. 237, 665–671.

Buschmann, M.D., Merzouki, A., Lavertu, M., Thibault, M., Jean, M., Darras, V., 2013. Chitosans for delivery of nucleic acids. Adv. Drug Deliv. Rev. 65 (9), 1234–1270. https://doi.org/10.1016/j.addr.2013.07.005.

Butcher, N.J., Mortimer, G.M., Minchin, R.F., 2016. Drug delivery: unravelling the stealth effect. Nat. Nanotechnol. 11 (4), 310–311. https://doi.org/10.1038/nnano.2016.6.

Campo, V.L., Kawano, D.F., Silva, D.B.d., Carvalho, I., 2009. Carrageenans: biological properties, chemical modifications and structural analysis—a review. Carbohydr. Polym. 77 (2), 167–180. https://doi.org/10.1016/j.carbpol.2009.01.020.

Cao, Z.T., Gan, L.Q., Jiang, W., Wang, J.L., Zhang, H.B., Zhang, Y., Wang, Y., Yang, X., Xiong, M., Wang, J., 2020. Protein binding affinity of polymeric nanoparticles as a direct Indicator of their pharmacokinetics. ACS Nano 14 (3), 3563–3575. https://doi.org/10.1021/acsnano.9b10015.

Carrstensen, H., Müller, R.H., Müller, B.W., 1992. Particle size, surface hydrophobicity and interaction with serum of parenteral fat emulsions and model drug carriers as parameters related to RES uptake. Clin. Nutr. 11 (5), 289–297. https://doi.org/10.1016/0261-5614(92)90006-C.

Casals, E., Pfaller, T., Duschl, A., Oostingh, G.J., Puntes, V., 2010. Time evolution of the nanoparticle protein corona. ACS Nano 4 (7), 3623–3632. https://doi.org/10.1021/nn901372t.

Chawla, P., Kumar, N., Bains, A., Dhull, S.B., Kumar, M., Kaushik, R., Punia, S., 2020. Gum arabic capped copper nanoparticles: synthesis, characterization, and applications. Int. J. Biol. Macromol. 146, 232–242. https://doi.org/10.1016/j.ijbiomac.2019.12.260.

Chen, X., Zhao, X., Gao, Y., Yin, J., Bai, M., Wang, F., 2018. Green synthesis of gold nanoparticles using carrageenan oligosaccharide and their in vitro antitumor activity. Mar. Drugs 16 (8). https://doi.org/10.3390/md16080277.

Cheung, R.C.F., Ng, T.B., Wong, J.H., Chan, W.Y., 2015. Chitosan: an update on potential biomedical and pharmaceutical applications. Mar. Drugs 13 (8), 5156–5186. https://doi.org/10.3390/md13085156.

Chourasia, M.K., Jain, S.K., 2004. Potential of guar gum microspheres for target specific drug release to colon. J. Drug Target. 12 (7), 435–442. https://doi.org/10.1080/10611860400006604.

Comba, S., Dalmazzo, D., Santagata, E., Sethi, R., 2011. Rheological characterization of xanthan suspensions of nanoscale iron for injection in porous media. J. Hazard. Mater. 185 (2–3), 598–605. https://doi.org/10.1016/j.jhazmat.2010.09.060.

Dasari, A., Guttena, V., 2016. Green synthesis, characterization, photocatalytic, fluorescence and antimicrobial activities of Cochlospermum gossypium capped Ag2S nanoparticles. J. Photochem. Photobiol. B Biol. 157, 57–69. https://doi.org/10.1016/j.jphotobiol.2016.02.002.

Deka, R., Sarma, S., Patar, P., Gogoi, P., Sarmah, J.K., 2020. Highly stable silver nanoparticles containing guar gum modified dual network hydrogel for catalytic and biomedical applications. Carbohydr. Polym., 248. https://doi.org/10.1016/j.carbpol.2020.116786.

Deshmukh, A.S., Aminabhavi, T.M., 2014. Pharmaceutical Applications of Various Natural Gums. Springer Science and Business Media LLC, pp. 1–30, https://doi.org/10.1007/978-3-319-03751-6_4-1.

Devi, P.R., Kumar, S.C., Selvamani, P., Subramanian, N., Ruckmani, K., 2015. Synthesis and characterization of Arabic gum capped gold nanoparticles for tumor-targeted drug delivery. Mater. Lett. 139, 241–244. https://doi.org/10.1016/j.matlet.2014.10.010.

Dhar, S., Maheswara Reddy, E., Shiras, A., Pokharkar, V., Prasad, B.L.V., 2008. Natural gum reduced/stabilized gold nanoparticles for drug delivery formulations. Chem. Eur. J. 14 (33), 10244–10250. https://doi.org/10.1002/chem.200801093.

Dhar, S., Murawala, P., Shiras, A., Pokharkar, V., Prasad, B.L.V., 2012. Gellan gum capped silver nanoparticle dispersions and hydrogels: cytotoxicity and in vitro diffusion studies. Nanoscale 4 (2), 563–567. https://doi.org/10.1039/c1nr10957j.

Diem, P.H.N., Thao, D.T.T., Phu, D.V., Duy, N.N., Quy, H.T.D., Hoa, T.T., Hien, N.Q., 2017. Synthesis of gold nanoparticles stabilized in dextran solution by gamma Co-60 ray irradiation and preparation of gold nanoparticles/dextran powder. J. Chem., 1–8. https://doi.org/10.1155/2017/6836375.

Dilbaghi, N., Kaur, H., Ahuja, M., Kumar, S., 2013. Evaluation of tropicamide-loaded tamarind seed xyloglucan nanoaggregates for ophthalmic delivery. Carbohydr. Polym. 94 (1), 286–291. https://doi.org/10.1016/j.carbpol.2013.01.054.

do Amaral Rodrigues, J., de Araújo, A.R., Pitombeira, N.A., Plácido, A., de Almeida, M.P., Veras, L.M.C., Delerue-Matos, C., Lima, F.C.D.A., Neto, A.B., de Paula, R.C.M., Feitosa, J.P.A., Eaton, P., Leite, J.R.S.A., da Silva, D.A., 2019. Acetylated cashew gum-based nanoparticles for the incorporation of alkaloid epiisopiloturine. Int. J. Biol. Macromol. 128, 965–972. https://doi.org/10.1016/j.ijbiomac.2019.01.206.

Dodi, G., Pala, A., Barbu, E., Peptanariu, D., Hritcu, D., Popa, M.I., Tamba, B.I., 2016. Carboxymethyl guar gum nanoparticles for drug delivery applications: preparation and preliminary in-vitro investigations. Mater. Sci. Eng. C 63, 628–636. https://doi.org/10.1016/j.msec.2016.03.032.

Dumitriu, 2005. Polysaccharides: Structural Diversity and Functional Versatility, second ed. vol. 1224 Marcel Dekker, CRC Press.

Ebrahiminezhad, A., Zare, M., Kiyanpour, S., Berenjian, A., Niknezhad, S.V., Ghasemi, Y., 2018. Biosynthesis of xanthangum-coated INPs by using *Xanthomonas campestris*. IET Nanobiotechnol. 12 (3), 254–258. https://doi.org/10.1049/iet-nbt.2017.0199.

Fam, S.Y., Chee, C.F., Yong, C.Y., Ho, K.L., Mariatulqabtiah, A.R., Tan, W.S., 2020. Stealth coating of nanoparticles in drug-delivery systems. Nano 10 (4). https://doi.org/10.3390/nano10040787.

Fleischer, C.C., Payne, C.K., 2014. Nanoparticle-cell interactions: molecular structure of the protein corona and cellular outcomes. Acc. Chem. Res. 47 (8), 2651–2659. https://doi.org/10.1021/ar500190q.

Frank, M.M., Fries, L.F., 1991. The role of complement in inflammation and phagocytosis. Immunol. Today 12 (9), 322–326. https://doi.org/10.1016/0167-5699(91)90009-I.

Gajalakshmi, B., Induja, S., Sivakumar, N., Raghavan, P.S., 2018. Guar gum stabilized copper oxide nanoparticles with enhanced thermal and antimicrobial properties. Asian J. Chem. 30 (5), 1099–1101. https://doi.org/10.14233/ajchem.2018.21200.

Ganeshkumar, M., Janani, M., Ponrasu, T., Suguna, L., 2018. Guar gum stabilized gold nanoparticles for colon cancer treatment. Clin. Oncol. 3, 153.

García-Ochoa, F., Santos, V.E., Casas, J.A., Gómez, E., 2000. Xanthan gum: production, recovery, and properties. Biotechnol. Adv. 18 (7), 549–579. https://doi.org/10.1016/S0734-9750(00)00050-1.

Goodarzi, N., Varshochian, R., Kamalinia, G., Atyabi, F., Dinarvand, R., 2013. A review of polysaccharide cytotoxic drug conjugates for cancer therapy. Carbohydr. Polym. 92 (2), 1280–1293. https://doi.org/10.1016/j.carbpol.2012.10.036.

Goyal, A.K., Garg, T., Rath, G., Gupta, U.D., Gupta, P., 2016. Chemotherapeutic evaluation of guar gum coated chitosan nanoparticle against experimental tuberculosis. J. Biomed. Nanotechnol. 12 (3), 450–463. https://doi.org/10.1166/jbn.2016.2180.

Gupta, A.P., Verma, D.K., 2014. Preparation and characterization of carboxymethyl guar gum nanoparticles. Int. J. Biol. Macromol. 68, 247–250. https://doi.org/10.1016/j.ijbiomac.2014.05.012.

Gustafson, H.H., Holt-Casper, D., Grainger, D.W., Ghandehari, H., 2015. Nanoparticle uptake: the phagocyte problem. Nano Today 10 (4), 487–510. https://doi.org/10.1016/j.nantod.2015.06.006.

Hadjesfandiari, N., Parambath, A., 2018. Stealth coatings for nanoparticles: polyethylene glycol alternatives. In: Engineering of Biomaterials for Drug Delivery Systems: Beyond Polyethylene Glycol. Elsevier Inc, pp. 345–361, https://doi.org/10.1016/B978-0-08-101750-0.00013-1.

Hajizadeh, S., Farhadi, K., Molaei, R., Forough, M., 2020. Silver nanoparticles-tragacanth gel as a green membrane for effective extraction and determination of capecitabine. J. Sep. Sci. 43 (13), 2666–2674. https://doi.org/10.1002/jssc.202000251.

Haque, S., Md, S., Sahni, J.K., Ali, J., Baboota, S., 2014. Development and evaluation of brain targeted intranasal alginate nanoparticles for treatment of depression. J. Psychiatr. Res. 48 (1), 1–12. https://doi.org/10.1016/j.jpsychires.2013.10.011.

Hassani, A., Mahmood, S., Enezei, H.H., Hussain, S.A., Hamad, H.A., Aldoghachi, A.F., Hagar, A., Doolaanea, A.A., Ibrahim, W.N., 2020. Formulation, characterization and biological activity screening of sodium alginate-gum Arabic nanoparticles loaded with curcumin. Molecules 25 (9). https://doi.org/10.3390/molecules25092244.

Hornig, S., Bunjes, H., Heinze, T., 2009. Preparation and characterization of nanoparticles based on dextran-drug conjugates. J. Colloid Interface Sci. 338 (1), 56–62. https://doi.org/10.1016/j.jcis.2009.05.025.

Iravani, S., 2020. Plant gums for sustainable and eco-friendly synthesis of nanoparticles: recent advances. Inorg. Nano-Metal Chem. 50 (6), 469–488. https://doi.org/10.1080/24701556.2020.1719155.

Jabri, T., Imran, M., Shafiullah, Rao, K., Ali, I., Arfan, M., Shah, M.R., 2018. Fabrication of lecithin-gum tragacanth muco-adhesive hybrid nano-carrier system for in-vivo performance of amphotericin B. Carbohydr. Polym. 194, 89–96. https://doi.org/10.1016/j.carbpol.2018.04.013.

Janes, K.A., Fresneau, M.P., Marazuela, A., Fabra, A., Alonso, M.J., 2001. Chitosan nanoparticles as delivery systems for doxorubicin. J. Control. Release 73 (2–3), 255–267. https://doi.org/10.1016/S0168-3659(01)00294-2.

Jiang, G., Park, K., Kim, J., Kim, K.S., Oh, E.J., Kang, H., Han, S.E., Oh, Y.K., Park, T.G., Hahn, S.K., 2008. Hyaluronic acid-polyethyleneimine conjugate for target specific intracellular delivery of siRNA. Biopolymers 89 (7), 635–642. https://doi.org/10.1002/bip.20978.

Joshy, K.S., Susan, M.A., Snigdha, S., Nandakumar, K., Laly, A.P., Sabu, T., 2018. Encapsulation of zidovudine in PF-68 coated alginate conjugate nanoparticles for anti-HIV drug delivery. Int. J. Biol. Macromol. 107, 929–937. https://doi.org/10.1016/j.ijbiomac.2017.09.078.

Kang, B., Opatz, T., Landfester, K., Wurm, F.R., 2015. Carbohydrate nanocarriers in biomedical applications: functionalization and construction. Chem. Soc. Rev. 44 (22), 8301–8325. https://doi.org/10.1039/c5cs00092k.

Kannan, R., Rahing, V., Cutler, C., Pandrapragada, R., Katti, K.K., Kattumuri, V., Robertson, J.D., Casteel, S.J., Jurisson, S., Smith, C., Boote, E., Katti, K.V., 2006. Nanocompatible chemistry toward fabrication of target-specific gold nanoparticles. J. Am. Chem. Soc. 128 (35), 11342–11343. https://doi.org/10.1021/ja063280c.

Kao, Y.J., Juliano, R.L., 1981. Interactions of liposomes with the reticuloendothelial system effects of reticuloendothelial blockade on the clearance of large unilamellar vesicles. Biochim. Biophys. Acta Gen. Subj. 677 (3–4), 453–461. https://doi.org/10.1016/0304-4165(81)90259-2.

Karagozlu, M.Z., Kim, S.K., 2014. Anticancer effects of chitin and chitosan derivatives. In: Advances in Food and Nutrition Research. 72. Academic Press Inc, pp. 215–225, https://doi.org/10.1016/B978-0-12-800269-8.00012-9.

Karakus, S., Ilgar, M., Tan, E., Sahin, Y.M., Tasaltin, N., Kilislioglu, A., 2019. The viscosity behaviour of PEGylated locust bean gum/rosin ester polymeric nanoparticles. In: Colloid Science in Pharmaceutical Nanotechnology. IntechOpen, https://doi.org/10.5772/intechopen.90248.

Karemore, M.N., Avari, J.G., 2019. In-situ gel of nifedipine for preeclampsia: optimization, in-vitro and in-vivo evaluation. J. Drug Deliv. Sci. Technol. 50, 78–89. https://doi.org/10.1016/j.jddst.2019.01.025.

Karemore, M.N., Bali, N.R., 2021. Gellan gum based gastroretentive tablets for bioavailability enhancement of cilnidipine in human volunteers. Int. J. Biol. Macromol. 174, 424–439. https://doi.org/10.1016/j.ijbiomac.2021.01.199.

Kattumuri, V., Katti, K., Bhaskaran, S., Boote, E.J., Casteel, S.W., Fent, G.M., Robertson, D.J., Chandrasekhar, M., Kannan, R., Katti, K.V., 2007. Gum arabic as a phytochemical construct for the stabilization of gold nanoparticles: in vivo pharmacokinetics and X-ray-contrast-imaging studies. Small 3 (2), 333–341. https://doi.org/10.1002/smll.200600427.

Katzbauer, B., 1998. Properties and applications of xanthan gum. Polym. Degrad. Stab. 59 (1–3), 81–84. https://doi.org/10.1016/s0141-3910(97)00180-8.

Kaur, H., Ahuja, M., Kumar, S., Dilbaghi, N., 2012. Carboxymethyl tamarind kernel polysaccharide nanoparticles for ophthalmic drug delivery. Int. J. Biol. Macromol. 50 (3), 833–839. https://doi.org/10.1016/j.ijbiomac.2011.11.017.

Khan, I., Saeed, K., Khan, I., 2019. Nanoparticles: properties, applications and toxicities. Arab. J. Chem. 12 (7), 908–931. https://doi.org/10.1016/j.arabjc.2017.05.011.

Khanna, V.K., 2012. Targeted delivery of nanomedicines. ISRN Pharmacol., 1–9. https://doi.org/10.5402/2012/571394.

Knop, K., Hoogenboom, R., Fischer, D., Schubert, U.S., 2010. Poly(ethylene glycol) in drug delivery: pros and cons as well as potential alternatives. Angew. Chem. Int. Ed. 49 (36), 6288–6308. https://doi.org/10.1002/anie.200902672.

Kong, J.M., Wong, C.V., Gao, Z.Q., Chen, X.T., 2008. Preparation of silver nanoparticles by microwave-hydrothermal technique. Synth. React. Inorg. Met.-Org. Nano-Met. Chem. 38 (2), 186–188. https://doi.org/10.1080/15533170801926218.

Kora, A.J., Arunachalam, J., 2012. Green fabrication of silver nanoparticles by gum tragacanth (astragalus gummifer): a dual functional reductant and stabilizer. J. Nanomater. 2012. https://doi.org/10.1155/2012/869765.

Kora, A.J., Sashidhar, R.B., Arunachalam, J., 2010. Gum kondagogu (*Cochlospermum gossypium*): a template for the green synthesis and stabilization of silver nanoparticles with antibacterial application. Carbohydr. Polym. 82 (3), 670–679. https://doi.org/10.1016/j.carbpol.2010.05.034.

Kumar, M.N.V.R., Muzzarelli, R.A.A., Muzzarelli, C., Sashiwa, H., Domb, A.J., 2004. Chitosan chemistry and pharmaceutical perspectives. Chem. Rev. 104 (12), 6017–6084. https://doi.org/10.1021/cr030441b.

Kumar, A., Aerry, S., Saxena, A., De, A., Mozumdar, S., 2012. Copper nanoparticulates in guar-gum: a recyclable catalytic system for the Huisgen [3 + 2]-cycloaddition of azides and alkynes without additives under ambient conditions. Green Chem. 14, 1298. https://doi.org/10.1039/c2gc35070j.

Kumar, P., Kumar, V., Kumar, R., Pruncu, C.I., 2020. Fabrication and characterization of ceftizoxime-loaded pectin nanocarriers. Nano 10 (8), 1–13. https://doi.org/10.3390/nano10081452.

Kumari, A.S., Maragoni, V., Dasari, A., Guttena, V., 2015. Green synthesis, characterization and catalytic activity of palladium nanoparticles by xanthan gum. Appl. Nanosci. 5 (3), 315–320. https://doi.org/10.1007/s13204-014-0320-7.

Lee, C., Ventola, M.S., 2012. The nanomedicine revolution: part 1: emerging concepts. P T 37 (9), 512–525.

Lee, Y.K., Choi, E.J., Webster, T.J., Kim, S.H., Khang, D., 2014. Effect of the protein corona on nanoparticles for modulating cytotoxicity and immunotoxicity. Int. J. Nanomedicine 10, 97–113. https://doi.org/10.2147/IJN.S72998.

Lemarchand, C., Gref, R., Couvreur, P., 2004. Polysaccharide-decorated nanoparticles. Eur. J. Pharm. Biopharm. 58 (2), 327–341. https://doi.org/10.1016/j.ejpb.2004.02.016.

Li, J., Huo, M., Wang, J., Zhou, J., Mohammad, J.M., Zhang, Y., Zhu, Q., Waddad, A.Y., Zhang, Q., 2012. Redox-sensitive micelles self-assembled from amphiphilic hyaluronic acid-deoxycholic acid conjugates for targeted intracellular delivery of paclitaxel. Biomaterials 33 (7), 2310–2320. https://doi.org/10.1016/j.biomaterials.2011.11.022.

Li, J., Zhai, J., Dyett, B., Yang, Y., Drummond, C.J., Conn, C.E., 2020. Effect of gum arabic or sodium alginate incorporation on the physicochemical and curcumin retention properties of liposomes. LWT-Food Sci. Technol. 139. https://doi.org/10.1016/j.lwt.2020.110571, 110571.

Liu, Z., Jiao, Y., Wang, Y., Zhou, C., Zhang, Z., 2008. Polysaccharides-based nanoparticles as drug delivery systems. Adv. Drug Deliv. Rev. 60 (15), 1650–1662. https://doi.org/10.1016/j.addr.2008.09.001.

Liu, Y., Qi, Q., Li, X., Liu, J., Wang, L., He, J., Lei, J., 2017. Self-assembled pectin-conjugated eight-arm polyethylene glycol-dihydroartemisinin nanoparticles for anticancer combination therapy. ACS Sustain. Chem. Eng. 5 (9), 8097–8107. https://doi.org/10.1021/acssuschemeng.7b01715.

Liu, Y., Rao, L., Zhang, H., Cen, Y., Cheng, K., 2018. Conjugation of vitamin E-TPGS and guar gum to carry borneol for enhancing blood–brain barrier permeability. J. Biomater. Appl. 33 (4), 590–598. https://doi.org/10.1177/0885328218799551.

Mahmoudi, M., Lynch, I., Ejtehadi, M.R., Monopoli, M.P., Bombelli, F.B., Laurent, S., 2011. Protein-nanoparticle interactions: opportunities and challenges. Chem. Rev. 111 (9), 5610–5637. https://doi.org/10.1021/cr100440g.

Martin, A., 2006. Textbook of Physical Pharmacy and Pharmaceutical Sciences, sixth ed. Lippincott Williams & Wilkins.

Martins, A.F., Facchi, S.P., Follmann, H.D.M., Pereira, A.G.B., Rubira, A.F., Muniz, E.C., 2014. Antimicrobial activity of chitosan derivatives containing N-quaternized moieties in its backbone: a review. Int. J. Mol. Sci. 15 (11), 20800–20832. https://doi.org/10.3390/ijms151120800.

McHugh, D.J., 2003. Carrageenan. In: A Guide to the Seaweed Industry: FAO Fisheries Technical Paper, pp. 61–72.

Mishra, B., Patel, B.B., Tiwari, S., 2010. Colloidal nanocarriers: a review on formulation technology, types and applications toward targeted drug delivery. Nanomedicine 6 (1), 9–24. https://doi.org/10.1016/j.nano.2009.04.008.

Mizrahy, S., Peer, D., 2012. Polysaccharides as building blocks for nanotherapeutics. Chem. Soc. Rev. 41 (7), 2623–2640. https://doi.org/10.1039/c1cs15239d.

Moghimi, S.M., Hamad, I., 2008. Liposome-mediated triggering of complement cascade. J. Liposome Res. 18 (3), 195–209. https://doi.org/10.1080/08982100802309552.

Moghimi, S.M., Patel, H.M., 1998. Serum-mediated recognition of liposomes by phagocytic cells of the reticuloendothelial system—the concept of tissue specificity. Adv. Drug Deliv. Rev. 32 (1–2), 45–60. https://doi.org/10.1016/S0169-409X(97)00131-2.

Muddineti, O.S., Kumari, P., Ajjarapu, S., Lakhani, P.M., Bahl, R., Ghosh, B., Biswas, S., 2016. Xanthan gum stabilized PEGylated gold nanoparticles for improved delivery of curcumin in cancer. Nanotechnology 27 (32). https://doi.org/10.1088/0957-4484/27/32/325101.

Müller, R.H., Wallis, K.H., Tröster, S.D., Kreuter, J., 1992. In vitro characterization of poly(methyl-methaerylate) nanoparticles and correlation to their in vivo fate. J. Control. Release 20 (3), 237–246. https://doi.org/10.1016/0168-3659(92)90126-c.

Nayak, A.K., Pal, D., 2011. Development of pH-sensitive tamarind seed polysaccharide-alginate composite beads for controlled diclofenac sodium delivery using response surface methodology. Int. J. Biol. Macromol. 49 (4), 784–793. https://doi.org/10.1016/j.ijbiomac.2011.07.013.

Nayak, A.K., Pal, D., 2017. Functionalization of tamarind gum for drug delivery. In: Thakur, V.K., Thakur, M.K. (Eds.), Functional Biopolymers. Springer International Publishing, pp. 25–56, https://doi.org/10.1007/978-3-319-66417-0_2.

Nazarzadeh Zare, E., Makvandi, P., Tay, F.R., 2019. Recent progress in the industrial and biomedical applications of tragacanth gum: a review. Carbohydr. Polym. 212, 450–467. https://doi.org/10.1016/j.carbpol.2019.02.076.

Necas, J., Bartosikova, L., 2013. Carrageenan: a review. Vet. Med. 58 (4), 187–205. https://doi.org/10.17221/6758-VETMED.

Nejatian, M., Abbasi, S., Azarikia, F., 2020. Gum tragacanth: structure, characteristics and applications in foods. Int. J. Biol. Macromol. 160, 846–860. https://doi.org/10.1016/j.ijbiomac.2020.05.214.

Ngo, D.H., Kim, S.K., 2014. Antioxidant effects of chitin, chitosan, and their derivatives. In: Advances in Food and Nutrition Research. vol. 73. Academic Press Inc, pp. 15–31, https://doi.org/10.1016/B978-0-12-800268-1.00002-0.

Nie, S., 2010. Understanding and overcoming major barriers in cancer nanomedicine. Nanomedicine 5 (4), 523–528. https://doi.org/10.2217/nnm.10.23.

Norman, M.E., Williams, P., Illum, L., 1992. Human serum albumin as a probe for surface conditioning (opsonization) of block copolymer-coated microspheres. Biomaterials 13 (12), 841–849. https://doi.org/10.1016/0142-9612(92)90177-P.

Novac, O., Lisa, G., Profire, L., Tuchilus, C., Popa, M.I., 2014. Antibacterial quaternized gellan gum based particles for controlled release of ciprofloxacin with potential dermal applications. Mater. Sci. Eng. C 35 (1), 291–299. https://doi.org/10.1016/j.msec.2013.11.016.

Oliveira, A.C.d.J., Araújo, A.R.d., Quelemes, P.V., Nadvorny, D., Soares-Sobrinho, J.L., Leite, J.R.S.d.A., da Silva-Filho, E.C., Silva, D.A.D., 2019. Solvent-free production of phthalated cashew gum for green synthesis of antimicrobial silver nanoparticles. Carbohydr. Polym. 213, 176–183. https://doi.org/10.1016/j.carbpol.2019.02.033.

Owens, D.E., Peppas, N.A., 2006. Opsonization, biodistribution, and pharmacokinetics of polymeric nanoparticles. Int. J. Pharm. 307 (1), 93–102. https://doi.org/10.1016/j.ijpharm.2005.10.010.

Pallavicini, P., Arciola, C.R., Bertoglio, F., Curtosi, S., Dacarro, G., D'Agostino, A., Ferrari, F., Merli, D., Milanese, C., Rossi, S., Taglietti, A., Tenci, M., Visai, L., 2017. Silver nanoparticles synthesized and coated with pectin: an ideal compromise for anti-bacterial and anti-biofilm action combined with wound-healing properties. J. Colloid Interface Sci. 498, 271–281. https://doi.org/10.1016/j.jcis.2017.03.062.

Palma, S.I.C.J., Carvalho, A., Silva, J., Martins, P., Marciello, M., Fernandes, A.R., del Puerto Morales, M., Roque, A.-C.A., 2015. Covalent coupling of gum arabic onto superparamagnetic iron oxide nanoparticles for MRI cell labeling: physicochemical and in vitro characterization. Contrast Media Mol. Imaging 10 (4), 320–328. https://doi.org/10.1002/cmmi.1635.

Pandey, S., Mishra, S.B., 2014. Catalytic reduction of p-nitrophenol by using platinum nanoparticles stabilised by guar gum. Carbohydr. Polym. 113, 525–531. https://doi.org/10.1016/j.carbpol.2014.07.047.

Paques, J.P., Sagis, L.M.C., van Rijn, C.J.M., van der Linden, E., 2014. Nanospheres of alginate prepared through w/o emulsification and internal gelation with nanoparticles of $CaCO_3$. Food Hydrocoll. 40, 182–188. https://doi.org/10.1016/j.foodhyd.2014.02.024.

Passirani, C., Barratt, G., Devissaguet, J.P., Labarre, D., 1998a. Interactions of nanoparticles bearing heparin or dextran covalently bound to poly(methyl methacrylate) with the complement system. Life Sci. 62 (8), 775–785. https://doi.org/10.1016/S0024-3205(97)01175-2.

Passirani, C., Barratt, G., Devissaguet, J.P., Labarre, D., 1998b. Long-circulating nanoparticles bearing heparin or dextran covalently bound to poly(methyl methacrylate). Pharm. Res. 15 (7), 1046–1050. https://doi.org/10.1023/A:1011930127562.

Patel, J., Maji, B., Moorthy, N.S.H.N., Maiti, S., 2020. Xanthan gum derivatives: review of synthesis, properties and diverse applications. RSC Adv. 10 (45), 27103–27136. https://doi.org/10.1039/d0ra04366d.

Patra, J.K., Das, G., Fraceto, L.F., Campos, E.V.R., Rodriguez-Torres, M.D.P., Acosta-Torres, L.S., Diaz-Torres, L.A., Grillo, R., Swamy, M.K., Sharma, S., Habtemariam, S., Shin, H.-S., 2018. Nano based drug delivery systems: recent developments and future prospects. J. Nanobiotechnol. 16, 1–33. https://doi.org/10.1186/s12951-018-0392-8.

Pitombeira, N.A.O., Veras Neto, J.G., Silva, D.A., Feitosa, J.P.A., Paula, H.C.B., De Paula, R.C.M., 2015. Self-assembled nanoparticles of acetylated cashew gum: characterization and evaluation as potential drug carrier. Carbohydr. Polym. 117, 610–615. https://doi.org/10.1016/j.carbpol.2014.09.087.

Pooja, D., Panyaram, S., Kulhari, H., Rachamalla, S.S., Sistla, R., 2014. Xanthan gum stabilized gold nanoparticles: characterization, biocompatibility, stability and cytotoxicity. Carbohydr. Polym. 110, 1–9. https://doi.org/10.1016/j.carbpol.2014.03.041.

Posocco, B., Dreussi, E., De Santa, J., Toffoli, G., Abrami, M., Musiani, F., Grassi, M., Farra, R., Tonon, F., Grassi, G., Dapas, B., 2015. Polysaccharides for the delivery of antitumor drugs. Materials 8 (5), 2569–2615. https://doi.org/10.3390/ma8052569.

Prasad, P.N., 2012. Introduction to Nanomedicine and Nanobioengineering, first ed. John Wiley & Sons, Inc.

Quito, E.M.P., Caro, R.R., Veiga, M.D., 2020. Carrageenan: drug delivery systems and other biomedical applications. Mar. Drugs 18, 583. https://doi.org/10.3390/md18110583.

Rahimi, M., Shafiei-Irannejad, V., Safa, D., K., & Salehi, R., 2018. Multi-branched ionic liquid-chitosan as a smart and biocompatible nano-vehicle for combination chemotherapy with stealth and targeted properties. Carbohydr. Polym. 196, 299–312. https://doi.org/10.1016/j.carbpol.2018.05.059.

Rampado, R., Crotti, S., Caliceti, P., Pucciarelli, S., Agostini, M., 2020. Recent advances in understanding the protein corona of nanoparticles and in the formulation of "stealthy" nanomaterials. Front. Bioeng. Biotechnol. 8, 166. https://doi.org/10.3389/fbioe.2020.00166.

Ranjbar-Mohammadi, M., Rahimdokht, M., Pajootan, E., 2019. Low cost hydrogels based on gum Tragacanth and TiO_2 nanoparticles: characterization and RBFNN modelling of methylene blue dye removal. Int. J. Biol. Macromol. 134, 967–975. https://doi.org/10.1016/j.ijbiomac.2019.05.026.

Rao, P.S., Ghosh, T.P., Krishna, S., 1946. Extraction and purification of tamarind seed polysaccharide. J. Sci. Ind. Res. 4, 705–713.

Rao, K., Imran, M., Jabri, T., Ali, I., Perveen, S., Shafiullah, Ahmed, S., Shah, M.R., 2017. Gum tragacanth stabilized green gold nanoparticles as cargos for Naringin loading: a morphological investigation through AFM. Carbohydr. Polym. 174, 243–252. https://doi.org/10.1016/j.carbpol.2017.06.071.

Rao, K., Roome, T., Aziz, S., Razzak, A., Abbas, G., Imran, M., Jabri, T., Gul, J., Hussain, M., Sikandar, B., Sharafat, S., Shah, M.R., 2018. Bergenin loaded gum xanthan stabilized silver nanoparticles suppress synovial inflammation through modulation of the immune response and oxidative stress in adjuvant induced arthritic rats. J. Mater. Chem. B 6 (27), 4486–4501. https://doi.org/10.1039/c8tb00672e.

Rastogi, L., Kora, A.J., Sashidhar, R.B., 2015. Antibacterial effects of gum kondagogu reduced/stabilized silver nanoparticles in combination with various antibiotics: a mechanistic approach. Appl. Nanosci. 5, 535–543. https://doi.org/10.1007/s13204-014-0347-9.

Riabov, V., Gudima, A., Kzhyshkowska, J., 2018. Macrophage-mediated foreign body responses. In: Vrana, N.E. (Ed.), Cell and Material Interface: Advances in Tissue Engineering, Biosensor, Implant, and Imaging Technologies. Taylor & Francis, p. 245.

Riehemann, K., Schneider, S.W., Luger, T.A., Godin, B., Ferrari, M., Fuchs, H., 2009. Nanomedicine—challenge and perspectives. Angew. Chem. Int. Ed. 48 (5), 872–897. https://doi.org/10.1002/anie.200802585.

Rivkin, I., Cohen, K., Koffler, J., Melikhov, D., Peer, D., Margalit, R., 2010. Paclitaxel-clusters coated with hyaluronan as selective tumor-targeted nanovectors. Biomaterials 31 (27), 7106–7114. https://doi.org/10.1016/j.biomaterials.2010.05.067.

Roser, M., Fischer, D., Kissel, T., 1998. Surface-modified biodegradable albumin nano- and microspheres. II: effect of surface charges on in vitro phagocytosis and biodistribution in rats. Eur. J. Pharm. Biopharm. 46 (3), 255–263. https://doi.org/10.1016/S0939-6411(98)00038-1.

Sagnella, S.M., Duong, H., Macmillan, A., Boyer, C., Whan, R., McCarroll, J.A., Davis, T.P., Kavallaris, M., 2014. Dextran-based doxorubicin nanocarriers with improved tumor penetration. Biomacromolecules 15 (1), 262–275. https://doi.org/10.1021/bm401526d.

Sahoo, R., Sahoo, S., Azizi, S., 2017. Tamarind seed polysaccharides and their nanocomposites for drug delivery: an economical, eco-friendly and novel approach. MJMS 2 (2), 32–40.

Salmaso, S., Caliceti, P., 2013. Stealth properties to improve therapeutic efficacy of drug nanocarriers. J. Drug Deliv., 1–19. https://doi.org/10.1155/2013/374252.

Sangnim, T., Limmatvapirat, S., Nunthanid, J., Sriamornsak, P., Sittikijyothin, W., Wannachaiyasit, S., Huanbutta, K., 2018. Design and characterization of clindamycin-loaded nanofiber patches composed of polyvinyl alcohol and tamarind seed gum and fabricated by electrohydrodynamic atomization. Asian J. Pharm. Sci. 13 (5), 450–458. https://doi.org/10.1016/j.ajps.2018.01.002.

Sashidhar, R.B., Raju, D., Karuna, R., 2015. Tree gum: gum kondagogu. In: Polysaccharides: Bioactivity and Biotechnology. Springer International Publishing, pp. 185–217, https://doi.org/10.1007/978-3-319-16298-0_32.

Schlenoff, J.B., 2014. Zwitteration: coating surfaces with zwitterionic functionality to reduce nonspecific adsorption. Langmuir 30 (32), 9625–9636. https://doi.org/10.1021/la500057j.

Senior, J.H., 1987. Fate and behavior of liposomes in vivo: a review of controlling factors. Crit. Rev. Ther. Drug Carrier Syst. 3 (2), 123–193.

Shanmuga, S., Singhal, M., Sen, S., 2015. Synthesis and characterization of carrageenan coated magnetic nanoparticles for drug delivery applications. Transl. Biomed. 6 (3), 1–5. https://doi.org/10.21767/2172-0479.100019.

Sharon, N., Lis, H., 1993. Carbohydrates in cell recognition. Sci. Am. 268 (1), 82–89. https://doi.org/10.1038/scientificamerican0193-82.

Sheng, Y., Liu, C., Yuan, Y., Tao, X., Yang, F., Shan, X., Zhou, H., Xu, F., 2009. Long-circulating polymeric nanoparticles bearing a combinatorial coating of PEG and water-soluble chitosan. Biomaterials 30 (12), 2340–2348. https://doi.org/10.1016/j.biomaterials.2008.12.070.

Sheorain, J., Mehra, M., Thakur, R., Grewal, S., Kumari, S., 2018. In vitro anti-inflammatory and antioxidant potential of thymol loaded bipolymeric (tragacanth gum/chitosan) nanocarrier. Int. J. Biol. Macromol. https://doi.org/10.1016/j.ijbiomac.2018.12.095.

Shinde, M., Bali, N., Rathod, S., Karemore, M., Salve, P., 2020. Effect of binary combinations of solvent systems on permeability profiling of pure agomelatine across rat skin: a comparative study with statistically optimized polymeric nanoparticles. Drug Dev. Ind. Pharm. 46 (5), 826–845. https://doi.org/10.1080/03639045.2020.1757697.

Soumya, R.S., Vineetha, V.P., Reshma, P.L., Raghu, K.G., 2013. Preparation and characterization of selenium incorporated guar gum nanoparticle and its interaction with H9c2 cells. PLoS One 8 (9). https://doi.org/10.1371/journal.pone.0074411.

Sugahara, S, Kajiki, M, Kuriyama, H, Kobayashi, T, 2008. Carrier Effects on Antitumor Activity and Neurotoxicity of AZ10992, a Paclitaxel–Carboxymethyl Dextran Conjugate, in a Mouse Model. Biological & Pharmaceutical Bulletin 31, 223–230. https://doi.org/10.1248/bpb.31.223.

Sworn, G., 2009. Gellan gum. In: Handbook of Hydrocolloids, second ed. Elsevier Inc, pp. 204–227, https://doi.org/10.1533/9781845695873.204.

Szekalska, M., Puciłowska, A., Szymańska, E., Ciosek, P., Winnicka, K., 2016. Alginate: current use and future perspectives in pharmaceutical and biomedical applications. Int. J. Polym. Sci. 2016. https://doi.org/10.1155/2016/7697031.

Tang, Y., Wang, X., Li, J., Nie, Y., Liao, G., Yu, Y., Li, C., 2019. Overcoming the reticuloendothelial system barrier to drug delivery with a "don't-eat-us" strategy. ACS Nano 13 (11), 13015–13026. https://doi.org/10.1021/acsnano.9b05679.

Telegeev, G, Kutsevol, N, Chumachenko, V, Naumenko, A, Telegeeva, P, Filipchenko, S, Harahuts, Yu, 2017. Dextran-Polyacrylamide as Matrices for Creation of Anticancer Nanocomposite. International Journal of Polymer Science, 1–9. https://doi.org/10.1155/2017/4929857.

Thau, L., Asuka, E., Mahajan, K., 2020. In: Pearls, S. (Ed.), Physiology, Opsonization. Stat Pearls, Treasure Island.

Torchilin, V.P., 1998. Polymer-coated long-circulating microparticulate pharmaceuticals. J. Microencapsul. 15 (1), 1–19. https://doi.org/10.3109/02652049809006831.

Tummalapalli, M., Deopura, B.L., Alam, M.S., Gupta, B., 2015. Facile and green synthesis of silver nanoparticles using oxidized pectin. Mater. Sci. Eng. C 50, 31–36. https://doi.org/10.1016/j.msec.2015.01.055.

Varvarà, P., Tranchina, L., Cavallaro, G., Licciardi, M., 2020. Preparation and characterization of gold nanorods coated with gellan gum and lipoic acid. Appl. Sci. 10 (23), 1–18. https://doi.org/10.3390/app10238322.

Vieira, S., Vial, S., Maia, F.R., Carvalho, M., Reis, R.L., Granja, P.L., Oliveira, J.M., 2015. Gellan gum-coated gold nanorods: an intracellular nanosystem for bone tissue engineering. RSC Adv. 5 (95), 77996–78005. https://doi.org/10.1039/c5ra13556g.

Wani, T.U., Raza, S.N., Khan, N.A., 2020. Nanoparticle opsonization: forces involved and protection by long chain polymers. Polym. Bull. 77 (7), 3865–3889. https://doi.org/10.1007/s00289-019-02924-7.

Wasiak, I., Kulikowska, A., Janczewska, M., Michalak, M., Cymerman, I.A., Nagalski, A., et al., 2016. Dextran nanoparticle synthesis and properties. PLoS ONE 11 (1), 1–17. https://doi.org/10.1371/journal.pone.0146237.

Wen, Y., Oh, J.K., 2014. Recent strategies to develop polysaccharide-based nanomaterials for biomedical applications. Macromol. Rapid Commun. 35 (21), 1819–1832. https://doi.org/10.1002/marc.201400406.

Wong, H.L., Wu, X.Y., Bendayan, R., 2012. Nanotechnological advances for the delivery of CNS therapeutics. Adv. Drug Deliv. Rev. 64 (7), 686–700. https://doi.org/10.1016/j.addr.2011.10.007.

Xue, D., Sethi, R., 2012. Viscoelastic gels of guar and xanthan gum mixtures provide long-term stabilization of iron micro- and nanoparticles. J. Nanopart. Res. 14 (11). https://doi.org/10.1007/s11051-012-1239-0.

Yuan, Z.X., Zhang, Z.R., Zhu, D., Sun, X., Gong, T., Liu, J., Luan, C.T., 2009. Specific renal uptake of randomly 50% A/-acetylated low molecular weight chitosan. Mol. Pharm. 6 (1), 305–314. https://doi.org/10.1021/mp800078a.

Zhang, X.G., Teng, D.Y., Wu, Z.M., Wang, X., Wang, Z., Yu, D.M., Li, C.X., 2008. PEG-grafted chitosan nanoparticles as an injectable carrier for sustained protein release. J. Mater. Sci. Mater. Med. 19 (12), 3525–3533. https://doi.org/10.1007/s10856-008-3500-8.

Zhang, L., Yu, F., Cole, A.J., Chertok, B., David, A.E., Wang, J., Yang, V.C., 2009. Gum arabic-coated magnetic nanoparticles for potential application in simultaneous magnetic targeting and tumor imaging. AAPS J. 11 (4), 693–699. https://doi.org/10.1208/s12248-009-9151-y.

Zhang, Z., Ortiz, O., Goyal, R., Kohn, J., 2014. Biodegradable polymers. In: Handbook of Polymer Applications in Medicine and Medical Devices. Elsevier Inc, pp. 303–335, https://doi.org/10.1016/B978-0-323-22805-3.00013-X.

Zhang, P., Zhao, S.R., Li, J.X., Hong, L., Raja, M.A., Yu, L.J., Liu, C.G., 2016. Nanoparticles based on phenylalanine ethyl ester-alginate conjugate as vitamin B2 delivery system. J. Biomater. Appl. 31 (1), 13–22. https://doi.org/10.1177/0885328216630497.

Zhao, X.J., Zhou, Z.Q., 2016. Synthesis and applications of pectin-based nanomaterials. Curr. Nanosci. 12 (1), 103–109. https://doi.org/10.2174/1573413711666150818224020.

Zia, K.M., Tabasum, S., Nasif, M., Sultan, N., Aslam, N., Noreen, A., Zuber, M., 2017. A review on synthesis, properties and applications of natural polymer based carrageenan blends and composites. Int. J. Biol. Macromol. 96, 282–301. https://doi.org/10.1016/j.ijbiomac.2016.11.095.

CHAPTER

2

Micro- and nanoscale drug delivery systems based on xanthan gum hydrogels

Ljiljana Djekic and Ana Ćirić

University of Belgrade, Faculty of Pharmacy, Belgrade, Serbia

2.1 Introduction

Xanthan gum (XG) is a natural extracellular polysaccharide (exopolysaccharide) produced by a gram-negative bacteria genus *Xanthomonas*. Since the early 1960s, XG has been produced on an industrial scale by the aerobic fermentation of glucose, sucrose, molasses, sugar cane, corn or other complex substrates, in the presence of the bacteria *Xanthomonas campestris* (El-Sawy et al., 2020; Hajikhani et al., 2019; Petri, 2015; Raafat et al., 2018). Some of the manufacturers of the commercially available XG are CP Kelco (United States), Jungbunzlauer Suisse AG (Switzerland), and Solvay S.A. (Belgium). XG for *in vivo* applications undergoes a strict purification procedure to achieve a high degree of purity (Han et al., 2012). It is a widely available nontoxic polymer approved by the FDA in 1969 as a food additive without any specific quantity limitations (Hajikhani et al., 2019; Petri, 2015). In addition, in Europe, it is listed in the European List of Permitted Food Additives (as E 415) and in accordance with the Joint WHO/FAO Expert Committee on Food Additives (JECFA), it is classified as Acceptable Daily Intake (ADI)-nonspecified additive nontoxic for human consumption (Gils et al., 2009). It has become one of the widely used pharmaceutical excipients which is highly functional, relatively inexpensive, and acceptable for different routes of administration. Moreover, XG is a biodegradable and bioadhesive excipient with good stability at different temperatures and under both acidic and alkaline conditions (El-Sawy et al., 2020; Gils et al., 2009; Kang et al., 2019; Morariu et al., 2018). In recent years, it has received great attention from researchers in the field of drug delivery (Ahmad et al., 2019; Cascone and Lamberti, 2020; Singhvi et al., 2019).

Micro- and Nanoengineered Gum-Based Biomaterials for Drug Delivery and Biomedical Applications
https://doi.org/10.1016/B978-0-323-90986-0.00007-8

XG is a branched polysaccharide that consists of a main linear chain of β-(1,4)-D-glucose (that is similar to cellulose backbone) and trisaccharide side chains consisting of two mannose residues with a glucuronic acid residue between them (in a 2:2:1 molar ratio). The mannose-(β-1,4) glucuronic acid-(β-1,2)-mannose side chains are attached to alternate glucose residues in the main chain by α-1,3 linkages (Fig. 2.1) (Abu Elella et al., 2020; El-Sawy et al., 2020; Kennedy et al., 2015; Kumar et al., 2018; Morariu et al., 2018; Raafat et al., 2018). D-mannose unit linked to the main chain contains an acetyl group at position O-6, while approximately half of the terminal D-mannose contains a pyruvic acid residue linked *via* keto group to the 4 and 6 positions, with an unknown distribution.

The molecular weight distribution of XG ranges from 2×10^6 to 20×10^6 g/mol, depending on the biosynthesis conditions and degree of a self-association between several individual chains (Petri, 2015). The native state of the polymer is a double-stranded, right-handed fivefold helix. XG is soluble in cold and hot water; however, it dissolves slowly and intensive agitation is required to avoid the formation of *fish-eyes* (a gelatinous layer of partially hydrated XG formed on the outside of the particle which prevents penetration of water towards the particle to complete swelling and dissolution of the polymer) (Kumar et al., 2018). The mesh size of the polymer can influence its dispersibility and dissolution rate, so commercially available XG are often milled to a predetermined mesh size in order to control these parameters (García-Ochoa et al., 2000). The aqueous solutions are stable over a wide range

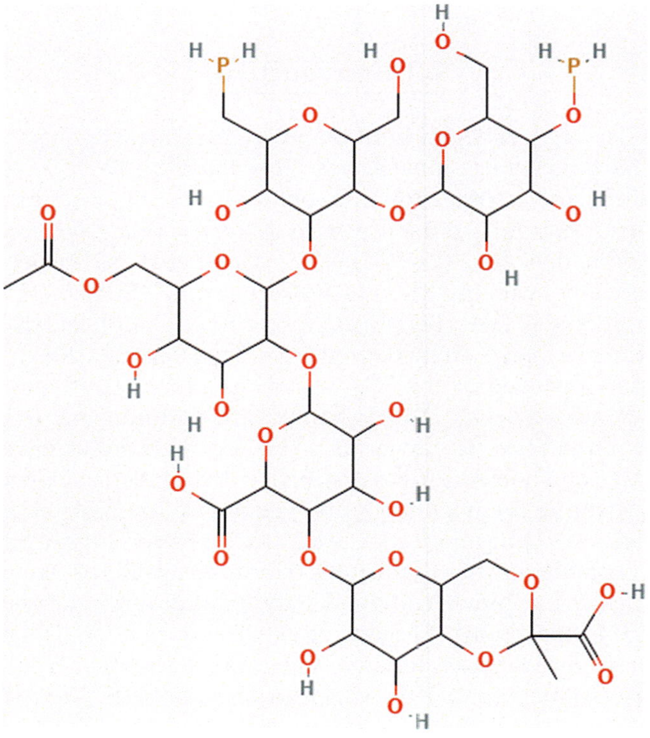

FIG. 2.1 Chemical structure of XG (https://pubchem.ncbi.nlm.nih.gov/compound/XC-Polymer).

of pH values (from 2 to 11), temperatures (up to 90 °C), and salt concentrations. In aqueous solutions at pH > 4.5 XG has polyanionic characteristics due to the presence of glucuronic acid and pyruvic acid groups in the side chain (Gils et al., 2009; Petri, 2015), thus physical interactions with positively charged (polycationic) polymers are possible. The common counter-ions in XG, stemming from the fermentation process, are sodium (Na^+), calcium (Ca^{2+}), or potassium (K^+) (Raafat et al., 2018). Therefore, as a polyelectrolyte, its conformation in the aqueous solution is a function of pH and ionic strength (Kennedy et al., 2015), but also it may be affected by temperature. At temperatures below the midpoint transition temperature (Tm ~ 40–50 °C) XG solutions exhibit a weak gel-like behavior and shear-thinning character when exposed to stress, because the ordered double helical strand structure forms three-dimensional (3D) networks (Ahmad et al., 2019; Iijima et al., 2007; Zhang et al., 2015). XG undergoes a transition from the ordered to the disordered (coil) conformation as well as to ordered (double helix) conformation as a function of temperature and ionic strength. In addition, salt stabilizes the ordered conformation, so at low ionic strength or high temperature, chains assume coil conformation, and on the other hand, under high ionic strength or low temperature, chains are arranged in helical conformation (Kumar et al., 2018; Morariu et al., 2018; Petri, 2015). XG in disordered form is available for catalysis of the main chain by fungal cellulases and it takes 2 days for complete biodegradation, while in ordered form the trisaccharide side chains of XG are likely to be a barrier to the common cellulases (Milas and Rinaudo, 1983). In an aqueous solution at 25 °C, the molecular conformation of the main chain of XG is extended, but in a partially ordered form (randomly broken helix), due to the two types of possible interactions between the chains: electrostatic repulsive interactions and attractive interactions by hydrogen bonds. Aqueous solutions of XG are pseudoplastic non-Newtonian fluids with a relatively high intrinsic viscosity due to its high molecular weight, helical structure and formation of a complex network of relatively weakly bound molecules by hydrogen bonds (Huang et al., 2018; Morariu et al., 2018). Rheological properties of XG aqueous solutions can be affected by the pyruvate and acetate content. Lower pyruvate and higher acetate contents result in lower viscosity of XG solution. In addition, XG solutions in deionized water have higher viscosity compared to those prepared with tap or groundwater, possibly due to the ionic strength differences between the water sources (García-Ochoa et al., 2000; Kumar et al., 2018; Singhvi et al., 2019). Although XG solutions show weak gel-like properties (Kang et al., 2019), a viscoelastic polymeric 3D network typical for hydrogels does not form, even at high polymer concentrations (Ahmad et al., 2019; Cascone and Lamberti, 2020; Huang et al., 2018; Kang et al., 2019). Despite a restriction for the formation of the rigid physically cross-linked homopolymeric hydrogels, XG has a large number of hydroxyl groups and carboxylic groups available for chemical cross-linking in presence of a suitable cross-linking agent to form such hydrogels. Moreover, it can form stable copolymeric hydrogels by conjugation with other polymers (polysaccharides) *via* physical interactions (e.g., hydrophobic, dipole-dipole and electrostatic interactions, hydrogen bonding) or by copolymerization with synthetic polymers in presence of appropriate cross-linking agents (El-Sawy et al., 2020; Kalia and Roy Choudhury, 2019). Additionally, XG can be chemically modified by carboxymethylation or different grafting methods (Badwaik et al., 2013). The combining of XG or corresponding chemical derivatives with other polymers represents a versatile strategy for tuning the desired properties of the hydrogels in accordance with the intended purpose (Gilbert et al., 2013; Vianna-Filho et al., 2013). There has been great interest in designing various XG-based physically cross-linked or chemically cross-linked homopolymeric and copolymeric hydrogels which could be utilized in the development of various drug delivery systems (bioadhesive/mucoadhesive semisolid hydrogels, film-forming hydrogels, beads, films, aerogels, microparticles, nanoparticles, injectable formulations, etc.) with different rheological and thermal properties, swelling capacity in aqueous media, and sensitivity to respond to the external stimuli (e.g., pH, temperature or specific molecules from the external environment) and release the drug substance. During the last few years, the scientists' focus has been expanded on the XG-based nanocomposite hydrogels, which represent hybrids of XG-based hydrogels and nanomaterials such as nanoparticles, microemulsions, and nanoemulsions. Their potential to enable significant improvements in stability and drug delivery performances compared to both hydrogels and nanocarriers is already recognized. This chapter describes a number of examples of biocompatible homopolymeric, copolymeric and nanocomposite hydrogels based on XG or modified XG, which are considered as drug carriers for oral, parenteral, and topical administration, with special reference to their preparation/synthesis, physicochemical characteristics, and capacity to enhance drug delivery.

2.2 Homopolymeric XG-based hydrogels

Homopolymeric hydrogels with XG as a single gelling agent in the system can be formulated by the introduction of suitable cross-linkers such as metal ions or small molecules with reactive end groups. The addition of a cross-linking agent increases the number of network connections between the XG molecules and thus restricts their movements and transforms the liquid polymer solution into the viscoelastic hydrogel (Kalia and Roy Choudhury, 2019). Cross-linkers must be nontoxic and biocompatible for the given route of administration. The number of such agents is limited, and some of them considered so far are trivalent iron ions (Fe^{3+}) (Kang et al., 2019), sodium trimetaphosphate (STMP) (Tao et al., 2016), and citric acid (CA) (Bueno et al., 2013; Bueno and Petri, 2014; Reddy and Yang, 2010).

Kang et al. have demonstrated that the cross-linking of XG with Fe^{3+} could be a simple and effective method for the preparation of layered hydrogels (Kang et al., 2019). Such hydrogels were prepared by soaking 0.01–0.07 g/mL XG aqueous solutions (poured into special molds with rectangular grooves or penicillin bottles) in aqueous iron(III) chloride solution (0.05 M) for 18 h. Trivalent iron ions integrate with the XG solution and coordinate with the carbonyl group (intra- or inter-) of the polymer molecules with a rigid and rod-like double-helical structure, to form a hydrogel network. Scanning electron microscopy (SEM) gives evidence of the uniform layered structure which was related to the gradual diffusion of Fe^{3+} through the XG solution from top to bottom and the formation of gel layers at the interface and subsequent permeation of Fe^{3+} into the XG solution under osmotic pressure to form the next layer of the hydrogel. For the obtained hydrogels, the enhanced mechanical strength was observed with the highest measured tensile strength of 62 MPa. In addition, they possessed good swelling property and the swelling degree decreased with increasing cross-linking degree and XG concentration. The iron has two oxidation states, so the developed hydrogels are recommended for further evaluation as potential redox-responsive carriers that enable drug release as a response to reducing agents present in surrounding biological environment. Moreover, with the incorporation of sodium lactate, the investigated hydrogels could exhibit reversible sol-gel conversion under ultraviolet (UV) light irradiation, which can also be a useful mechanism for controlling the delivery of the active substance.

Another approach for the formation of homopolymeric XG-based hydrogels is by cross-linking hydroxyl groups of the polymer with a nontoxic and water-soluble STMP in alkaline medium. Tao et al. monitored *in situ* cross-linking process by dynamical oscillation tests to evaluate the forming kinetics and mechanical stability of the XG-STMP hydrogels during the addition of an aqueous solution of STMP to 2%–5% (w/v) XG solutions in 0.1 M sodium hydroxide. The transition from hydrogen bonding to chemical cross-linking of the XG chains was observed at 37 °C. The structure of the lyophilized XG-STMP hydrogels was porous and interconnected. They were swollen with an aqueous solution of bovine serum albumin (BSA) as a model protein and after that rinsed with phosphate buffer saline (PBS, pH 7.4) and immersed into PBS media at 25 °C to assess the BSA release diffusion profile. BSA diffused gradually into the PBS medium in a sustained manner, and the release profiles showed a different trend with the increase of XG concentration in the investigated hydrogels (Tao et al., 2016).

CA is a cheap and nontoxic cross-linker (Reddy and Yang, 2010). XG hydrogels which were produced by condensation at 165 °C in the presence of CA (Fig. 2.2) showed homogeneous porous structure and practically no nanofibrils, higher cross-linking, gel content and charge densities (Fig. 2.3), in comparison with those prepared without CA where heterogeneous porous structure, lower gel content and nanofibrils (diameter $\sim$ 20–30 nm) were observed (Fig. 2.3) because intra- and intermolecular ester bonds are formed due to XG chains dehydration and trans-esterification (Fig. 2.2) (Bueno et al., 2013). The circular dichroism (CD) spectroscopic analysis of the conformation of cross-linked XG chains in the swollen networks prepared with CA showed that the polymer chains are coiled, but upon increasing the ionic strength they undergo a conformational transition to ordered state. Films and aerogels prepared from the XG hydrogels cross-linked with CA that were immersed in 0.1 M hydrochloric acid (pH 1) retain their stability and original swollen form for 6 months. The pH value of the aqueous media significantly affected the pores' morphology of the XG-CA hydrogel (Fig. 2.4). The pore size at pH 6.5 was decreased in comparison with those at pH 2. At pH $\sim$ 10 the pore size increased significantly and the morphological features of the XG-CA hydrogels (Bueno et al., 2013; Bueno and Petri, 2014).

2.3 Copolymeric XG-based hydrogels

Copolymeric hydrogels include other natural or synthetic polymers beside XG. The ordered conformation of XG in solution is available for hydrogen bonding, intermolecular electrostatic (ionic) interactions, van der Waals, and hydrophobic interactions with a second biocompatible polysaccharide (Berger et al., 2004; De Robertis et al., 2015; Sharma and Tiwari, 2020). Numerous novel drug delivery systems are based on physically (noncovalently) cross-linked polymers due to their low toxicity, biocompatibility, and ease of preparation (Berger et al., 2004). These drug delivery systems have been developed primarily in order to achieve controlled and predictable release of drugs with zero-order release kinetics (De Robertis et al., 2015). Synthetization of hydrogels from two polymers generally resulted in hybrid materials with improved mechanical, thermal, and microbiological stability, functionality, gelation behavior, and solubility in physiological fluids. Moreover, the chemical cross-linking of XG with both natural and synthetic polymers is a novel approach in development of the drug carriers with tailored drug delivery performances and therapeutical effectiveness as well to reduce

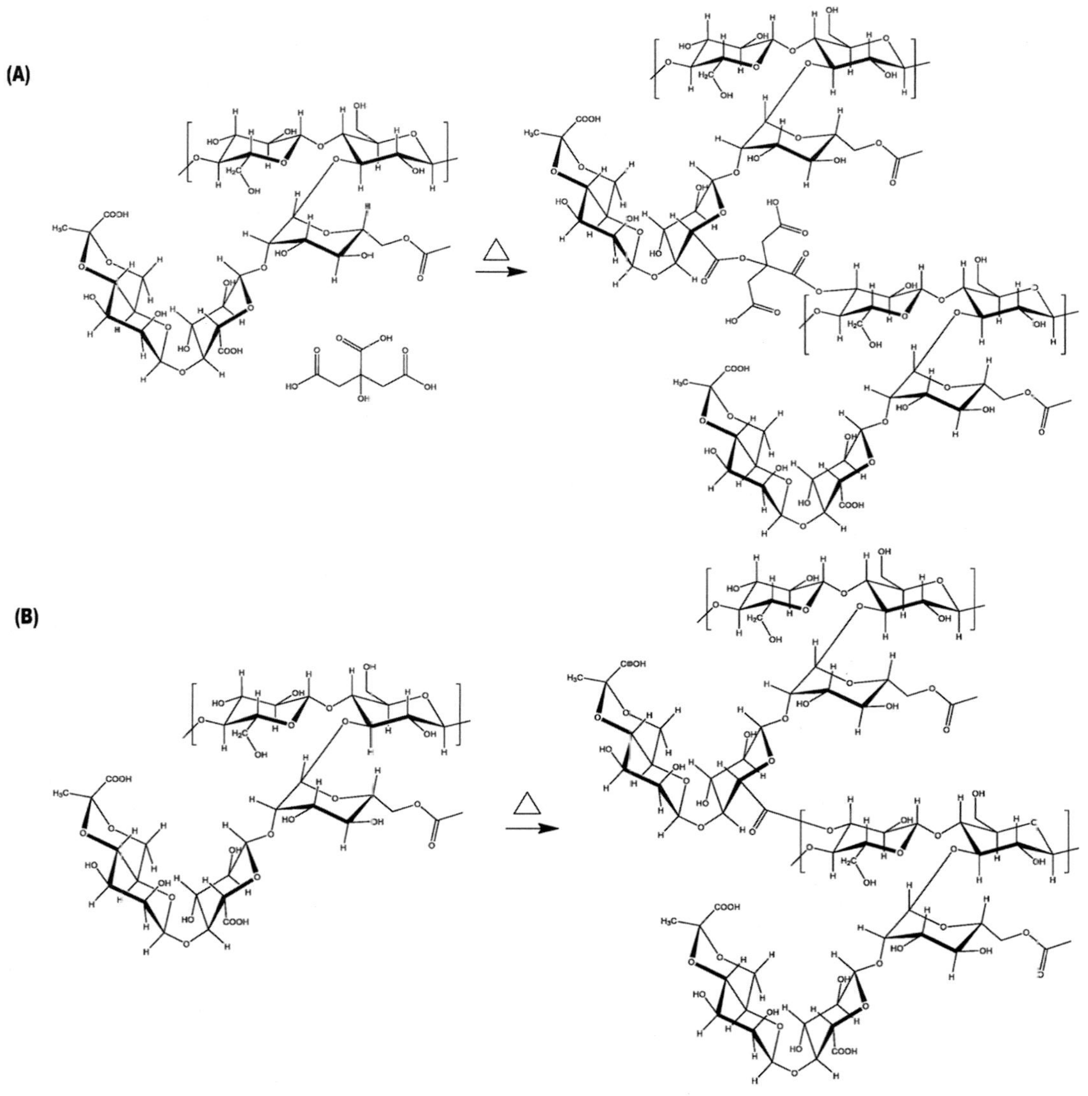

FIG. 2.2 Scheme of (A) XG cross-linking reaction with citric acid and (B) XG cross-linking reaction (dehydration). *With permission from Bueno, V.B., Bentini, R., Catalani, R.H., Petri, D.F.S., 2013. Synthesis and swelling behavior of xanthan-based hydrogels. Carbohyd. Polym. 92, 1091–1099.*

production costs (Hajikhani et al., 2019; Kalia and Roy Choudhury, 2019; Simões et al., 2020; Tao et al., 2016; Zheng et al., 2019a, b).

2.3.1 Physically cross-linked copolymeric XG-based hydrogels

The strongest noncovalent interactions in obtaining physically cross-linked copolymeric hydrogels are ionic interactions between oppositely charged polymers, resulting in the

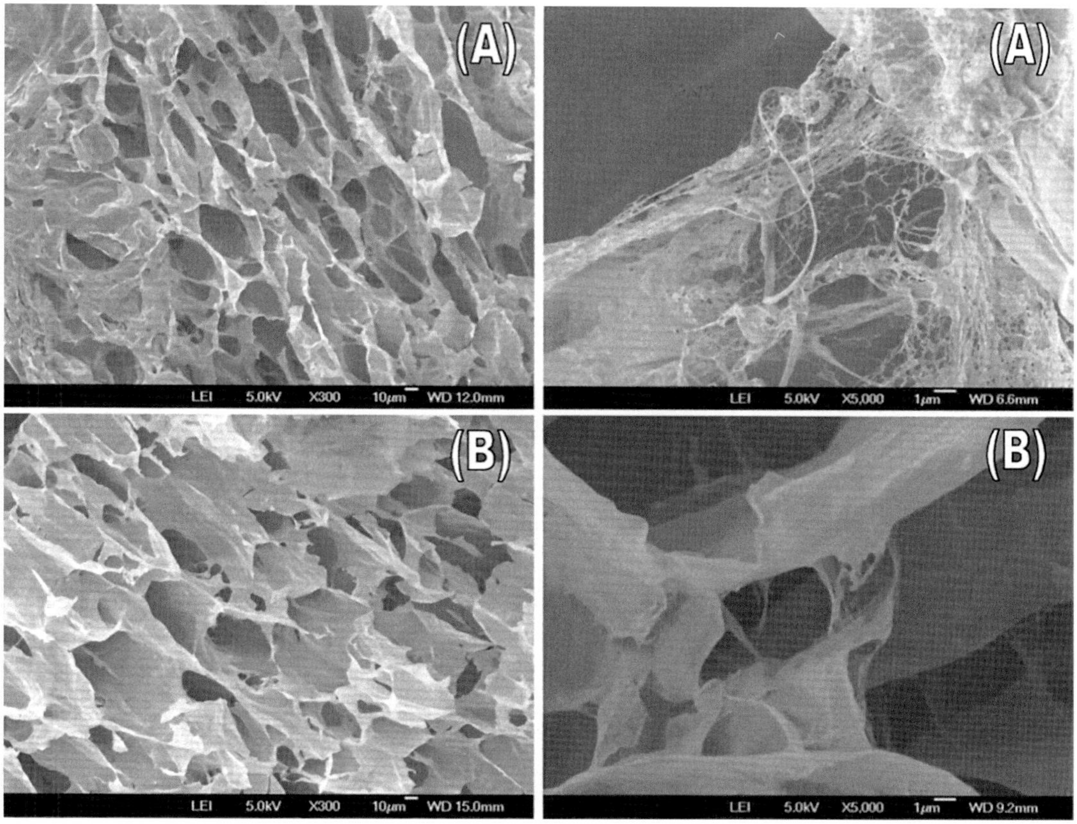

FIG. 2.3 SEM images obtained for (A) XG and (B) XG-CA hydrogels, each one in two different magnifications. *With permission from Bueno, V.B., Bentini, R., Catalani, R.H., Petri, D.F.S., 2013. Synthesis and swelling behavior of xanthan-based hydrogels. Carbohyd. Polym. 92, 1091–1099.*

formation of polyelectrolyte complexes (PECs). Although hydrogen, van der Waals, and hydrophobic interactions are weaker than ionic ones, they have also been investigated in the formulation of controlled-release drug delivery systems (Berger et al., 2004; Singhvi et al., 2019). All mentioned interactions are reversible and the obtained complexes are usually sensitive to external environmental stimuli, e.g., pH, temperature, ionic strength.

XG is a hydrophilic anionic polymer widely explored for the formulation of drug delivery systems based on PECs with chitosan (CH) (Luo and Wang, 2014). Many research groups synthesized CH/XG-based PEC hydrogels with promising pharmaceutical relevance. They are usually intended for the development of oral drug formulations due to their pH-dependent swelling and drug release (Merlusca et al., 2016; Nilesh et al., 2015), which will be discussed in more detail in Section 2.5 of this chapter. The interactions between polysaccharides and performances of the hydrogels are affected by several formulation factors including concentrations of the polymers and their molar ratio, pH and ionic strength of medium, temperature, and agitation. Magnin et al. applied rheological measurements to explore the complexation of CH and XG by the coacervation method. The kinetic curve clearly showed a classic sol-gel

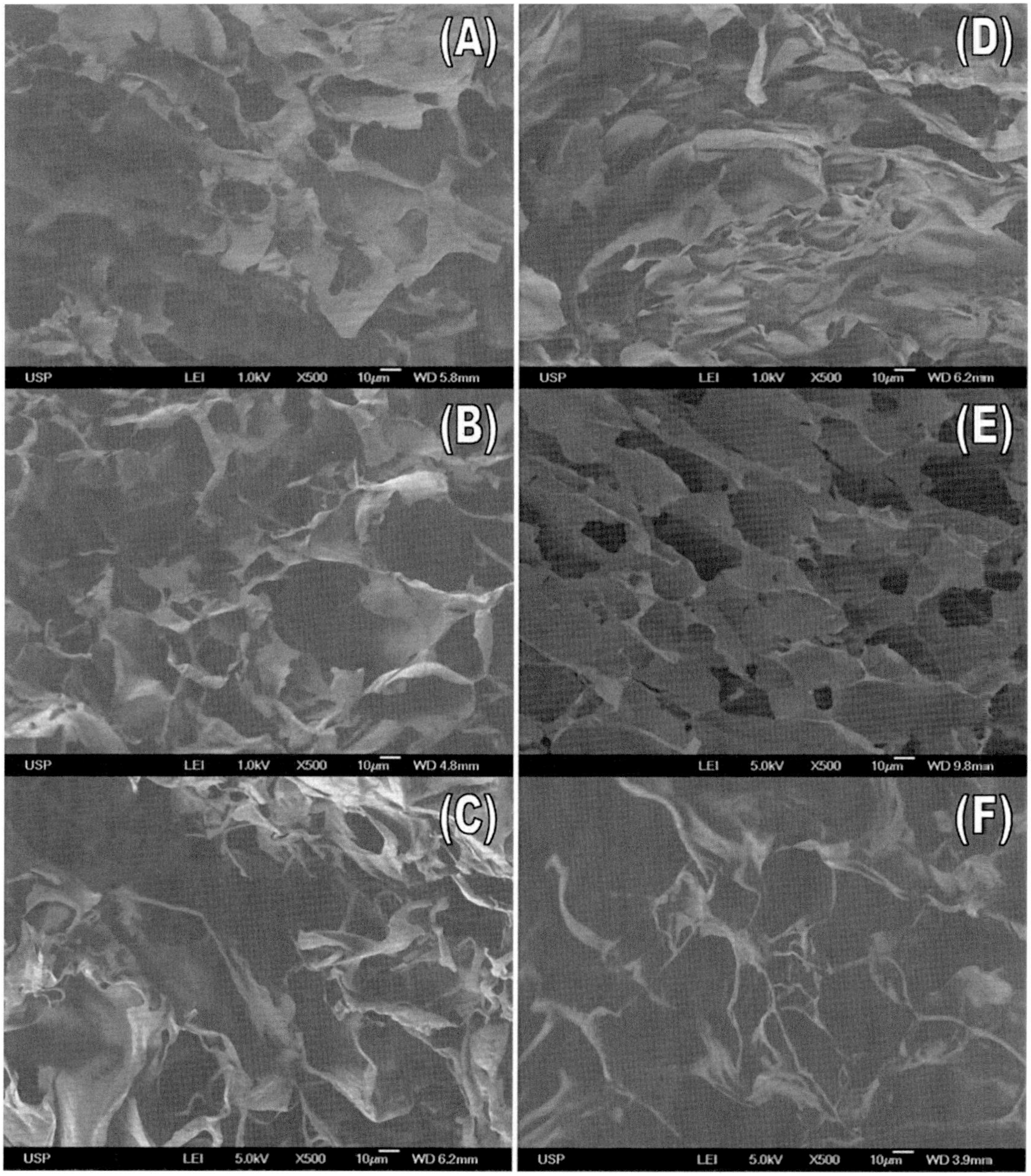

FIG. 2.4 SEM images of hydrogels. (A–C) XG hydrogel in acid, neutral and basic media, respectively; (D–F) XG-CA hydrogel in acid, neutral and basic media, respectively. *From Bueno, V.B., Bentini, R., Catalani, R.H., Petri, D.-F.S., 2013. Synthesis and swelling behavior of xanthan-based hydrogels. Carbohyd. Polym. 92, 1091–1099.*

transition and the organization of the hydrogel in a "quasiordered" network during the coacervation. The investigation has shown that the time of coacervation influenced the swelling degree and the rheological characteristics of the PECs. The PEC hydrogel obtained by the mixing of polymer dispersions have shown stable storage modulus and rheological data have proven its solid-like behavior. The subsequent cryopreparation of the PEC hydrogel helped to explain its structure. Microscopic investigations have shown that the hydrogel had a regular organization with a porous network composed of polymers' fibers. The size of pores has been estimated between 25 and 100 nm, and the diameter of fibers between 1 and 10 μm. The solid PEC particles, obtained by mechanical grinding of the freeze-dried bulk hydrogel, had a suitable diameter for pharmaceutical applications, so can be considered an appropriate matrix for drug incorporation. PEC hydrogels can also be formed *in situ* when the mixture of XG and CH powders gets in touch with the aqueous medium (Magnin et al., 2004). Dadou et al. (2020) developed and evaluated tablets based on CH and XG with controlled-release properties. Tablets were prepared by direct compression of the mixture of powdered XG, low or high molecular weight CH, and metoprolol succinate. When tablets came into contact with the dissolution medium, PECs started to form between the polymers on the surface of the tablets. The authors also showed that the interactions between XG and low molecular weight CH were more favorable than those between XG and high molecular weight CH, and that the complex with the most favorable interactions was formed when low molecular weight CH was used at 15% (Dadou et al., 2020). CH/XG PEC-based hydrogels have also been prepared by progressive acidification of XG solution containing homogenously dispersed CH particles in a noncharged state at pH values slightly higher than pK_a of this polymer. The following decrease of pH resulted in gradual charging of CH molecules, which led to the formation of electrostatic interactions between oppositely charged XG and CH. The molecules of polymers self-organized into supramolecular structures (i.e., PECs). Fibrillation or capsulation of polymer molecules depended on the charge and mixing ratio of XG and CH. The gelation of solutions took place in the case of fibrillar morphology. Where oppositely charged polymers self-organized into capsules, the hydrogel was not formed because of the absence of cross-links between particles with a diameter of few millimeters and a mechanically strong shell (Shchipunov et al., 2010).

Many formulation factors can influence the formation of CH/XG-based PECs and thereafter their use as drug carriers. Argin-Soysal et al. indicated that the cross-linking density between polymers in CH/XG-based PECs was less dependent on the concentration of CH solution than the concentration of XG solution and CH solution pH (Argin-Soysal et al., 2009). In a study by Iizhar Ahmed et al. (2014), it was indicated that the CH/XG mass ratio can also affect the physicochemical characteristics of PEC and the release profile of active substances (Iizhar Ahmed et al., 2014). They showed that the best sustained release during 12 h with the desired release rate of highly water-soluble isosorbide mononitrate was achieved for CH/XG mass ratio 1:2. In addition, the influence of the drug substance properties and its entrapment procedure on the complex formation should be considered. Very recently, Ćirić et al. studied the combined effect of pH adjusting agent type (hydrochloric, acetic, or lactic acid) and pH value (3.6, 4.6, and 5.6) on the formation of CH/XG-based PECs, their physicochemical characteristics, and influence on *in vitro* release kinetics of the hydrophobic drug ibuprofen. The results of conductivity and rheological measurements demonstrated the greatest interaction extent in PECs prepared with acetic acid at pH 3.6. Acid type and pH strongly influenced the yield and particle size, but did not have any influence on the residual moisture content of air-dried PECs. Differential scanning calorimetry (DSC) and Fourier-transform infrared spectroscopy (FTIR) analysis confirmed exclusively physical interactions between the polymers, while powder X-ray diffraction (PXRD) analysis showed the semicrystalline or amorphous structure of PECs, which also depended on the acid type and pH. PECs comprising acetic acid prepared at pH 4.6 and 5.6 had adequate rehydration ability in phosphate buffer pH 7.2 and they prolonged ibuprofen release up to 10 h at a PEC-to-drug mass ratio up to 1:2 (Ćirić et al., 2020). In the continuation of the research, the same research group examined the influence of the effect of entrapment procedure of ibuprofen, a poorly water-soluble drug, on physicochemical and drug release performances of CH/XG-based PECs to achieve controlled drug release as the ultimate goal (Ćirić et al., 2021). The formation of PECs for two drug entrapment procedures (before or after the mixing of polymers) at a pH of 4.6 or 5.6 and three CH/XG mass ratios (1:1, 1:2, and 1:3) was confirmed by a continuous decrease in conductivity and increased apparent viscosity and hysteresis values during the PECs formation. The most extensive cross-linking was shown when ibuprofen was added before the PECs formation at pH 4.6 and CH/XG mass ratio 1:1. On the other hand, PECs prepared at CH/XG mass ratios 1:2 and 1:3 had higher yields and drug entrapment efficiencies. DSC and FTIR analysis confirmed ibuprofen entrapment in PECs by establishing only physical interactions with polymers, while PXRD analysis demonstrated the partial disruption of its crystallinity. Drug entrapment highly influenced its release kinetics from PECs, and potentially optimal was considered the PEC prepared at pH 4.6 with ibuprofen entrapped before the mixing of polymers at CH/XG mass ratio 1:2, which provided controlled drug release by zero-order kinetics, high yield, and drug entrapment efficiency.

PECs can be prepared from the derivatives obtained by chemical modification of the polymers in order to minimize or eliminate their shortcomings and improve the drug delivery performances of the final hydrogel carrier. In a study by Hanna and Saad (2019), a polyelectrolyte hydrogel based on carboxymethyl xanthan gum (CMXG) and *N*-trimethyl chitosan (TMC) was synthesized and evaluated as a carrier for ciprofloxacin. TMC, as a partially quaternized derivative of CH, exhibits an enhanced water solubility over a wide pH range and better antimicrobial activity compared to CH. The aqueous solutions of the polymers obtained by etherification of hydroxyl groups of XG with carboxymethyl groups (carboxymethylation) were mixed under constant stirring for 2 h at room temperature to obtain the hydrogel. The dried hydrogel was combined with the drug solution. The ciprofloxacin encapsulation efficiency was increased up to 93.8 ± 2.1% by increasing the drug concentration up to 250 μg/mL. The FTIR analysis of TMC/CMXG hydrogel indicated the physical cross-linking by molecular interactions between the hydrophilic hydroxyl and carboxylate groups of CMXG and a quaternary amino group of TMC. Ciprofloxacin was dispersed within the hydrogel in a crystalline form and there was no significant interaction between the drug and the polyelectrolyte hydrogel. The surface of the drug-free dry hydrogel was rough and fibrous with many pores (Fig. 2.5), while the SEM microphotograph of the drug-loaded hydrogel showed that the drug was prone to agglomeration and the particles were distributed unevenly in the hydrogel matrix (Fig. 2.5). The extent of swelling of the prepared hydrogel in PBS (pH 7.4) at 37 °C increases gradually with increasing the immersion time up to the maximum of ~372 ± 8.0% after ~240 min, due to the presence of many hydrophilic groups along polymers chains. The increase in the rate of swelling of the hydrogel and drug loading efficiency increased the drug release rate up to 96.1 ± 1.8% for 150 min.

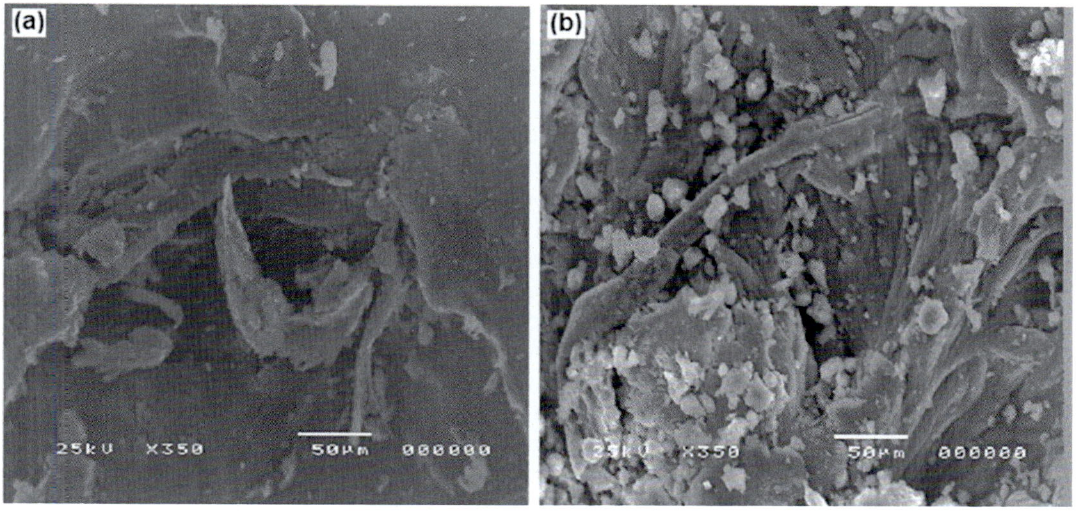

FIG. 2.5 SEM micrographs of (A) dry TMC/CMXG hydrogel and (B) ciprofloxacin loaded TMC/CMCX hydrogel (3.52%) at magnification ×350. *With permission from Hanna, D.H., Saad, G.R., 2019. Encapsulation of ciprofloxacin within modified xanthan gum-chitosan based hydrogel for drug delivery. Bioorg. Chem. 84, 115–124.*

The drug release was relatively fast and the release kinetics followed zero-order kinetics, so the prepared hydrogel can be considered promising for controlled drug release suitable for the achievement of high activity against gram-positive and gram-negative bacterial strains (Hanna and Saad, 2019). Carboxymethylation of XG was applied also for the achievement of controlled delivery of BSA and diltiazem (Maiti et al., 2009; Ray et al., 2010).

Different studies of XG-based copolymeric hydrogels with hydrocolloid gums (guar gum (Sandeep et al., 2014), acacia gum (Outuki et al., 2016)), cellulose derivatives (methylcellulose (MC) (Liu and Yao, 2015)), or cyclic oligosaccharides (Ahmad et al., 2019; Iijima et al., 2007; Zhang et al., 2015) have been investigated in order to elucidate the nature of the interactions established between the polymers and potential synergistic properties of the obtained systems. Sandeep et al. developed PEC-based microparticles of cationic guar gum and XG. Diclofenac sodium as a model drug was incorporated into the microparticles. DSC and FTIR studies confirmed the formation of PEC and the absence of chemical interactions between the drug and polymers, while SEM showed the rough surface of microparticles. Particle size varied between 294 and 300 μm, and the encapsulation efficiency of the drug was 96.47%. *In vitro* release studies indicated the extended drug release during 12 h. Compared with a diclofenac sodium solution, microparticles showed lower and prolonged drug release when subjected to *in vivo* pharmacokinetic evaluation in rabbits. Results of *in vitro* and *in vivo* studies indicated that developed microparticles can provide controlled release of diclofenac sodium after oral administration (Sandeep et al., 2014). XG is considered a convenient excipient for the spray-drying process (García-Ochoa et al., 2000). In the study of Outuki et al. microparticles loaded with *Eschweilera nana* extract were prepared by a spray-drying technique using a mixture of acacia gum and XG. The extract has strong antioxidant activity due to the presence of flavonoids, rutin, and hyperoside, but their low water solubility, bioavailability, and stability to environmental conditions limit their potential therapeutic application. The study showed that it was possible to prepare microparticles loaded with *E. nana* extract with a high encapsulation efficiency of rutin and hyperoside using the spray-drying technique. This may be associated with possible interactions between the polymers and the extract. All formulations were amorphous, hollow, and spherical with smooth surfaces. Thermal analysis revealed that the microencapsulation enabled thermal protection of the extract. The release of the microencapsulated rutin was slower than from the pure extract. The antioxidant activity was confirmed by the ability of the extract and the extract-loaded microparticles to decrease reactive oxygen species in the neutrophil respiratory burst (Outuki et al., 2016). Zhang et al. developed XG-based hydrogels with improved textural characteristics by physical crosslinking of XG with hydroxypropyl β-cyclodextrin (HPCD) at different XG/HPCD mass ratios (4:0, 4:1, 4:2, 4:3, 4:4, 4:5, and 4:6). The improvement of the texture of hydrogels was the result of the self-aggregation of XG helixes, as observed by optical and scanning electron microscopy. Rheological measurements showed that the addition of HPCD into XG hydrogel affected the dynamic behavior of the hydrogel system due to the enhancement of the supramolecular interactions. The XG/HPCD hydrogel with a mass ratio of 4:3 had a more elastic network than other XG/HPCD hydrogels, so was considered the most suitable for drug delivery. The hydrogel with anticancer drug methotrexate demonstrated effective controlled release, with a considerably higher amount released in simulated intestinal juice (pH 7.4) in comparison with simulated gastric juice (pH 1.2). The inclusion of the HPCD molecule in the system enhanced the binding of methotrexate. In addition, the hydrogel with antibiotic levofloxacin exhibited fair antibacterial effects. The supramolecular interactions in the network of hydrogels were mostly hydrogen bonds, as shown by FTIR spectroscopy. PXRD showed that the dried XG/HPCD hydrogels had amorphous characteristics that were found to be beneficial for the loading of drugs (Zhang et al., 2020). Combining XG with other polysaccharides also can be a useful strategy for the design of copolymeric hydrogels with the drug delivery potential based on their sensitivity to the temperature in the biological environment. Liu and Yao described thermosensitive shear-thinning injectable hydrogels based on XG and MC double networks. Such copolymeric hydrogels combine a good shear-thinning property of XG and a thermal gelation property of MC. They flow under shear stress at room temperature (before injection), but at body temperature (after injection) recover their mechanical properties and form high viscosity hydrogels and thus avoid leakage from the injection site. At 23 °C, the XG solution forms a weak gel with shear-thinning properties, while the MC solution is a low viscous liquid. An XG/MC blend solution was prepared by adding XG powder (1–3%) into MC solution (8–12%) in PBS pH 7.4, followed by stirring. The XG/MC blend contained only the XG network at 23 °C. The viscosity of the blend solution was lower than the XG solution because the MC concentration was much higher than the XG concentration and the XG 3D network was partly disturbed by the MC molecules. When the temperature increased to gelation temperature (37 °C), intermolecular hydrophobic interaction occurred and XG/MC blend solution transformed into hydrogel due to the formation of a thermoresponsive MC network, besides the XG network. Both the gelation temperature and gelation time of MC were decreased due to dehydration of the polymer caused by the carboxylic groups of XG and the sodium chloride from PBS. This double network structure is illustrated in Fig. 2.6. *In situ* gelation and formation of XG/MC hydrogel proved promising for sustained delivery of doxorubicin (Liu and Yao, 2015).

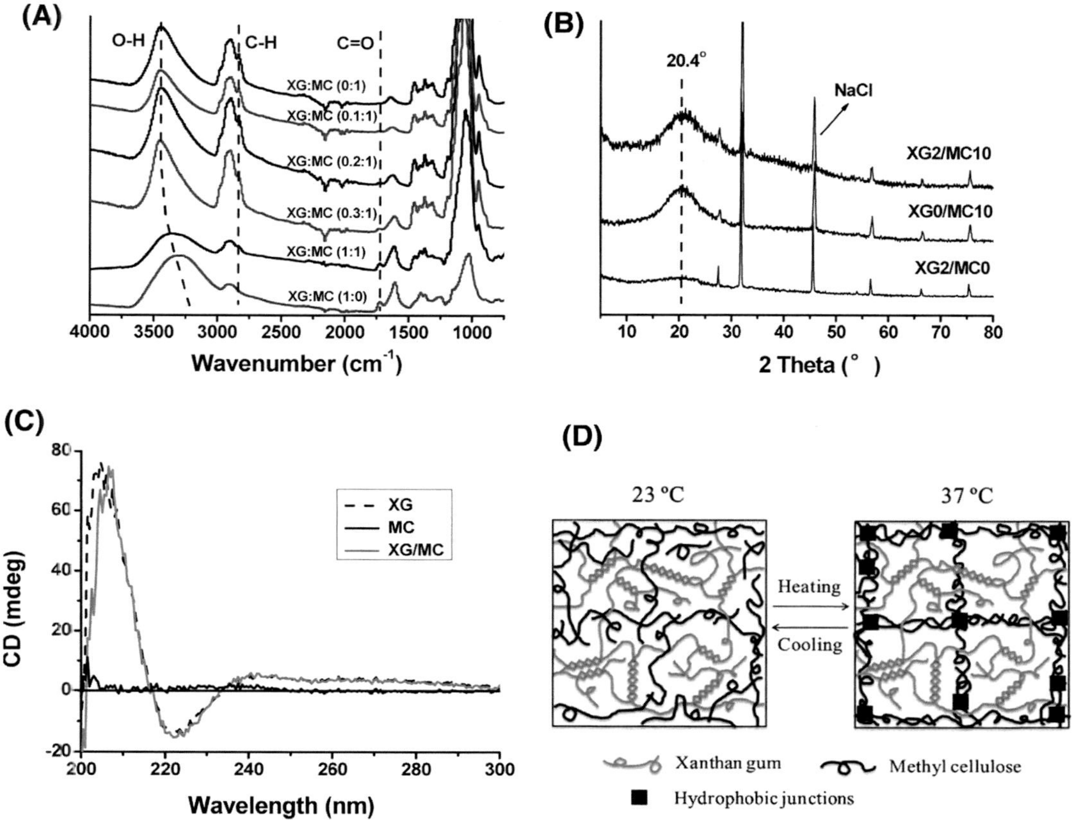

FIG. 2.6 (A) FTIR spectra of various freeze-dried XG/MC hydrogels; (B) XRD patterns of freeze-dried XG2/MC0, XG0/MC10 and XG2/MC10 hydrogels; (C) CD spectra of 2 mg/mL XG solution, 10 mg/mL MC solution, and XG/MC blend solution containing 2 mg/mL XG and 10 mg/mL MC measured at 23 °C; and (D) illustration of the gelation mechanism of XG/MC blend solution. *With permission from Liu, Z., Yao, P., 2015. Injectable thermo-responsive hydrogel composed of xanthan gum and methylcellulose double networks with shear-thinning property. Carbohyd. Polym. 132, 490–498.*

2.3.2 Chemically cross-linked copolymeric XG-based hydrogels

Chemical cross-linking for the formation of copolymer hydrogels with XG can be performed by various technological approaches. The literature describes different examples of covalent cross-linking of: XG with other polysaccharides (e.g., starch) in the presence of the chemical cross-linker (Shalviri et al., 2010; Simões et al., 2020); reactive XG derivatives (XG-aldehyde (XG-ALD)) with CH derivatives (Huang et al., 2018); and grafted XG derivatives with synthetic polymers (Abu Elella et al., 2020; Gils et al., 2009; Kulkarni and Sa, 2009; Sabaa et al., 2019; Simões et al., 2020; Trombino et al., 2019; Zheng et al., 2019a, b).

Shalviri et al. developed a film-forming hydrogel based on XG cross-linked with starch by using STMP (Fig. 2.7).

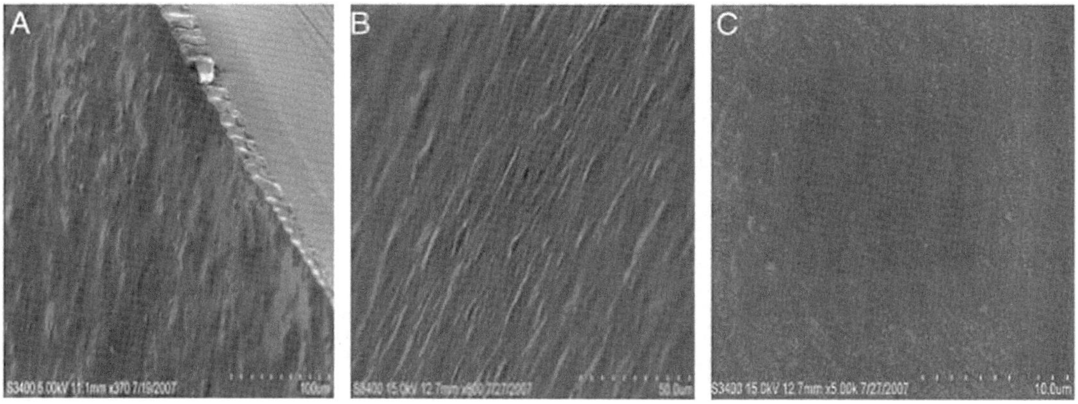

FIG. 2.7 SEM images of (A) surface (*bottom left corner*) and cross section (*top right corner*), (B) surface, and (C) cross section of cross-linked starch–xanthan gum film containing 10% xanthan gum and 5% STMP. *With permission from Shalviri, A., Liu, Q., Abdekhodaie, M.J., Wu, X.Y., 2010. Novel modified starch-xanthan gum hydrogels for controlled drug delivery: synthesis and characterization. Carbohyd. Polym. 79, 898–907.*

Starch is a homopolysaccharide composed of a mixture of two polymers: linear amylose (formed by D-glucose units linked by α-1,4-glycosidic bonds) and highly branched amylopectin with the molecules containing α-1,4-glycosidic bonds in the main chain and α-1,6 at branching points (Du et al., 2007). XG-starch hydrogels were prepared by mixing XG with the aqueous dispersion of starch and initiation of cross-linking by adding different amounts of STMP. The covalent cross-linking of XG and starch via STMP and formation of the hetero-polymer network was confirmed by FTIR and nuclear magnetic resonance (NMR) analysis. The obtained hydrogel was isolated by centrifugation, washed with water, and dried at room temperature to prepare films with an average thickness of 300–400 μm. The films were macroscopically homogeneous and transparent with few micropores. Increases were observed in the equilibrium swelling ratio and swelling rate of the hydrogels with increasing STMP and XG content, particularly when it was higher than 10%, likely due to the formation of monopolysaccharide phosphate species. The same trend was observed for the gel mesh size and drug permeability. The mesh sizes of the hydrogels were 2.84–6.74 nm (at pH 7.4) and hence large enough for the transport of small molecules as well as polypeptide and proteins. The drug permeability of the hydrogels was affected by drug charges. Permeability of the films for ibuprofen sodium and sodium salicylate in the anionic form (at pH 7.4) was significantly lower than their neutral forms (at pH 2) due to the electrostatic repulsion between the negatively charged drugs and the polymer network. Such film-forming carriers with selective drug permeability can be further investigated for the achievement of controlled release profiles of ionizable drugs (Shalviri et al., 2010).

Simões et al. prepared the starch/XG hydrogels in the presence of CA as a cross-linking agent and sodium hypophosphite (SHP) as a catalyst, and applied extrusion and thermopressing processes to obtain the hydrogel-based sheets (Fig. 2.8). The starch/XG hydrogels with different levels of CA (0–2.25 g/100 g polymer) were prepared in three stages. In the first stage, mixtures were prepared upon mixing of a glycerol solution of CA and SHP with the mixture of XG and starch, and afterward they were processed using a laboratory

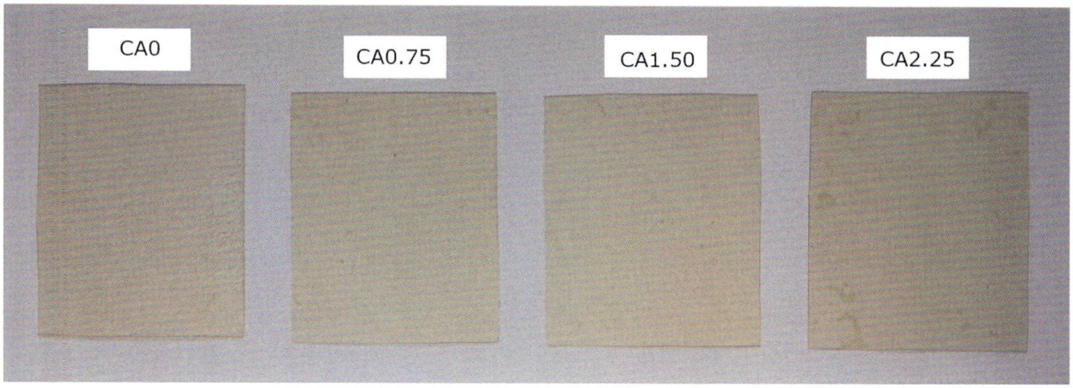

FIG. 2.8 Appearance of the starch/xanthan hydrogels. *With permission from Simões, B.M., Cagnin, C., Yamashita, F., Olivato, J.B., Garcia, P.S., de Oliveira, S.M., Eiras Grossmann, M.V., 2020. Citric acid as crosslinking agent in starch/xanthan gum hydrogels produced by extrusion and thermopressing. LWT, 125, 108950.*

single-screw extruder and the obtained profiles were pelletized. In the second stage, the pellets were reprocessed in the same equipment, resulting in partially homogeneous sheets. Finally, hydrogels were prepared from the extruded pieces pressurized in a hydraulic press at a temperature of 120 °C (Simões et al., 2020). Both extrusion and thermopressing (i.e., thermal and mechanical energy) can cause chemical and physical reactions within the exposed material (Ghosh Dastidar and Netravali, 2012; González-Seligra et al., 2017). The strength of starch/XG cross-linked hydrogels with CA was lower in comparison to noncross-linked hydrogels, due to acid hydrolysis of the polymer chains; however, CA-SHP increased the storage modulus of the hydrogels. In addition, extrusion promoted cross-linking between the components of the hydrogels, while thermopressing enabled producing homogeneous hydrogels with a smooth surface and good processability. Cross-linking strengthened the structure of the hydrogels, but decreased their swelling in water, in comparison with the control. CA affected the mechanical properties of the hydrogels, especially the increase in elongation promoted by the hydrolytic action, and ensured the preservation of swelled hydrogels' integrity. The obtained hydrogels can be further considered as potential drug carriers.

Huang et al. investigated the hydrogel obtained via cross-linking of XG-ALD and carboxymethyl-modified CH (CH-CM) (Huang et al., 2018). CH is a natural polysaccharide derived from the chitin shells of shrimp and other crustaceans (Illum, 1998). It is poorly soluble in physiological fluids, but this disadvantage can be overcome by its carboxymethylation. After mixing the solutions of XG-ALD and CH-CM, a stable 3D network was formed. The initial phase of the hydrogel formation was based mainly on hydrogen bonds (i.e., physical interactions), but a Schiff's base reaction between the aldehyde groups of XG-ALD and the amino groups of CH-CM was established very quickly and actually the chemical cross-linking occurred, as was confirmed by FTIR analysis. The cross-linking and morphology of the hydrogels depend on the polymers' mass ratio. The high density, small-sized pores, the quickest gelation time, and the highest swelling ratio to absorb large amounts of liquids were observed for the hydrogel with a XG-ALD/CH-CM mass ratio of 1%:0.33%, while the increase or decrease in the CH-CM content led to an irregular shape and morphology. The optimized XG-ALD/CH-CM hydrogel was viscoelastic solid with superior stability and self-healing property due to the dynamic equilibrium between the Schiff's base cross-linking and the individual polymer reactants. It released only a minor amount of a potent angiogenic factor (VEGF) in acid media due to the high degree of copolymerization and protective effect of the side chains of XG-ALD. Therefore, it can be considered as a controlled drug release carrier for oral administration. Additionally, it could be injected for abdominal wall reconstruction or it forms a gel *in situ* in an environment with much excretion or exudation, such as open wounds.

Physicochemical and drug delivery properties of XG-based hydrogels can be extended and optimized by grafting of XG with synthetic polymers, such as acrylamide (AM) (Zheng et al., 2019a), acrylic acid (AA) (Zheng et al., 2019b), methacrylic acid (MA) (Trombino et al., 2019), 2-hydroxyethyl methacrylate (HEMA) (Gils et al., 2009), and poly(N-vinyl imidazole) (PVI) (Abu Elella et al., 2020; Sabaa et al., 2019). Grafting of XG may enable new properties and functionalities, such as enhanced thermal stability and porosity, increased encapsulation efficiency, and swelling ability (Abu Elella et al., 2020; Gils et al., 2009; Kulkarni and Sa, 2009; Sabaa et al., 2019; Simões et al., 2020; Trombino et al., 2019; Zheng et al., 2019a; Zheng et al., 2019b). A recent study by Zheng et al. demonstrated the improvement of both thermal stability and adsorption capacity by grafting XG with AM and partly neutralized AA with trimethylolpropane triglycidyl ether (TTE) as a cross-linking agent. The introduction of amino and carboxyl groups by AM and AA enables rearrangement of the XG molecules with better regularity, leading to higher thermal stability and a more porous structure with a strong ability for water uptake and a pH-responsive swelling/deswelling property within the pH interval from 1.8 up to 7.2. The addition of TTE caused the growth of the grafted XG crystals with porous structure and enhancement of their thermal stability (Zheng et al., 2019a). Cross-linked XG grafts with PVI (XG-g-PVI) were synthesized via a free radical polymerization mechanism with various graft yields (0%–95%) and they were investigated as matrices for the incorporation of BSA (Abu Elella et al., 2020; Sabaa et al., 2019). PVI is a synthetic biodegradable water-soluble polymer that was used for grafting XG to form hydrogels suitable for protein delivery. X-ray diffractometry (XRD) confirmed the grafting process onto XG and the formation of XG-g-PVI by intermolecular H-bonding interactions between hydroxyl groups of XG chains and nitrogen atom on PVI chains. The surface morphology of the drug-unloaded graft hydrogels, examined by field emission scanning electron microscopy (FE-SEM), was heterogeneous, irregular, and highly porous, while the BSA-loaded graft hydrogel surface was rough and porous, filled with lustrous particles. This confirmed that BSA was diffused into the hydrogel matrix within many pores on its surface. FTIR spectrum and XRD patterns of BSA and BSA-loaded graft hydrogel indicated that it could encapsulate BSA through hydrogen bonding. BSA encapsulation efficiency increased with increasing the grafting percent due to the increased number of pores and irregular cavities and electrostatic interactions between the nitrogen atom in the PVI imidazole ring and carboxyl groups of BSA. In a study by Trombino et al., the hydrogel made of XG methacrylate (XG-MA) and *N,N′*-methylenebisacrylamide (MBAA) was obtained by free radical polymerization using ammonium persulfate (APS) as an initiation system. Esterification of XG with methacrylic acid was confirmed by FTIR while the presence of pendant methacrylic groups in XG-MA available for radical polymerization was detected by the ^{1}H NMR spectrum. The hydrogel was considered as a carrier for omega-3 polyunsaturated fatty acids (PUFAs). It was soaked in the aqueous dispersion of either docosahexaenoic acid (DHA) or its metabolic precursor, α-linolenic acid (ALA), and after filtration it was dried. The loadings of ALA and DHA were 71% and 57%, respectively, depending on the interactions between the polymeric matrix and the fatty acids. The PUFA-loaded hydrogels show a porous structure. The high hydrogel swelling degree at pH 7.4 suggested their suitability as potential candidates for adjuvant therapy of colorectal cancer (Trombino et al., 2019). Gils et al. described the procedure based on the free radical graft copolymerization method for synthesis of a superporous hydrogel through chemical cross-linking by graft copolymerization of HEMA and AA on to XG via a redox initiator system of APS and *N,N,N,N*-tetramethylethylenediamine (TMED) in the presence of MBAA (cross-linking agent), sodium bicarbonate (foaming agent), and polyoxyethylene/polyoxypropylene/polyoxyethylene triblock copolymer (foam stabilizer). HEMA is a biocompatible biomaterial with good chemical and hydrolytic stability suitable for the achievement of controlled release of drugs. Polymerization of AA with either hydrophilic or hydrophobic monomers in the presence of organic cross-linkers enables the formation of copolymers of tunable physicochemical properties. Thermogravimetric analysis showed the improved thermal stability of the grafted XG due to the strong bonding between the grafting polymer chains and XG matrices. The surface morphology of the obtained hydrogel was microporous and the connectivity between the pores enable fast swelling and deswelling kinetics of the hydrogels by convection. It was observed that increasing the HEMA/AA ratio (v/v) decreased the porosity of the hydrogel, and the highest water absorption was achieved for the hydrogel sample with a HEMA/AA ratio of 0.5. It was more porous and thus displayed a faster swelling rate. In addition, a high concentration of charged ionic groups in the hydrogel increases the swelling due to osmosis and charge repulsion. The swelling/deswelling behavior of the hydrogel was sensitive to environmental pH, which makes him a promising candidate for controlled drug delivery (Gils et al., 2009).

XG chains can be entrapped inside a cross-linked hydrophilic polymer to form an interpenetrating polymer network (IPN) (Giri et al., 2016) or semi-IPN (Hajikhani et al., 2019). Giri et al. designed the IPN hydrogel beads of casein and carboxymethyl-modified XG (XG-CM) by ionotropic gelation and covalent cross-linking method in an aqueous environment. The prepared beads were spherical with a rough surface. Their size ranged from 602.5 to 1075.45 μm and it was lower at the higher concentration of aluminum chloride ($AlCl_3$) solution (used as a medium for the preparation of the beads), lower concentration of the polymer, lower concentration of the model drug (theophylline), lower concentration of the cross-linker glutaraldehyde (GA), and longer exposure time. The drug entrapment efficiency was up to 85.12% and it increased as the concentration of $AlCl_3$ and glutaraldehyde decreased and the concentration of the polymer increased. Incorporation of the drug into the beads led to the transformation of a crystalline drug to an amorphous state. The hydrogels exhibited slow theophylline release in pH 1.2 KCl/HCl buffer with a maximum of $20.45 \pm 0.31\%$ released after 2 h. The presence of casein in the beads decreased the degree of swelling in pH 1.2 KCl/HCl buffer due to electrostatic repulsive forces between the protonated carboxyl groups, while at higher pH values (PBS pH 6.8), the swelling was significantly higher due to increased osmotic pressure inside the beads caused by ionization of carboxyl groups of the hydrogel. In semi-IPNs, one polymer network is independently synthesized or cross-linked in the presence of another one (Giri et al., 2016). XG has been used in combination with PAA in IPNs or semi-IPNs in order to decrease water solubility of PAA and premature drug release, as well as to preserve the pH sensitivity typical for this polyelectrolyte polymer (Ijaz et al., 2018). For this purpose, AA is polymerized in the presence of cross-linker agents to form hydrogels. Hajikhani et al. synthesized semi-IPNs of cross-linked AA and XG in the presence of *N,N′*-hexane-1,6-dilbisprop-2-enamide (HDE), or 1,4-butandioldimethacrylate (BDD) as the cross-linking agent (Fig. 2.9).

The semi-IPNs were synthesized through chemical cross-linking in a polymerization process using potassium persulfate as an initiator in the aqueous medium. The effects of two cross-linkers on the morphology, structure, and thermal stability of semi-IPNs were evaluated. BDD improved the thermal stability of semi-IPNs; however, HDE enables higher strength and porosity (from 20 to 200 μm) than BDD (<20 μm). The porosity of semi-IPNs increased with increasing the HDE content, while BDD-based semi-IPNs were more susceptible to crumbling with an accompanying sound effect with increasing the amount of this cross-linker. For such hydrogels are typical higher porosity, reduced mechanical properties, and higher capacity of water absorption. In addition, the higher swelling ratio of BDD-based semi-IPNs is related to the conversion of XG into the polymer network with the remaining HDE and the formation of a complex chemical network of the IPN type which reduces the swelling rate. On the other hand, the formation of the IPN prevents the removal of XG from the polymer network and thus enhances its stability. For the investigated semi-IPNs, increasing the amount of cross-linker decreased the swelling ratio in both cases. In addition, the swelling behaviors of the semi-IPNs were dependent on pH, making them suitable for use as drug delivery carriers (Hajikhani et al., 2019).

2.4 XG-based nanocomposite hydrogels

The hydrogels may act like a "host" with the large spaces within the 3D polymeric networks that can accommodate different nanomaterials as a "guest" to form nanocomposite hydrogels. Many nanocomposite hydrogels have been designed over the last two decades. Incorporation of different drug-loaded nanocarriers, such as metal or metal oxides nanoparticles and "soft" nanomaterials including liposomes, microemulsions, and nanoemulsions, into the polymeric network of the hydrogels, is becoming a promising strategy for the development of nanocomposite hydrogels which can be considered for drug delivery purposes. The nanocomposite hydrogels can be prepared from physically or chemically cross-linked polymer networks and the nanomaterials introduced into the hydrogel *in situ* or *ex situ*. The building of the nanocomposite hydrogels by combining the hydrogels and nanomaterials may result in a synergistic enhancement of physicochemical characteristics (e.g., mechanical strength, stiffness, elasticity, porosity, swelling capacity), physical stability, and drug delivery performances of the resulting hybrid system in comparison with the starting materials. Moreover, the potential risks of utilization of nanomaterials regarding human health and the environment can be diminished by combining them with biocompatible hydrogels (Ma et al., 2016; Praveen et al., 2015; Wahid et al., 2020).

Singh et al. prepared biodegradable nanocomposite hydrogel based on XG-g-PAA with *in situ* synthesized gold nanoparticles for controlled delivery of amoxicillin. For the synthesis of the hydrogel was applied the free radical graft polymerization method assisted by microwave

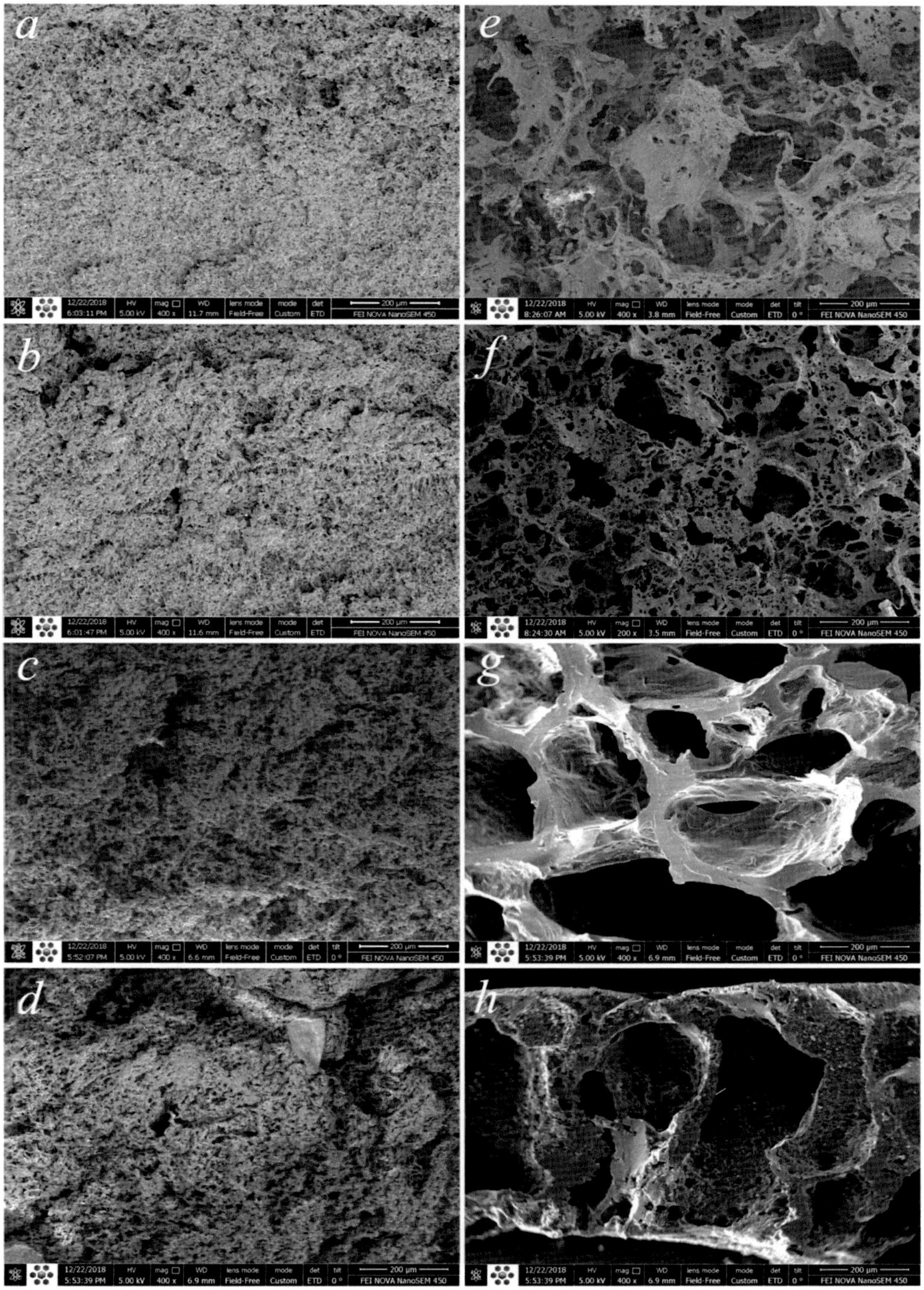

FIG. 2.9 SEM images of surface fracture of semi-IPN hydrogels: (A) BDOD-5; (B) BDOD-5; (C) BDOD-15; (D) BDOD-20; (E) MS-5; (F) MS-10; (G) MS-15; and (H) MS-20. *With permission from Hajikhani, M., Khanghahi, M.M., Shahrousvand, M., Mohammadi-Rovshandeh, J., Babaei, A., Khademi, S.M.H., 2019. Intelligent superabsorbents based on a xanthan gum/poly (acrylic acid) semi-interpenetrating polymer network for application in drug delivery systems. Int. J. Biol. Macromol. 139, 509–520.*

radiation. The formation of gold nanoparticles inside the polymer matrix has been observed from the change in color from colorless to pink-red, due to their surface plasmon resonance (SPR) effect. Grafting of AA with XG and synthesis of the nanoparticles inside polymer matrix took place without aggregation, the nanoparticles were uniformly dispersed, and the anisotropy of the nanocomposite hydrogel increased. In addition, DLS, XRD, and transmission electron microscopy (TEM) analysis revealed the crystalline structure and spherical shape of the nanoparticles with an average size of 75 ± 3 nm and Zeta potential of -33.5 mV. The grafting of PAA onto the XG backbone provided the formation of the porous mesh type structure, which was related to the high capacity for water absorption. Additionally, the incorporation of the nanoparticles into the polymer matrix leads to an increase in the size and specific surface area of the pores compared to native XG. Amoxicillin was loaded inside the nanocomposite by the interaction with hydroxyl groups of XG, compared to native XG. This study demonstrated that the grafting of XG with AA and incorporation of gold nanoparticles represents an effective strategy to improve the strength of both the polymer matrix itself and the linkage of drug molecules with the polymer. The obtained nanocomposite hydrogel could be considered a promising carrier to optimize the release profile of amoxicillin (Singh et al., 2020).

XG-polyvinyl alcohol (PVA)/zinc oxide (ZnO) nanocomposite hydrogels were prepared by a 60Co γ-ray irradiation-induced copolymerization process and their characterization for use as a wound dressing was performed by Raafat et al. (Raafat et al., 2018). Synthesis of the copolymeric hydrogels by gamma radiation is a simple technique that enables cross-linking without chemical cross-linking agents. Furthermore, it enabled the control over the degree of the cross-linking which affects the mechanical and swelling properties of the hydrogel. In an aqueous solution of XG exposed to gamma radiation, free radicals formed along the main chain due to radiolysis of water to OH- and H-radicals which abstract hydrogen from XG chains. Free radicals are also generated by breaking the weak bonds of the vinyl group on the AA moieties under ionization irradiation. Between these free radical points of the XG chains and the grafted moieties, covalent bonds are formed, resulting in cross-linking and formation of the hydrogels. A series of homogeneous pasty mixtures of XG and PVA with different compositions (50:50, 40:60, 30:70, 20:80, and 10:90, w/w) were cast in glass molds and irradiated using 60Co gamma rays. XG-PVA/ZnO nanocomposite hydrogels were prepared by mixing 0.5%, 1%, 2%, and 5% ZnO nanoparticles with 15% XG-PVA (30:70) to get a homogenous paste, then cast and irradiated at 30 kGy. Copolymerization of XG with PVA enhances the mechanical characteristics of the hydrogel, while the presence of ZnO nanoparticles (with the size of 15–25 nm) within the internal hydrogel network enables the formation of a homogenous porous structure, which is proven by SEM (Fig. 2.10). Adequate porosity of the nanocomposite hydrogel and the present ZnO nanoparticles control the fluid uptake ability (swelling), water retention capability, and water vapor transmission rate, and thus their prospective use in wound treatment.

Nanocomposite hydrogels composed of XG and AA containing magnesium oxide (MgO) nanoparticles were evaluated for oral delivery of anticancer drug methotrexate (El-Sawy et al., 2020). MgO nanoparticles with the size of 6.6 nm were synthesized using the sol-gel method. XG-AA/MgO nanocomposite hydrogels were prepared by an irradiation-induced copolymerization and cross-linking technique. Mixtures of XG and AA with different compositions (60:40, 50:50, 40:60, 30:70, 20:80, and 10:90, w/w) were irradiated at 20, 25, and

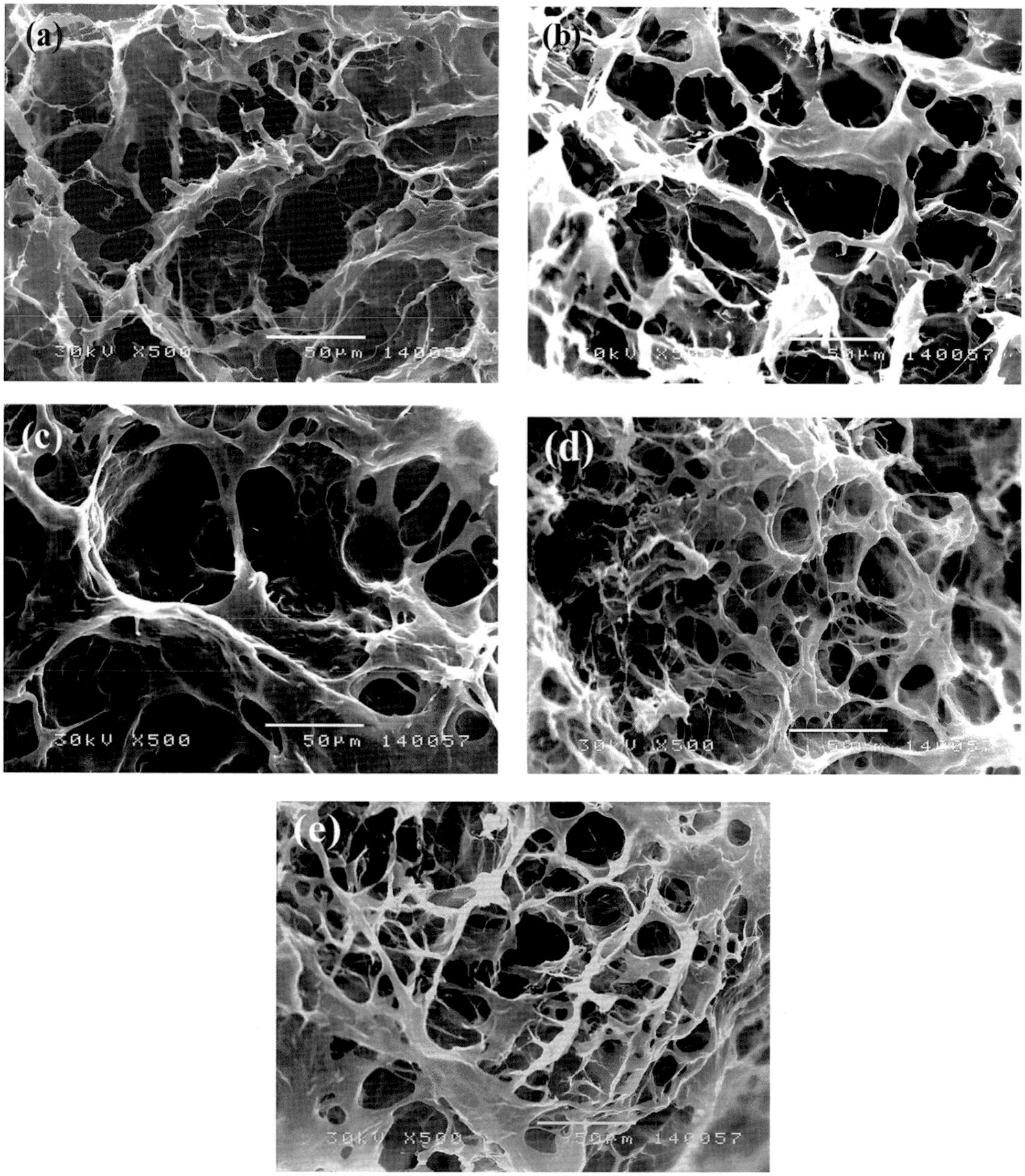

FIG. 2.10 SEM images of (XG-PVA) dressing hydrogel with different ZnO content: (A) 0%, (B) 0.5%, (C) 1%, (D) 2%, and (E) 5%. *With permission from Raafat, A.I., El-Sawy, N.M., Badawy, N.A., Mousa, E.A., Mohamed, A.M., 2018. Radiation fabrication of xanthan-based wound dressing hydrogels embedded ZnO nanoparticles: in vitro evaluation. Int. J. Biol. Macromol. 118, 1892–1902.*

30 kGy in small vials using 60Co gamma rays at a dose rate of 2 kGy/h, and the obtained cylindrical hydrogels were cut into disks, washed in excess water, and air-dried at room temperature up to constant weight. For the preparation of XG-AA/MgO nanocomposite hydrogels, 0.5%, 1%, and 3% MgO nanoparticles were added into the XG-AA mixture and then packed and irradiated under the same conditions as for nanoparticle-free hydrogels. FTIR analysis confirmed the incorporation of MgO nanoparticles within the XG-AA copolymer hydrogel. Their incorporation slightly decreases the gelation degree (94% and 96%) in comparison with XG-AA hydrogel. The swelling degree and network porosity were enhanced due to the presence of MgO nanoparticles with different sizes, morphologies, and surface charges, thus the expansion of the XG-AA hydrogel network and penetration of water were facilitated. In addition, the swelling behavior of XG-AA/MgO was pH-dependent (Fig. 2.11) and for all the investigated samples was Fickian at pH 1 and non-Fickian at pH 7. The described swelling behavior with a low swelling degree at low pH values is promising for use of XG-AA/MgO as a drug delivery carrier with a pH-responsive drug release profile upon oral administration.

In another study, Kennedy et al. combined XG with locust bean gum (LBG) to form thermally reversible hydrogels with silicon dioxide (SiO_2) nanoparticles with a diameter of 10–20 nm, which may be of interest for drug delivery. LBG is a nonionic polysaccharide derived from the seed of the carob tree. It consists of a linear backbone of (1,4)-β-D-mannose and every fourth mannose is substituted with a (1,6) α-D-galactose. In aqueous solutions, LBG is in a random coil conformation. XG and LBG interact to form a firm thermally reversible hydrogel. To prepare XG/LBG/SiO_2 hydrogels, dispersions of the nanoparticles in XG aqueous solution and LBG aqueous solution were mixed in a 1:1 ratio of XG to LBG to form a hydrogel with 1% (w/v) total polymer concentration and the nanoparticles content from 1% to 10%. The polymer-particle interactions with respect to nanoparticle concentration were analyzed by performing rheological characterization of viscoelastic parameters during the temperature ramping from 25 °C to 85 °C. It was observed that 10% of nanoparticles in the XG/LBG/SiO_2 hydrogel caused an increase in elasticity. Although it is well known that XG undergoes a conformational change at higher temperatures that impacts the rheological behavior, the complex modulus of the investigated nanocomposite hydrogels with 10% of the nanoparticles was relatively constant, which meant that the nanoparticles counteracted the effect of temperature increase on the material properties. SiO_2 particles are relatively nontoxic and can be used as carriers for poorly soluble drugs with high drug loading capacity (Kennedy et al., 2015).

Lu et al. developed the hydrogel based on XG-ALD and cross-linked with phosphatidylethanolamine (PE) liposomes. PE liposomes were prepared by sonication and extrusion. Aldehyde modification of XG was performed by oxidation of the vicinal hydroxyl groups in sugar ring into dialdehyde, with sodium periodate in aqueous solution at 80 °C, followed by dialysis and lyophilization. For the preparation of XG-ALD/PE liposomes hydrogel, ALD-XG was dissolved in PBS (0.1 M, pH 6.25) at 80 °C and mixed with the same volume of PE liposome dispersions (15, 20, or 30 mg/mL), at room temperature. XG-ALD/PE liposomes hydrogel formed quickly and the concentration of the XG-ALD solution and the PE liposomes dispersion did not significantly influence the gelation time. The establishment of the dynamic cross-linking balance within the hydrogel network and formation of the nanocomposite hydrogel was based on the Schiff base reaction between the aldehyde groups

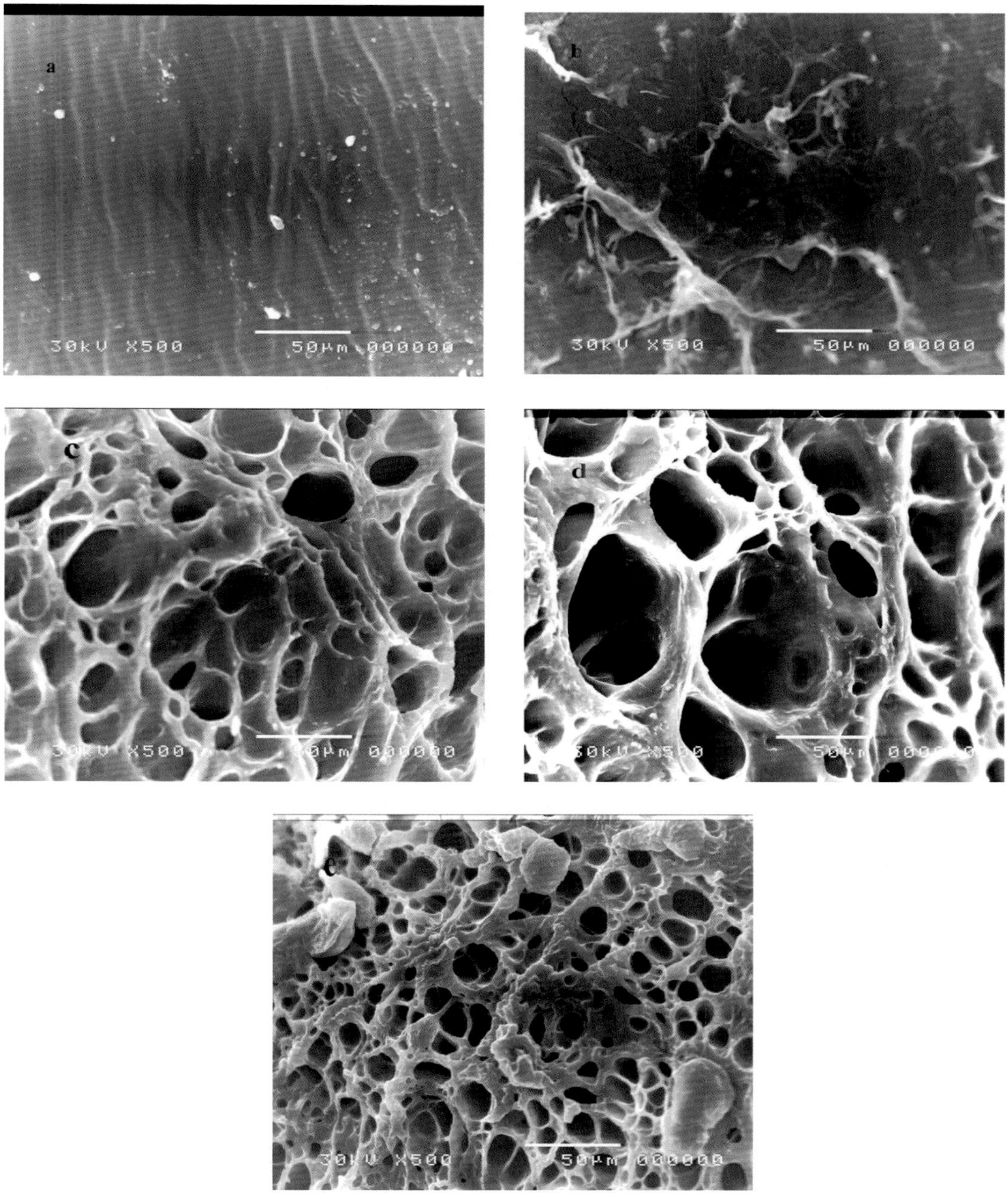

FIG. 2.11 SEM micrograph of (A) (XG-AA), (B) (XG-AA)/MgO 3% at pH 1, and (C) (XG-AA), (D) (XG-AA) MgO 0.5%, and (E) (XG-AA) MgO 3% at pH 7. Total feed solution (15%) and irradiation dose 30 kGy. *With permission from El-Sawy, N.M., Raafat, A.I., Badawy, N.A., Mohamed, A.M., 2020. Radiation development of pH-responsive (xanthan-acrylic acid)/MgO nanocomposite hydrogels for controlled delivery of methotrexate anticancer drug. Int. J. Biol. Macromol. 142, 254–264.*

of XG-ALD and the amino groups distributed around the outer shell of PE liposomes. The freeze-dried hydrogels have a porous structure with pore diameters in the range of 10–100 μm. The TEM image analysis of the hydrogel demonstrated the presence of the spherical liposomes with preserved integrity. Biodegradation of the ALD-XG/PE liposomes hydrogel was achieved by digestion of XG backbones in papain solution and the liposomes gradually released from the hydrogel. Due to the dynamic equilibrium of the Schiff base bonds, the hydrogel responds to various stimuli such as heat, pH variation, and histidine exposure, which could trigger the decomposition or/and regeneration of the hydrogel. The developed hydrogel can be considered as a multiresponsive, biocompatible, and injectable drug delivery carrier (Lu et al., 2012). Manconi et al. developed liposomes coated with CH/XG-based PEC (chitosomes) (Fig. 2.12) as a carrier for colon-specific delivery of protein

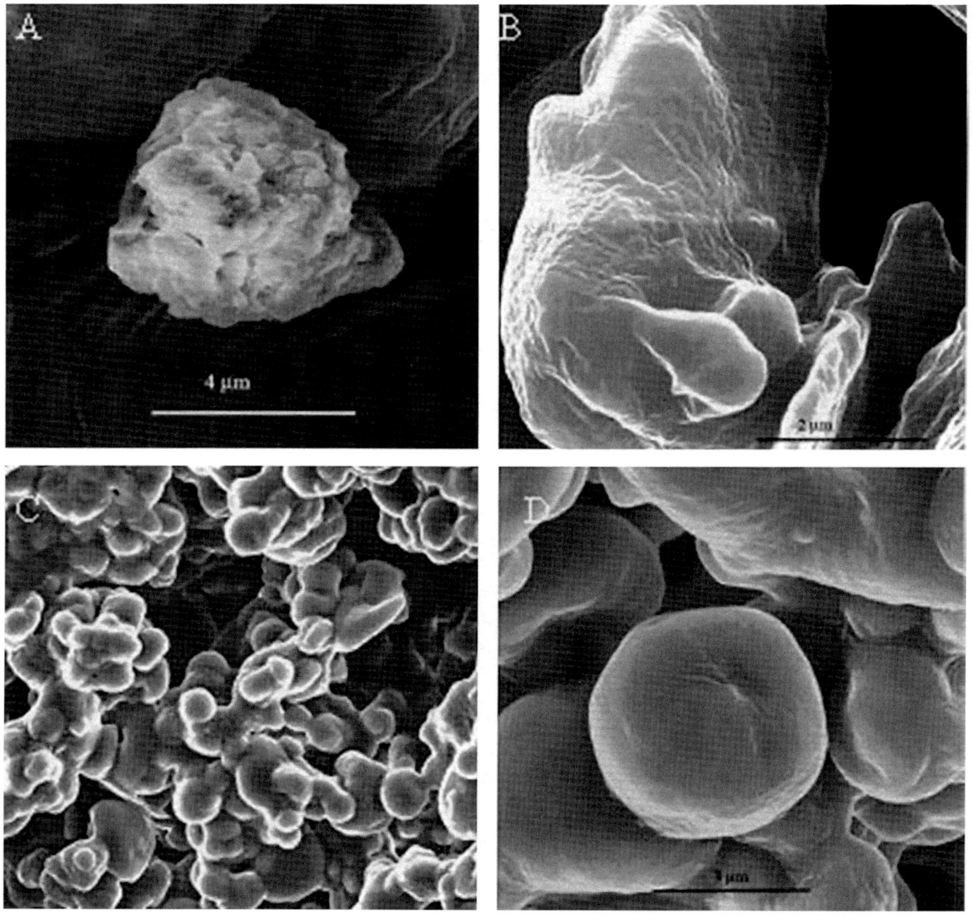

FIG. 2.12 SEM pictures of chitosomes: FD CH–XG 0.5/2.0 (A and B) and SD CH–XG 0.5/8.0 (C and D). *With permission from Manconi, M., Mura, S., Manca, M.L., Fadda, A.M., Dolz, M., Hernandez, M.J., Casanovas, A., Díez-Sales, O., 2010. Chitosomes as drug delivery systems for C-phycocyanin: Preparation and characterization. Int. J. Pharm. 392, 92–100.*

C-phycocyanin. Several CH/XG-microcomplexes were prepared at different CH/XG ratios and then freeze-dried or spray-dried. Rheological properties of the microcomplexes were studied to analyze the contribution of polymers in the reaction of microcomplexation and revealed a high influence of the XG amount on the microcomplex properties. Microcomplexes with elastic properties were considered convenient for the development of colon-specific drug carriers. It was established that spray-dried microcomplexes with CH/XG 0.5/8.0 mass ratio and freeze-dried microcomplexes with CH/XG 0.5/2.0 mass ratio are suitable for C-phycocyanin delivery. Chitosomes were prepared by coating C-phycocyanin-loaded liposomes with the CH/XG hydrogels using spray-drying or freeze-drying. Chitosomes prepared by a spray-drying technique using CH and XG in 0.5/8.0 mass ratio showed a regular surface and the release of the active ingredient by Fickian diffusion with a higher amount of drug released in simulated intestinal fluid (pH 7.4) in comparison with simulated gastric fluid (pH 1.2). The *in vitro* mucoadhesive study confirmed that the spray-drying method was advantageous to prepare C-phycocyanin-loaded chitosomes with excellent mucoadhesive properties for colonic drug delivery (Manconi et al., 2010).

The use of XG-based hydrogels in combination with oil-in-water microemulsions and nanoemulsions has also been described in the literature (Djekic et al., 2016; Koop et al., 2015; Mishra et al., 2018; Morsi et al., 2017). Oil-in-water microemulsions are thermodynamically stable transparent dispersions of oil nanodroplets in water, stabilized by a suitable surfactant and a cosurfactant. By combining microemulsions with hydrogels, nanocomposite systems can be obtained with preserved microemulsion properties (e.g., ease of manufacturing, long-term stability, enhanced drug loading capacity, and drug delivery potential) as well as optimized rheological properties and drug release kinetics. Optimization of the rheological properties and nanodroplet size of such hybrid systems can be promising for improving their application properties and topical drug delivery performances. The study of Djekic et al. demonstrates the combining of XG hydrogel with the nonionic oil-in-water microemulsion (M) to formulate a nanocomposite carrier for the percutaneous delivery of ibuprofen (Djekic et al., 2016). The 5% XG physical hydrogel was introduced into the drug-loaded microemulsion preconcentrate consisting of nonionic surfactants (Labrasol® and Solubilisant gamma®), the oil (isopropyl myristate), and ibuprofen, and after that water was added up to 100%, at the room temperature (Djekic et al., 2012). The final concentration of the polymer in the nanocomposite formulation ranged from 0.25% to 1.0%. All XG-containing microemulsions (HTM1-HTM4) were homogeneous, clear to slightly opalescent (Fig. 2.13), without traces of the undissolved drug, grittiness, or lumps.

Although it is well known that the consistency of XG aqueous solutions and hydrogels may change at temperatures above room temperature (García-Ochoa et al., 2000; Katzbauer, 1998), the consistency of the prepared systems depended only on the XG content. They were thick liquids at 0.5% of XG and soft semisolids at the higher investigated concentrations up to 1% and they were physically stable under ambient conditions as well as at 5 ± 3 °C and 40 ± 1 °C for 6 months. The oil-in-water microemulsion structure with the average droplet size (Z-ave) of 14.34 ± 0.98 nm was well preserved, as was demonstrated by conductivity and DLS measurements. The incorporation of XG hydrogel into the microemulsion resulted in the formation of the shear-thinning pseudoplastic systems with thixotropy. The apparent viscosity strongly depended on the polymer content, and the hysteresis area also increased exponentially with the XG concentration. XG maintained its intra- and intermolecular physical

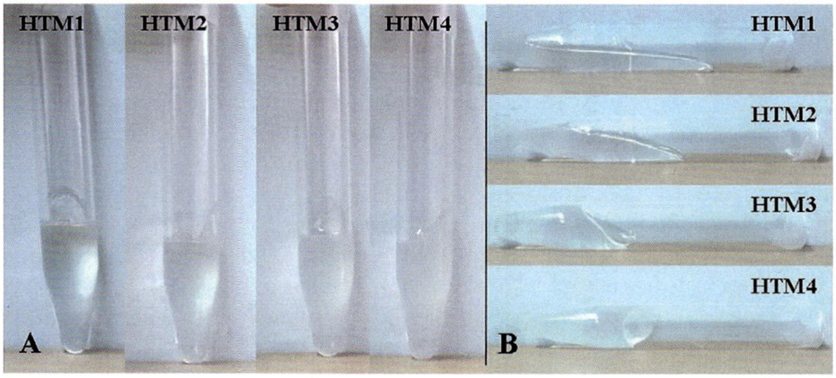

FIG. 2.13 The appearance of the samples prepared from the microemulsion with added xanthan gum in a concentration of 0.25% (HTM1), 0.50% (HTM2), 0.75% (HTM3), and 1.0% (HTM4), after 48 h from preparation: (A) in the upright cuvettes immediately after centrifugation test (an apparent positive meniscus in the cuvettes was formed during centrifugation) and (B) in cuvettes tilted at an angle of 90 degree for 5 min. *With permission from Djekic, L., Martinovic, M., Stepanović-Petrović, R., Micov, A., Tomić, M., Primorac, M., 2016. Formulation of hydrogel-thickened nonionic microemulsions with enhanced percutaneous delivery of ibuprofen assessed in vivo in rats. Eur. J. Pharm. Sci. 92, 255–265.*

interactions related to the weak hydrogel structure within the aqueous phase of the formulated nanocomposite systems, and a significant increase in the degree of interaction between the polymer molecules within the investigated concentration range was observed. The results of physicochemical characterization proved the formation of nanocomposite carrier comprising oil nanodroplets in XG hydrogel with the physical polymer network, the strength of which depended on the polymer concentration. In a similar study, Mishra et al. prepared nanocomposite carriers from XG hydrogel and oil-in-water microemulsion comprising diisopropyl adaptate, Cremophore®-EL, ethanol, and distilled water, and evaluated them as carriers for cutaneous administration of an antifungal drug liranaftate. The final concentrations of XG and the drug in the formulated system were 1.5% and 2%, respectively. The systems differed in terms of oil, surfactant, cosurfactant, and water content. The hybrid system was able to maintain the appearance and the structure of the oil-in-water microemulsion as well as the XG matrix. Physicochemical characterization revealed that all formulations were homogenous, transparent, colorless, and optically isotropic. The high conductivity (119–304 S cm^{-1}) confirmed the water-continuous structure. The average droplet size was 52.53–93.40 nm with a narrow size of distribution. The apparent viscosity of the nanocomposite carriers was suitable for the achievement of skin retention effect, which is desirable from the aspect of improving the percutaneous penetration of the drug (Mishra et al., 2018). *In situ* gelling nanoemulsions for ocular delivery of the antiglaucoma drug acetazolamide were developed by Morsi et al. (2017). The drug-loaded oil-in-water nanoemulsion composed of polysorbate 80 and/or Cremophor® EL as a surfactant, Transcutol® P or propylene glycol as a cosurfactant, and peanut oil was used to formulate ion-induced *in situ* gels when mixed with tear fluid for ocular administration. A small droplet size (11–15 nm) was observed in the nanoemulsion formulations. Although all nanoemulsion based *in situ* gelling formulations showed a high degree of transparency in the sol state, due to a low degree of light scattering by very small oil droplets and the polymer molecules, transparency was reduced by about 15% upon gelation. Physical cross-linking of the polymers reduces the clarity of the gel. The addition of XG (0.2%–0.6%) and gellan gum (0.3%) to the nanoemulsion significantly reinforced the viscosity and mucoadhesive force of *in situ* gel. This *in situ* gelling formulation thus presents a promising ocular drug delivery system that would provide less frequent administration and thus improve patient compliance for the treatment of glaucoma (Morsi et al., 2017). The observed relatively high viscosity of the XG/gellan based system was attributed to the anionic nature of XG and the electrostatic repulsions between the side chains that make its molecules extend (Khouryieh et al., 2007), as well as to the high molecular weight which increases the intermolecular association among polymer chains. In addition, the XG molecules align and physically cross-link via hydrogen bonding to form a weak hydrogel network, which would immobilize the free water and increase the viscosity. Additionally, the XG/gellan-based formulation showed good stability at all studied temperatures. Koop et al. combined XG with galactomannan (GM) (locust bean gum) from *Schizolobium parahybae* to prepare the nanocomposite hydrogel from the curcumin-loaded microemulsion for cutaneous administration. The oil-in-water microemulsion consisted of isopropyl myristate, lecithin, polysorbate 80, water, and curcumin. For the preparation of the hybrid hydrogel, XG and GM (1:1) were dispersed in distilled water at 1.25% total polysaccharide concentration and the microemulsion was added into the obtained hydrogel. XG forms physical hydrogels when mixed with the equivalent amounts of GM. The formation of such hydrogels involves synergistic interactions between two polysaccharides. The microemulsions were isotropic, transparent liquids with the mean apparent hydrodynamic diameter of the oil nanodroplets of 231.8 ± 7.6 nm. The hybrid hydrogel had the rheological behavior typical for weak hydrogels up to 45 days. Evaluation of the nanocomposite hydrogel by oscillatory rheometry at 32 °C (i.e., the temperature of the skin surface) showed the viscoelastic behavior that allows prolonged retention over the skin. Moreover, elastic recovery was observed immediately after the stress was removed. The mechanical spectra demonstrated the elastic behavior of the nanocomposite hydrogels and cross-linking between the molecules of XG and GM. The binary hydrogel network presented mechanical stress resistance for 45 days at the elevated temperature (45 °C). These nanocomposite carriers showed high curcumin encapsulation efficiency (99.5%) and sustained drug release for 12 h (Koop et al., 2015).

2.5 Drug delivery potential of xanthan gum hydrogels as nano- and microcarriers for different routes of administration

XG and XG-based materials have been investigated extensively for the preparation of controlled-release formulations for different routes of administration (e.g., matrix tablets for sustained and predictable drug release after oral administration, injectable shear-thinning hydrogels, hydrogels, and microsponges for cutaneous and percutaneous delivery of drugs, ophthalmic liquid dosage forms with prolonged retention in the precorneal area, mucoadhesive intranasal preparations, suppositories containing mucoadhesive microspheres) (Rowe et al., 2009; Singhvi et al., 2019). Recently, researchers have focused on the development of advanced hydrogels based on XG alone or in combination with other polymers, or modified/grafted-XG which are stimuli-responsive drug delivery systems able to respond to pH variation in different zones of the body, body temperature, or the presence of specific biological molecules, and release the drug load in a controlled manner (Singhvi et al., 2019). XG-based hydrogels are preferred as drug delivery systems for oral administration, but other routes (parenteral, (trans)dermal, ocular, nasal, rectal) are also considered.

2.5.1 Oral drug delivery

The advantages of the oral route of drug administration such as high patient compliance due to ease of administration, absence of sterility constraints, and versatility of the dosage form contribute to a permanently high interest in scientific research related to this area. Due to the high stability of XG-based hydrogels under acidic and neutral pH, the drug can be protected in the stomach as well as the stomach mucous can be protected by the encapsulation of the drug within such a carrier. This is a particularly useful strategy for the oral delivery of protein drugs and antibiotics that are unstable in the acidic environment of the stomach, and for cytotoxic drug substances. Furthermore, there are a large number of examples of the XG hydrogels-based carriers that are designed to achieve controlled (usually pH-dependent) and adjustable drug release. The drug release rate is usually limited in the stomach and increases in the small intestine or colon, i.e., in alkaline conditions where ester bonds are hydrolyzed (Petri, 2015). Targeted delivery to the colon is the important goal for the optimization of the treatment of inflammatory-based disorders and colorectal cancer (El-Sawy et al., 2020).

It has been demonstrated that PECs based on CH and XG have pH-sensitive swelling kinetics and thus a high potential for oral drug delivery. The aim of the investigation of Merlusca et al. was to incorporate neomycin sulfate into a hydrophilic XG/CH-based PEC to obtain a controlled-release oral drug delivery system suitable to protect the drug. FTIR analysis confirmed ionic interactions between the polymers as well as XG/CH/neomycin sulfate hydrogen bonding. Freeze-dried XG/CH complexes had a smooth surface with no apparent cracks and fractures, while the micrograph of the XG/CH/neomycin sulfate complex showed that the drug was adsorbed on the XG/CH complex in the form of distinct particles. The study showed a significantly higher swelling degree in the simulated intestinal fluid at pH 7.2 than in simulated gastric fluid at pH 2.0, indicating possible higher drug release in simulated intestinal fluid (Merlusca et al., 2016). An *in vivo* study on healthy Wistar rats showed the amelioration of the biochemical parameters (creatinine, urea, serum uric acid, and C-reactive protein levels) and histological changes of the colon due to reduced drug toxicity. In another study, the nanocomposite hydrogel based on XG-g-PAA with *in situ* synthesized gold nanoparticles was investigated for the controlled delivery of amoxicillin. Amoxicillin is a semisynthetic, orally absorbed antibiotic with very short residency time in the stomach due to its hydrolytic degradation in gastric acid (pH 1.2). The drug was loaded into nanocomposite hydrogel by swelling. The presence of the nanoparticles provided the strong linkage of drug molecules with polymer and increased the maximum drug loading efficiency up to ~85%, which was higher than in the nanoparticle-free sample. The release profiles of amoxicillin took place in accordance with the Korsmeyer–Peppas model. The presence of nanoparticles controlled the *in vitro* drug diffusion rate from the carrier and it was higher in alkaline medium (pH 9.2) in comparison to neutral (pH 7) and acidic (pH 2.4) media.

The nanocomposite was rated as biodegradable based on the results of the soil burial method for 75 days as well as FTIR and FE-SEM analysis results (Singh et al., 2020).

The oral delivery of different therapeutic proteins is limited by catalytic degradation in the stomach, poor permeability across the GIT mucosa due to hydrophilicity and large molecule size, and first pass metabolism in the liver (Sabaa et al., 2019). Polysaccharide-based hydrogels are recognized as a suitable option to design a carrier suitable for the avoidance of denaturation during encapsulation of different proteins. Sabaa et al. synthesized a hydrogel based on XG and PVI as a BSA carrier intended for controlled-release. The drug loading and encapsulation efficiency increased with increasing gelation time and loaded BSA concentration, while these parameters decreased with increasing the polymer concentration. The results of *in vitro* BSA release in phosphate buffer saline (pH 6.8) showed that an increase in the polymer concentrations led to an increase in BSA release. Kinetic studies of the *in vitro* release of BSA from the XG/PVI/BSA matrix followed non-Fickian and case II transport mechanisms. In addition, cytotoxicity results on normal cell lines (VERO cells) showed good biocompatibility of this hydrogel. Sodium dodecyl sulfate-polyacrylamide gel electrophoresis (SDS-PAGE) analysis confirmed that the structural integrity of BSA was not affected by the encapsulation or release conditions. For that reason, this hydrogel can be used as an efficient carrier for the oral delivery of proteins (Sabaa et al., 2019). Abu Elella et al. used cross-linked XG-g-PVI hydrogel as a biodegradable, pH-sensitive carrier for delivery of BSA in the intestines. BSA uptake and release from loaded hydrogel were performed in acidic (pH 1.2) and slightly alkaline (pH 7.4) solutions, within various time intervals (from 12 to 120 h) at 37 °C. The swelling rate and the accumulative BSA release percentage were faster in alkaline solution than in pH 1.2 medium. Different parameters (graft yield percentage (GY%) and MBA, BSA, and graft concentrations) affected loading, encapsulation efficiency (EE%), and *in vitro* release of BSA. EE% increased with the increase in GY%, BSA, and graft concentration and the decrease in MBA concentration, while the accumulative BSA release percentage increased with the GY% increase, and BSA, MBA, and graft concentrations decrease. At the least amount of BSA loaded (1 mg/mL) in the hydrogel matrix, a large pore fraction was formed and higher swelling and BSA release occurred in both media, while at the highest BSA load (3 mg/mL), the number of small pores was reduced and they shrank due to formation of the high crystalline domain. Such changes decreased the BSA release. The mechanism and kinetics of *in vitro* BSA release was studied by application of the Korsmeyer model for the analysis of the obtained BSA release profiles. The diffusion exponent values (n) were from 0.53 to 0.86; therefore, the release of BSA from the hydrogel matrix was sustained by a combination of diffusion and erosion. The BSA release rate gradually increased with the increase in the incubation time up to 120 h. The BSA release in pH 7.4 medium was faster than in pH 1.2 medium, although in both media followed non-Fickian release. All investigated BSA-loaded hydrogels were safe for the human lung normal cell line and showed good inhibitory effect against *Aspergillus niger* and *Staphylococcus aureus* (Abu Elella et al., 2020).

pH-responsive XG/AA nanocomposite hydrogels were used as a hydrogel matrix for MgO nanoparticles loaded with methotrexate (El-Sawy et al., 2020). The application of methotrexate in the treatment of cancer and autoimmunological diseases is related to several problems such as wide distribution and fast metabolism and excretion. Thus, it is a suitable candidate for the development of a controlled release drug delivery system. Incorporation of the drug-loaded MgO nanoparticles into XG/AA hydrogel enabled pH-depended release in the simulated stomach medium and in the simulated intestine medium (pH 7). The drug release profiles corresponded with the swelling degree of the nanocomposite hydrogels, which was low (30%–80%) in the simulated stomach medium, while it increased up to 900%–1600% in the simulated intestine medium. Therefore, the release and absorption of the cytotoxic drug were regulated by the swelling process and it was achieved mainly in the intestine, while it was minimal in the stomach. The incorporation of MgO nanoparticles into the hydrogel matrix enhanced the porosity of the nanocomposite hydrogel, increased the swelling degree, and prolonged the drug release.

Quercetin is a flavonoid with antioxidant and antiinflammatory properties, and is poorly absorbed when orally administered. Caddeo et al. prepared CH/XG-based microparticles to increase quercetin oral bioavailability and optimize its release in the colon. CH/XG-based hydrogel with incorporated quercetin was spray-dried to obtain microparticles. Microparticles were smooth and spherical, with a diameter of approximately 5 μm, and quercetin encapsulation efficiency of 63 ± 7%. They were mixed with lactose (lactose-to-polymers mass ratio 4:1), compressed into tablets, and coated with Eudragit® L100 to prevent degradation in acidic pH. Prepared tablets provided resistance to acidic conditions (13% of drug released at pH 2.0 after 2 h), allowing complete drug release in alkaline pH (7.4), mimicking the colonic environment. The release was controlled by non-Fickian diffusion of the dissolved drug out of the swollen polymeric tablets. Tablets did not disintegrate during the drug release test (24 h), which was explained by strong interactions between CH and XG in microparticles and built of a structured gel layer around the tablet core when they came into contact with the release medium. For that reason, microparticle-based tablets could be considered a promising dosage form for quercetin delivery to the colon in the oral therapy of inflammatory-based disorders (Caddeo et al., 2014).

You et al. prepared and characterized novel pH-sensitive konjac gum-XG-glycerol-sodium alginate hydrogel (KGM-XG-GLY-SA hydrogel) for the colon-specific delivery of drugs. A hydrophilic drug, hydrocortisone sodium succinate, was chosen as a model drug and successfully loaded into the hydrogel without the use of any organic solvent. The drug-loaded hydrogel exhibited good sustained release with minimal drug release at pH 1.2 (23.40% after 2 h), pH 6.8 (25.88% after 4 h), and significantly higher (70.20% after 4 h) at pH 7.4. It was assumed that the addition of glycerol into the hydrogel prevented the rapid dissolution of material at higher pH of the intestine. That ensured the controlled release of the entrapped drug. All the *in vitro* results showed that the KGM-XG-GLY-SA hydrogels had good colon-targeting characteristics. In addition, the *in vivo* study on laboratory rats or mice indicated that KGM-XG-GLY-SA hydrogels were nontoxic, reduced the spleen and thymus index, and had an obvious effect on the treatment of ulcerative colitis. Therefore, it can be considered that the developed hydrocortisone sodium succinate/KGM-XG-GLY-SA hydrogels, as colon-targeting vectors, might have great potential application in the treatment of ulcerative colitis (You et al., 2015).

Novel therapeutic strategies for the targeted delivery of antineoplastic drugs for the treatment of colorectal cancer include the utilization of the XG-based hydrogels as carriers with the potential to protect the drugs from oxidative degradation, improve their bioavailability and delivery to tissues, and reduce toxicity. Trombino et al. designed pH-sensitive XG-MA/MBAA-based hydrogels as carriers for PUFAs (DHA and ALA) with controlled swelling and drug release characteristics in the acidic (pH 1.2) to weakly acidic (pH 6.8) conditions of the upper gastrointestinal tract. On the other side, the carriers enabled the enhanced targeted PUFAs release in the colorectal segment due to large swelling degree in near-neutral pH 7.4. In the first stage, the burst effect occurred with the fast release of about 40% of the PUFAs from the outer surfaces of the carriers, whereas the portion of them enclosed in the internal spaces released more slowly. Such initially fast release of antineoplastics is favorable for tumor treatment. The antioxidant activity of the hydrogel was efficient to protect the PUFAs from lipid peroxidation for up to 2 h and increased significantly the ability of ALA to reduce HCT116 human colon adenocarcinoma cells growth *in vitro* in a time-dependent manner at all the concentrations used (5–50 μM) in comparison with free ALA. The XG-based hydrogel comprising 50 μM of DHA enhanced the HCT116 cells apoptosis; however, the enhancement was not significantly higher versus the already high antineoplastic effect of the free fatty acid. The comparison with the effects of the encapsulated and free forms of ALA and DHA pointed that the XG-based carriers themselves likely affected other biological processes (e.g., proliferation) and reduced HCT116 cell growth. The investigated hydrogel was degradable by colonic enzymes (Trombino et al., 2019).

2.5.2 Parenteral drug delivery

Shear-thinning hydrogels have attracted attention as parenteral drug delivery systems acceptable for injection into the different tissues. Such hydrogels are high viscosity systems that start to flow under shear stress (during administration) and recover their rheological properties at the injection site (Gutowska et al., 2001; Guvendiren et al., 2012; Lu et al., 2012). Therefore, they easily accommodate irregularly shaped compartments, remain at the desired position, and deliver the drug to the targeted tissues (Yu and Ding, 2008). The transition from the low viscosity polymer solution (sol) to the high viscosity hydrogel (gel) occurs upon removal of the stress, at the body temperature. Copolymeric hydrogels are particularly promising for optimization of the sol-gel transition temperature, biocompatibility and biodegradability of injectable formulations, the drug release profile, and rheological behavior to avoid leakage from the injection site.

Very recently, hydrogels based on XG and silk fibroin, using STP as a cross-linker, with injectable and self-healing properties, were prepared by Zhang et al. (2020). The injectability of prepared hydrogels was estimated by a three-stage model of oscillation-shear-oscillation, which was used to evaluate the destruction and regeneration of the hydrogels during the injection process. The hydrogels immediately recovered to the original storage modulus when shear was removed. The hydrogels were 3D printed into the self-supporting constructions comprising hydrogel fibers with connected porous structures. It was shown that fibers of hydrogels containing silk fibroin had a smaller width than formulation without silk fibroin. Rheological measurements indicated that silk fibroin-containing hydrogels formed rapidly and had more pronounced solid-like gel behavior than hydrogels without silk fibroin. Hydrogel structures were destroyed under a strain larger than critical strain but were rebuilt under a small strain within 120 s, suggesting their self-healing property. The incorporation of silk fibroin into hydrogels improved stiffness and retained recoverability. Due to the presence of carboxylate and phosphate groups in the hydrogel structure, they were able to absorb enough liquid electrolytes, which lead to effective ionic conductivity. Ion-conductive hydrogels with injectable, self-healing, controlled-release, and noncytotoxic properties were obtained (Zhang et al., 2020).

Liu and Yao prepared injectable hydrogels by blending XG and MC for sustained release and long-term delivery of doxorubicin. XG/MC blends were highly viscous solutions with shear-thinning properties at room temperature. The temperature change from 23 °C to 37 °C lead to the formation of thermoresponsive networks due to the presence of MC, via intermolecular hydrophobic interactions that caused XG/MC blend solutions to gel. The gelation times, transition temperatures, and storage moduli of the blends could be affected by XG and/or MC concentrations. With the increase of their concentrations, the storage moduli increased, while the gelation times and transition temperatures decreased. Both *in vitro* and *in vivo* studies demonstrated that the optimized polymer blend solution (2% of XG and 10% methylcellulose (XG2/MC10)) immediately recovered its high viscosity and rapidly formed hydrogel at body temperature after injection using a syringe. Doxorubicin was loaded into the hydrogel by mixing the drug with the optimized polymer blend solution at room temperature. The gelation, biocompatibility, and complete biodegradation after 36 days of the optimized hydrogel were demonstrated *in vivo*, by implantation of the hydrogel in rats (Fig. 2.14). The amount of the drug released *in vitro* depended on its concentration in the hydrogel (Liu and Yao, 2015).

The injectable hydrogel obtained via self-cross-linking of XG-ALD and CH-CM exhibited biocompatibility, biodegradability, self-healing characteristics, and antienzymatic hydrolysis (Huang et al., 2018). After injection, the liquid mixture of the polymers temporarily maintains a stable structure due to physical interactions (mostly hydrogen bonds) which are quickly replaced by the chemical cross-linking via the Schiff's base reaction. Therefore, the presence of the tissue exudates did not affect the hydrogels and they enabled controlled *in vitro* release of BSA for 10 h. Additionally, loading of angiogenic factor (VEGF) in these injectable hydrogels is considered promising in abdominal wall reconstruction.

2.5.3 Cutaneous and percutaneous drug delivery

XG-based hydrogels have been formulated for dermal (on the skin surface or into the deeper skin layers) and transdermal drug delivery. Their relatively high bioadhesion onto the skin surface, biocompatibility, biodegradability, and nontoxicity are well accepted for wound healing and (per)cutaneous drug delivery. The hydrophilicity of XG makes it convenient to formulate water-swellable hydrogels capable of absorbing and holding a large amount of water or tissue exudates while maintaining their physical performances and stability. The superporous hydrogel synthesized by graft copolymerization of HEMA and AA onto XG (Gils et al., 2009) can absorb quickly, swell, and retain aqueous solutions up to or more than 100 times their own weight due to highly porous structures that allow very fast penetration of water into the core of the dried matrix. Furthermore, such hydrogels are suitable bases for the immobilization of enzymes, being able to provide higher thermal stability. Negatively charged films based on XG hydrogels are proposed as carriers for positively charged lysozyme (Xu et al., 2016). The release of lysozyme was very limited at pH 7, and the films preserved the native structure of the enzyme and thus provided bactericidal activity which is promising for use in wound dressings. Wound dressings with ZnO nanoparticles incorporated in nanocomposite XG/PVA hydrogel with adequate porosity to control the fluid uptake ability (554–664%), water retention (1281–1603%), and water vapor transmission

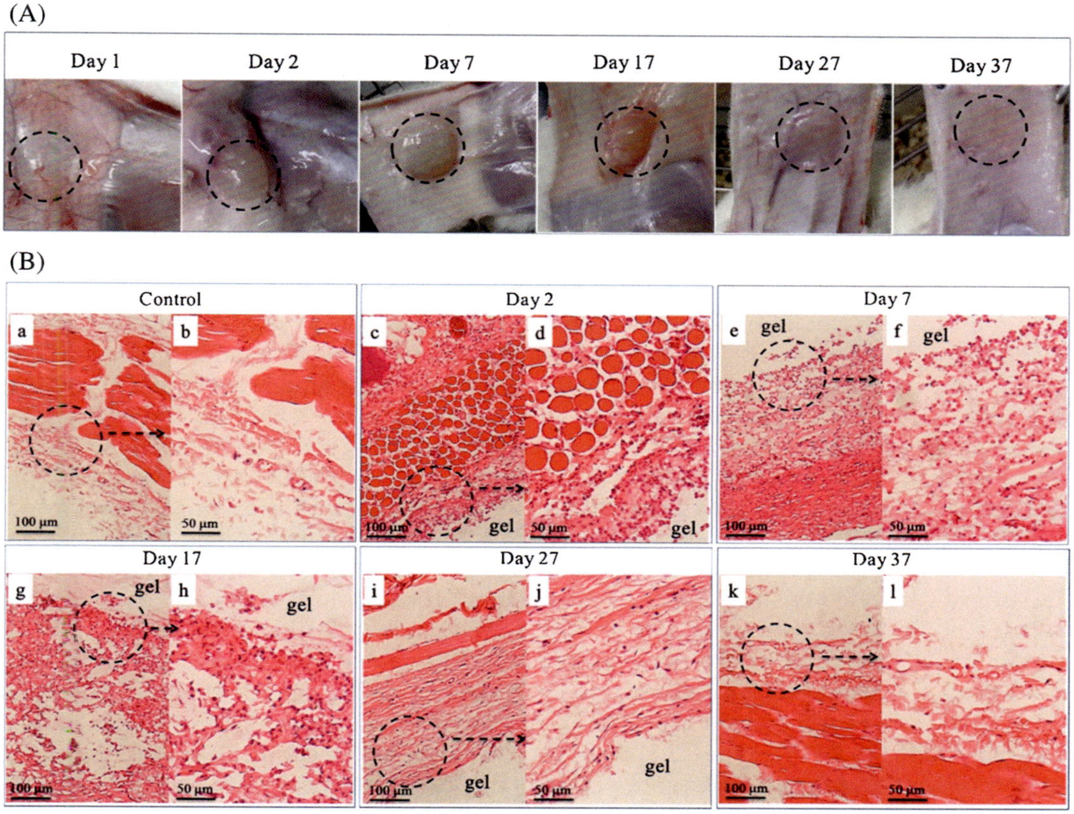

FIG. 2.14 (A) Photos of the hydrogels after injection of XG2/MC10 into the backside subcutaneous tissues of SD rats; the first photo (day 1) was taken at 30 min postinjection; (B) histological images of the tissues around the implanted XG2/MC10 hydrogel with H&E staining; the tissues were taken on day 2 (c and d), day 7 (e and f), day 17 (g and h), day 27 (i and j), and day 37 (k and l) postinjection. The gel labeled in the images indicates the location of remnant hydrogel. *With permission from Liu, Z., Yao, P., 2015. Injectable thermo-responsive hydrogel composed of xanthan gum and methylcellulose double networks with shear-thinning property. Carbohyd. Polym. 132, 490–498.*

rate (167–184 g/m^2 h), which was sufficient to keep a wound's surface moist and accelerate wound healing, were designed by Raafat et al. (2018). The dressings provide efficient antimicrobial activity against *Staphylococcus aureus*, *Escherichia coli*, and *Candida albicans*, due to the antibacterial properties of ZnO nanoparticles. Moreover, these hydrogels can be easily and completely removed from the fragile regenerated skin surface.

Curcumin-loaded XG/GM hydrogels combined with oil-in-water microemulsion were evaluated as *in situ* gel-forming wound healing drug delivery systems (Koop et al., 2015). The drug encapsulation efficiency was high (99.5%). Their biocompatibility was assessed by using the chorioallantoic membrane (*Gallus domesticus*). The hydrogels were nontoxic and thus safe for prolonged skin contact during the treatment. The *in vitro* cytotoxicity analysis suggested the absence of cutaneous membrane irritation. The *in vitro* release of curcumin was slow and followed the first-order kinetic profile, and it was related to the incorporation of the drug into the microemulsion nanodroplets as well as with the rheological properties of the hydrogels. The *in vitro* skin permeation analysis indicated that the level of curcumin at *stratum corneum*, epidermis, and dermis ranged from 2.15 to 2.50 g/mL. The suppression of topical inflammation on a mouse ear caused by the investigated hybrid hydrogels was 76.8%–63.2%, hence it was higher than the positive control. The nanocomposite hydrogels were recognized as potentially useful in the treatment of skin cancer and psoriasis.

Mishra et al. developed the nanocomposite hydrogel from oil-in-water microemulsion to enhance the permeation of liranaftate (2%) in the treatment of cutaneous fungal infection. The drug-loaded hydrogel and the saturated drug solution were compared *in vitro* and *ex vivo*. The optimized formulation was composed of Di-isopropyl adaptate (4.5%, w/w), Cremophor®-EL (30%, w/w), ethanol (10%, w/w), water (52%, w/w), and 1.5% (w/w) of XG. The permeation rate of liranaftate from the optimized formulation was higher in comparison with other investigated hydrogels and saturated drug solution. The optimized nanocomposite hydrogel was stable and enabled six times higher drug deposition capacity in the skin than saturated drug solution, due to its composition and structure, drug solubility enhancement, and modification of skin diffusivity by the microemulsion constituents. In addition, it did not cause any erythema in rats. The average zone of inhibition against *C. albicans* (25.52 ± 0.26 mm) was higher in comparison with saturated drug solution (13.44 ± 0.40 mm). The current study demonstrated the enhancement of cutaneous antifungal activity by the formulation of XG-based nanocomposite hydrogel (Mishra et al., 2018).

Maiti et al. prepared microsponges based on XG and ethylcellulose by the double emulsification-solvent evaporation technique. Prepared microsponges were air-dried and then dispersed in a carbomer gel base. The carrier was formulated using different drug/polymers mass ratios (0.4:1, 0.6:1, or 0.8:1) to achieve controlled delivery of diclofenac sodium to the skin. SEM showed the porous, spherical structure of the microsponges. An increase in the drug/polymers ratio increased their yield (79.1%–88.5%), drug entrapment efficiency (50.0%–64.1%), and mean diameter (181–255 μm). FTIR spectroscopy and DSC analyses demonstrated the chemically stable, amorphous nature of the drug in the microsponges. Microsponges with a high drug/polymers ratio exhibited the highest flux of the entrapped diclofenac sodium (0.0382 mg cm^{-2} h^{-1}), while the flux of the entrapped drug through excised rat skin decreased by 19.9% and 17.0% for the microsponges prepared at low and intermediate drug/polymers ratios, respectively. The drug permeation through rat skin followed Higuchi's diffusion kinetic model both for pure drug and diclofenac sodium entrapped into microsponges dispersed in the hydrogel base. The microsponges prepared at the lowest drug/polymers ratio exhibited the slowest drug permeation profile during 8 h and were considered the most suitable for controlled drug delivery. The gel containing optimized microsponges was comparable to the commercial gel formulation and did not show serious dermal reactions. For that reason, the microsponge system obtained at the lowest drug/polymers ratio could be considered a suitable carrier for the controlled release of diclofenac sodium to the skin (Maiti et al., 2011).

Djekic et al. developed nanocomposite hydrogels from oil-in-water microemulsion and XG with ibuprofen in a therapeutical concentration for percutaneous delivery (5%). A negative linear correlation was observed between the spread diameter and increasing concentration of XG and thixotropy of the investigated nanocomposite systems. Those prepared with the lower concentration of XG were more suitable for spreading on a larger skin area, while their thixotropy may prolong the retention at the site of application. The release of ibuprofen followed zero-order kinetics and the amount of ibuprofen released after 12 h ranged from $10.98 \pm 1.78\%$ to $22.80 \pm 4.00\%$. The release rate and the transmembrane flux decreased with increasing concentration of XG due to the viscosity-related reduction of the diffusion rate of the drug substance within the carrier and from the carrier in the acceptor medium. Additionally, the release rate was affected by the dynamic of the process of partition of ibuprofen within the colloidal structure of the nanocomposite carriers (i.e., between the oil droplets and the external aqueous phase comprising the polymer network). Therefore, the investigated nanocomposites possess characteristics of the sustained drug release carriers. The optimized nanocomposite formulation, which was prepared with 0.25% of XG, exhibited the highest spreadability (5.3 ± 0.5 cm) and ibuprofen release rate (1.65% h^{-1}). *In vivo* assessment of the optimized nanocomposite formulation for antihyperalgesic and antiedematous effects showed significantly higher efficacy in producing antihyperalgesic and at a lower extent antiedematous activity in prophylactic topical treatment protocol, in comparison with the referent carbomer hydrogel, while they were comparable in producing antihyperalgesic/antiedematous effects in the therapeutic protocol. The investigated nanocomposite formulation did not cause skin irritation *in vivo*. The developed nanocomposite formulation can be a promising drug delivery system for chronical topical treatment of osteoarthritis with fewer systemic side effects (Djekic et al., 2016).

XG hydrogel-based drug carriers could also be prepared to achieve transdermal drug delivery. The recent study of Surini et al. aimed to prepare coprocessed excipients of XG and cross-linked amylose with improved characteristics in comparison with single polymers as a matrix for a transdermal hydrogel with diclofenac sodium as a model drug. Four types of coprocessed excipients were prepared using two methods which combined cross-linking and coprocessing steps. Method A first involved the cross-linking (CL) of amylose (A) molecules using STMP (6% or 12%) as a cross-linker, and then the coprocessing (Co) of the polymers by mixing their aqueous dispersions and subsequent drying of the mixture in a drum dryer. Excipients labeled as Co-CLA6-XG and Co-CLA12-XG were obtained by this method. Method B involved the reversed sequence of steps, i.e., the coprocessing was performed first by mixing aqueous dispersions of amylose and XG and drying them in a drum dryer, and the subsequent step was the cross-linking of the coprocessed excipient with STMP (6% or 12%). Excipients labeled as CL6-Co-A-XG and CL12-Co-A-XG were obtained by this method. The obtained coprocessed excipients, containing amylose and XG in a 1:2 mass ratio, were used for the preparation of solid and compact transdermal hydrogels with diclofenac sodium. An *in vitro* penetration test was performed using Franz diffusion cells and rat skin membrane, while an *in vivo* penetration test was performed by applying the hydrogel to the abdominal skin of male Sprague-Dawley rats and measuring the plasma drug concentration. Both tests lasted 12 h. *In vitro* penetration results showed that the flux from Co-CLA6-XG, Co-CLA12-XG, CL6-Co-A-XG, and CL12-Co-A-XG transdermal hydrogels was $655.23 \pm 116.43\ \mu g\ cm^{-2}\ h^{-1}$, $569.08 \pm 26.58\ \mu g\ cm^{-2}\ h^{-1}$, $867.42 \pm 101.27\ \mu g\ cm^{-2}\ h^{-1}$, and $736.99 \pm 15.39\ \mu g\ cm^{-2}\ h^{-1}$, respectively. The percentage of cumulative drug penetration from all hydrogels through the rat skin *in vitro* was less than 50%, which was explained by a combination of the ability of coprocessed excipients to slow drug release and the characteristics of the diclofenac sodium (low partition coefficient). The measured areas under the curve for the Co-CLA6-XG, Co-CLA12-XG, CL6-Co-A-XG, and CL12-Co-A-XG transdermal hydrogels were 32.08 ± 5.40 μg mL^{-1} h, 34.27 ± 8.34 μg mL^{-1} h, 6.20 ± 2.90 μg mL^{-1} h, and 14.38 ± 2.38 μg mL^{-1} h, respectively. The results showed that the highest area under the curve was achieved by Co-CLA12-XG hydrogel. Areas under the curve for transdermal hydrogels CL6-Co-A-XG and CL12-Co-A-XG were significantly lower. These results indicated that the cross-linking degree of amylose and especially the method of preparation coprocessed excipient affected the drug release. For all those reasons, the authors considered that transdermal hydrogels of coprocessed excipients prepared by method A (Co-CLA6-XG and Co-CLA12-XG), as well as CL12-Co-A-XG, can be used as a transdermal matrix with controlled and prolonged drug release behavior (Surini et al., 2020).

Kulkarni and Sa grafted PAA to XG to obtain a polyanionic copolymer that is electroresponsive. The PAA-g-XG copolymer was synthesized by free radical polymerization under nitrogen atmosphere followed by alkaline hydrolysis. FTIR analysis confirmed the hydrolysis reaction, while the grafting reaction of PAA on XG was detected by the ^{1}H NMR technique. The PAA-g-XG hydrogels were considered as transdermal drug delivery carriers which are activated by an electric signal to provide on-demand ketoprofen release. The drug release was provided by deswelling of the previously swollen PAA-g-XG hydrogel when it was placed between a pair of electrodes. The deswelling of the hydrogel was observed in the vicinity of the anode as a response to the electric stimulation. The electric stimulus was switched "on" and "off" and the drug release followed a pulsated pattern. The drug release as a function of time depended on the applied electrical stimulus and concentration of the hydrogel. The deswelling of the hydrogel was increased with increased electrical stimulus and concentration of PAA-g-XG in the hydrogel. The ketoprofen-loaded PAA-g-XG hydrogel was used as a reservoir in the membrane-controlled drug delivery systems with PVA films as rate controlling membranes. The system was evaluated *in vitro* for drug permeation through excised rat abdominal skin. The applied electric stimulus caused the changes in the structure of the *stratum corneum* and cell structure, so ketoprofen permeation was enhanced under the electric stimulus in comparison to passive diffusion. The drug permeation was dependent upon the applied electric current strength and cross-link density of the PVA film (Kulkarni and Sa, 2009).

2.5.4 Other drug delivery routes

Mucoadhesion is an important aspect for the application of the hydrogels to mucous membranes of the eye, nose, or rectum, in order to achieve longer retention of the drug at the site of action or absorption and increase its bioavailability. *In situ* gelling systems are of special importance for the ocular and nasal administration of drugs. They are viscous liquids that undergo a sol-gel transition upon application, due to change in temperature, pH, or ionic strength. The formation of the viscous hydrogels with elastic properties allows precise drug administration and prolongs the attachment of the drug delivery system to the mucosal surface (Naik et al., 2020; Pathak, 2011; Robinson and Mlynek, 1995).

In a study by Shastri et al. (2010), an *in situ* thermoreversible and mucoadhesive hydrogel with antibiotic moxifloxacin hydrochloride was prepared using a combination of poloxamer 407 and poloxamer 188 with different mucoadhesive polymers, XG and sodium alginate.

Concentrations of mucoadhesive polymers varied from 0.3% to 0.7% and they were added to increase hydrogel strength and bioadhesion force and to increase the precorneal contact time and bioavailability of the drug. All prepared hydrogels were transparent, uniform in consistency, with the gelling temperature between 30 °C and 36 °C, and had good spreadability within a pH range between 6.8 and 7.4. A satisfactory bioadhesion (3298–4130 dyn/cm^2) on the sheep corneal surface and good hydrogel strength (95–128 s) were also observed. The rate of drug release decreased with an increase in the concentration of mucoadhesive polymers in the hydrogel formulation, but it was higher from the formulations containing XG (up to 70% after 12 h). The drug was released by Fickian and non-Fickian diffusion mechanisms. Prepared hydrogels provided greater bioadhesion force and hydrogel strength compared to poloxamer 407 or 188 themselves. Hydrogels containing sodium alginate had greater bioadhesion force, while XG-based hydrogels could release a higher amount of drug during 12 h (Shastri et al., 2010). Morsi et al. achieved the sustained release of acetazolamide from the nanocomposite hydrogels prepared from the oil-in-water nanoemulsion and XG/gellan and more intense and prolonged intraocular pressure lowering effect relative to that of commercial eye drops and oral tablet (Morsi et al., 2017). Besides the high drug solubilization capacity, good spreadability, and transcorneal penetration enhancement originating from the nanoemulsion, the promotion of the therapeutical effectiveness of the drug by the hybrid carrier was related to the ion-induced *in situ* gelling and mucoadhesivenes that increase the contact time between the drug delivery system and corneal surface. In this way, the rapid and extensive precorneal loss caused by the drainage and the high tear fluid turnover can be overcome (Naik et al., 2020).

Srivastava et al. developed thermoreversible and mucoadhesive *in situ* gels with reduced nasal mucociliary clearance to improve the local effect of the antioxidative and antiinflammatory polyherbal (*Moringa olifera* and *Embelia ribes*) extract in the treatment of allergic rhinitis. Gels were prepared by combining different concentrations (0.5%, 1%, or 1.5% w/v) of XG, hypromellose, or carbomer, with pluronic F127. The formulations, after administration into the nasal cavity, were rapidly transformed into viscous hydrogels at body temperature, which can diminish nasal mucociliary clearance and prolong the duration of action of the administered drug. The preparation of *in situ* nasal herbal gels by a combination of different concentrations of the polymers with pluronic F127 (10%, w/v) produced better and more effective drug delivery systems. The addition of the polymers in pluronic F127 increased the gel strength. Gel pH values were in the range between 5.2 and 5.9, which was appropriate for nasal administration. Formulations prepared with high concentrations of XG, hypromellose, or carbomer exhibited better mucoadhesion strength in comparison with pluronic F1277 (10%, w/v) itself. The spreadability of gels varied from 7.6 ± 0.21 to 11.7 ± 0.65 cm. Investigated *in situ* nasal herbal gels did not exhibit redness, edema, inflammation, and irritation during irritancy studies on mice, so were considered safe for use. The formulated mucoadhesive *in situ* gel systems are a promising approach for the intranasal delivery of polyherbal extracts for the treatment of allergic rhinitis (Srivastava et al., 2017).

Salunkhe et al. prepared suppositories containing mucoadhesive microspheres with granisetron hydrochloride (GH) using the combination of XG (0.1%–0.3%) and sodium alginate (5%–7%). Microspheres with GC were prepared by the emulsification method followed by cross-linking (cross-linking time 5, 10, or 20 min) with calcium chloride (7%, 9%, or 11%). The suppositories containing optimized microspheres (100 mg of GH, 5% of sodium alginate,

0.3% of XG, 7% of calcium chloride; 5 min cross-linking time) were prepared by the fusion method using hydrophilic (polyethylene glycol 1500, 4000, or 6000) or lipophilic (cocoa butter or suppository base Mayol W-45) base. The results showed that all prepared microspheres were spherical and had a size range of 23.56–36.76 μm and that the size increased with an increase in sodium alginate, XG, cross-linking agent, and drug concentration, as well as the cross-linking time. The drug encapsulation was in the range of 61.67–92.30%, and microspheres had satisfactory mucoadhesion. Mucoadhesion increased with an increase in polymer concentration but decreased at higher drug contents. The optimized microspheres had 83.67% mucoadhesion, 92.30% drug encapsulation, and showed extended release of GH during 4 h. The drug content in all suppositories was in the range of 93.20%–98.40%. Suppositories prepared with polyethylene glycol 4000 as a base were considered the best with extended GH release during 5 h with zero-order release kinetics, high drug content (98.4%), and optimum mechanical strength. These suppositories did not show any morphological changes in the rectal tissues of rats, indicating their safety. *In vivo* localization revealed satisfactory mucoadhesion of microspheres. The delivery of GH in the form of mucoadhesive suppositories can avoid the presystemic metabolism of the drug and may be considered an efficient alternative to the oral dosage forms and conventional suppositories (Salunkhe et al., 2014).

2.6 Concluding remarks

XG is a widely used pharmaceutical excipient that is gaining in importance due to its suitability to form different types of biocompatible homopolymeric, copolymeric, and nanocomposite hydrogels. Such XG-based hydrogels can be considered as promising microcarriers and nanocarriers with drug delivery capacity which can overcome the challenges of poor solubility, permeability through biological barriers, stability, and/or toxicological profile. To date, a number of different methods have been proposed to prepare XG-based hydrogels by physical or chemical cross-linking of the polymers. By adjustment of process and formulation parameters, it is possible to optimize both physicochemical and functional properties of the hydrogels and the drug delivery performances of the carrier. Combining XG with polymers of natural, semisynthetic, and synthetic origin represents an attractive strategy for creating pH-sensitive, thermosensitive, redox-sensitive, and electrosensitive carriers for controlled and targeted drug delivery. Most characterization studies of XG-based hydrogel drug carriers have been conducted for oral and (trans)dermal administration, but this versatile formulation approach has become increasingly important for other routes of application such as parenteral, ocular, nasal, and rectal.

References

Abu Elella, M.H., Sabaa, M.W., Hanna, D.H., Abdel-Aziz, M.M., Mohamed, R.R., 2020. Antimicrobial pH-sensitive protein carrier based on modified xanthan gum. J. Drug Deliv. Sci. Technol. 57, 101673.

Ahmad, S., Ahmad, M., Manzoor, K., Purwar, R., Ikram, S., 2019. A review on latest innovations in natural gums based hydrogels: preparations & applications. Int. J. Biol. Macromol. 136, 870–890.

Argin-Soysal, S., Kofinas, P., Lo, Y.M., 2009. Effect of complexation conditions on xanthan-chitosan polyelectrolyte complex gels. Food Hydrocoll. 23, 202–209.

Badwaik, H., Giri, T., Nakhate, K., Kashyap, P., Tripathi, D., 2013. Xanthan gum and its derivatives as a potential biopolymeric carrier for drug delivery system. Curr. Drug Deliv. 10, 587–600.

Berger, J., Reist, M., Mayer, J.M., Felt, O., Gurny, R., 2004. Structure and interactions in chitosan hydrogels formed by complexation or aggregation for biomedical applications. Eur. J. Pharm. Biopharm. 57, 35–52.

Bueno, V.B., Petri, D.F.S., 2014. Xanthan hydrogel films: molecular conformation, charge density and protein carriers. Carbohydr. Polym. 101, 897–904.

Bueno, V.B., Bentini, R., Catalani, R.H., Petri, D.F.S., 2013. Synthesis and swelling behavior of xanthan-based hydrogels. Carbohydr. Polym. 92, 1091–1099.

Caddeo, C., Nácher, A., Díez-Sales, O., Merino-Sanjuán, M., Fadda, A.M., Manconi, M., 2014. Chitosan-xanthan gum microparticle-based oral tablet for colon-targeted and sustained delivery of quercetin. J. Microencapsul. 31, 694–699.

Cascone, S., Lamberti, G., 2020. Hydrogel-based commercial products for biomedical applications: a review. Int. J. Pharm. 573, 118803.

Ćirić, A., Medarević, Đ., Čalija, B., Dobričić, V., Mitrić, M., Djekic, L., 2020. Study of chitosan/xanthan gum polyelectrolyte complexes formation, solid state and influence on ibuprofen release kinetics. Int. J. Biol. Macromol. 148, 942–955.

Ćirić, A., Medarević, Đ., Čalija, B., Dobričić, V., Rmandić, M., Barudžija, T., Malenović, A., Djekic, L., 2021. Effect of ibuprofen entrapment procedure on physicochemical and controlled drug release performances of chitosan/xanthan gum polyelectrolyte complexes. Int. J. Biol. Macromol. 167, 547–558.

Dadou, S.M., El-Barghouthi, M.I., Antonijevic, M.D., Chowdhry, B.Z., Badwan, A.A., 2020. Elucidation of the controlled-release behavior of metoprolol succinate from directly compressed xanthan gum/chitosan polymers: computational and experimental studies. ACS Biomater. Sci. Eng. 6, 21–37.

De Robertis, S., Bonferoni, M.C., Elviri, L., Sandri, G., Caramella, C., Bettini, R., 2015. Advances in oral controlled drug delivery: the role of drug-polymer and interpolymer non-covalent interactions. Expert Opin. Drug Deliv. 12, 441–453.

Djekic, L., Primorac, M., Filipic, S., Agbaba, D., 2012. Investigation of surfactant/cosurfactant synergism impact on ibuprofen solubilization capacity and drug release characteristics of nonionic microemulsions. Int. J. Pharm. 433 (1–2), 25–33. https://doi.org/10.1016/j.ijpharm.2012.04.070.

Djekic, L., Martinovic, M., Stepanović-Petrović, R., Micov, A., Tomić, M., Primorac, M., 2016. Formulation of hydrogel-thickened nonionic microemulsions with enhanced percutaneous delivery of ibuprofen assessed in vivo in rats. Eur. J. Pharm. Sci. 92, 255–265.

Du, X., Jia, N., Xu, S., Zhou, Y., 2007. Molecular structure of starch from *Pueraria lobata* (Willd.) Ohwi relative to kuzu starch. Starch/Staerke 59 (12), 609–613. https://doi.org/10.1002/star.200700604.

El-Sawy, N.M., Raafat, A.I., Badawy, N.A., Mohamed, A.M., 2020. Radiation development of pH-responsive (xanthan-acrylic acid)/MgO nanocomposite hydrogels for controlled delivery of methotrexate anticancer drug. Int. J. Biol. Macromol. 142, 254–264.

García-Ochoa, F., Santos, V.E., Casas, J.A., Gómez, E., 2000. Xanthan gum: production, recovery, and properties. Biotechnol. Adv. 18 (7), 549–579. https://doi.org/10.1016/S0734-9750(00)00050-1.

Ghosh Dastidar, T., Netravali, A.N., 2012. "Green" crosslinking of native starches with malonic acid and their properties. Carbohydr. Polym. 90 (4), 1620–1628. https://doi.org/10.1016/j.carbpol.2012.07.041.

Gilbert, L., Loisel, V., Savary, G., Grisel, M., Picard, C., 2013. Stretching properties of xanthan, carob, modified guar and celluloses in cosmetic emulsions. Carbohydr. Polym. 93 (2), 644–650. https://doi.org/10.1016/j.carbpol.2012.12.028.

Gils, P.S., Ray, D., Sahoo, P.K., 2009. Characteristics of xanthan gum-based biodegradable superporous hydrogel. Int. J. Biol. Macromol. 45 (4), 364–371. https://doi.org/10.1016/j.ijbiomac.2009.07.007.

Giri, T.K., Thakur, A., Tripathi, D.K., 2016. Biodegradable hydrogel bead of casein and modified xanthan gum for controlled delivery of theophylline. Curr. Drug Ther. 11 (2), 150–162. https://doi.org/10.2174/1574885511666160830123807.

González-Seligra, P., Guz, L., Ochoa-Yepes, O., Goyanes, S., Famá, L., 2017. Influence of extrusion process conditions on starch film morphology. LWT 84, 520–528. https://doi.org/10.1016/j.lwt.2017.06.027.

Gutowska, A., Jeong, B., Jasionowski, M., 2001. Injectable gels for tissue engineering. Anat. Rec. 263 (4), 342–349. https://doi.org/10.1002/ar.1115.

Guvendiren, M., Lu, H.D., Burdick, J.A., 2012. Shear-thinning hydrogels for biomedical applications. Soft Matter 8 (2), 260–272. https://doi.org/10.1039/c1sm06513k.

Hajikhani, M., Khanghahi, M.M., Shahrousvand, M., Mohammadi-Rovshandeh, J., Babaei, A., Khademi, S.M.H., 2019. Intelligent superabsorbents based on a xanthan gum/poly (acrylic acid) semi-interpenetrating polymer network for application in drug delivery systems. Int. J. Biol. Macromol. 139, 509–520.

Han, G., Wang, G., Zhu, X., Shao, H., Liu, F., Yang, P., Ying, Y., Wang, F., Ling, P., 2012. Preparation of xanthan gum injection and its protective effect on articular cartilage in the development of osteoarthritis. Carbohydr. Polym. 87 (2), 1837–1842. https://doi.org/10.1016/j.carbpol.2011.10.016.

Hanna, D.H., Saad, G.R., 2019. Encapsulation of ciprofloxacin within modified xanthan gum-chitosan based hydrogel for drug delivery. Bioorg. Chem. 84, 115–124.

Huang, J., Deng, Y., Ren, J., Chen, G., Wang, G., Wang, F., Wu, X., 2018. Novel in situ forming hydrogel based on xanthan and chitosan re-gelifying in liquids for local drug delivery. Carbohydr. Polym. 186, 54–63. https://doi.org/10.1016/j.carbpol.2018.01.025.

Iijima, M., Shinozaki, M., Hatakeyama, T., Takahashi, M., Hatakeyama, H., 2007. AFM studies on gelation mechanism of xanthan gum hydrogels. Carbohydr. Polym. 68 (4), 701–707. https://doi.org/10.1016/j.carbpol.2006.08.004.

Iizhar Ahmed, S., Ismail, A., Niveditha, P., 2014. Formulation and evaluation of polyelectrolyte complex-based matrix tablet of isosorbide mononitrate. Int. J. Pharm. Investig., 38. https://doi.org/10.4103/2230-973x.127739.

Ijaz, H., Tulain, U.R., Qureshi, J., 2018. Formulation and in vitro evaluation of pH-sensitive cross-linked xanthan gum-grafted acrylic acid copolymer for controlled delivery of perindopril erbumine (PE). Polym. Plast. Technol. Eng. 57 (5), 459–470. https://doi.org/10.1080/03602559.2017.1320722.

Illum, L., 1998. Chitosan and its use as a pharmaceutical excipient. Pharm. Res. 15 (9), 1326–1331. https://doi.org/10.1023/A:1011929016601.

Kalia, S., Roy Choudhury, A., 2019. Synthesis and rheological studies of a novel composite hydrogel of xanthan, gellan and pullulan. Int. J. Biol. Macromol. 137, 475–482. https://doi.org/10.1016/j.ijbiomac.2019.06.212.

Kang, M., Oderinde, O., Liu, S., Huang, Q., Ma, W., Yao, F., Fu, G., 2019. Characterization of xanthan gum-based hydrogel with Fe^{3+} ions coordination and its reversible sol-gel conversion. Carbohydr. Polym. 203, 139–147. https://doi.org/10.1016/j.carbpol.2018.09.044.

Katzbauer, B., 1998. Properties and applications of xanthan gum. Polym. Degrad. Stab. 59 (1–3), 81–84. https://doi.org/10.1016/s0141-3910(97)00180-8.

Kennedy, J.R.M., Kent, K.E., Brown, J.R., 2015. Rheology of dispersions of xanthan gum, locust bean gum and mixed biopolymer gel with silicon dioxide nanoparticles. Mater. Sci. Eng. C 48, 347–353. https://doi.org/10.1016/j.msec.2014.12.040.

Khouryieh, H.A., Herald, T.J., Aramouni, F., Alavi, S., 2007. Intrinsic viscosity and viscoelastic properties of xanthan/guar mixtures in dilute solutions: effect of salt concentration on the polymer interactions. Food Res. Int. 40 (7), 883–893. https://doi.org/10.1016/j.foodres.2007.03.001.

Koop, H.S., De Freitas, R.A., De Souza, M.M., Savi Jr., R., Silveira, J.L.M., 2015. Topical curcumin-loaded hydrogels obtained using galactomannan from Schizolobium parahybae and xanthan. Carbohydr. Polym. 116, 229–236. https://doi.org/10.1016/j.carbpol.2014.07.043.

Kulkarni, R.V., Sa, B., 2009. Electroresponsive polyacrylamide-grafted-xanthan hydrogels for drug delivery. J. Bioact. Compat. Polym. 24 (4), 368–384. https://doi.org/10.1177/0883911509104475.

Kumar, A., Rao, K.M., Han, S.S., 2018. Application of xanthan gum as polysaccharide in tissue engineering: a review. Carbohydr. Polym. 180, 128–144. https://doi.org/10.1016/j.carbpol.2017.10.009.

Liu, Z., Yao, P., 2015. Injectable thermo-responsive hydrogel composed of xanthan gum and methylcellulose double networks with shear-thinning property. Carbohydr. Polym. 132, 490–498.

Lu, H.D., Charati, M.B., Kim, I.L., Burdick, J.A., 2012. Injectable shear-thinning hydrogels engineered with a self-assembling dock-and-lock mechanism. Biomaterials 33 (7), 2145–2153. https://doi.org/10.1016/j.biomaterials.2011.11.076.

Luo, Y., Wang, Q., 2014. Recent development of chitosan-based polyelectrolyte complexes with natural polysaccharides for drug delivery. Int. J. Biol. Macromol. 64, 353–367. https://doi.org/10.1016/j.ijbiomac.2013.12.017.

Ma, Y.H., Yang, J., Li, B., Jiang, Y.W., Lu, X., Chen, Z., 2016. Biodegradable and injectable polymer-liposome hydrogel: a promising cell carrier. Polym. Chem. 7 (11), 2037–2044. https://doi.org/10.1039/c5py01773d.

Magnin, D., Lefebvre, J., Chornet, E., Dumitriu, S., 2004. Physicochemical and structural characterization of a polyionic matrix of interest in biotechnology, in the pharmaceutical and biomedical fields. Carbohydr. Polym. 55 (4), 437–453. https://doi.org/10.1016/j.carbpol.2003.11.013.

Maiti, S., Ray, S., Sa, B., 2009. Controlled delivery of bovine serum albumin from carboxymethyl xanthan microparticles xanthan microparticles. Pharm. Dev. Technol. 14 (2), 165–172. https://doi.org/10.1080/10837450802498878.

Maiti, S., Kaity, S., Ray, S., Sa, B., 2011. Development and evaluation of xanthan gum-facilitated ethyl cellulose microsponges for controlled percutaneous delivery of diclofenac sodium. Acta Pharma. 61 (3), 257–270. https://doi.org/10.2478/v10007-011-0022-6.

Manconi, M., Mura, S., Manca, M.L., Fadda, A.M., Dolz, M., Hernandez, M.J., Casanovas, A., Díez-Sales, O., 2010. Chitosomes as drug delivery systems for C-phycocyanin: preparation and characterization. Int. J. Pharm. 392, 92–100.

Merlusca, I.P., Plamadeala, P., Girbea, C., Popa, I.M., 2016. Xanthan-chitosan complex as a potential protector against injurious effects of neomycin. Cellul. Chem. Technol. 50, 577–583.

Milas, M., Rinaudo, M., 1983. Properties of the concentrated xanthan gum solutions. Polym. Bull. 10 (5–6), 271–273. https://doi.org/10.1007/BF00272235.

Mishra, B., Sahoo, S.K., Sahoo, S., 2018. Liranaftate loaded xanthan gum based hydrogel for topical delivery: physical properties and ex-vivo permeability. Int. J. Biol. Macromol. 107, 1717–1723. https://doi.org/10.1016/j.ijbiomac.2017.10.039.

Morariu, S., Bercea, M., Brunchi, C.E., 2018. Phase separation in xanthan solutions. Cellul. Chem. Technol. 52 (7–8), 569–576. http://www.cellulosechemtechnol.ro/pdf/CCT7-8(2018)/p.569-576.pdf.

Morsi, N., Ibrahim, M., Refai, H., El Sorogy, H., 2017. Nanoemulsion-based electrolyte triggered in situ gel for ocular delivery of acetazolamide. Eur. J. Pharm. Sci. 104, 302–314. https://doi.org/10.1016/j.ejps.2017.04.013.

Naik, J.B., Pardeshi, S.R., Patil, R.P., Patil, P.B., Mujumdar, A., 2020. Mucoadhesive micro-/nano carriers in ophthalmic drug delivery: an overview. BioNanoScience 10 (3), 564–582. https://doi.org/10.1007/s12668-020-00752-y.

Nilesh, K., Pravin, W., Jitendra, N., 2015. Development of floating chitosan-xanthan beads for oral controlled release of glipizide. Int. J. Pharm. Investig., 73. https://doi.org/10.4103/2230-973X.153381.

Outuki, P.M., de Francisco, L.M.B., Hoscheid, J., Bonifácio, K.L., Barbosa, D.S., Cardoso, M.L.C., 2016. Development of arabic and xanthan gum microparticles loaded with an extract of Eschweilera nana Miers leaves with antioxidant capacity. Colloids Surf. A Physicochem. Eng. Asp. 499, 103–112. https://doi.org/10.1016/j.colsurfa.2016.04.006.

Pathak, K., 2011. Mucoadhesion; a prerequisite or a constraint in nasal drug delivery? Int. J. Pharm. Investig. 1, 62–63.

Petri, D.F.S., 2015. Xanthan gum: a versatile biopolymer for biomedical and technological applications. J. Appl. Polym. Sci. 132 (23). https://doi.org/10.1002/app.42035.

Praveen, T., Jin, T.M., Abdul, K.A., James, Y.D., Jun, L.X., 2015. Nanoparticle-hydrogel composites: concept, design, and applications of these promising, multi-functional materials. Adv. Sci. https://doi.org/10.1002/advs.201400010, 1400010.

Raafat, A.I., El-Sawy, N.M., Badawy, N.A., Mousa, E.A., Mohamed, A.M., 2018. Radiation fabrication of xanthan-based wound dressing hydrogels embedded ZnO nanoparticles: in vitro evaluation. Int. J. Biol. Macromol. 118, 1892–1902.

Ray, S., Maiti, S., Sa, B., 2010. Polyethyleneimine-treated xanthan beads for prolonged release of diltiazem: in vitro and in vivo evaluation. Arch. Pharm. Res. 33 (4), 575–583. https://doi.org/10.1007/s12272-010-0412-1.

Reddy, N., Yang, Y., 2010. Citric acid cross-linking of starch films. Food Chem. 118 (3), 702–711. https://doi.org/10.1016/j.foodchem.2009.05.050.

Robinson, J.R., Mlynek, G.M., 1995. Bioadhesive and phase-change polymers for ocular drug delivery. Adv. Drug Deliv. Rev. 16 (1), 45–50. https://doi.org/10.1016/0169-409X(95)00013-W.

Rowe, R.C., Sheskey, P.J., Quinn, 2009. Handbook of Pharmaceutical Excipients. Pharmaceutical Press.

Sabaa, M.W., Hanna, D.H., Abu Elella, M.H., Mohamed, R.R., 2019. Encapsulation of bovine serum albumin within novel xanthan gum based hydrogel for protein delivery. Mater. Sci. Eng. C 94, 1044–1055. https://doi.org/10.1016/j.msec.2018.10.040.

Salunkhe, N.H., Jadhav, N.R., Mali, K.K., Dias, R.J., Ghorpade, V.S., Yadav, A.V., 2014. Mucoadhesive microsphere based suppository containing granisetron hydrochloride for management of emesis in chemotherapy. J. Pharm. Investig. 44 (4), 253–263. https://doi.org/10.1007/s40005-014-0123-6.

Sandeep, C., Deb, T.K., Moin, A., Shivakumar, H.G., 2014. Cationic guar gum polyelectrolyte complex micro particles. J. Young Pharm. 6 (4), 11–19. https://doi.org/10.5530/jyp.2014.4.3.

Shalviri, A., Liu, Q., Abdekhodaie, M.J., Wu, X.Y., 2010. Novel modified starch-xanthan gum hydrogels for controlled drug delivery: synthesis and characterization. Carbohydr. Polym. 79, 898–907.

Sharma, S., Tiwari, S., 2020. A review on biomacromolecular hydrogel classification and its applications. Int. J. Biol. Macromol. 162, 737–747. https://doi.org/10.1016/j.ijbiomac.2020.06.110.

Shastri, D.H., Prajapati, S.T., Patel, L.D., 2010. Design and development of thermoreversible ophthalmic in situ hydrogel of moxifloxacin HCL. Curr. Drug Deliv. 7 (3), 238–243. https://doi.org/10.2174/156720110791560928.

Shchipunov, Y., Sarin, S., Kim, I., Ha, C.S., 2010. Hydrogels formed through regulated self-organization of gradually charging chitosan in solution of xanthan. Green Chem. 12 (7), 1187–1195. https://doi.org/10.1039/b925138c.

Simões, B.M., Cagnin, C., Yamashita, F., Olivato, J.B., Garcia, P.S., de Oliveira, S.M., Eiras Grossmann, M.V., 2020. Citric acid as crosslinking agent in starch/xanthan gum hydrogels produced by extrusion and thermopressing. LWT 125, 108950.

Singh, J., Kumar, S., Dhaliwal, A.S., 2020. Controlled release of amoxicillin and antioxidant potential of gold nanoparticles-xanthan gum/poly (acrylic acid) biodegradable nanocomposite. J. Drug Deliv. Sci. Technol. 55. https://doi.org/10.1016/j.jddst.2019.101384.

Singhvi, G., Hans, N., Shiva, N., Kumar Dubey, S., 2019. Xanthan gum in drug delivery applications. In: Natural Polysaccharides in Drug Delivery and Biomedical Applications. Elsevier, pp. 121–144, https://doi.org/10.1016/B978-0-12-817055-7.00005-4.

Srivastava, R., Srivastava, S., Singh, S.P., 2017. Thermoreversible in-situ nasal gel formulations and their pharmaceutical evaluation for the treatment of allergic rhinitis containing extracts of Moringa olifera and Embelia ribes. Int. J. Appl. Pharm. 9 (6), 16–20. https://doi.org/10.22159/ijap.2017v9i6.18780.

Surini, S., Sicilia, A., Sofiani, R., Khoiriah, S., Indriatin, U.H., Sari, S.P., Harahap, Y., 2020. Development of transdermal dosage form using coprocessed excipients of xanthan gum and cross-linked amylose: in vitro and in vivo studies. Int. J. Appl. Pharm. 12 (1), 207–211. https://doi.org/10.22159/ijap.2020.v12s1.FF047.

Tao, Y., Zhang, R., Xu, W., Bai, Z., Zhou, Y., Zhao, S., Xu, Y., Yu, D., 2016. Rheological behavior and microstructure of release-controlled hydrogels based on xanthan gum crosslinked with sodium trimetaphosphate. Food Hydrocoll. 52, 923–933. https://doi.org/10.1016/j.foodhyd.2015.09.006.

Trombino, S., Serini, S., Cassano, R., Calviello, G., 2019. Xanthan gum-based materials for omega-3 PUFA delivery: preparation, characterization and antineoplastic activity evaluation. Carbohydr. Polym. 208, 431–440. https://doi.org/10.1016/j.carbpol.2019.01.001.

Vianna-Filho, R.P., Petkowicz, C.L.O., Silveira, J.L.M., 2013. Rheological characterization of O/W emulsions incorporated with neutral and charged polysaccharides. Carbohydr. Polym. 93 (1), 266–272. https://doi.org/10.1016/j.carbpol.2012.05.014.

Wahid, F., Zhao, X.J., Jia, S.R., Bai, H., Zhong, C., 2020. Nanocomposite hydrogels as multifunctional systems for biomedical applications: current state and perspectives. Compos. Part B Eng. 200. https://doi.org/10.1016/j.compositesb.2020.108208.

Xu, W., Li, Z., Jin, W., Li, P., Li, Y., Liang, H., Li, Y., Li, B., 2016. Structural and rheological properties of xanthan gum/lysozyme system induced by in situ acidification. Food Res. Int. 90, 85–90. https://doi.org/10.1016/j.foodres.2016.10.039.

You, Y.C., Dong, L.Y., Dong, K., Xu, W., Yan, Y., Zhang, L., Wang, K., Xing, F.J., 2015. In vitro and in vivo application of pH-sensitive colon-targeting polysaccharide hydrogel used for ulcerative colitis therapy. Carbohydr. Polym. 130, 243–253. https://doi.org/10.1016/j.carbpol.2015.03.075.

Yu, L., Ding, J., 2008. Injectable hydrogels as unique biomedical materials. Chem. Soc. Rev. 37 (8), 1473–1481. https://doi.org/10.1039/b713009k.

Zhang, F., Luan, T., Kang, D., Jin, Q., Zhang, H., Yadav, M.P., 2015. Viscosifying properties of corn fiber gum with various polysaccharides. Food Hydrocoll. 43, 218–227. https://doi.org/10.1016/j.foodhyd.2014.05.018.

Zhang, B., Wang, J., Li, Z., Ma, M., Jia, S., Li, X., 2020. Use of hydroxypropyl β-cyclodextrin as a dual functional component in xanthan hydrogel for sustained drug release and antibacterial activity. Colloids Surf. A Physicochem. Eng. Asp. 587. https://doi.org/10.1016/j.colsurfa.2019.124368.

Zheng, M., Lian, F., Xiong, Y., Liu, B., Zhu, Y., Miao, S., Zhang, L., Zheng, B., 2019a. The synthesis and characterization of a xanthan gum-acrylamide-trimethylolpropane triglycidyl ether hydrogel. Food Chem. 272, 574–579. https://doi.org/10.1016/j.foodchem.2018.08.083.

Zheng, M., Lian, F., Zhu, Y., Zhang, Y., Liu, B., Zhang, L., Zheng, B., 2019b. pH-responsive poly (xanthan gum-g-acrylamide-g-acrylic acid) hydrogel: preparation, characterization, and application. Carbohydr. Polym. 210, 38–46. https://doi.org/10.1016/j.carbpol.2019.01.052.

CHAPTER

3

Chitosan-based nanoengineered drug delivery system

Sreejan Manna[a], *Aishik Banerjee*[a], *Sougata Jana*[b,c], *and Manas Bhowmik*[d]

[a]Department of Pharmaceutical Technology, Brainware University, Kolkata, West Bengal, India [b]Department of Pharmaceutics, Gupta College of Technological Sciences, Asansol, West Bengal, India [c]Department of Health and Family Welfare, Directorate of Health Services, Kolkata, West Bengal, India [d]Department of Pharmaceutical Technology, Jadavpur University, Kolkata, West Bengal, India

3.1 Introduction

In the field of pharmaceutical research, targeting of therapeutic agents play an important role to improve the efficacy of the system. Over the past few decades, nanoparticles (NPs) are widely used for tissue-specific drug targeting. The structural, dimensional, and surface properties of NPs are very useful for in vivo localization of drugs (Ibrahim et al., 2019; Patra et al., 2018). Nanoengineered particles are generally described as small solid colloidal particles having the size range of 1–1000 nm (Jana et al., 2013; Jeevanandam et al., 2018; Manna et al., 2016). Drugs can be easily encapsulated in a nanoparticulate core which can also protect the drug from various environmental and chemical degradation enhancing the in vivo stability. Nanoparticle-based drug delivery systems are efficient for delivering large and diversified molecules that are either hydrophobic or amphiphile in nature (Singh and Lillard, 2009; Yang et al., 2015). Other advantages of the nanoengineered drug delivery system include controlled release of therapeutics at the site of action, improved bioavailability, and modifiable surface characteristics (Devarajan and Sonavane, 2007; Gandhi et al., 2014; Gelperina et al., 2005; Mudshinge et al., 2011). Nanoparticles are widely used as carriers for drugs and biological agent, genetic delivery, delivery of antigen and vaccines, diagnostic agents, etc. (Mandal et al., 2018; Rizvi and Saleh, 2018; Shaw et al., 2017; Yetisgin et al., 2020). The pharmacokinetics and therapeutic index of a drug molecule can be increased significantly

Micro- and Nanoengineered Gum-Based Biomaterials for Drug Delivery and Biomedical Applications
https://doi.org/10.1016/B978-0-323-90986-0.00010-8

when the drugs are loaded into nanoparticles through physical encapsulation and chemical conjugation (Jana et al., 2016; Zhang et al., 2008).

Following the current trend in pharmaceutical research, biopolymers have gained considerable attention from researchers, particularly in synthesis of hydrogel forming and particulate drug carriers. Biopolymers are naturally occurring compounds, generally materialized form various natural sources including plants, animals, bacteria, and fungi (Ahsan et al., 2018; Sun et al., 2020; Torres et al., 2019; Yadav et al., 2015). Being biodegradable and biocompatible in nature, natural polymers have shown promising application in the fields of drug delivery and biomedical application. Chitosan is a naturally occurring polysaccharide synthesized from deacetylation of chitin (Ahsan et al., 2018; Elieh-Ali-Komi and Hamblin, 2016; Mohebbi et al., 2019). Apart from being used as an excipient in various drug delivery systems, chitosan has been reported to be used for widespread applications including raw materials for dietary supplement (Illum, 1998), as a penetration enhancer of drug molecules (Mohammed et al., 2017), wound dressing materials (Park et al., 2018), etc. This linear amino polysaccharide can also be used in the treatment of damaged oral cavity tissues due to its regenerative inductive properties (Ahmad et al., 2017; Jayakumar et al., 2010; Wieckiewicz et al., 2017). It has found potential application as an adsorbent for dye removal (Vakili et al., 2014), and as polymeric scaffolds to be used in skin, bone, and cartilage healing therapy (Oryan and Sahvieh, 2017; Rodríguez-Vázquez et al., 2015). Over the years, chitosan-based nanomaterials have been investigated by scientists for delivering therapeutic moieties into bio-targets (Ali and Ahmed, 2018; Li et al., 2018; Naskar et al., 2019). In this chapter, we have described chitosan-based nanoengineered carriers for delivering drug molecules in biological systems.

3.1.1 Source and chemistry of chitosan

Chitin is the second most naturally occurring polysaccharide primarily obtained from the exoskeletons of crab, shrimp, lobster, cuttlefish, etc. (Nishimura et al., 1991; Rinaudo, 2006). The ratio of glucosamine unit to *N*-acetyl glucosamine unit present in the structure of chitosan determines the degree of deacetylation (varying between 60% and 100%), as shown in Fig. 3.1. Depending on the degree of deacetylation, the properties of chitosan can vary (Layek and Mandal, 2020). The structural backbone of chitosan forms a long unbranched structure consisting β-(1–4) glycosidic linkages (Austin et al., 1981; Elieh-Ali-Komi and Hamblin, 2016; Synowiecki and Al-Khateeb, 2003; Younes and Rinaudo, 2015). Nitrogen can be present in the chitosan from 5% to 8%, depending upon the extent of the deacetylation process (Dutta et al., 2002). Chitosan can react with keto acids and be followed by sodium borohydride to produce nonprotic amino groups (Hanna Rosli et al., 2020). Apart from that, at room temperature it can also form aldimines and ketimines on reaction with aldehydes and ketones, respectively (Dutta et al., 2004). A pK_a of 6.5 is observed for the primary amines of chitosan, which can result in protonation in acidic aqueous solution. It solubilizes in acidic solvents, but remains insoluble in neutral as well as alkaline media (Cheung et al., 2015). Due to the presence of amino groups, chitosan can behave as a mucoadhesive polymer, forming an electrostatic bond with mucin (Singla and Chawla, 2001). The excellent biodegradability, safety, hydrogel forming ability, and mucoadhesive property of chitosan makes it a perfect candidate for a drug carrier (Ali and Ahmed, 2018; Felt et al., 1998; Hamedi et al., 2018; Iqbal Hassan Khan et al., 2019; Jana et al., 2015; Jana and Sen, 2017; Prabaharan, 2008).

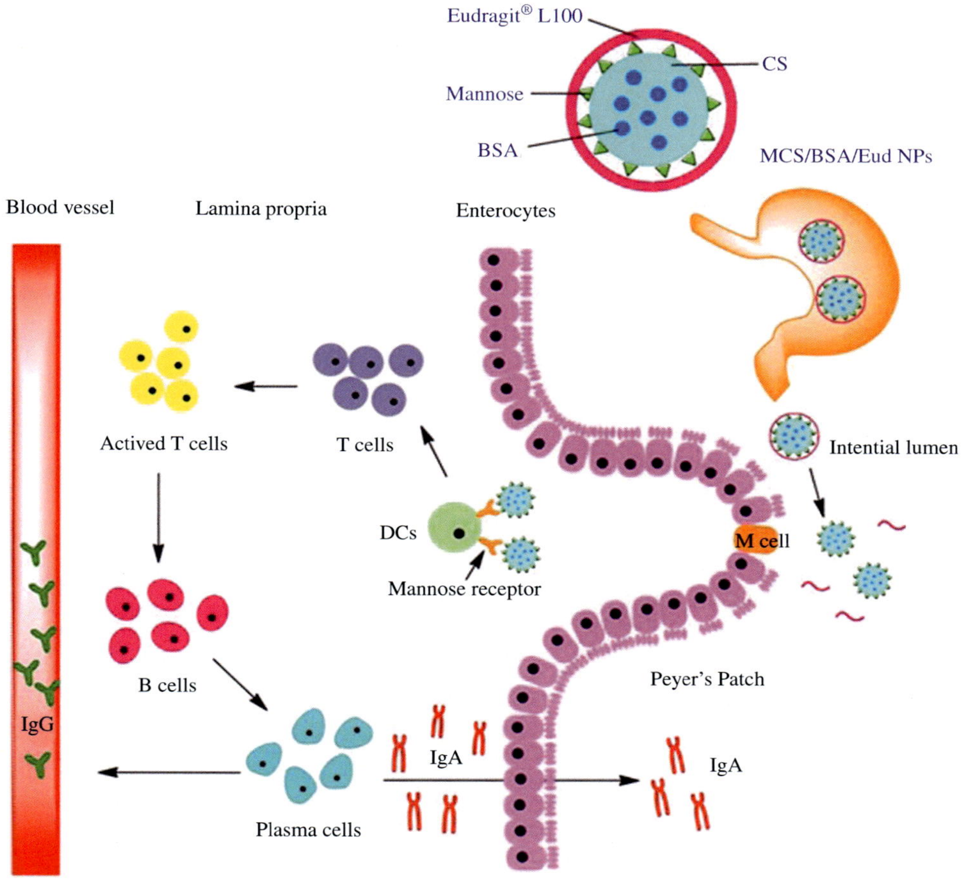

FIG. 3.1 Schematic diagram representing mucosal and systemic immune responses through oral delivery of chitosan-based NPs. *From Xu, B., Zhang, W., Chen, Y., Xu, Y., Wang, B., Zong, L., 2018. Eudragit® L100-coated mannosylated chitosan nanoparticles for oral protein vaccine delivery. Int. J. Biol. Macromolecules 113, 534–542. https://doi.org/10.1016/j.ijbiomac.2018.02.016.*

3.2 Drug delivery applications of chitosan-based nanoengineered systems

Various chitosan-based nanoengineered therapeutic carriers are summed up in Table 3.1.

3.2.1 Intestine targeting

Zare et al. developed chitosan-based novel nanocarriers to enhance the oral bioavailability of doxorubicin. Doxorubicin was entrapped in chitosan-tripolyphosphate NPs synthesized by an ionic gelation technique. An average particle size was reported of about 150 ± 10 nm. Intestinal permeation of doxorubicin was evaluated using rat intestine, which indicated an improved permeation of the drug (Zare et al., 2018). Cerchiara et al. have

TABLE 3.1 Drug delivery applications of chitosan-based nanoengineered carriers.

Sl. no.	Applications	Polymers used	Therapeutic agents	References
1	Intestinal targeting	Chitosan, tripolyphosphate	Doxorubicin	Zare et al. (2018)
		Chitosan, tripolyphosphate	Vancomycin	Cerchiara et al. (2015)
		Chitosan	Resveratrol	Iglesias et al. (2019)
		Chitosan, fucoidan	Berberine	Wu et al. (2014)
		Mannosylated chitosan, tripolyphosphate, Eudragit L100	Bovine serum albumin	Xu et al. (2018)
		Chitosan	Rifamixin	Kumar and Newton (2017)
		Chitosan, pectin	Curcumin	Alkhader et al. (2017)
2	Brain targeting	Chitosan, poly(lactic-*co*-glycolic acid), sialic acid	Curcumin	Kuo et al. (2019)
		Lactoferrin conjugated *N*-trimethyl chitosan, polylactide-*co*-glycoside	Huperizine A	Meng et al. (2018)
		Chitosan, l-valine	Saxagliptin	Fernandes et al. (2018)
		Chitosan	Cyclovirobuxine D	Hanmei et al. (2018)
		Chitosan, D-α-tocopherol polyethylene glycol 1000 succinate	Docetaxel	Agrawal et al. (2017)
		Chitosan	Rutin	Ahmad et al. (2017)
		Chitosan, hydroxypropyl-β-cyclodextrin	Scutellarin	Liu and Ho (2017)
3	Hepatic targeting	Chitosan, biotin, avidin	Trans-resveratrol	Bu et al. (2013)
		Chitosan, Fe@C	Carboplatin	Li et al. (2009)
		Polymetformin, penetratin, DSPE-PEG 2000	Interleukin-22	Zai et al. (2019)
		Carboxymethyl chitosan, phenylboronic acid pinacol ester, active targeted ligand CD147	Doxorubicin	Qu et al. (2018)
		Galactosylated *O*-carboxymethyl chitosan, stearic acid	Doxorubicin	Guo et al. (2013)
		Chitosan, carbopol, folic acid	Gemcitabine	Viota et al. (2013)
		Galactosylated chitosan	Norcantharidin	Wang et al. (2010)

TABLE 3.1 Drug delivery applications of chitosan-based nanoengineered carriers—Cont'd

Sl. no.	Applications	Polymers used	Therapeutic agents	References
4	Pulmonary targeting	Chitosan, AS1411 aptamer	Methotrexate	Guo et al. (2018)
		Oleic acid conjugated chitosan	Paclitaxel and Quercetin	Liu et al. (2017)
		Glycol chitosan	–	Yhee et al. (2017)
		Glycol chitosan, fluorophore cyanine 5.5	–	On et al. (2020)
		Chitosan, poly(lactic-*co*-glycolic acid)	Catechin hydrate	Ahmad et al. (2020)
		Chitosan, hyaluronic acid	Cyanin 3 labeled siRNA	Zhang et al. (2019)
		Chitosan, hyaluronic acid	Everolimus	Chiesa et al. (2018)
5	Transdermal targeting	Chitosan, poly(*N*-isopropylacrylamide-*co*-acrylic acid)	Clindamycin phosphate and tretinoin	Shamsi et al. (2017)
		Chitosan, lecithin	Baicalein	Dong et al. (2020)
		Chitosan, albumin	Aceclofenac	Jana et al. (2014)
		Chitosan, poly(lactide-*co*-glycolide), lecithin	Betamethasone valerate	Özcan et al. (2013)
		Carboxymethyl chitosan, mesoporous silica, pullulan	Colchicine	Mohamed et al. (2020)
		Chitosan, alginate	Benzoyl peroxide	Friedman et al. (2013)
		Chitosan, nicotinamide	Tacrolimus	Yu et al. (2018)
		Chitosan, sodium hyaluronate	Quercetin	Jeon et al. (2015)
6	Ocular targeting	Chitosan, sulfobutylether-β-cyclodextrin	Honokiol	Deng et al. (2018)
		Chitosan, HPMC E5	Sparfloxacin	Ambhore et al. (2016)
		Chitosan, sodium tripolyphosphate	Rosmarinic acid	da Silva et al. (2016)
		Chitosan, poly(lactic-*co*-glycolic acid), tripolyphosphate, hyaluronic acid	Ranibizumab	Elsaid et al. (2016)
		Chitosan, tripolyphosphate	Ketorolac tromethamine	Fathalla et al. (2016)
		Chitosan, hyaluronic acid, hydroxypropyl methyl cellulose	Ceftazidime	Silva et al. (2017)
		Chitosan, poly(lactic acid)	Amphotericin B	Zhou et al. (2013)
		Chitosan	Triamcinolone acetonide	Cheng et al. (2019)

Continued

TABLE 3.1 Drug delivery applications of chitosan-based nanoengineered carriers—Cont'd

Sl. no.	Applications	Polymers used	Therapeutic agents	References
7	Renal targeting	Catechol-derived low molecular weight chitosan	Autofluorescent doxorubicin	Qiao et al. (2014)
		Chitosan	siRNA	Yang et al. (2015)
		Chitosan	siRNA	Gao et al. (2014)
		Chitosan	Allopurinol	Kandav et al. (2019)
		Chitosan	Curcumin	Anwar et al. (2020)
		Chitosan	Metformin	Wang et al (2021)

synthesized chitosan-based nanocarriers for colon-targeted delivery of vancomycin. An ionotropic gelation technique was employed to develop chitosan-tripolyphosphate NPs followed by lyophilization or spray drying. The antibacterial efficacy was determined using *S. aureus*, which demonstrated excellent bactericidal activity. The developed formulations showed good water uptake along with pH-triggered drug release pH 7.4, resulting in an approximately 10-fold increase in vancomycin release (Cerchiara et al., 2015). Iglesias et al. have synthesized chitosan-based nanocarriers for delivering resveratrol, which is used in the therapy of inflammatory bowel diseases, Crohn's disease, and ulcerative colitis. The functionality of this drug complex is based on the sustained colonic release of resveratrol for long-lasting mucoadhesive drug depots. The release pattern of resveratrol in simulated colon conditions for 48 h indicates a potential difference in drug release rate between other formulations and chitosan nanocomposites (Iglesias et al., 2019).

Chitosan- and fucodian-based nanocarriers were developed by Wu et al. for oral administration of berberine. The developed nanocomplex was reported for high drug loading efficiency. A relatively slow drug release was observed in acidic pH, where as a rapid release of berberine was noted in pH 7.4, indicating the pH-dependent drug release of developed NPs. Study reports suggested the efficacy of the system in restoring the intestinal barrier function for damaged intestinal epithelial layers (Wu et al., 2014). Xu et al. synthesized mannosylated chitosan- and tripolyphosphate-based NPs by an ionic gelation technique for oral protein vaccine delivery. Bovine serum albumin was incorporated in nanoparticulate core with maximum entrapment efficiency of $90.38 \pm 9.12\%$. The developed NPs were further coated with Eudragit L100 for intestine-specific delivery. A strong mucosal IgA and IgG antibody responses were produced after oral administration of the NPs, as shown in Fig. 3.1 (Xu et al., 2018). Rifamixin-loaded chitosan NPs were developed by an ionic gelation technique for effective colon targeting to treat inflammatory bowel disease. The zeta potential was reported to be 37.79, confirming the stability of NPs in nanosuspension. A restricted rifamixin release in the upper gastrointestinal tract was observed while the maximum drug was released in the colonic environment (Kumar and Newton, 2017). Mucoadhesive chitosan-pectin NPs were synthesized by Alkhader et al. to deliver curcumin in colonic conditions. The majority of curcumin release was observed in a pectinase-enriched environment. After exposure in various media, the SEM images showed a retained NP matrix in an acidic environment and a distorted matrix in a pectinase-enriched environment (Alkhader et al., 2017).

3.2.2 Brain targeting

Kuo et al. have synthesized curcumin-loaded chitosan-poly(lactic-*co*-glycolic acid) (PLGA) NPs suitably modified by sialic acid for penetrating the blood brain barrier (BBB). An increase in chitosan concentration showed steady release of curcumin from synthesized NPs. The immunochemical staining test performed on endothelial cells of a human brain confirmed the positive penetration across the BBB. The immunofluorescence images exhibited inhibiting the proliferation of brain cancer stem cells and U87MG cells (Kuo et al., 2019). Meng et al. developed huperizine A-loaded NPs synthesized by using lactoferrin conjugated *N*-trimethyl chitosan and polylactide-*co*-glycoside. The NPs were developed using Box-Behnken design by an emulsion-solvent evaporation technique. Reduced toxicity in the 16HBE cell line was observed in comparison to huperizine A solution. The cellular uptake study demonstrated higher accumulation of NPs in SH-SY5Y and 16HBE cells. An increased fluorescence intensity and prolonged residence time was reported by an in vivo imaging study (Meng et al., 2018). Fernandes et al. developed an chitosan-I-valine conjugate which was employed to synthesize NPs to deliver saxagliptin, an extremely hydrophilic drug used in the treatment of Alzheimer's disease. Fluorescent dye and Rhodamine B were incorporated in the nanoengineered system for evaluating the BBB permeability. The in vivo study performed in rats showed a lower C_{max} than pure saxagliptin. The brain uptake study demonstrated 53 ng/mL of the drug compared to no detectable saxagliptin after 24 h of administration (Fernandes et al., 2018).

Hanmei et al. have investigated the efficacy of an intranasal route for delivering cyclovirobuxine D to the brain. The chitosan-based NPs were developed by following an ionotropic gelation technique. The entrapment efficiency was reported as 62.82 ± 2.59% with a polydispersity index of 0.19 ± 0.01. The average sizes of NPs were reported within 235.37 ± 12.71 nm. A sustained release of cyclovirobuxine D (88.03 ± 2.30%) was observed after 24 h. An in vivo study confirmed a higher AUC at the brain, indicating the efficacy of intranasal administration of a chitosan-based nanoparticulate system (Hanmei et al., 2018). Agrawal et al. developed docetaxel-loaded nanocarriers containing chitosan and D-α-tocopherol polyethylene glycol 1000 succinate. An in vitro study including C6 glioma cells demonstrated cytotoxicity along with increased cellular uptake. The in vivo study indicated an enhancement in relative bioavailability compared to a marketed formulation of docetaxel (Agrawal et al., 2017). Ahmad et al. have investigated the efficacy of chitosan NPs to deliver an antioxidant agent, rutin, which can minimize the risk of ischemic disease. The particle size was reported at <100 nm with a zeta potential of 31.04, as shown in Fig. 3.2. The brain targeting potential was found to be 93.00% ± 5.69% after intranasal administration, indicating increased bioavailability of rutin in respect to intravenous administration. The study findings confirmed improved histopathology and neurobehavioral activity in cerebral ischemic rats (Ahmad et al., 2017). Another study to treat cerebrovascular ischemia involving scutellarin-loaded chitosan-HP-β-CD-based nanoparticles synthesized by Liu et al. via ionotropic cross-linking technology. The in vivo study performed in C57BL mice showed increased drug concentration in the brain, indicating the brain targeting ability of the nanoparticulate system. The efficacy of the developed NPs were evaluated for both the oral

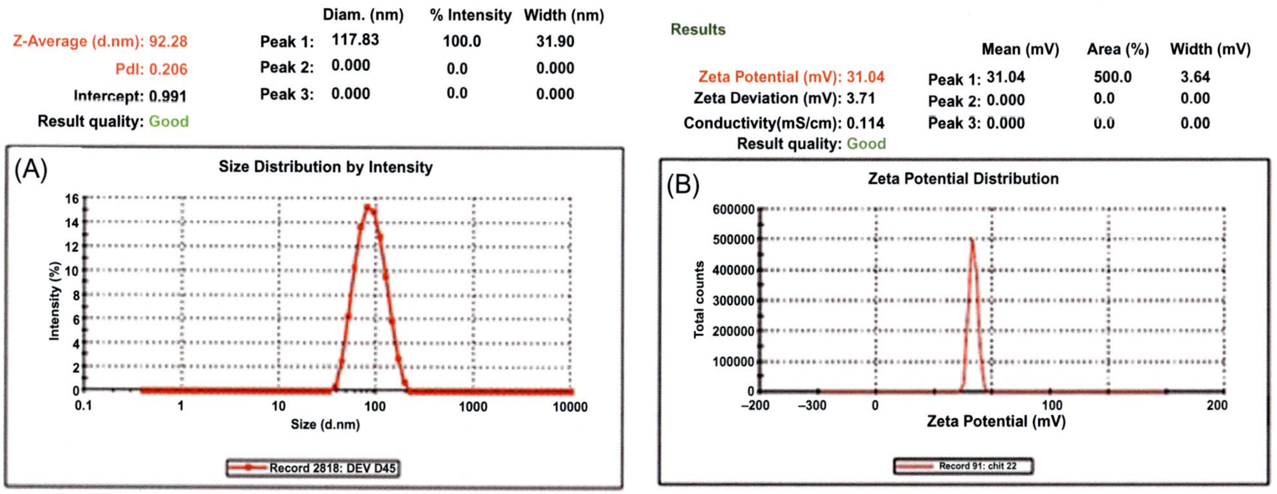

FIG. 3.2 Particle size distribution (A) and zeta potential (B) of chitosan NPs. *From Ahmad, N., Ahmad, R., Naqvi, A.A., Alam, M.A., Ashafaq, M., Samim, M., Iqbal, Z., Ahmad, F.J., 2016. Rutin-encapsulated chitosan nanoparticles targeted to the brain in the treatment of cerebral ischemia. Int. J. Biol. Macromolecules, 91, 640–655. https://doi.org/10.1016/j.ijbiomac.2016.06.001.*

and intranasal route, but the study reports suggested that the nasal route had more potential for treating cerebrovascular ischemia (Liu and Ho, 2017).

3.2.3 Hepatic targeting

Biotin- and avidin-modified chitosan NPs were developed by Bu et al. to achieve tumor-specific liver targeting. Trans-resveratrol-loaded NPs were reported to accumulate in liver following parenteral administration. Pharmacokinetic profiling revealed that a maximum liver targeting index of 2.70 was achieved. An inhibitory study performed on HepG2 cells suggested improved anticancer activity in respect to unmodified NPs and trans-resveratrol solution (Bu et al., 2013). Li et al. developed chitosan-based magnetic NPs loaded with carboplatin-Fe@C for liver targeting. A reverse microemulsion technique was employed for synthesizing the magnetic nanocages. A sustained drug release of 91% was reported after 120 h. An in vivo study demonstrated an increase in tumor temperature within 10 min of parenteral administration (Li et al., 2009). Self-assembled nanocomplexes were synthesized by electrostatic interaction to incorporate interleukin-22 for the therapy of nonalcoholic fatty liver disease. The NPs showed reduced hepatic steatosis with enhanced metabolic syndrome in a mice model. The study's results demonstrated activation of Nrf2/SOD1 and STAT3/Erk1/2 signaling transductions as well as modulating in vivo lipid metabolism (Zai et al., 2019). Carboxymethyl chitosan-based pH-sensitive nanomicelles were developed for tumor-specific hepatic delivery of doxorubicin. The nanomicelles showed complete and rapid release of doxorubicin in 5.3 pH. The cellular uptake study indicated that modification of CD147 has improved cellular internalization. An in vivo study indicated enhanced antiproliferative property with reduced side effects (Qu et al., 2018). Guo et al. have synthesized self-assembled nanoparticles based on galactosylated *O*-carboxymethyl chitosan grafted stearic acid conjugate for doxorubicin delivery to the liver. This conjugate self-assembles with a diameter of 160 nm by probe sonication in an aqueous medium. Galactosylated *O*-carboxymethyl chitosan grafted stearic acid conjugates demonstrated excellent potential to be applied in the treatment of cancer (Guo et al., 2013). Viota et al. have developed magnetic NPs for delivering a chemotherapeutic agent gemcitabine to tumoral tissue. The developed NPs were coated with layers of chitosan and poly (acrylic acid), and were finally coated with folic acid. Gemcitabine was adsorbed on the surface of NPs. An in vitro study exhibited the efficacy of developed magnetic NPs to deliver the drug to the liver (Viota et al., 2013). A drug used in the therapy of hepatocarcinoma, norcantharidin, was incorporated in galactosylated chitosan-based nanocarriers. An ionic cross-linking technique was employed to develop the NPs. Strong cytotoxicity was reported against hepatic carcinoma cells with excellent in vitro cellular uptake. An in vivo study performed in mice having H22 hepatic tumors showed better tumor inhibition than free norcantharidin (Wang et al., 2010).

3.2.4 Pulmonary targeting

Guo et al. developed a multifunctional nanocarrier through electrostatic interaction of a chitosan-fluorescent gold nanocluster and AS1411 aptamer. Methotrexate was incorporated in the developed NPs via hydrophobic interaction. The drug release study indicated a pH-dependent release of methotrexate. An in vitro study demonstrated selective uptake of NPs by cancer cells along with improved anticancer efficacy in A549 lung cancer cells. An in vivo study performed in a mice model exhibited tumor-specific accumulation of drug inhibiting tumor growth (Guo et al., 2018). Liu et al. developed an oleic acid conjugated chitosan-based novel nanocarriers system for lung targeting of chemotherapeutic drugs. Paclitaxel and quercetin were incorporated in NPs, which were further polymerized into polymeric microspheres for obtaining suitable particle size. An in vivo study revealed prolonged circulation time. The biodistribution study exhibited lung-specific accumulation of drugs, indicating the targeting ability of the system (Liu et al., 2017). Yhee et al. have investigated the efficacy of glycol chitosan NPs in lung targeting to treat idiopathic pulmonary fibrosis (IPF). The cellular uptake of glycol chitosan NPs in human lung fibroblast was investigated, which demonstrated an increased cellular uptake of NPs. The study findings suggested that glycol chitosan NPs can interact with the collagen matrix and have the potential as a drug carrier system for targeting fibrotic lung fibroblasts (Yhee et al., 2017). On et al. developed cyanin 5.5 or fluorophore conjugated theranostic glycol chitosan NPs to compare tumor targeting ability in VX2 tumoral rabbit and mouse models. The spherical NPs demonstrated good stability in aqueous condition for up to 8 days. Exceptional tumor accumulation with prolonged biodistribution was observed in both rabbit and mouse models. The precise tumor targeting facilitated successful image-guided surgery for tumoral rabbit lung (On et al., 2020). Catechin hydrate-loaded chitosan-coated PLGA NPs were developed by Ahmad et al. for improving pulmonary bioavailability. A solvent evaporation technique was employed to fabricate the chitosan-based NPs studied for anticancer efficacy on H1299 lung cells. SEM and TEM study revealed smooth and spherical-shaped NPs, as shown in Fig. 3.3. The drug release study demonstrated that a cancer microenvironment triggered release of catechin hydrate. The intranasal administration using rat model showed an improved C_{max} in the lung compared to the oral and intravenous routes (Ahmad et al., 2020). Hyaluronic acid-decorated chitosan NPs were synthesized by Zhang et al. for potential lung targeting of siRNAs. Cyanin 3 labeled

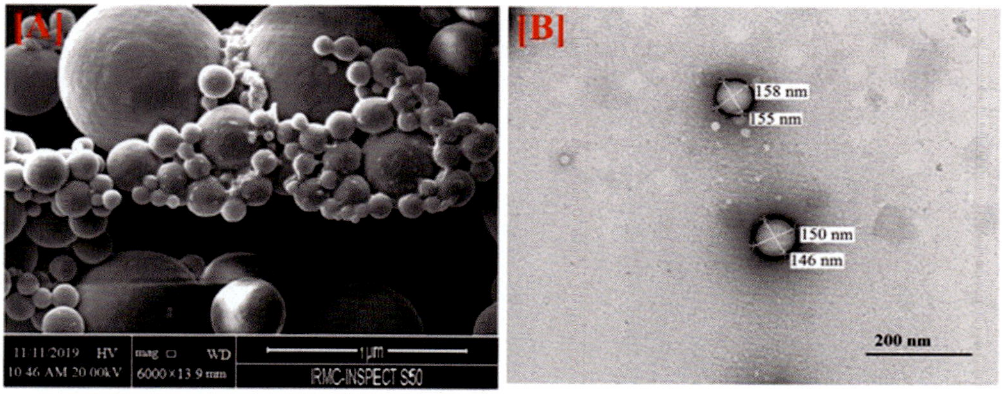

FIG. 3.3 Scanning electron microscopic images (A) and transmission electron microscopy images (B) of chitosan-coated PLGA NPs. *From Ahmad, N., Ahmad, R., Alrasheed, R.A., Almatar, H.M.A., Al-Ramadan, A.S., Buheazah, T.M., AlHomoud, H.S., Al-Nasif, H.A., Alam, M.A., 2020. A chitosan-PLGA based catechin hydrate nanoparticles used in targeting of lungs and cancer treatment. Saudi J. Biol. Sci. 27 (9), 2344–2357. https://doi.org/10.1016/j.sjbs.2020.05.023.*

siRNA was characterized using a tumoral mice model and in A549 cells. The study's report revealed effective targeting in A549 cells through CD44 receptor and inhibition in cell proliferation. An in vivo study demonstrated a similar outcome in comparison to unmodified siRNA-loaded NPs (Zhang et al., 2019). Another study conducted by Chiesa et al. showed the efficacy of hyaluronic acid-modified chitosan NPs for pulmonary targeting of everolimus. The conjugation of hyaluronic acid aided the targeting of NPs into CD44 cells. A sustained release up to 24 h was observed meeting the therapeutic requirements for chronic lung allograft dysfunction (Chiesa et al., 2018).

3.2.5 Transdermal targeting

Shamsi et al. developed chitosan-poly(*N*-isopropylacrylamide-*co*-acrylic acid)-based NPs for transdermal applications. Clindamycin phosphate and tretinoin were incorporated using microfluid technology. The developed NPs showed a sustained release with minimum inhibitory as well as bactericidal concentration. An in vivo study performed in rats and rabbits showed no erythema or skin sensitivity with better drug permeation (Shamsi et al., 2017). Dong et al. synthesized baicalein-loaded NPs containing chitosan and lecithin for improving transdermal drug permeability. A solvent injection technique was used to develop the NPs by self-assembled interaction between the polymers. The ex vivo and in vivo study demonstrated prolonged retention and efficient penetration of baicalein without showing skin irritation (Dong et al., 2020). Jana et al. synthesized chitosan- and albumin-based NPs by heat coagulation technique for sustained aceclofenac delivery. The NPs were further incorporated to a carbopol 940-based aqueous gel suitable for topical application. The in vivo study was performed by using carrageenan-treated rats which demonstrated improved swelling inhibition compared to marketed aceclofenac formulation over 4 h (Jana et al., 2014). Chitosan was used in combination with poly(lactide-*co*-glycolide) and lecithin to synthesize nanoparticles for topical administration of steroids. A steroidal drug, betamethasone valerate, was incorporated in the nanoparticle core. The permeation study revealed a 1.58-fold increase in drug concentration in epidermis for chitosan-based NPs, in respect to PLGA NPs. A suitable 10% *w*/w chitosan gel was developed to facilitate topical application of prepared NPs. The pharmacodynamic profiling indicated antiinflammatory properties (Özcan et al., 2013). Mohamed et al. have incorporated colchicine through mesoporous silica NPs containing oxidized pullulan and carboxymethyl chitosan. The developed composites were used for treating cotton fabric to prepare transdermal patches. The ex vivo colchicines permeation was determined, which demonstrated an extended release of drug for up to 24 h with enhanced flux. Fluorescein was used for confocal laser microscopy to evaluate the skin permeation further. Improved locomotor activity was observed in mono-iodoacetate-induced osteoarthritis model of rats (Mohamed et al., 2020). Friedman et al. investigated the efficacy of chitosan-based NPs for the treatment of cutaneous pathogen. The immunological and antimicrobial properties of chitosan-alginate NPs were evaluated in combination with benzoyl peroxide, an antiacne drug. Superior antimicrobial activity was reported against *Propionibacterium acnes* compared with benzoyl peroxide alone. The toxicity study revealed a reduced toxicity profile in eukaryotic cells (Friedman et al., 2013). Chitosan-nicotinamide NPs were developed by Yu et al. for transdermal application of tacrolimus which is used in the treatment of atopic dermatitis. The use of nicotinamide showed a positive impact on drug entrapment. The in vivo skin permeation study showed significantly increased drug permeation for chitosan- and nicotinamide-based NPs. This highly efficacious system facilitated a reduction of dose for the treatment of atopic dermatitis (Yu et al., 2018). Multilayered liposomes were developed using layer-by-layer technology for topical delivery of quercetin. Cationic chitosan and anionic sodium hyaluronate were used for liposome synthesis. Transmission electron microscopy confirmed the formation of a spherical polyelectrolyte complex. The quercetin release from liposomal core was sustained with increasing numbers of bilayers following the Korsmeyer–Peppas model. In vitro skin permeation results demonstrated similar skin permeability for positively and negatively charged particles (Jeon et al., 2015).

3.2.6 Ocular targeting

Honokiol-loaded chitosan-sulfobutylether-β-cyclodextrin NPs were developed by Deng et al. for retinal neovascularization therapy. Sulfobutylether-β-cyclodextrin was used to form an inclusion complex facilitating the incorporation of insoluble honokiol in chitosan NPs synthesized by an ionotropic gelation technique. An in vivo study indicated improved ocular bioavailability along with good ophthalmic tolerability. The maximum honokiol concentration was found to improve by 1.65 times in respect to honokiol suspension (Deng et al., 2018). A sparfloxacin-loaded ophthalmic nanosuspension was prepared by Ambhore et al. with HPMC E5 and chitosan using a solvent diffusion technique. The in vitro drug release study confirmed sustained release of sparfloxacin for up to 9 h from a chitosan-based formulation compared to 6 h for HPMC nanosuspension. The study findings indicated excellent biocompatibility, ocular tolerance, and in vivo antimicrobial efficacy of the developed sparfloxacin nanosuspension (Ambhore et al., 2016). The efficacy of chitosan NPs developed by an ionic gelation technique using sodium tripolyphosphate was investigated by da Silva et al. The mucoadhesive nature of the NPs showed an increased ocular retention time, facilitating higher ocular penetration of an antioxidant rosmarinic acid. A study conducted on a human corneal cell line demonstrated no cytotoxic effect (da Silva et al., 2016). Chitosan NPs were synthesized and incorporated in poly(lactic-*co*-glycolic acid)-based microparticles for sustained release of ranibizumab. The chitosan-tripolyphosphate-hyaluronic acid NPs were entrapped in microparticles by using a modified w/o/w emulsification technique. Enhanced antiangiogenic activity was observed for the prepared formulations. The developed nanoengineered system also exhibited improved drug loading and better release profile (Elsaid et al., 2016). An ionic gelation technique was employed for synthesizing chitosan-based NPs for ocular delivery of ketorolac tromethamine. The in vitro ketorolac tromethamine release demonstrated significant variation ($P < 0.05$) compared to ketorolac tromethamine solution. The mucoadhesion study revealed a good adhesion property. Excised porcine eyeballs were used for ex vivo corneal permeation which indicated increased ocular retention time (Fathalla et al., 2016). Silva et al. developed a chitosan- and hyaluronic acid-based mucoadhesive nanoparticulate system for delivering an antibiotic agent, ceftazidime. The NPs were incorporated in an eye drop containing 0.75% (w/v) hydroxypropyl methyl cellulose. The mucoadhesive property of the NPs facilitated the increase in ocular residence time. The in vitro drug release study showed a prolonged release of ceftazidime from chitosan-based nanocarriers (Silva et al., 2017). Poly(lactic acid)-grafted chisotan was used to synthesize self-aggregated nanocarriers for ophthalmic delivery of amphotericin B. A dialysis technique was used to develop the nano-vehicles having a mean size of 200 nm. The in vitro study indicated sustained release ability. An ocular irritation study was performed in rabbits and demonstrated no detectable irritation on application. Amphotericin B showed similar efficacy against *Candida albicans* as free amphotericin B (Zhou et al., 2013). Chitosan-coated liposomal delivery was developed by Cheng et al. for ocular targeting of triamcinolone acetonide to treat macular edema. The cellular uptake was investigated using a fluorescent dye. Clinical safety was established by determining intraocular pressure and corneal thickness. A preclinical study conducted in a rat model showed excellent efficacy in reducing retinal edema. The toxicity study was performed in a cell culture, which indicated the nontoxic nature of chitosan-coated liposomal eye drops (Cheng et al., 2019).

3.2.7 Renal targeting

Qiao et al. have investigated the kidney targeting of a chitosan-based nanocomplex developed by coordination driven assembly. Autofluorescent doxorubicin was incorporated in the nanocomplex for determining cellular uptake and targeting ability. The developed nanodevice exhibited satisfactory stability in a physiological environment with pH-triggered drug release. A tissue distribution study demonstrated a renal uptake rate of (r(e)) 25.6, indicating tissue-specific accumulation in the kidney (Qiao et al., 2014). The efficacy of chitosan-based siRNA NPs was investigated for knockdown of COX-2 to prevent kidney injury induced by UUO (unilateral ureteral obstruction). The accumulation of chitosan-siRNA NPs in an obstructed kidney was confirmed by confocal microscopy and optical imaging methods. A study conducted in mice demonstrated diminishing of UUO after parenteral administration of developed NPs. Decreased COX-2 immunoreactivity with marginal tubular damage was reported in treated UUO mice (Yang et al., 2015). Another study conducted by Gao et al., involving chitosan-siRNA NPs, demonstrated that siRNA is accumulated in proximal tubule epithelial cells for more than 48 h. When the developed NPs were administered in a mice model, with megalin gene knockout, they were distributed in cells which expressed megalin. Knockdown of AQP1 (aquaporin 1) up to 50% was achieved through i.v. administration of chitosan-AQP1 siRNA in the mice model (Gao et al., 2014). Kandav et al. have synthesized allopurinol-incorporated chitosan-coated magnetic NPs for effective kidney targeting to treat nephrolithiasis. The magnetic NPs were developed by a chemical coprecipitation technique. Enhanced kidney uptake of 19.07-fold was reported for allopurinol in comparison to pure drug after 2 h of administration. Studies conducted in mice showed significant ($P < 0.01$) efficacy, estimated by serum uric acid level and urine pH (Kandav et al., 2019). Anwar et al. developed curcumin-loaded chitosan NPs to minimize renal toxicity caused by cypermethrin. A solvent displacement technique was used to synthesize NPs. An in vivo study performed in rabbits suggested that the developed NPs are efficacious against the oxidative damage and biochemical alterations in the kidney caused by the intake of cypermethrin (Anwar et al., 2020). Metformin-incorporated chitosan NPs were developed by Wang et al. as therapeutic options for polycystic kidney disease. The orally administered

NPs showed good mucoadhesion with considerable permeability. The in vivo study in a murine model indicated good tolerance on repeated administration and reduced cyst burden in respect to free drug. Creatinine and blood urea nitrogen level were reported to be identical to those of the untreated group, which indicates good biocompatibility health (Wang et al., 2021).

3.3 Conclusions

The drug delivery application of chitosan in the development of nanoengineered drug carriers has been summarized in this chapter. As chitosan is a polysaccharide-based biopolymer, its abundant availability in nature has propelled its application in pharmaceutical and biomedical research. Apart from having excellent biodegradability, the mucoadhesive nature of chitosan has drawn researchers' attention for enhancing the retention time and improving bioavailability. Among various drug delivery systems, chitosan-based nanoengineered drug carriers are one of the most investigated delivery techniques for selective targeting of therapeutic molecules. Continuous advancement has been reported in the field of nanotechnology, which promises to reduce risks associated in therapy and to develop a more efficacious delivery system.

References

Agrawal, P., Singh, R.P., Sonali, Kumari, L., Sharma, G., Koch, B., Rajesh, C.V., Mehata, A.K., Singh, S., Pandey, B.L., Meuthu, M.S., 2017. TPGS-chitosan cross-linked targeted nanoparticles for effective brain cancer therapy. Mater. Sci. Eng. C 74, 167–176. https://doi.org/10.1016/j.msec.2017.02.008.

Ahmad, M., Manzoor, K., Singh, S., Ikram, S., 2017. Chitosan centered bionanocomposites for medical specialty and curative applications: a review. Int. J. Pharm. 529 (1–2), 200–217. https://doi.org/10.1016/j.ijpharm.2017.06.079.

Ahmad, N., Ahmad, R., Alrasheed, R.A., Almatar, H.M.A., Al-Ramadan, A.S., Buheazah, T.M., AlHomoud, H.S., Al-Nasif, H.A., Alam, M.A., 2020. A chitosan-PLGA based catechin hydrate nanoparticles used in targeting of lungs and cancer treatment. Saudi J. Biol. Sci. 27 (9), 2344–2357. https://doi.org/10.1016/j.sjbs.2020.05.023.

Ahsan, S.M., Thomas, M., Reddy, K.K., Sooraparaju, S.G., Asthana, A., Bhatnagar, I., 2018. Chitosan as biomaterial in drug delivery and tissue engineering. Int. J. Biol. Macromol. 110, 97–109. https://doi.org/10.1016/j.ijbiomac.2017.08.140.

Ali, A., Ahmed, S., 2018. A review on chitosan and its nanocomposites in drug delivery. Int. J. Biol. Macromol. 109, 273–286. https://doi.org/10.1016/j.ijbiomac.2017.12.078.

Alkhader, E., Billa, N., Roberts, C.J., 2017. Mucoadhesive chitosan-pectinate nanoparticles for the delivery of curcumin to the Colon. AAPS PharmSciTech 18 (4), 1009–1018. https://doi.org/10.1208/s12249-016-0623-y.

Ambhore, N.P., Dandagi, P.M., Gadad, A.P., 2016. Formulation and comparative evaluation of HPMC and water soluble chitosan-based sparfloxacin nanosuspension for ophthalmic delivery. Drug. Deliv. Transl. Res. 6 (1), 48–56. https://doi.org/10.1007/s13346-015-0262-y.

Anwar, M., Muhammad, F., Akhtar, B., Ur Rehman, S., Saleemi, M.K., 2020. Nephroprotective effects of curcumin loaded chitosan nanoparticles in cypermethrin induced renal toxicity in rabbits. Environ. Sci. Pollut. Res. Int. 27 (13), 14771–14779. https://doi.org/10.1007/s11356-020-08051-5.

Austin, P.R., Brine, C.J., Castle, J.E., Zikakis, J.P., 1981. Chitin: new facets of research. Science 212 (4496), 749–753. https://doi.org/10.1126/science.7221561.

Bu, L., Gan, L.C., Guo, X.Q., Chen, F.Z., Song, Q., Zhao, Q., Gou, X.J., Hou, S.X., Yao, Q., 2013. Trans-resveratrol loaded chitosan nanoparticles modified with biotin and avidin to target hepatic carcinoma. Int. J. Pharm. 452 (1–2), 355–362. https://doi.org/10.1016/j.ijpharm.2013.05.007.

Cerchiara, T., Abruzzo, A., Di Cagno, M., Bigucci, F., Bauer-Brandl, A., Parolin, C., Vitali, B., Gallucci, M.C., Luppi, B., 2015. Chitosan based micro- and nanoparticles for colon-targeted delivery of vancomycin prepared by alternative processing methods. Eur. J. Pharm. Biopharm. 92, 112–119. https://doi.org/10.1016/j.ejpb.2015.03.004.

Cheng, T., Li, J., Cheng, Y., Zhang, X., Qu, Y., 2019. Triamcinolone acetonide-chitosan coated liposomes efficiently treated retinal edema as eye drops. Exp. Eye Res. 188, 107805. https://doi.org/10.1016/j.exer.2019.107805.

Cheung, R.C.F., Ng, T.B., Wong, J.H., Chan, W.Y., 2015. Chitosan: an update on potential biomedical and pharmaceutical applications. Mar. Drugs 13 (8), 5156–5186. https://doi.org/10.3390/md13085156.

Chiesa, E., Dorati, R., Conti, B., Modena, T., Cova, E., Meloni, F., Genta, I., 2018. Hyaluronic acid-decorated chitosan nanoparticles for CD44-targeted delivery of everolimus. Int. J. Mol. Sci. 19 (8). https://doi.org/10.3390/ijms19082310.

da Silva, S.B., Ferreira, D., Pintado, M., Sarmento, B., 2016. Chitosan-based nanoparticles for rosmarinic acid ocular delivery- -In vitro tests. Int. J. Biol. Macromol. 84, 112–120. https://doi.org/10.1016/j.ijbiomac.2015.11.070.

Deng, F., Hu, W., Chen, H., Tang, Y., Zhang, L., 2018. Development of a chitosan-based nanoparticle formulation for ophthalmic delivery of honokiol. Curr. Drug. Deliv. 15 (4), 594–600. https://doi.org/10.2174/1567201814666170419113933.

Devarajan, P.V., Sonavane, G.S., 2007. Preparation and in vitro/in vivo evaluation of gliclazide loaded Eudragit nanoparticles as a sustained release carriers. Drug Dev. Ind. Pharm. 33 (2), 101–111. https://doi.org/10.1080/03639040601096695.

Dong, W., Ye, J., Wang, W., Yang, Y., Wang, H., Sun, T., Gao, L., Liu, Y., 2020. Self-assembled lecithin/chitosan nanoparticles based on phospholipid complex: a feasible strategy to improve entrapment efficiency and transdermal delivery of poorly lipophilic drug. Int. J. Nanomedicine 15, 5629–5643. https://doi.org/10.2147/IJN.S261162.

Dutta, P.K., Ravikumar, M.N.V., Dutta, J., 2002. Chitin and chitosan for versatile applications. J. Macromol. Sci. Polym. Rev. 42 (3), 307–354. https://doi.org/10.1081/MC-120006451.

Dutta, P.K., Duta, J., Tripathi, V.S., 2004. Chitin and chitosan: chemistry, properties and applications. J. Sci. Ind. Res. 63 (1), 20–31.

Elieh-Ali-Komi, D., Hamblin, M.R., 2016. Chitin and chitosan: production and application of versatile biomedical nanomaterials. Int. J. Adv. Res. 4 (3), 411–427.

Elsaid, N., Jackson, T.L., Elsaid, Z., Alqathama, A., Somavarapu, S., 2016. PLGA microparticles entrapping chitosan-based nanoparticles for the ocular delivery of ranibizumab. Mol. Pharm. 13 (9), 2923–2940. https://doi.org/10.1021/acs.molpharmaceut.6b00335.

Fathalla, Z.M.A., Khaled, K.A., Hussein, A.K., Alany, R.G., Vangala, A., 2016. Formulation and corneal permeation of ketorolac tromethamine-loaded chitosan nanoparticles. Drug. Dev. Ind. Pharm. 42 (4), 514–524. https://doi.org/10.3109/03639045.2015.1081236.

Felt, O., Buri, P., Gurny, R., 1998. Chitosan: a unique polysaccharide for drug delivery. Drug Dev. Ind. Pharm. 24 (11), 979–993.

Fernandes, J., Ghate, M.V., Basu Mallik, S., Lewis, S.A., 2018. Amino acid conjugated chitosan nanoparticles for the brain targeting of a model dipeptidyl peptidase-4 inhibitor. Int. J. Pharm. 547 (1–2), 563–571. https://doi.org/10.1016/j.ijpharm.2018.06.031.

Friedman, A.J., Phan, J., Schairer, D.O., Champer, J., Qin, M., Pirouz, A., Blecher-Paz, K., Oren, A., Liu, P.T., Modlin, R.L., Kim, J., 2013. Antimicrobial and anti-inflammatory activity of chitosan-alginate nanoparticles: a targeted therapy for cutaneous pathogens. J. Investig. Dermatol. 133 (5), 1231–1239. https://doi.org/10.1038/jid.2012.399.

Gandhi, A., Jana, S., Sen, K.K., 2014. In-vitro release of acyclovir loaded Eudragit RLPO® nanoparticles for sustained drug delivery. Int. J. Biol. Macromol. 67, 478–482. https://doi.org/10.1016/j.ijbiomac.2014.04.019.

Gao, S., Hein, S., Dagnæs-Hansen, F., Weyer, K., Yang, C., Nielsen, R., Christensen, E.I., Fenton, R.A., Kjems, J., 2014. Megalin-mediated specific uptake of chitosan/siRNA nanoparticles in mouse kidney proximal tubule epithelial cells enables AQP1 gene silencing. Theranostics 4 (10), 1039–1051. https://doi.org/10.7150/thno.7866.

Gelperina, S., Kisich, K., Iseman, M.D., Heifets, L., 2005. The potential advantages of nanoparticle drug delivery systems in chemotherapy of tuberculosis. Am. J. Respir. Crit. Care Med. 172 (12), 1487–1490. https://doi.org/10.1164/rccm.200504-613PP.

Guo, H., Zhang, D., Li, C., Jia, L., Liu, G., Hao, L., Zheng, D., Shen, J., Li, T., Guo, Y., Zhang, Q., 2013. Self-assembled nanoparticles based on galactosylated O-carboxymethyl chitosan-graft-stearic acid conjugates for delivery of doxorubicin. Int. J. Pharm. 458 (1), 31–38. https://doi.org/10.1016/j.ijpharm.2013.10.020.

Guo, X., Zhuang, Q., Ji, T., Zhang, Y., Li, C., Wang, Y., Li, H., Jia, H., Liu, Y., Du, L., 2018. Multi-functionalized chitosan nanoparticles for enhanced chemotherapy in lung cancer. Carbohydr. Polym. 195, 311–320. https://doi.org/10.1016/j.carbpol.2018.04.087.

Hamedi, H., Moradi, S., Hudson, S.M., Tonelli, A.E., 2018. Chitosan based hydrogels and their applications for drug delivery in wound dressings: a review. Carbohydr. Polym. 199, 445–460. https://doi.org/10.1016/j.carbpol.2018.06.114.

Hanmei, W., Sisi, L., Jiabao, W., Lei, Y., Ning, J., Qing, W., Yang, Y., 2018. A novel delivery method of Cyclovirobuxine D for brain-targeting: chitosan coated nanoparticles loading cyclovirobuxine D by intranasal administration. J. Nanosci. Nanotechnol., 5274–5282. https://doi.org/10.1166/jnn.2018.15371.

Hanna Rosli, N.A., Loh, K.S., Wong, W.Y., Mohamad Yunus, R., Khoon Lee, T., Ahmad, A., Chong, S.T., 2020. Review of chitosan-based polymers as proton exchange membranes and roles of chitosan-supported ionic liquids. Int. J. Mol. Sci. 21 (2). https://doi.org/10.3390/ijms21020632.

Ibrahim, K., Khalid, S., Idrees, K., 2019. Nanoparticles: properties, applications and toxicities. Arab. J. Chem., 908–931. https://doi.org/10.1016/j.arabjc.2017.05.011.

Iglesias, N., Galbis, E., Díaz-Blanco, M.J., Lucas, R., Benito, E., De-Paz, M.V., 2019. Nanostructured chitosan-based biomaterials for sustained and colon-specific resveratrol release. Int. J. Mol. Sci. 20 (2). https://doi.org/10.3390/ijms20020398.

Illum, L., 1998. Chitosan and its use as a pharmaceutical excipient. Pharm. Res. 15 (9), 1326–1331. https://doi.org/10.1023/A:1011929016601.

Iqbal Hassan Khan, M., An, X., Dai, L., Li, H., Khan, A., Ni, Y., 2019. Chitosan-based polymer matrix for pharmaceutical excipients and drug delivery. Curr. Med. Chem. 26 (14), 2502–2513. https://doi.org/10.2174/0929867325666180927100817.

Jana, S., Sen, K.K., 2017. Chitosan—locust bean gum interpenetrating polymeric network nanocomposites for delivery of aceclofenac. Int. J. Biol. Macromol. 102, 878–884. https://doi.org/10.1016/j.ijbiomac.2017.04.097.

Jana, S., Maji, N., Nayak, A.K., Sen, K.K., Basu, S.K., 2013. Development of chitosan-based nanoparticles through inter-polymeric complexation for oral drug delivery. Carbohydr. Polym. 98 (1), 870–876. https://doi.org/10.1016/j.carbpol.2013.06.064.

Jana, S., Manna, S., Nayak, A.K., Sen, K.K., Basu, S.K., 2014. Carbopol gel containing chitosan-egg albumin nanoparticles for transdermal aceclofenac delivery. Colloids Surf. B: Biointerfaces 114, 36–44. https://doi.org/10.1016/j.colsurfb.2013.09.045.

Jana, S., Laha, B., Maiti, S., 2015. Boswellia gum resin/chitosan polymer composites: controlled delivery vehicles for aceclofenac. Int. J. Biol. Macromol. 77, 303–306. https://doi.org/10.1016/j.ijbiomac.2015.03.029.

Jana, S., Sen, K.K., Gandhi, A., 2016. Alginate based nanocarriers for drug delivery applications. Curr. Pharm. Des. 22 (22), 3399–3410. https://doi.org/10.2174/1381612822666160510125718.

Jayakumar, R., Prabaharan, M., Nair, S.V., Tamura, H., 2010. Novel chitin and chitosan nanofibers in biomedical applications. Biotechnol. Adv. 28 (1), 142–150. https://doi.org/10.1016/j.biotechadv.2009.11.001.

Jeevanandam, J., Barhoum, A., Chan, Y.S., Dufresne, A., Danquah, M.K., 2018. Review on nanoparticles and nanostructured materials: history, sources, toxicity and regulations. Beilstein J. Nanotechnol. 9 (1), 1050–1074. https://doi.org/10.3762/bjnano.9.98.

Jeon, S., Yoo, C.Y., Park, S.N., 2015. Improved stability and skin permeability of sodium hyaluronate-chitosan multilayered liposomes by layer-by-layer electrostatic deposition for quercetin delivery. Colloids Surf. B: Biointerfaces 129, 7–14. https://doi.org/10.1016/j.colsurfb.2015.03.018.

Kandav, G., Bhatt, D.C., Jindal, D.K., 2019. Targeting kidneys by superparamagnetic allopurinol loaded chitosan coated nanoparticles for the treatment of hyperuricemic nephrolithiasis. Daru 27 (2), 661–671. https://doi.org/10.1007/s40199-019-00300-4.

Kumar, J., Newton, A.M.J., 2017. Rifaximin–chitosan nanoparticles for inflammatory bowel disease (IBD). Recent Patents Inflamm. Allergy Drug Discov. 11 (1), 41–52. https://doi.org/10.2174/1872213X10666161230111226.

Kuo, Y.C., Wang, L.J., Rajesh, R., 2019. Targeting human brain cancer stem cells by curcumin-loaded nanoparticles grafted with anti-aldehyde dehydrogenase and sialic acid: colocalization of ALDH and CD44. Mater. Sci. Eng. C 102, 362–372. https://doi.org/10.1016/j.msec.2019.04.065.

Layek, B., Mandal, S., 2020. Natural polysaccharides for controlled delivery of oral therapeutics: a recent update. Carbohydr. Polym. 230. https://doi.org/10.1016/j.carbpol.2019.115617.

Li, F.-r., Yan, W.-h., Guo, Y.-h., Qi, H., Zhou, H.-x., 2009. Preparation of carboplatin-Fe@C-loaded chitosan nanoparticles and study on hyperthermia combined with pharmacotherapy for liver cancer. Int. J. Hyperthermia 25 (5), 383–391. https://doi.org/10.1080/02656730902834949.

Li, J., Cai, C., Li, J., Li, J., Li, J., Sun, T., Wang, L., Wu, H., Yu, G., 2018. Chitosan-based nanomaterials for drug delivery. Molecules 23 (10). https://doi.org/10.3390/molecules23102661.

Liu, S., Ho, P.C., 2017. Intranasal administration of brain-targeted HP-β-CD/chitosan nanoparticles for delivery of scutellarin, a compound with protective effect in cerebral ischaemia. J. Pharm. Pharmacol. 69 (11), 1495–1501. https://doi.org/10.1111/jphp.12797.

Liu, K., Chen, W., Yang, T., Wen, B., Ding, D., Keidar, M., Tang, J., Zhang, W., 2017. Paclitaxel and quercetin nanoparticles co-loaded in microspheres to prolong retention time for pulmonary drug delivery. Int. J. Nanomedicine 12, 8239–8255. https://doi.org/10.2147/IJN.S147028.

Mandal, D., Shaw, T.K., Dey, G., Pal, M.M., Mukherjee, B., Bandyopadhyay, A.K., Mandal, M., 2018. Preferential hepatic uptake of paclitaxel-loaded poly-(D-L-lactide-co-glycolide) nanoparticles—a possibility for hepatic drug targeting: pharmacokinetics and biodistribution. Int. J. Biol. Macromol. 112, 818–830. https://doi.org/10.1016/j.ijbiomac.2018.02.021.

Manna, S., Lakshmi, U.S., Racharla, M., Sinha, P., Kanthal, L.K., Kumar, S.P.N., 2016. Bioadhesive HPMC gel containing gelatin nanoparticles for intravaginal delivery of tenofovir. J. Appl. Pharm. Sci. 6 (8), 22–29. https://doi.org/10.7324/JAPS.2016.60804.

Meng, Q., Wang, A., Hua, H., Jiang, Y., Wang, Y., Mu, H., Wu, Z., Sun, K., 2018. Intranasal delivery of Huperzine a to the brain using lactoferrin-conjugated N-trimethylated chitosan surface-modified PLGA nanoparticles for treatment of Alzheimer's disease. Int. J. Nanomedicine 13, 705–718. https://doi.org/10.2147/IJN.S151474.

Mohamed, A.L., Elmotasem, H., Salama, A.A.A., 2020. Colchicine mesoporous silica nanoparticles/hydrogel composite loaded cotton patches as a new encapsulator system for transdermal osteoarthritis management. Int. J. Biol. Macromol. 164, 1149–1163. https://doi.org/10.1016/j.ijbiomac.2020.07.133.

Mohammed, M.A., Syeda, J.T.M., Wasan, K.M., Wasan, E.K., 2017. An overview of chitosan nanoparticles and its application in non-parenteral drug delivery. Pharmaceutics 9 (4). https://doi.org/10.3390/pharmaceutics9040053.

Mohebbi, S., Nezhad, M.N., Zarrintaj, P., Jafari, S.H., Gholizadeh, S.S., Saeb, M.R., Mozafari, M., 2019. Chitosan in biomedical engineering: a critical review. Curr. Stem Cell Res. Ther. 14 (2), 93–116. https://doi.org/10.2174/1574888X13666180912142028.

Mudshinge, S.R., Deore, A.B., Patil, S., Bhalgat, C.M., 2011. Nanoparticles: emerging carriers for drug delivery. Saudi Pharm. J. 19 (3), 129–141. https://doi.org/10.1016/j.jsps.2011.04.001.

Naskar, S., Koutsu, K., Sharma, S., 2019. Chitosan-based nanoparticles as drug delivery systems: a review on two decades of research. J. Drug Target. 27 (4), 379–393. https://doi.org/10.1080/1061186X.2018.1512112.

Nishimura, S.I., Kohgo, O., Kurita, K., Kuzuhara, H., 1991. Chemospecific manipulations of a rigid polysaccharide: syntheses of novel chitosan derivatives with excellent solubility in common organic solvents by regioselective chemical modifications. Macromolecules 24 (17), 4745–4748. https://doi.org/10.1021/ma00017a003.

On, K.C., Rho, J., Yoon, H.Y., Chang, H., Yhee, J.Y., Yoon, J.S., Jeong, S.Y., Kim, H.K., Kim, K., 2020. Tumor-targeting glycol chitosan nanoparticles for image-guided surgery of rabbit orthotopic VX2 lung cancer. Pharmaceutics 12 (7), 1–14. https://doi.org/10.3390/pharmaceutics12070621.

Oryan, A., Sahvieh, S., 2017. Effectiveness of chitosan scaffold in skin, bone and cartilage healing. Int. J. Biol. Macromol. 104, 1003–1011. https://doi.org/10.1016/j.ijbiomac.2017.06.124.

Özcan, İ., Azizoğlu, E., Şenyiğit, T., Özyazici, M., Özer, O., 2013. Comparison of PLGA and lecithin/chitosan nanoparticles for dermal targeting of betamethasone valerate. J. Drug Target. 21 (6), 542–550. https://doi.org/10.3109/1061186X.2013.769106.

Park, J.U., Song, E.H., Jeong, S.H., Song, J., Kim, H.E., Kim, S., 2018. Chitosan-based dressing materials for problematic wound management. In: Advances in Experimental Medicine and Biology. vol. 1077. Springer New York LLC, pp. 527–537, https://doi.org/10.1007/978-981-13-0947-2_28.

Patra, J.K., Das, G., Fraceto, L.F., Campos, E.V.R., Rodriguez-Torres, M.D.P., Acosta-Torres, L.S., Diaz-Torres, L.A., Grillo, R., Swamy, M.K., Sharma, S., Habtemariam, S., Shin, H.S., 2018. Nano based drug delivery systems: recent developments and future prospects 10 Technology 1007 Nanotechnology 03 Chemical Sciences 0306 Physical Chemistry (incl. Structural) 03 Chemical Sciences 0303 Macromolecular and Materials Chemistry 11 Medical and Health Sciences 1115 Pharmacology and Pharmaceutical Sciences 09 Engineering 0903 Biomedical Engineering Prof Ueli Aebi, Prof Peter Gehr. J. Nanobiotechnol. 16 (1). https://doi.org/10.1186/s12951-018-0392-8.

Prabaharan, M., 2008. Review paper: chitosan derivatives as promising materials for controlled drug delivery. J. Biomater. Appl. 23 (1), 5–36. https://doi.org/10.1177/0885328208091562.

Qiao, H., Sun, M., Su, Z., Xie, Y., Chen, M., Zong, L., Gao, Y., Li, H., Qi, J., Zhao, Q., Gu, X., Ping, Q., 2014. Kidney-specific drug delivery system for renal fibrosis based on coordination-driven assembly of catechol-derived chitosan. Biomaterials 35 (25), 7157–7171. https://doi.org/10.1016/j.biomaterials.2014.04.106.

Qu, C., Li, J., Zhou, Y., Yang, S., Chen, W., Li, F., You, B., Liu, Y., Zhang, X., 2018. Targeted delivery of doxorubicin via CD147-mediated ROS/pH dual-sensitive nanomicelles for the efficient therapy of hepatocellular carcinoma. AAPS J. 20 (2). https://doi.org/10.1208/s12248-018-0195-8.

Rinaudo, M., 2006. Chitin and chitosan: properties and applications. Prog. Polym. Sci. 31 (7), 603–632. https://doi.org/10.1016/j.progpolymsci.2006.06.001.

Rizvi, S.A.A., Saleh, A.M., 2018. Applications of nanoparticle systems in drug delivery technology. Saudi Pharm. J. 26 (1), 64–70. https://doi.org/10.1016/j.jsps.2017.10.012.

Rodríguez-Vázquez, M., Vega-Ruiz, B., Ramos-Zúñiga, R., Saldaña-Koppel, D.A., Quiñones-Olvera, L.F., 2015. Chitosan and its potential use as a scaffold for tissue engineering in regenerative medicine. Biomed. Res. Int. 2015. https://doi.org/10.1155/2015/821279.

Shamsi, M., Zahedi, P., Ghourchian, H., Minaeian, S., 2017. Microfluidic-aided fabrication of nanoparticles blend based on chitosan for a transdermal multidrug delivery application. Int. J. Biol. Macromol. 99, 433–442. https://doi.org/10.1016/j.ijbiomac.2017.03.013.

Shaw, T.K., Mandal, D., Dey, G., Pal, M.M., Paul, P., Chakraborty, S., Ali, K.A., Mukherjee, B., Bandyopadhyay, A.K., Mandal, M., 2017. Successful delivery of docetaxel to rat brain using experimentally developed nanoliposome: a treatment strategy for brain tumor. Drug Deliv. 24 (1), 346–357. https://doi.org/10.1080/10717544.2016.1253798.

Silva, M.M., Calado, R., Marto, J., Bettencourt, A., Almeida, A.J., Gonçalves, L.M.D., 2017. Chitosan nanoparticles as a mucoadhesive drug delivery system for ocular dministration. Mar. Drugs 15 (12). https://doi.org/10.3390/md15120370.

Singh, R., Lillard, J.W., 2009. Nanoparticle-based targeted drug delivery. Exp. Mol. Pathol. 86 (3), 215–223. https://doi.org/10.1016/j.yexmp.2008.12.004.

Singla, A.K., Chawla, M., 2001. Chitosan: some pharmaceutical and biological aspects—an update. J. Pharm. Pharmacol. 53 (8), 1047–1067. https://doi.org/10.1211/0022357011776441.

Sun, B., Wang, W., He, Z., Zhang, M., Kong, F., Sain, M., 2020. Biopolymer substrates in buccal drug delivery: current status and future trend. Curr. Med. Chem. 27 (10), 1661–1669. https://doi.org/10.2174/0929867325666181001114750.

Synowiecki, J., Al-Khateeb, N.A., 2003. Production, properties, and some new applications of chitin and its derivatives. Crit. Rev. Food Sci. Nutr. 43 (2), 145–171. https://doi.org/10.1080/10408690390826473.

Torres, F.G., Troncoso, O.P., Pisani, A., Gatto, F., Bardi, G., 2019. Natural polysaccharide nanomaterials: an overview of their immunological properties. Int. J. Mol. Sci. 20 (20). https://doi.org/10.3390/ijms20205092.

Vakili, M., Rafatullah, M., Salamatinia, B., Abdullah, A.Z., Ibrahim, M.H., Tan, K.B., Gholami, Z., Amouzgar, P., 2014. Application of chitosan and its derivatives as adsorbents for dye removal from water and wastewater: a review. Carbohydr. Polym. 113, 115–130. https://doi.org/10.1016/j.carbpol.2014.07.007.

Viota, J.L., Carazo, A., Munoz-Gamez, J.A., Rudzka, K., Gómez-Sotomayor, R., Ruiz-Extremera, A., Salmerón, J., Delgado, A.V., 2013. Functionalized magnetic nanoparticles as vehicles for the delivery of the antitumor drug gemcitabine to tumor cells. Physicochemical in vitro evaluation. Mater. Sci. Eng. C 33 (3), 1183–1192. https://doi.org/10.1016/j.msec.2012.12.009.

Wang, Q., Zhang, L., Hu, W., Hu, Z.H., Bei, Y.Y., Xu, J.Y., Wang, W.J., Zhang, X.N., Zhang, Q., 2010. Norcantharidin-associated galactosylated chitosan nanoparticles for hepatocyte-targeted delivery. Nanomedicine 6 (2), 371–381. https://doi.org/10.1016/j.nano.2009.07.006.

Wang, J., Chin, D., Poon, C., Mancino, V., Pham, J., Li, H., Ho, P.-Y., Hallows, K.R, Chung, E.J., 2021. Oral delivery of metformin by chitosan nanoparticles for polycystic kidney disease. J. Control. Release 329, 1198–1209. https://doi.org/10.1016/j.jconrel.2020.10.047.

Wieckiewicz, M., Boening, K.W., Grychowska, N., Paradowska-Stolarz, A., 2017. Clinical application of chitosan in dental specialities. Mini-Rev. Med. Chem. 17 (5), 401–409. https://doi.org/10.2174/1389557516666160418123054.

Wu, S.J., Don, T.M., Lin, C.W., Mi, F.L., 2014. Delivery of berberine using chitosan/fucoidan-taurine conjugate nanoparticles for treatment of defective intestinal epithelial tight junction barrier. Mar. Drugs 12 (11), 5677–5697. https://doi.org/10.3390/md12115677.

Xu, B., Zhang, W., Chen, Y., Xu, Y., Wang, B., Zong, L., 2018. Eudragit® L100-coated mannosylated chitosan nanoparticles for oral protein vaccine delivery. Int. J. Biol. Macromol. 113, 534–542. https://doi.org/10.1016/j.ijbiomac.2018.02.016.

Yadav, P., Yadav, H., Shah, V.G., Shah, G., Dhaka, G., 2015. Biomedical biopolymers, their origin and evolution in biomedical sciences: a systematic review. J. Clin. Diagn. Res. 9 (9), 21–25. https://doi.org/10.7860/JCDR/2015/13907.6565.

Yang, D., Ma, P., Hou, Z., Cheng, Z., Li, C., Lin, J., 2015. Current advances in lanthanide ion (Ln3+)-based upconversion nanomaterials for drug delivery. Chem. Soc. Rev. 44 (6), 1416–1448. https://doi.org/10.1039/c4cs00155a.

Yetisgin, A.A., Cetinel, S., Zuvin, M., Kosar, A., Kutlu, O., 2020. Therapeutic nanoparticles and their targeted delivery applications. Molecules 25 (9). https://doi.org/10.3390/molecules25092193.

Yhee, J.Y., Yoon, H.Y., Kim, H., Jeon, S., Hergert, P., Im, J., Panyam, J., Kim, K., Nho, R.S., 2017. The effects of collagen-rich extracellular matrix on the intracellular delivery of glycol chitosan nanoparticles in human lung fibroblasts. Int. J. Nanomedicine 12, 6089–6105. https://doi.org/10.2147/IJN.S138129.

Younes, I., Rinaudo, M., 2015. Chitin and chitosan preparation from marine sources. Structure, properties and applications. Mar. Drugs 13 (3), 1133–1174. https://doi.org/10.3390/md13031133.

Yu, K., Wang, Y., Wan, T., Zhai, Y., Cao, S., Ruan, W., Wu, C., Xu, Y., 2018. Tacrolimus nanoparticles based on chitosan combined with nicotinamide: enhancing percutaneous delivery and treatment efficacy for atopic dermatitis and reducing dose. Int. J. Nanomedicine 13, 129–142. https://doi.org/10.2147/IJN.S150319.

Zai, W., Chen, W., Wu, Z., Jin, X., Fan, J., Zhang, X., Luan, J., Tang, S., Mei, X., Hao, Q., Liu, H., Ju, D., 2019. Targeted Interleukin-22 gene delivery in the liver by polymetformin and penetratin-based hybrid nanoparticles to treat nonalcoholic fatty liver disease. ACS Appl. Mater. Interfaces 11 (5), 4842–4857. https://doi.org/10.1021/acsami.8b19717.

Zare, M., Samani, S.M., Sobhani, Z., 2018. Enhanced intestinal permeation of doxorubicin using chitosan nanoparticles. Adv. Pharm. Bull. 8 (3), 411–417. https://doi.org/10.15171/apb.2018.048.

Zhang, L., Gu, F.X., Chan, J.M., Wang, A.Z., Langer, R.S., Farokhzad, O.C., 2008. Nanoparticles in medicine: therapeutic applications and developments. Clin. Pharmacol. Ther. 83 (5), 761–769. https://doi.org/10.1038/sj.clpt.6100400.

Zhang, Wenhua, Xu, Wenhua, Lan, Yu, He, Xuliang, Liu, Kaibin, Liang, Ye, 2019. Antitumor effect of hyaluronic-acid-modified chitosan nanoparticles loaded with siRNA for targeted therapy for non-small cell lung cancer. Int. J. Nanomedicine 14, 5287–5301. https://doi.org/10.2147/IJN.S203113.

Zhou, Wenjun, Wang, Yuanyuan, Jian, Jiuying, Song, Shengfang, 2013. Self-aggregated nanoparticles based on amphiphilic poly(lactic acid)-grafted-chitosan copolymer for ocular delivery of amphotericin B. Int. J. Nanomedicine 8, 3715–3728. https://doi.org/10.2147/IJN.S51186.

CHAPTER

4

Pectin-based micro- and nanomaterials in drug delivery

De-Qiang Li[a,b], *Feng Xu*[b], and *Jun Li*[a]

[a]College of Chemistry and Chemical Engineering, Xinjiang Agricultural University, Urumchi, Xinjiang, P.R. China [b]Beijing Key Laboratory of Lignocellulosic Chemistry, Beijing Forestry University, Beijing, P.R. China

Abbreviations

ADH	adipic acid dihydrazide
BMEP	bis[2-methacryloyloxy] ethyl phosphate
BSA	bull serum albumin
CA	citric acid
ChCl	choline chloride
DCC	dicyclohexylcarbodiimide
DDS	drug delivery system
DE	degree of esterification
DES	deep eutectic solvent
DL	double layer
EE	encapsulation efficiency
EMI	electromagnetic induction
Gal A	D-galacturonic acid
GLU	glutaraldehyde
HFCs	human dermal fibroblast cells
HG	homogalacturonan
HM	high-methoxyl
HNT	halloysite nanotube
LM	low-methoxyl
MMP	matrix metalloproteinase
MMT	montmorillonite
NHS	N-hydroxysuccinimide
NorPEC	norbornene-functionalized pectin
QP	quaternized pectin

Micro- and Nanoengineered Gum-Based Biomaterials for Drug Delivery and Biomedical Applications
https://doi.org/10.1016/B978-0-323-90986-0.00015-7

RFAE radio-frequency assisted extraction
RGD peptide ligand CGGGGRGDSP
RG-I rhamnogalacturonan-I
RG-II rhamnogalacturonan-II
Rha rhamnose
SCF simulated colon fluid
SGF simulated gastric fluid
SLNs solid lipid nanoparticles
SLR ratio of liquid to solid
SSIF simulated small intestinal fluid
TEOS tetraethyl orthosilicate
VEGF vascular endothelial growth factor
XGA xylogalacturonan

4.1 Introduction

Drug delivery system (DDS) is a device that contained therapeutic substances. It significantly promotes efficacy and safety by controlling the release rate, time, and sites in the human body. Natural and man-made polymers have been employed to prepare carriers for DDS. Natural macromolecules have attracted more and more attention, because of the advantages of potential health benefits, environmental friendliness, biodegradation, sustainability, nontoxicity, etc.

Pectin, heterogeneously branched polysaccharides that exist in primary cell walls and intercellular regions of higher plants, has nowadays attracted increasing interest for the application in the DDS, due to the hydration, ion exchange, and rheological properties. The chemical structure surely plays a key role in these physicochemical properties. However, the structure of pectin is hard to determine due to the structural changes during its isolation, storage, and processing. In general, the application of pectin in DDS mainly depends on its surface functional groups. Hydroxyl, amide, and carboxyl functional groups are the major functional groups in the pectin chain (Fig. 4.1).

The ubiquitous functional groups of pectin provide conditions for their application in DDS. The –COOH of the pectin molecule could exist in different forms of $-COO^-$, –COOH, and $-COOH_2^+$ under different pH conditions. Thus, we could composite pectin with different feedstocks that present the opposite charge by adjusting the pH value of the medium. Some materials, including chitosan (Cheikh et al., 2019; Giacomazza et al., 2014; Liu et al., 2019; Luo and Wang, 2014; Luppi et al., 2010; Shitrit et al., 2019), casein (Chang et al., 2017; Li et al., 2013; Wusigale et al., 2020), soy protein (Jin et al., 2018, 2019; Sun et al., 2014), and whey protein (Assadpour et al., 2017; Gharehbeglou et al., 2019; Lutz et al., 2009; Raei et al., 2017; Serrano-Cruz et al., 2013; Souza et al., 2012), are widely used to offer a positive charge, yielding

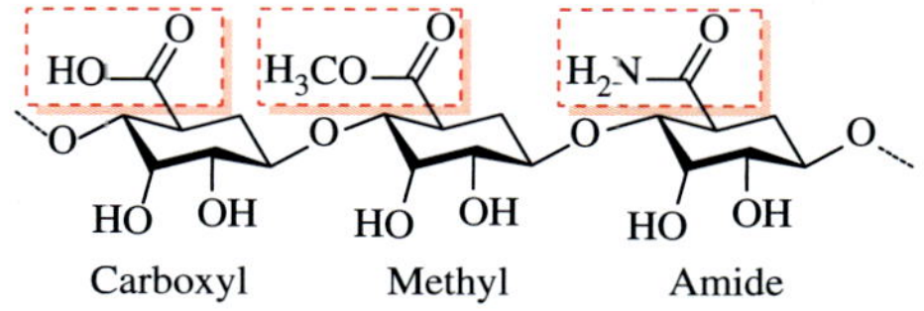

FIG. 4.1 Different functional groups in the pectin molecules.

electrostatic interaction. In addition to electrostatic interaction, the ionic cross-linking method is another important form; the electron-rich groups could coordinate with polyvalent metal ions such as Ca^{2+} and Zn^{2+}, described as an "egg-box" model (Basak and Bandyopadhyay, 2014; Braccini and Pérez, 2001; Cao et al., 2020a). Compared with the physical cross-linking strategies, chemical cross-linking methods are also widely used based on the abundant hydroxyl, carboxyl, formylamino, and methyl ester groups, resulting in changing properties of pectin including hydrophilia and mechanical strength (Maior et al., 2008; Renard and Jarvis, 1999a, b; Yamada and Shiiba, 2015). All the abovementioned strategies including physical and chemical reactions provide potentials for forming hydrogel, film, microsphere, etc., that are applied in the DDS.

Herein, we retrieved "pectin" and "drug delivery" in the "web of science", and the results were shown in Fig. 4.2. It is obvious that the study of pectin-based DDS is a special-interest research topic. Following investigation, pectin has revealed several advantages in biomedical applications including prebiotic effect, hypoglycemic effect, hypocholesterolemia effect, and anticancer effect. Thus, pectin will attract more and more attention.

4.2 Properties of pectin

Pectin is resistant to hydrolysis by digestive enzymes and therefore belongs to the dietary fiber class (Chang et al., 2017; Li et al., 2013; Wusigale et al., 2020). Thus, pectin-based materials exhibit some health benefits. Pectin has been widely used in the food industry due to its gelling properties. As is widely known, the chemical structure plays a decisive role. Unfortunately, we can only be sure that pectin is present in different polymeric forms including homogalacturonan (HG, accounting for about 65% of pectins in plants), rhamnogalacturonan-I (RG-I, accounting for 20%–35% of pectins in plants), xylogalacturonan (XGA, <10% of pectins

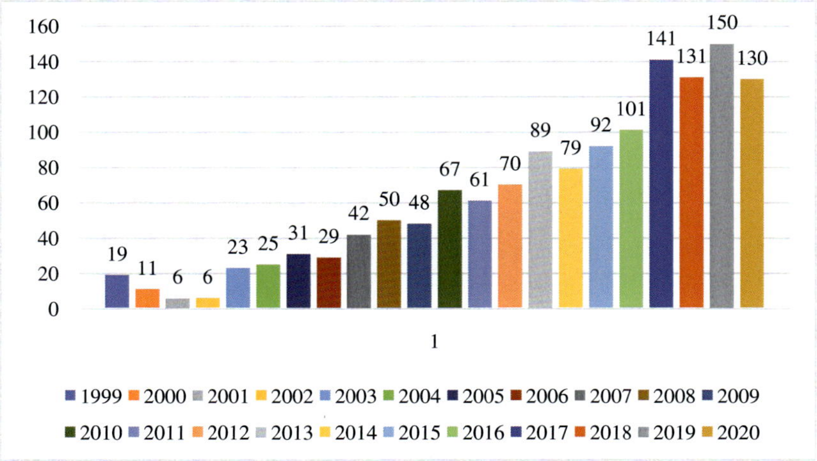

FIG. 4.2 Quantity of papers published per year (since 1999).

in plants), and rhamnogalacturonan-II (RG-II, <10% of pectins in plants). HG consists of linear chains of D-galacturonic acid (Gal A), some of them methylated, in some cases, acetylated, and covalently α-1,4-linked. RG-I is a branched Gal A with α-1,2-linked rhamnose (Rha) residues which can be substituted with β-1,4-galactan, branched arabinan, and/or arabinogalactan side chains. XGA is substituted with xylose residues by β-1,4 linkages. RG-II is composed of 12 types of glycosyl residues, such as glucuronic acid, aceric acid, apiose, and fucose, linked together by at least 22 different glycosidic bonds (Cuervo et al., 2014).

The content of methyl groups directly determines the gelling ability of pectin (Jacob et al., 2020; Muhoza et al., 2019). It can be expressed by the degree of esterification (DE), which is the ratio of the methylated galacturonic acid residues to the total of galacturonic acid units present in the pectin molecules. According to the DE value, pectin can be classified into high-methoxyl pectin (HMP, 60%–75% of DE value) and low-methoxyl pectin (LMP, 20%–40%). Molecular size is another factor that significantly influences the solubleness of pectin, as well as its gelling properties. In addition, the solubility of pectin in aqueous media is affected by several parameters, such as counterion nature, ionic strength, pH value, and temperature (Chang et al., 2017; Li et al., 2013; Wusigale et al., 2020). The inter- and/or intra-chain associations will lead to gelation once the pectin is solubilized. In this process, numerous hydroxyl groups and free, methylated, or amidated carboxyl groups could form hydrogen bonds, hydrophobic interactions, or ionic bonds (Thibault and Ralet, 2003). All the forces influence the gelling, rheological property, and hydration, resulting in different performances of pectin-based DDS.

HM pectin with high soluble solids contents, generally higher than 55%, could form gelation in a concentrated acidic medium with high sugar concentration (Abboud et al., 2020). The high concentration of sugar promotes interchain interaction rather than chain-solvent interactions. The H^+ significantly restrains the –COOH changes to $-COO^-$, and consequently reduces electrostatic repulsion between molecular chains. Thus, hydrogen bonds between the undissociated carboxyl and secondary alcohol groups could form, and hydrophobic interactions between methoxyl groups exist; the two actions combine to align molecular helices.

LM pectin could form gels in the presence of polyvalent metal ions without sugar. In addition, the pH value ranges from 2.0 to 6.0, which is much higher than that for HM pectin (Thibault and Ralet, 2003). The well-known "egg-box" model can explain the gelation mechanism between pectin molecules and polyvalent metal ions (Braccini and Pérez, 2001). The formation of gelation is due to the junction zones between the "smooth" regions of different pectin chains through metal cation bridges, in which dissociated carboxyl groups and hydroxyl groups are certain to play a pivotal role. When the pH is below the pK_a of galacturonic acid, undissociated carboxyl groups predominate, and more H-bonds form in the gel network (Chang et al., 2017; Li et al., 2013; Wusigale et al., 2020). As a result, the gel presents rigid and nonshear reversible properties. Oppositely, dissociated carboxyl groups favor ionic cross-linking, which leads to a spreadable and shear reversible gel network (Choi et al., 2018; Li et al., 2017; Zhang et al., 2020a). Additionally, the gel from LMP is sensitive to the chemical structure and composition of pectin, such as the forms of carboxyl, and the amounts and distribution of substituents (Feng et al., 2019; Jacob et al., 2020; Khotimchenko, 2020; Li et al., 2016, 2020; Norcino et al., 2018; Shi et al., 2020; Synytsya et al., 2020; Zaitseva et al., 2020).

4.3 Pectin extraction

Commercial pectin is produced mainly from apple pomace and citrus peels, both by-products from juice or cider manufacturing. In addition, sugar beet pulp (Gómez et al., 2016; Li et al., 2015; Yapo et al., 2007), sunflower head residues (Ezzati et al., 2020; Tan et al., 2020), passion fruit (Liew et al., 2016), and pomelo (Liew et al., 2019; Wandee et al., 2019) are also selected as the feedstocks to extract pectin. The occurrence and proportions of pectin in different sources are variable among plant species (Gentilini et al., 2014). Due to variations in these parameters, pectin from different sources shows different molecule sizes and DE values, which significantly influences the properties of pectin. Meanwhile, the extraction process also significantly influences the structural and technological characteristics of pectin (Ezzati et al., 2020; Li et al., 2015, 2019; Liew et al., 2019). Pectin is generally extracted in an acidic medium at high temperatures. However, the severe extraction processes are detrimental to the chemical structure of pectin, resulting in deesterification and depolymerization (Ezzati et al., 2020; Zhang et al., 2020a, b), and the corresponding hydrolytic process of HG is shown in Fig. 4.3. Thus, emerging technologies have been developed to isolate pectin from the abovementioned feedstocks in order to minimize the damage of pectin chemical structure, and avoid environmental contamination.

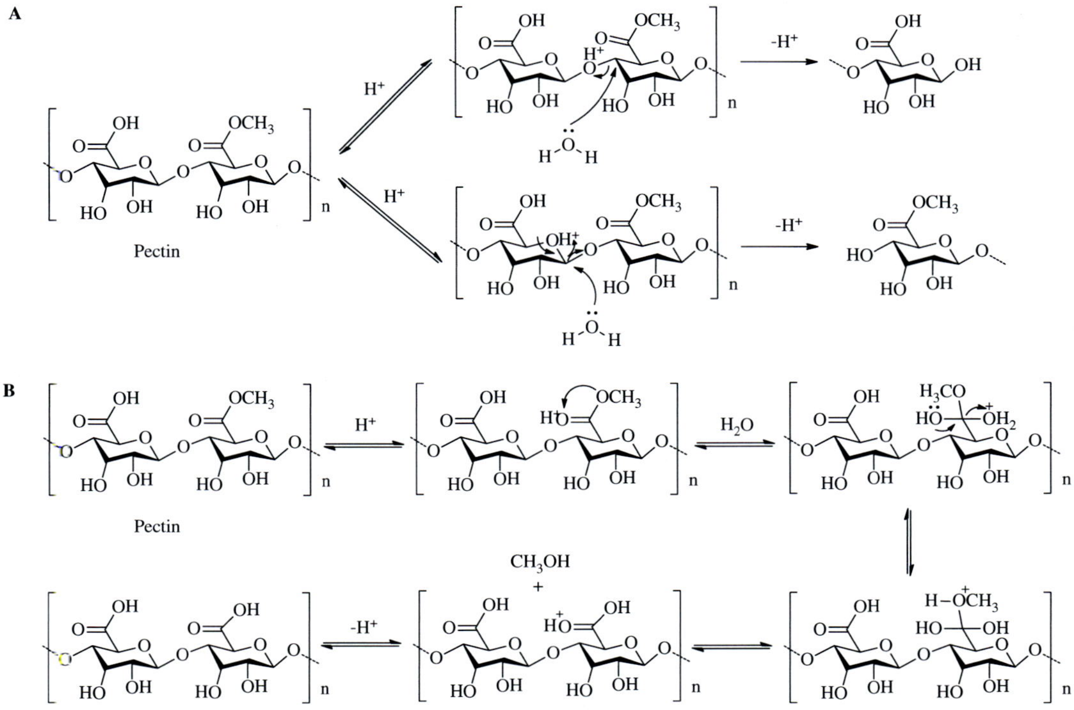

FIG. 4.3 Hydrolytic process of HG: (A) depolymerization and (B) de-esterification.

4.3.1 Traditional extraction process

Heating extraction in an acidic medium is the traditional method used in industry for pectin extraction from agro-industrial waste. Mineral acids were employed as the catalyst for hydrolysis reactions. To obtain pectin with less structural damage, organic acids including tartaric, citric, or lactic acid has be used as alternatives. In this process, the type and concentration of acid, extraction temperature, and extraction time can influence not only the yield, but also the chemical structure of pectin. The extraction process is generally performed in a pH range of 1–3 for 0.5–3 h with a temperature range of 80–100 °C. Industrially, extracted pectin is typically separated using hydraulic presses or a centrifugation process; afterward, the extract is filtered and concentrated. Thus, the ratio of liquid to solid (SLR) is also an important factor that influences the yield of pectin and energy consumption of the extraction process.

Pectin extraction from sugar beet pulp by citric acid (CA) under different conditions was studied using Box–Behnken design for four independent variables (pH, extraction temperature, extraction time, and SLR) (Li et al., 2015). The yield ranged from 6.3% to 23.0%, and the content of protein varied from 1.5% to 4.5%. The yield increased as the pH decreased of the extraction medium, extending time and advancing temperature, and an opposite relationship of effects between variables and content of protein was obtained. The total carbohydrate was essentially the same, while the content of Gal A, Rha, Ara, Xyl, Man, Glc, and Gal presented higher values in the condition of lower pH and longer extraction time. Moreover, the isolated pectin from sugar beet pulp could prepare steady oil-in-water emulsions for more than 30 days. To describe accurately the influence of extraction variables on the DE of pectin, the effects of different extractants on the DE were studied (Li et al., 2019). Sulfuric acid and citric acid were the representatives of a strong and weak acid, respectively. The DE of pectin extracted by sulfuric acid (DEs) ranged from 28.13 to 43.02% and that by citric acid (DEc) ranged from 26.98% to 43.78%. In addition, the values of DEs were generally higher than DEc values.

Zhang et al. have compared five extraction methods, including hydrothermal, acidic, alkaline, Viscozyme® L, and FoodPro® CBL extraction to obtain high yield leaf pectin with higher viscosity and better gelling properties (Zhang et al., 2020a, b). Results showed that alkaline extraction presented the highest pectin yield (9.2%), with galacturonic acid content of 41.6%, average molecular weight of 3.34×10^5 Da, and particle size of 296 nm. The pectin extracted by alkali presented a strong shear thinning property with a viscosity of about 90 mPas, which was four times higher than those by other processes. The gel rigidity of alkali-extracted pectin was10 times as those of other pectins including commercial citrus pectin. Pectin extracted by Viscozyme® L presented highest yield (8.5%), and contents of rhamnogalacturonan I and rhamnogalacturonan II. However, it appeared poor viscosity and gelling properties due to the hydrolysis of the side chains.

4.3.2 Emerging technologies for pectin extraction

Ultrasound-, microwave-, radio frequency-, electromagnetic-, and enzyme-assisted extraction were employed to obtain pectin from various feedstocks. Generally, ultrasonic wave (16 Hz to 16 kHz) consists of mechanical vibrations, which can be used for extraction, drying, emulsification, and homogenization, as well as for microorganism and enzyme inactivation, among others. For the extraction process, ultrasonic wave could induce the collapse of cavitation bubbles near cell walls, and therefore enhance solvent entrance into the cells and intensification of the mass transfer. Microwave-assisted extraction is another important method for pectin isolation. Microwave belongs to nonionizing radiation with a frequency range of 300 MHz to 300 GHz. Microwave generates heat after contacting with the cellular polar compounds. The resultant heat induces ionic conduction and dipoles rotation in the feedstocks.

Influences of power and time of microwave-assisted extraction, SLR, and pH value of the sulfuric acid solution on molecular weight were studied by response surface methodology (Li et al., 2012). The pH values exhibited a significant effect on molecular weight, which varied from 7.8×10^3 to 315.3×10^3 Da. Compared with the pectin extracted by conventional heating (Yapo et al., 2007), of which weight-average molar mass values ranged widely from 20,200 to 90,100 Da, pectin extracted by the microwave-assisted heat process presented a higher molecular weight which is beneficial to the gelling properties. To obtain the influence of the extraction process, pectin was also extracted from black carrots pomace in hot acidic water (pH 2.5) using the microwave at 110 °C for 5 min, ultrasound at 70 °C for 30 min, and conventional heating at 110 °C for 90 min (Sucheta et al., 2020). The conventional heating extraction process presented the highest yield (22%), followed by microwave- and ultrasound-assisted treatments. Moreover, pectin from the traditional heating process presented maximum retention of anthocyanins (1213 mg/L), phenolics (1832 mg/L), and antioxidant activity (180 μM/mL). The microwave-assisted extraction process exhibited the highest efficiency among the three approaches, which is beneficial to the development of functional food. Meanwhile, pectin extracted by the microwave-assisted method showed excellent physicochemical properties such as water-holding capacity and particle size. In addition, pectin from the conventional heating process and microwave-assisted extraction process also evinced higher DE values and Gal A contents than that from the ultrasound, while presented poorer thermal stability.

Liew et al. have studied the influences of extractor concentration, extraction temperature, and time on the yield of pectin and DE by the acidic and enzymatic extraction methods (Liew et al., 2016). Citric acid and celluclast were selected as environmentally friendly extraction agents. The yield and DE of pectin extracted by traditional method were 7.16% and 71.02% under optimized condition of extraction time of 102 min with citric acid concentration of 0.19% (w/w) at 75 °C. Meanwhile, pectin extracted by enzymatic extraction methods gave pectin yield of 7.12%, and DE of 85.45% with a celluclast concentration of 1.67% (w/w) at 61.11 °C. Overall, the enzymatic extraction method presented greater capability to isolate HM pectin.

Radio-frequency assisted extraction (RFAE) is another novel technology and has been employed to extract pectin from jackfruit peel (Naik et al., 2020). The optimum RFAE conditions were found to be an RF time of 61.50 min, the ratio of liquid to solid 20.63:1 (mL/g), and pH 2.61. The pectin obtained under the optimum conditions presented a yield and DE of 29.40% and 65.07%, respectively. Additionally, the conventional method was employed, and the corresponding yield was 12.37%, which was much lower than that from the RFAE process. However, the pectin extracted by the two methods showed nearly pseudoplastic flow behavior. Thus, RFAE of pectin under acidic conditions is an efficient and environmentally friendly method.

Electromagnetic induction (EMI) was employed to extract pectin. Zouambia et al. have studied electromagnetic induction and conventional heating for extracting pectin from citrange (Zouambia et al., 2017). It was found that the pectin extracted by EMI for 30 min presented a similar yield (24%, w/w) to that by conventional heating process for 90 min. Meanwhile, the pectin extracted by two methods presented similar compositions and physicochemical properties including galacturonic acid content (29.10%–29.40%) and DE (61.00%–62.50%). Thus, the EMI heating process is an efficient method without influencing the chemical composition and the physicochemical properties of the pectin.

In addition to emerging technologies, new solvents were also used for pectin extraction. Deep eutectic solvents (DESs), a series of eutectic mixtures of Bronsted–Lewis bases and acids in a certain ratio, have been considered as green solvent alternatives to conventional solvents and have attracted considerable attention. Recently, they have been prepared and applied to various areas of chemistry. Extraction of pectin by DES is a novel approach. Shafie et al. (2019) have studied the influences of variables on the extraction yield. Choline chloride (ChCl) and citric acid (CA) monohydrate were employed as the hydrogen-bond receptor and donor, respectively. The results indicated the optimal conditions were as follows: percentage of DES was 3.74% (w/v); molar ratio of DES components was 1:1; and the extraction time and temperature were 2.5 h and 80 °C, respectively. Benvenutti et al. have used six natural DESs including pairings of ChCl and propylene glycol, ChCl and CA, ChCl and malic acid, CA and glucose, CA and propylene glycol, and betaine and CA (Benvenutti et al., 2020). The corresponding results showed that CA and glucose were the most effective pairing for pectin extraction. Although the advantages of DESs have been demonstrated, many difficulties still need to be overcome. DESs generally presented high viscosity due to H-bonding, which significantly influences the mass transfer in the extraction process. In addition, the higher viscosity caused difficulties in the isolation of pectin from extract, which may result in unforeseen added costs.

4.4 Modification of pectin

Due to the drawbacks of rapid hydration, swelling, and erosion which were opposite from drug delivery application, pectin has been modified to change its properties in the past decades. The numerous hydroxyl and carboxyl groups make pectin capable of preparing a broad spectrum of derivatives which have been widely applied to form desirable materials in the field of drug delivery systems. Generally, pectin modification can be divided into substitution (esterification, amidation, etherification, acetalation, etc.), chain propagation (cross-linking and/or grafting) and depolymerization (oxidation, acid and enzymatic hydrolysis, β-elimination, and mechanical degradation).

4.4.1 Esterification

Pectin could react with various anhydrides and fatty acids, and corresponding products were obtained with different properties including gelling, swelling, emulsion, and thermostability. Thus, the alkylation modification supports the basic for the application in drug delivery and biomedicine.

The esterification of pectin was generally carried out in the presence of anhydrides. Succinic, acetic, and maleic anhydrides are the familiar chemicals which were employed to modify pectin (Leroux et al., 2003; Renard and Jarvis, 1999a, b; Wang et al., 2019). When the reaction substrates were fatty acids, they needed to be translated to corresponding anhydrides. Luca Monfregola et al. developed a free solvent process for pectin modification (Monfregola et al., 2011). In this process, pectin reacted with palmitic, oleic, and linoleic acids anhydride in the presence of K_2CO_3 via mechanical milling in an agate mortar, and modified pectin was obtained at increasing contents of palmitic, oleic, and linoleic acids. The modification process presented weight loss which was similar to the degradation path of pure pectin. In addition, a decreasing mass of evaporating water on increasing the fatty acid concentration, in particular for the palmitic acid modification, indicated reduced water sorption by the modified powders. Kamnev et al. performed a similar modification procedure in which oleic, linoleic, and palmitic acids were collected as the substrates to modify apple peel pectin (Kamnev et al., 2015). In a study published by Luca Monfregola et al., naturally occurring oils including oleic acid, linoleic acid, and hydroxyl-oleic acid were used to functionalize pectin (Monfregola et al., 2011). In the abovementioned process, dicyclohexylcarbodiimide (DCC) was employed to produce fatty acids anhydrides. Choi et al. have also modified pectic polysaccharide by fatty acid (succinic acid); in their study, acetic anhydride was used as a dehydrant to obtain succinic acid anhydrides (Choi et al., 2018).

Compared to fatty acids and anhydride, halohydrocarbon was also used for pectin modification. Pappas et al. prepared pectin derivatives by alkylation of tetrabutylammonium pectate with benzyl bromide and decyl bromide in DMSO (Pappas et al., 2004). Renard and Jarvis prepared methyl-esterified pectin using CH_3I at room temperature for 24 h (Renard and Jarvis, 1999a, b). *N*-alkylation of pectin could be performed with alkyl bromide in a heterogeneous system (e.g., isopropyl alcohol) (Zheng et al., 2015). The Fischer esterification was also performed to synthesize thiolated pectin via reaction of pectin and mercaptopropionic acid in the presence of HCl (Arora et al., 2016). EDC/NHS was used to catalyze reaction of esterification reaction between pectin and carboxylic acid for thiolated pectin preparation (Cheewatanakornkool et al., 2017). Finally, organic compounds contained glycidyl could reacted with the -COOH of pectin molecules under acidic conditions to obtain the carboxylic ester (Maior et al., 2008; Renard and Jarvis, 1999a, b; Yamada and Shiiba, 2015).

4.4.2 Amidation

Amidation of pectin was generally referred to the reaction between –COOH ($COOCH_3$) in a pectin molecule and amino compound. As early as 1995, Dongowski treated esterified pectin by aqueous NH_4OH in cold EtOH for different times (Dongowski, 1995). Recently, Deshmukha performed a similar operation by changing the solvent to isopropyl alcohol (Deshmukh et al., 2020). Later, Sinitsya et al. reported a universal approach in which partially amidated pectin derivatives (*N*-alkyl pectinamides) were prepared from highly methoxylated pectin by treatment with different primary amines in methanol (Sinitsya et al., 2000). This reaction system was used for the reaction of different types of amino compounds such as aliphatic amine, aromatic amine, alcohol amine, and enamine. In the decades that followed, several teams followed the approach to obtain amidated pectin (Bera and Kumar, 2018; Li et al., 2016; Mishra et al., 2008). In particular, amino compounds could capture protons from water molecules to form ammonium salts. The remaining OH^- can react with –COOH in pectin molecules to form $–COO^-$. Thus, a by-product named ammonium carboxylate salts was formed by electrostatic bonding.

Fajardo et al. have prepared pectin–NH_2 by functionalizing pectin with primary amine groups in the presence of EDC (Fajardo et al., 2012). The authors firstly reduced the degree of esterification of pectin by NaOH to improve the capacity of functionalization. Then, pectin and ethylenediamine were mixed and dissolved in water followed by adding EDC as catalyst. The pH condition of reaction system was adjusted to 5 by HCl solution. Recently, Arora developed a new process through activating -COOH in pectin molecules using EDC-NHS and then reacted with free amine groups (Arora et al., 2017). Briefly, EDC/NHS (1:1, mol:mol) was added into pectin solution. The mixture was stirring for a certain time to activate the carboxyl group, followed by adding an amino compound. The pH condition was a significant variable which needed to be fixed at 8 .

4.4.3 Silanization

Silanization could be carried out by adding silicane into pectin suspension or solution. Rachini has treated hemp fibers which contained pectin. Briefly, γ-aminopropyltriethoxysilane (APS) was added into a hemp fiber suspension, in an 80/20 ethanol/water mixture, by stirring at 120 °C (Rachini et al., 2009). Yamada and Shiiba prepared a water-insoluble pectin-inorganic composite material by mixing pectin and a silane coupling reagent, bis(3-trimethoxysilylpropyl)amine (Yamada and Shiiba, 2015). The pectin was also mixed with tetraethyl orthosilicate to obtain pectin-silica gel (Vityazev et al., 2017).

4.4.4 Etherification

The etherification reaction was based on the hydroxyl of pectin molecules. Generally, the ring-opening reaction of an epoxy group is the common approach. Under alkaline condition, the epoxy group could react with hydroxyl to form an ether (Ahn et al., 2017). As a comparison, the epoxy group could form an ester with –$COOCH_3$ by a transesterification reaction under acidic conditions (Maior et al., 2008). Chloroacetic acid was also collected as the substrate which could react with pectin in ethanol-NaOH solution at 80 °C under constant stirring for 1 h. An acetal group was special ethers from the view of structure. It could be formed by the reaction between the aldehyde group and vicinal diol of pectin molecule (Chang et al., 2017; Li et al., 2013; Wusigale et al., 2020). In this respect, glutaraldehyde was most widely used due to its relatively low toxicity (Wang et al., 2019).

4.4.5 Oxidization

Galacturonic acid of pectin has an o-diols which can be selectively oxidized to dialdehyde by periodate (Gupta et al., 2013). The obtained aldehyde present higher reactivity than hydroxyl and carboxyl in the pectin molecules. Thus, the oxidization provided various new approaches to form pectin-based materials. Recently, we have obtained oxidized pectin which was further cross-linked by chitosan to obtain an injectable and self-healing hydrogel (Li et al.,

2020). Similarly, Ahadi et al. have prepared a hydrogel scaffold of silk fibroin/oxidized pectin (SF/OP) based on the Schiff's base mechanism (Ahadi et al., 2019). In addition, the formyl groups provided reaction sites for preparation of prodrug such as pectin-adriamycin conjugates.

4.4.6 Grafting and cross-linking reaction

Pectin could be modified by grafting through a free radical reaction. The widely used monomers include acrylamides, acrylates, alkyl acrylates, and alkyl acrylamides. Persulfate, ammonium ceric sulfate, and ammonium persulfate-bisulfite are commonly used as initiators of graft reactions. To obtain mechanical properties, *N,N′*-methylenebisacrylamide (MBA) was generally used as a cross-linker. Babaladimath and Badalamoole developed a supermagnetic polysaccharide-based nanocomposite gel via graft copolymerization of 2-acrylamido-2-methylpropanesulfonic acid on pectin by using ammonium peroxodisulfate as an initiator and MBA as a cross-linker (Babaladimath and Badalamoole, 2018a, b). Modification of pectin was also carried out by grafting of vinyl monomers, i.e., tert-butyl acrylate and methacrylamide, in a homogeneous aqueous phase using ceric ammonium nitrate as the initiator (Raj et al., 2019). The synthesized pectin-based materials were also obtained without using any initiators and cross-linking agents. Bhuyan et al. have synthesized pectin-*N,N*-dimethylacrylamide (DMAA) hydrogel by gamma radiation without adding any accessory ingredient (Bhuyan et al., 2018). A similar approach was employed to prepare pectin-polyacrylamide hybrid materials (Abou El Fadl and Maziad, 2015). All the grafting reactions could significantly improve the molecular weight of pectin, which influenced the application potential. The single cross-linking reaction can also increase the molecular weight. Ionically cross-linked pectin-based micro- and nanomaterials are commonly used, which was formed in the presence of Ca^{2+}, Zn^{2+}, Fe^{3+}, etc., based on the "egg-box" model (Shahzad et al., 2019). In addition, a chemical cross-linking reaction can also be performed based on the reactions mentioned above. Succinic anhydride (Wang et al., 2019), glutaric dialdehyde (Leroux et al., 2003; Renard and Jarvis, 1999a, b; Wang et al., 2019), EDC/NHS (Wang et al., 2016), and epichlorohydrin (Semdé et al., 2003) have been collected as effective cross-linkers.

4.4.7 Electrostatic interaction

The carboxyl groups in pectin molecules can dissociate into $-COO^-$ in the solution when the pH value is above the pK_a of galacturonic acid (around 3.6). Thus, the dissociated pectin can form composite with positively charged molecules via electrostatic attraction. (Li et al., 2020; Shitrit et al., 2019).

Chitosan, the only cationic polysaccharide in nature, can also dissociate in the near-neutral and acididc solution . Thus, pectin and chitosan molecules could self-composite under appropriate conditions (Cheikh et al., 2019; Işıklan and Tokmak, 2019). When the two polysaccharides were left in a solution with pH value below 3.6, the carboxyl groups in pectin existed as $-COOH$, and amino groups of the chitosan molecule changed to $-NH_3^+$, which was beneficial to the dissolution. In this case, no electrostatic interaction exists between two polysaccharides, and the mixture existed as a solution. When the pH value increased, the carboxyl groups gradually changed to $-COO^-$, and electrostatic interaction will appear. As the pH of the solution continues to rise, the ammonium will lose their protons and become amino groups, and the electrostatic interaction between the two macromolecules will disappear. Based on the above analysis, pectin/chitosan-based composite materials have been widely studied due to their easy operation in the last decades (Cheikh et al., 2019; Giacomazza et al., 2014; Liu et al., 2019; Luo and Wang, 2014; Luppi et al., 2010; Shitrit et al., 2019).

Proteins are amphoteric compounds that exhibit different acidity or alkalinity under different conditions. Thus, various proteins were also collected to induce pectin forming composite materials. Casein (Chang et al., 2017; Li et al., 2013; Wusigale et al., 2020), soy protein (Feng et al., 2019; Jin et al., 2018, 2019; Sun et al., 2014), and whey protein (Assadpour et al., 2017; Gharehbeglou et al., 2019; Lutz et al., 2009; Raei et al., 2017; Serrano-Cruz et al., 2013; Souza et al., 2012) are widely used to composite with pectin to form microbeads, hydrogel, film, etc. Pereira et al. reported cell-instructive pectin-based hydrogels cross-linked via thiol-norbornene photo-click chemistry for skin tissue engineering (Pereira et al., 2018). They treated pectin with tetrabutylammonium salt which could be adsorbed on the surface of pectin molecule chains by electrostatic interaction. Thus, the surface coating induced the solubilization of pectin in organic solvents, such as DMSO.

4.4.8 Others

In addition to the reaction types listed above, some other ways can be used to achieve the purpose of pectin modification. Acid and alkali can also be used for the treatment of pectin. Generally, the treatment process was similar to extraction due to the same mechanism. From the view of chemical structure, acid and alkali modification can significantly induce the depolymerization of pectin chains and reduce content of neutral sugars (Rosenbohm et al., 2003), which was beneficial to the adsorption of pectin in blood and the small intestine, and extended the scope of biomedical and biopharmaceutical applications as a result (Maxwell et al., 2012; Takei et al., 2011). A similar result can be achieved with treatment by enzymes such as pectin methylesterase, polygalacturonase (Marić et al., 2018). Enzyme-catalyzed oxidative cross-linking of feruloyl groups can promote gelation of sugar beet pectin (Takei et al., 2011). Horseradish peroxidase and laccase were also employed to treat pectin by oxidative cross-linking of feruloyl groups, which can promote gelation of pectin (Zaidel et al., 2013). Compared with chemical and biological treatment, physical treatment may be more environmentally friendly and present advantages including easy operation, low-cost, free reagents, etc. In recent decades, microwave treatment, ultrasonic treatment, and UV irradiation treatment have been developed. The properties of pectin treated by the physical method are mainly reflected in the decrease of molecular weight (Anuar et al., 2012).

4.5 Pectin in biomedical applications

4.5.1 Prebiotic effect

Nowadays, the relationships with microbial populations through our digestive system has been gradually discovered and has received more and more attention due to the direct understanding and requirements of human health (Ezzati et al., 2020; Tan et al., 2020). In the past decades, nutrition developed rapidly as a result. Probiotics need to be highlighted here. Therein, *Faecalibacterium prausnitzii* and *Roseburia intestinalis*, along with other species in the digestive tract, present beneficial effects to colon cancer and ulcerative colitis (Gómez et al., 2016; Singh et al., 2020). Acetate and butyrate, short-chain fatty acids derived from the intestinal fermentation of pectins, exert a beneficial effect on the digestibility and fermentation of pectin (Ferreira-Lazarte et al., 2019; Tan and Nie, 2020). Probiotics may feed on pectins and produce secondary metabolites which inhibit the growth of nonprobiotic. Gómez et al. isolated pectic oligosaccharides from sugar beet pulp and lemon peel wasters, which promoted joint populations of *bifidobacteria* and *lactobacilli* from 19% up to 34% and 29%, respectively (Gómez et al., 2016). Moreover, the results indicated that low molecular weight and degree of methylation were beneficial to prebiotic capacity.

4.5.2 Hypoglycemic effect

In addition to prebiotic activity, pectin can also contribute to a reduction of glucose intolerance in diabetics, even lowering the blood cholesterol level (Brouns et al., 2012). This may be due to the fact that higher food viscosity caused by dietary fibers decreases the speed of the gastrointestinal transit, lowering the intestinal absorption rate of glucose as a result. Khramova et al. reported that pectin gel in acidic gastric medium increases the rheological properties of gastric digesta and reduces glycemic response in mice (Khramova et al., 2019). Previous studies have shown that pectin can influence the in vitro digestibility of starch, which was achieved by increasing the viscosity of digesta and inhibiting amylase activity by intermolecular interactions (Qi et al., 2018; Zaitseva et al., 2020). To understand whether unextracted pectin could influence in vitro digestibility of in situ starch in vegetables, Bai et al. studied relations between digestibility and structures of pumpkin starches and pectins (Bai et al., 2020). Results included: (1) pumpkin starch existed as high proportion of long amylopectin chains; (2) pumpkin pectin exhibited wide diversity in molecular structure; and (3) structural parameters of both pectin (especially DE value) and starch worked together to affect the in vitro digestibility.

4.5.3 Hypocholesterolemia effect

In terms of the hypocholesterolemia effect, dietary fibers (pectin included) contributed to the reduction of cholesterol levels; and have promoted their application to reduce the risk of cardiovascular disease. Some in vivo studies confirm to some extent that the physicochemical characteristics of pectin influences the metabolism of cholesterol. More importantly, the chemical structure of pectin was the major factor that reduced the cholesterol levels in plasma and liver when high methoxyl pectins are consumed. Zhu and coworkers studied the effects of hawthorn (*Crataegus pinnatifida Bunge*) pectin and its hydrolyzates on the cholesterol homeostasis of hamsters that were fed high-cholesterol diets (Zhu et al., 2015, 2017). The results indicated that haw pectin pentasaccharide was more effective than haw pectin and hydrolyzates on decreasing the body weight gain (by 38.2%), liver weight (by 16.4%), and plasma and hepatic total cholesterol (TC; by 23.6% and 27.3%, respectively) of hamsters (Zhu et al., 2015).

Moreover, the authors have proven that haw pectin penta-oligogalacturonide (HPPS) played an important role in improvement of cholesterol metabolism and promotion of the conversion of cholesterol to bile acids (BA) in mice that were fed a high-cholesterol diet (Zhu et al., 2017). Therein, HPPS significantly decreased plasma and hepatic TC levels but increased plasma high-density lipoprotein cholesterol and apolipoprotein A-I levels, compared to a high-cholesterol diet. The analysis results of bile acids showed that HPPS significantly decreased hepatic and small intestine BA levels, but increased the gallbladder BA levels, and finally decreased the total BA pool size, compared to HCD. In addition, Gunness et al. revealed that pectin can reduce plasma cholesterol in pigs, but had different effects on triglycerides and bile acids (Gunness et al., 2021).

4.5.4 Anticancer effect

Recently, pectin from various resources and its derivatives have been proven to have antiproliferative effects on various cancer cell lines, such as colon cancer, prostate cancer, breast cancer, and pancreatic cancer, due to their special chemical structure (Bergman et al., 2010; Li et al., 2010). Generally, pectin with higher neutral sugar content presents higher bioactivity due to the fact that galactan side-chains on pectin can bind to and inhibit the pro-metastatic protein galectin-3, resulting in the suppression of cancer cell proliferation, aggregation, adhesion, and metastasis. From this point of view, the depolymerized pectin was beneficial to inhibit galectin-3. Maxwell et al. reported that modified sugar beet pectin induces apoptosis of colon cancer cells via an interaction with the neutral sugar side chains (Maxwell et al., 2016). They treated pectin by alkali and found that increasing content of RG-I could significantly increase pectin bioactivity. However, the DE seemed to have only a slight influence on anticancer activity. In terms of pectin structure inhibiting cancer cells, Zhang et al. obtained water-soluble polysaccharides from *notopterygium incisum* roots and separated these into two homogeneous fractions noted as arabinogalactan (AG) and HG fraction, which were covalently linked with AG and RG-II domains. The results indicated that the AG domain exhibited stronger binding avidity to galectins including galectin-1, galectin-3, galectin-7, and galectin-8 than the RG-II and HG domain, while oligogalacturonides did not. (Choi et al., 2018; Li et al., 2017; Zhang et al., 2020a). Citrus pectin was also treated by UV/H_2O_2 to obtain MCP4 and MCP10 under pH conditions of 4 and 10, respectively (Cao et al., 2020b). The differences between two series of modified pectin occurred at branching and methoxylation degree, which caused higher antiinflammatory activity of MCP10 in inhibiting the NF-κB expression and producing pro-inflammatory factors TNF-α and IL-1β of THP-1 cells stimulated by lipopolysaccharide, as well as a stronger inhibitory effect on Caco-2 cell proliferation. There are still many other studies in this field that we cannot elaborate on here (Hu et al., 2020; Pedrosa et al., 2020; do Prado et al., 2019a; Wu et al., 2020).

4.6 Pectin-based hybrid materials in drug delivery

Due to the versatile modification approaches and anticancer bioactivity (Hu et al., 2020; Pedrosa et al., 2020; do Prado et al., 2019b; Wu et al., 2020), pectin is one of the most suitable feedstocks that used in DDS application (Zaitseva et al., 2020). Additionally, pectin presents some uncommon advantages including peculiar gelling mechanisms under acidic conditions, biological degradation in the digestive tract, biocompatibility, etc. (Basak and Bandyopadhyay, 2014; Braccini and Pérez, 2001; Cao et al., 2020a). These reasons are why pectin has been used in DDS applications for more than 70 years. Generally, pectin-based hybrid materials could be obtained by cross-linking and/or grafting reactions (Cheikh et al., 2019; Giacomazza et al., 2014; Liu et al., 2019; Luo and Wang, 2014; Luppi et al., 2010; Shitrit et al., 2019). Metal salts and bifunctional organic molecules can be used as crosslinkers for the preparation of pectin-based materials, so-called ionically and chemically cross-linked materials. LM pectin can form a hydrogel in the presence of Ca^{2+}, Zn^{2+}, etc., as explained by the well-known "egg-box" models (Braccini and Pérez, 2001; Munjeri et al., 1998a, b; Murray and Finland, 1946; Musabayane et al., 2003; Wakerly et al., 1997). However, the obtained ionically cross-linked hydrogel demonstrated undesirable performances such as low strength, toughness, and ion exchange in aqueous solution, because of the nature of labile coordination bonds (Abedini et al., 2018; Assifaoui et al., 2013; Khotimchenko, 2020). In this respect, chemical and ionic-chemical combined cross-linking strategies may improve the abovementioned properties due to the stability of covalent bond and controlled rate of covalent-binding reaction. Generally, glutaraldehyde (GLU) (Cui et al., 2017), succinic anhydride (Wang et al., 2019), maleic anhydride (Leroux et al., 2003; Renard and Jarvis, 1999a, b; Wang et al., 2019), EDC/NHS (Cheewatanakornkool et al., 2017; Cui et al., 2017), and epichlorohydrin (Semdé et al., 2003) are the widely used reagents in the chemical cross-linking strategies.

4.6.1 Ionic cross-linking strategies

Pectin-based drug delivery via ionic cross-linking strategies has been reported as early as 1946 (Murray and Finland, 1946). When calcium salts are added into the aqueous solution of pectin, coordination bonds formed between pectin chains and Ca^{2+}, giving rise to a gel network in a dilute solution. It is well known as the "egg-box" model, as shown in Fig. 4.4. With increasing concentrations of pectin and/or metal ions, the number of fractal flocs of pectin increased simultaneously with the cross-linking density, which caused significant increases in the values of the bulk elastic modulus (Basak and Bandyopadhyay, 2014). The transformation of performance provided the potential for drug delivery applications.

The "one-pot" method, mixing pectin and calcium salt in a solution, was the mainstream for preparing the pectin-based film, bulk, tablet, etc., with micro- and nano-structure. Wakerly et al. studied amidated pectin as a potential carrier for colonic drug delivery (Wakerly et al., 1997). Increasing the dosage of calcium as a cross-linking agent could significantly increase the viscosities of amidated pectin gels to a maximum value. However, the gel viscosity presented a decreasing trend with the further addition of calcium ions, which caused a reduction in drug release ratio. Besides, drug release rate was faster for pectin with less amidation in the presence of calcium ions, which may be related to matrix erosion. In addition, Semdé et al. combined HM pectin or calcium pectinates with commercially available aqueous polymer dispersions for colon-specific drug delivery (Semdé et al., 1998).

The "two-step" method was that dropwise adding droplet contained pectin and drug model into calcium-ion-containing solutions. Calcium pectinate gel microbeads were

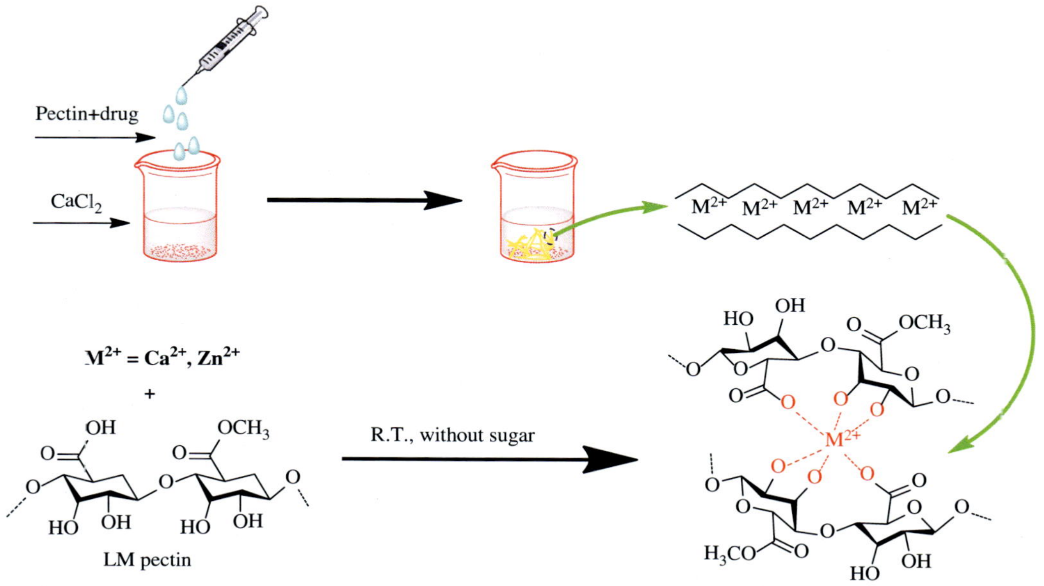

FIG. 4.4 Formation of a calcium pectate gel network.

prepared by extruding a bull serum albumin (BSA)-loaded pectin solution into an agitated calcium chloride solution (Sriamornsak, 1998). The mean diameter of the obtained microbeads was related to the DE value of pectin and cross-linking time. The influence of BSA dosage on the mean diameter of microbeads seemed to be unclear. The results of the drug release study suggested promising calcium-pectinate-based oral delivery of protein for colonic delivery.

It has been shown that the drug released from calcium pectinate beads could be divided into two stages: (1) solvent permeated into the calcium pectinate network, which caused swelling in the dissolution medium, and (2) ion exchange happened between calcium and potassium and/or sodium ions from the dissolution medium, and disintegration occurred subsequently. As a result, the ionically cross-linked pectin carriers cannot effectively protect the drug from the release medium. Silica-coated calcium pectinate beads were fabricated for colonic drug delivery (Assifaoui et al., 2013). The calcium pectinate beads were immersed in a prehydrolyzed tetraethyl orthosilicate (TEOS) solution at a pH of 2 for 30 min. After filtration, the wet beads were introduced in a tris(hydroxymethyl)aminomethane/HCl solution (pH 7.6) to induce silica condensation. According to the characterizations, weak interactions occurred between the silica layer and the organic beads. The existence of silica coating could work well as a barrier to block water uptake and reduce the swelling ratio of the beads as a result.

4.6.2 Covalent cross-linking strategies

To overcome the disadvantage of ionically cross-linked pectin-based DDS, covalent cross-linking strategies are the primary routes due to the strength and controlled reaction of the covalent bond. Generally, covalent cross-linking strategies were performed based on esterification, amidation, etherification, acetalation, and chain propagation of pectin.

Succinic anhydride and glutaric dialdehyde were collected as cross-linking reagents to prepare chemically cross-linked pectin microcarriers for loading diclofenac sodium (Wang et al., 2019). The encapsulation efficiency (EE) could reach 78.81%. However, the cumulative release rates of the sample PG1 were 3.04%, 3.66%, and 79.43% in simulated gastric fluid (SGF), small intestinal fluid (SSIF), and colon fluid (SCF), respectively, after a 12 h release. The ultra-low cumulative release rate under acidic conditions may be caused by the disparate existence of free carboxyl groups under different pH conditions. When the DDS was put into the acidic buffer solution, the carboxyl groups existed as the $-COOH_2{}^+$ and/or $-COOH$ which was averse to forming a pore structure on the surface. Oppositely, the carboxyl group shifted to $-COO^-$ in the alkaline medium. The existence of $-COO^-$ significantly increased the hydrophilicity of the carrier surface and contributed to the pore structure formation, which led to the drug release. A similar cross-linking strategy can also be found in the literature, as reported by Cui et al. (2017).

Click chemistry has been considered as biocompatible reactions that were designed to conjoin substrates to biomolecules (Li et al., 2021). Cell-instructive pectin hydrogels were obtained via thiol-norbornene photo-click chemistry for skin tissue engineering (Pereira et al., 2018). The norbornene-functionalized pectin (NorPEC) was synthesized through the reaction with carbic anhydride in dimethyl sulfoxide, followed by adding into NaCl solution containing the photoinitiator VA-086, the peptide cross-linker CGPQG↓IWGQC, and the cell-adhesive peptide ligand CGGGGRGDSP (RGD peptide). Due to the cleavage ability of matrix metalloproteinase (MMP) on CGPQG↓IWGQC, the cell-loading hydrogels were evaluated for MMP-responsiveness. The dosage of MMP significantly influenced the cross-linking density, elastic modulus, and swelling ratio. The RGD peptide, an adhesive unit in the pectin-based network, promoted cell adhesion and spreading within the hydrogels. Fibroblasts assay presented obvious differences in the spreading and elongation with different dosages of RGD. Specifically, the low concentration of RGD (0.5 mM) contributed to the heterogeneous spreading in the inner-gel network.

Chitosan has been considered to be unsuitable for DDS application (Kean and Thanou, 2010). Cross-linking with other polysaccharides that are biocompatible may be an ideal approach. Carbodiimide cross-linking chemistry has been employed to synthesize a pectin-chitosan conjugate, which was obtained by attaching primary amine in chitosan molecules to the carboxyl group of pectin in the presence of dicyclohexylcarbodiimide and N-hydroxysuccinimide ester (DCC/NHS) under anhydrous conditions (Tian et al., 2020). The present study compared biocompatibility of pectin, chitosan, and their conjugates on the Hela cell line and showed acceptable results. To enhance the stability of solid lipid nanoparticles (SLNs), pectin was employed for surface modification by cross-linking with GLU and EDC/NHS (Wang et al., 2016). SLNs were firstly prepared by combining the solvent-diffusion and hot homogenization techniques. Pectin was further coated on the surface of NaCas-emulsified SLNsfor preparing double-layer (DL) SLNs via electrostatic interaction. Finally, the DLSLNs were chemically cross-linked by GLU or EDC/NHS. The particle size of DLSLNs before cross-linking was around 320 nm, while the particle size of DLSLNs cross-linked by GLU and EDC/NHS was about 500 nm and below 300 nm, respectively. Furthermore, the zeta potential presented an obvious decrease after cross-linking. Adjacent hydroxyl groups could be selectively oxidized to dialdehyde by periodate. The obtained 2,3-dialdehyde pectin could react with chitosan. Oxidized pectin/chitosan/γ-Fe_2O_3 ternary hydrogel system was prepared via Schiff's base reaction (Li et al., 2020). To promote the targeting of DDS, nano γ-Fe_2O_3 was introduced into the hybrid gel network, which was loaded on the surface of hydrogel, and the particle size was around 0.25 μm. *S*-shaped magnetic hysteresis loops were obtained over the applied magnetics with an MS value of 4.86 emu/g. All the hydrogel-based DDS presented pH- and thermo-sensitive properties in the release medium and persistently released 5-FU for more than 12 h. The addition of γ-Fe_2O_3 was not only beneficial to the target but also enhanced the anticancer property without cytotoxicity. Oxidized pectin was also collected to cross-link with gelatin which contained lysine amino groups (Shi et al., 2020). The oxidized pectin was first treated by electrospinning to fabricate fibriform pectin, which was subsequently cross-linked with gelatin. The cells could grow into the inner of gelatin-crosslinked pectin nanofiber mats rather than on the surface, suggesting that the obtained hybrid fibers hold great potential for soft tissue regeneration. Furthermore, the maximum tensile strength and ultimate tensile strain could reach 2.3 MPa and 15%, respectively.

Cross-linking with synthetic polymers was an optional approach to improve the performances of pectin including mechanical strength, drug loading capacity, swelling, and erosion rates. In situ synthesized pectin-polymer hybrid bulks have been developed, such as poly (methyl vinyl ether-*co*-maleic acid)-pectin-based hydrogel (gel, film, and microneedles) (Demir et al., 2017), pectin-conjugated eight-arm polyethylene glycol-dihydroartemisinin nanoparticles (Liu et al., 2017), pectin-*g*-poly(2-acrylamido-2-methyl-1-propane sulfonic acid) silver nanocomposite hydrogel beads (Babaladimath and Badalamoole, 2018a, b), and pectin-eight-arm polyethylene glycol-drug conjugates (Liu et al., 2018). Eswaramma et al. reported a potential dual responsive hydrogel network (PPAD) from pectin, poly((2-dimethylamino)ethyl methacrylate)), and a phosphate cross-linker bis[2-methacryloyloxy] ethyl phosphate (BMEP) by a simple free radical polymerization (Eswaramma et al., 2017). High pH and thermo-responsiveness were obtained by analyzing the results of swelling experiments. The values of network parameters indicated that the mesh size and molecular weight between two cross-links significantly influenced the structure of the PPAD hydrogel network. Additionally, in vitro release kinetics followed non-Fickian diffusion and fitted well with the Higuchi model.

4.6.3 Dual cross-linking strategies

Ionic cross-linking and covalent cross-linking strategies have proved to be feasible in the preparation of DDS. However, some performances still need to be improved, such as the potential biotoxicity of the excess and/or unreacted chemicals. Thus, dual cross-linking strategies, a combination of ionic cross-linking and covalent cross-linking methods, can not only reduce the dosage of reagents but also achieve the ideal properties of pectin-based drug delivery.

Tan et al. prepared carboxymethyl sago pulp/pectin hydrogel beads by ionic cross-linking and further cross-linked them by electron beam irradiation for a colon-targeted drug (Tan et al., 2016). Carboxymethyl sago pulp/pectin hydrogel beads could keep an intact structure in a swelling medium at pH 7.4 for 24 h. The swelling degree increased as the pH value increased, indicated pH-sensitivity. In addition, the EE was about 50%. The cumulative release rate was less than 9% at SGF. Furthermore, the hydrogel-based DDS could persistently release the drug for more than 30 h.

Vityazeva et al. prepared pectin–glycerol gel beads by Ca^{2+}-cross-linking and then cross-linking by esterification in the presence of HCl (Vityazev et al., 2020). The LM pectin could form gels with glycerol in an acidified water-ethanol environment without heating, while DM pectin cannot, indicating the importance of free carboxyl in pectin molecules. In addition, adding the dosage of glycerol significantly improved the rheological and textural properties. In SGF, the antiswelling performance of chemically crosslinked beads was greater than that of Ca^{2+}-crosslinked, indicating the strength of the covalent bond.

Cui et al. studied the effects of the cross-linking method on the performance of pectin-based nanofibers. Mono-crosslinking was the "one-pot" method mentioned above, while dual crosslinking is achieved by employing GLU or adipic acid dihydrazide (ADH) as crosslinker after Ca^{2+} crosslinking. GLU cross-linking improved their mechanical strength moderately but did not inhibit their degradation, while ADH cross-linking improved their mechanical strength and slowed down their degradation dramatically. The biocompatibility assay did not indicate obvious cytotoxicity of the obtained fibers. In addition, ADH cross-linked pectin fibers presented cell adhesion and proliferation performances (Cui et al., 2017).

4.7 Pectin-based composite materials in drug delivery applications

4.7.1 Pectin-organic polymer composite materials

Generally, the surface negative charge of pectin was the major factor for the preparation of pectin-based composite materials. Natural macromolecule, which contained a certain amount of -NH_2, could support the positive charge for the composite approach. Chitosan, gelation, and casein are commonly used components.

Though chitosan has been widely studied in terms of drug delivery, it has been considered to be unsuitable for this (Kean and Thanou, 2010). Thus, how to reduce the biotoxicity of chitosan is worth studying. Pectin was the ideal partner that could solve the biotoxicity of chitosan. Additionally, electrostatic interactions between the two polysaccharides were responsible for increasing water resistance and mechanical performance of chitosan/pectin blend films, which provided an approach for modulating physical properties including hydrophilia and mechanical property. Physically cross-linked pectin/chitosan nanogel composite hydrogel with shear-thinning performance was obtained (Shitrit et al., 2019). When the temperature was below 42 °C, the pectin and chitosan nanogels could form hydrogel, which presented shear-thinning performance and self-healing properties due to the electrostatic interaction between pectin and chitosan chains. Furthermore, increasing the concentration of chitosan nanogels significantly increased the viscosity and Young's modulus, while decreasing equilibrium swelling, which suggested a relation between cross-linking density and nanogel content. Bio-based chitosan/pectin composite films were prepared by solution casting (Norcino et al., 2018). Polyelectrolyte complexes were formed through electrostatic interaction between $-COO^-$ in pectin and $-NH^{3+}$ in chitosan molecules. In addition, the blend systems changed from Newtonian to pseudoplastic gel-like systems. When pectin and chitosan were mixed at a mass ratio of 75:25, pectin-chitosan gel formation occurred. Mucoadhesive vaginal film containing fluconazole was obtained from chitosan and pectin via electrostatic interaction (Bera and Kumar, 2018; Li et al., 2016; Mishra et al., 2008). Compared with the common process, glycerol was employed as a plasticizer, and DMSO was selected as a permeation enhancer. The results of the optimized batch showed drug diffusion of 79% in 24 h, a tensile strength of 8.7 N mm^{-2}, elongation of 10%, moisture content of 2.5, and swelling index of 2.2. In vitro antifungal activity indicated noninterference on *Lactobacillus* action. In terms of in vivo vaginal irritation study, there was no obvious erythema or inflammation at the vaginal site of healthy female rabbits after a 3-day assay. Zhu et al. prepared pectin/chitosan beads containing porous starch embedded with doxorubicin hydrochloride (Zhu et al., 2015, 2017). Porous starch acted as an effective adsorbent to load doxorubicin hydrochloride, and pectin/chitosan was the coating in the presence of Ca^{2+}. The in vitro simulated digestion demonstrated the effectiveness, as only a 13.80% release ratio was observed in the upper gastrointestinal tract, whereas release ratios of 17.56% and 67.04% were obtained for pectin/porous starch-based and pectin-based DDS, respectively.

Pectin with different DE was collected to composite with gelatin for cinnamaldehyde delivery (Muhoza et al., 2019). The coacervates could form with high yield, good size, and morphology at pH of 4.23 for gelatin/HM pectin with a ratio of 3:1, and at 4.37 for gelatin/LM pectin with a ratio of 6:1. Gelatin molecules exhibited a conformational change after the complexation occurred by electrostatic interaction. Microencapsulation contained cinnamaldehyde from gelatin and pectin via complex coacervation presented a significant increase in degradation temperature, from 180 to 220 °C to 350–400 °C. In addition, the gelatin/high methyl pectin composite revealed a promising encapsulation efficiency (EE) of 89.2% and controlled release performance. Gelatin-pectin-biphasic calcium phosphate composite coated BMP-2 and VEGF was fabricated for bone healing (Amirian et al., 2015). A pectin-type B gelatin polyelectrolyte complex was also developed for curcumin delivery (Shih et al., 2018).

Generally, the negatively charged molecules were not suitable to composite with pectin due to electrostatic repulsion. To overcome the undesirable interaction, the positive-charged polymers could be used as a "bridge" to form ternary complexes. To improve the storage and digestion stability of curcumin emulsions, soy protein-pectin-phenolic acids ternary nanocomplexes were obtained by ultrasonication (Jin et al., 2019). Ferulic acid, ellagic acid, and tannic acid afforded distinguishing influences on the physical and structural properties of ternary complexes. The emulsifying property of the obtained three complexes from high to low were soy protein-pectin-ferulic acid complex, protein-pectin-ellagic acid, and soy protein-pectin-tannic acid complex. In addition, the protein-pectin-ferulic acid complex could store curcumin for 30 days and presented a slow release of curcumin during simulated digestion, which was due to the interaction between components including hydrogen bonding, electrostatic interactions, and intermolecular stacking effects. Jin et al. prepared soy protein/pectin/tea polyphenol ternary nanoparticles via photocatalysis (Jin et al., 2018). In the process, the tea polyphenol was spontaneously adsorbed on the surface of soy protein due to hydrogen bonds and hydrophobic interactions. The loss of fluorescence intensity indicated the layer-by-layer structure of ternary nanoparticles, in which pectin acted as the inner core, surrounding by soy protein and then tea polyphenol. In situ gelling particles were obtained from an alginate-pectin blend for ANXA1 N-terminal peptide Ac2–26 loading (Gaudio et al.,

2020). In this study, Ac2–26 also acted as a crosslinker that could link pectin and alginate. The EE was up to 83%. The obtained particles presented a burst effect within 3 h, followed by a sustained release for more than 24 h.

4.7.2 Pectin-inorganic composite materials

Pectin-based composites have the potential for drug delivery. However, the surface wettability of pectin/natural macromolecules composites is still the disadvantage that significantly influences mechanical strength in simulated body fluid. The addition of inorganic nanomaterials can adjust the surface wettability of pectin-based materials, which changes the mutual effect between the composites and simulated body fluid as a result. However, there are still few studies on pectin-inorganic composite materials.

Montmorillonite (MMT) was a commonly used inorganic material to composite with natural macromolecules. However, the nonocclusive lamellar structure restricted the potential in sustained drug release. To deal with this limitation effectively, quaternized pectin (QP) was introduced to coat MMT for delivering 5-FU (Meng et al., 2020). The cytotoxicity assay proved the great biocompatibility of QP-MMT composite film. The XRD characterization indicated that an intercalation reaction occurred in the blending process. The optimal drug delivery showed high EE (36.50%) and loading efficiency (80.30%). In addition, the intercalated DDS presented greater sustained-release performance, with sustained-release time above 8 h, than pure pectin-based drug delivery in all simulated release mediums. MMT, an excellent adsorbent for loading metformin, was also used as a core which was coated by chitosan and pectin/chitosan composite shell (Rebitski et al., 2020). The pectin coating could ensure stability in the stomach, and chitosan could present mucoadhesive to attain controlled release performance in the intestinal tract. The chitosan and pectin core-shell beads encapsulating the metformin–clay intercalation compound was more efficient due to higher EE of around 90%, in contrast to 60% reached in the chitosan-based formulation. A similar study was performed by Cheikh et al., who chose diclofenac as the drug model (Cheikh et al., 2019). The formation of a polyelectrolyte complex from HM pectin, chitosan, and clay resulted in a decrease in the crystallinity and thermal stability of the drug. The results indicated that the obtained nanocomposite was suitable to load anionic drugs for colon-targeted delivery. Diethanolamine-grafted HM pectin-arabic gum-modified MMT composites were prepared for intragastric ziprasidone HCl delivery based on the floating and mucoadhesion mechanisms (Bera and Kumar, 2018; Li et al., 2016; Mishra et al., 2008). The optimal complex presented EE of 61% and the release ratio of 52% after 8 h release. Kinetics studies demonstrated Fickian diffusion, which was also proven in the published data (Bera et al., 2019). Halloysite nanotube (HNT), a kaolin group clay mineral with a spurious tubular structure, has attracted much attention due to its various advantages including biocompatibility, hollow tubular morphology, high surface area, and modifiable surface. Jamshidzadeh et al. developed three-ply nanocarriers based on HNT sandwiched by chitosan/pectin layers with biocompatible and pH-responsive performances for controlled release of phenytoin sodium (Jamshidzadeh et al., 2020). The composite carriers presented greater loading capacity of 34.6 mg/g than that of pure HNT (18.3 mg/g). In vitro release studies showed that the obtained nanocomposites presented a low release of phenytoin sodium in the SGF as a contrast to a more release ratio in the simulated intestinal fluid.

Carbon-based materials have been collected to form DDS due to the biocompatibility and high specific surface area. Graphene seemed to be an all-purpose material that could blend with various feedstocks to change their performance. Pectin-coated chitosan-graphene oxide nanocomposites were prepared for delivering DsiRNA effectively into cells (Haliza et al., 2017). Herein, pectin acted as a compatibilization agent to protect nanocomposites from erosion in the stomach and small intestine. In addition, the electrostatic interaction among pectin, chitosan, and graphene oxide was the major driving force for nanocomposite forming. The mean particle size of the obtained nanocomposites was 554.5 ± 124.6 nm. TEM analysis revealed that the shape of chitosan-graphene oxide composite became irregular after pectin coating. The nanocomposites presented high EE of $92.6 \pm 3.9\%$. Additionally, the DsiRNA loaded nanocomposites could selectively inhibit the growth of the colon cancer cell line (Caco-2 cells) and were able to decrease VEGF level significantly. Hussien et al. prepared a pectin-magnetic graphene oxide composite for paclitaxel delivery (Hussien et al., 2018). A cytotoxicity assay showed that the composite carrier was cytocompatible, with more than 80% relative cell viability. In addition, the existence of pectin could increase cytocompatibility. The drug release assessment presented a pH-responsive release. A nano-carbon sphere (NCS) was also employed as the feedstock to form a pectin-inorganic composite due to being nontoxic. To obtain a well-interacted complex, the NCS was first modified by 3-aminopropyltriethoxysilane. Then, the modified NCS was mixed with pectin to prepare a pectin-modified NCS composite film in the presence of Ca^{2+}. A cytotoxicity assay indicated great biocompatibility of the composite carriers, with cell viability above 80% after culturing for 72 h. The EE was in the range of 30.1%–52.6%. As a comparison, the pectin-Ca^{2+} film showed the lowest EE value. The results of simulated drug release in vitro showed that all the composite DDS was greater in release properties than calcium pectinate-based DDS. In addition, the optimal sample presented cumulative release rates of 32.17%, 22.77%, and 63.89% in SGF, SSIF, and SCF, respectively.

Metal particles and metal oxide particles were also collected to blend with pectin. Pectin-capped gold nanoparticles (PT-capped AuNPs) were synthesized in situ by mixing pectin and $HAuCl_4$ in different proportions (de Almeida et al., 2020). Then the obtained pectin-capped Au nanoparticles were added into a chitosan solution. Pectin acted as a reductant in the process for preparing pectin-Au nanoparticles, which can be inferred from the preparation of pectin-based silver nanoparticles (Kodoth et al., 2019). In addition, the obtained PT-capped AuNPs presented negative zeta potential, which indicated the cladding of pectin. Thus, the second function of pectin was to combine with chitosan via electrostatic attraction. Pectin-coated gold nanoparticles (GNPs) were also prepared as a candidate for curcumin drug delivery (Khodashenas et al., 2020). Though the formation process was carried out by simple blending, the obtained curcumin entrapment efficiencies were all ultra-high—at least 96%. In addition, the values of cumulative release ratio indicated excellent sustained-release performance under the condition in both an acidic and an alkaline simulated medium. The results of the antibacterial test revealed that the DDS and curcumin appeared an similar performance. Chitosan/pectin/ZnO porous films were prepared by loading ZnO in the lyophilized pectin/chitosan composite (Soubhagya et al., 2020). ZnO was obtained in situ from zinc acetate dihydrate in the presence of NaOH. The obtained porous films presented swelling degree and water retention capability in the ranges of 189%–465% and 230%–390%, respectively. MTT assay was performed to calculate the biocompatibility of chitosan/pectin/ZnO films, which revealed biocompatibility against the primary human dermal fibroblast cells (HFCs).

4.8 Conclusions

Pectin-based DDSs have been developed for more than 70 years. Several series of DDSs have been fabricated by physical, ionic, and covalent cross-linking methods. All the strategies have their advantages and disadvantages, limited by the nature of chemical reactions and physical actions. However, the pectin-based DDS with real controlled-release performance is still worthy of attention. Furthermore, studies on pectin can not only be focused on the DDSs including oral DDS, transdermal DDS, and transnasal DDS. The injectable hydrogel, micro-, and nano-spheres may also need to be developed due to the advantages of biocompatibility, biodegradability, prebiotic effect, hypoglycemic effect, hypocholesterolemia effect, anticancer effect, etc.

References

Abboud, K.Y., Iacomini, M., Simas, F.F., Cordeiro, L.M.C., 2020. High methoxyl pectin from the soluble dietary fiber of passion fruit peel forms weak gel without the requirement of sugar addition. Carbohydr. Polym. 246, 116616.

Abedini, F., Ebrahimi, M., Roozbehani, A.H., Domb, A.J., Hosseinkhani, H., 2018. Overview on natural hydrophilic polysaccharide polymers in drug delivery. Polym. Adv. Technol. 29 (10), 2564–2573.

Abou El Fadl, F.I., Maziad, N.A., 2015. Radiation syntheses of pectin/acrylamide (PEC/PAM) and pectin/diethylaminoethylmethacrylate (PEC/DEAMA) hydrogels as drug delivery systems. J. Radioanal. Nucl. Chem. 303 (1), 623–630.

Ahadi, F., Khorshidi, S., Karkhaneh, A., 2019. A hydrogel/fiber scaffold based on silk fibroin/oxidized pectin with sustainable release of vancomycin hydrochloride. Eur. Polym. J. 118, 265–274.

Ahn, S., Halake, K., Lee, J., 2017. Antioxidant and ion-induced gelation functions of pectins enabled by polyphenol conjugation. Int. J. Biol. Macromol. 101, 776–782.

Amirian, J., Linh, N.T.B., Min, Y.K., Lee, B.T., 2015. Bone formation of a porous gelatin-pectin-biphasic calcium phosphate composite in presence of BMP-2 and VEGF. Int. J. Biol. Macromol. 76, 10–24.

Anuar, N.K., Wong, T.W., Taib, M.N., 2012. Microwave modified non-crosslinked pectin films with modulated drug release. Pharm. Dev. Technol. 17 (1), 110–117.

Arora, V., Sood, A., Shah, J., Kotnala, R.K., Jain, T.K., 2016. Synthesis and characterization of thiolated pectin stabilized gold coated magnetic nanoparticles. Mater. Chem. Phys. 173, 161–167.

Arora, V., Sood, A., Shah, J., Kotnala, R.K., Jain, T.K., 2017. Synthesis and characterization of pectin-6-aminohexanoic acid-magnetite nanoparticles for drug delivery. Mater. Sci. Eng. C 80, 243–251.

Assadpour, E., Jafari, S.M., Maghsoudlou, Y., 2017. Evaluation of folic acid release from spray dried powder particles of pectin-whey protein nano-capsules. Int. J. Biol. Macromol. 95, 238–247.

Assifaoui, A., Bouyer, F., Chambin, O., Cayot, P., 2013. Silica-coated calcium pectinate beads for colonic drug delivery. Acta Biomater. 9 (4), 6218–6225.

Babaladimath, G., Badalamoole, V., 2018a. Pectin-graft-poly(2-acrylamido-2-methyl-1-propane sulfonic acid) silver nanocomposite hydrogel beads: evaluation as matrix material for sustained release formulations of ketoprofen and antibacterial assay. J. Polym. Res. 25 (9), 202.

Babaladimath, G., Badalamoole, V., 2018b. Magnetic nanoparticles embedded in pectin-based hydrogel for the sustained release of diclofenac sodium. Polym. Int. 67 (8), 983–992.

Bai, Y., Zhang, M., Chandra Atluri, S., Chen, J., Gilbert, R.G., 2020. Relations between digestibility and structures of pumpkin starches and pectins. Food Hydrocoll. 106, 105894.

Basak, R., Bandyopadhyay, R., 2014. Formation and rupture of Ca^{2+} induced pectin biopolymer gels. Soft Matter 10 (37), 7225–7233.

Benvenutti, L., Sanchez-Camargo, A.d.P., Zielinski, A.A.F., Ferreira, S.R.S., 2020. NADES as potential solvents for anthocyanin and pectin extraction from *Myrciaria cauliflora* fruit by-product: in silico and experimental approaches for solvent selection. J. Mol. Liq. 315, 113761.

Bera, H., Kumar, S., 2018. Diethanolamine-modified pectin based core-shell composites as dual working gastroretentive drug-cargo. Int. J. Biol. Macromol. 108, 1053–1062.

Bera, H., Abbasi, Y.F., Yoke, F.F., Seng, P.M., Kakoti, B.B., Ahmmed, S.K.M., Bhatnagar, P., 2019. Ziprasidone-loaded arabic gum modified montmorillonite-tailor-made pectin based gastroretentive composites. Int. J. Biol. Macromol. 129, 552–563.

Bergman, M., Djaldetti, M., Salman, H., Bessler, H., 2010. Effect of citrus pectin on malignant cell proliferation. Biomed. Pharmacother. 64 (1), 44–47.

Bhuyan, M.M., Okabe, H., Hidaka, Y., Dafader, N.C., Rahman, N., Hara, K., 2018. Synthesis of pectin-N, N-dimethyl acrylamide hydrogel by gamma radiation and application in drug delivery (in vitro). J. Macromol. Sci. A 55 (4), 369–376.

Braccini, I., Pérez, S., 2001. Molecular basis of Ca^{2+}-induced gelation in alginates and pectins: the egg-box model revisited. Biomacromolecules 2 (4), 1089–1096.

Brouns, F., Theuwissen, E., Adam, A., Bell, M., Berger, A., Mensink, R.P., 2012. Cholesterol-lowering properties of different pectin types in mildly hyper-cholesterolemic men and women. Eur. J. Clin. Nutr. 66 (5), 591–599.

Cao, L., Lu, W., Mata, A., Nishinari, K., Fang, Y., 2020a. Egg-box model-based gelation of alginate and pectin: a review. Carbohydr. Polym. 242, 116389.

Cao, J., Yang, J., Wang, Z., Lu, M., Yue, K., 2020b. Modified citrus pectins by UV/H_2O_2 oxidation at acidic and basic conditions: structures and in vitro anti-inflammatory, anti-proliferative activities. Carbohydr. Polym. 247, 116742.

Chang, C., Wang, T., Hu, Q., Luo, Y., 2017. Zein/caseinate/pectin complex nanoparticles: formation and characterization. Int. J. Biol. Macromol. 104, 117–124.

Cheewatanakornkool, K., Niratisai, S., Manchun, S., Dass, C.R., Sriamornsak, P., 2017. Thiolated pectin–doxorubicin conjugates: synthesis, characterization and anticancer activity studies. Carbohydr. Polym. 174, 493–506.

Cheikh, D., García-Villén, F., Majdoub, H., Zayani, M.B., Viseras, C., 2019. Complex of chitosan pectin and clay as diclofenac carrier. Appl. Clay Sci. 172, 155–164.

Choi, Y.R., Lee, Y.K., Chang, Y.H., 2018. Structural and rheological properties of pectic polysaccharide extracted from *Ulmus davidiana* esterified by succinic acid. Int. J. Biol. Macromol. 120, 245–254.

Cuervo, A., Gueimonde, M., Margolles, A., González, S., 2014. Pectin: dietary sources, properties and health benefits. In: Pectin: Chemical Properties, Uses and Health Benefits. Nova Science Publishers, Inc, pp. 17–26.

Cui, S., Yao, B., Gao, M., Sun, X., Gou, D., Hu, J., Zhou, Y., Liu, Y., 2017. Effects of pectin structure and crosslinking method on the properties of crosslinked pectin nanofibers. Carbohydr. Polym. 157, 766–774.

de Almeida, D.A., Sabino, R.M., Souza, P.R., Bonafé, E.G., Venter, S.A.S., Popat, K.C., Martins, A.F., Monteiro, J.P., 2020. Pectin-capped gold nanoparticles synthesis in-situ for producing durable, cytocompatible, and superabsorbent hydrogel composites with chitosan. Int. J. Biol. Macromol. 147, 138–149.

Demir, Y.K., Metin, A.Ü., Şatıroğlu, B., Solmaz, M.E., Kayser, V., Mäder, K., 2017. Poly (methyl vinyl ether-co-maleic acid)—pectin based hydrogel-forming systems: gel, film, and microneedles. Eur. J. Pharm. Biopharm 117, 182–194.

Deshmukh, R., Harwansh, R.K., Paul, S.D., Shukla, R., 2020. Controlled release of sulfasalazine loaded amidated pectin microparticles through Eudragit S 100 coated capsule for management of inflammatory bowel disease. J. Drug Deliv. Sci. Technol. 55, 101495.

do Prado, S.B.R., Shiga, T.M., Harazono, Y., Hogan, V.A., Raz, A., Carpita, N.C., Fabi, J.P., 2019a. Migration and proliferation of cancer cells in culture are differentially affected by molecular size of modified citrus pectin. Carbohydr. Polym. 211, 141–151.

do Prado, S.B.R., Santos, G.R.C., Mourão, P.A.S., Fabi, J.P., 2019b. Chelate-soluble pectin fraction from papaya pulp interacts with galectin-3 and inhibits colon cancer cell proliferation. Int. J. Biol. Macromol. 126, 170–178.

Dongowski, G., 1995. Influence of pectin structure on the interaction with bile acids under in vitro conditions. Z. Lebensm. Unters. Forsch. 201 (4), 390–398.

Eswaramma, S., Reddy, N.S., Rao, K.S.V.K., 2017. Phosphate crosslinked pectin based dual responsive hydrogel networks and nanocomposites: development, swelling dynamics and drug release characteristics. Int. J. Biol. Macromol. 103, 1162–1172.

Ezzati, S., Ayaseh, A., Ghanbarzadeh, B., Heshmati, M.K., 2020. Pectin from sunflower by-product: optimization of ultrasound-assisted extraction, characterization, and functional analysis. Int. J. Biol. Macromol. 165, 776–786.

Fajardo, A.R., Lopes, L.C., Pereira, A.G.B., Rubira, A.F., Muniz, E.C., 2012. Polyelectrolyte complexes based on pectin–NH2 and chondroitin sulfate. Carbohydr. Polym. 87 (3), 1950–1955.

Feng, L., Jia, X., Zhu, Q., Liu, Y., Li, J., Yin, L., 2019. Investigation of the mechanical, rheological and microstructural properties of sugar beet pectin/soy protein isolate-based emulsion-filled gels. Food Hydrocoll. 89, 813–820.

Ferreira-Lazarte, A., Moreno, F.J., Cueva, C., Gil-Sánchez, I., Villamiel, M., 2019. Behaviour of citrus pectin during its gastrointestinal digestion and fermentation in a dynamic simulator (simgi®). Carbohydr. Polym. 207, 382–390.

Gaudio, P., Amante, C., Civale, R., Bizzarro, V., Petrella, A., Pepe, G., Campiglia, P., Russo, P., Aquino, R.P., 2020. In situ gelling alginate-pectin blend particles loaded with Ac2-26: a new weapon to improve wound care armamentarium. Carbohydr. Polym. 227, 115305.

Gentilini, R., Munarin, F., Petrini, P., Tanzi, M.C., 2014. Pectin gels for biomedical application. In: Pectin: Chemical Properties, Uses and Health Benefits. Nova Science Publishers Inc, pp. 1–15.

Gharehbeglou, P., Jafari, S.M., Hamishekar, H., Homayouni, A., Mirzaei, H., 2019. Pectin-whey protein complexes vs. small molecule surfactants for stabilization of double nano-emulsions as novel bioactive delivery systems. J. Food Eng. 245, 139–148.

Giacomazza, D., Sabatino, M.A., Catena, A., Leone, M., San Biagio, P.L., Dispenza, C., 2014. Maltose-conjugated chitosans induce macroscopic gelation of pectin solutions at neutral pH. Carbohydr. Polym. 114, 141–148.

Gómez, B., Gullón, B., Yáñez, R., Schols, H., Alonso, J.L., 2016. Prebiotic potential of pectins and pectic oligosaccharides derived from lemon peel wastes and sugar beet pulp: a comparative evaluation. J. Funct. Foods 20, 108–121.

Gunness, P., Zhai, H., Williams, B.A., Zhang, D., Gidley, M.J., 2021. Pectin and mango pulp both reduce plasma cholesterol in pigs but have different effects on triglycerides and bile acids. Food Hydrocoll. 112, 106369.

Gupta, B., Tummalapalli, M., Deopura, B.L., Alam, M.S., 2013. Functionalization of pectin by periodate oxidation. Carbohydr. Polym. 98 (1), 1160–1165.

Haliza, K., Iqbal, M.A.M.C., Nursyafiqah, M., Ying, N.L., Adhwa, M.B.P.A., 2017. Cell growth inhibition effect of DsiRNA vectorised by pectin-coated chitosan-graphene oxide nanocomposites as potential therapy for colon cancer. J. Nanomater. 2017, 4298218.

Hu, S., Kuwabara, R., Beukema, M., Ferrari, M., de Haan, B.J., Walvoort, M.T.C., de Vos, P., Smink, A.M., 2020. Low methyl-esterified pectin protects pancreatic β-cells against diabetes-induced oxidative and inflammatory stress via galectin-3. Carbohydr. Polym. 249, 116863.

Hussien, N.A., Işıklan, N., Türk, M., 2018. Pectin-conjugated magnetic graphene oxide nanohybrid as a novel drug carrier for paclitaxel delivery. Artif. Cells Nanomed. Biotechnol. 46 (1), 264–273.

Işıklan, N., Tokmak, Ş., 2019. Development of thermo/pH-responsive chitosan coated pectin-graft-poly(N,N-diethyl acrylamide) microcarriers. Carbohydr. Polym. 218, 112–125.

Jacob, E.M., Borah, A., Jindal, A., Pillai, S.C., Yamamoto, Y., Maekawa, T., Kumar, D.N.S., 2020. Synthesis and characterization of citrus-derived pectin nanoparticles based on their degree of esterification. J. Mater. Res. 35 (12), 1514–1522.

Jamshidzadeh, F., Mohebali, A., Abdouss, M., 2020. Three-ply biocompatible pH-responsive nanocarriers based on HNT sandwiched by chitosan/pectin layers for controlled release of phenytoin sodium. Int. J. Biol. Macromol. 150, 336–343.

Jin, B., Zhou, X., Liu, Y., Li, X., Mai, Y., Liao, Y., Liao, J., 2018. Physicochemical stability and antioxidant activity of soy protein/pectin/tea polyphenol ternary nanoparticles obtained by photocatalysis. Int. J. Biol. Macromol. 116, 1–7.

Jin, B., Zhou, X., Zhou, S., Liu, Y., Guan, R., Zheng, Z., Liang, Y., 2019. Influence of phenolic acids on the storage and digestion stability of curcumin emulsions based on soy protein-pectin-phenolic acids ternary nano-complexes. J. Microencapsul. 36 (7), 622–634.

Kamnev, A.A., Calce, E., Tarantilis, P.A., Tugarova, A.V., Luca, S.D., 2015. Pectin functionalised by fatty acids: diffuse reflectance infrared Fourier transform (DRIFT) spectroscopic characterisation. J. Mol. Struct. 1079, 74–77.

Kean, T., Thanou, M., 2010. Biodegradation, biodistribution and toxicity of chitosan. Adv. Drug Deliv. Rev. 62 (1), 3–11.

Khodashenas, B., Ardjmand, M., Baei, M.S., Rad, A.S., Akbarzadeh, A., 2020. Conjugation of pectin biopolymer with au-nanoparticles as a drug delivery system: experimental and DFT studies. Appl. Organomet. Chem. 34 (6), e5609.

Khotimchenko, M., 2020. Pectin polymers for colon-targeted antitumor drug delivery. Int. J. Biol. Macromol. 158, 1110–1124.

Khramova, D.S., Vityazev, F.V., Saveliev, N.Y., Burkov, A.A., Beloserov, V.S., Martinson, E.A., Litvinets, S.G., Popov, S.V., 2019. Pectin gelling in acidic gastric condition increases rheological properties of gastric digesta and reduces glycaemic response in mice. Carbohydr. Polym. 205, 456–464.

Kodoth, A.K., Ghate, V.M., Lewis, S.A., Prakash, B., Badalamoole, V., 2019. Pectin-based silver nanocomposite film for transdermal delivery of donepezil. Int. J. Biol. Macromol. 134, 269–279.

Leroux, J., Langendorff, V., Schick, G., Vaishnav, V., Mazoyer, J., 2003. Emulsion stabilizing properties of pectin. Food Hydrocoll. 17 (4), 455–462.

Li, Y., Niu, Y., Wu, H., Sun, Y., Li, Q., Kong, X., Liu, L., Mei, Q., 2010. Modified apple polysaccharides could induce apoptosis in colorectal cancer cells. J. Food Sci. 75 (8), 224–229.

Li, D.Q., Jia, X., Wei, Z., Liu, Z.Y., 2012. Box-Behnken experimental design for investigation of microwave-assisted extracted sugar beet pulp pectin. Carbohydr. Polym. 88 (1), 342–346.

Li, X., Fang, Y., Phillips, G.O., Al-Assaf, S., 2013. Improved sugar beet pectin-stabilized emulsions through complexation with sodium caseinate. J. Agric. Food Chem. 61 (6), 1388–1396.

Li, D.Q., Du, G.M., Jing, W.W., Li, J.F., Yan, J.Y., Liu, Z.Y., 2015. Combined effects of independent variables on yield and protein content of pectin extracted from sugar beet pulp by citric acid. Carbohydr. Polym. 129, 108–114.

Li, C., Nie, H., Chen, Y., Xiang, Z.-Y., Li, J.-B., 2016. Amide pectin: a carrier material for colon-targeted controlled drug release. J. Appl. Polym. Sci. 133 (29), 43697.

Li, D.q., Wang, J., Guo, Z.g., Li, J., Shuai, J., 2017. Pectin gels cross-linked by Ca^{2+}: an efficient material for methylene blue removal. J. Mol. Liq. 238, 36–42.

Li, J., Zhang, L., Li, J., Li, D., 2019. Comparative studies of combined influence of variables on the esterification degree of pectin extracted by sulfuric acid and citric acid. Adv. Polym. Technol. 2019, 6313241.

Li, D., Wang, Sh., Meng, Y., Guo, Z., Cheng, M., Li, J., 2021. Fabrication of self-healing pectin/chitosan hybrid hydrogel via Diels-Alder reactions for drug delivery with high swelling property, pH-responsiveness, and cytocompatibility. Carbohyd. Polym. 268, 118244.

Li, D.q., Wang, S.y., Meng, Y.j., Li, J.f., Li, J., 2020. An injectable, self-healing hydrogel system from oxidized pectin/chitosan/γ-Fe_2O_3. Int. J. Biol. Macromol. 164, 4566–4574.

Liew, S.Q., Chin, N.L., Yusof, Y.A., Sowndhararajan, K., 2016. Comparison of acidic and enzymatic pectin extraction from passion fruit peels and its gel properties. J. Food Process Eng. 39 (5), 501–511.

Liew, S.Q., Teoh, W.H., Yusoff, R., Ngoh, G.C., 2019. Comparisons of process intensifying methods in the extraction of pectin from pomelo peel. Chem. Eng. Process. Process Intensif. 143, 107586.

Liu, Y., Qi, Q., Li, X., Liu, J., Wang, L., He, J., Lei, J., 2017. Self-assembled pectin-conjugated eight-arm polyethylene glycol-dihydroartemisinin nanoparticles for anticancer combination therapy. ACS Sustain. Chem. Eng. 5 (9), 8097–8107.

Liu, Y., Liu, K., Li, X., Xiao, S., Zheng, D., Zhu, P., Li, C., Liu, J., He, J., Lei, J., Wang, L., 2018. A novel self-assembled nanoparticle platform based on pectin-eight-arm polyethylene glycol-drug conjugates for co-delivery of anticancer drugs. Mater. Sci. Eng. C 86, 28–41.

Liu, C., Tan, Y., Xu, Y., McCleiments, D.J., Wang, D., 2019. Formation, characterization, and application of chitosan/pectin-stabilized multilayer emulsions as astaxanthin delivery systems. Int. J. Biol. Macromol. 140, 985–997.

Luo, Y., Wang, Q., 2014. Recent development of chitosan-based polyelectrolyte complexes with natural polysaccharides for drug delivery. Int. J. Biol. Macromol. 64, 353–367.

Luppi, B., Bigucci, F., Abruzzo, A., Corace, G., Cerchiara, T., Zecchi, V., 2010. Freeze-dried chitosan/pectin nasal inserts for antipsychotic drug delivery. Eur. J. Pharm. Biopharm. 75 (3), 381–387.

Lutz, R., Aserin, A., Wicker, L., Garti, N., 2009. Double emulsions stabilized by a charged complex of modified pectin and whey protein isolate. Colloids Surf. B: Biointerfaces 72 (1), 121–127.

Maior, J.F.A.S., Reis, A.V., Muniz, E.C., Cavalcanti, O.A., 2008. Reaction of pectin and glycidyl methacrylate and ulterior formation of free films by reticulation. Int. J. Pharm. 355 (1), 184–194.

Marić, M., Grassino, A.N., Zhu, Z., Barba, F.J., Brnčić, M., Rimac Brnčić, S., 2018. An overview of the traditional and innovative approaches for pectin extraction from plant food wastes and by-products: ultrasound-, microwaves-, and enzyme-assisted extraction. Trends Food Sci. Technol. 76, 28–37.

Maxwell, E.G., Belshaw, N.J., Waldron, K.W., Morris, V.J., 2012. Pectin-an emerging new bioactive food polysaccharide. Trends Food Sci. Technol. 24 (2), 64–73.

Maxwell, E.G., Colquhoun, I.J., Chau, H.K., Hotchkiss, A.T., Waldron, K.W., Morris, V.J., Belshaw, N.J., 2016. Modified sugar beet pectin induces apoptosis of colon cancer cells via an interaction with the neutral sugar side-chains. Carbohydr. Polym. 136, 923–929.

Meng, Y.j., Wang, S.y., Guo, Z.w., Cheng, M.m., Li, J., Li, D.q., 2020. Design and preparation of quaternized pectin-montmorillonite hybrid film for sustained drug release. Int. J. Biol. Macromol. 154, 413–420.

Mishra, R.K., Datt, M., Pal, K., Banthia, A.K., 2008. Preparation and characterization of amidated pectin based hydrogels for drug delivery system. J. Mater. Sci. Mater. Med. 19 (6), 2275–2280.

Monfregola, L., Bugatti, V., Amodeo, P., De Luca, S., Vittoria, V., 2011. Physical and water sorption properties of chemically modified pectin with an environmentally friendly process. Biomacromolecules 12 (6), 2311–2318.

Muhoza, B., Xia, S., Cai, J., Zhang, X., Duhoranimana, E., Su, J., 2019. Gelatin and pectin complex coacervates as carriers for cinnamaldehyde: effect of pectin esterification degree on coacervate formation, and enhanced thermal stability. Food Hydrocoll. 87, 712–722.

Munjeri, O., Collett, J.H., Fell, J.T., Sharma, H.L., Smith, A.M., 1998a. In vivo behavior of hydrogel beads based on amidated pectins. Drug Deliv. 5 (4), 239–241.

Munjeri, O., Hodza, P., Osim, E.E., Musabayane, C.T., 1998b. An investigation into the suitability of amidated pectin hydrogel beads as a delivery matrix for chloroquine. J. Pharm. Sci. 87 (8), 905–908.

Murray, R., Finland, M., 1946. Pectin adjuvant for oral penicillin. Proc. Soc. Exp. Biol. Med. 62 (2), 240–242.

Musabayane, C.T., Munjeri, O., Matavire, T.P., 2003. Transdermal delivery of chloroquine by amidated pectin hydrogel matrix patch in the rat. Ren. Fail. 25 (4), 525–534.

Naik, M., Rawson, A., Rangarajan, J.M., 2020. Radio frequency-assisted extraction of pectin from jackfruit (Artocarpus heterophyllus) peel and its characterization. J. Food Process Eng. 43 (6), e13389.

Norcino, L.B., de Oliveira, J.E., Moreira, F.K.V., Marconcini, J.M., Mattoso, L.H.C., 2018. Rheological and thermo-mechanical evaluation of bio-based chitosan/pectin blends with tunable ionic cross-linking. Int. J. Biol. Macromol. 118, 1817–1823.

Pappas, C.S., Malovikova, A., Hromadkova, Z., Tarantilis, P.A., Ebringerova, A., Polissiou, M.G., 2004. Determination of the degree of esterification of pectinates with decyl and benzyl ester groups by diffuse reflectance infrared Fourier transform spectroscopy (DRIFTS) and curve-fitting deconvolution method. Carbohydr. Polym. 56 (4), 465–469.

Pedrosa, L.d.F., Lopes, R.G., Fabi, J.P., 2020. The acid and neutral fractions of pectins isolated from ripe and overripe papayas differentially affect galectin-3 inhibition and colon cancer cell growth. Int. J. Biol. Macromol. 164, 2681–2690.

Pereira, R.F., Barrias, C.C., Bártolo, P.J., Granja, P.L., 2018. Cell-instructive pectin hydrogels crosslinked via thiol-norbornene photo-click chemistry for skin tissue engineering. Acta Biomater. 66, 282–293.

Qi, X., Al-Ghazzewi, F.H., Tester, R.F., 2018. Dietary fiber, gastric emptying, and carbohydrate digestion: a mini-review. Starch/Staerke 70, 1700346.

Rachini, A., Le Troedec, M., Peyratout, C., Smith, A., 2009. Comparison of the thermal degradation of natural, alkali-treated and silane-treated hemp fibers under air and an inert atmosphere. J. Appl. Polym. Sci. 112 (1), 226–234.

Raei, M., Shahidi, F., Farhoodi, M., Jafari, S.M., Rafe, A., 2017. Application of whey protein-pectin nano-complex carriers for loading of lactoferrin. Int. J. Biol. Macromol. 105, 281–291.

Raj, M., Savaliya, R., Joshi, S., Raj, L., Keharia, H., 2019. Biodegradability, thermal, chemical, mechanical and morphological behavior of LDPE/pectin and LDPE/modified pectin blend. Polym. Bull. 76 (10), 5173–5195.

Rebitski, E.P., Darder, M., Carraro, R., Ruiz-Hitzky, E., 2020. Chitosan and pectin core-shell beads encapsulating metformin-clay intercalation compounds for controlled delivery. New J. Chem. 44 (24), 10102–10110.

Renard, C.M.G.C., Jarvis, M.C., 1999a. Acetylation and methylation of homogalacturonans 1: optimisation of the reaction and characterisation of the products. Carbohydr. Polym. 39 (3), 201–207.

Renard, C.M.G.C., Jarvis, M.C., 1999b. Acetylation and methylation of homogalacturonans 2: effect on ion-binding properties and conformations. Carbohydr. Polym. 39 (3), 209–216.

Rosenbohm, C., Lundt, I., Christensen, T.M.I.E., Young, N.W.G., 2003. Chemically methylated and reduced pectins: preparation, characterisation by 1H NMR spectroscopy, enzymatic degradation, and gelling properties. Carbohydr. Res. 338 (7), 637–649. https://doi.org/10.1016/S0008-6215(02)00440-8.

Semdé, R., Amighi, K., Pierre, D., Devleeschouwer, M.J., Moës, A.J., 1998. Leaching of pectin from mixed pectin/insoluble polymer films intended for colonic drug delivery. Int. J. Pharm. 174 (1–2), 233–241.

Semdé, R., Moës, A.J., Devleeschouwer, M.J., Amighi, K., 2003. Synthesis and enzymatic degradation of epichlorohydrin cross-linked pectins. Drug Dev. Ind. Pharm. 29 (2), 203–213.
Serrano-Cruz, M.R., Villanueva-Carvajal, A., Morales Rosales, E.J., Ramírez Dávila, J.F., Dominguez-Lopez, A., 2013. Controlled release and antioxidant activity of Roselle (*Hibiscus sabdariffa* L.) extract encapsulated in mixtures of carboxymethyl cellulose, whey protein, and pectin. LWT-Food Sci. Technol. 50 (2), 554–561.
Shafie, M.H., Yusof, R., Gan, C.Y., 2019. Deep eutectic solvents (DES) mediated extraction of pectin from *Averrhoa bilimbi*: optimization and characterization studies. Carbohydr. Polym. 216, 303–311.
Shahzad, A., Khan, A., Afzal, Z., Umer, M.F., Khan, J., Khan, G.M., 2019. Formulation development and characterization of cefazolin nanoparticles-loaded cross-linked films of sodium alginate and pectin as wound dressings. Int. J. Biol. Macromol. 124, 255–269.
Shi, X., Cui, S., Song, X., Rickel, A.P., Sanyour, H.J., Zheng, J., Hu, J., Hong, Z., Zhou, Y., Liu, Y., 2020. Gelatin-crosslinked pectin nanofiber mats allowing cell infiltration. Mater. Sci. Eng. C 112, 110941.
Shih, F.Y., Su, I.J., Chu, L.L., Lin, X., Kuo, S.C., Hou, Y.C., Chiang, Y.T., 2018. Development of pectin-type B gelatin polyelectrolyte complex for curcumin delivery in anticancer therapy. Int. J. Mol. Sci. 19 (11), 3625.
Shitrit, Y., Davidovich-Pinhas, M., Bianco-Peled, H., 2019. Shear thinning pectin hydrogels physically cross-linked with chitosan nanogels. Carbohydr. Polym. 225, 115249.
Singh, R.P., Prakash, S., Bhatia, R., Negi, M., Singh, J., Bishnoi, M., Kondepudi, K.K., 2020. Generation of structurally diverse pectin oligosaccharides having prebiotic attributes. Food Hydrocoll. 108, 105988.
Sinitsya, A., Čopíková, J., Prutyanov, V., Skoblya, S., Machovič, V., 2000. Amidation of highly methoxylated citrus pectin with primary amines. Carbohydr. Polym. 42 (4), 359–368.
Soubhagya, A.S., Moorthi, A., Prabaharan, M., 2020. Preparation and characterization of chitosan/pectin/ZnO porous films for wound healing. Int. J. Biol. Macromol. 157, 135–145.
Souza, F.N., Gebara, C., Ribeiro, M.C.E., Chaves, K.S., Gigante, M.L., Grosso, C.R.F., 2012. Production and characterization of microparticles containing pectin and whey proteins. Food Res. Int. 49 (1), 560–566.
Sriamornsak, P., 1998. Investigation of pectin as a carrier for oral delivery of proteins using calcium pectinate gel beads. Int. J. Pharm. 169 (2), 213–220.
Sucheta, Misra, N.N., Yadav, S.K., 2020. Extraction of pectin from black carrot pomace using intermittent microwave, ultrasound and conventional heating: kinetics, characterization and process economics. Food Hydrocoll. 102, 105592.
Sun, Q., Wang, F., Han, D., Zhao, Y., Liu, Z., Lei, H., Song, Y., Huang, X., Li, X., Ma, A., Yuan, G., Li, X., Yang, Z., 2014. Preparation and optimization of soy protein isolate-high methoxy pectin microcapsules loaded with *Lactobacillus delbrueckii*. Int. J. Food Sci. Technol. 49 (5), 1287–1293.
Synytsya, A., Poučková, P., Zadinová, M., Troshchynska, Y., Štětina, J., Synytsya, A., Saloň, I., Král, V., 2020. Hydrogels based on low-methoxyl amidated citrus pectin and flaxseed gum formulated with tripeptide glycyl-L-histidyl-L-lysine improve the healing of experimental cutting wounds in rats. Int. J. Biol. Macromol. 165, 3156–3168.
Takei, T., Sugihara, K., Ijima, H., Kawakami, K., 2011. In situ gellable sugar beet pectin via enzyme-catalyzed coupling reaction of feruloyl groups for biomedical applications. J. Biosci. Bioeng. 112 (5), 491–494.
Tan, H., Nie, S., 2020. Deciphering diet-gut microbiota-host interplay: investigations of pectin. Trends Food Sci. Technol. 106, 171–181.
Tan, H.L., Tan, L.S., Wong, Y.Y., Muniyandy, S., Hashim, K., Pushpamalar, J., 2016. Dual crosslinked carboxymethyl sago pulp/pectin hydrogel beads as potential carrier for colon-targeted drug delivery. J. Appl. Polym. Sci. 133 (19), 43416.
Tan, J., Hua, X., Liu, J., Wang, M., Liu, Y., Yang, R., Cao, Y., 2020. Extraction of sunflower head pectin with superfine grinding pretreatment. Food Chem. 320, 126631.
Thibault, J.-F., Ralet, M.-C., 2003. Physico-chemical properties of pectins in the cell walls and after extraction. In: Voragen, F., Schols, H., Visser, R. (Eds.), Advances in Pectin and Pectinase Research. Springer Netherlands, pp. 91–105.
Tian, L., Singh, A., Singh, A.V., 2020. Synthesis and characterization of pectin-chitosan conjugate for biomedical application. Int. J. Biol. Macromol. 153, 533–538.
Vityazev, F.V., Fedyuneva, M.I., Golovchenko, V.V., Patova, O.A., Ipatova, E.U., Durnev, E.A., Martinson, E.A., Litvinets, S.G., 2017. Pectin-silica gels as matrices for controlled drug release in gastrointestinal tract. Carbohydr. Polym. 157, 9–20.

Vityazev, F.V., Khramova, D.S., Saveliev, N.Y., Ipatova, E.A., Burkov, A.A., Beloserov, V.S., Belyi, V.A., Kononov, L.-O., Martinson, E.A., Litvinets, S.G., Markov, P.A., Popov, S.V., 2020. Pectin-glycerol gel beads: preparation, characterization and swelling behaviour. Carbohydr. Polym. 238 (15), 116166.

Wakerly, Z., Fell, J., Attwood, D., Parkins, D., 1997. Studies on amidated pectins as potential carriers in colonic drug delivery. J. Pharm. Pharmacol. 49 (6), 622–625.

Wandee, Y., Uttapap, D., Mischnick, P., 2019. Yield and structural composition of pomelo peel pectins extracted under acidic and alkaline conditions. Food Hydrocoll. 87, 237–244.

Wang, T., Ma, X., Lei, Y., Luo, Y., 2016. Solid lipid nanoparticles coated with cross-linked polymeric double layer for oral delivery of curcumin. Colloids Surf. B: Biointerfaces 148, 1–11.

Wang, S.y., Li, J., Zhou, Y., Li, D.q., Du, G.m., 2019. Chemical cross-linking approach for prolonging diclofenac sodium release from pectin-based delivery system. Int. J. Biol. Macromol. 137, 512–520.

Wu, D., Zheng, J., Hu, W., Zheng, X., He, Q., Linhardt, R.J., Ye, X., Chen, S., 2020. Structure-activity relationship of citrus segment membrane RG-I pectin against Galectin-3: the galactan is not the only important factor. Carbohydr. Polym. 245, 116526.

Wusigale, Liang, L., Luo, Y., 2020. Casein and pectin: structures, interactions, and applications. Trends Food Sci. Technol. 97, 391–403.

Yamada, M., Shiiba, S., 2015. Preparation of pectin-inorganic composite material as accumulative material of metal ions. J. Appl. Polym. Sci. 132 (24), 42056.

Yapo, B.M., Robert, C., Etienne, I., Wathelet, B., Paquot, M., 2007. Effect of extraction conditions on the yield, purity and surface properties of sugar beet pulp pectin extracts. Food Chem. 100 (4), 1356–1364.

Zaidel, D.N.A., Chronakis, I.S., Meyer, A.S., 2013. Stabilization of oil-in-water emulsions by enzyme catalyzed oxidative gelation of sugar beet pectin. Food Hydrocoll. 30 (1), 19–25.

Zaitseva, O., Khudyakov, A., Sergushkina, M., Solomina, O., Polezhaeva, T., 2020. Pectins as a universal medicine. Fitoterapia 146, 104676.

Zhang, C., Zhu, X., Zhang, F., Yang, X., Ni, L., Zhang, W., Liu, Z., Zhang, Y., 2020a. Improving viscosity and gelling properties of leaf pectin by comparing five pectin extraction methods using green tea leaf as a model material. Food Hydrocoll. 98, 105246.

Zhang, M., Zu, H., Zhuang, X., Yu, Y., Wang, Y., Zhao, Z., Zhou, Y., 2020b. Structural analyses of the HG-type pectin from notopterygium incisum and its effects on galectins. Int. J. Biol. Macromol. 162, 1035–1043.

Zheng, X.-F., Lian, Q., Yang, H., Zhu, H., 2015. Alkyl pectin: hydrophobic matrices for controlled drug release. J. Appl. Polym. Sci. 132 (3), 41302.

Zhu, R.-G., Sun, Y.-D., Li, T.-P., Chen, G., Peng, X., Duan, W.-B., Zheng, Z.-Z., Shi, S.-L., Xu, J.-G., Liu, Y.-H., Jin, X.-Y., 2015. Comparative effects of hawthorn (*Crataegus pinnatifida* Bunge) pectin and pectin hydrolyzates on the cholesterol homeostasis of hamsters fed high-cholesterol diets. Chem. Biol. Interact. 238, 42–47.

Zhu, R.-G., Sun, Y.-D., Hou, Y.-T., Fan, J.-G., Chen, G., Li, T.-P., 2017. Pectin penta-oligogalacturonide reduces cholesterol accumulation by promoting bile acid biosynthesis and excretion in high-cholesterol-fed mice. Chem. Biol. Interact. 272, 153–159.

Zouambia, Y., Youcef Ettoumi, K., Krea, M., Moulai-Mostefa, N., 2017. A new approach for pectin extraction: electromagnetic induction heating. Arab. J. Chem. 10 (4), 480–487.

CHAPTER

5

Gellan gum nanoparticles in drug delivery

Ana Letícia Rodrigues Costa and Lucimara Gaziolla de la Torre
Department of Materials and Bioprocess Engineering, School of Chemical Engineering, University of Campinas, Campinas, Brazil

5.1 Introduction

Conventionally, excipients are widely used as inert additives that provide the appropriate dosage form, safety, and efficacy in administering pharmaceutical formulations to patients (Haywood and Glass, 2011; Goswami and Naik, 2014). Nowadays, novel excipients have been tailored and developed for the needs of different delivery routes (transdermal, pulmonary, oral, ocular) in order to improve the drug potency by changing its pattern of release, absorption, distribution, metabolism, and elimination (Heer, 2013; Dwarakanadha Reddy and Swarnalatha, 2010). However, regulatory restrictions have hampered innovation, since new excipients must comply with many regulatory restrictions on their safety, quality, and functionality and follow strict guidelines for good manufacturing practices (GMP) (Heer, 2013; Milivojevic et al., 2019). Thus, the demand for natural polymers instead of the most common synthetic or semisynthetic excipients is currently increasing in the pharmaceutical formulation due to their low toxicity, widespread availability, cost-effectiveness, and inherent biocompatibility (Kulkarni et al., 2005; Gopal et al., 2017).

Natural polysaccharides are hydrophilic carbohydrate polymers with high molecular weight, composed of repeated units of monosaccharides linked by glycosidic bonds (Sah et al., 2016). They are produced commercially from plants (e.g., pectin, cellulose, starch), animals (e.g., chitosan, chitin, glycosaminoglycan), microorganisms (e.g., dextran, xanthan gum, gellan gum), and algae (e.g., agar, alginate, and carrageenan), and widely used in food, cosmetic, paper, textiles, and other industries as viscous enhancing, water-holding, emulsifying, coating, encapsulating, gelling, and other agents (Tako, 2015; Zhang et al., 2013). Specifically, natural polysaccharides present the gelation capacity as one essential property that can be explored in the pharmaceutical and biomedical fields. Gels are a result of suitable

Micro- and Nanoengineered Gum-Based Biomaterials for Drug Delivery and Biomedical Applications
https://doi.org/10.1016/B978-0-323-90986-0.00009-1

physical or chemical cross-links among polysaccharide chains, which lead to the formation of a soft material that combines the properties of solids and fluids. These gels can retain large amounts of water or biological fluids while maintaining their structure; in addition, their porosity contributes to high drug loading capacity and swelling properties, enabling the controlled release of the drug (Zhang et al., 2016).

With the advent of nanotechnology, polysaccharides have been widely applied in nanoparticle delivery systems since they offer advantages for site-specific and/or time-controlled delivery, and reduce toxic side effects of drugs. Polysaccharide-based nanoparticles are composed of hydrogel particulate entities with nanometric size, usually below 1000 nm, so they have the properties of gels and the advantages of nanotechnology at the same time (Zhang et al., 2016; Liu et al., 2008). In addition, nanogels are reported to be an ideal drug delivery system due to their ease of manufacture, high stability, uniform particle size distribution, and stimuli-responsive nature that are unprecedented for common pharmaceutical nanocarriers (Chacko et al., 2012; Oh et al., 2007).

Among all natural polysaccharides, gellan gum (GG) is increasingly attracting attention to many different dosage forms. In the conventional dosage forms, GG has been investigated as a tablet binder, granulating agent, or coating material, but those applications are less attractive. Thus, relatively modern formulations (beads, films, microspheres/microcapsules, and nanoparticles) have been developed to alter drug pharmacokinetics and improve GG's bioavailability through delayed or/and sustained release, site-specific release, and receptor-targeted delayed-release (Milivojevic et al., 2019; Builders and Attama, 2011).

This chapter deals specifically with the current approaches to synthesize GG-based nanoparticles as a drug delivery system, describing their properties and characteristics for the drug administration by different routes.

5.2 Gellan gum

Gellan gum (GG) is an anionic exopolysaccharide produced by aerobic fermentation of the bacterium *Sphingomonas elodea*, composed of a linear unit of tetrasaccharides (1,3-β-D-glucose, 1,4-β-D-glucuronic, 1,4-β-D-glucose, 1,4-α-L-rhamnose) containing a free carboxyl side group and a molecular weight of about 500 kDa (Giavasis et al., 2000; Bacelar et al., 2016; Kang and Pettitt, 1993). The Food and Drug Administration approved its use as a food additive in 1992 due to its proven biocompatibility and low cytotoxicity (Osmałek et al., 2014). The natural form of GG is named high acyl or acetylate gellan gum because it has two acyl substituents: acetate and glycerate. Such substituents can be removed from the repetitive structure by alkali treatment. Thus, in its deacetylate state, GG is called low acyl or deacetylate gellan gum (Fig. 5.1), which is commercially available in powder form and is the most commonly used in the pharmaceutical and biomedical fields (Osmałek et al., 2014; Kuo et al., 1986; Nitta et al., 2010; Mao et al., 2000). In addition, it is a low-cost and high-reproducible biomaterial manufactured on an industrial scale, which also presents favorable physicochemical, mechanical, and functional properties such as abundance in nature, nontoxicity, easy modification, rapid gelation in the presence of cations, adjustable gel elasticity, mucoadhesiveness, thermal stability, acid reliability, and high transparency (Giavasis et al., 2000; Bajaj et al., 2006; Milas et al., 1990; Quinn et al., 1993).

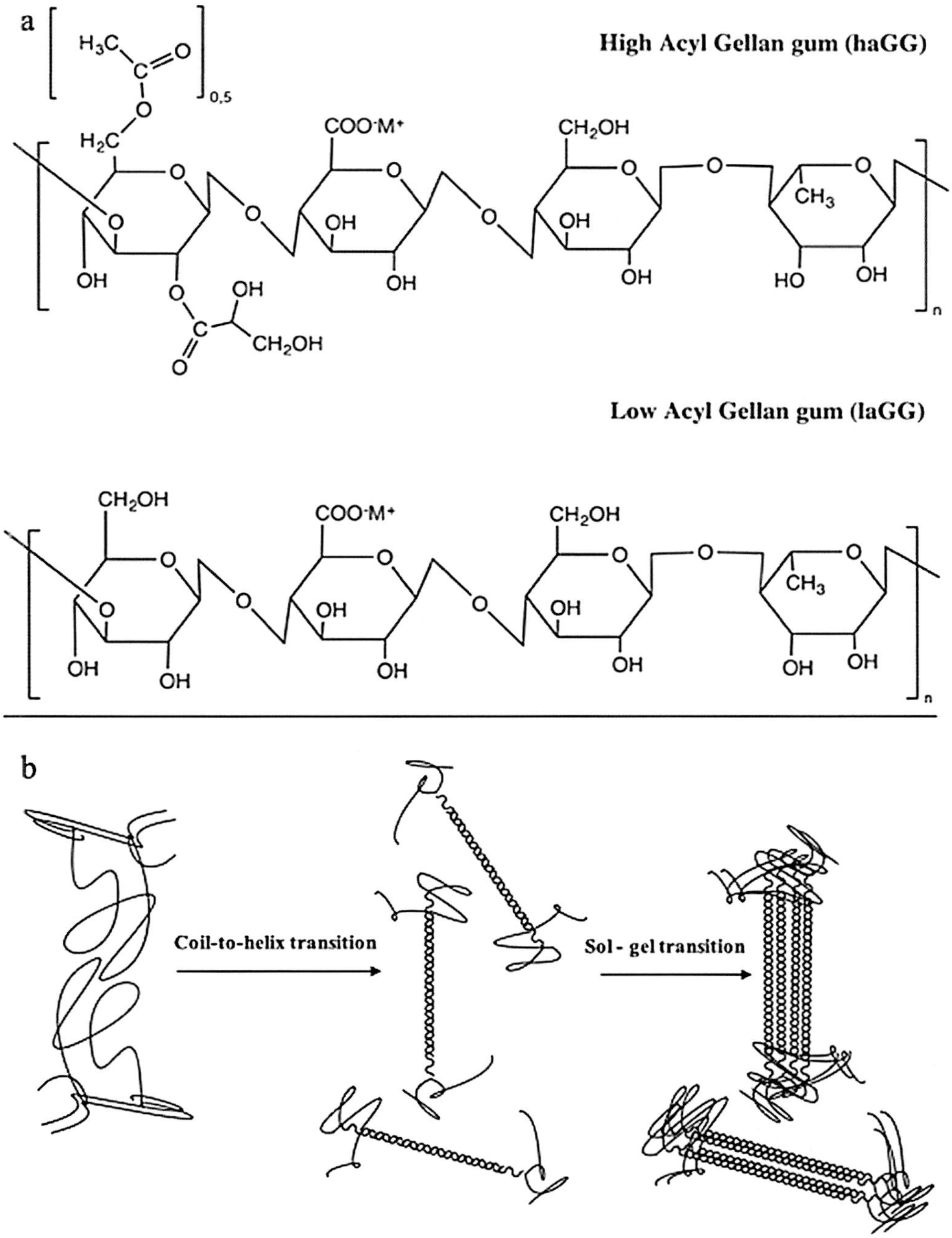

FIG. 5.1 (A) Chemical structure of high acyl gellan gum and low acyl gellan gum and (B) coil-to-helix and sol-gel transition of gellan gum. *Data from Palumbo, F.S., Federico, S., Pitarresi, G., Fiorica, C., Giammona, G., 2020. Gellan gum-based delivery systems of therapeutic agents and cells. Carbohydr. Polym. 229. https://doi.org/10.1016/j.carbpol.2019.115430.*

The gelling mechanism of GG has two stages. Initially, the disordered chains are organized by forming a double-helical junction zone (coil-to-helix transition), which is a thermally reversible transition (Fig. 5.1B). After the coil-to-helix transition followed by cooling (30–50 °C), hydrophobic double helix segments may interact to form a three-dimensional network (sol-gel transition) mediated by hydrogen bonding with water and complexation with cations (Moritaka et al., 1995; Grasdalen and Smidsrød, 1987; Miyoshi et al., 1994). The properties of GG hydrogels are dependent on many factors, including the content of acyl groups and polysaccharide concentration, type and concentration of cations, pH, presence of other hydrocolloids, and cooling conditions (Bajaj et al., 2007; Morris et al., 2012; Vilela et al., 2015; Santos and Cunha, 2018; Picone and Cunha, 2011; Yamamoto and Cunha, 2007; Jay et al., 1998). For example, GG gels formed by divalent cations are stronger and formed at much lower ion concentrations than those produced by monovalent ones. In the latter, gelation occurs due to the screening of the electrostatic repulsion between ionized carboxylate groups on the GG chains. In contrast, divalent cations induce gelation and aggregation via chemical bonding by direct bridges between double helix pairs, in addition to the screening effects (Bacelar et al., 2016; Morris et al., 2012; Costa et al., 2017; Santos et al., 2020; Pires Vilela et al., 2011; Santos and Cunha, 2019; Singh et al., 2004). In addition, it is known that the gelled structures obtained from GG are stronger and less permeable than those obtained from other polysaccharides, and can be formed in lower concentrations of polysaccharide and the gelling agent when compared to κ-carrageenan and alginate (Chan et al., 2009; Perrechil et al., 2011). GG hydrogels can also be thermo-responsive (Graham et al., 2019) and susceptible to enzymatic degradation *in vivo* by lysozyme, amylase, and trypsin, resulting in loss of mass (Palumbo et al., 2020). All GG characteristics can be widely explored in different drug delivery formulations, including nanoparticles for sustained and controlled drug delivery.

5.3 Production and characterization of gellan gum nanoparticles

The strategy of compartmentalizing drugs into individual nanometric subunits may favor the occurrence of dynamic biological interactions, leading to better cellular uptake, extended half-life in the systemic circulation, and better control of drug targeting and specificity, since drug delivery becomes strongly influenced by the system carrier characteristics and not only by drug features (Kumar et al., 2016; Prezotti et al., 2020; Qian et al., 2013). Polysaccharide-based nanoparticles are three-dimensional hydrogel networks in nanoscale size, formed by physical or chemical cross-linking polysaccharide chains (Musazzi et al., 2018; Kumari and Badwaik, 2019). These nanocarrier systems have unique potentials due to the combination of the hydrogel properties (low interfacial tension and stability with biological fluids, versatility, flexibility, biocompatibility, swelling, and high-water content) with the advantages of nanotechnology (high carrier capacity, high stability against premature chemical/enzymatic degradation, modulation of physicochemical, and drug release properties, being actively or passively targeted to the desired site) (Qian et al., 2013; D'Arrigo et al., 2014).

Numerous natural polysaccharides have been used to design nanoparticles, such as chitosan, alginate, dextran, pectin, hyaluronic acid, xanthan, and gellan gum. In addition, a wide range of molecular weights and a significant number of functional groups (hydroxyl, amino, and carboxyl groups) in polysaccharides' backbones for chemical modification lead to the variations in nanoparticles properties such as mechanical strength, surface charge, shape, porosity, amphiphilicity, and degradability, enabling them to be implemented in a wide range of drug delivery systems (Kumari and Badwaik, 2019; Basu et al., 2015; Garg et al., 2019).

As aforementioned, polysaccharide-based nanoparticles can be synthesized by physical (amphiphilic and electrostatic association) or chemical (disulfide, amine-based, and photo-induced) cross-linking methods (Kumari and Badwaik, 2019). However, techniques for their preparation are limited. Specifically, GG can form nanoparticle delivery carriers through of the self-assembly process, ionic cross-linking, polyelectrolyte complexation, and heat cross-linking, as discussed below.

5.3.1 Production of gellan gum nanoparticles

Self-assembly process

Recently, the use of amphiphilic polymer has attracted much attention for developing nanoparticles capable of incorporating and delivering hydrophobic drugs to enhance their therapeutic efficiency (by improving solubility) and reduce their cytotoxic side effects (Basu et al., 2015; Maiti et al., 2014). Amphiphilic polymers, including the natural polysaccharides, can be constructed by grafting hydrophobic moieties into the hydrophilic polymer backbone or producing block polymers that lead to alternating hydrophilic and hydrophobic segments (Varshosaz et al., 2018). In the latter, amphiphilic polymers can form self-assembly nanomicelle structures in water, presenting the inner hydrophobic core that could favor an efficient drug loading capacity and the outer hydrophilic shell that reduces its interaction with serum proteins and prolongs its circulation time in the blood (Zhang et al., 2016; Lai et al., 2012). The hydrophobic core allows drug loading by hydrophobic or electrostatic interactions and determines the drug release rate and other pharmacokinetics parameters (Cho et al., 2009). In general, the drug-loaded stability in polymer-based nanomicelles is high due to the slower dissociation rate of these systems than surfactant-based nanomicelles (Varshosaz et al., 2018); however, the core-shell structure formed by hydrophobic interaction is unstable in the systemic circulation (Kumari and Badwaik, 2019). Nanoparticles obtained from amphiphilic polymers synthesized by grafting hydrophobic molecules into the polymer chains are more stable in the human body than those formed by block polymers. Their self-assembly process allows inter- or intramolecular interactions between several hydrophobic groups through various points of cross-linking to form nanohydrogels instead of forming a typical nanomicelle structure (Varshosaz et al., 2018; Akiyoshi et al., 1993).

GG has been used as a hydrophilic polysaccharide for the design of amphiphilic polymer in both construction approaches. A long alkyl chain (C18) and poly (ethylene glycol) (PEG) were used as a hydrophobic block to produce amphiphilic polymers by conjugating C18 and PEG chains to the GG backbone via etherification reaction (Maiti et al., 2014; Sadhukhan et al., 2015). In contrast, the esterification reaction was used to link hydrophobic moieties, e.g., cholesterol and also lipophilic drugs, to the carboxylic groups of GG backbone (Musazzi et al., 2018; D'Arrigo et al., 2014; D'Arrigo et al., 2012; Manconi et al., 2018b; Manconi et al., 2018a). In both methods, the solubility of the GG-amphiphilic polysaccharide in the aqueous environment is altered, resulting in the formation of GG-nanomicelles or GG-nanohydrogels by a self-assembling process (Manconi et al., 2018a).

The loading of drugs into the GG-nanomicelles is usually performed by the solvent evaporation method. The hydrophobic drug dissolved in an organic solvent (ethanol or chloroform) is added gradually to the aqueous dispersion of GG, which occurs under continuous agitation. Thus, the linking of drug molecules to the hydrophobic blocks of the amphiphilic GG occurs simultaneously with the nanomicelle self-assembly process. At the end of this process, the dispersion is filtered to remove excess insoluble drug (Maiti et al., 2014; Sadhukhan et al., 2015). When lipophilic drugs are chemically conjugated to GG chains, the drug loading process also occurs simultaneously with the nanohydrogel self-assembly process and under a high energy level, which is supplied to the system by ultrasound (bath sonicator for 30 min) (D'Arrigo et al., 2012, 2014). For example, the prednisolone moiety attached to GG chains via a short hydrocarbon chain acts simultaneously as a promoter of the nanohydrogels formation and a pharmacological agent. However, in some amphiphilic polymers, the hydrophobic moiety is not the drug but the site where the lipophilic drug will be bound. Cholesterol is the most chosen hydrophobic molecule for biomedical applications, as it can promote the formation of GG-nanohydrogels without pharmacological activity (D'Arrigo et al., 2012). In this case, nanohydrogels are first prepared by autoclaving (121 °C for 30 min) or ultra-sonication (tip-sonicated for 200 cycles) processes. The drug loading process into the preformed nanohydrogels also occurs through autoclave (121 °C for 20 min) or ultra-sonication (tip-sonicated for 15–25 cycles) processes, but at lower energy levels (shorter autoclave times or sonication cycles) (Musazzi et al., 2018; D'Arrigo et al., 2012, 2014; Manconi et al., 2018a, b). Autoclaving proved to be a more efficient process for the production of smaller and more homogeneously dispersed GG-nanohydrogels compared with those produced by ultrasound, which was probably caused by a difference in the physicochemical interactions of GG chains and the aqueous medium (Manconi et al., 2018a, b).

Ultrasonication has been used as an efficient process to reduce the molecular weight of GG, since the native GG did not lead to the formation of the nanohydrogels due to its high molecular weight and stiffness (D'Arrigo et al., 2012). The mechanism of GG depolymerization by ultrasonication is not entirely clear; however, it is known that intense mechanical forces associated with the cavitation phenomenon promoted the hemolytic chemical bond cleavage of the GG chains, without causing changes in the chemical structure of the polysaccharide (D'Arrigo et al., 2012; Heymach and Jost, 1968; Taghizadeh and Bahadori, 2009). In addition, the reduction in the molecular weight of native GG was strongly associated with the ultrasonication time. The depolymerization strategy led to the achievement of a more adequate and reliable polymer system, which also improved the size and morphology of the formed GG-nanohydrogels (D'Arrigo et al., 2012). Other less common processes can be used to reduce the molecular weight of GG, as the microfluidizer and high-pressure homogenizer. In these high-energy processes, shear forces are applied to disrupt the polymeric bonds for the production of the depolymerized polysaccharide (Musazzi et al., 2018; Costa et al., 2018).

Ionic cross-linking

Ionic cross-linking (or ionotropic gelation) is a relatively simple technique due to its easy experimental procedures and mild preparation conditions. Therefore, it is considered an eco-friendly technique since it generally does not rely on the use of toxic solvents (Liu et al., 2008;

Jătariu et al., 2011; Patil et al., 2010). Ionotropic cross-linking involves the ionic interaction between cationic and anionic polysaccharides and their counterions, originating a hydrogel network (Kuo and Ma, 2001; Berger et al., 2004). Since different polysaccharides began to be used (i.e., alginate, pectin, gellan gum, chitosan) as drug delivery systems, the ionic cross-linking technique has been widely explored for this purpose.

The preparation of GG-based nanoparticles by the ionic cross-linking method is only emerging since traditional ionotropic gelation techniques by mixing or dripping polysaccharide dispersions in saline solutions lead to the production of macro- or microsized GG-hydrogels. Currently, emulsification techniques using high energy levels have been employed to obtain nanosized GG-particles (Vigneshwaran et al., 2019; Pokharkar et al., 2015; Jayaprakash et al., 2014). In addition, smaller sizes have been achieved where double emulsions are prepared instead of single ones. In general, the emulsification cross-linking method consists of forming a primary water-in-oil emulsion by mixing the aqueous dispersion of GG in a nonpolar phase, composed of an organic solvent and an anionic surfactant using an ultrasound device. The most used surfactant is dioctyl sodium sulfosuccinate (AOT). The secondary water-in-oil-in-water emulsion is prepared by sonicating the primary emulsion into an aqueous polyvinyl alcohol solution (PVA). Aqueous saline solution should be added gradually to the secondary emulsion with continuous stirring in order to promote the ionic interaction between calcium ions and the negatively ionized carboxylate groups on the GG chains and thus form nanogels. At the end of this process, nanohydrogels are recovered by centrifugation and washed to remove PVA (Pokharkar et al., 2015; Jayaprakash et al., 2014).

Hydrophilic and hydrophobic drugs were loaded in GG-nanohydrogels produced by the emulsification cross-linking method, presenting both encapsulation efficiency above 70%. The peculiarity of this technique is that depending on the polarity of the drug, it must be mixed with the GG dispersion or the nonpolar phase prior to the first emulsification process (Pokharkar et al., 2015; Jayaprakash et al., 2014). Other factors can also influence the nanoparticles preparation process and properties, including the GG molecular weight, ionic strength, pH, concentration, and cross-linking agent ratio to polysaccharide (Koukaras et al., 2012; Gan et al., 2005).

The other approach to the production of GG-nanohydrogel is the spray gelation method based on ionic cross-linking. This approach is very similar to the ionotropic gelation technique by dripping; however, it allows the production of nanosized structures by nebulization. Based on this, nanohydrogels were obtained from mixtures of GG and pectin by employing air-jet nebulization to produce submicrometric droplets sprayed on the cross-linking solution aluminum chloride ($AlCl_3$). In addition, a poorly water-soluble drug was successfully loaded into pectin-GG-nanohydrogels, although it has previously been solubilized in the aqueous dispersion. High encapsulation efficiency (>80%) was associated with the effective entrapment of the drug in the cross-linked polymer network. The low drug solubility in the aqueous medium also prevented the drug from leaching into the aqueous $AlCl_3$ cross-linking solution (Prezotti et al., 2020).

Polyelectrolyte complexation

Polyelectrolytes are polymers carrying charged ionizable groups in their repeating units (Dobrynin and Rubinstein, 2005). Thus, both cationic and anionic polysaccharides can form polyelectrolyte complexation through strong electrostatic interactions between charged groups of at least two oppositely charged partners without any cross-linking agent (Sarika et al., 2015; Lu et al., 2012). As with the ionic cross-linking method, the advantage of polyelectrolyte complexation is its cost-effectiveness, and its simplicity of technique, since the formation of intermolecular electrostatic interactions occurs through simple mixing of the oppositely charged polysaccharides (Hamman, 2010).

GG was investigated for the preparation of polyelectrolyte complex nanoparticles in different ratios of GG to chitosan and GG to polyethyleneimine (PEI) (Kumar et al., 2016; Goyal et al., 2011; Abioye et al., 2015). As GG carries negative charges when dissolved in water, chitosan and PEI are promising choices for use as positively charged groups to create electrostatic interactions between oppositely charged chains and thus produce nanoparticles. Chitosan is the only natural, cationic, water-soluble, nontoxic, and biocompatible polysaccharide. In addition, it has three types of functional groups: an amino group, and two primary and secondary hydroxyl groups at the C-2, C-3, and C-6 positions, respectively, which are promising reactive sites for electrostatic interactions (Gómez et al., 2006; Liu et al., 2008; Zou et al., 2016). On the other hand, the very high positive charge density of PEI appears to be the main cause of its significant toxicity, limiting its use as a drug and gene delivery system *in vivo*. PEI is a promising candidate as a vector due to its relatively high transfection level in various target organs for several delivery routes. This polymer was complexed with negatively charged GG to partially neutralize its excess positive charge to form GG-PEI nanoparticles (Goyal et al., 2011; Erbacher et al., 2004), to overcome PEI toxicity for *in vivo* applications.

Polyelectrolyte complex nanoparticles' production technique is based on the dripping of one polysaccharide dispersed on the other with an opposite charge under continuous agitation. In addition, to load GG-chitosan nanoparticles, the drug could be mixed with a dispersion of chitosan or chemically associated with chitosan chains before the dripping process (Kumar et al., 2016; Abioye et al., 2015). After that, centrifugation, filter purification, or/and lyophilization processes were applied to obtain drug-loaded nanogels dispersed in water or dried (Kumar et al., 2016; Goyal et al., 2011; Abioye et al., 2015).

The use of chemically modified GG associated with the aforementioned polyelectrolyte complexation technique has been proposed to obtain nanoparticles for specific biomedical applications (Mehnath et al., 2019; Novac et al., 2014). Quaternized GG was synthesized by grafting N-(3-chloro-2-hydroxypropyl)-trimethyl ammonium chloride into the GG primary hydroxyl groups by nucleophilic substitution, in the presence of alkali and under specific reaction conditions. Quaternized GG-chitosan nanohydrogels formed showed retention of quaternary ammonium moieties' antibacterial activity, an essential feature in treating skin infections (Novac et al., 2014). GG maleate was recently formed by the addition of free radical polymerizable groups through esterification reactions, followed by grafting natural silk sericin over the maleate-GG surface. Maleate-GG-silk sericin-chitosan nanohydrogels loaded with rifampicin and pyrazinamide were used to overcome the problems associated with tuberculosis therapy (Mehnath et al., 2019).

Heat cross-linking

GG-based nanofibers have been developed as a novel gastroretentive drug delivery system due to their high surface area to volume ratio and the mucoadhesive nature of GG, which improves drug efficacy and release (Vashisth et al., 2017). GG-nanofibers were produced by the electrospun technique using polyvinyl alcohol (PVA) as a supporting polymer and water as a solvent since both polymers are water-soluble. PVA can decrease the repulsive forces of the highly negatively charged GG dispersion, allowing its electrospinning. The polysaccharides dispersion is electrospun in a high voltage environment through a needle and travels toward the grounded collector during the process. The electrospinning process parameters, including applied voltage, dispersion flow rate, and tip-collector distance, can be easily controlled depending on the desired application, while the physicochemical properties of electrospun nanofibers rely only on the chosen polysaccharide characteristics. (Vashisth et al., 2015; Vashisth and Pruthi, 2016).

Specifically for the production of GG-PVA nanofibers by electrospun technique, a longer tip-collector distance was desirable, as the water takes a long time to evaporate compared to other volatile electrospinning solvents (Vashisth and Pruthi, 2016). Like most other natural polysaccharides-based nanofibers, the GG-PVA nanofibers presented poor mechanical properties and insufficient resistance in the aqueous environment (Frenot and Chronakis, 2003). However, to enhance nanofibers' stability, some cross-linking methods have been applied, such as chemical, heat, and ionic cross-linking. Among all these methods, heat cross-linking is the most efficient method to improve the physicochemical properties of GG-nanofibers and their potential for application as a drug delivery system (Vashisth et al., 2017; Vashisth and Pruthi, 2016). Heat cross-linking induces the release of water molecules linked to polymeric chains, which increases hydrophobic inter- and intramolecular interactions, leading to the formation of a three-dimensional network between polysaccharide chains. Since both polysaccharides are thermostable, the GG-PVA nanofibers submitted to the heat cross-linking method (at 150 °C for 15 min) presented an enhancement in stability and mechanical strength, swelling properties, as well as biocompatibility compared to noncross-linked nanofibers (Vashisth et al., 2017).

Fig. 5.2 shows the different production strategies for the synthesis of GG nanoparticles via self-assembly process, ionic cross-linking, polyelectrolyte complexation, and heat cross-linking. In addition, Section 5.4 of this chapter provides more information about the design and characteristics of GG nanoparticles developed for different biomedical applications, including information on the type of drugs loaded and the administration routes.

5.3.2 Characterization of gellan gum nanoparticles

Numerous analytical techniques have been used to determine the properties of GG nanoparticles such as size, polydispersity, zeta potential, drug entrapment efficiency, *in vitro/ex vivo/in vivo* drug release, swelling behavior, critical aggregation concentration, Fourier transform infrared spectroscopy (FTIR), differential scanning calorimetry, nuclear magnetic resonance, microscopy techniques, and others. Thus, several analyses can be applied to evaluate nanoparticles depending on their specific biological application. This topic describes the most commonly used tests for characterizing GG nanoparticles, including FTIR, morphology, zeta potential, size, polydispersity, drug entrapment efficiency, and drug release *in vitro*.

FTIR spectroscopy is a suitable technique used to analyze the functional groups of polysaccharides by absorption spectroscopy in the infrared region. Intermolecular interactions

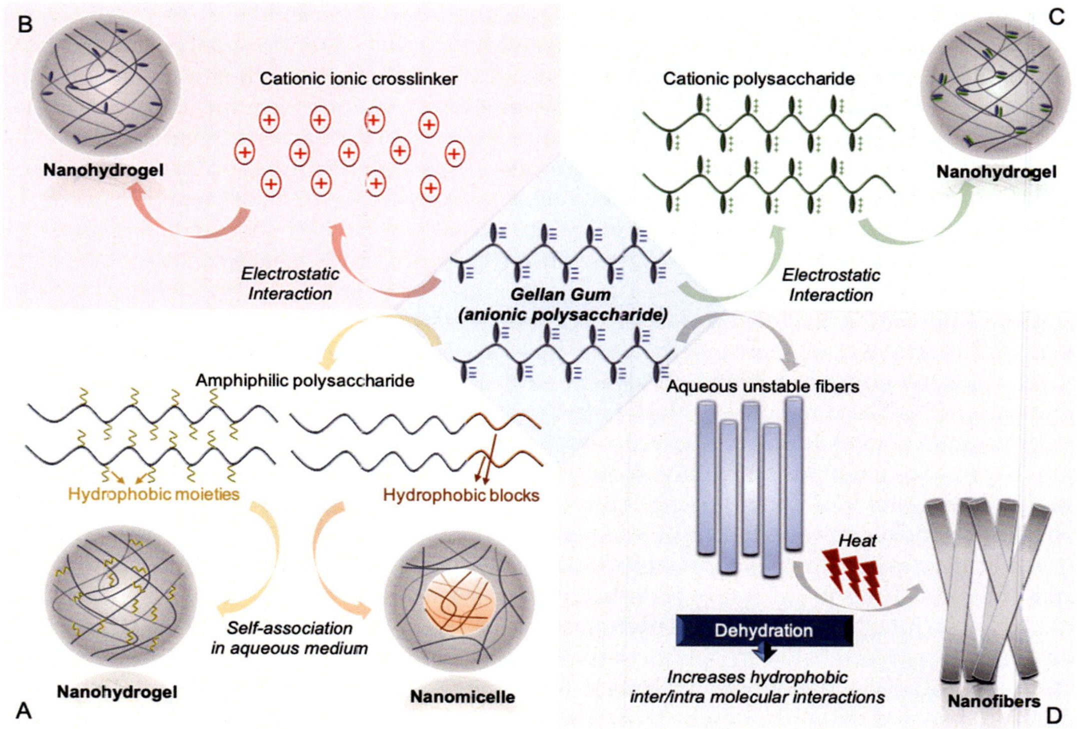

FIG. 5.2 Schematic view of gellan gum-based nanoparticle synthesis. (A) Self-assembly process to produce nanomicelle and nanohydrogel; (B) ionic cross-linking; (C) polyelectrolyte complexation; and (D) heat cross-linking.

affect the stretching or vibration of functional groups in GG chains, which can be detected by FTIR analysis (Vigneshwaran et al., 2019; Mehnath et al., 2019). By analyzing the FTIR spectrum, chemical changes in the GG chains by quaternary ammonium salt and maleic anhydride were confirmed (Mehnath et al., 2019; Novac et al., 2014), as well as the synthesis of amphiphilic GG by chemical conjugation of GG with long alkyl chain (C18), poly (ethylene glycol) (PEG), and hydrophobic moieties (Musazzi et al., 2018; Sadhukhan et al., 2015; Maiti et al., 2014). Nanoparticles were also analyzed in the FTIR instrument to investigate the chemical interactions between the entrapped drug and the polymeric matrix (Kumar et al., 2016; Vigneshwaran et al., 2019; Pokharkar et al., 2015) and confirm the presence of different polymers (cationic and anionic) in nanoparticles produced by polyelectrolyte complexation (Goyal et al., 2011; Dahiya et al., 2017).

The shape and morphological characteristics of nanoparticles have been observed by various microscopy techniques, including atomic force microscopy (AFM) (Garg et al., 2019; Goyal et al., 2011; Sehgal et al., 2017), scanning electron microscopy (SEM) (Prezotti et al., 2020; Vigneshwaran et al., 2019; Novac et al., 2014), high resolution scanning electron microscopy (HR-SEM) (Mehnath et al., 2019), transmission electron microscopy (TEM) (Kumar et al., 2016; Pokharkar et al., 2015; Mehnath et al., 2019; Dahiya et al., 2017), and cryogen electron transmission microscopy (cryo-TEM) (Sadhukhan et al., 2015). Usually, GG nanoparticles are spherical or almost spherical, especially those produced by the self-assembly or polyelectrolyte complexation (Kumar et al., 2016; Prezotti et al., 2020; Garg et al., 2019; Sadhukhan et al., 2015; Goyal et al., 2011; Mehnath et al., 2019; Maiti et al., 2014; Dahiya et al., 2017). However, ionic cross-linking nanoparticles have shown aggregation and irregular shape with a slightly rough surface (Vigneshwaran et al., 2019; Pokharkar et al., 2015).

The aggregation of nanoparticles due to interparticle van der Waals attractions can impair the long-term stability of systems. However, nanoparticles with high surface charge can prevent aggregation by electrostatic stabilization. Zeta potential is a measure of the surface charge of dispersed particulate systems by electrophoretic mobility. Dispersions with zeta potential values greater than +25 mV or less than −25 mV can guarantee the stability of nanoparticles during storage (Manconi et al., 2018a; Maiti et al., 2014). Due to the presence of negatively charged carboxyl groups in GG structure, nanoparticles produced from this polysaccharide might exhibit negative zeta potential (values varying between <−7 and −70 mV). However, they may also have positive zeta potential (<+5 to +70 mV) depending on the other ingredients added and the techniques applied to GG nanoparticles production (Table 5.1). It is already proven that the surface charge is one of the most critical factors that can influence the GG nanoparticles biodistribution, their interaction with biological barriers, and organ accumulation. Despite the unfavorable interactions between the negatively charged particles and the cell membrane (also negatively charged), it is believed that their internalization occurs through nonspecific binding and clustering of the nanoparticles at the fewer cationic sites on the plasma membrane and their subsequent endocytosis (D'Arrigo et al., 2012).

Size is also an important parameter to determine the interaction of nanoparticles with the cell membrane and their penetration when crossing physiological barriers, depending on the tissue, the target site, and the circulation (Kumari et al., 2010). Generally, nanometric carriers also comprise submicro particles below 1000 nm in size and various morphologies, including nanomicelles, nanohydrogels, nanofibers, etc. The submicron size of nanoparticles offers several advantages over microparticles, including relatively higher intracellular uptake (Pinto Reis et al., 2006). The degree of uniformity of the particle size distribution can be determined by measuring the polydispersity index. This parameter can also indicate nanoparticle aggregation and its stability during storage. Samples with a broad size distribution have polydispersity index values greater than 0.7, while for monodisperse particles, these values can vary from 0.01 to 0.5–0.7 (Sreeram et al., 2008). Both size distribution and polydispersity index are usually determined by photon correlation spectroscopic analysis, which analyzes the fluctuations in the intensity of the light backscattered by the dispersed particles (Sreeram et al., 2008; Manca et al., 2014). As shown in Table 5.1, GG nanoparticles have a wide range of size and polydispersity index values. These parameters were highly dependent on the GG molecular mass, the nanoparticle production process, the chosen ingredients, their proportions in the formulation, and the type of encapsulated drug and its interactions with the matrix.

Controlling the size and polydispersity of GG nanoparticles and other properties, such as porosity, mechanical stability, swelling, and response to different physical stimuli, offers a wide of possibilities for the regulation of drug entrapment efficiency and its release kinetic (Milivojevic et al., 2019). The *in vitro* drug diffusion is the most common and simple test to elucidate the release kinetics of the drugs from the loaded GG nanoparticles. The drug release

TABLE 5.1 Brief description of the process of obtaining the chemically modified gellan gum (GG), the formation process of GG nanoparticles and their characterization in terms of size, polydispersity index (PI), zeta potential, and drug entrapping efficiency (DEE).

Nanoparticle	Chemical modification of GG	GG nanoparticles formation process	Size, PI	Zeta potential	DEE	References
Hydrogel	GG–prednisolone or GG-cholesterol: obtained by dissolving GG in *N*-methyl-2-pyrrolidone (NMP) and by adding Br–butyric-prednisolone or Br–butyric-cholesterol dissolved in NMP. The reaction occurred under specific conditions	*Self-assembly*: GG–prednisolone or GG-cholesterol were dispersed in filtered bidistilled water and the suspensions were sonicated for 30 min in an ultrasonic bath sonicator	GG-prednisolone: 320 nm, PI = 0.274	GG-prednisolone: −20.7 mV	–	Palumbo et al. (2020)
			GG-cholesterol: 340 nm, PI = 0.311	GG-cholesterol: −26.4 mV		
			GG-prednisolone: 285 nm, PI = 0.214	GG-prednisolone: −20.7 mV		Jay et al. (1998)
			GG-cholesterol: 340 nm, PI = 0.199			
		Self-assembly: GG-cholesterol was dispersed in phosphate-buffered saline and the suspension was: (a) tip-sonicated for 200 cycles, 5 s on/2 s off, at 13 μm of probe amplitude, with a high intensity ultrasonic or (b) autoclaved at 121 °C for 30 min	∼30–120 nm	∼−50 mV	∼45%	Kumar et al. (2016)
			∼350 nm, PI <0.30	∼−30 mV	∼37%	Prezotti et al. (2020)

	GG-cholesterol: obtained by linking cholesterol (esterified with 4-bromobutyric acid) to the polymer chains in *N*-methyl pyrrolidone.	*Self-assembly:* GG-cholesterol or GG-riboflavin were dispersed in distilled water and the suspensions were autoclaved for 20 min at 121 °C and 1.10 bar. The resulting suspension was withdrawn and centrifuged	GG-cholesterol: 177 nm	~−40 mV in both cases	GG-cholesterol: 282 μg/mg	33
	GG-riboflavin: obtained by linking riboflavin tetrabutyrate (functionalized with 1,6-dibromohexane) to the polymer chains		GG-riboflavin: 178 nm		GG-riboflavin: 301 μg/mg	
Micelles	A long alkyl chain (C18) was grafted onto GG polysaccharide by etherification reaction	*Self-assembly:* GG-nanomicelles structures were synthesized by the solvent evaporation method	371–750 nm	−67.2 to −48.3 mV	>95%	Picone and Cunha (2011)
	Chlorinated poly (ethylene glycol) was grafted onto GG polysaccharide by etherification reaction		426.8–912.6 nm, PI ≤0.761	−28 mV	18.86%	Graham et al. (2019)
Hydrogel	*Quaternized GG*: synthesized by grafting N-(3-chloro-2-hydroxypropyl)-trimethyl ammonium chloride onto GG primary hydroxyl groups by nucleophilic substitution, in the presence of alkali, under specific reaction conditions	*Polyelectrolyte complexation*: chitosan (cationic polymer) was added dropwise to the quaternized GG (anionic polymer) using a peristaltic pump at a flow rate of 1 mL/min under stirring	5–46 nm	–	–	Vigneshwaran et al. (2019)
	Maleate-GG: synthesized by esterification, adding of free radical polymerizable groups in the presence of acetone. Natural silk sericin was also grafted onto the maleate-GG surface to produce sericin functionalized maleate-GG	*Polyelectrolyte complexation:* maleate-GG-silk serin (anionic polymer) was added dropwise to the chitosan solution (cationic polymer) under stirring	160–180 nm, PI = 0.26	−6.48 mV	~0.2% to 0.9%	Berger et al. (2004)

Continued

TABLE 5.1 Brief description of the process of obtaining the chemically modified gellan gum (GG), the formation process of GG nanoparticles and their characterization in terms of size, polydispersity index (PI), zeta potential, and drug entrapping efficiency (DEE)—cont'd

Nanoparticle	Chemical modification of GG	GG nanoparticles formation process	Size, PI	Zeta potential	DEE	References
Hydrogel	–	*Polyelectrolyte complexation*: a preheated solution of branched polyethylenimine (PEI, cationic polymer) was added dropwise a preheated dispersion of GG (anionic polymer) with continuous stirring	122–322 nm	+21.3 to +31.5 mV	–	Taghizadeh and Bahadori (2009)
		Polyelectrolyte complexation: the ternary nanogels were prepared by incorporating ibuprofen-chitosan nanoconjugate to the GG added with propylene glycol under continuous stirring	14.15–677.82 nm	–	–	Costa et al. (2018)
		Polyelectrolyte complexation: GG (anionic polymer) was added to the chitosan dispersion (cationic polymer) with continuous stirring	155.7 nm	+32.1 mV	–	Morris et al. (2012)
		Ionic cross-linking: GG nanoparticles were prepared using the emulsion cross-linking technique. Double emulsions (w/o/w) were produced by sonication, using an anionic surfactant (Aerosol OT™, AOT) and polyvinyl alcohol. Subsequently, aqueous calcium chloride solution (cross-linker) was added to the double emulsion under stirring	125–900 nm	−52 to −35 mV	4%–98%	Cho et al. (2009)
			180 nm, PDI 0.128	−22.6 mV	45.73%–80.07%	Lai et al. (2012)

Hydrogel		*Ionic cross-linking*: GG nanoparticles were synthesized by the solvent evaporation emulsification method. GG was dropwise to the acetone solution to produce single emulsions (w/o). Subsequently, aqueous calcium chloride solution (cross-linker) was added to the double emulsion under stirring	2–3 nm, 100–200 nm, and 400–900 nm	−18.1 mV	–	Varshosaz et al. (2018)
		Polyelectrolyte complexation and ionic cross-linking: GG and xanthan were encapsulated with dexamethasone disodium phosphate-chitosan nanoparticles and calcium chloride was added as a cross-linker. Subsequently, nanoparticles were poured inside the porous baghdadite scaffold	<100 nm	–	25%	Lu et al. (2012)
		Polyelectrolyte complexation and ionic cross-linking: GG was added dropwise to the chitosan–$MgCl_2$ solution under continuous gentle stirring at room temperature	250–720 nm	+61 to +72.9 mV	91.85%	Sarika et al. (2015)
		Polyelectrolyte complexation and ionic cross-linking: nanogels were obtained from mixtures of GG and pectin by employing air-jet nebulization to produce submicrometric droplets that were sprayed on the cross-linking solution of $AlCl_3$	337 nm, PDI 0.973	+5.87 mV	82.05%	Vilela et al. (2015)

Continued

TABLE 5.1 Brief description of the process of obtaining the chemically modified gellan gum (GG), the formation process of GG nanoparticles and their characterization in terms of size, polydispersity index (PI), zeta potential, and drug entrapping efficiency (DEE)—cont'd

Nanoparticle	Chemical modification of GG	GG nanoparticles formation process	Size, PI	Zeta potential	DEE	References
Fibers		*Heat cross-linking*: GG-nanofibers were produced by the electrospun technique using polyvinyl alcohol (PVA) as a supporting polymer and water as a solvent. Subsequently, GG-PVA nanofibers were submitted to heat cross linking method (at 150 °C for 15 min)	Average fiber diameter from 25 to 40 nm	–	78.47%	Pokharkar et al. (2015)

behavior is usually obtained following the dialysis method using a molecular weight cut-off membrane. The nanoparticle can be dispersed in phosphate-buffered saline buffer or other simulate fluids such as simulated body fluid and simulated gastric or intestinal fluids, and then placed on a dialysis membrane. At predetermined time intervals, suitable aliquots are withdrawn and then assessed by spectrometry or chromatography methods (Garg et al., 2019; Vigneshwaran et al., 2019; Maiti et al., 2014; Dahiya et al., 2017). The drug release profile depends, besides the factors mentioned, on the initial amount of drug-loaded into the nanoparticle. A low drug entrapment efficiency can be a limiting attribute to carry drugs in nanoparticles since the expected biological effects may not be achieved. In general, drug entrapment efficiency is determined by calculating the free drug concentration in the supernatant compared to the total drug. The free drug in the formulation can be separated by centrifugation or dialysis and the collected supernatant analyzed by spectrometry or chromatography methods (Manconi et al., 2018a, b; Pokharkar et al., 2015; Dahiya et al., 2017). The values of drug entrapping efficiency (DEE), zeta potential, polydispersity index (PI), size of the GG nanoparticles, and a brief description of their formation process are summarized in Table 5.1.

5.4 Gellan gum nanoparticles (GG nanoparticles) in drug delivery

5.4.1 GG nanoparticles in oral drug delivery

The oral drug administration is considered the preferred route due to its ease of administration and convenience, which results in greater patient compliance than invasively administered drug delivery systems. In addition, oral delivery formulations must provide effective drug concentrations for a long time in systemic circulation and safely (Milivojevic et al., 2019; Ma and Williams, 2018). The advantages of using nanoparticles compared to conventional oral delivery forms include increased drug mucoadhesion and retention in the gastrointestinal tract (GIT) as well as overcoming limitations associated with the drug release at the wrong place due to its premature absorption and/or metabolization (Prezotti et al., 2020; Gamboa and Leong, 2013). In general, an oral drug delivery system based on nanoparticles should retain suitable stability in the GIT to achieve absorption in the intestines, adhere to mucus membranes, and have suitable permeability to be absorbed into the circulatory system (Ma and Williams, 2018).

GG nanoparticles have been widely studied for oral delivery of drugs with hydrophilic or lipophilic characteristics (D'Arrigo et al., 2012, 2014; Dahiya et al., 2017) or poor stability toward enzymatic content of the GIT (Prezotti et al., 2020; Sadhukhan et al., 2015) as well as for the development of multiple drugs and gastroretentive/mucoadhesive systems, mainly for delivery and/or retention of drugs at specific GIT sites (Prezotti et al., 2020; D'Arrigo et al., 2012, 2014; Vashisth et al., 2017). The choice of GG in these applications is based on its properties, including biocompatibility, mucoadhesiveness, and resistance to the acidic environment and enzymes (i.e., pectinase, amylase, lipase, etc.) of the GIT (Prezotti et al., 2020; Yang et al., 2013).

Recently, GG-nanohydrogel was developed by a combination of polyelectrolyte complexation (chitosan and gellan gum) and ionic gelation technique (magnesium dichloride ($MgCl_2$)

used as cross-linker) for oral delivery of the hydrophilic drug, epigallocatechin gallate (EGCG). EGCG is an antioxidant compound known to act against many diseases without exhibiting any side effects. The combination of GG with chitosan improves encapsulation efficiency and mucoadhesiveness of EGCG-loaded nanohydrogel and the sustained drug release behavior compared to pure EGCG (Dahiya et al., 2017).

GG-nanohydrogel and other GG nanoparticles have also gained attention as a vehicle for lipophilic drugs (even multiple drugs) to overcome the difficulties associated with low drug solubility and biological barriers to oral drug absorption. For example, GG-nanomicelles formed by conjugating the poly (ethylene glycol) chain to the GG backbone have been described as promising systems for increasing drug solubility due to the combined hydrophobic/hydrophilic domains. Simvastatin is an antihyperlipidemic drug with poor water solubility and low oral bioavailability due to extensive first-pass metabolism in the intestinal gut and liver. *In vivo* experiments showed that simvastatin-loaded poly (ethylene oxide)-*g*-GG nanomicelles performed better than the pure drug in reducing total cholesterol, low-density lipoprotein, and triglyceride levels in hyperlipidemic rabbits, which was mainly attributed to the improvement in drug solubility (Sadhukhan et al., 2015). In another study, GG-nanohydrogels were designed for the simultaneous delivery of both anticancer and antiinflammatory drugs (D'Arrigo et al., 2014). Prednisolone, an antiinflammatory drug poorly soluble in water, was chemically cross-linking with GG carboxylic groups to act as a hydrophobic portion promoting their self-assembling process, while paclitaxel was physically encapsulated into GG-nanohydrogel. Prednisolone-paclitaxel-loaded nanogels overcame solubility difficulties of prednisolone and improved drugs uptake into the cells, enhancing the bioavailability and *in vitro* cytotoxic effect on tumor cells due to the synergistic effect of both drugs (D'Arrigo et al., 2012, 2014).

GG nanoparticles have also been developed for the preferential release of encapsulated drugs in the upper (gastric) or lower (intestinal) parts of the GIT (Prezotti et al., 2020; Vashisth et al., 2017). Nanoparticles based on GG-pectin mixture were developed for colon-targeted release of resveratrol. Resveratrol-loaded GG-pectin nanohydrogel promoted effective *in vitro* modulation of drug release rates and favored its retention in rat intestinal tissue, desirable attributes to target resveratrol to the colon (Prezotti et al., 2020). Although GG is a promising material for drug delivery to the lower parts of the GIT (due to its resistance to acidic pH and gastric enzymes), GG nanoparticles have been explored as an alternative to acting as a novel gastroretentive drug delivery system. The relative mucoadhesive nature and superior biocompatibility of GG-based nanofibers compared to other forms (hydrogel and films) allowed the ofloxacin-loaded GG-polyvinyl alcohol nanofibers to remain in rat gastric mucosa for a long time. The gastroretentive property and sustained release of the drug from these nanofibers can improve the pharmacokinetics of the dosage form in the GIT compared to the pure drug (Vashisth et al., 2017).

5.4.2 GG nanoparticles in topical (dermal and transdermal) drug delivery

Topical delivery includes dermal (local) drug administration for treating skin diseases and transdermal (systematic) when the skin is used as the site for drug administration.

Transdermal delivery is a noninvasive drug administration and an excellent alternative to oral administration, reducing the systematic side effects and providing sustained delivery in lower drug exposure levels. In addition, drug bioavailability is improved by avoiding gastrointestinal and hepatic first-pass metabolic degradation, and may enhance the safety of the treatments when the orally administrated drug causes gastrointestinal adverse effects and hepatic damage (Milivojevic et al., 2019; Musazzi et al., 2018; Bajaj et al., 2011). Despite these advantages, transdermal drug administration remains a challenge since human skin is a natural protective barrier of low absorption, mainly in its upper layers. The upper skin layers, denominated *stratum corneum*, are composed of flattened dead cells and arranged as "bricks" of hydrophilic proteins embedded in a hydrophobic lipid "mortar" (Abioye et al., 2015; Heisig et al., 1996; Carmona-Moran et al., 2016). Thus, in order to increase the drug permeation through the skin, some strategies have been used (Salim et al., 2012; Carafa et al., 2004; Mahdi et al., 2016; Li et al., 2001), including the use of penetration enhancers (such as dimethyl sulfoxide (DMSO), isopropyl alcohol, ethanol, and propylene glycol) in the drug formulation (Williams and Barry, 2012; Stott et al., 1998) and the loading of drugs into nanogels.

The drug-loaded nanogel penetration mechanism through the skin is not fully understood but may be related to the physicochemical stability of the nanostructure, the loss of its integrity with skin depth, and a copenetration of its components (Rancan et al., 2016; Šmejkalová et al., 2017). Thus, the drug permeation rate is mainly modulated by the delivery systems, resulting in the composition and concentration of the gelling polymer (Abioye et al., 2015; Carmona-Moran et al., 2016; Asbill and Bumgarner, 2007). In addition, it is known that its nanoscale structure improves the drug concentration in the dermal layer, which can also favor the permeation rate of loaded drugs (Vigneshwaran et al., 2019).

Recently, delivery systems based on GG-nanohydrogels have been developed for topical drug administration in healing of wounds (Manconi et al., 2018a, b), skin infections (Kumar et al., 2016; Novac et al., 2014), and antiinflammatory applications (Abioye et al., 2015). GG nanoparticles were obtained from: (i) GG in the native state (Vigneshwaran et al., 2019); (ii) GG chemically modified by reaction with quaternary ammonium, cholesterol, or riboflavin groups (Musazzi et al., 2018; Manconi et al., 2018a, b; Novac et al., 2014); or (iii) GG ionic gelation using chitosan or branched polyethylenimine as cationic polymers (Goyal et al., 2011; Abioye et al., 2015).

Ex vivo release studies on pig skin have evidenced the potential of chitosan-ibuprofen-GG nanohydrogels (produced by drug-polymer nanoassembly and ionic gelation) to increase the transdermal delivery of antiinflammatory ibuprofen by improving skin penetration/retention, percutaneous drug release through particle size reduction, and conversion of the crystalline drug into amorphous particles (Abioye et al., 2015). In another study, branched polyethyleneimine (PEI) was shown to interact electrostatically with GG acid to form nanogels that were used for gene delivery. Plasmid DNA-loaded PEI-GG nanohydrogels showed the best transfection efficiency in tested cell lines (HEK293, HeLa, and HepG2 cells) and primary mouse skin cells (keratinocytes) compared with polyethyleneimine, Lipofectamine, and other commercial transfection agents, and also exhibited minimum cytotoxicity (Goyal et al., 2011).

In vitro transdermal release tests have confirmed the ability of quarternized GG-chitosan nanohydrogels and cholesterol- or riboflavin-conjugated GG-nanohydrogels to sustain the release of the antibiotic ciprofloxacin and the antiinflammatory piroxicam, respectively, through the dermal layers (Musazzi et al., 2018; Manconi et al., 2018a, b; Novac et al., 2014). In particular, piroxicam has been proposed to treat nonmelanoma skin cancers, but it is a very poorly permeable compound. Grafting hydrophobic moieties (cholesterol or riboflavin) in GG chains promotes the self-assembly process, which results in the formation of the GG-nanohydrogel capable of incorporating the piroxicam into nanogels by hydrophobic interactions between drug molecules and hydrophobic polymer zones. In the topical delivery, the hydrophobic moieties of the cholesterol- or riboflavin-conjugated GG-nanohydrogels interact by cohesive forces with the lipidic matrix of the upper skin layers (*stratum corneum*) and then gradually disassemble, thus diffusing in the epidermis layers. (Musazzi et al., 2018).

The potential of GG nanoparticles in wound healing was evaluated by producing GG-cholesterol nanohydrogels loaded with the antiinflammatory baicalin. The bacailin-loaded GG-cholesterol nanohydrogel supports 3 T3 fibroblast cell growth and counteracts the toxic effect induced by hydrogen peroxide in cells. In addition, topical administration of the nanohydrogel onto TPA-injured mouse skin led to complete wound healing. Probably, the skin repair occurs due to several factors; among them the nanogel acts as a moisturizing agent in the *stratum corneum*, which leads to the drug accumulation in the epidermis layers, followed by its diffusion in the dermis where it can counteract oxidative stress, regulating the process of inflammation (Manconi et al., 2018a).

5.4.3 GG nanoparticles in ophthalmic and intranasal drug delivery

Ophthalmic drug delivery is the most challenging administration route due to reduced intraocular bioavailability and high drug dilution and elimination rates caused by limited corneal permeability, rapid tear renewal, and blinking (Osmałek et al., 2014; Pijls et al., 2005). Thus, to achieve a therapeutic effect similar to conventional oral delivery systems, frequent applications are mandatory, leading to problems with patient compliance or drug overdosing (Gupta et al., 2010; Sanzgiri et al., 1993). Despite these disadvantages, ocular delivery is one of the most desirable drug administration routes due to its easy preparation/application and high patient convenience (Milivojevic et al., 2019). Thus, several drug delivery systems have been developed to increase ocular bioavailability by prolonging drug residence time and penetration, including in situ gels, liposomes, nanoemulsions, nanosuspensions, and nanogels.

GG has been successfully applied as an ocular delivery system due to its properties, including transparency, thickening, and gelling (Singh et al., 2009). In many cases, GG is used as a component of complex ocular formulations with other polymers, such as alginate, carboxymethyl cellulose, carrageenan, and chitosan to produce micro/macrogels with appropriate gel strength, mucoadhesiveness, and prolonged drug retention/permeation at the corneal site (Gupta et al., 2010; Kesavan et al., 2010; Luaces-Rodríguez et al., 2017). Ocular formulation based on GG-nanoparticles is rare; however, it is known that appropriate particle size and narrow size distribution may also ensure less irritation and blurred vision (Guinedi et al., 2005).

Thus, GG nanoparticles were prepared using the emulsion cross-linking technique and an anionic surfactant (Aerosol OT™, AOT) for ophthalmic delivery of the antibiotic doxycycline hydrochloride (DXY). Minimum inhibitory concentration assays have shown that DXY-loaded GG-AOT nanohydrogels inhibited bacterial growth at very low concentrations than that of the pure drug. In addition, *ex vivo* diffusion tests have suggested that these nanoparticles can release the drug for a sustained period of time. GG-AOT nanohydrogels also produce minor irritation and opacity in rabbit eyes, confirming their safety in treating eye infections (Pokharkar et al., 2015).

Conventionally, the drug administration through the nasal cavity is used to treat local diseases, such as rhinitis, sinusitis, and nasal infections and congestion (Illum, 2003; Abbas and Marihal, 2014; Bhise et al., 2008). Recently, intranasal drug delivery has attracted considerable attention as an alternative route to systemic therapy due to its numerous advantages such as relatively large surface area and high blood flow, thus achieving rapid absorption and avoiding first-pass metabolism and a harsh GIT environment. However, the mucociliary clearance is the primary barrier for drug delivery by this route, hindering drug absorption and retention (Hao et al., 2016; Hosny and Hassan, 2014; Devkar et al., 2014).

GG has been generally adopted as an in situ gelling matrix for intranasal formulations to enhance the nasal mucosal permeability, improving drug bioavailability and the residence time (Patil and Sawant, 2009). These improvements are achieved by administering low viscosity aqueous solutions based on polymers in spray into the nasal cavity. The solution will turn into a gel on the nasal mucosa under physiological conditions by different trigger mechanisms such as pH, ionic strength, and temperature (Saindane et al., 2013; Hosny, 2009). The in situ nasal gels are currently combined with nanotechnology delivery systems, such as nanosized microemulsion (Hosny and Hassan, 2014), nanosuspension (Hao et al., 2016; Saindane et al., 2013), and cubosomes (Patil et al., 2019), in order to enhance their pharmacodynamic activity and mucoadhesiveness.

The relatively new approach to nasal drug delivery is the dry intranasal nanoparticle systems, including those developed from GG. They are based on the in situ transition from a dry powder to a swollen gel when the dry nanoparticles come into contact with the nasal fluid (Mehnath et al., 2019; Maiti et al., 2014). GG-nanomicelles were formed by grafting long alkyl chains (C18) onto the GG polysaccharide and then loaded with budesonide, a synthetic antiinflammatory corticosteroid used to treat allergic rhinitis. Budesonide-loaded GG-nanomicelles exhibited a more extended dissolution profile in simulated nasal fluid (pH 5.5) and prolonged *in vitro* drug release potential, which could avoid repetitive application of conventional nasal sprays and provide long-term relief to patients with allergic rhinitis symptoms (Maiti et al., 2014). Recently, maleate GG-silk sericin-chitosan nanohydrogels loaded with rifampicin and pyrazinamide were developed to overcome the hepatotoxicity problems associated with these drugs used in tuberculosis therapy. Hepatotoxicity is a side effect caused by the action of the drug on hepatocytes rather than alveolar macrophages cells that harbor *Mycobacterium tuberculosis*. The proposed nasal sustained-release system was composed of multifunctional biomaterials loaded with a high drug amount. This composition generated a stimuli-responsive nanosystem and effectively binds with tuberculosis cells, penetrates them, and releases both drugs, leading to disruption of the cell membrane (Mehnath et al., 2019).

5.4.4 GG nanoparticles in other drug delivery routes

GG has been widely applied as an in situ injectable nanoparticle-loaded hydrogel system for local delivery of drugs in osteomyelitis and osteoporosis treatment and cartilage regeneration applications (Posadowska et al., 2015, 2016a, b; Kouhi et al., 2020). In situ injectable hydrogels are applied as a minimally invasive injection in the defect region, then the formed hydrogel quickly adapts its shape to the site, establishing an efficient integration with the host tissue (Oliveira et al., 2008).

Less common formulations based on GG nanoparticles have been developed for intravenous administration and scaffolds implants (Garg et al., 2019; Goyal et al., 2011; Sehgal et al., 2017). Heparin was anchored to GG chains, and the usnic acid-loaded heparin-GG nanohydrogels were prepared (Garg et al., 2019). Heparin has been used for tumor targeting due to its capability to bind specifically to the receptors present in the tumor cell, whereas usnic acid is an anticancer drug that presents limited therapeutic use due to its high hepatotoxicity. Usnic acid-loaded heparin-GG nanohydrogels were administered through intravenous route in albino rats. The usnic acid trapping in nanoparticles has been shown to increase its antitumor activity and significantly reduce its hepatotoxicity (Garg et al., 2019; Santos et al., 2005). In addition, heparin-GG nanohydrogels have sustained the drug release, which may be recognized due to hydrophilic heparin coating on the GG, causing long circulation (Garg et al., 2019).

Branched polyethyleneimine (PEI)-GG nanohydrogels used as topical gene delivery have also been tested as intravenous gene delivery systems. After intravenous injection of plasmid DNA-loaded PEI-GG nanogels, naked DNA, or polyethyleneimine, it was observed that nanogels were the best delivery agent among the three, which may be due to the presence of the GG portion that prevents interactions with blood serum proteins, increasing blood circulation time and resulting in more expression of the target gene (Goyal et al., 2011).

A more complex drug delivery system based on a tissue-engineered scaffold was developed to provide sustained delivery of an osteoinductive drug, dexamethasone disodium phosphate (DXP). DXP-encapsulated chitosan nanoparticles (DXP-CN) using nanostructured gellan gum and xanthan hydrogel (GX) were incorporated in porous baghdadite ($Ca_3ZrSi_2O_9$) scaffolds, a zirconia-modified calcium silicate ceramic. Fig. 5.3 shows a schematic diagram of the preparation and optical images of a nanostructured DXP-CN-GX-baghdadite scaffold. The optimization of the cross-linker and GG-xanthan concentrations led to a homogeneous distribution of hydrogel coating within baghdadite scaffolds, which resulted in sustained delivery of DXP (78 ± 6% over 5 days) compared with free DXP loaded in uncoated baghdadite scaffolds (92 ± 8% release in 1 h) (Sehgal et al., 2017).

Table 5.2 summarizes most of the currently available publications related to the application of GG nanoparticles in drug delivery.

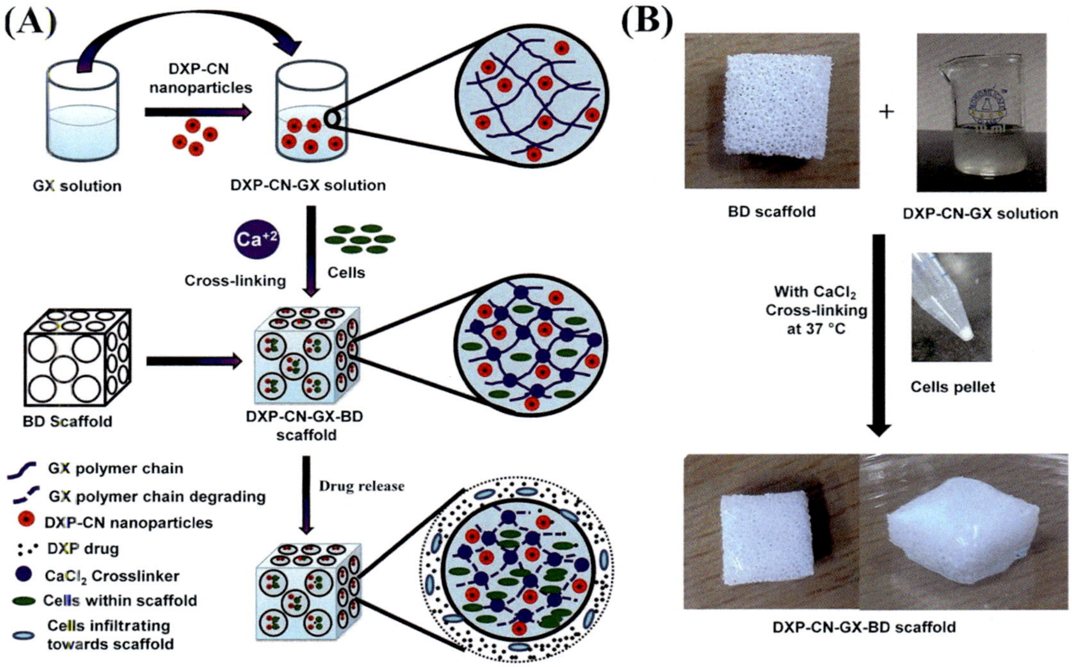

FIG. 5.3 (A) Schematic diagram showing the preparation of nanostructured dexamethasone disodium phosphate (DXP)-chitosan nanoparticles (CN)-gellan-xanthan (GX)-baghdadite (BD) scaffold. (B) Optical images of baghdadite scaffold before and after coating with DXP–CN–GX hydrogel. *From Sehgal, R.R., Roohani-Esfahani, S.I., Zreiqat, H., Banerjee, R., 2017. Nanostructured gellan and xanthan hydrogel depot integrated within a baghdadite scaffold augments bone regeneration. J. Tissue Eng. Regen. Med. 11 (4), 1195–1211. https://doi.org/10.1002/term.2023.*

TABLE 5.2 Gellan gum nanoparticles (GG) as a drug delivery system.

Administration	Polymers	Active ingredient	Application	References
Topical	Quaternized GG/chitosan	Ciprofloxacin	Antibiotic	Vigneshwaran et al. (2019)
	GG/chitosan	Ibuprofen	Antiinflammatory	Costa et al. (2018)
	GG/chitosan	Ketoconazole	Antifungal	Morris et al. (2012)
	GG/cholesterol	Baicalin	Wound healing	Kumar et al. (2016)
	GG/cholesterol	Baicalin	Wound healing	Prezotti et al. (2020)
	GG/cholesterol/riboflavin	Piroxicam	Cutaneous pathologies, antiinflammatory	Picone and Cunha (2011)
	GG	Pyridoxine	Diabetic peripheral neuropathy treatment	Varshosaz et al. (2018)

Continued

TABLE 5.2 Gellan gum nanoparticles (GG) as a drug delivery system—cont'd

Administration	Polymers	Active ingredient	Application	References
Topical/ intravenous	GG/ polyethylenimine	Plasmid DNA	Nonviral gene vector (gene therapy)	Taghizadeh and Bahadori (2009)
Oral	GG	Prednisolone	Antiinflammatory	Palumbo et al. (2020)
	GG	Paclitaxel, prednisolone	Anticancer, antiinflammatory	Jay et al. (1998)
	GG/di-octyl sodium sulfosuccinate	Raloxifene	Breast cancer and osteoporosis treatment	Cho et al. (2009)
	GG/poly(ethylene glycol)	Simvastatin	Antihyperlipidemic	Graham et al. (2019)
	GG/chitosan	Epigallocatechin gallate	Antioxidant, antibacterial	Sarika et al. (2015)
	GG/pectin	Ketoprofen	Antiinflammatory	Vilela et al. (2015)
Oral (nanofibers)	GG/polyvinyl alcohol	Ofloxacin	Antibiotic	Pokharkar et al. (2015)
Ophthalmic	GG	Doxycycline hydrochloride	Antibiotic	Lai et al. (2012)
Intranasal	Maleate-GG/silk sericin/chitosan	Rifampicin, pyrazinamide	Tuberculosis therapy	Berger et al. (2004)
	Long alkyl chains (C18)-grafted GG	Budesonide	Budesonide	Pires Vilela et al. (2011)
Intravenous	Heparin modified GG	Usnic acid	Anticancer	Santos et al. (2020)
Scaffolds implant	GG/chitosan/ xanthan	Dexamethasone disodium phosphate	Bone regeneration	Lu et al. (2012)

5.5 Conclusions

Among the numerous nano delivery systems, polysaccharide-based nanoparticles have gained increasing attention in the field of drug delivery, mainly due to the ease of manufacture, high stability, high water-holding capacity, adjustable chemical and mechanical properties, ability to incorporate both hydrophilic and hydrophobic drugs, and feasibility of administration by different routes. This chapter specifically discussed the preparation and characterization (such as size, polydispersity index, zeta potential, and drug entrapping efficiency) of gellan gum (GG)-nanoparticles and their application in the field of drug delivery. As can be seen, the ability of GG to be transformed into nanoparticles by different techniques offers a promising strategy for combining polysaccharide properties and advantages of nanotechnology, such as site-specific and/or time-controlled delivery of drugs. GG is biodegradable, nontoxic, with thermal resistance and resistance to acid medium, can form gels, and it is mucoadhesive. These characteristics make this polysaccharide a versatile excipient to be used as nanocarriers for different drug administration routes (topical, oral, nasal, ophthalmic, and others). Despite these advantages, GG's full potential as a nanosystem for drug delivery has not yet been explored. There are relatively few scientific reports describing the use of GG nanoparticles in pharmaceutical formulations compared to other polysaccharides (such as alginate) and dosage forms (such as tablet binder, granulating agent, or coating material). This chapter has pointed out the most current paths in the synthesis of GG nanoparticles, their potentialities, and encourages further studies.

Acknowledgments

Ana Letícia Rodrigues Costa Lelis is grateful for the financial support of Coordination for the Improvement of Higher Education Personnel (CAPES) (grant # 88887.473207/2020-00) and São Paulo Research Foundation (FAPESP) (grant # 2020/02313-0). We also thank CAPES finance code 001. Lucimara Gaziolla de la Torre thanks the National Council for Scientific and Technological Development (CNPq) (productivity grant 302212/2019-1).

References

Abbas, Z., Marihal, S., 2014. Gellan gum-based mucoadhesive microspheres of almotriptan for nasal administration: formulation optimization using factorial design, characterization, and in vitro evaluation. J. Pharm. Bioallied Sci. 6, 267–277.

Abioye, A.O., Issah, S., Kola-Mustapha, A.T., 2015. Ex vivo skin permeation and retention studies on chitosan-ibuprofen-gellan ternary nanogel prepared by in situ ionic gelation technique—a tool for controlled transdermal delivery of ibuprofen. Int. J. Pharm. 490, 112–130.

Akiyoshi, K., Deguchi, S., Moriguchi, N., Yamaguchi, S., Sunamoto, J., 1993. Self-aggregates of hydrophobized polysaccharides in water. Formation and characteristics of nanoparticles. Macromolecules 26, 3062–3068.

Asbill, C.S., Bumgarner, G.W., 2007. Transdermal drug delivery. In: Pharmaceutical Manufacturing Handbook: Production and Processes. Wiley, pp. 793–807.

Bacelar, A.H., Silva-Correia, J., Oliveira, J.M., Reis, R.L., 2016. Recent progress in gellan gum hydrogels provided by functionalization strategies. J. Mater. Chem. B 4, 6164–6174.

Bajaj, I.B., Saudagar, P.S., Singhal, R.S., Pandey, A., 2006. Statistical approach to optimization of fermentative production of gellan gum from *Sphingomonas paucimobilis* ATCC 31461. J. Biosci. Bioeng. 102, 150–156.

Bajaj, I.B., Survase, S.A., Saudagar, P.S., Singhal, R.S., 2007. Gellan gum: fermentative production, downstream processing and applications. Food Technol. Biotechnol. 45, 341–354.

Bajaj, S., Whiteman, A., Brandner, B., 2011. Transdermal drug delivery in pain management. Contin. Educ. Anaesth. Crit. Care Pain 11, 39–43.

Basu, A., Kunduru, K.R., Abtew, E., Domb, A.J., 2015. Polysaccharide-based conjugates for biomedical applications. Bioconjug. Chem. 26, 1396–1412.

Berger, J., et al., 2004. Structure and interactions in covalently and ionically crosslinked chitosan hydrogels for biomedical applications. Eur. J. Pharm. Biopharm. 57, 19–34.

Bhise, S., Yadav, A., Avachat, A., Malayandi, R., 2008. Bioavailability of intranasal drug delivery system. Asian J. Pharm. 2, 201–215.

Builders, P.F., Attama, A.A., 2011. Functional properties of biopolymers for drug delivery applications. In: Biodegradable Materials: Production, Properties and Applications. Nova Science Pub Inc, pp. 103–154.

Carafa, M., Marianecci, C., Lucania, G., Marchei, E., Santucci, E., 2004. New vesicular ampicillin-loaded delivery systems for topical application: characterization, in vitro permeation experiments and antimicrobial activity. J. Control. Release 95, 67–74.

Carmona-Moran, C.A., et al., 2016. Development of gellan gum containing formulations for transdermal drug delivery: component evaluation and controlled drug release using temperature responsive nanogels. Int. J. Pharm. 509, 465–476.

Chacko, R.T., Ventura, J., Zhuang, J., Thayumanavan, S., 2012. Polymer nanogels: a versatile nanoscopic drug delivery platform. Adv. Drug Deliv. Rev. 64, 836–851.

Chan, E.S., Lee, B.B., Ravindra, P., Poncelet, D., 2009. Prediction models for shape and size of Ca-alginate macrobeads produced through extrusion-dripping method. J. Colloid Interface Sci. 338, 63–72.

Cho, J.K., Park, W., Na, K., 2009. Self-organized nanogels from pullulan-g-poly(L-lactide) synthesized by one-pot method: physicochemical characterization and in vitro doxorubicin release. J. Appl. Polym. Sci. 113, 2209–2216.

Costa, A.L.R., Gomes, A., Ushikubo, F.Y., Cunha, R.L., 2017. Gellan microgels produced in planar microfluidic devices. J. Food Eng. 209, 18–25.

Costa, A.L.R., Gomes, A., Tibolla, H., Menegalli, F.C., Cunha, R.L., 2018. Cellulose nanofibers from banana peels as a Pickering emulsifier: high-energy emulsification processes. Carbohydr. Polym. 194, 122–131.

D'Arrigo, G., et al., 2012. Self-assembled gellan-based nanohydrogels as a tool for prednisolone delivery. Soft Matter 8, 11557–11564.

D'Arrigo, G., Navarro, G., Di Meo, C., Matricardi, P., Torchilin, V., 2014. Gellan gum nanohydrogel containing anti-inflammatory and anti-cancer drugs: a multi-drug delivery system for a combination therapy in cancer treatment. Eur. J. Pharm. Biopharm. 87, 208–216.

Dahiya, S., Rani, R., Kumar, S., Dhingra, D., Dilbaghi, N., 2017. Chitosan-gellan gum bipolymeric nanohydrogels—a potential nanocarrier for the delivery of epigallocatechin Gallate. Bionanoscience 7, 508–520.

Devkar, T.B., Tekade, A.R., Khandelwal, K.R., 2014. Surface engineered nanostructured lipid carriers for efficient nose to brain delivery of ondansetron HCl using *Delonix regia* gum as a natural mucoadhesive polymer. Colloids Surf. B: Biointerfaces 122, 143–150.

Dobrynin, A.V., Rubinstein, M., 2005. Theory of polyelectrolytes in solutions and at surfaces. Prog. Polym. Sci. 30, 1049–1118.

Dwarakanadha Reddy, P., Swarnalatha, D., 2010. Recent advances in novel drug delivery systems. Int. J. PharmTech Res. 1, 316–326.

Erbacher, P., et al., 2004. Genuine DNA/polyethylenimine (PEI) complexes improve transfection properties and cell survival. J. Drug Target. 12, 223–236.

Frenot, A., Chronakis, I.S., 2003. Polymer nanofibers assembled by electrospinning. Curr. Opin. Colloid Interface Sci. 8, 64–75.

Gamboa, J.M., Leong, K.W., 2013. In vitro and in vivo models for the study of oral delivery of nanoparticles. Adv. Drug Deliv. Rev. 65, 800–810.

Gan, Q., Wang, T., Cochrane, C., McCarron, P., 2005. Modulation of surface charge, particle size and morphological properties of chitosan-TPP nanoparticles intended for gene delivery. Colloids Surf. B: Biointerfaces 44, 65–73.

Garg, A., et al., 2019. Heparin appended ADH-anionic polysaccharide nanoparticles for site-specific delivery of usnic acid. Int. J. Pharm. 557, 238–253.

Giavasis, I., Harvey, L.M., McNeil, B., 2000. Gellan gum. Crit. Rev. Biotechnol. 20, 177–211.

Gómez, L., Ramírez, H.L., Neira-Carrillo, A., Villalonga, R., 2006. Polyelectrolyte complex formation mediated immobilization of chitosan-invertase neoglycoconjugate on pectin-coated chitin. Bioprocess Biosyst. Eng. 28, 387–395.

Gopal, P.N.V., Murthy, K.V.R., Rao, K.R.S.S., 2017. Gellan gum—multifunctional excipient in drug delivery. PHARMANEST Int. J. Adv. Pharm. Sci. 8, 1–10.

Goswami, S., Naik, S., 2014. Natural gums and its pharmaceutical application. J. Sci. Innov. Res. 3, 112–121.

Goyal, R., et al., 2011. Gellan gum blended PEI nanocomposites as gene delivery agents: evidences from in vitro and in vivo studies. Eur. J. Pharm. Biopharm. 79, 3–14.

Graham, S., Marina, P.F., Blencowe, A., 2019. Thermoresponsive polysaccharides and their thermoreversible physical hydrogel networks. Carbohydr. Polym. 207, 143–159.

Grasdalen, H., Smidsrød, O., 1987. Gelation of gellan gum. Carbohydr. Polym. 7, 371–393.

Guinedi, A.S., Mortada, N.D., Mansour, S., Hathout, R.M., 2005. Preparation and evaluation of reverse-phase evaporation and multilamellar niosomes as ophthalmic carriers of acetazolamide. Int. J. Pharm. 306, 71–82.

Gupta, H., Velpandian, T., Jain, S., 2010. Ion-and pH-activated novel in-situ gel system for sustained ocular drug delivery. J. Drug Target. 18, 499–505.

Hamman, J.H., 2010. Chitosan based polyelectrolyte complexes as potential carrier materials in drug delivery systems. Mar. Drugs 8, 1305–1322.
Hao, J., et al., 2016. Fabrication of an ionic-sensitive in situ gel loaded with resveratrol nanosuspensions intended for direct nose-to-brain delivery. Colloids Surf. B: Biointerfaces 147, 376–386.
Haywood, A., Glass, B.D., 2011. Pharmaceutical excipients—where do we begin? Aust. Prescr. 34, 112–114.
Heer, D., 2013. Novel excipients as different polymers: a review. J. Drug Deliv. Ther. 3, 202–207.
Heisig, M., Lieckfeldt, R., Wittum, G., Mazurkevich, G., Lee, G., 1996. Non steady-state descriptions of drug permeation through stratum corneum. I. The biphasic brick-and-mortar model. Pharm. Res. 13, 421–426.
Heymach, G.J., Jost, D.E., 1968. The alteration of molecular weight distributions of polymers by ultrasonic energy. J. Polym. Sci. C Polym. Symp. 25, 145–153.
Hosny, K.M., 2009. Preparation and evaluation of thermosensitive liposomal hydrogel for enhanced transcorneal permeation of ofloxacin. AAPS PharmSciTech 10, 1336–1342.
Hosny, K.M., Hassan, A.H., 2014. Intranasal in situ gel loaded with saquinavir mesylate nanosized microemulsion: preparation, characterization, and in vivo evaluation. Int. J. Pharm. 475, 191–197.
Illum, L., 2003. Nasal drug delivery—possibilities, problems and solutions. J. Control. Release 87, 187–198.
Jătariu, A.N., Popa, M., Curteanu, S., Peptu, C.A., 2011. Covalent and ionic co-cross-linking-an original way to prepare chitosan-gelatin hydrogels for biomedical applications. J. Biomed. Mater. Res. A 98, 342–350.
Jay, A.J., Colquhoun, I.J., Ridout, M.J., et al., 1998. Analysis of structure and function of gellans with different substitution patterns. Carbohydr. Polym. 35, 179–188.
Jayaprakash, S., Santhiagu, A., Jasemine, S., 2014. Optimization of process variables for the preparation of gellan gum-raloxifene nanoparticles using statistical design. Am. J. Pharm. Health Res. 2, 34–49.
Kang, K.S., Pettitt, D.J., 1993. Xanthan, gellan, welan, and rhamsan. In: Industrial Gums, third ed. Academic Press, pp. 341–397 (Chapter 13).
Kesavan, K., Nath, G., Pandit, J.K., 2010. Preparation and in vitro antibacterial evaluation of gatifloxacin mucoadhesive gellan system. DARU J. Pharm. Sci. 18, 237–246.
Kouhi, M., Varshosaz, J., Hashemibeni, B., Sarmadi, A., 2020. Injectable gellan gum/lignocellulose nanofibrils hydrogels enriched with melatonin loaded forsterite nanoparticles for cartilage tissue engineering: fabrication, characterization and cell culture studies. Mater. Sci. Eng. C 115, 111114.
Koukaras, E.N., Papadimitriou, S.A., Bikiaris, D.N., Froudakis, G.E., 2012. Insight on the formation of chitosan nanoparticles through ionotropic gelation with tripolyphosphate. Mol. Pharm. 9, 2856–2862.
Kulkarni, G.T., Gowthamarajan, K., Dhobe, R.R., Yohanan, F., Suresh, B., 2005. Development of controlled release spheriods using natural polysaccharide as release modifier. Drug Deliv. 12, 201–206.
Kumar, S., Kaur, P., Bernela, M., Rani, R., Thakur, R., 2016. Ketoconazole encapsulated in chitosan-gellan gum nanocomplexes exhibits prolonged antifungal activity. Int. J. Biol. Macromol. 93, 988–994.
Kumari, L., Badwaik, H.R., 2019. Polysaccharide-based nanogels for drug and gene delivery. In: Polysaccharide Carriers for Drug Delivery. Elsevier Ltd, pp. 497–557.
Kumari, A., Yadav, S.K., Yadav, S.C., 2010. Biodegradable polymeric nanoparticles based drug delivery systems. Colloids Surf. B: Biointerfaces 75, 1–18.
Kuo, C.K., Ma, P.X., 2001. Ionically crosslinked alginate hydrogels as scaffolds for tissue engineering: part 1. Structure, gelation rate and mechanical properties. Biomaterials 22, 511–521.
Kuo, M.S., Mort, A.J., Dell, A., 1986. Identification and location of l-glycerate, an unusual acyl substituent in gellan gum. Carbohydr. Res. 156, 173–187.
Lai, Y., et al., 2012. A novel micelle of coumarin derivative monoend-functionalized PEG for anti-tumor drug delivery: in vitro and in vivo study. J. Drug Target. 20, 246–254.
Li, J., Kamath, K., Dwivedi, C., 2001. Gellan film as an implant for insulin delivery. J. Biomater. Appl. 15, 321–343.
Liu, Z., Jiao, Y., Wang, Y., Zhou, C., Zhang, Z., 2008. Polysaccharides-based nanoparticles as drug delivery systems. Adv. Drug Deliv. Rev. 60, 1650–1662.
Lu, X., et al., 2012. Polyelectrolyte complex nanoparticles of amino poly(glycerol methacrylate)s and insulin. Int. J. Pharm. 423, 195–201.
Luaces-Rodríguez, A., et al., 2017. Cysteamine polysaccharide hydrogels: study of extended ocular delivery and biopermanence time by PET imaging. Int. J. Pharm. 528, 714–722.
Ma, X., Williams, R.O., 2018. Polymeric nanomedicines for poorly soluble drugs in oral delivery systems: an update. J. Pharm. Investig. 48, 61–75.

Mahdi, M.H., Conway, B.R., Mills, T., Smith, A.M., 2016. Gellan gum fluid gels for topical administration of diclofenac. Int. J. Pharm. 515, 535–542.
Maiti, S., Chakravorty, A., Chowdhury, M., 2014. Gellan co-polysaccharide micellar solution of budesonide for allergic anti-rhinitis: an in vitro appraisal. Int. J. Biol. Macromol. 68, 241–246.
Manca, M.L., et al., 2014. Improvement of quercetin protective effect against oxidative stress skin damages by incorporation in nanovesicles. Colloids Surf. B: Biointerfaces 123, 566–574.
Manconi, M., et al., 2018a. Preparation of gellan-cholesterol nanohydrogels embedding baicalin and evaluation of their wound healing activity. Eur. J. Pharm. Biopharm. 127, 244–249.
Manconi, M., et al., 2018b. Nanodesign of new self-assembling core-shell gellan-transfersomes loading baicalin and in vivo evaluation of repair response in skin. Nanomed. Nanotechnol. Biol. Med. 14, 569–579.
Mao, R., Tang, J., Swanson, B.G., 2000. Texture properties of high and low acyl mixed gellan gels. Carbohydr. Polym. 41, 331–338.
Mehnath, S., et al., 2019. Sericin-chitosan doped maleate gellan gum nanocomposites for effective cell damage in mycobacterium tuberculosis. Int. J. Biol. Macromol. 122, 174–184.
Milas, M., Shi, X., Rinaudo, M., 1990. On the physicochemical properties of gellan gum. Biopolymers 30, 451–464.
Milivojevic, M., Pajic-Lijakovic, I., Bugarski, B., Nayak, A.K., Hasnain, M.S., 2019. Gellan gum in drug delivery applications. In: Natural Polysaccharides in Drug Delivery and Biomedical Applications. Elsevier Inc, pp. 145–186.
Miyoshi, E., Takaya, T., Nishinari, K., 1994. Gel-sol transition in gellan gum solutions. I. Rheological studies on the effects of salts. Food Hydrocoll. 8, 505–527.
Moritaka, H., Nishinari, K., Taki, M., Fukuba, H., 1995. Effects of pH, potassium chloride, and sodium chloride on the thermal and rheological properties of Gellan gum gels. J. Agric. Food Chem. 43, 1685–1689.
Morris, E.R., Nishinari, K., Rinaudo, M., 2012. Gelation of gellan—a review. Food Hydrocoll. 28, 373–411.
Musazzi, U.M., et al., 2018. Gellan nanohydrogels: novel nanodelivery systems for cutaneous administration of piroxicam. Mol. Pharm. 15, 1028–1036.
Nitta, Y., Takahashi, R., Nishinari, K., 2010. Viscoelasticity and phase separation of aqueous Na-type gellan solution. Biomacromolecules 11, 187–191.
Novac, O., Lisa, G., Profire, L., Tuchilus, C., Popa, M.I., 2014. Antibacterial quaternized gellan gum based particles for controlled release of ciprofloxacin with potential dermal applications. Mater. Sci. Eng. C 35, 291–299.
Oh, J.K., Siegwart, D.J., Matyjaszewski, K., 2007. Synthesis and biodegradation of nanogels as delivery carriers for carbohydrate drugs. Biomacromolecules 8, 3326–3331.
Oliveira, J.T., et al., 2008. Injectable gellan gum hydrogels as supports for cartilage tissue engineering applications: in vitro characterization and initial in vivo studies. In: 8th World Biomaterials Congress.
Osmałek, T., Froelich, A., Tasarek, S., 2014. Application of gellan gum in pharmacy and medicine. Int. J. Pharm. 466, 328–340.
Palumbo, F.S., Federico, S., Pitarresi, G., Fiorica, C., Giammona, G., 2020. Gellan gum-based delivery systems of therapeutic agents and cells. Carbohydr. Polym. 229, 115430.
Patil, S.B., Sawant, K.K., 2009. Development, optimization and in vitro evaluation of alginate mucoadhesive microspheres of carvedilol for nasal delivery. J. Microencapsul. 26, 432–443.
Patil, J.S., Kamalapur, M.V., Marapur, S.C., Kadam, D.V., 2010. Ionotropic gelation and polyelectrolyte complexation: the novel techniques to design hydrogel particulate sustained, modulated drug delivery system: a review. Dig. J. Nanomater. Biostruct. 5, 241–248.
Patil, R.P., Pawara, D.D., Gudewar, C.S., Tekade, A.R., 2019. Nanostructured cubosomes in an in situ nasal gel system: an alternative approach for the controlled delivery of donepezil HCl to brain. J. Liposome Res. 29, 264–273.
Perrechil, F.A., Sato, A.C.K., Cunha, R.L., 2011. κ-Carrageenan-sodium caseinate microgel production by atomization: critical analysis of the experimental procedure. J. Food Eng. 104, 123–133.
Picone, C.S.F., Cunha, R.L., 2011. Influence of pH on formation and properties of gellan gels. Carbohydr. Polym. 84, 662–668.
Pijls, R.T., et al., 2005. Studies on a new device for drug delivery to the eye. Eur. J. Pharm. Biopharm. 59, 283–288.
Pinto Reis, C., Neufeld, R.J., Ribeiro, A.J., Veiga, F., 2006. Nanoencapsulation I. Methods for preparation of drug-loaded polymeric nanoparticles. Nanomedicine 2, 8–21.
Pires Vilela, J.A., Cavallieri, Â.L.F., Lopes da Cunha, R., 2011. The influence of gelation rate on the physical properties/structure of salt-induced gels of soy protein isolate-gellan gum. Food Hydrocoll. 25, 1710–1718.
Pokharkar, V., Patil, V., Mandpe, L., 2015. Ocular-engineering of polymer-surfactant nanoparticles of doxycycline hydrochloride for ocular drug delivery.Pdf. Drug Deliv. 22, 955–968.

Posadowska, U., et al., 2015. Injectable nanoparticle-loaded hydrogel system for local delivery of sodium alendronate. Int. J. Pharm. 485, 31–40.

Posadowska, U., Brzychczy-Wloch, M., Pamula, E., 2016a. Injectable gellan gum-based nanoparticles-loaded system for the local delivery of vancomycin in osteomyelitis treatment. J. Mater. Sci. Mater. Med. 27, 1–9.

Posadowska, U., et al., 2016b. Injectable hybrid delivery system composed of gellan gum, nanoparticles and gentamicin for the localized treatment of bone infections. Expert Opin. Drug Deliv. 13, 613–620.

Prezotti, F.G., et al., 2020. Oral nanoparticles based on gellan gum/pectin for colon-targeted delivery of resveratrol. Drug Dev. Ind. Pharm. 46, 236–245.

Qian, Z.Y., Fu, S.Z., Feng, S.S., 2013. Nanohydrogels as a prospective member of the nanomedicine family. Nanomedicine 8, 161–164.

Quinn, F.X., Hatakeyama, T., Yoshida, H., Takahashi, M., Hatakeyama, H., 1993. The conformational properties of gellan gum hydrogels. Polym. Gels Netw. 1, 93–114.

Rancan, F., et al., 2016. Effects of thermoresponsivity and softness on skin penetration and cellular uptake of polyglycerol-based nanogels. J. Control. Release 228, 159–169.

Sadhukhan, S., Bakshi, P., Datta, R., Maiti, S., 2015. Oral-poly(ethylene oxide)-g-gellan polysaccharide nanocarriers for controlled gastrointestinal delivery of simvastatin.pdf. J. Appl. Polym. Sci. 132, 42399.

Sah, S.K., Tiwari, A.K., Shrivastava, B., Bairwa, R., Bishnoi, N., 2016. Natural gums emphasized grafting technique: applications and perspectives in floating drug delivery system. Asian J. Pharm. 10, 72–83.

Saindane, N.S., Pagar, K.P., Vavia, P.R., 2013. Nanosuspension based in situ gelling nasal spray of carvedilol: development, in vitro and in vivo characterization. AAPS PharmSciTech 14, 189–199.

Salim, N., Basri, M., Rahman, M.B.A., Abdullah, D.K., Basri, H., 2012. Modification of palm kernel oil esters nanoemulsions with hydrocolloid gum for enhanced topical delivery of ibuprofen. Int. J. Nanomedicine 7, 4739–4747.

Santos, T.P., Cunha, R.L., 2018. Role of process variables on the formation and in vitro digestion of gellan gels. Carbohydr. Polym. 192, 111–117.

Santos, T.P., Cunha, R.L., 2019. In vitro digestibility of gellan gels loaded with jabuticaba extract: effect of matrix-bioactive interaction. Food Res. Int. 125, 108638.

Santos, N.P., et al., 2005. Usnic acid-loaded nanocapsules: an evaluation of cytotoxicity. J. Drug Deliv. Sci. Technol. 15, 355–361.

Santos, T.P., Costa, A.L.R., Michelon, M., Costa, L.P., Cunha, R.L., 2020. Development of a microfluidic route for the formation of gellan-based microgels incorporating jabuticaba (*Myrciaria cauliflora*) extract. J. Food Eng. 276, 109884.

Sanzgiri, Y.D., et al., 1993. Gellan-based systems for ophthalmic sustained delivery of methylprednisolone. J. Control. Release 26, 195–201.

Sarika, P.R., Pavithran, A., James, N.R., 2015. Cationized gelatin/gum arabic polyelectrolyte complex: study of electrostatic interactions. Food Hydrocoll. 49, 176–182.

Sehgal, R.R., Roohani-Esfahani, S.I., Zreiqat, H., Banerjee, R., 2017. Nanostructured gellan and xanthan hydrogel depot integrated within a baghdadite scaffold augments bone regeneration. J. Tissue Eng. Regen. Med. 11, 1195–1211.

Singh, B.N., Trombetta, L.D., Kim, K.H., 2004. Biodegradation behavior of gellan gum in simulated colonic media. Pharm. Dev. Technol. 9, 399–407.

Singh, S.R., et al., 2009. L-carnosine: multifunctional dipeptide buffer for sustained-duration topical ophthalmic formulations. J. Pharm. Pharmacol. 61, 733–742.

Šmejkalová, D., et al., 2017. Hyaluronan polymeric micelles for topical drug delivery. Carbohydr. Polym. 156, 86–96.

Sreeram, K.J., Nidhin, M., Indumathy, R., Nair, B.U., 2008. Synthesis of iron oxide nanoparticles of narrow size distribution on polysaccharide templates. Bull. Mater. Sci. 31, 93–96.

Stott, P.W., Williams, A.C., Barry, B.W., 1998. Transdermal delivery from eutectic systems: enhanced permeation of a model drug, ibuprofen. J. Control. Release 50, 297–308.

Taghizadeh, M.T., Bahadori, A., 2009. Degradation kinetics of poly (vinyl-pyrrolidone) under ultrasonic irradiation. J. Polym. Res. 16, 545–554.

Tako, M., 2015. The principle of polysaccharide gels. Adv. Biosci. Biotechnol. 6, 22–36.

Varshosaz, J., Taymouri, S., Ghassami, E., 2018. Supramolecular self-assembled nanogels a new platform for anticancer drug delivery. Curr. Pharm. Des. 23, 5242–5260.

Vashisth, P., Pruthi, V., 2016. Synthesis and characterization of crosslinked gellan/PVA nanofibers for tissue engineering application. Mater. Sci. Eng. C 67, 304–312.

Vashisth, P., Kumar, N., Sharma, M., Pruthi, V., 2015. Biomedical applications of ferulic acid encapsulated electrospun nanofibers. Biotechnol. Rep. 8, 36–44.

Vashisth, P., et al., 2017. Ofloxacin loaded gellan/PVA nanofibers—synthesis, characterization and evaluation of their gastroretentive/mucoadhesive drug delivery potential. Mater. Sci. Eng. C 71, 611–619.

Vigneshwaran, R., Sankar, M., Vigneshkumar, S., 2019. Synthesis and characterization of gellan gum-pyridoxine nanoparticle for treatment of localized peripheral neuropathy. Int. J. Eng. Adv. Technol. 9, 576–580.

Vilela, J.A.P., Perrechil, F.D.A., Picone, C.S.F., Sato, A.C.K., Da Cunha, R.L., 2015. Preparation, characterization and in vitro digestibility of gellan and chitosan-gellan microgels. Carbohydr. Polym. 117, 54–62.

Williams, A.C., Barry, B.W., 2012. Penetration enhancers. Adv. Drug Deliv. Rev. 64, 128–137.

Yamamoto, F., Cunha, R.L., 2007. Acid gelation of gellan: effect of final pH and heat treatment conditions. Carbohydr. Polym. 68, 517–527.

Yang, F., Xia, S., Tan, C., Zhang, X., 2013. Preparation and evaluation of chitosan-calcium-gellan gum beads for controlled release of protein. Eur. Food Res. Technol. 237, 467–479.

Zhang, N., Wardwell, P.R., Bader, R.A., 2013. Polysaccharide-based micelles for drug delivery. Pharmaceutics 5, 329–352.

Zhang, H., Zhai, Y., Wang, J., Zhai, G., 2016. New progress and prospects: the application of nanogel in drug delivery. Mater. Sci. Eng. C 60, 560–568.

Zou, P., et al., 2016. Advances in characterisation and biological activities of chitosan and chitosan oligosaccharides. Food Chem. 190, 1174–1181.

CHAPTER

6

Gum kondagogu as a potential material for micro- and nanoparticulate drug delivery

Rimpy and Munish Ahuja

Drug Delivery Research Laboratory, Department of Pharmaceutical Sciences, Guru Jambheshwar University of Science and Technology, Hisar, Haryana, India

6.1 Introduction

Carbohydrate polymers derived from nature can be classified on the basis of their origin as plant polysaccharides (e.g., kondagogu, tragacanth, ghatti, karaya, arabic, pectin, locust bean, and guar gum), microbial polysaccharides (e.g., gellan, dextran, xanthan, pullulan, and glycan gum) and seaweed polysaccharide (e.g., agar, carrageenan, and alginate). All these polysaccharides possess versatile characteristics such as biodegradability, biocompatibility, and easy availability with excellent thickening, gelling, and emulsifying properties. Due to these advantages, usages of natural gums is increasing around the world in every field from pharmaceutical, food, or other packaging to environmental, agricultural, or chemical engineering (Mohammadinejad et al., 2020). Properties of plant-based gums like strong hydrophilicity and low cytotoxicity make them suitable candidates for nano-based drug delivery system or other biomedical applications. Gum kondagogu is one such plant polysaccharide which has been explored by various researchers for drug delivery applications during the last decade (Ahmad et al., 2019).

6.1.1 Gum kondagogu

Gum kondagogu is an exudate obtained from the bark of the Indian tree *"Cochlospermum gossypium"* belonging to the family *"Bixaceae."* This deciduous, small to medium-sized tree is found distributed widely across India from Garhwal Himalayas to West Bengal, and Central

Micro- and Nanoengineered Gum-Based Biomaterials for Drug Delivery and Biomedical Applications
https://doi.org/10.1016/B978-0-323-90986-0.00006-6

India to Deccan Peninsula. The major portion of the gum is collected by the tribal people from the forests of Andhra Pradesh under the aegis of Girijan Co-operative Corporation, which holds the monopoly rights over the collection and sale of gum kondagogu. Gum kondagogu is reported to possess thermogenic, sedative, and anodyne properties, and has been traditionally used for the treatment of cough, diarrhea, and dysentery. Apart from its medicinal value, it has been used for its emulsifying, gelling, and stabilizing properties in calico printing, paper, ice cream, and cigar industries for a relatively long time (Singh and Singh, 2013).

Proximate analysis of gum kondagogu unveiled that it contains on percentage weight basis an average of 80% of total fiber, 15.25% of moisture, 7.3% of ash, 6.3% of protein content, 2.2% of lipids, and traces of tannins (Vinod and Sashidhar, 2010). The carbohydrate component in the gum comprises acidic sugars (63%) and neutral sugars (37%). The D-glucuronic acid (19 26% mol), β-D-galactouronic acid (13.22 mol%), and α-D-galactouronic acid (11.22 mol%) account for the acidic components, while the neutral sugars consist mainly of rhamnose (12.85% mol), arabinose (2.52% mol), mannose (8.3% mol), α-D-glucose (2.48% mol) and β-D-glucose (2.52% mol), and few traces of galactose and fructose (Vinod et al., 2008; Vinod and Sashidhar, 2009, 2010). The amino acids profiling of gum kondagogu carried out by GC–MS analysis after hydrolysis of the protein components established that gum kondagogu contains an average of the following amino acids in μg per gram of gum: aspartic acid (72.2), methionine (44.2), proline (42.4), glutamic acid (34.2), tyrosine (32.8), alanine (32.2), threonine (30.4), tryptophan (3.8), valine (7.2), glycine (5.0), and leucine (3.8) (Vinod et al., 2010). The fatty acid composition of lipids present in gum kondagogu has also been analyzed using GC–MS. The gum was found to contain 84% and 16% of saturated and unsaturated fatty acids, respectively. The contents in μg per gram of various fatty acid methyl esters present in the gum kondagogu were found to be stearic acid (25.4), palmitic acid (20.2), lauric acid (10.2), erucic acid (9.9), palmitoleic acid (7.9), lignoceric acid (6.2), behenic acid (5.2), oleic acid (5.1), myristic acid (4.9), capric acid (2.5), linoleic acid (1.8), arachidic acid (1.1), and γ-linolenic acid (0.8) (Janaki and Sashidhar, 1998). In addition, the elemental analysis of gum revealed that it is composed of 17.5, 15.2, and 2.5 mg per gram of calcium, potassium, and sodium, respectively, as major elements while the magnesium, silicon, sulfur, chlorine, manganese, iron, copper, strontium, barium, and aluminum are present in small quantities (Vinod et al., 2010).

Gum kondagogu has properties similar to the gum obtained from *Sterculia urens* (karaya gum); as a result, the Food and Agricultural Organization (FAO) in 1991 listed the gum obtained from *Cochlospermum* species (family *Bixaceae*) under karaya gum (FAO, 1991). To establish its separate identity and exploit its applications, various researchers have carried out morphological, physicochemical, structural, and rheological characterization of gum kondagogu.

Chemical composition and structure

Structural elucidation studies by chemical and spectroscopic methods were carried out to establish the structure of gum kondagogu (Fig. 6.1). The results of FTIR spectroscopic studies established that it contains hydroxyl, acetyl, carbonyl, and carboxyl as major functional groups. Furthermore, structural elucidation was carried out using ^{1}H, ^{13}C NMR, and GC–MS studies on the gum samples, which were hydrolyzed and subjected to Smith degradation analysis; this revealed that gum kondagogu is a heteropolysaccharide of substituted

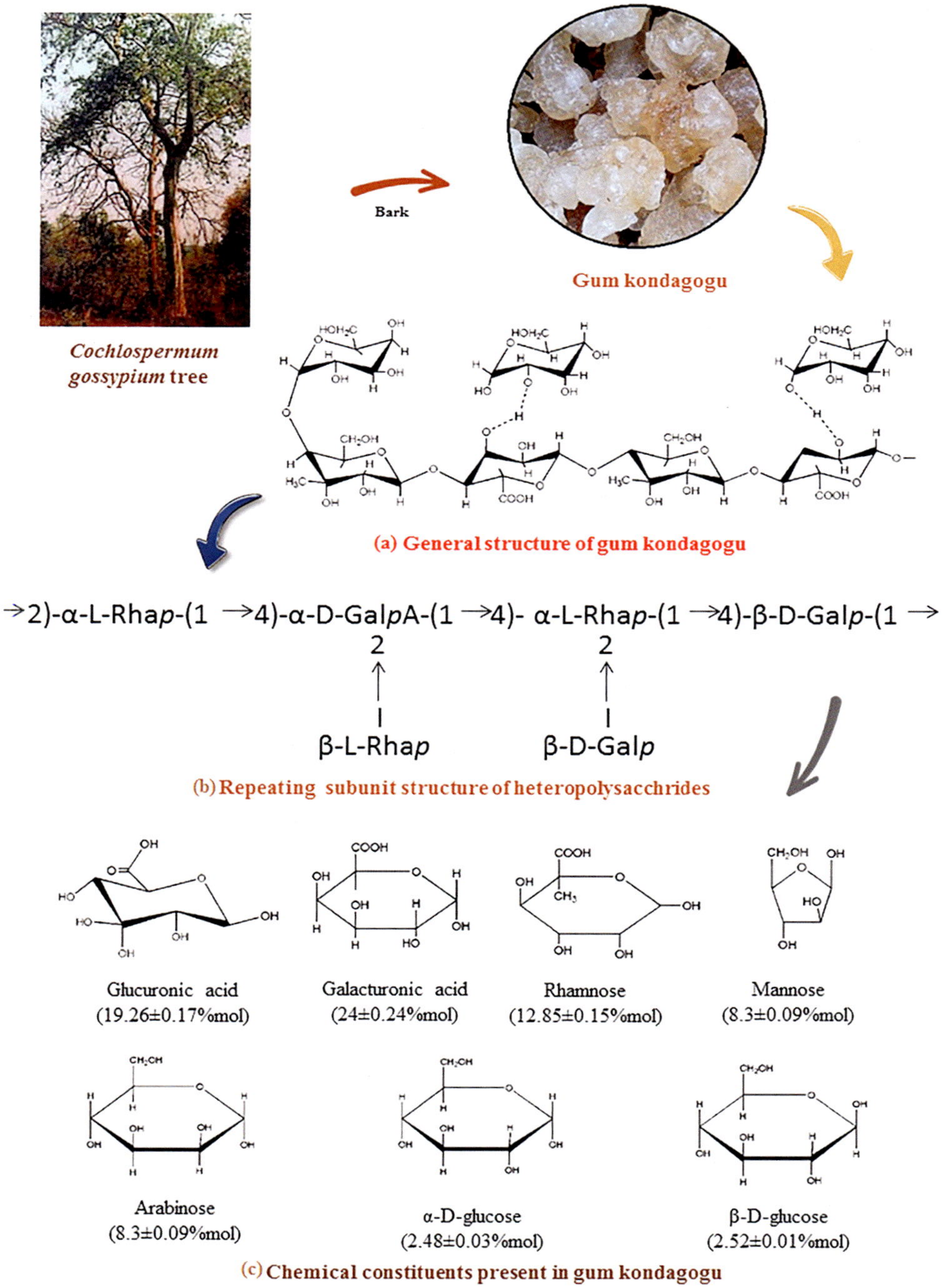

FIG. 6.1 (A) General structure of gum kondagogu with their chemical constituents, (B) repeating subunit structure of heteropolysaccharides, and (C) chemical constituents present in gum kondagogu. *Panels (A) and (B): adapted from Rastogi, L., Sashidhar, R.B., Karunasagar, D., Arunachalam, J., 2014. Gum kondagogu reduced/stabilized silver nanoparticles as direct colorimetric sensor for the sensitive detection of Hg2+ in aqueous system. Talanta, 118, 111–117. https://doi.org/10.1016/j.talanta.2013.10.012.*

rhamanogalactouronan type comprising of (1→2) β-D-Gal p, (1→6) β-D-Gal p, (1→4) β-D-Glc p A, 4-O-Me-α-D-Glc p A, and (1→2) α-L-Rha (Vinod et al., 2008).

Physicochemical properties

The native gum is obtained in the form of tears. The gum is usually purified by dissolving in water and removing the undissolved impurities. The dissolved gum is freeze-dried and used. Surface morphological analysis by scanning electron microscopy revealed that the native gum has irregularly shaped particles. The intrinsic viscosity, weight average molecular mass, and zeta potential of the native gum are reported to be 32 dL/g, 7230 kDa, and −23.4 mV, respectively. Furthermore, the solution conformational analysis showed that gum kondagogu shows a semiflexible chain. Gum kondagogu has high water binding capacity of 35 mL/g. Rheological studies conducted on gum kondagogu solutions indicated that at gum concentrations >0.6% (w/v), it shows Newtonian and shear-thinning behavior (Sashidhar et al., 2015; Srivastava et al., 2018; Vegi et al., 2009).

6.2 Modifications of gum kondagogu

Natural gums are highly used and recommended polymers because of their various advantages as mentioned earlier, but their properties like high viscosity, uncontrolled hydration, batch-to-batch variability, microbial contamination and instability, fall in viscosity on storage poor mechanical strength, and less stability at high temperature or physiological conditions compared to the synthetic polymers limit their applications (Chaudhary and Pawar, 2014). In order to overcome these drawbacks and improve the functionality of natural gums, various researchers tend to modify the functional properties of natural gums by chemical derivatization or physical processing (Rana et al., 2011). This section describes various modifications carried out on gum kondagogu and the changes in the physicochemical properties brought about as a result of these modifications.

6.2.1 Carboxymethylation

The carboxymethylation reaction is one of the favorable and extensively used modification techniques all around the world due to easier and milder processing and reliable usage. As the name indicates, the carboxymethyl group is attached to the primary or secondary group of the polysaccharide chain. The modification reaction involves Williamson's ether synthesis in which etherification of the compound is carried out in a two-step reaction. Basically, in the first step, the reaction of sodium hydroxide with hydroxyl groups on the polysaccharide chain results in the generation of alkoxide. This is followed by a nucleophilic substitution reaction of monochloroacetate with alkoxide in the second step, leading to substitution of the carboxymethyl group on the polymer backbone (Fig. 6.2A) (de Nooy et al., 2000). Kumar and Ahuja (2012) synthesized the carboxymethyl derivative of gum kondagogu by reacting gum kondagogu (1.25%, w/v) in aqueous sodium hydroxide with an aqueous monochloroacetic acid solution (75%, w/v) at 70 °C for 30 min. The reaction yielded carboxymethyl substituted gum kondagogu with a degree of substitution of 0.2. The incorporation of carboxymethyl

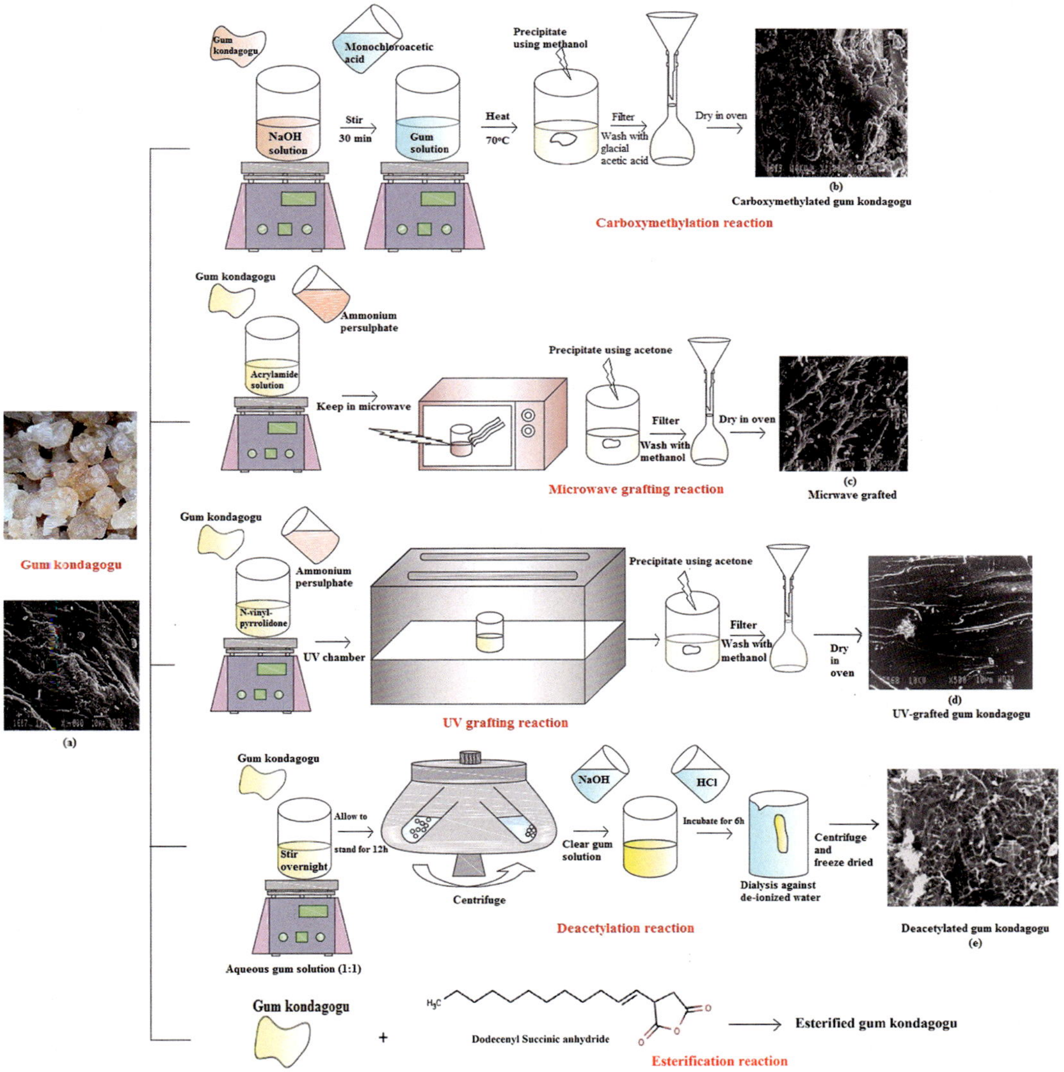

FIG. 6.2 Different types of modification reaction performed on gum kondagogu with their micrographical images including native gum (A), carboxymethylated (B), microwave grafted (C), UV-grafted (D), and deacetylated gum kondagogu (E). *Panels A and D: adapted from Malik, S., Kumar, A., Ahuja, M., 2012. Synthesis of gum kondagogu-g-poly (N-vinyl-2-pyrrolidone) and its evaluation as a mucoadhesive polymer. Int. J. Biol. Macromol. 51 (5), 756–762. https://doi.org/10.1016/j.ijbiomac.2012.07.009; panel B: adapted from Kumar, A., Ahuja, M., 2012. Carboxymethyl gum kondagogu: Synthesis, characterization and evaluation as mucoadhesive polymer. Carbohydr. Polym. 90 (1), 637–643. https://doi.org/10.1016/j.carbpol.2012.05.089; panel C: adapted from Malik, S., Ahuja, M., 2011. Gum kondagogu-g-poly (acrylamide): Microwave-assisted synthesis, characterisation and release behaviour. Carbohydr. Polym. 86 (1), 177–184. https://doi.org/10.1016/j.carbpol.2011.04.027; panel E: adapted from Vinod, V.T.P., Sashidhar, R.B., Suresh, K.I., Rama Rao, B., Vijaya Saradhi, U.V.R., Prabhakar Rao, T., 2008. Morphological, physico-chemical and structural characterization of gum kondagogu (*Cochlospermum gossypium*): a tree gum from India. Food Hydrocolloids, 2 2(5), 899–915. https://doi.org/10.1016/j.foodhyd.2007.05.006.*

functionality on the backbone polysaccharide chain imparts an anionic character on it. The columbic repulsion between the backbone chains of gum kondagogu prevents their aggregation, leading to a fall in the gum's apparent viscosity. Furthermore, it was observed that carboxymethylation of gum kondagogu increased its degree of crystallinity and made its surface rougher. The increased anionic character of the gum alters swelling and ionic-gelling behavior, which results in the modification of release characteristics (Kumar and Ahuja, 2012). The carboxymethylated gum has also been observed to have a higher bioadhesive property than the native gum. Apart from the increased aqueous solubility, carboxymethyl modification has been reported to improve biological activity such as the antimicrobial, immunomodulatory, and antitumor activity of polysaccharides. The exact mechanism of improved biological activities on the incorporation of carboxymethyl functionality in the polysaccharides is not yet elucidated; however, it is suggested to be due to the change in chain conformation, electrostatic interaction with receptors and proteins (Chakka and Zhou, 2020).

6.2.2 Grafting

Graft copolymerization is one of the most commonly employed polymer modification methods used to prepare hybrid polymers from natural and synthetic polymers. Usually, the vinyl monomers are graft copolymerized on the polysaccharide backbone chain with the formation of graft copolymers with tunable properties. Graft copolymerization has been used to modify the swelling, pH responsiveness, mucoadhesiveness, and release characteristics of natural gums.

Grafting of gum kondagogu was also done using acrylamide and NVP under the microwave and UV radiation, respectively (Fig. 6.2B and C). Malik and Ahuja (2011) carried out the graft modification of acrylamide on gum kondagogu using the microwave-assisted technique. Aqueous solutions of gum kondagogu (2%–3%, w/v) containing ammonium persulfate (10–40 mmol) as a redox initiator were irradiated in a domestic microwave oven at 20%–40% microwave power for 30–120 s. The pendant hydroxyl (–OH) groups of gum kondagogu are cleaved due to dielectric heating on exposure to microwave irradiation, resulting in the formation of gum kondagogu macroradicals. The reaction between the gum kondagogu macroradicals with acrylamide free radicals results in a series of free radical chain reactions which culminate in the formation of gum kondagogu-g-poly(acrylamide) and polyacrylamide homopolymer. By varying the microwave power, exposure time, and concentrations of gum, vinyl monomer, and redox initiator, one can optimize the reaction to maximize the formation of graft copolymer and minimize the formation of homopolymer. Grafting of acrylamide on gum kondagogu was observed to increase the rate of swelling and erosion of the gum matrices, which the authors speculated might be due to a porous matrix. The modification of swelling behavior translated into the modification of release behavior, as was observed in faster release of diclofenac from the matrix tablets prepared using graft copolymer, compared to tablets prepared using gum kondagogu (Malik and Ahuja, 2011). In another study, Malik et al. (2012) carried out grafting of N-vinyl-2-pyrrolidone on gum kondagogu employing a UV radiation-initiated free radical chain reaction. In this case, gum kondagogu macroradicals formed by the abstraction of hydrogen ions by ammonium persulfate and also by cleavage of hydroxyl bond by UV radiation reacted with pyrrolidone free radicals, which finally led to the formation of gum kondagogu-g-poly(vinyl pyrrolidone). The grafting of N-vinyl-pyrrolidone, a mucoadhesive polymer on gum kondagogu, resulted in improving the mucoadhesive property, which was further utilized in designing the mucoadhesive buccal patches of metronidazole (Malik et al., 2012).

6.2.3 Deacetylation

A number of natural gums and polysaccharides such as acetan, arabic, chitin, gellan, karaya, konjac, and kondagogu are found to be acetylated in their native state. Acetylated gum usually has properties different from deacetylated gum. Native gum is reported to lose the acetyl content on aging and on exposure to increased temperature, humidity, and particle size reduction (Srivastava et al., 2018). Deacetylation has been observed to modify the rheological behavior and gelling characteristics of gums. Gum kondagogu has high acetyl content; it is deacetylated by treatment with alkali hydroxides or ammonia followed by neutralization with acid. The deacetylated product thus obtained is purified by dialysis to remove the unreacted reagents, followed by lyophilization to get the purified deacetylated gum (Fig. 6.2D). Deacetylation was observed to change the surface morphology of gum kondagogu with the change in particle shape from polyhedral to fibrillar particles. The acetyl groups in the gum were observed to be essential for maintaining the structural integrity of gum. The difference in particle size and surface area is reported to influence the hydration behavior, intrinsic viscosity, and molecular mass of the gum. The intrinsic viscosity, molecular mass, and zeta potential of the deacetylated gum kondagogu were determined to be 59 dL/g, 36,100 kDa, and −37.4 mV, respectively (Vinod et al., 2008). The higher intrinsic viscosity and molar mass of deacetylated gum kondagogu were attributed to the higher solubilization and gelation of the deacetylated gum, which in turn led to higher intrinsic viscosity and molar mass as compared to the native gum. Furthermore, thermogravimetric analysis of native and deacetylated gum kondagogu pointed to the higher thermal stability of the deacetylated gum.

6.2.4 Esterification

The presence of reactive functional groups such as hydroxyl, carboxyl, and amine in natural polysaccharides paves the way for their reaction with acids and their derivatives. Esterification reactions between natural polysaccharides and carboxylic acid anhydride have been employed to introduce the charged groups in uncharged or partially charged natural polysaccharides. Carboxylic acid anhydrides such as octenyl succinic anhydride, dodecenyl succinic anhydride, and octadecenyl succinic anhydride have been employed to modify natural polysaccharides (Fig. 6.2E). It has been observed that modifying polysaccharides with carboxylic acid anhydrides leads to an improvement in their emulsifying properties (Li et al., 2016). Dodecenyl succinic anhydride has been used to impart the hydrophobic properties on gum arabic, konjac, inulin, starch, and gum karaya (Soni et al., 2019). Gum kondagogu has also been modified by reacting with dodecenyl succinic anhydride under alkaline conditions (pH of 8.5) at 25 °C for 7 h, followed by neutralization with acid and lyophilization. The modified gum kondagogu was observed to have antibacterial properties.

Conclusively, it can be inferred that modification of gum kondagogu using reactions like carboxymethylation, grafting, deacetylation, and esterification makes it a promising, desirable, and better candidate with enhanced properties for further usage in various applications.

6.3 Applications of gum kondagogu as a microparticulate and nanoparticulate carrier

Microparticles are particles of micrometer dimensions, which may be matrix systems (microspheres) or coated systems (microcapsules). A number of synthetic and natural polymers have been used to prepare microparticles for applications in drug delivery, cosmetics, enzyme immobilization, agriculture, textiles, etc. (Lombardo and Villares, 2020). Natural polysaccharides, because of their biocompatibility, biodegradability, and easy availability, are the materials of choice in designing microparticles. Alginate, pectin, chitosan, and cellulose derivatives are the most frequently used polysaccharides in the fabrication of microparticles due to their ionic gelling behavior (Wang et al., 2006). However, when the size of particles is reduced to nanometer dimensions, a quantum change appears in the physicochemical and biological properties of materials. From the drug delivery point of view, the permeability, cellular uptake, and in vivo distribution and circulation of the drugs can be altered by engineering them into nanoparticles. Polysaccharides, because of their unique properties as described above, have emerged as versatile materials for nanoparticulate drug delivery (Saravanan et al., 2012). This section gives an overview of the research work carried out to explore the applications of gum kondagogu in micro- and nanoparticulate formulations.

6.3.1 Microparticles

Gum kondagogu, being an anionic gum, interacts with cations, but the interaction is not strong enough to result in the formation of microspheres or beads. Studies involving the use of gum kondagogu have therefore either used the derivative of gum kondagogu with improved gelling characteristics or used gum kondagogu in combination with other natural polymers. Naidu et al. (2009) prepared microparticles between gum kondagogu and chitosan using a coacervation-phase separation method based on interaction between the NH_3^+ group of chitosan and the $-COO^-$ of gum kondagogu (Fig. 6.3). They prepared diclofenac sodium-loaded polyelectrolyte complex microparticles, and found that the interaction between the gum kondagogu and chitosan in 4:1–5:1 (w/w) was optimal, providing microparticles 175 μm in size, having diclofenac entrapment of 85% and 90% yield. They also observed that the strong interaction between the protonated amino group of chitosan and carboxylate group of gum kondagogu resulted in a tight network in an acidic environment (pH 1.2) leading to slower swelling and release of diclofenac, while at pH 6.8 due to deprotonation of amino group of chitosan the matrix gets loosened releasing the drug. The results of pharmacokinetic studies conducted on rats showed a 2.2- to 2.7-fold increase in elimination half-life ($t_{1/2}$) and 5.3- to 5.8-fold increase in relative bioavailability of diclofenac from the polyelectrolyte complex microparticles compared to the pure drug. Both chitosan and gum kondagogu are

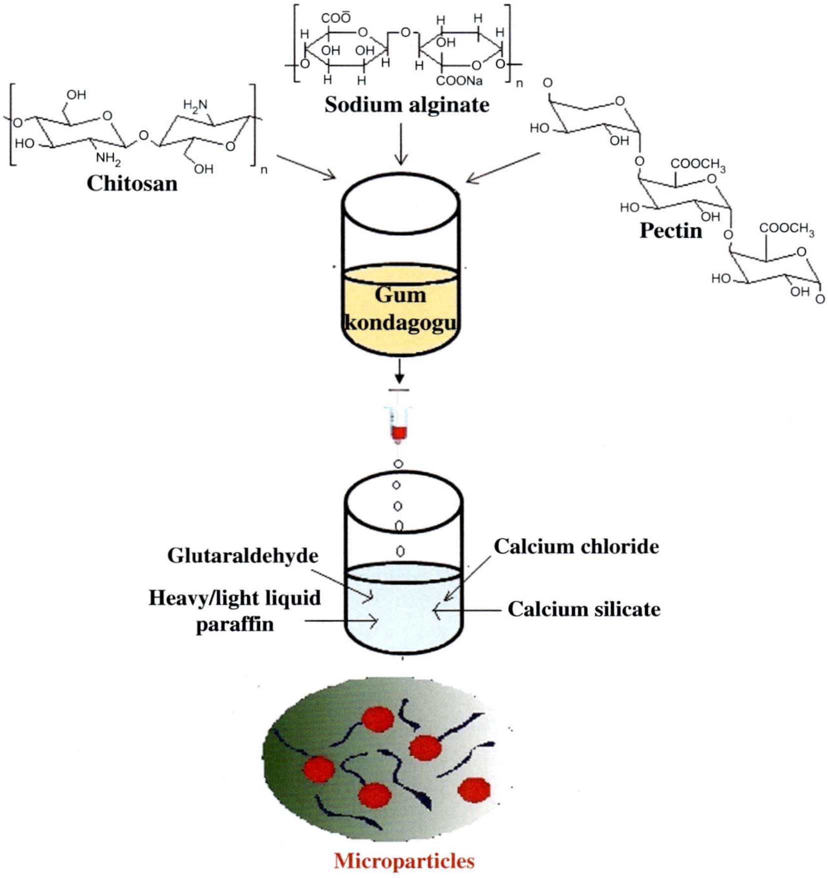

FIG. 6.3 Different approaches using formation of microparticles.

mucoadhesive in nature. There might also be some role of bioadhesion, which the authors have not probed (Naidu et al., 2009).

Krishna and Murthy (2010) explored the mucoadhesive nature of gum kondagogu. They formulated glipizide-loaded composite microcapsules of gum kondagogu and sodium alginate using both ionotropic gelation and emulsification ionotropic gelation techniques. Aqueous solutions containing about 3% each of gum kondagogu and sodium alginate and containing glipizide in a polymer-drug ratio of 2:1 to 1:1 were ionically gelled by reacting with calcium chloride (15%, w/v) solutions. In the ionotropic gelation method, the drug-containing polymer solution was extruded into a cross-linker solution using a #18G needle syringe, while in the emulsion gelation method the ionic cross-linking reaction was carried out in heavy liquid paraffin as a medium under stirring. Spherical, free-flowing microcapsules with an average size of 920 μm and microencapsulation efficiency of 76%–93% were obtained. However, the study does not provide any data to describe the difference in particle size and microencapsulation efficiency of microcapsules obtained from the two cross-linking techniques. The microcapsules with a coat:core ratio of 2:1 were able to sustain the release of the drug for 8 h following zero-order kinetics. Furthermore, microcapsules tested by in vitro wash-off test showed bioadhesion for 8 h. In vivo pharmacodynamic evaluation of glipizide microcapsules in rabbits showed sustained hypoglycemic effects for 14 h (Krishna and Murthy, 2010).

A similar study involving comparative evaluation of gliclazide-loaded mucoadhesive composite microcapsules of sodium alginate with xanthan or karaya gum or gum kondagogu ionically cross-linked with calcium chloride was carried out by Mankala et al. (2011). The study explores both ionic gelation and emulsification ionic gelation methods. Free-flowing, spherical microcapsules with microencapsulation efficiency of 57%–61%, which released gliclazide for up to 12 h following zero-order kinetics and non-Fickian release mechanism, were reported. The study concluded that the formulation of sodium alginate and xanthan gum microcapsules prepared by the ionic gelation method was the best. However, the in vitro release profile of gliclazide from the various formulations regardless of method of preparation did not appear to be significantly different. The study does not provide adequate mucoadhesive data to support the authors' assertion (Mankala et al., 2011).

Bera et al. (2017) prepared the gastroretentive microparticulate system of a combination of low methoxy pectin and gum kondagogu reinforced with mesoporous silica and ionically cross-linked with Zn^{2+} for delivery of flurbiprofen. To achieve gastroretention, flurbiprofen solutions in low-density oils, viz., light liquid paraffin or olive or sunflower low oils, were encapsulated into the pectin-gum kondagogu matrices. The effects of polymer ratio, type of oil, and calcium silicate incorporation in the matrices were studied using one factor at a time. Initially, an emulsion was prepared by homogenizing an oil phase (2.5%, v/v) containing flurbiprofen and an aqueous phase comprising of pectin and gum kongagogu (5:1 to 1:1) with or without calcium silicate (0.5%, w/v) at 5000 rpm for 15 min. The flurbiprofen-loaded microparticulate beads were obtained by adding the emulsion into aqueous zinc acetate (5%, w/v) using a syringe with a 21G needle. The spherical beads had an average diameter of 1.55–1.87 mm, a density of 0.477–0.761 g/cm^3, and encapsulation efficiency of 45%–87%. The beads showed floating lag times of $\leq$3 s with a buoyancy of 15%–84% at the end of 7 h. The study reported that the size of beads increased with increasing the concentrations of gum kondagogu in beads. The beads prepared using sunflower oil showed higher entrapment, which was attributed to higher partitioning of flurbiprofen in sunflower oil. Furthermore, the incorporation of calcium silicate in the bead improved the drug entrapment, which was suggested to be due to additional internal gelation of matrices by Ca^{2+} ions. The study concluded that the flurbiprofen-loaded emulgel beads prepared using sunflower oil and pectin-gum kondagogu (3:1) and calcium silicate as a matrix reinforcer provided sustained drug delivery with combined mucoadhesivity and buoyancy (Bera et al., 2017).

Carboxymethyl derivative of gum kondagogu, which has a more anionic character than the native gum, was found to interact strongly with Ca^{2+} ions, forming ionically gelled beads. Kumar and Ahuja (2012) explored this interaction to formulate metformin-loaded Ca^{+2}-cross-linked beads of carboxymethyl gum kondagogu. The spherical, mucoadhesive beads were prepared by extruding an aqueous solution of carboxymethyl gum kondagogu (2.5%, w/v) through a #18G needle into calcium chloride solutions (5%–20%, w/v). Increasing the cross-linker concentration increased the percentage drug entrapment and reduced the burst release of the metformin. The beads showed >80% ex vivo bioadhesion at 24 h. Furthermore, the beads released metformin in a sustained manner for 24 h, releasing the drug by Fickian diffusion following zero-order kinetics (Kumar and Ahuja, 2012).

Apart from ionic cross-linking, chemical cross-linking with glutaraldehyde has been employed to formulate acyclovir-loaded gum kondagogu microspheres (Mankala et al., 2011). The microspheres were prepared by the emulsion cross-linking method. An aqueous solution of gum kondagogu containing acyclovir was emulsified in an oil phase comprising a mixture of heavy and light liquid paraffin (50:50) with Tween 80 and Span 80 as emulsifying agents. The cross-linking of the internal gum kondagogu phase was achieved by adding glutaraldehyde. It was observed in the study that increasing the concentration of gum kondagogu from 1.5% to 3.0% (w/v) resulted in an increase in microparticle size from 56 to 88 μm, while increasing the concentration of acyclovir from 15% to 30% (w/w) led to an increase in percentage drug entrapment from 68% to 77%. The microspheres provided sustained release of acyclovir over 48 h, following Higuchi's square root kinetics.

6.3.2 Polyelectrolyte nanoparticles

The interaction between the anionic carboxymethyl gum kondagogu and cationic chitosan was observed to result in opalescent solutions to precipitates depending upon the concentrations of the two polyelectrolytes (Kumar and Ahuja, 2013). This interaction was optimized by varying the concentrations of carboxymethyl gum kondagogu (0.01%–0.1%, w/v) and chitosan (0.05%–0.2%, w/v) using response surface methodology to prepare ofloxacin-loaded nanoparticles (Fig. 6.4). Optimized polyelectrolyte nanoparticles with particle size 285 nm and 63% drug entrapment were observed to be ovoid-shaped and had antibacterial activity comparable to the equivalent of ofloxacin solution against *Micrococcus luteus* and *Pseudomonas aeruginosa*. Moreover, the polyelectrolyte nanoparticles were able to sustain the drug release over 24 h releasing the drug, following Higuchi's square root kinetics.

6.3.3 Metal nanoparticles

Metals and metal ions have been used for biological applications for ages. In the Ayurvedic and Siddha systems of medicine, oxides of gold, silver, zinc, iron, copper, etc. have been used as "*Bhasmas*" for treatment of various diseases and as "*Rasyanas*" for rejuvenation (Pal et al., 2014). The unique optical, surface, and biological properties endowed on metal nanoparticles compared to bulk metals have enabled their use in wide-ranging applications such as biosensors, bioremediation, cancer therapy, catalysis, photography, drug delivery, targeting, etc. A number of top-down and bottom-up methods have been used for the synthesis of metal nanoparticles (Jamkhande et al., 2019). The top-down methods include laser ablation, mechanical milling, sputtering, chemical etching, etc. The bottom-up methods commonly employed are the sol-gel method, chemical vapor deposition, pyrolysis, electrodeposition, chemical and hydrothermal reduction, spinning, and a host of biological methods (use of microbes and plants). Among the various methods, biological or green synthesis methods are currently attracting worldwide attention. Green synthesis methods are environmentally

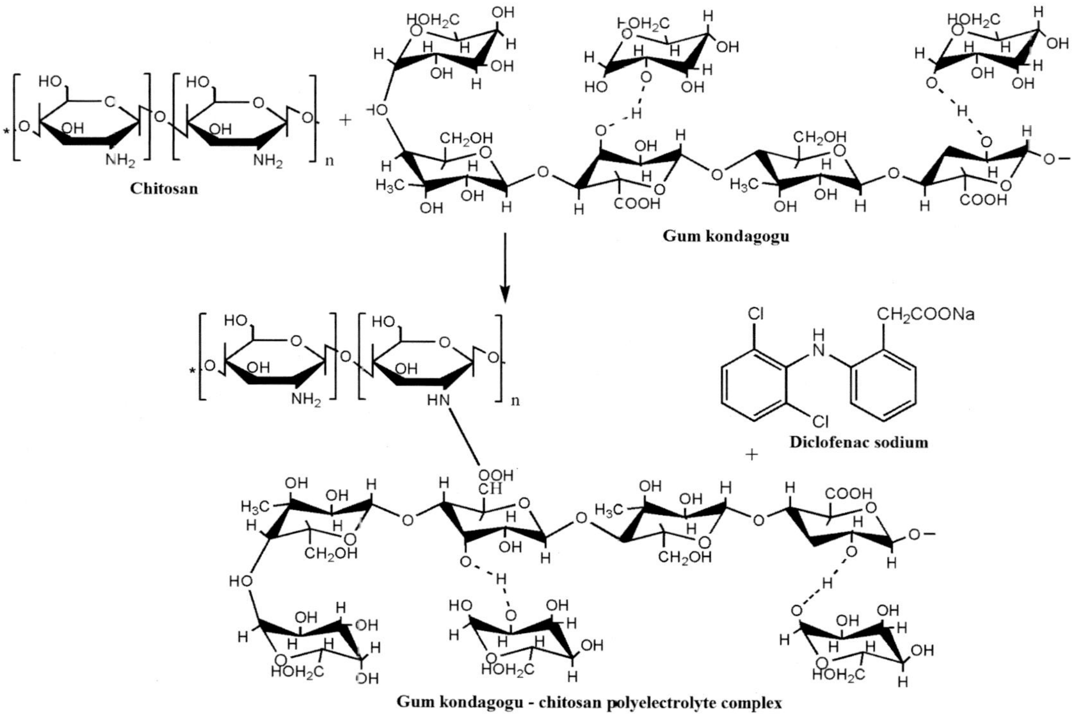

FIG. 6.4 Polyelectrolyte complex of chitosan and gum kondagogu. *Adapted from Naidu, V.G.M., Madhusudhana, K., Sashidhar, R.B., Ramakrishna, S., Khar, R.K., Ahmed, F.J., Diwan, P.V., 2009. Polyelectrolyte complexes of gum kondagogu and chitosan, as diclofenac carriers. Carbohyd. Polym. 76 (3), 464–471. https://doi.org/10.1016/j.carbpol.2008.11.010.*

friendly as they avoid the use of toxic chemicals, are economical. Among green synthesis methods, the use of natural gums-induced reduction of metal ions to metal nanoparticles has drawn considerable attention. There are a number of advantages afforded by natural gums in the synthesis of nanoparticles such as their solubility in water avoiding hazardous organic solvents, biodegradability, and biocompatibility, economy, and easy availability. In addition to their reductant action, natural gums also act as capping agents for stabilizing the nanoparticles. This section reviews the various studies employing gum kondagogu for the synthesis of metal nanoparticles.

6.3.4 Gold nanoparticles (AuNPs)

Gold was known by some as the "elixir of life" in ancient times. Its nanoparticles formulation in the medicinal field is considered an effective candidate for targeted drug delivery, gene delivery, and diagnosis and treatment of diseases especially cancer, biosensing, imaging, molecular therapeutics, etc. (Arvizo et al., 2010; Cai and Yao, 2013; Cai et al., 2008; Zeng et al., 2011). The properties of gold nanoparticles, such as high biocompatibility, excellent affinity to bind with receptors, and noncytotoxicity, make them a reliable candidate as medicinal therapeutic agents (Dhar et al., 2008; Tiwari et al., 2011). The combination of gold with different compounds like gold-silica or gold-colloidal nanospheres and gold nanoshells are popular products for cancer treatment. Green synthesized gold nanoparticles are prepared mainly using the metal salt solution reduction method, which is not only nontoxic, economic, and environmental friendly, but also prepared easily without undue hassle. Different sources of reducing agents such as plant extracts, essential oil, fruits, bacteria, fungi, and polyphenols have been employed to synthesize gold nanoparticles (Gurunathan et al., 2014; Thakor et al., 2011). Among natural gums, xanthan gum (Pooja et al., 2014), katira gum (Maity et al., 2012), gum arabic (Wu and Chen, 2012), and gum kondagogu are being used to produce gold nanoparticles (Reddy et al., 2015; Santos et al., 2014).

Vinod et al. (2011) utilized gum kondagogu as a reducing, capping, and stabilizing agent in gold nanoparticle synthesis. The synthesis was carried out by heating the reaction mixture of chloroauric acid and gum kondagogu at 75 °C at 250 rpm for 1 h under different pH conditions (pH 4–12). It was observed that the reduction of Au^{3+} to Au^{o} could be achieved at a pH of 10. The formation of gold nanoparticles at a pH of 10 was explained by the fact that at this pH, the glucose ring opens by abstraction of α-proton of the ring, resulting in oxidation of glucose to gluconic acid by Au^{3+} ion with its concomitant reduction to Au^{o} (Vinod et al., 2011). In a similar study (Reddy et al., 2015), the reduction of chloroauric acid by gum kondagogu was carried out by heating their solutions in an autoclave at 120 °C for 15 min. The effects of varying concentrations of chloroauric acid (0.1–1 mM) and gum kondagogu (0.1%–0.5%, w/v) on nanoparticle synthesis revealed that the amount of gold nanoparticles increased by increasing the concentrations of chloroauric acid as well as gum kondagogu. The study reported that gum kondagogu-stabilized gold nanoparticles were discreet, spherical in shape, had an average size of 12 nm, and were stable to changes in pH and the presence of salt. Furthermore, the nanoparticles showed potent antibacterial activity against gram-positive and gram-negative bacteria. A similar research work studied the effect of autoclaving time (0–60 min) on the synthesis of gum kondagogu-capped gold nanoparticles (Selvi et al., 2017) (Fig. 6.5). It was observed that increasing the autoclaving time up to 30 min increases the quantity of gold nanoparticles as monitored by measuring the intensity of absorbance at 534 nm, but beyond this time a decrease in quantity occurs; this was attributed to the agglomeration of gold nanoparticles. The study also compared the effect of temperatures (25–95 °C) on the stability of gum kondagogu-capped gold nanoparticles, but the researchers did not find any change in the intensity of absorbance at 534 nm with variation in temperature. The nanoparticles synthesized were spherical in shape with a smooth surface and an average size of 4–12 nm. Furthermore, the gum kondagogu-capped gold nanoparticles exhibited antiproliferative activity in B16F10 melanoma cell line, which was suggested to be due to induction of apoptosis.

In a similar study, carboxymethyl gum kondagogu was employed to reduce chloroauric acid to gold nanoparticles, and for stabilizing the synthesized nanoparticles (Seku et al., 2019). The study evaluated the effect of varying the concentrations of carboxymethyl gum kondagogu (0.1%–1%, w/v) and chloroauric acid (0.1–1 mM), and the effect of autoclaving time (5–30 min) on nanoparticle synthesis by varying one factor at a time. As expected, increasing the concentrations of the reactants or autoclaving time enhanced the synthesis of gold nanoparticles. The spherical, crystallite nanoparticles of average size 11 nm and zeta potential −19 mV were found to be stable in the pH range of 5–12 but showed aggregation at pH

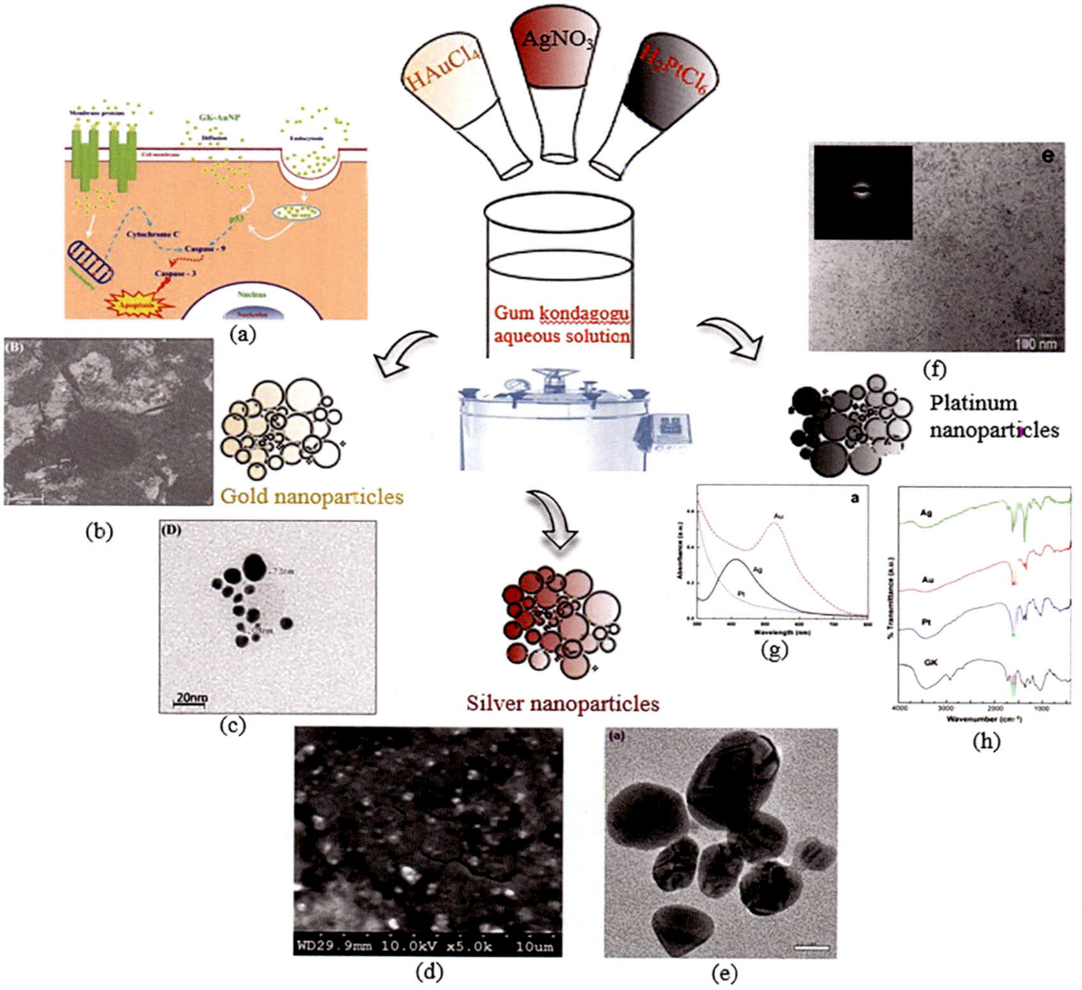

FIG. 6.5 Preparation of gold, silver, and platinum nanoparticles with their respective characterization images which includes schematic representation of synthesis (A), SEM (B), and TEM image (C) of gold nanoparticles; SEM (D) and TEM image (E) of silver nanoparticles; and TEM (F), UV (G), and IR spectra (H) of platinum nanoparticles. *Panels A–C: adapted from Selvi, S.K., Mahesh, J., Sashidhar, R.B., 2017. Bioactive carbohydrates and dietary fibre anti-proliferative activity of gum kondagogu* (Cochlospermum gossypium)*—gold nanoparticle constructs on B16F10 melanoma cells: an in vitro model. Bioact. Carbohydr. Diet. Fibre, 11, 38–47. https://doi.org/10.1016/j.bcdf.2017.07.002; panels D, F–H: adapted from Vinod, V.T.P., Saravanan, P., Sreedhar, B., Devi, D.K., Sashidhar, R.B., 2011. A facile synthesis and characterization of Ag, Au and Pt nanoparticles using a natural hydrocolloid gum kondagogu* (Cochlospermum gossypium)*. Colloids Surf. B: Biointerfaces, 83 (2), 291–298. https://doi.org/10.1016/j.colsurfb.2010.11.035; panel E: adapted from Kora, A.J., Sashidhar, R.B., Arunachalam, J., 2010. Gum kondagogu* (Cochlospermum gossypium)*: a template for the green synthesis and stabilization of silver nanoparticles with antibacterial application. Carbohydr. Polym. 82 (3), 670–679. https://doi.org/10.1016/j.carbpol.2010.05.034.*

2. Aggregation of nanoparticles was also observed on addition of sodium chloride at concentrations ≥2 M. Furthermore, the nanoparticles showed antibacterial activity against *Escherichia coli* and *Bacillus subtilis* that was comparable to the antibiotic ampicillin.

Gum kondagogu-capped gold nanoparticles, conjugated with folic acid (an active targeting ligand) and fluorescein isothiocyanate (a fluorescent dye), were evaluated for targeting and imaging applications in breast and lung cancer cells (Kumar et al., 2018). The carboxylic groups of folic acid were esterified with the hydroxyl groups of gum kondagogu-capped gold nanoparticles employing carbodiimide chemistry reactions. The folate-conjugated nanoparticles were labeled with fluorescein by the reaction between the carbonyl functional groups of gum kondagogu with the hydroxyl group of fluorescein using the same carbodiimide chemistry reaction. The folate-coupled and fluorescent-labeled gum kondagogu gold nanoparticles had an average particle size of 37 nm with a zeta potential of −23.7 mV. Furthermore, the nanoparticles showed biocompatibility with A549 and MCF-7 cells. However, there was no uptake of nanoparticles in A549 cells which lack folate receptors, while MCF-7 cells having folate receptors showed the uptake of nanoparticles.

6.3.5 Silver nanoparticles (AgNPs)

Silver nanoparticles have found wide applications in many fields such as in catalysis, energy conversion, optics electronics, photochemistry, sensing, and medicine compared to other bulk metals (Desireddy et al., 2013). Silver possesses good antimicrobial action and has been employed for wound healing formulations. Various natural and synthetic polymers like chitosan, heparin, cellulose, acacia gum, carboxymethyl curdlan, gellan gum, gellan gum, starch, *Lysiloma acapulcensis*, poly(ethylene glycol), and poly-(N-vinyl-2-pyrrolidone) have been investigated for the green synthesis of silver nanoparticles (Garibo et al., 2020). Gum kondagogu, which contains an abundance of hydroxyl, carboxyl, and carbonyl functional groups, is reported to complex with silver ions and reduce it to elemental silver. Apart from the polysaccharides, the proteins and tannins present in the gum also act as capping agents, preventing the agglomeration of silver nanoparticles (Kora and Sashidhar, 2014). Kora et al. (2010) prepared silver nanoparticles by autoclaving the silver nitrate solution in the presence of gum kondagogu at 121 °C and 15 psi (Fig. 6.5). The study observed that reducing the size of gum powder from 300 to 38 μm improves the efficiency of silver nanoparticle synthesis; this might be due to the increased dissolution of gum. Similar results were seen on increasing the concentration of gum from 0.1% to 0.5% (w/v) and increasing the autoclaving time from 10 to 60 min. These results were explained to be due to the increased reduction capacity of gum because of the greater availability of hydroxyl functionality and time for carrying out the reduction. The researchers were able to synthesize monodispersed, spherical silver nanoparticles of average size around 3 nm by autoclaving 1 mM of silver nitrate with 0.5% (w/v) of gum kondagogu at 121 °C for 60 min. The study reported the antibacterial activity of synthesized nanoparticles against gram-positive and gram-negative bacterial strains (Kora et al., 2010). In another study, the silver nanoparticles of size 4.5 nm prepared as above were evaluated for antibacterial, antibiofilm, and cytotoxic activity. The study reported the minimal inhibitory concentration of silver nanoparticles against *Staphylococcus aureus*, *Pseudomonas aeruginosa*, and *E. coli* to be 10, 5, and 2 μg/mL, respectively. Furthermore, the study showed antibiofilm activity of silver nanoparticles at the concentration of 2 μg/mL. The antibacterial and antibiofilm activity was attributed to the formation of higher reactive oxygen species and bacterial cell membrane damage induced by silver nanoparticles. However, the nanoparticles beyond the concentration of 2.5 μg/mL were found to be cytotoxic as measured by MTT assay on human cervical (HeLa) cell lines, which was suggested to be due to smaller particle size. To overcome the limitation of cytotoxicity of silver nanoparticles, the researchers evaluated the antibacterial potential of gum kondagogu reduced silver nanoparticles in combination with the antibiotics, viz., streptomycin or gentamicin or ciprofloxacin, by determining the fractional inhibitory concentration of different combinations (Rastogi et al., 2014). The result indicated that the combination of silver nanoparticles and streptomycin is synergistic while the combination of silver nanoparticles with gentamicin or ciprofloxacin was found to be additive. The study suggested that a combination of antibiotics and silver nanoparticles at noncytotoxic levels (i.e., 1 μg/mL) may be used for antibacterial applications with decreased potential for antibiotic resistance. Silver nanoparticles for antifungal application were prepared by reducing silver nitrate by gum kondagogu solutions at varying pH (5.0 or 5.5 or 6.0) and autoclaving for 15–165 min. The results revealed that the pH of 6.0 and autoclaving time of 120 min are optimal for the preparation of silver nanoparticles. The spherical nanoparticles of size 3.6 nm exhibited a minimal inhibitory concentration of 4.5 and 3.5 μg/mL against *Aspergillus parasiticus* and *Aspergillus flavus*, respectively. The antifungal activity of nanoparticles was attributed to the damage to the fungal mycelial membrane and oxidative stress (Malkapur et al., 2017).

The green synthetic methods of gum kondagogu-induced reduction of silver ions into silver nanoparticles used heating in an autoclave; a method involving heating under milder conditions of 45 °C has also been explored (Vinod et al., 2011). In this study, the effect of varying pH of gum kondagogu solution from 3 to 7 on the synthesis of silver nanoparticles at 45 °C under stirring (250 rpm) and in a nitrogen atmosphere was investigated. The study reported that the gum kondagogu completely reduced the silver ions into silver nanoparticles at a pH of 5. However, the reduction process was slow and complete reduction took 12 h. The plausible mechanism suggested was that the silver ions were reduced extracellularly in gum solution followed by the precipitation on the gum kondagogu matrices. Monodispersed nanoparticles of average size 5.5 nm, which did not aggregate even after 6 months of storage, were obtained.

A different method for synthesis of silver nanoparticles using carboxymethyl gum kondagogu as a capping and reducing agent and utilizing microwave heating has been reported (Seku et al., 2018a). The study involved exposure of reaction mixture comprising silver nitrate and aqueous solutions of carboxymethyl gum kondagogu to microwave radiation of 750 W for 50–90 s. The study found that increasing the concentration of carboxymethyl gum kondagogu from 0.1% to 0.5% (w/v) led to a decrease in particle size of nanoparticles, while increasing the concentration of silver nitrate from 0.1% to 0.5% (w/v) provided more number of nanoparticles. Furthermore, the nanoparticles were observed to be of spherical shape with an average size of 9 nm, showing good free radical scavenging and antibacterial activity.

Apart from elemental silver, its sulfide, i.e., Ag_2S, has been used as quantum dots for fluorescent labeling of biological objects. Silver sulfide was prepared in situ by adding sodium sulfide (0.1 M) to the reaction mixture of gum kondagogu (0%–1%, w/v) and silver nitrate

(0.1 M) with stirring under dark conditions, followed by autoclaving the reaction mixture at 121 °C, 15 psi for 1 h (Ayodhya and Veerabhadram, 2016). Silver sulfide nanoparticles of 25 nm with a zeta potential of −30 mV and spherical shape were obtained. The gum kondagogu-capped silver sulfide nanoparticles showed higher antibacterial activity in comparison with the uncapped silver sulfide nanoparticles against the strains of *E. coli*, *P. aeruginosa*, *S. aureus*, and *B. thirugiensis*.

6.3.6 Platinum nanoparticles (PtNPs)

A general method used to produce the metallic nanoparticles is the addition of a relevant reducing agent in metal salt. Various natural gum, plant extracts, microorganisms, egg yolk, algae, fungi, bovine serum albumin, and sheep milk are also already being used as reducing/stabilizing agents to produce noble platinum nanoparticles (Fahmy et al., 2020). Platinum nanoparticles of gum kondagogu were also prepared by using this method. Vinod et al. (2011) prepared the nanoparticles of three metals in one study, i.e., gold, silver, and platinum. All three nanoparticles were analyzed completely and compared to distinguish their stability behavior. Gold and silver nanoparticles have already been discussed in the above sections. However, platinum nanoparticles were prepared using a hexachloroplatinic acid (H_2PtCl_6) reagent. Basically, a certain amount of platinic acid and gum kondagogu was mixed at different pH (4, 5, 6, 8, and 10) and autoclaved for 15 min at 15 psi. The final optimum batch of platinum nanoparticles was found to produce at pH 8 at 121 °C. The formation of a black color aqueous solution confirms the production of platinum nanoparticles (Vinod et al., 2011). Fig. 6.5 shows the UV, FTIR, and TEM photographs of gum kondagogu-based platinum nanoparticles. In addition, it was noticed that the size of platinum nanoparticles was the smallest (3–10 nm diameter) compared to other metallic nanoparticles. This smaller size of nanoparticles leads to better biomedical activity due to greater surface area. Therefore, the results of the whole study indicated that platinum nanoparticulate formulation is a better option for biomedical applications rather than gold and silver nanoparticles. In another investigation, it was stated that reaction time, temperature, pH, and rate of adding required entity control the shape and size of green synthesized platinum nanoparticles (Fahmy et al., 2020). Platinum is employed as a catalyst in many electrochemical and chemical reactions but the agglomeration of platinum nanoparticles reduce their catalytic performance, which can be prevented by depositing the nanoparticles on the support materials. Deacteylated gum kondagogu was used to reduce $PtCl_4$ and graphene oxide simultaneously. The platinum nanoparticles deposited on reduced graphene oxide support were evaluated further as green catalysts (Venkateshaiah et al., 2019).

6.3.7 Copper nanoparticles (CuNPs)

Copper, an extremely conductive and noble metal, is very cheap, economic, and environmentally friendly. Nanoparticles prepared using copper metals are highly conductive in nature, with excellent catalytic and antibacterial activity. Several important properties such as better chemical stability, bioavailability, and intense plasmon resonance of copper nanoparticles are reported in the literature. Green synthesis of copper nanoparticles using natural and synthetic polymers such as gum arabic (Chawla et al., 2020), tree gum (Ramalechume et al., 2020), gum acacia (Dwivedi et al., 2021), guar gum (Kumar et al., 2012), etc., has already been reported. Gum kondagogu-based copper nanoparticles were synthesized using a two-stage chemical reduction method (Suresh et al., 2016a, b). Gum kondagogu was employed as a stabilizer, ascorbic acid as an antioxidant agent, hydrazine hydrate as the reducing agent, and copper sulfate as the precursor. Firstly, three different solutions—gum kondagogu and ascorbic acid, copper sulfate, and hydrazine hydrate and sodium hydroxide—were prepared in distilled water separately. Then the prepared gum solution was dropped in copper sulfate solution with stirring followed by the addition of hydrazine hydrate alkaline solution. The prepared blue color mixture eventually turned a brown-black color after continuous stirring for 1 h. The product was finally washed with methanol to obtain gum kondagogu-capped copper nanoparticles. Similarly to other nanoparticles, these nanoparticles also had spherical shapes with polycrystalline nature and an average diameter of 15 nm. The nanoparticles exhibited biofilm inhibition activity against *Klebsiella pneumonia*. In another reported study, the same authors prepared copper nanoparticles by adopting the above procedure, but they used copper nitrate instead of copper sulfate. However, in this study, photoluminescence spectra were observed additionally. It was concluded from the whole study that due to the presence of copper and natural gum, photoluminescence emission and intensity are increased and exhibit improved behavior (Suresh et al., 2016a). Seku et al. (2018b), using a similar method, synthesized copper nanoparticles employing the carboxymethyl derivative of gum kondagogu as a stabilizer and copper chloride as the precursor. The spherical-shaped copper nanoparticles of average size 14 nm were observed to be stable for at least 30 days in the air. The obtained metallic nanoformulation was found to have antibacterial and antifungal properties (Seku et al., 2018b).

6.3.8 Palladium nanoparticles (PdNPs)

Palladium is another recognized catalyst used for fine chemical and pharmaceutical product synthesis. Gum kondagogu have been explored as the bio-reducing agent for the synthesis of palladium nanoparticles. Aqueous solutions of kondagogu and palladium chloride were autoclaved at 121 °C, 15 psi for 30 min. The effect of variation in pH, concentrations of $PdCl_2$, and gum were observed in the study. Spherical, crystalline palladium nanoparticles of average size 6.5 nm and zeta potential −22.1 mV were obtained at the optimum gum concentration of 0.5%, palladium chloride −1 mM, and at pH 7. Additionally, these synthesized nanoparticles were reported to have better catalytic activity at even low concentrations for the reduction of 4-nitrophenol. With this, it was claimed that green synthesized gum kondagogu-based palladium nanoparticles can be employed as a commercial product in the remediation of toxic chemicals (Rastogi et al., 2015). The same authors in another similar study reported the different activity of the nanoformulation. Palladium nanoparticles were synthesized using gum kondagogu as a reducing agent and $PdCl_2$ following the above method. The study explored the possibility of substituting peroxidase with a gum kondagogu-palladium nanocomposite for the determination of glucose. The study observed a good affinity of the palladium nanocomposite toward 3,3,5,5-tetramethylbenizidine and hydrogen peroxide, which was utilized to develop a colorimetric assay method for the determination of glucose in human serum. The study concluded that the stable gum kondagogu-palladium nanocomposite has the potential for use in clinical diagnostics as a substitute for peroxidase enzyme (Rastogi et al., 2017).

6.3.9 Bimetallic nanoparticles

Bimetallic nanoparticles, also called alloy nanoparticles, are formed by simultaneous reduction of two metal ions. Bimetallic nanoparticles of Fe-Co, Fe-Ni, Au-Ag, Fe-Ni, etc., have been used for preparing nanoparticles with unique properties such as good chemical stability, high magnetic saturation, and high X-ray absorptivity. Subbiah and Beedu (2018) synthesized the Au-Ag bimetallic nanoparticles by simultaneous reduction of chloroauric acid (1 mM) and silver nitrate (1 mM) using gum kondagogu (0.5%, w/v) by autoclaving at 121 °C at 15 psi for 15 min. The bimetallic nanoparticles were observed to be spherically shaped with a smooth surface with an average diameter of 1–12 nm and zeta potential of −24.3 mV. Furthermore, the Au-Ag nanoparticles demonstrated antiproliferative activity against the B16F10 melanoma cell lines; this was shown to be due to the upregulation of the apoptotic p-53, caspase-3, and caspase-9 genes, and downregulation of the antiapoptotic Bcl-2 and Bcl-x(K) genes (Subbiah and Beedu, 2018).

6.3.10 Metal oxide nanoparticles

Oxides of metals such as iron, titanium, silicon, cerium, and zinc have been explored widely because of their unique properties for diverse biomedical applications such as biosensors, drug targeting, photoablation therapy, and bioimaging (McNamara and Tofail, 2017). Among the various metallic nanoparticles, iron oxide nanoparticles are popular nanocarriers to deliver multifunctional drugs in a controlled manner. These engineered nanoparticles have great advantages like good solubility and biocompatibility, unique physicochemical and biological properties, chemical stability, high surface area, small bandgap (2.1 eV), low toxicity, etc. (Iqbal et al., 2019). They also have extensive applications in different areas such as antigen detection, enzymes, vaccines, pathogens, radiology, sensors, biomedicine, and diagnostics (Ruan et al., 2018; Shah and Dobrovolskaia, 2018; Zhao et al., 2018). Natural polymer or plant extracts like *Araucaria heterophylla*, *Azadirachta indica* (Samrot et al., 2020), almond gum (Jaison et al., 2020), gum acacia (Mohamed et al., 2020), gum ghatti (Kulal and Badalamoole, 2020), and many more have been used for green synthesis of superparamagnetic iron oxide nanoparticles. Gum kondagogu has also been used in green synthesis of iron oxide nanoparticles (Raju et al., 2016). The reaction mixture consisting of aqueous solution of gum kondagogu (0.2% or 0.4%, w/v) and 1 mM ferrous sulfate was autoclaved at 121 °C for various time intervals (15–90 min). The increase in the concentration of gum was observed to increase the efficiency of nanoparticle synthesis. A similar effect was observed on increasing the autoclaving time up to 60 min, but beyond that, no significant increase in the efficiency of nanoparticle synthesis was found. The study reported the synthesis of circular, monodispersed nanoparticles of size 2–6 nm and zeta potential −20.7 mV. The synthesized nanoparticles were observed to facilitate the water uptake capacity as well as the seed germination property in agricultural biomass.

Titanium dioxide (TiO_2) is another versatile metal oxide that finds applications in skin care, cosmetics, and pharmaceutical products. It is used to confer whiteness to toothpaste, paint, paper, inks, and plastics products, and to promote photosynthesis and nitrogen metabolism in food plants (Raju et al., 2016; Saranya et al., 2018). Because of the aforementioned diverse applications, TiO_2 nanoparticles are one of the favorable nanoparticles formulations which can be used to design the environmental friendly sustainable pharmaceutical formulation. Saranya et al. (2018) carried out green synthesis of titanium oxide nanoparticles using gum kondagogu as a template. Titanium oxysulfate (0.1 M), a precursor of titanium oxide, was reacted with gum kondagogu (0.5%, w/v) at 90–95 °C under stirring, followed by centrifugation, washing, and drying of the product. The Ti^{3+}-gum kondagogu complex thus obtained was calcined at 500 °C, leaving behind the TiO_2 nanoparticles. The spherical, monodispersed nanoparticles of size 8–13 nm were also evaluated for photocatalytic activity. Furthermore, it was concluded that TiO_2 nanoparticles of gum kondagogou can be good candidates for dye effluent and water purification treatment (Saranya et al., 2018). Apart from this, no other research is reported in the literature that emphasizes the preparation of gum kondagogu-based TiO_2 nanoparticles, which indicates that there is still more scope to explore the advantage of TiO_2 nanoparticles on gum kondagogu or other similar gums.

6.3.11 Electrospun nanofibers

Electrospinning is the easiest and simple method of producing nanofibers. Electrospun nanofibers are light in weight, possess high porosity, and have a surface area in the range of 1–100 m^2/g. In addition, electrospun nanofibers have excellent mechanical strength and good permeability. Furthermore, electrospinning offers flexibility, the potential to scale up, and cost-effectiveness as well as the capacity to spin an extensive range of polymers. In biomedical applications, electrospun nanofibers have been used in wound healing, tissue engineering, and drug delivery. Padil et al. (2015) evaluated deacetylated gum kondagogu and its blend with polyvinyl alcohol in various ratios ranging from gum kondagogu/polyvinyl alcohol (100/0 to 0/100) for electrospinning. The deacetylated gum kondagogu solutions alone could not be electrospun due to their higher viscosities. The solutions of gum kondagogu/polyvinyl alcohol in 30/70 had viscosity appropriate to produce electrospun fibers with uniform linear morphology and diameter of 150 nm. It was suggested that the hydroxyl groups of polyvinyl alcohol form hydrogen bonds with the carboxyl and hydroxyl groups of gum kondagogu, reducing their viscosity and enabling their electrospinning into nanofibers with a uniform diameter. The nanofibers were also treated with methane plasma, which increased the BET surface area and contact angle of the nanofibers. The study further evaluated the plasma-treated nanofiber membranes for extraction of gold, silver, and platinum nanoparticles by adsorption. In a similar study, electrospun composite nanofibers of activated carbon, polyvinyl alcohol, and gum kondagogu were prepared and evaluated for adsorbing the toxic metals. In yet another study, a comparative evaluation of electrospun nanofibers of polyvinyl alcohol or polyethylene oxide with gum arabic or karaya or kondagogu was carried out (Thekkae et al., 2015). The study reported that blends of natural gums with polyethylene oxide were not suitable for electrospinning as they provided nonuniform fibers, while blends with polyvinyl alcohol provided nanofibers with uniform diameter and smooth surface (Padil et al., 2016). The nanofibers were hydrophobically modified by plasma treatment. Among the nanofibers of different gums, plasma-treated nanofibers electrospun using gum kondagogu showed the highest contact angle, BET surface area, and degree of stability. However, no report on biomedical applications of gum kondagogu-based electrospun nanofibers could be found. However, considering the properties of gum kondagogu-based electrospun fibers, there exists great potential in pharmaceutical applications of gum kondagogu-based electrospun nanofibers.

6.4 Conclusions

Gum kondagogu was initially used as a cheap substitute for gum karaya. A number of studies have been carried out to characterize its structural, physicochemical, and biological properties, to establish its separate identity. Various scientific studies utilizing the gelling, thickening, emulsifying, and bioadhesive properties of gum kondagogu and its derivatives have established their role as a polymer for modifying the release behavior of medicinal agents. Much work has been directed toward formulating mucoadhesive microparticles and for synthesizing metal nanoparticles using gum kondagogu as a template, but only little effort has been devoted to exploring its use in nanometric carriers for drugs. Gum kondagogu has excellent potential for use as biopolymer in electrospinning nanofibers. So far, the electrospun nanofibers based on gum kondagogu have been tested for bioremediation of toxic metals. There is a lot of scope in designing electrospun nanofiber-based wound dressings and membranes for drug delivery applications. In conclusion, the potential of gum kondagogu in nanoparticulate drug delivery has not been fully explored.

References

Ahmad, S., Ahmad, M., Manzoor, K., Purwar, R., Ikram, S., 2019. A review on latest innovations in natural gums based hydrogels: preparations & applications. Int. J. Biol. Macromol. 136, 870–890. https://doi.org/10.1016/j.ijbiomac.2019.06.113.

Arvizo, R., Bhattacharya, R., Mukherjee, P., 2010. Gold nanoparticles: opportunities and challenges in nanomedicine. Expert Opin. Drug Deliv. 7 (6), 753–763. https://doi.org/10.1517/17425241003777010.

Ayodhya, D., Veerabhadram, G., 2016. Biology antimicrobial activities of *Cochlospermum gossypium* capped Ag_2S nanoparticles. J. Photochem. Photobiol. B 157, 57–69. https://doi.org/10.1016/j.jphotobiol.2016.02.002.

Bera, H., Nadimpalli, J., Kumar, S., Vengala, P., 2017. Kondogogu gum-Zn+2-pectinate emulgel matrices reinforced with mesoporous silica for intragastric furbiprofen delivery. Int. J. Biol. Macromol. 104, 1229–1237. https://doi.org/10.1016/j.ijbiomac.2017.07.027.

Cai, H., Yao, P., 2013. In situ preparation of gold nanoparticle-loaded lysozyme-dextran nanogels and applications for cell imaging and drug delivery. Nanoscale 5 (7), 2892–2900. https://doi.org/10.1039/c3nr00178d.

Cai, W., Gao, T., Hong, H., Sun, J., 2008. Applications of gold nanoparticles in cancer nanotechnology. Nanotechnol. Sci. Appl. 1, 17–32. https://doi.org/10.2147/nsa.s3788.

Chakka, V.P., Zhou, T., 2020. Carboxymethylation of polysaccharides: synthesis and bioactivities. Int. J. Biol. Macromol. 165, 2425–2431. https://doi.org/10.1016/j.ijbiomac.2020.10.178.

Chaudhary, P.D., Pawar, H.A., 2014. Recently investigated natural gums and mucilages as pharmaceutical excipients: an overview. J. Pharm. 2014, 1–9. https://doi.org/10.1155/2014/204849.

Chawla, P., Kumar, N., Bains, A., Dhull, S.B., Kumar, M., Kaushik, R., Punia, S., 2020. Gum arabic capped copper nanoparticles: synthesis, characterization, and applications. Int. J. Biol. Macromol. 146, 232–242. https://doi.org/10.1016/j.ijbiomac.2019.12.260.

de Nooy, A.E.J., Rori, V., Masci, G., Dentini, M., Crescenzi, V., 2000. Synthesis and preliminary characterisation of charged derivatives and hydrogels from scleroglucan. Carbohydr. Res. 324, 116–126. https://doi.org/10.1016/S0008-6215(99)00286-4.

Desireddy, A., Conn, B.E., Guo, J., Yoon, B., Barnett, R.N., Monahan, B.M., Kirschbaum, K., Griffith, W.P., Whetten, R.-L., Landman, U., Bigioni, T.P., 2013. Ultrastable silver nanoparticles. Nature 501 (7467), 399–402. https://doi.org/10.1038/nature12523.

Dhar, S., Maheswara Reddy, E., Shiras, A., Pokharkar, V., Prasad, B L.V., 2008. Natural gum reduced/stabilized gold nanoparticles for drug delivery formulations. Chem. Eur. J. 14 (33), 10244–10250. https://doi.org/10.1002/chem.200801093.

Dwivedi, L.M., Shukla, N., Baranwal, K., Gupta, S., Siddique, S., Singh, V., 2021. Gum acacia modified Ni doped CuO nanoparticles: an excellent antibacterial material. J. Clust. Sci. 32 (1), 209–219. https://doi.org/10.1007/s10876-020-01779-7.

Fahmy, S.A., Preis, E., Bakowsky, U., Azzazy, H.M.E.S., 2020. Platinum nanoparticles: green synthesis and biomedical applications. Molecules 25 (21). https://doi.org/10.3390/molecules25214981.

FAO, 1991. Compendium of Food Additive Specifications. Food Agriculture Organisation.

Garibo, D., Borbón-Nuñez, H.A., de León, J.N.D., García Mendoza, E., Estrada, I., Toledano-Magaña, Y., Tiznado, H., Ovalle-Marroquin, M., Soto-Ramos, A.G., Blanco, A., Rodríguez, J.A., Romo, O.A., Chávez-Almazán, L.A., Susarrey-Arce, A., 2020. Green synthesis of silver nanoparticles using Lysiloma acapulcensis exhibit high-antimicrobial activity. Sci. Rep. 10 (1). https://doi.org/10.1038/s41598-020-69606-7.

Gurunathan, S., Han, J.W., Park, J.H., Kim, J.H., 2014. A green chemistry approach for synthesizing biocompatible gold nanoparticles. Nanoscale Res. Lett. 9 (1), 1–11. https://doi.org/10.1186/1556-276X-9-248.

Iqbal, J., Abbasi, B.A., Ahmad, R., Shahbaz, A., Zahra, S.A., Kanwa, S., Munir, A., Rabbani, A., Mahmood, T., 2019. Biogenic synthesis of green and cost effective iron nanoparticles and evaluation of their potential biomedical properties. J. Mol. Struct. 1199, 1–13. https://doi.org/10.1016/j.molstruc.2019.126979.

Jaison, D., Chandrasekaran, G., Mothilal, M., 2020. pH-sensitive natural almond gum hydrocolloid based magnetic nanocomposites for theragnostic applications. Int. J. Biol. Macromol. 154, 256–266. https://doi.org/10.1016/j.ijbiomac.2020.03.103.

Jamkhande, P.G., Ghule, N.W., Bamer, A.H., Kalaskar, M.G., 2019. Metal nanoparticles synthesis: an overview on methods of preparation, advantages and disadvantages, and applications. J. Drug Deliv. Sci. Technol. 53. https://doi.org/10.1016/j.jddst.2019.101174.

Janaki, B., Sashidhar, R.B., 1998. Physico-chemical analysis of gum kondagogu (*Cochlospermum gossypium*): a potential food additive. Food Chem. 61 (1–2), 231–236. https://doi.org/10.1016/S0308-8146(97)00089-7.

Kora, A.J., Sashidhar, R.B., 2014. Biogenic silver nanoparticles synthesized with rhamnogalacturonan gum: antibacterial activity, cytotoxicity and its mode of action. Arab. J. Chem. 11 (3), 313–323. https://doi.org/10.1016/j.arabjc.2014.10.036.

Kora, A.J., Sashidhar, R.B., Arunachalam, J., 2010. Gum kondagogu (*Cochlospermum gossypium*): a template for the green synthesis and stabilization of silver nanoparticles with antibacterial application. Carbohydr. Polym. 82 (3), 670–679. https://doi.org/10.1016/j.carbpol.2010.05.034.

Krishna, R.R., Murthy, T.E.G.K., 2010. Preparation and evaluation of mucoadhesive microcapsules of glipizide formulated with gum kondagogu: in vitro and in vivo. Acta Pharm. Sci. 52 (3), 335–344. http://www.actapharmasciencia.org/actapharmasciencia.org/pdf/2010_vol52_no3_11.pdf.

Kulal, P., Badalamoole, V., 2020. Efficient removal of dyes and heavy metal ions from waste water using gum ghatti-graft-poly(4-acryloylmorpholine) hydrogel incorporated with magnetite nanoparticles. J. Environ. Chem. Eng. 8 (5). https://doi.org/10.1016/j.jece.2020.104207.

Kumar, A., Ahuja, M., 2012. Carboxymethyl gum kondagogu: synthesis, characterization and evaluation as mucoadhesive polymer. Carbohydr. Polym. 90 (1), 637–643. https://doi.org/10.1016/j.carbpol.2012.05.089.

Kumar, A., Ahuja, M., 2013. Carboxymethyl gum kondagogu-chitosan polyelectrolyte complex nanoparticles: preparation and characterization. Int. J. Biol. Macromol. 62, 80–84. https://doi.org/10.1016/j.ijbiomac.2013.08.035.

Kumar, A., Aerry, S., Saxena, A., De, A., Mozumdar, S., 2012. Copper nanoparticulates in guar-gum: a recyclable catalytic system for the Huisgen [3 + 2]-cycloaddition of azides and alkynes without additives under ambient conditions. Green Chem. 14 (5), 1298–1301. https://doi.org/10.1039/c2gc35070j.

Kumar, S.S.D., Mahesh, A., Antoniraj, M.G., Rathore, H.S., Houreld, N.N., Kandasamy, R., 2018. Cellular imaging and folate receptor targeting delivery of gum kondagogu capped gold nanoparticles in cancer cells. Int. J. Biol. Macromol. 109, 220–230. https://doi.org/10.1016/j.ijbiomac.2017.12.069.

Li, J., Hu, X., Li, X., Ma, Z., 2016. Effects of acetylation on the emulsifying properties of Artemisia sphaerocephala Krasch. polysaccharide. Carbohydr. Polym. 144, 531–540. https://doi.org/10.1016/j.carbpol.2016.02.039.

Lombardo, S., Villares, A., 2020. Engineered multilayer microcapsules based on polysaccharides nanomaterials. Molecules 25 (19). https://doi.org/10.3390/molecules25194420.

Maity, S., Kumar Sen, I., Sirajul Islam, S., 2012. Green synthesis of gold nanoparticles using gum polysaccharide of *Cochlospermum religiosum* (katira gum) and study of catalytic activity. Phys. E: Low-Dimens. Syst. Nanostruct. 45, 130–134. https://doi.org/10.1016/j.physe.2012.07.020.

Malik, S., Ahuja, M., 2011. Gum kondagogu-g-poly (acrylamide): microwave-assisted synthesis, characterisation and release behaviour. Carbohydr. Polym. 86 (1), 177–184. https://doi.org/10.1016/j.carbpol.2011.04.027.

Malik, S., Kumar, A., Ahuja, M., 2012. Synthesis of gum kondagogu-g-poly(N-vinyl-2-pyrrolidone) and its evaluation as a mucoadhesive polymer. Int. J. Biol. Macromol. 51 (5), 756–762. https://doi.org/10.1016/j.ijbiomac.2012.07.009.

Malkapur, D., Devi, M.S., Rupula, K., Sashidhar, R.B., 2017. Biogenic synthesis, characterisation and antifungal activity of gum kondagogu-silver nano bio composite construct: assessment of its mode of action. IET Nanobiotechnol. 11 (7), 866–873. https://doi.org/10.1049/iet-nbt.2017.0043.

Mankala, S.K., Nagamalli, N.K., Raprla, R., Kommula, R., 2011. Preparation and characterization of mucoadhesive microcapsules of gliclazide with natural gums. Stamford J. Pharm. Sci. 4 (1), 38–48. https://doi.org/10.3329/sjps.v4i1.8865.

McNamara, K., Tofail, S.A.M., 2017. Nanoparticles in biomedical applications. Adv. Phys. 2 (1), 54–88. https://doi.org/10.1080/23746149.2016.1254570.

Mohamed, A.A., Mahmoud, G.A., ElDin, M.R.E., Saad, E.A., 2020. Synthesis and properties of (gum acacia/polyacryamide/SiO_2) magnetic hydrogel nanocomposite prepared by gamma irradiation. Polym. Plast. Technol. Mater. 59 (4), 357–370. https://doi.org/10.1080/25740881.2019.1647240.

Mohammadinejad, R., Kumar, A., Ranjbar-Mohammadi, M., Ashrafizadeh, M., Han, S.S., Khang, G., Roveimiab, Z., 2020. Recent advances in natural gum-based biomaterials for tissue engineering and regenerative medicine: a review. Polymers 12 (1). https://doi.org/10.3390/polym12010176.

Naidu, V.G.M., Madhusudhana, K., Sashidhar, R.B., Ramakrishna, S., Khar, R.K., Ahmed, F.J., Diwan, P.V., 2009. Polyelectrolyte complexes of gum kondagogu and chitosan, as diclofenac carriers. Carbohydr. Polym. 76 (3), 464–471. https://doi.org/10.1016/j.carbpol.2008.11.010.

Padil, V.V.T., Stuchlík, M., Černík, M., 2015. Plasma modified nanofibres based on gum kondagogu and their use for collection of nanoparticulate silver, gold and platinum. Carbohydr. Polym. 121, 468–476. https://doi.org/10.1016/j.carbpol.2014.11.074.

Padil, V.V.T., Senan, C., Wacławek, S., Černík, M., 2016. Electrospun fibers based on Arabic, karaya and kondagogu gums. Int. J. Biol. Macromol. 91, 299–309. https://doi.org/10.1016/j.ijbiomac.2016.05.064.

Pal, D., Sahu, C.K., Haldar, A., 2014. Bhasma: the ancient Indian nanomedicine. J. Adv. Pharm. Technol. Res. 5 (1), 4–12. https://doi.org/10.4103/2231-4040.126980.

Pooja, D., Panyaram, S., Kulhari, H., Rachamalla, S.S., Sistla, R., 2014. Xanthan gum stabilized gold nanoparticles: characterization, biocompatibility, stability and cytotoxicity. Carbohydr. Polym. 110, 1–9. https://doi.org/10.1016/j.carbpol.2014.03.041.

Raju, D., Mehta, U.J., Beedu, S.R., 2016. Biogenic green synthesis of monodispersed gum kondagogu (*Cochlospermum gossypium*) iron nanocomposite material and its application in germination and growth of mung bean (*Vigna radiata*) as a plant model. IET Nanobiotechnol. 10 (3), 141–146. https://doi.org/10.1049/iet-nbt.2015.0112.

Ramalechume, C., Shamili, P., Krishnaveni, R., Swamidoss, C.M.A., 2020. Synthesis of copper oxide nanoparticles using tree gum extract, its spectral characterization, and a study of its anti- bactericidal properties. Mater. Today Proc. 33, 4151–4155. https://doi.org/10.1016/j.matpr.2020.06.587.

Rana, V., Rai, P., Tiwary, A.K., Singh, R.S., Kennedy, J.F., Knill, C.J., 2011. Modified gums: approaches and applications in drug delivery. Carbohydr. Polym. 83 (3), 1031–1047. https://doi.org/10.1016/j.carbpol.2010.09.010.

Rastogi, L., Sashidhar, R.B., Karunasagar, D., Arunachalam, J., 2014. Gum kondagogu reduced/stabilized silver nanoparticles as direct colorimetric sensor for the sensitive detection of Hg^{2+} in aqueous system. Talanta 118, 111–117. https://doi.org/10.1016/j.talanta.2013.10.012.

Rastogi, L., Beedu, S.R., Kora, A.J., 2015. Facile synthesis of palladium nanocatalyst using gum kondagogu (*Cochlospermum gossypium*): a natural biopolymer. IET Nanobiotechnol. 9 (6), 362–367. https://doi.org/10.1049/iet-nbt.2014.0055.

Rastogi, L., Karunasagar, D., Sashidhar, R.B., Giri, A., 2017. Peroxidase-like activity of gum kondagogu reduced/stabilized palladium nanoparticles and its analytical application for colorimetric detection of glucose in biological samples. Sensors Actuators B Chem. 240, 1182–1188. https://doi.org/10.1016/j.snb.2016.09.066.

Reddy, G.B., Madhusudhan, A., Ramakrishna, D., Ayodhya, D., Venkatesham, M., Veerabhadram, G., 2015. Green chemistry approach for the synthesis of gold nanoparticles with gum kondagogu: characterization, catalytic and antibacterial activity. J. Nanostruct. Chem. 5, 185–193. https://doi.org/10.1007/s40097-015-0149-y.

Ruan, H., Wu, X., Yang, C., Li, Z., Xia, Y., Xue, T., Shen, Z., Wu, A., 2018. A supersensitive CTC analysis system based on triangular silver nanoprisms and SPION with function of capture, enrichment, detection, and release. ACS Biomater. Sci. Eng. 4 (3), 1073–1082. https://doi.org/10.1021/acsbiomaterials.7b00825.

Samrot, A.V., Bhavya, K.S., Angalene, J.L.A., Roshini, S.M., Preethi, R., Steffi, S.M., Raji, P., Kumar, S.S., 2020. Utilization of gum polysaccharide of *Araucaria heterophylla* and *Azadirachta indica* for encapsulation of cyfluthrin loaded super paramagnetic iron oxide nanoparticles for mosquito larvicidal activity. Int. J. Biol. Macromol. 153, 1024–1034. https://doi.org/10.1016/j.ijbiomac.2019.10.232.

Santos, S.A.O., Pinto, R.J.B., Rocha, S.M., Marques, P.A.A.P., Neto, C.P., Silvestre, A.J.D., Freire, C.S.R., 2014. Unveiling the chemistry behind the green synthesis of metal nanoparticles. ChemSusChem 7 (9), 2704–2711. https://doi.org/10.1002/cssc.201402126.

Saranya, S.K.S.K.S., Padil, V.V.T., Senan, C., Pilankatta, R., Saranya, S.K.K., George, B., Wacławek, S., Černík, M., 2018. Green synthesis of high temperature stable anatase titanium dioxide nanoparticles using gum kondagogu: characterization and solar driven photocatalytic degradation of organic dye. Nano 8 (12). https://doi.org/10.3390/nano8121002.

Saravanan, P., Vinod, V.T.P., Sreedhar, B., Sashidhar, R.B., 2012. Gum kondagogu modified magnetic nano-absorbant—an efficient protocol for removal of various toxic metal ions. Mater. Sci. Eng. C 32, 581–586. https://doi.org/10.1016/j.msec.2011.12.015.

Sashidhar, R.B., Raju, D., Karuna, R., 2015. Tree gum: gum kondagogu. In: Polysaccharides: Bioactivity and Biotechnology. Springer International Publishing, pp. 185–217, https://doi.org/10.1007/978-3-319-16298-0_32.

Seku, K., Ganapuram, B.R., Pejjai, B., Kotu, G.M., Narasimha, G., 2018a. Hydrothermal synthesis of copper nanoparticles, characterization and their biological applications. Int. J. Nano Dimens. 9 (1), 7–14.

Seku, K., Reddy, B., Babu, G., Kishore, P., Kadimpati, K., Golla, N., 2018b. Microwave assisted synthesis of silver nanoparticles and their application in catalytic, antibacterial and antioxidant activities. J. Nanostruct. Chem. 8 (2), 179–188. https://doi.org/10.1007/s40097-018-0264-7.

Seku, K., Gangapuram, B.R., Pejjai, B., Hussain, M., Hussaini, S.S., Golla, N., Kadimpati, K.K., 2019. Eco-friendly synthesis of gold nanoparticles using carboxymethylated gum *Cochlospermum gossypium* (CMGK) and their catalytic and antibacterial applications. Chem. Pap. 73 (7), 1695–1704. https://doi.org/10.1007/s11696-019-00722-z.

Selvi, S.K., Mahesh, J., Sashidhar, R.B., 2017. Ioactive carbohydrates and dietary fibre anti-proliferative activity of gum kondagogu (*Cochlospermum gossypium*)—gold nanoparticle constructs on B16F10 melanoma cells: an in vitro model. Bioact. Carbohydr. Diet. Fibre 11, 38–47. https://doi.org/10.1016/j.bcdf.2017.07.002.

Shah, A., Dobrovolskaia, M.A., 2018. Immunological effects of iron oxide nanoparticles and iron-based complex drug formulations: therapeutic benefits, toxicity, mechanistic insights, and translational considerations. Nanomedicine 14 (3), 977–990. https://doi.org/10.1016/j.nano.2018.01.014.

Singh, J., Singh, B., 2013. Gum Kondagogu – a natural exudate from Cochlospermum gossypium. Botanica 62–63, 115–123.

Soni, N., Shah, N.N., Singhal, R.S., 2019. Dodecenyl succinylated guar gum hydrolysate as a wall material for microencapsulation: synthesis, characterization and evaluation. J. Food Eng. 242, 133–140. https://doi.org/10.1016/j.jfoodeng.2018.08.030.

Srivastava, S., Chowdhury, A.R., Sharma, S.C., 2018. Control of deacetylation in gum karaya on storage for quality retention. Int. J. Bioresour. Sci. 5 (2), 1–5. https://doi.org/10.30954/2347-939655.02.2018.2.

Subbiah, K.S., Beedu, S.R., 2018. Biogenic synthesis of biopolymer-based Ag–Au bimetallic nanoparticle constructs and their anti-proliferative assessment. IET Nanobiotechnol. 12 (8), 1047–1055. https://doi.org/10.1049/iet-nbt.2018.5135.

Suresh, Y., Annapurna, S., Bhikshamaiah, G., Singh, A.K., 2016a. Green luminescent copper nanoparticles. IOP Conf. Ser. Mater. Sci. Eng. 149 (1). https://doi.org/10.1088/1757-899X/149/1/012187.

Suresh, Y., Annapurna, S., Singh, A.K., Chetana, A., Pasha, C., Bhikshamaiah, G., 2016b. Characterization and evaluation of anti-biofilm effect of green synthesized copper nanoparticles. Mater. Today Proc. 3 (6), 1678–1685. https://doi.org/10.1016/j.matpr.2016.04.059.

Thakor, A.S., Jokerst, J., Zavaleta, C., Massoud, T.F., Gambhir, S.S., 2011. Gold nanoparticles: a revival in precious metal administration to patients. Nano Lett. 11 (10), 4029–4036. https://doi.org/10.1021/nl202559p.

Thekkae, V., Vinod, P., Stanislaw, W., Miroslav, C., 2015. Activated Carbon Nanofibres from Gum Kondagogu for Remediation of Toxic Metals. Nanocon, Brno, Czech Republic, EU, pp. 1–8.

Tiwari, P.W., Vig, K., Dennis, V.A., Singh, S.R., 2011. Functionalized gold nanoparticles and their biomedical applications. Nano 1 (1), 31–63. https://doi.org/10.3390/nano1010031.

Vegi, G.M.N., Sistla, R., Srinivasan, P., Beedu, S.R., Khar, R.K., Diwan, P.V., 2009. Emulsifying properties of gum kondagogu (*Cochlospermum gossypium*), a natural biopolymer. J. Sci. Food Agric. 89, 1271–1276. https://doi.org/10.1002/jsfa.3568.

Venkateshaiah, A., Silvestri, D., Ramakrishnan, R.K., Wacławek, S., Padil, V.V.T., Černík, M., Varma, R.S., 2019. Gum kondagoagu/reduced graphene oxide framed platinum nanoparticles and their catalytic role. Molecules 24 (20). https://doi.org/10.3390/molecules24203643.

Vinod, V.T.P., Sashidhar, R.B., 2009. Solution and conformational properties of gum kondagogu (*Cochlospermum gossypium*)—a natural product with immense potential as a food additive. Food Chem. 116 (3), 686–692. https://doi.org/10.1016/j.foodchem.2009.03.009.

Vinod, V.T.P., Sashidhar, R.B., 2010. Surface morphology, chemical and structural assignment of gum kondagogu (*Cochlospermum gossypium* DC.): an exudate tree gum of India. Indian J. Nat. Prod. Resour. 1 (2), 181–192. http://nopr.niscair.res.in/bitstream/123456789/9825/1/IJNPR%201(2)%20181-192.pdf.

Vinod, V.T.P., Sashidhar, R.B., Suresh, K.I., Rama Rao, B., Vijaya Saradhi, U.V.R., Prabhakar Rao, T., 2008. Morphological, physico-chemical and structural characterization of gum kondagogu (*Cochlospermum gossypium*): a tree gum from India. Food Hydrocoll. 22 (5), 899–915. https://doi.org/10.1016/j.foodhyd.2007.05.006.

Vinod, V.T.P., Sashidhar, R.B., Sarma, V.U.M., Raju, S.S., 2010. Comparative amino acid and fatty acid compositions of edible gums kondagogu (*Cochlospermum gossypium*) and karaya (*Sterculia urens*). Food Chem. 123 (1), 57–62. https://doi.org/10.1016/j.foodchem.2010.03.127.

Vinod, V.T.P., Saravanan, P., Sreedhar, B., Devi, D.K., Sashidhar, R.B., 2011. A facile synthesis and characterization of Ag, Au and Pt nanoparticles using a natural hydrocolloid gum kondagogu (*Cochlospermum gossypium*). Colloids Surf. B: Biointerfaces 83 (2), 291–298. https://doi.org/10.1016/j.colsurfb.2010.11.035.

Wang, W., Liu, X., Xie, Y., Zhang, H., Yu, W., Xiong, Y., Xie, W., Ma, X., 2006. Microencapsulation using natural polysaccharides for drug delivery and cell implantation. J. Mater. Chem. 16 (32), 3252–3267. https://doi.org/10.1039/b603595g.

Wu, C.C., Chen, D.H., 2012. Spontaneous synthesis of gold nanoparticles on gum arabic-modified iron oxide nanoparticles as a magnetically recoverable nanocatalyst. Nanoscale Res. Lett. 7. https://doi.org/10.1186/1556-276X-7-317.

Zeng, S., Yong, K.T., Roy, I., Dinh, X.Q., Yu, X., Luan, F., 2011. A review on functionalized gold nanoparticles for biosensing applications. Plasmonics 6 (3), 491–506. https://doi.org/10.1007/s11468-011-9228-1.

Zhao, Y., Zhao, X., Cheng, Y., Guo, X., Yuan, W., 2018. Iron oxide nanoparticles-based vaccine delivery for cancer treatment. Mol. Pharm. 15 (5), 1791–1799. https://doi.org/10.1021/acs.molpharmaceut.7b01103.

CHAPTER

7

Gum-based nanoparticles in cancer therapy

Maria John Newton Amaldoss[a,b,c] *and Reeta*[d]

[a]Australian Centre for Nanomedicine, UNSW Sydney, Sydney, NSW, Australia [b]Adult Cancer Program, Lowy Cancer Research Centre, UNSW Sydney, Sydney, NSW, Australia [c]Prince of Wales Clinical School, UNSW Sydney, Sydney, NSW, Australia [d]St. Soldier Institute of Pharmacy, Jalandhar, Punjab, India

7.1 Introduction

In recent years, human life has been positively impacted by nanotechnology (Patil et al., 2008). The potential benefits of nanotechnology extend to fields such as medicine (Soares et al., 2018), mechanical (Gu et al., 2010), electrical (Contreras et al., 2017), communication (Hao et al., 2020; Abdel-Mottaleb et al., 2011), solar energy (Abdel-Mottaleb et al., 2011), oxide fuel battery (Fan et al., 2018), cosmetic, clothes (Raj et al., 2012; Gharehaghaji, 2019), agriculture (Prasad et al., 2017), etc. Nanotechnology offers unlimited benefits to the medical field. Pharmaceutically it provides significant advantages in solubility, stability, and efficacy. Nanomedicine, often known as therapeutic nanoparticles, is defined as "very specialised molecular medical interventions for treating disease or regenerating damaged tissues like bone, muscle, or nerve." To be accurate, they should be less than 100 nm; nonetheless, this dimension is more diffuse (Jeevanandam et al., 2018; Webster, 2006). However, nanoparticles may not be a collective term to describe the various nano-drug delivery systems such as nanosphere, nanorods, nanosuspension, nanocrystals, and nano-vesicular systems such as nano lipid carriers and liposomes. Most pharmaceutical nano-drug delivery systems are prepared using polymers of natural and synthetic origin (Patra et al., 2018). Nanoparticles prepared using synthetic and natural polymers are called polymeric nanoparticles (Madkour, 2019). These polymeric nanoparticles are usually made from biodegradable and non-biodegradable polymers. In this category, natural gums belong to natural polymeric nanoparticles, which have a significant role in drug delivery, pharmaceutical product development, and therapy (Ngwuluka et al., 2014).

Micro- and Nanoengineered Gum-Based Biomaterials for Drug Delivery and Biomedical Applications
https://doi.org/10.1016/B978-0-323-90986-0.00013-3

Drug delivery and therapy are the two main functions of therapeutic nanoparticles. Nanoparticle-based targeted drug delivery systems are very efficient in delivering drugs for many challenging diseases such as cancer, Alzheimer's disease, Parkinson's disease, diseases related to the gastrointestinal tract (GIT) and viral and bacterial infections (Kabanov and Gendelman, 2007). Conventional pharmaceutical products such as parenteral, tablets, capsules, solutions, suspensions, emulsions, and ointments are known for their limitations such as high dose requirements, low bioavailability, first-pass effect, patient intolerance, and stability issues (Gupta et al., 2013; Wen et al., 2015). Conventional pharmaceutical products are associated with fluctuations in the plasma drug level and cannot provide sustained or targeted effects, so novel drug delivery systems emerge in the market (Sheehan et al., 2012). The next generation of sophisticated drug delivery technologies is attracted to nanoparticles. Many nanotechnology-based drug delivery systems are in the pipeline to reach the market few of them are already approved by US Food and Drug Administration (USFDA) (Patra et al., 2018). The nanoparticles demonstrate properties entirely different from their bulk materials, especially in physical and chemical properties (Jeevanandam et al., 2018; Schwirn et al., 2014; Khan et al., 2019). In the pharmaceutical nanoparticle category, gum-based nanoparticles are emerging as an efficient nano-drug delivery system that delivers the drug to various sites for multiple diseases (Shariatinia, 2019; Mohammadinejad et al., 2020). Since natural gums are derived from natural sources, they demonstrate a significant advantage over synthetic polymer-based drug delivery systems in terms of safety and biocompatibility (Mohammadinejad et al., 2020).

7.2 Principal natural gums in pharmaceutical applications

Gums have diverse applications as thickeners, emulsifiers, viscosity enhancers, sweeteners, etc. (Choudhary and Pawar, 2014). Modern pharmaceutical applications include binders and drug release modifiers. However, most of the gums are required in very high concentrations to function as drug release modifiers (Choudhary and Pawar, 2014). They are also demonstrating their high swellability, solubility at acidic pH conditions which also impacts their pharmaceutical properties (Rizwan et al., 2017). As a result, modification of the physicochemical features of gums is required for their regular use. Gums are modified by functional group derivatization, polymer grafting, and ion crosslinking. These are beneficial in optimizing the medicinal effects and pharmaceutical applications of gums (Patel et al., 2020).

Several characteristics that distinguish gums from other natural polymers, including their hydrophilicity toward water to create a gel, their limited calories generation and absorption, their capacity to form a film, and their ability to operate as a bulk forming agent in the large intestine. Crosslinking, electrostatic and ionic interaction, hydrophobicity, inter- and intramolecular hydrogen bonding, and chemical modification all contribute to the film-forming property. (Mudgil et al., 2014; Thombare et al., 2016) (Table 7.1). Gums used in biomedical applications are obtained from one of the sources listed below.

- *Marine*: agar, carrageenans, alginic acid, laminarin.
- *Microbial*: glycan, dextran, gellan gum, dextran; xanthan, guardian, pullulan, zanflo.
- *Plant*: *seed gums*—guar gum, locust bean gum, starch, amylase, cellulose); *tree exudates*—gum arabica, gum ghatti, gum karaya, gum tragacanth, khaya and albizia gums; *tuber*—konjac, potato starch; *extracts*—pectin larch gum, gums.

TABLE 7.1 Principal pharmaceutical natural gums and composition.

Gum	Source	Composition	References
Gum arabic	*Acacia Senegal* (*Leguminosae*)	Galactose, arabinose, rhamnose, glucuronic acid, 4-*O*-methyl glucuronic acid	Ali et al. (2009)
Xanthan gum	*Xanthomonas campestris*	Glucose, mannose, glucuronic acid	Rosalam and England (2006)
Alginates	*Laminaria hyperborea*, *Ascophyllum nodosum*, and *Macrocystis pyrifera*	Alternating blocks of β-D-mannuronic acid (M) and its C-5 epimer α-L-guluronic acid (G) repeating units linked by a 1,4-linkage	Szekalska et al. (2016)
Gellan gum	*Sphingomonas elodea*	D-glucose D-glucuronic acid, and L-rhamnose	Sworn (2009)
Dextrin	*Leuconostoc mesenteroides lectobacteriaceae*	D-glucose units connected by alpha-(1–6) linkages	Khalikova et al. (2005)
Cashew gum	*Anacardium occidentale* (*Anacardiaceae*)	Mannose, galactose, glucuronic acid	Animesh et al. (2012)
Gum ghatti	*Anogeissus latifolia* (*Combretaceae*)	L-arabinose, D-galactose, D-mannose, D-xylose, D-glucuronic acid	Deshmukh et al. (2012)
Guar gum	*Cyamopsis etragonolobus* (*Leguminosae*)	β-D-mannose monomers, single sugar side-chains of α-D-galactose	Mudgil et al. (2014)
Gum karaya	*Sterculia urens* (*Sterculiaceae*)	D-galactose, L-rhamnose, D-galacturonic acid	BeMiller (2019)
Gum kondagogu	*Cochlospermum gossypium* (*Bixaceae*)	Arabinose, galactose, rhamnose, mannose, α-D-glucose, and sugar acids like D-glucuronic acid and D-galacturonic acid	Vinod et al. (2008a)
Gum tragacanth	*Astragalus gummifer* (*Astragalus*)	D-galactose, L-fucose, D-xylose, L-arabinose, L-rhamnose	Nazarzadeh Zare et al. (2019)
Tamarind gum	*Tamarindus indica* (*Leguminoseae*)	β-(1,4)-D-glucan backbone with α-(1,6)-D-xylose branches that are partially substituted with β-(1,2)-D-galactose [2,3]	Khushbu and Warkar (2020)
Konjac gum	*Amorphophallus konjac* (*Araceae*)	D-glucose and D-mannose	Khan and Marya (2019)

7.2.1 Xanthan gum

Xanthan gum is a high molecular weight polysaccharide, and is also known as corn sugar gum. Xanthan gum contains approximately 1.25% of pyruvic acid (Singhvi et al., 2019). It is soluble in hot or cold water and produces a viscous solution at low concentrations. Xanthan gum is an effective thickener and emulsifier in pharmaceutical formulation. It has pseudoplastic properties that are very useful in preparing toothpaste and ointment to hold their shape. The majority of drugs are compatible with Xanthan gum. Xanthan gum is an excellent suspending agent for antispasmodics topically for esophageal spasm. The gum demonstrates good resistance to acidic and alkaline conditions when mixed with another polymer such as HPMC to modify the drug release profile (Vinay et al., 2017).

7.2.2 Alginates

Alginate is a water-soluble linear polysaccharide derived from β-D-mannuronic acid (M) and α-L-guluronic acid (G) found in brown algae and certain bacterial strains (e.g., *Pseudomonas aeruginosa*) (Szekalska et al., 2016). Alginate, also known as alginic acid, is an anionic polymer found in the cell wall of algae and produced by bacteria such as Pseudomonas Azobacteria. After polymerization, D-mannuronic acid is converted enzymatically to L-guluronic acid, which differs at C5. While their polymers are structurally similar, their confirmation polymer is significantly different. Alginate is a biocompatible and biodegradable mucoadhesive carrier for a wide variety of pharmaceutical molecules (Dheer et al., 2017). Alginate hydrogels are used in wound healing, drug delivery, and tissue engineering (Lee and Mooney, 2012). Crosslinkers enable controlled delivery of drugs and biological molecules from alginate gels, which are widely used in cell transplantation and tissue engineering.

7.2.3 Guar gum

Guar gum is a polysaccharide called galactomannan containing sugar galactose and mannose. Guar gum is derived from the Indian Cluster Bean (*Cyamopsis tetragonoloba*). It is composed of a 1,4-mannose backbone that is partially substituted with 1,6 galactose side chains (Aronson, 2016). It is hydrophilic, and swells in cold water to form a viscous colloidal dispersion and solution. The gelling property of guar gum controls the drug release from the dosage form. Guar gum is a colon-targeting polymer, and the glycosidic linkage of guar gum demonstrates degradation in the large intestine by the activity of colonic microbiota which aids the drug release (Aminabhavi et al., 2014).

7.2.4 Tamarind gum

Tamarind seed polysaccharide is a useful excipient for a wide range of pharmaceutical applications (Durai et al., 2012; Newton et al., 2015). The structure of tamarind seed polysaccharide has been proposed as a carbohydrate composed of a main chain or β-D-(1,4)-linked glucopyranosyl units with a side chain consisting of a single xylopyranosyl unit attached to every second, third, and fourth D-glucopyranosyl unit through an α-D-(1–6) linkage. One D-galactopyranosyl unit is attached to one of the xylopyranosyl units through a β-D-(1–2) linkage (Sharma et al., 2014a). Tamarind gum (TG) has potential uses in the pharmaceutical and cosmetic industries. Tamarind seed polysaccharide is unique in its high drug holding capacity, increased swelling index, and high thermal stability, especially for various novel drug delivery systems (Shukla et al., 2018). Tamarind seed polysaccharide is also used as a thickener, binder, release retardant, modifier, suspending agent, viscosity enhancer, and emulsifying agent, and as a carrier for novel drug delivery systems in oral, buccal, colon, ocular systems, nanofabrication, and wound dressing (Lovegrove et al., 2017).

7.2.5 Tragacanth gum

The exudate of *Astragalus* is a long curving string or ribbon of rapidly hardening gum (Mohammadifar et al., 2006). Gum *tragacanth* is an effective suspending agent for many pharmaceutical products. It is used as a suspending agent and increases the external phase's viscosity and prevents undissolved material sedimentation. Therefore, gum tragacanth is used as a suspending agent for high-density insoluble powders (Kulkarni and Shaw, 2016). Glycerite of tragacanth is used as a tablet binder. Tragacanth has an active component polysaccharide. The main fraction, bassorin (60%–70%), contains methyl ether groups and only swells in water, forming gel particles, and contains cellulose and proteins. Gum tragacanth is also used as a rheological modified in food preparations (BeMiller, 2019).

7.2.6 Cashew gum

Cashew gum is one of the flexible natural polysaccharides. The occurrence of glucuronic acid with a pK_a value of <3.5 makes cashew gum act as a polyanion at a $pH > 4$ (Ologunagba et al., 2017). Due to these anionic properties, cashew gum interacts readily with cation. Cashew gum is an acidic polysaccharide containing metallic cation ions such as Na^+, K^+, Mg^{2+}, and Ca^{2+}. Cashew gum is typically used as a thickening, gelling, and stabilizing agent (Ribeiro et al., 2016; Animesh et al., 2012). This gum also has the potential to be used as a tablet binder, superabsorbent hydrogen gelling agent, and viscosity enhancer. Cashew gum is used as a surfactant for microencapsulation, and is compatible with many pharmaceutical excipients.

7.2.7 Gum karaya

Gum karaya has a galacturonic-type partially acetylated ramified structure and is commercially obtained from magnesium salt (Padil et al., 2018). It is branched and has a large molecular mass of around 1.6×10^7 Da (Raj et al., 2021) and comprises 55%–60% of neutral monosaccharide such as galactose and rhamnose, 8% of acetyl groups, and 37%–40% of galacturonic and glucuronic acids. In this main gum, a chain is formed by -D-galacturonic acid and L-rhamnose units. Side chains are linked to the main chain by 1,2-D-galactose bonds or 1,3-D-glucuronic bonds for galacturonic acid. Gum karaya, sometimes known as sterculia gum, is the dried exudation of the *Sterculia urens* tree and other *Sterculia* species. Gum karaya is used as an emulsifier, stabilizer, and thickening agent in the pharmaceutical and food industries. The gum's viscosity increases linearly with the concentration, and at 4% concentration, it acts as a rubber gel-like paste because it has excellent adhesive properties (Featherstone, 2015).

7.2.8 Gum ghatti

Gum ghatti is the amorphous translucent exudate of *Anogeissus latifolia* (Roxb). The bark is used as a medicinal agent for skin diseases such as sores, boils, and itching, snake and scorpion bites, stomach diseases, colic, cough, and diarrhea (Parvathi et al., 2009). The gum is also used for antioxidant and wound healing activity (Shafie et al., 2020). As mentioned in the standard specification for gum ghatti by the 29th JECFA (1985), a water-soluble fraction (approximately 90%) has a molecular weight of about 12,000 Da. Gum ghatti has some physical properties that are superior to other gum products (Thombare et al., 2018), including greater acid resistance, salt tolerance, and oil binding capacity. It is a cost-effective emulsifier with excellent emulsification properties, even at considerably lower concentrations.

7.2.9 Konjac gum

Konjac mannan is a type of glucomannan derived from the tuber of the konjac plant, more specifically the *Amorphophallus konjac* species found in Southeast Asia. Konjac is a high molecular weight, water-soluble, and neutral polysaccharide (Fang and Wu, 2004). The molecules are rich in hydroxyl groups that quickly dissolve in water, leading to increased viscosity that forms a thick hydrocolloid in low concentrations (Li and Nie, 2016); this makes hydrocolloids economically useful in the manufacture of foods and as a pharmaceuticals excipient. Konjac gum dietary fiber has a long history in food and traditional Chinese medicine (Zalewski et al., 2015). Konjac gum has good film-forming ability, biocompatibility, biodegradability, and gelation performance, which is one of its most prominent features (Chen et al., 2016a). Konjac gum has been widely used in food (Jian et al., 2016; Zhang et al., 2016), pharmaceutical carriers, tissue scaffolds, and absorbing materials (Chen et al., 2016a; Fan et al., 2016).

7.2.10 Kondagogu gum

Kondagogu is a high molecular weight complex acetylated gum; it is an anionic polysaccharide, consisting of D-galacturonic acid, D-galactose, and L-rhamnose. Kondagogu is widely used to prepare hydrophilic matrix tablets because of its high water swellability, nontoxicity, and low cost (Vinod et al., 2008b). Unlike other water-soluble gums, it does not dissolve in water but absorbs it to form a viscous colloidal solution. Kondagogu demonstrates a swelling property in water to the extent that 3%–4% sol will produce a gel with a uniform smoothness and texture (Vinod et al., 2008a). Gum kondagogu swells by absorbing water and forms thixotropic gels. It is similar to gum karaya in physicochemical properties in terms of intrinsic viscosity and water-binding capacity (Malik and Ahuja, 2011).

7.2.11 Dextran

Dextran is colloidal in nature, hydrophilic, and water-soluble. Its solubility is due to the 95% linear linkage, making it more water-soluble and more suitable for various biomedical applications (Varshosaz, 2012). Dextran is biocompatible, and therefore is widely reported as a polymeric carrier in novel drug delivery systems (Dheer et al., 2017). Dextran is used for drug delivery in dextran hydrogels, micelles, and core-shell type nanoparticles (Dheer et al., 2017; Jeong et al., 2011). The biocompatibility and biodegradability of dextran cause it to be used in drug delivery systems most often to decrease inflammatory response and stimulate wound healing and perfect skin rehabilitation (Ishak et al., 2016). Dextran can be modified chemically, and the glycoconjugates can be derived through reactions such as etherification, esterification, amidation, and oxidation. Because of their instability in circulation, the efficacy of a number of therapeutically active proteins and peptides is severely limited. Covalent modification of the protein or enzyme with hydrophilic polymers like dextran or PEG has proven to be effective. Both in vitro and in vivo, these conjugates are more stable than the native protein (Varshosaz, 2012).

7.2.12 Biosafety and general advantages and disadvantages

The main advantage of GNPs is their biocompatibility and hydrophilicity. GNPs can carry the loaded drug moiety to the target site safely and effectively. The biodegradability of GNPs offers tissue compatibility and minimal tissue toxicity. Several gums are in the category of Generally Regarded As Safe (GRAS). A three-generation reproductive test was conducted on albino rats including males and females (Fiume et al., 2016). The rats were fed with 0, 0.25, and 0.5 g/kg body weight of xanthan gum. Ten males and 20 females were used in the first-generation group, and 20 males and 20 females were used for the next two generations. The successive generation was from the weanlings. The survival and reproductive parameters showed no significant difference between the treated and control parenteral rats. The parenterally treated rats' body weight was slightly decreased compared to the control group of each generation. There is no difference in the developmental parameters, and malformation between test, control, and offspring. This study indicated the safety profile of natural gums. An analysis was performed to evaluate the repeated dosing toxicity of partially hydrolyzed guar gum in rats and salmonella's mutagenic potential. The guar gum polysaccharide derived from the Indian origin was used for this study. This gum is used as a thickener and emulsion stabilizer worldwide in the food industry. The results revealed that no significant toxicity was observed at the amount of gum given. Under current use and concentration criteria, the Cosmetic Ingredient Review (CIR) Expert Panel decided that 34 microbial polysaccharide gums are safe for use in cosmetic formulations. Microbial polysaccharide gums have been utilized as an emulsion stabilizer, binder, viscosity-increasing agent, and skin conditioner. Based on existing animal and clinical data, the Panel determined the safety of the product. Parenterally given polysaccharides appear to be biotransformed in both animal and human studies. These big molecules, on the other hand, appear to be poorly absorbed via the skin, with minimal bioavailability. The CIR Expert Panel concluded that topical treatment of these substances would not induce systemic effects based on toxicological data from other exposure routes. Acute inhalation of xanthan gum and gellan gum, as well as a 4-week inhalation of beta-glucan in guinea pigs, caused no harm. There were no acute or long-term pulmonary consequences from xanthan gum powder exposure. The evidence for systemic toxicity, irritation, sensitization, and other effects of microbial polysaccharide gums was assessed by the Panel. Several microbial polysaccharide gums examined in oral studies exhibited no systemic toxicity, are not reproductive or developmental toxicants, are not genotoxic, and are not irritants or sensitizers, according to the Panel.

The advantages and limitations of gum-based nanoparticles are listed below from the various literature (Patra et al., 2018; Padil et al., 2018; Deshmukh and Aminabhavi, 2021; Dhar et al., 2008; Aakash et al., 2019; Torres et al., 2019; Salatin and Jelvehgari, 2017):

- increase the stability of the loaded drug;
- increase the shelf life of the loaded drug;
- better solubility of the drug;
- high drug loading capacity;
- less chemical interaction with loaded drug;

- particulate size and surface characterization can be manipulated easily to achieve passive or active targeting;
- increase therapeutic efficacy and reduce side effects;
- varieties of polymers are available for the matrix of choice;
- site specificity can be achieved by attaching ligands for targeting;
- GNPs can be delivered safely through different route such as oral, nasal, parenteral, and intraocular;
- multipurpose is possible by modifying the polymeric characteristic and surface properties, which helps perform therapy, drug delivery, and imaging simultaneously;
- highly suitable for IV delivery without embolism;
- formulation development is rapid and large-scale production is possible; and
- low immunogenicity and toxicity.

The limitations are as follows:

- chances of aggregation of a particle in the body;
- size being too small may lead to aggregation of particle and particle growth;
- predetermined drug loading may not be possible due to the smaller size;
- loss of drug loading due to the physical changes of the particles in a biological environment;
- microbial growth during storage and handling; and
- poor stability in varying environmental conditions.

7.3 Method of preparation of GNPs

After choosing a gum for nanoparticle synthesis, the right method is crucial for developing successful nanoparticles. The favorably reported and successful strategies are described in this section. Most of the preparation process applied to the polysaccharide-based nanoparticle synthesis is also suitable for gum with a slight modification to be ideal for the particular gum used for the synthesis. The selection of surfactant and cosurfactant plays a crucial role in the successful formation of GNPs. The following factors are considered critical for the synthesis of nanoparticles of ideal qualities:

- size of nanoparticles/microparticles required;
- properties of the drug to be loaded such as solubility, pK_a, and stability;
- surface charge required on the nanoparticles;
- types of tissue/cells to be penetrated or treated;
- degree of biodegradability required;
- biocompatibility and less toxicity;
- chemical compatibility between drug, excipients, and gums used;
- type of drug release mechanism required;
- number of therapeutic and nontherapeutic candidates loaded; and
- amount of the total loading on the nanoparticles.

7.3.1 Solvent evaporation (SE)

Solvent evaporation (SE) is the method first employed for the preparation of nanoparticles. The SE technique involves a volatile solvent such as chloroform, dichloromethane, and ethyl acetate. The polymer solution is prepared in the suitable concentration in a suitable solvent such as ethyl acetate, which replaces the chloroform and dichloromethane (Desgouilles et al., 2003). In this method, the polymer dissolves in the appropriate solvent. If the drug is hydrophobic, it can dissolve in the same solvent to develop a polymer-drug solution. If the drug is hydrophilic, it can be dissolved in the aqueous phase separately. The polymer-solvent solution is then added into the aqueous solution containing the surfactant or emulsifying solution to form an o/w emulsion. Once the emulsion is formed and stable, the organic solvent is evaporated under reduced pressure or by stirring or by rota-evaporator (Hernández-Giottonini et al., 2020). If needed, secondary approaches such as ultrasonication or high-speed homogenization can be employed to reduce particle size (Mulia et al., 2019). An ion-sensitive polymer of gellan gum ophthalmic in situ gel was loaded with nanoparticles (Kesarla et al., 2016). The moxifloxacin nanoparticles were separated via freeze-drying. The nanoparticles demonstrated the diffusion-controlled drug release from the nanoparticles for 12 h.

7.3.2 Nanoprecipitation

This method is based on the principle of the solvent evaporation method (Barreras-Urbina et al., 2016). This method involves precipitation of polymer from organic solution, and then polymer organic solution is diffused into the aqueous media, leading to formation of nanoparticles. Briefly, the polymer is dissolved in water-immiscible solvent with intermediate polarity, and then this phase is injected into the aqueous solution containing a stabilizer, which is usually a surfactant (Gericke et al., 2020). The polymer is deposited in the interphase between water and organic solution, and organic solution causes fast diffusion of solvent and leads to a colloidal suspension. The solvent displacement technique is used to precipitate nanoparticles in small volumes of nontoxic oil introduced into the organic phase, leading to nanoparticles' precipitation. This approach is appropriate for lipophilic drugs but not for hydrophilic drugs, as hydrophilic drugs dissolve in water and cannot be adequately entrapped in the oil phase (Shen et al., 2017). Guar gum nanoparticles were prepared by nanoprecipitation and the polymer cross-linking method. The model drug lipase and crystal violet were used in the study. It was found that the formation of nanoparticles is influenced by the molecular mass of galactomannan, solvent, cross-linkers, and agitation. The nanoparticles' size measured by the laser light scattering was 20–50 nm, and the polydispersity index was 0.1–0.4 (Soumya et al., 2010).

7.3.3 Emulsification solvent diffusion

This method is a modified solvent evaporation technique. The polymer plays an essential role in the emulsification technique (Rosca et al., 2004; Kumar et al., 2012; Baimark and Srisuwan, 2013). The polymer is dissolved in a suitable solvent such as propylene carbonate, benzyl alcohol. The solvent should be water-miscible so that the polymer or solvent mixture can be saturated with water to attain a thermodynamic equilibrium of both the liquids. The diffusion of the solvent dispersed phase has to be promoted by the excessive addition of water; the organic solvent is partially miscible with water. The polymer's water-saturated phase now emulsified in a suitable aqueous solution containing a stabilizer leads to the diffusion of solvent to external phase and subsequent formation of nanoparticles. In a final step, the solvent can be removed by the filtration or solvent evaporation by a rota-evaporator.

Alginate nanoparticles are prepared alone and mostly in combination with chitosan and other synthetic polymers (Choukaife et al., 2020). Alginate is one of the most explored polymers in the synthesis of microparticles and nanoparticles. Alginate nanocarriers are prepared in the form of nano aggregates, nanocapsules, nanospheres, etc. Alginate nanoparticles are prepared by two primary methods: complexation and water/oil emulsification. Chen et al. (2017) studied the structural and rheological characterization of hydrophobic alginate nanoparticles in an aqueous solution. Amphiphilic polymers that form a self-assembled structure in aqueous media have been investigated for various diagnostic and therapeutic applications. In this research, hydrophobic alginate with different molecular weight was synthesized. The system of Ugi-Alg was analyzed by ^{1}H NMR spectrophotometer. Methods such as dynamic light scattering electron spin resonance experiment and transmission electron microscopy were used to characterize the alginate nanoparticles. The research demonstrated that mid-MW Ugi-Alg (MW-2.8×10^5 g/mL) produces stable nanoparticles with less sodium concentration.

7.3.4 Dialysis

Dialysis is a simple and effective method for the preparation of nanosized particles. In this method, the polymer of choice is dissolved in an organic solvent and kept inside the dialysis tube containing a specific molecule weight cutoff membrane (Gericke et al., 2020). The precise molecular weight cutoff helps the nanoparticles of specific particle size to move through it, while particles larger than this size are retained. This displaces the solvent inside the membrane and is followed by progressive aggregation of the polymer. Due to the lower solubility, the formation of a suspension of nanoparticles can be achieved by precipitation (Chronopoulou et al., 2009). It is assumed that the dialysis membrane method follows the similar mechanisms of nanoprecipitation supported by dialysis. A dialysis membrane technique based on the semipermeable membrane technique allows the passive transport of solvent to slow down polymer solution mixing with nonsolvent (Jeon et al., 2000).

Nanoprecipitation and dialysis were employed for the formation of acetylated cashew gum. The nanoparticle prepared by the dialysis showed greater stability than the nanoparticles synthesized by the nanoprecipitation method (Dias et al., 2016). The nanoparticles were loaded with the diclofenac diethylamine. The drug loading by both the methods was 60%. The cytotoxicity assay revealed no significant toxicity toward the cell viability. The prepared nanoparticles demonstrated a controlled release profile. Moreover, the transdermal penetration was 90 for the drug delivered by these type of nanoparticles.

7.3.5 Ionic gelation

This method involves the mixing of two aqueous phases. The first aqueous phase contains polymer solutions such as chitosan, and the second aqueous phase includes gum arabic. The chitosan and gum arabic were used for the synthesis of insulin nanoparticles. This is the reaction between the materials of oppositely charged ions which is positively charged chitosan vs the negatively charged gum arabic that leads to the formation of the gel; therefore it is called ionic gelation (Naeye et al., 2011).

Katuwavila et al. (2016) developed a drug delivery system of Fe^{2+} using an alginate biopolymer. Ion-loaded alginate nanoparticles were prepared to control the ionic gelation technique. FTIR and thermogravimetric analysis confirmed the successful loading of Fe^{2+} ions. Electron energy loss spectroscopy confirmed the Fe^{2+} loading of 70%, which was optimized at 0.06% Fe (*W*/*V*). The nanoparticles prepared are in the range 15–30 nm with a zeta potential of −38 mV, demonstrating their stability. Xu et al. (2018) developed xanthan gum and lysozyme nanoparticles of like charges. In this study, xanthan gum and lysozyme's interaction revealed that the nanoparticles' size was between 1.7 and 108 nm. The xanthan gum lysozyme nanoparticles have a spherical shape.

7.3.6 Supercritical fluid technology (SFT)

Supercritical fluid technology (SFT) is an alternate method for the preparation of polymeric nanoparticles. SFT requires the fluid to remain at single-phase regardless of pressure or temperature above its critical temperature. Usually, supercritical fluid (SC Co_2) is widely used in this method due to mild critical conditions. The most common techniques used in SFT are supercritical antisolvent (SAS) method and the rapid expansion of supercritical solution (RESS). The SAS process is usually done by the liquid solvent such as methanol which is completely miscible with SC Co_2. The dissolving fluid must be micronized because the solute is insoluble in the supercritical fluid (SCF), which leads to the precipitation of solute as nanoparticles when the solvent power of SCF dramatically decreases (Franco and De Marco, 2021). RESS is the modified method in which expansion of the supercritical solution into a liquid solution occurs (Sun et al., 2005). The RESS process for the nanoparticle production was studied, and it differs from the SAS process. In RESS, the drug or material which needs to be formed as nanoparticles (solute) is dissolved in the SCF, and then this solution is allowed to expand through a small orifice to the area of low pressure; this causes a sudden reduction in the pressure and leads to the precipitation of the solute as nanoparticles. In the RESS technique, no organic solvent is used, since this leads to the formation of microscale product rather than nanoscale. To overcome these disadvantages, a new SFT was introduced, known as RESOIV. In this technique (Pathak et al., 2019), the liquid solvent suppresses the particle growth in the expansion jet, leading to nanosized products' formation.

7.4 Characterization techniques for GNPs

7.4.1 X-ray diffraction

X-ray diffraction (XRD) analysis was carried out to detect the crystallinity of the pure drug. The XRD patterns of the pure drug need to be recorded first. Later the spectrum of major excipients used in the synthesis needs to be identified, and the gums used to prepare nanoparticles. The significant peaks of all the ingredients used in the synthesis of nanoparticles need to be identified. The dummy nanoparticles are first prepared without the drug loading, and the XRD spectrum of the dummy nanoparticle recorded. Later, the spectrum of drug-loaded nanoparticles has to be compared with the dummy nanoparticles for the significant peaks and any significant shift in the intensity and the position of the peaks. If the spectrum of nanoparticles does not clearly indicate the peaks, it is recommended to study the drug's interaction with the other ingredients used by making a physical mixture. The comparison of drug and optimized nanoparticles loaded with the drug was usually studied by using Cu monochromatic radiation ($\lambda = 1.5406^\circ$A). A synchrotron light beam in a wavelength of 1.5406 Å at 40 kV and 40 mA , were used as the X-ray source. The (Si (Li) PSD) detector detects the scattering intensity. The diffractograms recorded for 2Θ were between 20 and 75 (Rimple and Newton, 2018). XRD provides information about the crystallinity of the drug and any changes in the crystallinity, which leads to the amorphousness due to the interaction with the gums.

7.4.2 Fourier transform infra-red spectroscopy (FTIR)

The study of polymer compatibility plays a significant role in the development of a stable dosage form. The FTIR spectrum of a pure drug alone and a pure drug with various excipients is investigated (Bruker alpha 2000, German). The samples were analyzed in the range of 4000–400 cm^{-1} wavenumber and percentage transmittance. The samples were prepared by using a KBr pellet technique of 2 mg sample in 200 mg KBr (Ebrahiminezhad et al., 2019). The experiments were repeated to ensure reproducibility. The significant peaks of the drug were identified in the mixture of the sample. This data reveals the possible interaction between the active pharmaceutical ingredients with gums used and the other excipients used in the study (Reeta et al., 2020). If the drug's significant peaks are not affected by mixing with the other ingredient of the formula, this confirms the absence of interaction and any chemical change. The developed nanoparticles formula or the nanoparticles needed to be studied by FTIR and analyzed for the significant peaks of the key ingredients.

7.4.3 Particle size analysis and surface charge by dynamic light scattering

The mean particle size polydispersity index (PDI) and zeta potential of nanoparticles can be analyzed using the dynamic light scattering (DLS) technique. The instrument zeta sizer provided with the condition at 25 °C using a scattering angle of 90 degree (e.g., Zetasizer, model ZS, Malvern Instruments, UK) (Reeta et al., 2020) was used for the study. Approximately 1 mL of the dilute solution or suspension containing nanoparticle was filled in the instrument's cuvette and analyzed. The mean diameter of particles present in the system will be recorded as the *Z* average; the nanoparticles' homogeneity can be known from the PDI value (Chacko and Newton, 2019). A similar instrument was used for the determination of the zeta potential of the particles in solution or suspension (Kumar and Newton, 2016). The zeta potential of prepared nanoparticles was also analyzed under the same conditions as that fixed for the particles' size analysis. For the zeta potential determination, the nanoparticles suspension was filled with the specially designed device. The samples for zeta potential study was diluted appropriately and sonicated, then kept in a low conductivity zeta cell to meet the instrumental requirements, and then the values were recorded. The instrument provides the zeta potential of the nanoparticles with the indication of charge, either positive or negative. A zeta potential greater than ±25 indicates a greater degree of stability. There no other method to determine the surface charge of the nanoparticles. However, zeta potential is not an indicator of actual charge on the surface of the nanoparticles. It is an indicator of the charge of the nanoparticles under the applied external electric field.

7.4.4 Transmission electron microscopy (TEM)

Transmission electron microscopy (TEM), a form of analysis, is used to study the physical feature, shape, and surface morphology of nanoparticles. Since TEM is contrast-based, there is a possibility of detecting the entrapped drug molecules. These molecules usually appear as a contrast point behind the particle's transparent outer layer (Patra et al., 2018). In this method, the samples were placed in a copper grid coated with carbon. Previously the samples were diluted appropriately with HPLC grade distilled water or solvent in which the polymer or drug is less soluble and subjected to sonication for 3 min (Gontard et al., 2014). TEM provides information such as morphological, compositional, and crystallographic (Kumar et al., 2019) details. In contrast, HRTEM gives more detailed information about the nanoparticles' features such as defects, stacking faults, dislocation, precipitates, and grain boundaries (Titus et al., 2019).

7.4.5 Atomic force microscopy (AFM)

Atomic force microscopy (AFM) is an ultra-high-resolution technique employed in the characterization of nanoparticles. This method is based on the particle's physical scanning in nanosized range using the probe tip of atomic scale. TEM, HRTEM, and AFM provide information on the size, morphology, and crystal structure of the nanomaterials (Mourdikoudis et al., 2018). The analysis was done by running the machine in the noncontact topping mode. Thus, the samples to be analyzed were drop coated on glass substance. The nanoparticles were characterized by analyzing the pattern obtained on surface topography and by analyzing the AFM data (Klapetek et al., 2011). This method provides a topographical map of the sample based on forces that occur between the tip and a surface of the sample. The samples are usually scanned in contact or noncontact mode. In contact mode, topographically a map is generated by tapping the probe on the surface of the sample. In noncontact mode, the probe has hover over the sample without touching it. The advantages of AFM are that it can image a nonconducting sample without any pretreatment (Rao et al., 2007). It is useful in imaging delicate, biological, and pharmaceutical nanoparticles. AFM provides the most accurate size report which does not require any mathematical treatment. AFM techniques provide a picture which helps us to understand biological conditions.

7.4.6 Laser confocal imaging microscopy

Laser confocal imaging microscopy has a wide variety of applications in studying the biological interaction of the nanoparticles. In general, it is used to detect the cellular uptake of

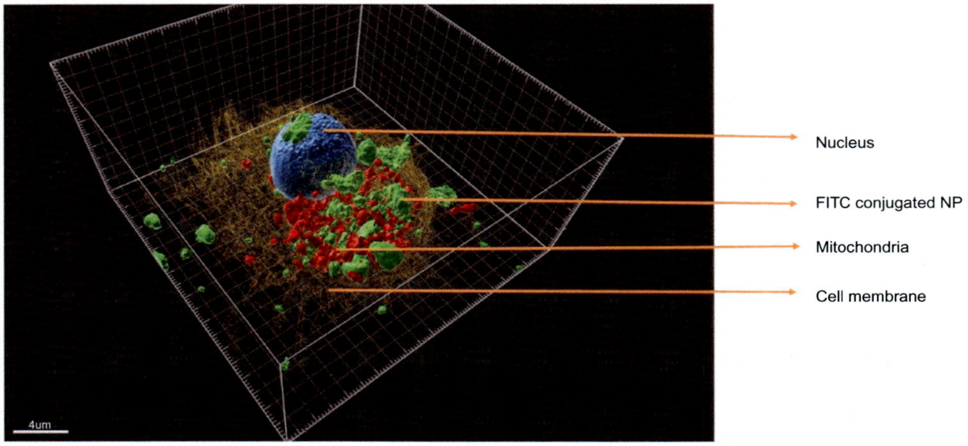

FIG. 7.1 Laser confocal microscopy imaging for the detection of cellular uptake of nanoparticles in fibrosarcoma cell line. The different organelles of the cells are stained by different dyes: *multitracker deep red* for mitochondrial staining, DII for membrane staining, and Hoechst-33,342 for DNA labeling. FITC conjugated nanoparticles appear in *green*.

the nanoparticles (Rossetti et al., 2013). Other applications include interaction with a biological system, depth of penetration, route of internalization in skin, quantification of skin penetration, and morphological analysis of the tissues. The method for detecting cellular uptake of the nanoparticles is described as follows. In this method, the nanoparticles were tagged with a suitable fluorescent tag such as fluorescein-5-isothiocyanate (FITC). In this method, the cells were seeded at 1×10^6 cells per 10 mm wells in 35 mm glass bottom dishes, after which the cells were allowed to adhere in an incubator with 5% CO_2 in air at 37 °C for 24 h. The nanoparticles were tagged with FITC using a suitable linking agents such as silanes or through the EDC–NHS (1-ethyl-3-(3-dimethylaminopropyl)carbodiimide hydrochloride, Fluorescein (5/6-carboxyfluorescein succinimidyl ester) chemistry. A suspension of FITC-conjugated nanoparticles at a concentration of 100 μg/mL and a volume of 500 L was administered to each petridish and incubated with 5% CO_2 in air at 37 °C for at least 4–12 h. After incubation, the cells were washed twice gently with DPBS to remove excess noninternalized nanoparticles present on the cell's surfaces. The cell organelles then needed to be stained by different dyes, including deep red for staining, DII for membrane staining, and Hoechst-33,342 for DNA labeling. The 4% paraformaldehyde solution was used to fix the cells. The cells then were imaged by confocal laser scanning microscopy (e.g., Zeiss LSM 880 or Leica SP8 DLS). The extent of cellular uptake can be determined by using the cells without nanoparticle treatment (Fig. 7.1).

7.4.7 In vitro drug release by membrane diffusion technique

The drug release from the GNPs can be studied by using the membrane diffusion technique. Membrane diffusion techniques consist of two components. The first is the donor compartment, which is typically an open-ended tube with a dialysis membrane attached to one end. Nanoparticles are typically introduced into this compartment and maintained vertically inside the receptor compartment. The receptor compartment contains medium simulating bodily fluids, and collects the drug released from the donor compartment. The dialysis membrane has a pore size of 2.4 nm. Approximately the molecular weight cutoff of 12,000–14,000 was used (Thatipamula et al., 2011; Dudhipala and Gorre, 2020). Previously, the membrane was soaked in HPLC distilled water for 12 h before the study to enable the pores in the membrane to allow diffusion of molecules (Bhupinder and Newton, 2016). This membrane was then tided into an open-ended test tube on one side. Another side remained open to introduce GNPs, or in the form of dispersion or emulsion or the solid particle dispersed in the double-distilled water. This setup was then kept vertical in the beaker containing media of simulated body fluid. Based on the requirement of experimental conditions media mimicking body fluids such as gastric fluid or intestinal fluids (Kumar and Newton, 2016), or simulated nasal fluid (Reeta et al., 2020), simulated tear fluid (Rimple and Newton, 2018), or plasma were used. The simulated fluid of 100 ml was used in the receptor compartment for nanoparticles based drug release study. The temperature was maintained at 37 ± 5 °C throughout the study. An aliquot of 1–5 mL of samples was withdrawn at various time intervals from the receptor compartment. The system was replenished with fresh media every time and samples were analyzed by spectrophotometrically by UV or HPLC.

7.4.8 In vivo imaging studies

The nanoparticle imaging can be done by using computed tomography (CT) of X-rays, magnetic resonance imaging (MRI), and positron emission tomography (PET) (Key and Leary, 2014). MRI provides anatomical. Physiological, and molecular information, whereas CT provides anatomical information, and PET provides physiological and molecular information. However, each method has its own limitations. In animal studies, fluorescent imaging provides fundamental information about the nanoparticles' performance in vivo. In this method, mice can be used for microscopic in vivo imaging by using suitable fluorescent tags (Choi et al., 2015). This can be done by administrating the nanoparticles in suspension form (approximately 200–300 μL) through the tail vein. Previously the animal should be anesthetized with a suitable anesthetic agent and bioimaging can be carried out on the animal. This is also described as functional bioimaging. In this method, the nanoparticles were tagged with a suitable fluorescent probe such as quantum dots and dispersed in a suitable vehicle to form the suspension. The near infrared luminescent nanoparticles are already in the use of biological imaging by using quantum dots (Cai et al., 2006; Yong et al., 2009). Nanoparticles in the form of a suspension were injected intravenously through the tail vein and imaged using a suitable radiation source that excites the sample. The microscopic images of brain or cancer cell imaging of the organ or cells can be studied by this method. The whole imaging process involving primary organs, cells, blood vessels, and tissues can be studied using this method. However, the single imaging technique cannot provide correct information about nanoparticle performance in vivo. Therefore, the other techniques such as MRI, CT, and PET need to be combined to obtain the most reliable data on the nanoparticle performance in vivo (Key and Leary, 2014). The sensitivity and specificity of the information can be improved by using exogenous contrast agents, and each imaging technique has its own limitation is spatial resolution, sensitivity, and tissue penetration depth.

7.5 General biomedical applications of gums

7.5.1 Controlling drug release

Gums can be used as an osmotic drug delivery systems for controlling the drug release. A monolithic omotic system was formed with two orifices on both side surfaces. The water-insoluble naproxen was incorporated as the drug. Gum arabic is used as an osmotic agent and an expanding agent. The drug release was significantly controlled and followed zero-order drug release for 12 h. It is indicated that these osmotic drug delivery systems could control drug delivery (Lu et al., 2003).

In another study (Ofori-Kwakye et al., 2012), the mechanical storing properties of cashew gum were investigated. The study results revealed that cashew gum was acidic, and its acidity is not changed by increasing the concentration and storage time. The binding property of cashew gum was compared with that of acacia. It was observed that the disintegration time of the tablet increased with an increase in the concentration of cashew gum. The swelling index of the gum was not affected by the pH of the media used. The films prepared by using cashew gum with HPMC and CMC were smooth, uniform, and transparent. This study revealed that the 7.5% (w/v) cashew gum enhanced the mechanical strength of paracetamol tablets. The film coating does not affect the drug release properties of the tablets.

A gum obtained from the almond tree was studied for its binding property in tablet formulations. The drug release was retarded with increased almond gum concentration as compared to synthetic gum, and the release mechanism followed non-Fickian diffusion. Almond gum was found to be useful for the preparation of uncoated tablet dosage form (Sarojini et al., 2010). The sustained release properties of guar gum, xanthan gum, and gum tragacanth were evaluated alone and in combination with the polyvinylpyrrolidone K90 (Iqbal et al., 2011). The drug release profile of the tablets was analyzed in comparison with the commercial tablets. Diclofenac sodium was used as a model drug, and the drug release profile was studied for 12 h. The results revealed that the drug release pattern follows the Higuchi model and super case-II and non-Fickian diffusion.

Moreover, in vitro studies on healthy human volunteers using a nonblinded cross-study revealed that the bioavailability of the formulation containing gum tragacanth and PVP·K30 was 0.91. The researchers carried out in vitro *and* in vivo release studies. Both these studies showed that gum tragacanth could be used to sustain the release of the drug.

Gelatin-acacia based tolnaftate microcapsules and microspheres were prepared using coacervation and emulsion-solvent evaporation methods. X-ray diffraction analysis shows the crystalline drug presence in the microcapsules and microspheres. A stability study of the formulation was performed and found that the drug was stable in microspheres and microcapsules formulation for about 6 months (Dash, 1997). Naturally occurring gum was used as mucoadhesive buccal drug delivery systems (Park and Munday, 2004). The drug concentration was maintained between 0% and 50% of xanthan gum, karaya gum, and guar gum used for this study. Among these, guar gum demonstrated poor mucoadhesive property, karaya gum demonstrated better mucoadhesive properties, and karaya gum formulation demonstrated zero-order drug release; however, more than 50% (w/w) is required to provide a suitable sustained release. Xanthan gum showed superior mucoadhesive properties and demonstrated zero-order drug release pattern.

7.5.2 Modifying rheological property of pharmaceuticals and food products

Gums are widely utilized as viscosity enhancers and rheological modifiers. Determining the rheological properties is essential for optimizing food thickeners, stabilizers, and emulsifiers in both pharmaceuticals and food preparations. The sugar content in the gums directly influences the shear-thinning properties of the gums (Vinod et al., 2008a). The rheological properties of the pharmaceuticals and food items can be modified using different concentrations and types of polymers. Rheological properties modification is advantageous in pharmaceutical suspensions and emulsions. In diets, rheological properties need to be modified for swallowing, as the swallowing ability varies significantly in different dysphagia patients. A study conducted in Korea evaluated the different concentrations of the gums for their suitability to be used as a rheological modifier in diets. The study was conducted on the two commercially available gums in Korea, such as food thickener of xanthan gum and guar gum mixture. The study results demonstrated their high shear-thinning properties as all concentrations were of 1.0%–3.0%. The flow behavior of the six species of the Iranian tragacanth dispersions was investigated in different temperatures and ionic strengths in the concentration range of 0.05%–1.5% (w/w) for their shear-thinning properties (Balaghi et al., 2010). The results revealed that all the gum dispersions have shear-thinning properties and can used as natural alternatives to the commercially available hydrocolloids.

7.5.3 Drug delivery carrier

Natural gums can also be modified to make products for drug delivery systems, making them a viable alternative to synthetic excipients. There is a lot of potential for gums and mucilage derived from plants to be used as effective drug delivery carriers. Moreover, the natural gums of polysaccharides are not digested by the upper gastrointestinal tract due to the nonavailability of polysaccharide-digesting enzymes, therefore the gum-based nanoparticles are found as efficient alternatives for synthetic polymethacrylate, usually used for colon-targeted polymers (Newton et al., 2011). Colon-targeted drug delivery systems are highly recommended in diseases such as inflammatory bowel disease (IBD) (Newton et al., 2011; Newton and Lakshmanan, 2020). The coated colon-targeted drug delivery systems available in the market are mostly dependent on the polymethacrylate derivatives. The gum-based polymers pectin, chitosan, tamarind gum, guar gum, and xanthan gum can be used in colon-targeted delivery successfully (Kumar and Newton, 2015). The pH-dependent synthetic drug delivery polymers may fail to release the drug at the colon due to premature drug delivery. Sometimes the enteric coating created by using the synthetic polymers may not get dissolved in the predetermined manner, which affects therapeutic efficacy (Jatinder and Newton, 2015).

In a study (Fosu et al., 2016), crude cashew gum was purified and characterized in terms of some physicochemical properties. Later, ibuprofen matrix tablet formulations containing ~200 mg ibuprofen were prepared with varying ratios of cashew and xanthan gum by direct compression. The study results revealed that the blend of xanthan gum and cashew gum was found to be an effective colonic drug delivery carrier. In a study (Dhamija et al., 2017), formulations of compression-coated tablets of 5-aminosalicylic acid were prepared and further evaluated for colon targeting ability. The drug 5-aminosalicylic acid is degraded in the stomach in the presence of gastric pH, so colon targeting is the best way to achieve a therapeutic effect. Initially, the drug's core tablet was prepared using microcrystalline cellulose as diluents; it was then compression-coated using various concentrations of *Prunus amygdalus* gum (almond gum). Almond gum successfully delivered the drug in the colon.

Tamarind gum, okra gum, and chitosan-compressed tablets were prepared for chronotherapeutic colonic drug delivery (Newton et al., 2015). Propranolol HCl was incorporated in the formulation. Propranolol HCl was delivered in a chronotherapeutic manner. The tablets were prepared by direct compression. Propranolol HCl undergoes hepatic metabolism, which causes poor bioavailability. The matrix tablet of propranolol HCl tablets was successfully delivered to the colon in a time-dependent manner. Carbopol was used as an auxiliary polymer to modify the drug release, and physicochemical characteristics of the tablet formulation. Among all the colon-targeted tablets, tamarind gum demonstrated superior drug release characteristics. Recently, tamarind gum has been employed as a colon targeted drug delivery system for rifaximin as a chronotherapeutic drug delivery system (Amaldoss et al., 2021). The findings indicate that tamarind gum can be successfully synthesized into nanoparticles for colonic drug delivery. This study shows that tamarind gum nanoparticles loaded with rifaximin can deliver the maximum amount of drugs to the colon. The nanoparticle compositions were stable, with a zeta potential that was substantially higher. In rats, the best formulation provided considerable protection from TNBS-induced colitis. In comparison to the untreated TNBS group, the produced nanoparticles were able to improve healthy biochemical parameters.

Alginate nanoparticles were also used as a potential carrier for rifampicin, isoniazid, and pyrazinamide (Ahmad et al., 2006). Alginate nanoparticles prepared by the cationic gelation method were administered to M-tuberculosis H37Rv-infected mice. They demonstrated complete bacterial clearance from the organ compared to 45 conventional dosages of orally administered free drugs (Kaur et al., 2016). Alginate gum nanoparticles were used for various applications such as drug carrier, gene transfection, novel drug delivery, antitubercular therapy, and targeted drug delivery for cancer. Marci et al. (2016) developed trimethyl chitosan sodium alginate nanoparticles by the ionotropic gelation technique for targeted drug delivery. The additional polymer chitosan with alginate contributes to the positive surface charges that prolong encapsulated molecules' contact time with the epithelium. Chitosan alginate nanoparticles of nifedipine were prepared by Li et al. (2008) as a carrier for many molecules such as protein, DNA, insulin, and cyclosporin-A. Calcium alginate nanoparticles can be prepared by inducing the gelation of calcium ions. Calcium alginate nanoparticles were also used in the delivery of various molecules.

Dextran nanoparticles were prepared with PLGA polymer and used for the delivery of doxorubicin to the doxorubicin-resistant HuCC-T1 cells. The antiproliferation study revealed that the nanoparticles showed a superior antitumor effect compared to the free doxorubicin. The results revealed that the doxorubicin hardly penetrated the tumor cells' membranes, whereas the doxorubicin-loaded nanoparticles were effectively engulfed by the tumor cells (Jeong et al., 2011). Similarly, paclitaxel-carboxymethyl dextran conjugate was administered to xenograft mice bearing colon carcinoma resistant to paclitaxel shown the tumor regression (Sugahara et al., 2007) (Fig. 7.2).

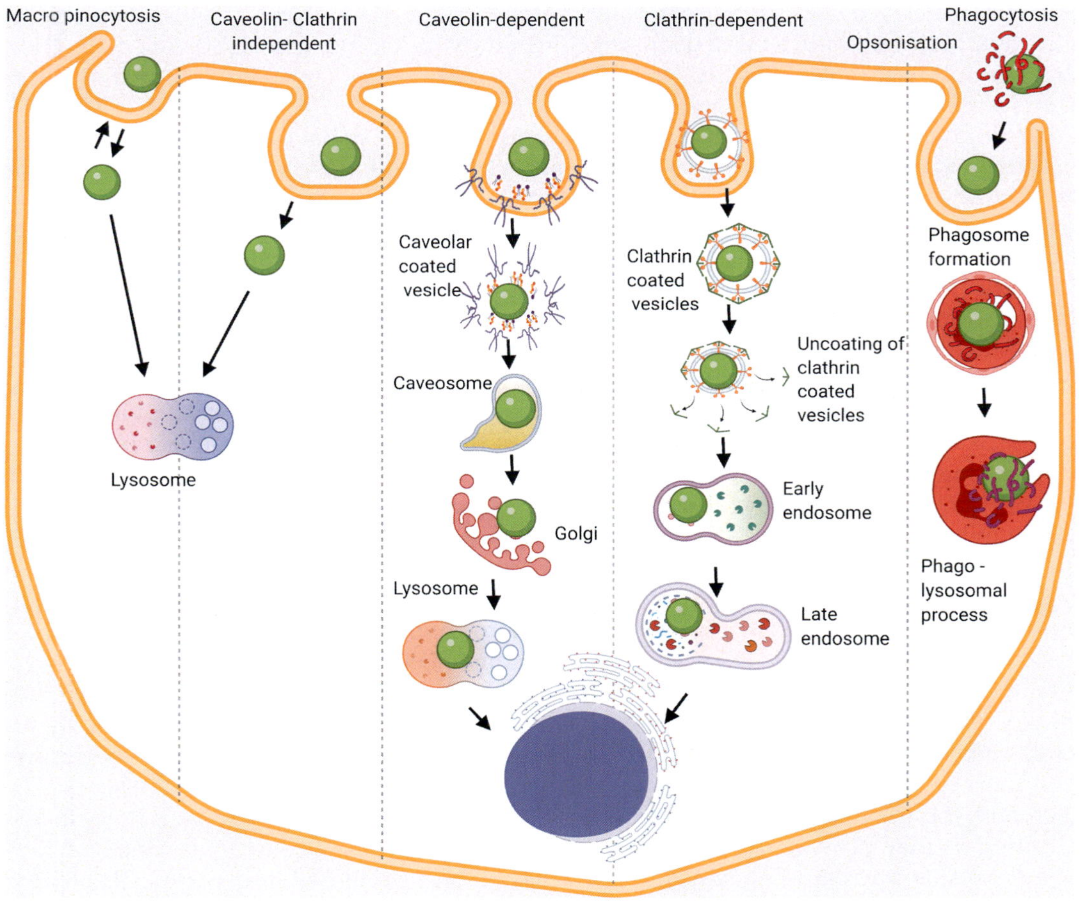

FIG. 7.2 Cellular uptake mechanisms of nanoparticles. (A) Pinocytosis involves the internalization of small molecules in the size range of 50–1000 nm. (B) Clathrin caveolae-independent endocytosis occurs in the absence of clathrin and caveolin. Folic acid conjugated with nanoparticles follows this pathway. (C) In caveolae-dependent endocytosis, the caveolin vesicles form multicaveolar structures of caveosomes which fuse with early endosomes in a bidirectional way. Cell signaling and transcytosis follow this pathway. (D) In clathrin-dependent endocytosis, a specific ligand in extracellular fluids interacts with the receptor on the cell membrane and forms a ligand-receptor complex, which is attracted to locations of clathrin abundance in the cell membrane. (E) Phagocytosis involves engulfment of larger molecules and solid particles in the size range of 100–10,000 nm.

7.5.4 Tissue engineering and regeneration

Tissue engineering is the process that deals with the regeneration of tissues or replacing malfunctioning organs and tissues (Mano et al., 2007). This process is done by using the scaffold materials, cells, and bioactive molecules. There are different strategies available for the use of three-dimensional porous scaffolds and hydrogel matrices for distinct tissue engineering strategies. The polymer of natural origin is advantageous in tissue engineering as its properties match with the extracellular matrix.

The gums of polysaccharides include uronic acid-containing polysaccharides and proteins (Stephen et al., 1990). Gums are important in tissue engineering because of their water solubility and high viscosity, as well as their gel-forming, thickening, and stabilizing properties. High solubility, biocompatibility, and biodegradability make gums suitable materials for scaffold formation in tissue engineering.

Hydrophilic moieties containing hydroxyl and polar functional groups enable hydrogen bonding, which plays a major role in the film-forming properties of the gums. Gums have the unique properties such as hydrophilicity to form a thick solution, less contribution of calories as they are not digested in the upper GIT, and film-forming properties due to their intermolecular forces such as cross-linking, electrostatic, hydrophobic, ionic interaction, and hydrogen bonding. Gums also offer flexibility in cross-linking functional groups from other moieties (Mudgil et al., 2014; Thombare et al., 2016). Polysaccharides gums such as dextran, pullulan, alginate, guar gum, gellan, and xanthan gum are widely used in tissue engineering. Gum-based nanoparticles were tested for tissue engineering of blood vessels, myocardium, heart valve, bone, and partial or tracheal cartilage. Gum-based nanoparticles are also experimented in vertebral discs, skin, liver, skeleton muscles, and urinary bladder.

As gellan gums are anionic linear microbial polysaccharides, they possess desirable biocompatibility and biodegradability is an added advantage. To cite an example, gellan gum has been used for vascularization due to its natural advantages of good temperature sensitivity and divalent ion response in tissue engineering. Functionalized gellan gum with collagen type 1 was fabricated to encapsulate bone marrow-derived mesenchymal stem cells for vascularization (Chen et al., 2018). This study combined collagen with the gellan gum as the latter demonstrates a high phase transition temperature which is disadvantage in encapsulating scaffolds (Grasdalen and Smidsrød, 1987; Singh and Kim, 2005). Collagen is a significant component of the extracellular matrix which contains RGD (arginine-glycine-aspartic) sequences, which are highly helpful in the adhesion and proliferation of stem cells. Therefore, there is a natural advantage in using gellan gum-collagen composites in tissue engineering as they provide improved mechanical properties for tissue engineering and provide a suitable microenvironment for cell adhesion and proliferation in stem cell therapy. The overall results confirmed that collagen-functionalized gellan gums provided improved pore size, which helps in developing the good structure for cell survival and proliferation. Therefore, this gellan-collagen system proved its potential in 3D printing live cell scaffolds for vasculogenic differentiation (Table 7.2).

TABLE 7.2 General application of gum-based nanoparticles and microparticles.

Natural polymers	Category	Pharmaceutical applications	References
Pectin	Polysaccharide	Colon cancer Therapy	Cheng et al. (2013)
		Reducing blood cholesterol	
		Removing lead and mercury from the gastrointestinal tract and respiratory organs	
Alginate	Polysaccharide	Efficient gene transfection	Lee and Mooney (2012)
		Delivery of a variety of low molecular weight drugs	
		Protein delivery	
Guar gum	Polysaccharide	Food additive	Manjunath et al. (2016)
		Cardio, gut, and dental protective	
		Antimicrobial agent	
		Antiinflammatory agent and anticoagulant	
		Shelf-life enhancer	
		Drug delivery agent	
		Sensor and tumor imaging	
Xanthan gum	Polysaccharide	Recommended for colon-associated diseases	Kumar et al. (2018a) and Schnizlein et al. (2020)
		Delivery of protein and peptide drugs	
Okra gum		Okra mucilage showed more antibacterial activity	Mohammadi et al. (2018) and Nayak et al. (2018)
		Okra gum is one of the plant-derived polysaccharides	
		Cross-linked alginate for use as controlled drug-releasing matrices	
Locust bean gum	Protein-based	Biocompatible, bioabsorbable, biodegradable	Braz et al. (2018)
		Mucoadhesive	
		Stabilizer, emulsifier, and gelling agent	
Karaya gum	Polysaccharide	Used in the synthesis of metal nanoparticles	Pooja et al. (2015)
		Adhesive in leak-proof sealing rings for postsurgical drainage pouches or ostomy bags	
		Coarse gum particles are very effective as a bulk laxative	
		Neither digested nor absorbed in the human ingestion channel	
		Applied on dental plates as an adhesive	

Continued

TABLE 7.2 General application of gum-based nanoparticles and microparticles—cont'd

Natural polymers	Category	Pharmaceutical applications	References
Tamarind gum	Polysaccharide	Hydrophilic gel-forming property	Nayak and Pal (2018) and Durai et al. (2012)
		Mucoadhesive properties	
		Novel drug delivery system for oral, buccal, colon, and ocular systems, nanofabrication	
		Wound dressing	
		Cosmetics	

7.6 Gum-based nanoparticles in cancer therapy

Polymeric nanoparticles are biodegradable and not accumulate in the body, which makes them free from severe toxicities and a highly suitable carrier for anticancer drugs (Kumar and Newton, 2017). Biocompatibility, cytocompatibility, and the absence of an immune response are all advantages of natural gum-based nanoparticles (Kenry and Liu, 2018). Several anticancer drugs such as paclitaxel, 5-fluorouracil, and doxorubicin were successfully prepared as polymeric nanoparticles drug delivery systems (De Jong and Borm, 2008). The anticancer agent is usually toxic to normal and healthy cells with a high volume of distribution. This causes serious side effects in cancer chemotherapy. The use of nanoparticles in tumor cell toxicity is advantageous due to their high specificity. This reduces the high dose-related side effects (Jin et al., 2020). Gum-based nanoparticles, due to their smaller size, can penetrate through various biological barriers and small capillaries to deliver the anticancer agent at the specific site. Drug delivery is made efficient by polymeric gum nanoparticles because they avoid the reticular endothelium system and prevent the drug from premature inactivation. Gum-based nanoparticles can also carry DNA, protein, peptides, and other low molecular weight compounds for cancer diagnostic treatments (Yu et al., 2012) (Fig. 7.3).

7.6.1 Gum-based nanoparticles in cancer drug delivery

Natural gum-based nanoparticles cause a minimal toxicological issue which is advantageous over synthetic polymeric nanoparticles (Patra et al., 2018). Gum-based nano- and microparticles have the ability to modify drug release and drug loading. Gum-based nano- and microparticles usually contain a matrix system where nanoparticles have the internal core surrounded by a polymeric membrane. This makes gum-based nano- and microparticles ideal drug nanocarriers for cancer therapy, drug delivery, and vaccines, and carriers for biomolecules such as DNA, and the delivery of antibiotics (Swierczewska et al., 2016). Drug delivery to cancer cells is heavily limited by decreased drug, efflux pumps, increased drug metabolism, physiological barriers, and various cellular mechanisms (Sriraman et al., 2014). Many drugs are poorly soluble and have low bioavailability, meaning they can be

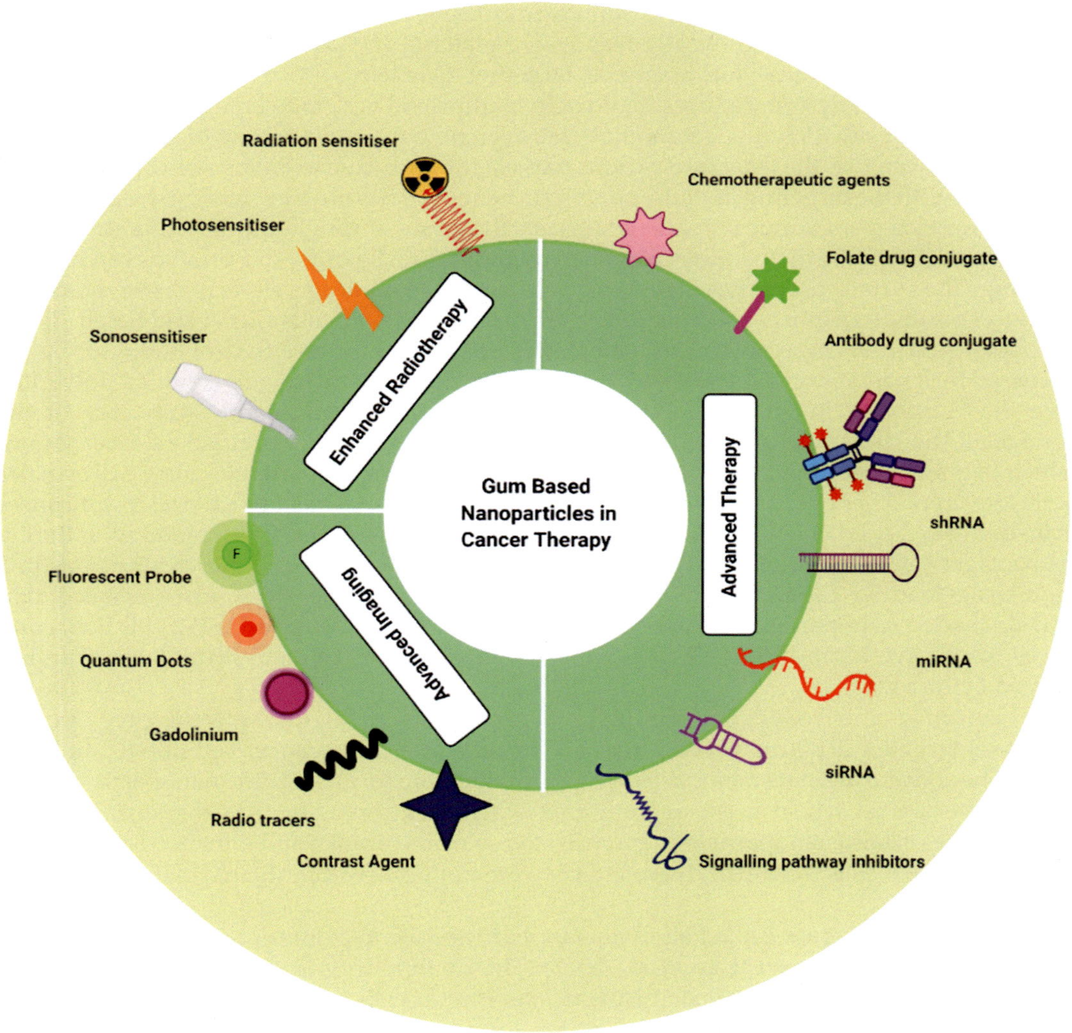

FIG. 7.3 Role of gum-based nanoparticles in cancer therapy. (A) GNPs are used in advanced therapy for drug delivery (chemotherapeutic agent), targeting (folate drug conjugate and antibodies), gene silencing through siRNA, and inhibiting the cancer signals by specifically delivering the signaling pathway inhibitors. (B) GNPs capable of enhancing theragnostic properties of imaging probes such as fluorescent probes, quantum dots, and contrast agents by specifically delivering at the target site. (C) GNPs deliver the photosensitizers and radio sensitizers at the cellular level, which enables site-specific radio therapy at a specific tissue or location which reduces radiation-mediated side effects at the nontargeted locations.

readily cleared from the body by the reticular endothelium system (Gigliobianco et al., 2018). In other words, the efficiency of a drug used for a longer time is limited by dose-dependent side effects (Peper, 2009). The site-specificity of nanoparticles can be achieved by attaching the targeting ligands such as antibodies, peptides, nucleic acids such as RNA, and other biomolecules (Friedman et al., 2013). Stimuli-responsive nanoparticles can be guided to the target by incorporating magnetically responsive material on nanoparticles' surface such as magnetite and maghemite for magnetic targeting (Medeiros et al., 2011). The targeted nano-drug delivery system reduces the toxicity and provides a more efficient drug distribution at the target site. Polymeric nanoparticles overcome the resistance of biological barriers and efficiently deliver the drug into various parts of the body due to their smaller particle size. Combined with the targeting ligands, polymeric nanoparticles are a promising drug carrier to various restricted anatomical regions such as the brain through the blood-brain barrier (Blanco et al., 2015). Targeted molecules such as antibodies, ligands, and proteins can be used to change the surface properties of nanoparticles for targeted drug delivery. These molecules can successfully transport cancer medicines to cancer tissue or cells (Rizvi and Saleh, 2018).

Cisplatin-loaded tragacanth gum nanogels with hydrophilic and hydrophobic core were prepared by a nanoemulsion process (Verma et al., 2020). These nanogels exhibit a polygonal core-shell structure in which cisplatin is embedded in the tragacanth gum core of the nanogels. The drug release profile of cisplatin demonstrated a sustained release pattern which helps in extending the circulation time of the cisplatin. Overall, the unique properties of the developed nanogel of tragacanth gum demonstrated a promising material in anticancer drug delivery. Saeed et al. (2013) developed alginate-based nanoparticles loaded with the cytotoxic agent ICD-85 venom-derived peptides. Sodium alginate nanoparticles demonstrate high drug loading capacity and sustained release profile. The study also revealed that ICD-85-loaded nanoparticles demonstrated significant cytotoxicity of HEP-2 cell line, compared to the free form of ICD-85. This confirms the potential of nano-drug delivery in improved cytotoxicity.

pH-sensitive and reduction-responsive nanoparticles of alginates were prepared for docetaxel-targeted drug delivery to treat HT-29 cells of a colon cancer cell line (Chiu et al., 2020). The sodium alginate nanoparticles were thiolated to form disulfide nanoparticles, then fluorescein-labeled wheat germ agglutinin (fWGA) was conjugated on the surface of the nanoparticles. These nanoparticles demonstrated selective uptake and enhanced cytotoxicity effects on HT-29 cells; however, these effects were not observed in the normal mouse fibroblast L929.

Similar disulfide cross-linked alginate nanoparticles were synthesized for the delivery of doxorubicin and paclitaxel (Gao et al., 2017). The double drug-loaded nanoparticles were shown to have significant cytotoxicity toward the HepG2 cells and HeLa cells; however, they improved the cell survival and growth of the liver normal cells (L-O2). The nanoparticles demonstrated the selective uptake in the cancer cells; moreover, the cardiotoxicity of the doxorubicin was very low when it was administered through the alginate nanoparticles.

Carboxymethyl cellulose dextran nanoparticles were prepared for the doxorubicin delivery. The carboxymethyl dextran was conjugated with lithocholic acid (Thambi et al., 2014). The lithocholic acid and carboxymethyl dextran matrix was loaded with doxorubicin. The disulfide bond linked the lithocholic acid with the nanoparticles to aid the drug release in the presence of GSH. These developed nanoparticles demonstrated higher toxicity to SCC7 cells, and were superior to nanoparticles without the disulfide bond. The nanoparticles effectively delivered the nanoparticles to the nuclei of SCC7 cells. These nanoparticles also demonstrated better biodistribution compared to the simple doxorubicin loaded nanoparticles.

Table 7.3 lists examples of nanoparticles from the gums used as efficient cancer drug delivery carriers.

TABLE 7.3 Gum-based nanoparticles in cancer therapy and drug delivery.

The natural polymeric material used	Drug/agent linked with the NP	Cancer type/pathway/ type of pathway	Key results and conclusions	References
Alginate nanoparticles	Oligopeptide	Caco-2 cells	Oligopeptide–alginate nanoparticles as a melittin carrier and the possibility of their usage as a carrier for peptide-based drugs. Oligopeptide-side chain in alginate offers a specific binding site for melittin and is effective in cancer chemotherapy	Wattanakul et al. (2019)
Alginate-based FGF-2 nanoparticle	Recombinant FGF-2	ERK1/2 signaling pathway	The growth-inhibitory and cytotoxic effects of nanoparticle-mediated intracellular delivery of FGF-2 may be broadly applicable to cancer cells	Miao et al. (2020)
Gum arabic-based gallic acid nanoparticles	Gallic acid with NF-kappaB	RAS-ERK and the PI3K/ AKT signaling pathways	This study also revealed several antineoplastic properties. The gum arabic enhanced the selective uptake in hepatic cells, and enhanced cytotoxicity in breast cancer cells	Hassani et al. (2020)
Dextran-based polymericles	Doxorubicin/ siRNA delivery	Inhibition of ABCB1 (MDR1)	Lipid-modified dextran-based polymeric nanoparticles are a promising platform for siRNA delivery. Nanocarriers loaded with MDR1 siRNA are a potential treatment strategy for reversing MDR in osteosarcoma	Susa et al. (2010)
Gum arabic-encapsulated gold nanoparticles	Gold	Apoptotic pathway and reduction in the levels of the inflammatory mediator TNF-α and the angiogenesis inducer VEGF	Considerable cytotoxicity against A549 cells. The treatment of lung tumor-bearing mice. Laser exposure enhanced the apoptotic pathway, and the elevations in cytochrome-c, death receptor five, and the subsequent upregulation of caspase-3 confirmed the enhancement. Reduction in the levels of the inflammatory mediator TNF-α and the angiogenesis inducer VEGF	Gamal-Eldeen et al. (2017)
Gum arabic and maltodextrin-loaded nanoparticles	Epigallocatechin-3-gallate	Apoptosis of Du145 prostate cancer cells	Reducing the cell viability and inducing apoptosis of Du145 prostate cancer cells. The clonogenic assay demonstrated that encapsulation of epigallocatechin-3-gallate enhanced its inhibitory effect on cell proliferation (10%–20%)	Rocha et al. (2011)

Continued

TABLE 7.3 Gum-based nanoparticles in cancer therapy and drug delivery—cont'd

The natural polymeric material used	Drug/agent linked with the NP	Cancer type/pathway/type of pathway	Key results and conclusions	References
Gum arabic-conjugated gold nanoparticles	Au nanoparticles	Induced cancer cell apoptosis by activation of death receptors DR5 and caspase-3	GA-AuNPs, with or without laser irradiation, induced cancer cell apoptosis	Gamal-Eldeer et al. (2014)
Tamarind gum semiinterpenetrating nanocomposites	Erlotinib	Induced apoptosis	Erlotinib-loaded formulation suppressed a549 cell proliferation and induced apoptosis more effectively than a new drug	Bera et al. (2020)
Konjac glucomannan microsphere	Anti-mir-31 oligonucleotide and curcumin	Suppressing BMP and tgfβ signaling pathways	The released PS-mir-31i/Cur from the microsphere was mucus-penetrating, efficiently passing through the colonic mucus layer, and allowed Cur and mir-31i to specifically target colon tumor cells with the guide of CD133 targeting peptides. Consequently, rectal delivery of sokgm-PS-mir-31i/Cur microspheres suppressed tumor growth	Zhao et al. (2020)
Dextran nanoparticles	Cisplatin-loaded LHRH	Metastasis/breast cancer	Significantly enhanced the antitumor and antimetastasis efficacy, compared to the nontargeted nanoparticles. These results suggest that nanoparticles show great potential for targeted chemotherapy of metastatic breast cancer	Li et al. (2015)
Alginate-sulfate nanoparticles	Plasmid DNA	Human breast cancer	Results show that the developed AlgS-Ca^{2+}-plasmid DNA NPs may be used as an effective nonviral carrier for plasmid DNA	Goldshtein et al. (2019)
Acetylated dextran microparticles	cyclic GMP-AMP	Targeted immunotherapy enhances natural killer cell and $CD8^{+}$ T cell	Treatment results in NK and T cell-dependent antitumor immune response	Watkins-Schulz et al. (2019)

7.6.2 Nucleic acid-conjugated GNPs for cancer therapy

Nucleic acid conjugation is a technique for efficiently expressing the linked encoded protein in an appropriate host or host cell. Gene therapy has the potential to enhance the anticancer effect with minimal adverse reactions. The primary small noncoding RNAs such as microRNAs (miRNAs), small interfering RNAs (siRNAs), and Piwi-interacting RNAs (piRNAs) (Baulcombe, 2004; Carthew and Sontheimer, 2009) are predominantly used in this strategy. These noncoding RNA play an essential role in epigenetic regulation of gene expression. Because of its ability to knock down the targeted genes and in experiments against cancer in animal models, the siRNA method seems appealing (Sayour et al., 2015). However, siRNA delivery is challenging due to the complex biological microenvironment, degradation of the biomolecules, and low site-specificity; all these contributed to the poor therapeutic effect (Sayour et al., 2015; Chen et al., 2016b). The siRNA delivery also causes side effects such as cross-reaction between nucleic acids and drug, and nucleic acid and somatic cells. Therefore, there is a need for a smart vector that successfully delivers the drugs into the site of action to overcome these challenges. In the recent decade, substantial research has shown that siRNA distribution as a method is successful against a variety of malignancies (Oh and Park, 2009; Schiffelers et al., 2004; Luo et al., 2017). Seventy percent of the genes' delivery in clinical trials are done by the viral vectos such as retroviruses, lentviruses, and adenoviruses. Despite the fact that viral vectors are the preferred choice in modern gene therapy, they have drawbacks such as carcinogenesis, immunogenicity, broad tropism, limited DNA packaging capacity, and vector production difficulty.

There is a need for the alternative to viral vectors to be less toxic, less immunogenic, and easy to fabricate (Wang et al., 2015; Nafissi et al., 2014; Yin et al., 2014). Nanomaterial plays a significant role in the delivery of siRNA. The nanomaterials-based delivery of siRNA has been found to be an effective method in cancer treatment (Mei et al., 2019). A variety of nano delivery carriers are available for nucleic acid delivery: liposomes, polymeric nanoparticles, inorganic nanoparticles, vesicular nano-sized carriers, solid lipid nanoparticles, and lipid nanocarriers (Conde et al., 2015; Ragelle et al., 2014). Due to their escape from the degenerative endolysosomal compartment and into the cytoplasm, nanoparticles loaded with plasmid DNA are also an effective gene delivery technology. This leads to the intracellular uptake of nanoparticles which escape the endolysosomal system and release DNA at sustained release pattern (Miki et al., 2017).

Natural polymer-based nanoparticles are successful in effective and safe delivery of nucleic acids. Xu et al. (2014) Prepared the redox responsive gelatine nanoparticles to deliver combination of wt-p53 expressing plasmid DNA and gemcitabine to treat pancreatic cancer. Gene therapy, in combination with a cytotoxic agent, was found suitable for improving pancreatic cancer therapy. The redox responsive thiolated gelatin nanoparticle systems efficiently deliver the drug in the glutathione-medicated reducing intracellular environment. This is successfully used for the site-specific wt-p53 expressing plasmid DNA and gemcitabine delivery by targeting the epidermal growth factor receptor. Efficiency studies were performed in subcutaneous adenocarcinoma-bearing mice; demonstrated improved in vivo targeting. The study results revealed that the gelatin nanoparticles safely delivered the drug and plasmid DNA, which significantly improved the gene-drug combination treatment's performance compared to the gene or drug alone treated groups. Thiolated gelatin nanoparticles loaded with wt-p53 plasmid or gemcitabine effectively induced cell apoptosis, which was confirmed from the increase of mRNA levels, protein apoptotic biomarker expression, and a significant decrease in antiapoptotic transcription factors.

The reverse microemulsion was used to encapsulate the enhanced fluorescent protein and encoded plasmid using chitosan and alginate (You et al., 2006). These biocompatible and biodegradable nanoparticles were used to encapsulate plasmid DNA for gene delivery via the cell endocytosis pathway. Another study (Yang et al., 2010) revealed that chitosan alginate nanoparticles successfully increase the transfection rate. A study on chitosan-alginate-pAcGFP1-C1 plasmid complex nanoparticles evaluated the efficiency of ultrasound on the gene transfection in subcutaneous tumors. The results revealed that these nanoparticles achieved efficient gene transfection with the support of ultrasound.

Table 7.4 summarizes the combination of gum-based nanoparticles in nucleic acid delivery.

7.6.3 GNPs in combination with inorganic nanoparticles for cancer therapy

There is an increasing concern for biocompatibility, stability, and long-term toxicities of INPs that require more research to assure their safe and effective use (Shah and Dobrovolskaia, 2018). In recent years, significant research efforts have been made to prepare natural polymer-based inorganic nanocomposites. Incorporating inorganic nanoparticles into the matrix of natural polymers offers substantial advantages in their performance by facilitating properties such as superparamagnetic, electron and photon transport and surface plasmon resonance effect (Li et al., 2010).

Improving the physical performance of inorganic nanoparticles by tuning the various properties is quite expensive and time-consuming, therefore using natural polymers such as gums is an economical yet practical approach to improve their performance. The natural polymers can be incorporated into inorganic nanoparticles by various methods such as dip coating, spin coating, film casting, and printing; the inorganic nanoparticles can also be synthesized by using natural gums as another way of incorporating natural polymers into inorganic nanoparticles. These particles are also called nanocomposites, and the nanocomposites prepared by this method have shown improvements in various properties such as optical, mechanical, electrical, and magnetic properties (Breiner and Mark, 1998; Rong et al., 2001). The combination of inorganic and organic properties also offers the generation of new functions. The nanocomposite was first synthesized with gold nanoparticles, which showed improved optical properties such as dichroism (Caseri, 2000, 2008). Later, other inorganic metals such as Ag, Pt, Fe, Pd, Rh, Cu Ce, Si, and Ti were also prepared as nanocomposites and their various advantages in cancer therapy were explored.

Gum arabic-coated radioactive gold nanoparticles (GA-^{198}AuNPs) demonstrated many advantages over traditional brachytherapy in treating prostate cancer (Axiak-Bechtel et al., 2014). GA-^{198}AuNPs are advantageous in homogenous dose distribution and higher dose-rate irradiation. The green synthesis of gold nanoparticles with a gum kondagogu capping was carried out. The anionic gum-capped kondagogu-gold nanoparticles enabled the conjugation of folic acid and fluorescein isothiocyanate (FITC) to generate fluorescently labelled (GNP F2) GNPs. GNP F2 were also evaluated for targeted delivery in both folate receptor-positive (MCF-7) and folate receptor-negative (A549) cells. This study revealed that folate-conjugated kondagogu nanoparticles are effective nanocarriers for targeted drug delivery and cancer cellular imaging through the folate receptor conjugation (Kumar et al., 2018b).

PEGylated gold nanoparticles stabilized with xanthan gum were prepared to deliver curcumin to treat cancer (Muddineti et al., 2016). This approach leads to the formation of spherical AuNPs with a mean hydrodynamic diameter of 80 ± 3 nm. They have optimized

TABLE 7.4 Gum-based nanoparticles conjugated with nucleic acids.

Gum-based nanoparticles	Nucleic acid loaded/ interfered	Potential target and cancer type/cells	Key results and conclusions	References
PEI alginate nanoparticles	siRNA	Inhibit tumor lymphangiogenesis and lymphatic metastasis	siRNA delivered with PEI-alginate nanoparticles can effectively inhibit differentiation and lymphangiogenesis. Inhibiting VEGFR-3 signaling with siRNA/nanocomplexes is an effective therapy for suppressing tumor lymphangiogenesis and lymphatic metastasis	Li et al. (2014)
Dextran-based polymeric nanoparticle	Mir-199a-3p	Inhibit the growth and proliferation of osteosarcoma cells	Dextran nanoparticles could deliver both mir-199a-3p and let-7a into osteosarcoma cell lines (KHOS and U-2OS) successfully. Dextran nanoparticles loaded with mirs efficiently downregulated the expression of target proteins and effectively inhibited the growth and proliferation of osteosarcoma cells	Zhang et al. (2015)
Dextran-based nanoparticles	SiRNA	Target myeloid cells of the liver	Dextran-based nanoparticles were taken up by cells of the myeloid lineage without compromising their viability. In vivo, empty and siRNA-carrying dextran-based nanoparticles were distributed to the liver. Serum parameters indicated no in vivo toxicity	Foerster et al. (2016)
Gellan gum-PEI nanocomposites	siRNA	In vitro (primary keratinocytes, HEK293, HeLa, and HepG2 cells) and in vivo (*Drosophila melanogaster*)	The nanocomposites exhibited negligible toxicity in vitro (primary keratinocytes, HEK293, HeLa, and HepG2 cells) and in vivo (*D. melanogaster*) compared to PEI or lipofectamine	Zhao et al. (2020)
Gum arabic-encapsulated gold nanoparticles	miR-210 and miR-21	CAL-127 cells	The expression of hypoxia-regulating miRNAs (hypoxia miR-210 and miR-21), regulators of HIF-1α, is high in OTSCC. Gum arabic-encapsulated gold nanoparticles (GA-AuNPs) show promising modality in cancer treatment	Gamal-Eldeen et al. (2021)
Xanthan gum-stabilized gold nanoparticles	siRNA/ shRNA	Mannose receptor overexpressing cancer cells	Self-targeted drug carriers for the delivery of chemotherapeutic agents/ siRNA/shRNA	Pooja and Sistla (2019)
Chitosan-alginate-DNA nanoparticles	pAcGFP1-C1 plasmid	Transfection of cells and subcutaneous tumors	Chitosan-alginate-DNA nanoparticles protect the transgene from DNase I degradation. The transgene product expression could be enhanced efficiently in tumor tissues	Yang et al. (2010)

drug loading. In this reaction, ascorbic acid and PEG-xanthan gum were used as a reducing agent and stabilizing agents, respectively. Previously the curcumin was prepared as a solid dispersion with the help of polyvinyl pyrrolidone k 30. The developed nanoparticles demonstrated the higher cellular uptake and cytotoxicity in the B16F10 murine melanoma cancer cell line. The overall results demonstrated that the developed nanoparticles were very efficient compared to the free curcumin in treating melanoma cancer cells.

The synthesis of environmentally friendly gold nanoparticles was done by using xanthan gum as a reducing agent (Pooja et al., 2014). In this synthesis, xanthan gum was used as a reducing agent as well as a stabilizing agent. The developed nanoparticles were loaded with the anticancer drug doxorubicin. The study evaluated the suitability of the gum-based nanoparticles in the delivery of doxorubicin HCl. The nanoparticle loaded with the doxorubicin was found to be an efficient drug delivery carrier. The overall results confirmed that the nanoparticles were three times more effective in cytotoxicity toward cytotoxicity in A549 cells than the free doxorubicin (Fig. 7.4; Table 7.5).

7.6.4 Cancer theragnostic applications

Theragnostics is the performance of therapeutic and diagnosis simultaneously by using a single formulation (Swierczewska et al., 2016). Nanoparticles are a versatile candidate for multifunctionality in a single device. They are highly suitable for theragnostic applications due to their targeting ability and capacity of delivering multiple candidates in a single platform (Swierczewska et al., 2016; Xie et al., 2010). Nanoparticles also offer the advantage of delivering the drug through external stimuli or by the external stimuli-guided drug delivery. Externally controlled drug delivery can precisely deliver the predetermined amount of drug at the site of action. Generally, two different types of nanoparticles are used in the bioimaging technique: luminescent nanoprobe for optical imaging and magnetic nanoparticles for MRI scan (Geppert and Himly, 2021). Polymeric gold nanoparticles used in the fluorescent probe are useful to detect cancer and as a biomarker (Chinen et al., 2015). Polymeric nanoparticles are also used to detect viral or bacterial DNA (Seyfoori et al., 2021). Several molecular imaging techniques are available in which gum-based polymeric nanoparticles are employed; optical imaging, magnetic resonance imaging, ultrasonic imaging positron emission tomography, and fluorescent laser confocal microscopy are some of the well-known imaging techniques (Price, 2012).

New-generation gold nanoparticles were prepared by Kattumuri et al. using gum arabic (Kattumuri et al., 2007). The AuNPs were stabilized by gum arabic and analyzed for their in vitro and in vivo pharmacokinetics on pigs to understand the organ-specific drug delivery of gum arabic-conjugated gold nanoparticles. The X-ray CT contrast technique was used for molecular imaging. This method synthesized the biocompatible nanoparticles for therapeutic and diagnostic applications. The developed AuNPs demonstrated the multiple advantages such as site-specific drug delivery and photoactive agent for optical imaging, which aids CT imaging and X-ray based cancer therapy.

Polyarabic acid coated magnetic nanoparticles demonstrated superior imaging, therapeutic, and biocompatibility properties (Patitsa et al., 2017). Polyarabic acid is a water-soluble polysaccharide derived from acacia, which is highly biocompatible. The polyarabic acid

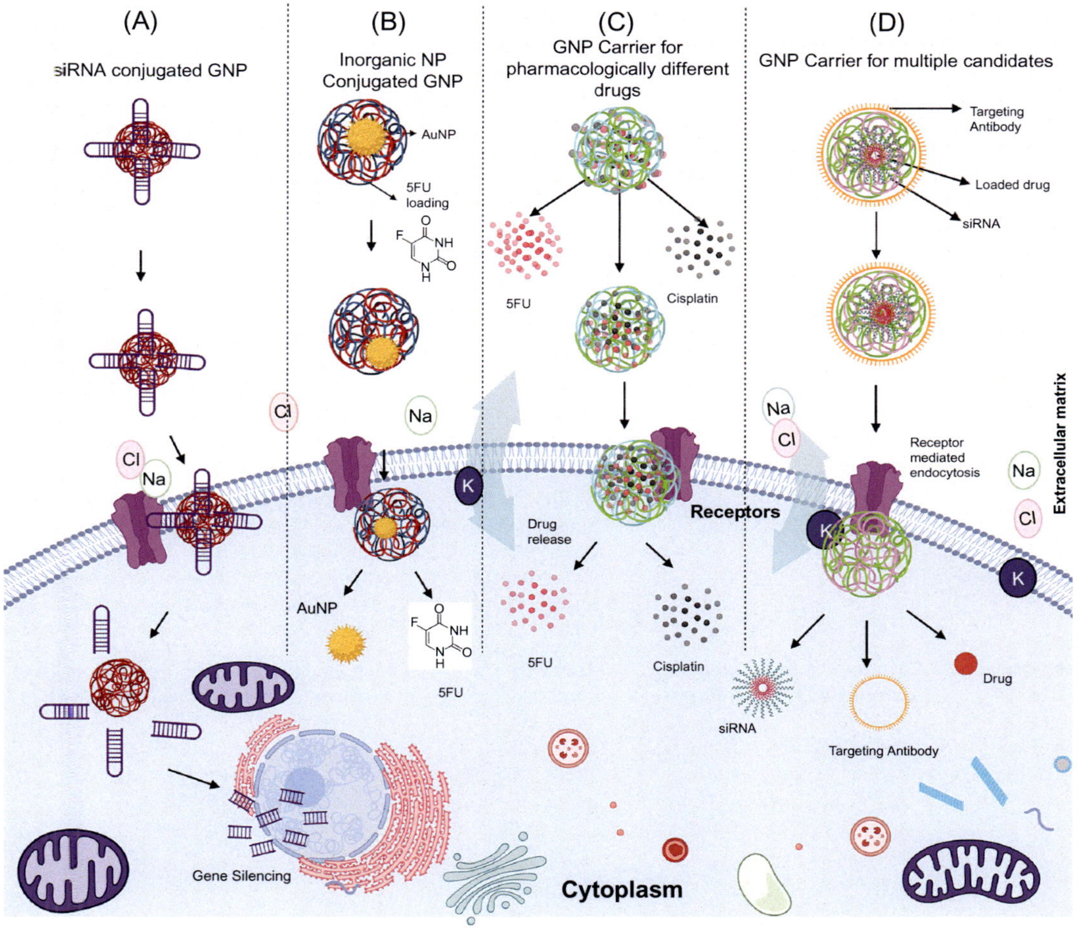

FIG. 7.4 Cancer drug delivery mechanisms of gum-based nanoparticles: (A) SiRNA drug delivery by GNP by conjugation for cancer therapy; (B) inorganic nanoparticles drug delivery by GNPs by coating or conjugation; (C) GNPs loaded with different drugs for drug delivery; and (D) GNPs work as a versatile single carrier for therapeutic agents and targeting ligands such as drugs, antibodies, and siRNA.

coated iron oxide nanoparticle was loaded with doxorubicin. The nanoparticles demonstrated enhanced penetration through the cell membrane, higher drug loading, and controlled drug release behavior. This nanoparticle also demonstrated minimal in vivo toxicity, due to the conjugation with the arabic acid. The developed nanoparticles have comparable contrast properties with commercial contrast agents. Overall, the polyarabic acid coating is advantageous in drug delivery and MRI applications.

Water-dispersible spherical nanoparticles of guar gum were prepared by the acid hydrolysis technique (Ghosh et al., 2015). These nanoparticles were further functionalized with fluorescein isothiocyanate through the hydroxy-propyl amine spacer group. These

TABLE 7.5 Gums in combination with inorganic nanoparticles.

Inorganic nanomaterial	Strategy	Natural polymer used	Cancer type/cells	Key results and conclusions	References
Gold nanoparticles	Encapsulation	Gum arabic	A549 cells/lung cancer	GA-AuNPs significantly cytotoxicity against A549 cells. Gum arabic-conjugated AuNPs treatment and laser exposure enhanced the apoptotic pathway of lung tumor-bearing mice, the elevations in cytochrome-c, caspase-3, and a significant reduction in the levels of the inflammatory mediator TNF-α	Gamal-Eldeen et al. (2017)
Copper nanoparticles	Reduction reaction	Cashew gum	4T1 LUC (4T1 mouse mammary tumor cell line) and NIH 3T3 cells (murine fibroblast cells)	Viability cell assays for CG-CuNPs at (0.250 mM) inhibited by 70% the growth of 4T1 LUC and NIH 3T3 cells over a 24 h period	Amorim et al. (2019)
Iron oxide	Layer-by-layer self-assembly	Alginate/chitosan	MDA-MB-231/breast cancer	Magnetic field aided effective targeted delivery of curcumin. The fluorescence-activated cell sorting (FACS) assay indicated that MDA-MB-231 cells treated with alginate/chitosan demonstrated a 3–6-fold increase in uptake	Song et al. (2018)
Gold nanoparticles	Covalent conjugation and esterification reaction	Alginate	C6 glioma and MCF-7/breast cancer	Alginate conjugated AuNPs exhibited enhanced cytotoxic effect against C6 glioma and MCF-7 cancer cell lines. Highly hemocompatible	Dey et al. (2016)
Silver nanoparticles	Polyelectrolyte complex formation	Alginate-chitosan	U87MG (human glioblastoma)	Alginate-chitosan conjugated silver nanoparticles generated ROS in cells causing mitochondrial dysfunction, and DNA damage leads to apoptosis inU87MG cells glioblastoma. Induce apoptosis in cells at a very low concentration	Sharma et al. (2014b)

TABLE 7.5 Gums in combination with inorganic nanoparticles—cont'd

Inorganic nanomaterial	Strategy	Natural polymer used	Cancer type/cells	Key results and conclusions	References
Magnetic iron oxide nanoparticles	Conjugation	Gum arabic	9L glioma cells	Gum arabic successfully stabilized MNP through a facile procedure. GA-MNP product possessed several major reactive functional groups for easy linking with biofunctional compounds. Higher cellular uptake by tumor cells and accumulated in brain tumors 12-fold higher than the normal brain	Zhang et al. (2009)
Gold nanoparticles	Conjugation	Gum kondagogu	A549 and MCF-7 cells	The in vitro viability study indicated a high level of biocompatibility in A549 and MCF-7 cells. F2-GNPs are an effective nanocarrier for targeted delivery, cellular imaging, and diagnostic purposes	Kumar et al. (2018b) and Sharma et al. (2014b)
Gold nanoparticles	Conjugation	Gum arabic (220)	(MDA-MB-231)/(PANC-1)/(PC-3)	Nanoparticles work as an anticancer agent against human breast (MDA-MB-231), pancreatic (PANC-1), and prostate (PC-3) cancers. Gum arabic provided a protein matrix support for enhanced trans-resveratrol loading onto the surface of the AuNPs	Thipe et al. (2019)

nanoparticles are efficient in bioimaging and biomolecular interaction as a pH-sensing probe in the subcellular environment. The nanoparticles' fluorescent intensity can be modulated by pH changes, and the fluorescent intensity was significantly augmented in the pH range of 6–8 h, which helps sense pH in the subcellular environemnt.

7.7 Conclusions

Gum-based nanoparticles are a versatile nanodrug delivery system that can be used for a variety of therapeutic and nontherapeutic applications. Due to their biocompatibility, GNPs outperform other nanoformulations. They are establishing themselves as the next-generation nanocarrier, redefining the dosage form, dosing, and pharmacokinetics of a variety of therapeutic compounds and xenobiotics. GNPs excel in cancer therapy due to their high drug loading capacity and ability to accommodate numerous candidates on a single platform. When paired with biomolecules such as siRNA, they are highly effective in cancer therapy. Their interaction with inorganic nanoparticles has recently been shown to be quite beneficial in terms of increasing the biocompatibility of inorganic nanoparticles. However, several research topics require additional exploration, and future research will provide additional information on the in vivo profile, toxicological profile, biodistribution, and clearance of these compounds.

References

Aakash, D., Neeraj, R., Ashok, K., Rimmy, N., Prabodh, C.S., Arun, K.S., 2019. Prospective of natural gum nanoparticulate against cardiovascular disorders. Curr. Chem. Biol. 13 (3), 197–211.

Abdel-Mottaleb, M.S.A., Byrne, J.A., Chakarov, D., 2011. Nanotechnology and solar energy. Int. J. Photoenergy 2011, 194146.

Ahmad, Z., Pandey, R., Sharma, S., Khuller, G.K., 2006. Alginate nanoparticles as antituberculosis drug carriers: formulation development, pharmacokinetics and therapeutic potential. Indian J. Chest Dis. Allied Sci. 48 (3), 171–176.

Ali, B.H., Ziada, A., Blunden, G., 2009. Biological effects of gum arabic: a review of some recent research. Food Chem. Toxicol. 47 (1), 1–8.

Amaldoss, M.J.N., Najar, I.A., Kumar, J., Sharma, A., 2021. Therapeutic efficacy of rifaximin loaded tamarind gum polysaccharide nanoparticles in TNBS induced IBD model Wistar rats. Rep. Pract. Oncol. Radiother. 26 (5), 712–729. https://doi.org/10.5603/RPOR.a2021.0100.

Aminabhavi, T.M., Nadagouda, M.N., Joshi, S.D., More, U.A., 2014. Guar gum as platform for the oral controlled release of therapeutics. Expert Opin. Drug Deliv. 11 (5), 753–766.

Amorim, A., Mafud, A.C., Nogueira, S., Jesus, J.R., Araújo, A.R., Plácido, A., et al., 2019. Copper nanoparticles stabilized with cashew gum: antimicrobial activity and cytotoxicity against 4T1 mouse mammary tumor cell line. J. Biomater. Appl. 34 (2), 188–197.

Animesh, K., Afrasim, M., Shruthi, R., Ayaz, A., Hosakote, G.S., 2012. Cashew gum a versatile hydrophyllic polymer: a review. Curr. Drug Ther. 7 (1), 2–12.

Aronson, J.K. (Ed.), 2016. Guar gum. In: Meyler's Side Effects of Drugs, sixteenth ed. Elsevier, Oxford, p. 634.

Axiak-Bechtel, S.M., Upendran, A., Lattimer, J.C., Kelsey, J., Cutler, C.S., Selting, K.A., et al., 2014. Gum arabic-coated radioactive gold nanoparticles cause no short-term local or systemic toxicity in the clinically relevant canine model of prostate cancer. Int. J. Nanomedicine 9, 5001–5011.

Baimark, Y., Srisuwan, Y., 2013. Preparation of polysaccharide-based microspheres by a water-in-oil emulsion solvent diffusion method for drug carriers. Int. J. Polym. Sci. 2013, 761870.

Balaghi, S., Mohammadifar, M.A., Zargaraan, A., 2010. Physicochemical and rheological characterization of gum Tragacanth exudates from six species of Iranian Astragalus. Food Biophys. 5 (1), 59–71.

Barreras-Urbina, C.G., Ramírez-Wong, B., López-Ahumada, G.A., Burruel-Ibarra, S.E., Martínez-Cruz, O., Tapia-Hernández, J.A., et al., 2016. Nano- and micro-particles by nanoprecipitation: possible application in the food and agricultural industries. Int. J. Food Prop. 19 (9), 1912–1923.

Baulcombe, D., 2004. RNA silencing in plants. Nature 431 (7006), 356–363.

BeMiller, J.N., 2019. Gum arabic and other exudate gums. In: JN, B.M. (Ed.), Carbohydrate Chemistry for Food Scientists, third ed. AACC International Press, pp. 313–321 (Chapter 16).

Bera, H., Abbasi, Y.F., Lee Ping, L., Marbaniang, D., Mazumder, B., Kumar, P., et al., 2020. Erlotinib-loaded carboxymethyl temarind gum semi-interpenetrating nanocomposites. Carbohydr. Polym. 230, 115664.

Bhupinder, K., Newton, M.J., 2016. Impact of pluronic F-68 vs Tween 80 on fabrication and evaluation of acyclovir SLNs for skin delivery. Recent Pat. Drug Deliv. Formul. 10 (3), 207–221.

Blanco, E., Shen, H., Ferrari, M., 2015. Principles of nanoparticle design for overcoming biological barriers to drug delivery. Nat. Biotechnol. 33 (9), 941–951.

Braz, L., Grenha, A., Corvo, M.C., Lourenço, J.P., Ferreira, D., Sarmento, B., et al., 2018. Synthesis and characterization of locust bean gum derivatives and their application in the production of nanoparticles. Carbohydr. Polym. 181, 974–985.

Breiner, J., Mark, J., 1998. Preparation, structure, growth mechanisms and properties of siloxane composites containing silica, titania or mixed silica–titania phases. Polymer 39 (22), 5483–5493.

Cai, W., Shin, D.W., Chen, K., Gheysens, O., Cao, Q., Wang, S.X., et al., 2006. Peptide-labeled near-infrared quantum dots for imaging tumor vasculature in living subjects. Nano Lett. 6 (4), 669–676.

Carthew, R.W., Sontheimer, E.J., 2009. Origins and mechanisms of miRNAs and siRNAs. Cell 136 (4), 642–655.

Caseri, W., 2000. Nanocomposites of polymers and metals or semiconductors: historical background and optical properties. Macromol. Rapid Commun. 21 (11), 705–722.

Caseri, W., 2008. Inorganic nanoparticles as optically effective additives for polymers. Chem. Eng. Commun. 196 (5), 549–572.

Chacko, A.C., Newton, A.M.J., 2019. Synthesis and characterization of valacyclovir HCl hybrid solid lipid nanoparticles by using natural oils. Recent Pat. Drug Deliv. Formul. 13 (1), 46–61.

Chen, J., Zhang, W., Li, X., 2016a. Adsorption of Cu (II) ion from aqueous solutions on hydrogel prepared from Konjac glucomannan. Polym. Bull. 73 (7), 1965–1984.

Chen, J., Guo, Z., Tian, H., Chen, X., 2016b. Production and clinical development of nanoparticles for gene delivery. Mol. Ther. Methods Clin. Dev. 3, 16023.

Chen, K., Li, J., Feng, Y., He, F., Zhou, Q., Xiao, D., et al., 2017. Structural and rheological characterizations of nanoparticles of environment-sensitive hydrophobic alginate in aqueous solution. Mater. Sci. Eng. C Mater. Biol. Appl. 70 (Pt 1), 617–627.

Chen, H., Zhang, Y., Ding, P., Zhang, T., Zan, Y., Ni, T., et al., 2018. Bone marrow-derived mesenchymal stem cells encapsulated in functionalized gellan gum/collagen hydrogel for effective vascularization. ACS Appl. Bio Mater. 1 (5), 1408–1415.

Cheng, H., Zhang, Z., Leng, J., Liu, D., Hao, M., Gao, X., et al., 2013. The inhibitory effects and mechanisms of rhamnogalacturonan I pectin from potato on HT-29 colon cancer cell proliferation and cell cycle progression. Int. J. Food Sci. Nutr. 64 (1), 36–43.

Chinen, A.B., Guan, C.M., Ferrer, J.R., Barnaby, S.N., Merkel, T.J., Mirkin, C.A., 2015. Nanoparticle probes for the detection of cancer biomarkers, cells, and tissues by fluorescence. Chem. Rev. 115 (19), 10530–10574. https://doi.org/10.1021/acs.chemrev.5b00321.

Chiu, H.I., Ayub, A.D., Mat Yusuf, S.N.A., Yahaya, N., Abd Kadir, E., Lim, V., 2020. Docetaxel-loaded disulfide cross-linked nanoparticles derived from thiolated sodium alginate for colon cancer drug delivery. Pharmaceutics 12 (1), 38.

Choi, M., Kwok, S.J.J., Yun, S.H., 2015. In vivo fluorescence microscopy: lessons from observing cell behavior in their native environment. Physiology (Bethesda) 30 (1), 40–49.

Choudhary, P.D., Pawar, H.A., 2014. Recently investigated natural gums and mucilages as pharmaceutical excipients: an overview. J. Pharm. 2014, 204849.

Choukaife, H., Doolaanea, A.A., Alfatama, M., 2020. Alginate nanoformulation: influence of process and selected variables. Pharmaceuticals (Basel) 13 (11), 335.

Chronopoulou, L., Fratoddi, I., Palocci, C., Venditti, I., Russo, M.V., 2009. Osmosis based method drives the self-assembly of polymeric chains into micro- and nanostructures. Langmuir 25 (19), 11940–11946.

Conde, J., Ambrosone, A., Hernandez, Y., Tian, F., McCully, M., Berry, C.C., et al., 2015. 15 years on siRNA delivery: beyond the state-of-the-art on inorganic nanoparticles for RNAi therapeutics. Nano Today 10 (4), 421–450.

Contreras, J.E., Rodriguez, E.A., Taha-Tijerina, J., 2017. Nanotechnology applications for electrical transformers—a review. Electr. Power Syst. Res. 143, 573–584.

Dash, A.K., 1997. Determination of the physical state of drug in microcapsule and microsphere formulations. J. Microencapsul. 14 (1), 101–112.

De Jong, W.H., Borm, P.J.A., 2008. Drug delivery and nanoparticles: applications and hazards. Int. J. Nanomed. 3 (2), 133–149. https://doi.org/10.2147/ijn.s596.

Desgouilles, S., Vauthier, C., Bazile, D., Vacus, J., Grossiord, J.-L., Veillard, M., et al., 2003. The design of nanoparticles obtained by solvent evaporation: a comprehensive study. Langmuir 19 (22), 9504–9510.

Deshmukh, A.S., Aminabhavi, T.M., 2021. Pharmaceutical applications of various natural gums natural gums. In: Ramawat, K.G., Mérillon, J.-M. (Eds.), Polysaccharides: Bioactivity and Biotechnology. Springer International Publishing, Cham, pp. 1–30.

Deshmukh, A.S., Setty, C.M., Badiger, A.M., Muralikrishna, K.S., 2012. Gum ghatti: a promising polysaccharide for pharmaceutical applications. Carbohydr. Polym. 87 (2), 980–986.

Dey, S., Sherly, M.C.D., Rekha, M.R., Sreenivasan, K., 2016. Alginate stabilized gold nanoparticle as multidrug carrier: evaluation of cellular interactions and hemolytic potential. Carbohydr. Polym. 136, 71–80.

Dhamija, K., Arora, V., Kulkarni, G., Vandana, 2017. Formulation of *Prunus Amygdalus* (Family: Rosaceae) gum based compression coated tablet for colon targeted delivery of 5-aminosalicylic acid. Novel Appr. Drug Des. Devel. 2 (5), 67–70.

Dhar, S., Reddy, E.M., Shiras, A., Pokharkar, V., Prasad, B.L.V., 2008. Natural gum reduced/stabilized gold nanoparticles for drug delivery formulations. Chem. Eur. J. 14 (33), 10244–10250.

Dheer, D., Arora, D., Jaglan, S., Rawal, R.K., Shankar, R., 2017. Polysaccharides based nanomaterials for targeted anti-cancer drug delivery. J. Drug Target. 25 (1), 1–16.

Dias, S.F.L., Nogueira, S.S., de França, D.F., Guimarães, M.A., de Oliveira Pitombeira, N.A., Gobbo, G.G., et al., 2016. Acetylated cashew gum-based nanoparticles for transdermal delivery of diclofenac diethyl amine. Carbohydr. Polym. 143, 254–261.

Dudhipala, N., Gorre, T., 2020. Neuroprotective effect of ropinirole lipid nanoparticles enriched hydrogel for Parkinson's disease: in vitro, ex vivo, pharmacokinetic and pharmacodynamic evaluation. Pharmaceutics 12 (5), 448.

Durai, R., Rajalakshmi, G.R., Joseph, J., Kanchalochana, S., Hari, V., 2012. Tamarind seed polysaccharide: a promising natural excipient for pharmaceuticals. Int. J. Green Pharm. 6, 270.

Ebrahiminezhad, A., Moeeni, F., Taghizadeh, S.-M., Seifan, M., Bautista, C., Novin, D., et al., 2019. Xanthan gum capped ZnO microstars as a promising dietary zinc supplementation. Foods 8 (3), 88.

Fan, L., Yi, J., Tong, J., Zhou, X., Ge, H., Zou, S., et al., 2016. Preparation and characterization of oxidized konjac glucomannan/carboxymethyl chitosan/graphene oxide hydrogel. Int. J. Biol. Macromol. 91, 358–367.

Fan, L., Zhu, B., Su, P.-C., He, C., 2018. Nanomaterials and technologies for low temperature solid oxide fuel cells: recent advances, challenges and opportunities. Nano Energy 45, 148–176.

Fang, W., Wu, P., 2004. Variations of Konjac glucomannan (KGM) from Amorphophallus konjac and its refined powder in China. Food Hydrocoll. 18 (1), 167–170.

Featherstone, S., 2015. 8—Ingredients used in the preparation of canned foods. In: Featherstone, S. (Ed.), A Complete Course in Canning and Related Processes, fourteenth ed. Woodhead Publishing, Oxford, pp. 147–211.

Fiume, M.M., Heldreth, B., Bergfeld, W.F., Belsito, D.V., Hill, R.A., Klaassen, C.D., et al., 2016. Safety assessment of microbial polysaccharide gums as used in cosmetics. Int. J. Toxicol. 35 (1 Suppl), 5S–49S.

Foerster, F., Bamberger, D., Schupp, J., Weilbächer, M., Kaps, L., Strobl, S., et al., 2016. Dextran-based therapeutic nanoparticles for hepatic drug delivery. Nanomedicine 11 (20), 2663–2677.

Fosu, M.-A., Ofori-Kwakye, K., Kuntworbe, N., Bonsu, M.A., 2016. Investigation of blends of cashew and xanthan gums as a potential carrier for colonic delivery of ibuprofen. Int. J. PharmTech Res. 9 (7), 369–380.

Franco, P., De Marco, I., 2021. Nanoparticles and nanocrystals by supercritical CO_2-assisted techniques for pharmaceutical applications: a review. Appl. Sci. 11 (4), 1476.

Friedman, A.D., Claypool, S.E., Liu, R., 2013. The smart targeting of nanoparticles. Curr. Pharm. Des. 19 (35), 6315–6329.

Gamal-Eldeen, A.M., Moustafa, D., El-Daly, S.M., Katti, K.V., 2014. P0131 efficacy of gum arabic-conjugated gold nanoparticles as a photothermal therapy for lung cancer: in vitro and in vivo approaches. Eur. J. Cancer 50, e46.

Gamal-Eldeen, A.M., Moustafa, D., El-Daly, S.M., Abo-Zeid, M.A.M., Saleh, S., Khoobchandani, M., et al., 2017. Gum arabic-encapsulated gold nanoparticles for a non-invasive photothermal ablation of lung tumor in mice. Biomed. Pharmacother. 89, 1045–1054.

Gamal-Eldeen, A.M., Baghdadi, H.M., Afifi, N.S., Ismail, E.M., Alsanie, W.F., Althobaiti, F., et al., 2021. Gum arabic-encapsulated gold nanoparticles modulate hypoxamiRs expression in tongue squamous cell carcinoma. Mol. Cell. Toxicol. 17, 111–121.

Gao, C., Tang, F., Zhang, J., Lee, S.M., Wang, R., 2017. Glutathione-responsive nanoparticles based on a sodium alginate derivative for selective release of doxorubicin in tumor cells. J. Mater. Chem. B 5 (12), 2337–2346.

Geppert, M., Himly, M., 2021. Iron oxide nanoparticles in bioimaging—an immune perspective. Front. Immunol. 12, 688927. https://doi.org/10.3389/fimmu.2021.688927.

Gericke, M., Schulze, P., Heinze, T., 2020. Nanoparticles based on hydrophobic polysaccharide derivatives—formation principles, characterization techniques, and biomedical applications. Macromol. Biosci. 20 (4), 1900415.

Gharehaghaji, A.A., 2019. Chapter 18—Nanotechnology in sport clothing. In: Subic, A. (Ed.), Materials in Sports Equipment, second ed. Woodhead Publishing, pp. 521–568.
Ghosh, S.K., Abdullah, F., Mukherjee, A., 2015. Fabrication and fluorescent labeling of guar gum nanoparticles in a surfactant free aqueous environment. Mater. Sci. Eng. C Mater. Biol. Appl. 46, 521–529.
Gigliobianco, M.R., Casadidio, C., Censi, R., Di Martino, P., 2018. Nanocrystals of poorly soluble drugs: drug bioavailability and physicochemical stability. Pharmaceutics 10 (3), 134.
Goldshtein, M., Shamir, S., Vinogradov, E., Monsonego, A., Cohen, S., 2019. Co-assembled Ca^{2+} alginate-sulfate nanoparticles for intracellular plasmid DNA delivery. Mol. Ther. Nucleic Acids 16, 378–390.
Gontard, L.C., Knappett, B.R., Wheatley, A.E.H., Chang, S.L.Y., Fernández, A., 2014. Impregnation of carbon black for the examination of colloids using TEM. Carbon 76, 464–468.
Grasdalen, H., Smidsrød, O., 1987. Gelation of gellan gum. Carbohydr. Polym. 7 (5), 371–393.
Gu, L.Z., Hong, Q., Xiang, C.J., 2010. The application of nanotechnology for mechanical manufacturing. Key Eng. Mater. 447-448, 86–90.
Gupta, S., Kesarla, R., Omri, A., 2013. Formulation strategies to improve the bioavailability of poorly absorbed drugs with special emphasis on self-emulsifying systems. ISRN Pharm. 2013, 848043.
Hao, H., Hui, D., Lau, D., 2020. Material advancement in technological development for the 5G wireless communications. Nanotechnol. Rev. 9 (1), 683–699.
Hassani, A., Azarian, M.M.S., Ibrahim, W.N., Hussain, S.A., 2020. Preparation, characterization and therapeutic properties of gum arabic-stabilized gallic acid nanoparticles. Sci. Rep. 10 (1), 17808.
Hernández-Giottonini, K.Y., Rodríguez-Córdova, R.J., Gutiérrez-Valenzuela, C.A., Peñuñuri-Miranda, O., Zavala-Rivera, P., Guerrero-Germán, P., et al., 2020. PLGA nanoparticle preparations by emulsification and nanoprecipitation techniques: effects of formulation parameters. RSC Adv. 10 (8), 4218–4231.
Iqbal, Z., Khan, R., Nasir, F., Khan, J.A., Rashid, A., Khan, A., et al., 2011. Preparation and in-vitro in-vivo evaluation of sustained release matrix diclofenac sodium tablets using PVP-K90 and natural gums. Pak. J. Pharm. Sci. 24 (4), 435–443.
Ishak, R.A., Osman, R., Awad, G.A., 2016. Dextran-based nanocarriers for delivery of bioactives. Curr. Pharm. Des. 22 (22), 3411–3428.
Jatinder, K., Newton, A.M.J., 2015. IBD modern concepts, nano drug delivery and patents: an update. Recent Pat. Nanomed. 5 (2), 122–145.
Jeevanandam, J., Barhoum, A., Chan, Y.S., Dufresne, A., Danquah, M.K., 2018. Review on nanoparticles and nanostructured materials: history, sources, toxicity and regulations. Beilstein J. Nanotechnol. 9, 1050–1074.
Jeon, H.-J., Jeong, Y.-I., Jang, M.-K., Park, Y.-H., Nah, J.-W., 2000. Effect of solvent on the preparation of surfactant-free poly(DL-lactide-co-glycolide) nanoparticles and norfloxacin release characteristics. Int. J. Pharm. 207 (1), 99–108.
Jeong, Y.I., Kim, D.H., Chung, C.W., Yoo, J.J., Choi, K.H., Kim, C.H., et al., 2011. Doxorubicin-incorporated polymeric micelles composed of dextran-b-poly(DL-lactide-co-glycolide) copolymer. Int. J. Nanomedicine 6, 1415–1427.
Jian, W., Wu, H., Wu, L., Wu, Y., Jia, L., Pang, J., et al., 2016. Effect of molecular characteristics of Konjac glucomannan on gelling and rheological properties of Tilapia myofibrillar protein. Carbohydr. Polym. 150, 21–31.
Jin, C., Wang, K., Oppong-Gyebi, A., Hu, J., 2020. Application of nanotechnology in cancer diagnosis and therapy—a mini-review. Int. J. Med. Sci. 17 (18), 2964–2973. https://doi.org/10.7150/ijms.49801.
Kabanov, A.V., Gendelman, H.E., 2007. Nanomedicine in the diagnosis and therapy of neurodegenerative disorders. Prog. Polym. Sci. 32 (8–9), 1054–1082.
Kattumuri, V., Katti, K., Bhaskaran, S., Boote, E.J., Casteel, S.W., Fent, G.M., et al., 2007. Gum arabic as a phytochemical construct for the stabilization of gold nanoparticles: in vivo pharmacokinetics and X-ray-contrast-imaging studies. Small 3 (2), 333–341.
Katuwavila, N.P., Perera, A.D., Dahanayake, D., Karunaratne, V., Amaratunga, G.A., Karunaratne, D.N., 2016. Alginate nanoparticles protect ferrous from oxidation: potential iron delivery system. Int. J. Pharm. 513 (1–2), 404–409.
Kaur, M., Garg, T., Narang, R.K., 2016. A review of emerging trends in the treatment of tuberculosis. Artif. Cells Nanomed. Biotechnol. 44 (2), 478–484.
Kenry, Liu, B., 2018. Recent advances in biodegradable conducting polymers and their biomedical applications. Biomacromolecules 19 (6), 1783–1803. https://doi.org/10.1021/acs.biomac.8b00275.
Kesarla, R., Tank, T., Vora, P.A., Shah, T., Parmar, S., Omri, A., 2016. Preparation and evaluation of nanoparticles loaded ophthalmic in situ gel. Drug Deliv. 23 (7), 2363–2370.

Key, J., Leary, J.F., 2014. Nanoparticles for multimodal in vivo imaging in nanomedicine. Int. J. Nanomedicine 9, 711–726.
Khalikova, E., Susi, P., Korpela, T., 2005. Microbial dextran-hydrolyzing enzymes: fundamentals and applications. Microbiol. Mol. Biol. Rev. 69 (2), 306–325.
Khan, H., Marya, 2019. Chapter 3.28—Konjac (Amorphophallus konjac). In: Nabavi, S.M., Silva, A.S. (Eds.), Nonvitamin and Nonmineral Nutritional Supplements. Academic Press, pp. 307–312.
Khan, I., Saeed, K., Khan, I., 2019. Nanoparticles: properties, applications and toxicities. Arab. J. Chem. 12 (7), 908–931.
Khushbu, Warkar, S.G., 2020. Potential applications and various aspects of polyfunctional macromolecule-carboxymethyl tamarind kernel gum. Eur. Polym. J. 140, 110042.
Klapetek, P., Valtr, M., Nečas, D., Salyk, O., Dzik, P., 2011. Atomic force microscopy analysis of nanoparticles in non-ideal conditions. Nanoscale Res. Lett. 6 (1), 514.
Kulkarni, V.S., Shaw, C., 2016. Chapter 5—Use of polymers and thickeners in semisolid and liquid formulations. In: Kulkarni, V.S., Shaw, C. (Eds.), Essential Chemistry for Formulators of Semisolid and Liquid Dosages. Academic Press, Boston, pp. 43–69.
Kumar, J., Newton, A.M.J., 2015. IBD modern concepts, nano drug delivery and patents: an update. Curr. Nanomed. 5 (2), 122–145.
Kumar, J., Newton, A.M.J., 2016. Colon targeted rifaximin nanosuspension for the treatment of inflammatory bowel disease (IBD). Anti-Inflammatory Anti-Allergy Agents Med. Chem. 15 (2), 101–117.
Kumar, S., Dilbaghi, N., Saharan, R., Bhanjana, G., 2012. Nanotechnology as emerging tool for enhancing solubility of poorly water-soluble drugs. BioNanoScience 2 (4), 227–250.
Kumar, A., Rao, K.M., Han, S.S., 2018a. Application of xanthan gum as polysaccharide in tissue engineering: a review. Carbohydr. Polym. 180, 128–144.
Kumar, S.S.D., Mahesh, A., Antoniraj, M.G., Rathore, H.S., Houreld, N.N., Kandasamy, R., 2018b. Cellular imaging and folate receptor targeting delivery of gum kondagogu capped gold nanoparticles in cancer cells. Int. J. Biol. Macromol. 109, 220–230.
Kumar, J., Newton, A.M.J., 2017. Rifaximin—chitosan nanoparticles for inflammatory bowel disease (IBD). Recent Pat. Inflamm. Allergy Drug Discov. 11 (1), 41–52. https://doi.org/10.2174/1872213X10666161230111226.
Kumar, P.S., Pavithra, K.G., Naushad, M., 2019. Chapter 4—Characterization techniques for nanomaterials. In: Thomas, S., EHM, S., Kalarikkal, N., Oluwafemi, S.O., Wu, J. (Eds.), Nanomaterials for Solar Cell Applications. Elsevier, pp. 97–124.
Lee, K.Y., Mooney, D.J., 2012. Alginate: properties and biomedical applications. Prog. Polym. Sci. 37 (1), 106–126.
Li, J.-M., Nie, S.-P., 2016. The functional and nutritional aspects of hydrocolloids in foods. Food Hydrocoll. 53, 46–61.
Li, P., Dai, Y.-N., Zhang, J.-P., Wang, A.-Q., Wei, Q., 2008. Chitosan-alginate nanoparticles as a novel drug delivery system for nifedipine. Int. J. Biomed. Sci. 4 (3), 221–228.
Li, S., Meng Lin, M., Toprak, M.S., Kim, D.K., Muhammed, M., 2010. Nanocomposites of polymer and inorganic nanoparticles for optical and magnetic applications. Nano Rev. 1 (1), 5214.
Li, T., Wang, G.-D., Tan, Y.-Z., Wang, H.-J., 2014. Inhibition of lymphangiogenesis of endothelial progenitor cells with VEGFR-3 siRNA delivered with PEI-alginate nanoparticles. Int. J. Biol. Sci. 10 (2), 160–170.
Li, M., Tang, Z., Zhang, Y., Lv, S., Li, Q., Chen, X., 2015. Targeted delivery of cisplatin by LHRH-peptide conjugated dextran nanoparticles suppresses breast cancer growth and metastasis. Acta Biomater. 18, 132–143.
Lovegrove, A., Edwards, C.H., De Noni, I., Patel, H., El, S.N., Grassby, T., et al., 2017. Role of polysaccharides in food, digestion, and health. Crit. Rev. Food Sci. Nutr. 57 (2), 237–253.
Lu, E.-X., Jiang, Z.-Q., Zhang, Q.-Z., Jiang, X.-G., 2003. A water-insoluble drug monolithic osmotic tablet system utilizing gum arabic as an osmotic, suspending and expanding agent. J. Control. Release 92 (3), 375–382.
Luo, X., Wang, W., Dorkin, J.R., Veiseh, O., Chang, P.H., Abutbul-Ionita, I., et al., 2017. Poly(glycoamidoamine) brush nanomaterials for systemic siRNA delivery in vivo. Biomater. Sci. 5 (1), 38–40.
Madkour, L.H., 2019. Chapter 13—Nanoparticle and polymeric nanoparticle-based targeted drug delivery systems. In: Madkour, L.H. (Ed.), Nucleic Acids as Gene Anticancer Drug Delivery Therapy. Academic Press, pp. 191–240.
Malik, S., Ahuja, M., 2011. Gum kondagogu-g-poly (acrylamide): microwave-assisted synthesis, characterisation and release behaviour. Carbohydr. Polym. 86 (1), 177–184.
Manjunath, M., Anjali, Gowda, D.V., Kumar, P., Srivastava, A., Osmani, R.A., et al., 2016. Guar gum and its pharmaceutical and biomedical applications. Adv. Sci. Eng. Med. 8 (8), 589–602.
Mano, J.F., Silva, G.A., Azevedo, H.S., Malafaya, P.B., Sousa, R.A., Silva, S.S., et al., 2007. Natural origin biodegradable systems in tissue engineering and regenerative medicine: present status and some moving trends. J. R. Soc. Interface 4 (17), 999–1030.

Marci, L., Meloni, M.C., Maccioni, A.M., Sinico, C., Lai, F., Cardia, M.C., 2016. Formulation and characterization studies of trimethyl chitosan/sodium alginate nanoparticles for targeted drug delivery. ChemistrySelect 1 (4), 669–674.
Medeiros, S.F., Santos, A.M., Fessi, H., Elaissari, A., 2011. Stimuli-responsive magnetic particles for biomedical applications. Int. J. Pharm. 403 (1), 139–161.
Mei, Y., Wang, R., Jiang, W., Bo, Y., Zhang, T., Yu, J., et al., 2019. Recent progress in nanomaterials for nucleic acid delivery in cancer immunotherapy. Biomater. Sci. 7 (7), 2640–2651.
Miao, T., Little, A.C., Aronshtam, A., Marquis, T., Fenn, S.L., Hristova, M., et al., 2020. Internalized FGF-2-loaded nanoparticles increase nuclear ERK1/2 content and result in lung cancer cell death. Nanomaterials (Basel) 10 (4), 612.
Miki, D., Zhu, P., Zhang, W., Mao, Y., Feng, Z., Huang, H., et al., 2017. Efficient generation of diRNAs requires components in the posttranscriptional gene silencing pathway. Sci. Rep. 7 (1), 301.
Mohammadi, H., Kamkar, A., Misaghi, A., 2018. Nanocomposite films based on CMC, okra mucilage and ZnO nanoparticles: physico mechanical and antibacterial properties. Carbohydr. Polym. 181, 351–357.
Mohammadifar, M.A., Musavi, S.M., Kiumarsi, A., Williams, P.A., 2006. Solution properties of targacanthin (water-soluble part of gum tragacanth exudate from Astragalus gossypinus). Int. J. Biol. Macromol. 38 (1), 31–39.
Mohammadinejad, R., Kumar, A., Ranjbar-Mohammadi, M., Ashrafizadeh, M., Han, S.S., Khang, G., et al., 2020. Recent advances in natural gum-based biomaterials for tissue engineering and regenerative medicine: a review. Polymers 12 (1), 176.
Mourdikoudis, S., Pallares, R.M., Thanh, N.T.K., 2018. Characterization techniques for nanoparticles: comparison and complementarity upon studying nanoparticle properties. Nanoscale 10 (27), 12871–12934.
Muddineti, O.S., Kumari, P., Ajjarapu, S., Lakhani, P.M., Bahl, R., Ghosh, B., et al., 2016. Xanthan gum stabilized PEGylated gold nanoparticles for improved delivery of curcumin in cancer. Nanotechnology 27 (32), 325101.
Mudgil, D., Barak, S., Khatkar, B.S., 2014. Guar gum: processing, properties and food applications—a review. J. Food Sci. Technol. 51 (3), 409–418.
Mulia, K., Safiera, A., Pane, I.F., Krisanti, E.A., 2019. Effect of high speed homogenizer speed on particle size of polylactic acid. J. Phys. Conf. Ser. 1198 (6), 062006.
Naeye, B., Deschout, H., Röding, M., Rudemo, M., Delanghe, J., Devreese, K., et al., 2011. Hemocompatibility of siRNA loaded dextran nanogels. Biomaterials 32 (34), 9120–9127.
Nafissi, N., Alqawlaq, S., Lee, E.A., Foldvari, M., Spagnuolo, P.A., Slavcev, R.A., 2014. DNA ministrings: highly safe and effective gene delivery vectors. Mol. Ther. Nucleic Acids. 3 (6), e165.
Nayak, A.K., Pal, D., 2018. Functionalization of tamarind gum for drug delivery. In: Thakur, V.K., Thakur, M.K. (Eds.), Functional Biopolymers. Springer International Publishing, Cham, pp. 25–56.
Nayak, A.K., Ara, T.J., Saquib Hasnain, M., Hoda, N., Inamuddin, 2018. Chapter 32—Okra gum–alginate composites for controlled releasing drug delivery. In: Asiri, A.M., Mohammad, A. (Eds.), Applications of Nanocomposite Materials in Drug Delivery. Woodhead Publishing, pp. 761–785.
Nazarzadeh Zare, E., Makvandi, P., Tay, F.R., 2019. Recent progress in the industrial and biomedical applications of tragacanth gum: a review. Carbohydr. Polym. 212, 450–467.
Newton, A.M.J., Lakshmanan, P., 2020. Comparative efficacy of chitosan, pectin based mesalamine colon targeted drug delivery systems on TNBS-induced IBD model rats. Anti-Inflammatory Anti-Allergy Agents Med. Chem. 19 (2), 113–127. https://doi.org/10.2174/1871523018666190118112230.
Newton, A.M.J., Prabhakaran, L., Kabilan, P., 2011. Drug deliveries to colonic region. Indian Pharm. 9, 23.
Newton, A.M.J., Indana, V.L., Kumar, J., 2015. Chronotherapeutic drug delivery of tamarind gum, chitosan and okra gum controlled release colon targeted directly compressed propranolol HCl matrix tablets and in-vitro evaluation. Int. J. Biol. Macromol. 79, 290–299.
Ngwuluka, N.C., Ochekpe, N.A., Aruoma, O.I., 2014. Naturapolyceutics: the science of utilizing natural polymers for drug delivery. Polymers 6 (5), 1312–1332.
Ofori-Kwakye, K., Amekyeh, H., El-Duah, M., Kipo, S.L., 2012. Mechanical and tablet coating properties of cashew tree (Anacardium occidentale L) gum-based films. Asian J. Pharm. Clin. Res. 5 (4), 62–68.
Oh, Y.K., Park, T.G., 2009. siRNA delivery systems for cancer treatment. Adv. Drug Deliv. Rev. 61 (10), 850–862.
Ologunagba, M.O., Jain, S., Thanki, K., Suresh, S., Furtado, S., Azubuike, C.P., Silva, O.B., 2017. Extraction and characterization of the gum exudate of Anacardium occidentale for its potential as an excipient in drug delivery systems. Trop. J. Nat. Prod. Res. 1 (2), 76–83. https://doi.org/10.26538/tjnpr/v1i2.6.
Padil, V.V.T., Wacławek, S., Černík, M., Varma, R.S., 2018. Tree gum-based renewable materials: sustainable applications in nanotechnology, biomedical and environmental fields. Biotechnol. Adv. 36 (7), 1984–2016.

Park, C.R., Munday, D.L., 2004. Evaluation of selected polysaccharide excipients in Buccoadhesive tablets for sustained release of nicotine. Drug Dev. Ind. Pharm. 30 (6), 609–617.

Parvathi, K., Ramesh, C., Krishna, V., Paramesha, M., Kuppast, I., 2009. Hypolipidemic activity of gum ghatti of Anogeissus latifolia. Pharmacogn. Mag. 5 (19), 11.

Patel, J., Maji, B., Moorthy, N.S.H.N., Maiti, S., 2020. Xanthan gum derivatives: review of synthesis, properties and diverse applications. RSC Adv. 10 (45), 27103–27136.

Pathak, C., Vaidya, F.U., Pandey, S.M., 2019. Chapter 3—Mechanism for development of nanobased drug delivery system. In: Mohapatra, S.S., Ranjan, S., Dasgupta, N., Mishra, R.K., Thomas, S. (Eds.), Applications of Targeted Nano Drugs and Delivery Systems. Elsevier, pp. 35–67.

Patil, M., Mehta, D.S., Guvva, S., 2008. Future impact of nanotechnology on medicine and dentistry. J. Indian Soc. Periodontol. 12 (2), 34–40.

Patitsa, M., Karathanou, K., Kanaki, Z., Tzioga, L., Pippa, N., Demetzos, C., et al., 2017. Magnetic nanoparticles coated with polyarabic acid demonstrate enhanced drug delivery and imaging properties for cancer theranostic applications. Sci. Rep. 7 (1), 775.

Patra, J.K., Das, G., Fraceto, L.F., Campos, E.V.R., Rodriguez-Torres, M.D.P., Acosta-Torres, L.S., et al., 2018. Nano based drug delivery systems: recent developments and future prospects. J. Nanobiotechnol. 16 (1), 71.

Peper, A., 2009. Aspects of the relationship between drug dose and drug effect. Dose-Response 7 (2), 172–192.

Pooja, D., Sistla, R., 2019. Design of eco-friendly gold nanoparticles for cancer treatment. In: Dinesh Kumar, L. (Ed.), RNA Interference and Cancer Therapy: Methods and Protocols. Springer New York, New York, NY, pp. 215–221.

Pooja, D., Panyaram, S., Kulhari, H., Rachamalla, S.S., Sistla, R., 2014. Xanthan gum stabilized gold nanoparticles: characterization, biocompatibility, stability and cytotoxicity. Carbohydr. Polym. 110, 1–9.

Pooja, D., Panyaram, S., Kulhari, H., Reddy, B., Rachamalla, S.S., Sistla, R., 2015. Natural polysaccharide functionalized gold nanoparticles as biocompatible drug delivery carrier. Int. J. Biol. Macromol. 80, 48–56.

Prasad, R., Bhattacharyya, A., Nguyen, Q.D., 2017. Nanotechnology in sustainable agriculture: recent developments, challenges, and perspectives. Front. Microbiol. 8, 1014.

Price, J.C., 2012. Molecular brain imaging in the multimodality era. J. Cereb. Blood Flow Metab. 32 (7), 1377–1392. https://doi.org/10.1038/jcbfm.2012.29.

Ragelle, H., Riva, R., Vandermeulen, G., Naeye, B., Pourcelle, V., Le Duff, C.S., et al., 2014. Chitosan nanoparticles for siRNA delivery: optimizing formulation to increase stability and efficiency. J. Control. Release 176, 54–63.

Raj, S., Jose, S., Sumod, U.S., Sabitha, M., 2012. Nanotechnology in cosmetics: opportunities and challenges. J. Pharm. Bioallied Sci. 4 (3), 186–193.

Raj, V., Lee, J.-H., Shim, J.-J., Lee, J., 2021. Recent findings and future directions of grafted gum karaya polysaccharides and their various applications: a review. Carbohydr. Polym. 258, 117687. https://doi.org/10.1016/j.carbpol.2021.117687.

Rao, A., Schoenenberger, M., Gnecco, E., Glatzel, T., Meyer, E., Brändlin, D., et al., 2007. Characterization of nanoparticles using atomic force microscopy. J. Phys. Conf. Ser. 61, 971–976.

Reeta, R.M., John, M., Newton, A., 2020. Fabrication and characterisation of lavender oil and plant phospholipid based sumatriptan succinate hybrid nano lipid carriers. J. Pharm. Biomed. Sci. 6 (1), 91–104. https://doi.org/10.18502/pbr.v6i1.3430.

Ribeiro, A.J., de Souza, F.R.L., Bezerra, J.M., Oliveira, C., Nadvorny, D., Monica, F., et al., 2016. Gums' based delivery systems: review on cashew gum and its derivatives. Carbohydr. Polym. 147, 188–200.

Rimple, Newton, M.J., 2018. Impact of ocular compatible lipoids and castor oil in fabrication of brimonidine tartrate nanoemulsions by 33 full factorial design. Recent Patents Inflamm. Allergy Drug Discov. 12 (2), 169–183.

Rizvi, S.A.A., Saleh, A.M., 2018. Applications of nanoparticle systems in drug delivery technology. Saudi Pharm. J. 26 (1), 64–70.

Rizwan, M., Yahya, R., Hassan, A., Yar, M., Azzahari, A.D., Selvanathan, V., et al., 2017. pH sensitive hydrogels in drug delivery: brief history, properties, swelling, and release mechanism, material selection and applications. Polymers (Basel) 9 (4), 137.

Rocha, S., Generalov, R., MDC, P., Peres, I., Juzenas, P., Coelho, M.A., 2011. Epigallocatechin gallate-loaded polysaccharide nanoparticles for prostate cancer chemoprevention. Nanomedicine (Lond.) 6 (1), 79–87.

Rong, M.Z., Zhang, M.Q., Zheng, Y.X., Zeng, H.M., Friedrich, K., 2001. Improvement of tensile properties of nano-SiO_2/PP composites in relation to percolation mechanism. Polymer 42 (7), 3301–3304.

Rosalam, S., England, R., 2006. Review of xanthan gum production from unmodified starches by Xanthomonas comprestris sp. Enzym. Microb. Technol. 39 (2), 197–207.

Rosca, I.D., Watari, F., Uo, M., 2004. Microparticle formation and its mechanism in single and double emulsion solvent evaporation. J. Control. Release 99 (2), 271–280.

Rossetti, F.C., Depieri, L.V., Bentley, M., 2013. Confocal laser scanning microscopy as a tool for the investigation of skin drug delivery systems and diagnosis of skin disorders. In: Confocal Laser Microscopy—Principles and Applications in Medicine, Biology, and the Food Sciences. IntechOpen, pp. 99–140.

Saeed, M., Abbas Zare, M., Ali, S., Nasser Mohammadpour, D., Saman, S., Mehrasa Rahimi, B., 2013. Preparation and characterization of sodium alginate nanoparticles containing ICD-85 (venom derived peptides). Int. J. Innov. Appl. Stud. 4 (3), 534–542.

Salatin, S., Jelvehgari, M., 2017. Natural polysaccharide based nanoparticles for drug/gene delivery. Pharm. Sci. 23 (2), 84–94.

Sarojini, S., Kunam, D.S., Manavalan, R., Jayanthi, B., 2010. Effect of natural almond gum as a binder in the formulation of diclofenac sodium tablets. Int. J. Pharm. Sci. Res. 1 (3), 55–60.

Sayour, E.J., Sanchez-Perez, L., Flores, C., Mitchell, D.A., 2015. Bridging infectious disease vaccines with cancer immunotherapy: a role for targeted RNA based immunotherapeutics. J. Immunother. Cancer 3, 13.

Schiffelers, R.M., Ansari, A., Xu, J., Zhou, Q., Tang, Q., Storm, G., et al., 2004. Cancer siRNA therapy by tumor selective delivery with ligand-targeted sterically stabilized nanoparticle. Nucleic Acids Res. 32 (19), e149.

Schnizlein, M.K., Vendrov, K.C., Edwards, S.J., Martens, E.C., Young, V.B., 2020. Dietary xanthan gum alters antibiotic efficacy against the murine gut microbiota and attenuates Clostridioides difficile colonization. mSphere 5 (1), e00708-19.

Schwirn, K., Tietjen, L., Beer, I., 2014. Why are nanomaterials different and how can they be appropriately regulated under REACH? Environ. Sci. Eur. 26 (1), 4.

Seyfoori, A., Shokrollahi Barough, M., Mokarram, P., Ahmadi, M., Mehrbod, P., Sheidary, A., Madrakian, T., Kiumarsi, M., Walsh, T., McAlinden, K.D., Ghosh, C.C., Sharma, P., Zeki, A.A., Ghavami, S., Akbari, M., 2021. Emerging advances of nanotechnology in drug and vaccine delivery against viral associated respiratory infectious diseases (VARID). Int. J. Mol. Sci. 22 (13). https://doi.org/10.3390/ijms22136937.

Shafie, N.A., Suhaili, N.A., Taha, H., Ahmad, N., 2020. Evaluation of antioxidant, antibacterial and wound healing activities of Vitex pinnata. F1000Res 9, 187.

Shah, A., Dobrovolskaia, M.A., 2018. Immunological effects of iron oxide nanoparticles and iron-based complex drug formulations: therapeutic benefits, toxicity, mechanistic insights, and translational considerations. Nanomedicine 14 (3), 977–990.

Shariatinia, Z., 2019. Chapter 2—Pharmaceutical applications of natural polysaccharides. In: Hasnain, M.S., Nayak, A.K. (Eds.), Natural Polysaccharides in Drug Delivery and Biomedical Applications. Academic Press, pp. 15–57.

Sharma, M., Mondal, D., Mukesh, C., Prasad, K., 2014a. Preparation of tamarind gum based soft ion gels having thixotropic properties. Carbohydr. Polym. 102, 467–471.

Sharma, S., Chockalingam, S., Sanpui, P., Chattopadhyay, A., Ghosh, S.S., 2014b. Silver nanoparticles impregnated alginate-chitosan-blended nanocarrier induces apoptosis in human glioblastoma cells. Adv. Healthc. Mater. 3 (1), 106–114.

Sheehan, J.J., Reilly, K.R., Fu, D.-J., Alphs, L., 2012. Comparison of the peak-to-trough fluctuation in plasma concentration of long-acting injectable antipsychotics and their oral equivalents. Innov. Clin. Neurosci. 9 (7–8), 17–23.

Shen, S., Wu, Y., Liu, Y., Wu, D., 2017. High drug-loading nanomedicines: progress, current status, and prospects. Int. J. Nanomedicine 12, 4085–4109.

Shukla, A.K., Bishnoi, R.S., Kumar, M., Fenin, V., Jain, C.P., 2018. Applications of tamarind seeds polysaccharide-based copolymers in controlled drug delivery: an overview. Asian J. Pharm. Pharmacol. 4 (1), 23.

Singh, B.N., Kim, K.H., 2005. Effects of divalent cations on drug encapsulation efficiency of deacylated gellan gum. J. Microencapsul. 22 (7), 761–771.

Singhvi, G., Hans, N., Shiva, N., Kumar Dubey, S., 2019. Chapter 5—Xanthan gum in drug delivery applications. In: Hasnain, M.S., Nayak, A.K. (Eds.), Natural Polysaccharides in Drug Delivery and Biomedical Applications. Academic Press, pp. 121–144.

Soares, S., Sousa, J., Pais, A., Vitorino, C., 2018. Nanomedicine: principles, properties, and regulatory issues. Front. Chem. 6, 360.

Song, W., Su, X., Gregory, D.A., Li, W., Cai, Z., Zhao, X., 2018. Magnetic alginate/chitosan nanoparticles for targeted delivery of curcumin into human breast cancer cells. Nanomaterials (Basel) 8 (11), 907.

Soumya, R.S., Ghosh, S., Abraham, E.T., 2010. Preparation and characterization of guar gum nanoparticles. Int. J Biol. Macromol. 46 (2), 267–269.

Sriraman, S.K., Aryasomayajula, B., Torchilin, V.P., 2014. Barriers to drug delivery in solid tumors. Tissue Barriers 2, e29528.

Stephen, A., Churms, S., Vogt, D., 1990. Methods Plant Biochemistry. Elsevier.

Sugahara, S.-I., Kajiki, M., Kuriyama, H., Kobayashi, T.-R., 2007. Complete regression of xenografted human carcinomas by a paclitaxel–carboxymethyl dextran conjugate (AZ10992). J. Control. Release 117 (1), 40–50.

Sun, Y.-P., Meziani, M.J., Pathak, P., Qu, L., 2005. Polymeric nanoparticles from rapid expansion of supercritical fluid solution. Chem. Eur. J. 11 (5), 1366–1373.

Susa, M., Iyer, A.K., Ryu, K., Choy, E., Hornicek, F.J., Mankin, H., et al., 2010. Inhibition of ABCB1 (MDR1) expression by an siRNA nanoparticulate delivery system to overcome drug resistance in osteosarcoma. PLoS One 5 (5), e10764.

Swierczewska, M., Han, H.S., Kim, K., Park, J.H., Lee, S., 2016. Polysaccharide-based nanoparticles for theranostic nanomedicine. Adv. Drug Deliv. Rev. 99 (Pt A), 70–84.

Sworn, G., 2009. Chapter 9—Gellan gum. In: Phillips, G.O., Williams, P.A. (Eds.), Handbook of Hydrocolloids, second ed. Woodhead Publishing, pp. 204–227.

Szekalska, M., Puciłowska, A., Szymańska, E., Ciosek, P., Winnicka, K., 2016. Alginate: current use and future perspectives in pharmaceutical and biomedical applications. Int. J. Polym. Sci. 2016, 7697031.

Thambi, T., You, D.G., Han, H.S., Deepagan, V.G., Jeon, S.M., Suh, Y.D., et al., 2014. Bioreducible carboxymethyl dextran nanoparticles for tumor-targeted drug delivery. Adv. Healthc. Mater. 3 (11), 1829–1838.

Thatipamula, R., Palem, C., Gannu, R., Mudragada, S., Yamsani, M., 2011. Formulation and in vitro characterization of domperidone loaded solid lipid nanoparticles and nanostructured lipid carriers. Daru 19 (1), 23–32.

Thipe, V.C., Panjtan Amiri, K., Bloebaum, P., Raphael Karikachery, A., Khoobchandani, M., Katti, K.K., et al., 2019. Development of resveratrol-conjugated gold nanoparticles: interrelationship of increased resveratrol corona on anti-tumor efficacy against breast, pancreatic and prostate cancers. Int. J. Nanomedicine 14, 4413–4428.

Thombare, N., Jha, U., Mishra, S., Siddiqui, M.Z., 2016. Guar gum as a promising starting material for diverse applications: a review. Int. J. Biol. Macromol. 88, 361–372.

Thombare, N., Mate, C., Thamilarasi, K., Srivastava, S., Chowdhury, A.R., 2018. Physico-chemical characterization and microbiological evaluataion of gum ghatti as potential food additive. Multilogic Sci. 8, 316–319.

Titus, D., James Jebaseelan Samuel, E., Roopan, S.M., 2019. Chapter 12—Nanoparticle characterization techniques. In: Shukla, A.K., Iravani, S. (Eds.), Green Synthesis, Characterization and Applications of Nanoparticles. Elsevier, pp. 303–319.

Torres, F.G., Troncoso, O.P., Pisani, A., Gatto, F., Bardi, G., 2019. Natural polysaccharide nanomaterials: an overview of their immunological properties. Int. J. Mol. Sci. 20 (20), 5092.

Varshosaz, J., 2012. Dextran conjugates in drug delivery. Expert Opin. Drug Deliv. 9 (5), 509–523.

Verma, C., Negi, P., Pathania, D., Anjum, S., Gupta, B., 2020. Novel Tragacanth gum-entrapped lecithin nanogels for anticancer drug delivery. Int. J. Polym. Mater. Polym. Biomater. 69 (9), 604–609.

Vinay, S.V., Kalyani, S., Hemant, R.B., 2017. Xanthan gum a versatile biopolymer: current status and future prospectus in hydro gel drug delivery. Curr. Chem. Biol. 11 (1), 10–20.

Vinod, V.T.P., Sashidhar, R.B., Sarma, V.U.M., Vijaya Saradhi, U.V.R., 2008a. Compositional analysis and rheological properties of gum kondagogu (Cochlospermum gossypium): a tree gum from India. J. Agric. Food Chem. 56 (6), 2199–2207.

Vinod, V., Sashidhar, R., Suresh, K., Rao, B.R., Saradhi, U.V., Rao, T.P., 2008b. Morphological, physico-chemical and structural characterization of gum kondagogu (*Cochlospermum gossypium*): a tree gum from India. Food Hydrocoll. 22 (5), 899–915.

Wang, H., Jiang, Y., Peng, H., Chen, Y., Zhu, P., Huang, Y., 2015. Recent progress in microRNA delivery for cancer therapy by non-viral synthetic vectors. Adv. Drug Deliv. Rev. 81, 142–160.

Watkins-Schulz, R., Tiet, P., Gallovic, M.D., Junkins, R.D., Batty, C., Bachelder, E.M., et al., 2019. A microparticle platform for STING-targeted immunotherapy enhances natural killer cell- and CD8 + T cell-mediated anti-tumor immunity. Biomaterials 205, 94–105.

Wattanakul, K., Imae, T., Chang, W.-W., Chu, C.-C., Nakahata, R., Yusa, S.-I., 2019. Oligopeptide-side chained alginate nanocarrier for melittin-targeted chemotherapy. Polym. J. 51 (8), 771–780.

Webster, T.J., 2006. Nanomedicine: what's in a definition? Int. J. Nanomed. 1 (2), 115–116. https://doi.org/10.2147/nano.2006.1.2.115.

Wen, H., Jung, H., Li, X., 2015. Drug delivery approaches in addressing clinical pharmacology-related issues: opportunities and challenges. AAPS J. 17 (6), 1327–1340.

Xie, J., Lee, S., Chen, X., 2010. Nanoparticle-based theranostic agents. Adv. Drug Deliv. Rev. 62 (11), 1064–1079.

Xu, J., Singh, A., Amiji, M.M., 2014. Redox-responsive targeted gelatin nanoparticles for delivery of combination wt-p53 expressing plasmid DNA and gemcitabine in the treatment of pancreatic cancer. BMC Cancer 14 (1), 75.

Xu, W., Li, B., Li, J., Huang, L., Liu, H., Zhu, D., et al., 2018. Rheological and spectral analysis of xanthan gum/lysozyme system during nanoparticle fabrication. Int. J. Food Sci. Technol. 53 (11), 2595–2601.

Yang, S.-J., Chang, S.-M., Tsai, K.-C., Chen, W.-S., Lin, F.-H., Shieh, M.-J., 2010. Effect of chitosan-alginate nanoparticles and ultrasound on the efficiency of gene transfection of human cancer cells. J. Gene Med. 12 (2), 168–179.

Yin, H., Kanasty, R.L., Eltoukhy, A.A., Vegas, A.J., Dorkin, J.R., Anderson, D.G., 2014. Non-viral vectors for gene-based therapy. Nat. Rev. Genet. 15 (8), 541–555.

Yong, K.T., Roy, I., Ding, H., Bergey, E.J., Prasad, P.N., 2009. Biocompatible near-infrared quantum dots as ultrasensitive probes for long-term in vivo imaging applications. Small 5 (17), 1997–2004.

You, J.-O., Liu, Y.-C., Peng, C.-A., 2006. Efficient gene transfection using chitosan-alginate core-shell nanoparticles. Int. J. Nanomedicine 1 (2), 173–180.

Yu, M.K., Park, J., Jon, S., 2012. Targeting strategies for multifunctional nanoparticles in cancer imaging and therapy. Theranostics 2 (1), 3–44.

Zalewski, B.M., Chmielewska, A., Szajewska, H., 2015. The effect of glucomannan on body weight in overweight or obese children and adults: a systematic review of randomized controlled trials. Nutrition 31 (3), 437–442.e2.

Zhang, L., Yu, F., Cole, A.J., Chertok, B., David, A.E., Wang, J., et al., 2009. Gum arabic-coated magnetic nanoparticles for potential application in simultaneous magnetic targeting and tumor imaging. AAPS J. 11 (4), 693–699.

Zhang, L., Lyer, A.K., Yang, X., Kobayashi, E., Guo, Y., Mankin, H., et al., 2015. Polymeric nanoparticle-based delivery of microRNA-199a-3p inhibits proliferation and growth of osteosarcoma cells. Int. J. Nanomedicine 10, 2913–2924.

Zhang, T., Li, Z., Wang, Y., Xue, Y., Xue, C., 2016. Effects of konjac glucomannan on heat-induced changes of physicochemical and structural properties of surimi gels. Food Res. Int. 83, 152–161.

Zhao, R., Du, S., Liu, Y., Lv, C., Song, Y., Chen, X., et al., 2020. Mucoadhesive-to-penetrating controllable peptosomes-in-microspheres co-loaded with anti-miR-31 oligonucleotide and curcumin for targeted colorectal cancer therapy. Theranostics 10 (8), 3594–3611.

CHAPTER

8

Gum-based micro- and nanobiomaterials in gene delivery

M.R. Rekha

Division of Biosurface Technology, Biomedical Technology Wing Sree Chitra Tirunal Institute for Medical Sciences & Technology, Thiruvananthapuram, Kerala, India

8.1 Introduction

Natural gums belong to the category of polysaccharides with varying composition and properties based on the origin of source. The main monomers are glucose, galactose and mannose along with other monosaccharides such as arabinose, xylose, rhamnose, etc. The natural gum sources range from microbes to plants to the animal kingdom. Owing to their appropriate properties such as nontoxicity, biocompatibility, and biodegradability, gums are widely explored for various biomedical applications. The various natural gums include acacia, arabic, guar, gellan, ghatti, karaya, konjac, locust bean tamarind, tragacanth, and xanthan. Recently gums have attracted attention in the food and pharmaceutical industries as potential raw materials for various applications. This chapter deals exclusively with gene delivery systems based on various gums and their derivatives.

8.2 Classification

Natural gums are long chain polysaccharides containing one or more types of monosaccharide units such as arabinose, galactose, glucose, mannose, xylose, arabinose, rhamnose, or fucose, or uronic acids such as gulucuronic acid, galacturonic acid, mannuronic acid, L-guluronic acid, etc. (Cui and Wang, 2005). In nature, they are found in various sources such as plants, algae, and microbes, and polysaccharides form viscous solution in an aqueous medium, as shown in Fig. 8.1 (Suhail et al., 2019). The chemical composition and chemical structure of these polysaccharides varies with respect to the source from which they have been derived, and also according to the extraction methods. The gum from plant sources is

Micro- and Nanoengineered Gum-Based Biomaterials for Drug Delivery and Biomedical Applications
https://doi.org/10.1016/B978-0-323-90986-0.00019-4

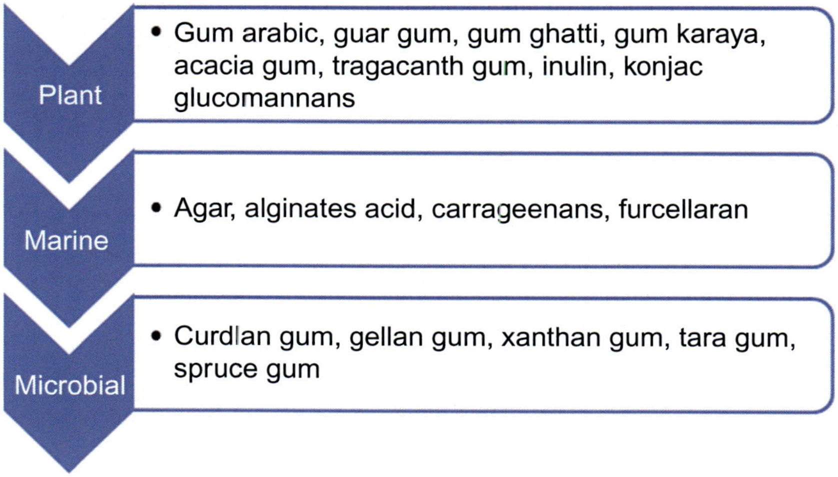

FIG. 8.1 Classification of natural gums.

obtained from trees, seed coatings, etc., marine origin gum is mainly from algae, and that from microbes is produced by a fermentation process. Gums are thus basically polysaccharides, and these include both nonionic and ionic categories (Suhail et al., 2019).

8.3 Biomedical application of gums

Biopolymers have been of wide use in the pharmaceutical industry for decades and recently natural gums have also been receiving wide attention owing to their numerous applications as binders, thickening agents, emulsifiers, suspending agents, disintegrants, gelling agents, etc. (Alonso-Sande et al., 2009; Vipul et al., 2013). The main reasons for this interest include their nontoxic nature, easy availability, low cost, biocompatibility, etc. In addition, gums, being polysaccharides, can be easily modified by various conjugation reactions. Through these reactions, various molecules of interest can be introduced into the polymeric chain, modulate the characteristics, and design suitable carriers for therapeutic molecule delivery applications. Though natural gums are being widely explored for conventional pharmaceutical applications, recently a surge in research on developing novel drug delivery systems and tissue engineering scaffolds has been seen (Fig. 8.2). This chapter focuses solely on the gene delivery systems derived from natural gums.

8.4 Nonviral gene delivery

Gene therapy is the introduction of genetic material as a medicine into the target cell as a replacement against a faulty gene or delivering an absent gene to cure or slow down the progression of a disease (Nikunj and Verma, 2000). Gene therapy is mediated by

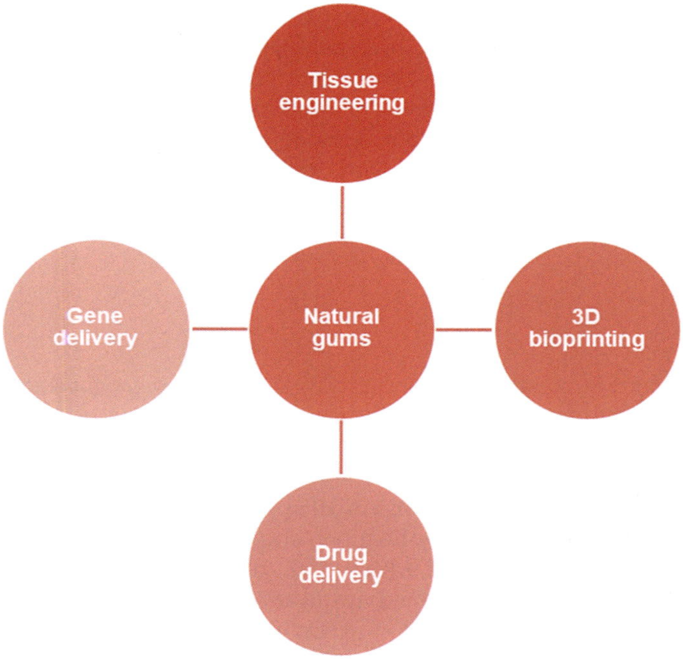

FIG. 8.2 Major biomedical applications of natural gums.

delivering appropriate nucleic acid (DNA or RNA) into cells for the purpose of enhancing or inhibiting gene expression using suitable vectors, and is called transfection (De Smedt et al., 2000). The two broad groups of vectors employed for gene delivery are viral and nonviral vectors. Viral vectors are highly efficient transfecting agents but their application is limited owing to immunogenicity, toxicity, low encapsulation capacity, etc. As a result, nonviral vectors have been widely investigated, and among these polymeric vectors hold special attention owing to their high availability, biocompatibility, low toxicity, good loading capacity, ability to design and engineer suitable carriers, etc. The wider scope of polymer chemistry provides an opportunity to synthesize more effective gene delivery vehicles while maintaining biocompatibility, facile manufacturing, and robust and safe formulation. Among the polymeric vectors, biopolymers are considered to be a better choice owing to their nontoxicity, biocompatibility, biodegradability, easy modification, etc.

8.5 Natural gum-based gene delivery vectors

As mentioned earlier, naturally derived gums are broadly classified into groups based on their source and its properties. The main gums that will be discussed in this chapter are as follows: gellan gum, xanthan and dextran (which are of microbial origin), alginates (which are of marine origin), and other gums including guar, karaya, tragacanth, ghatti, etc. The genetic material is the medicine in gene therapy and is anionic in nature. Nonviral vectors are mostly cationic in nature and form compact nanosized polyplexes by ionic interactions. However, most of the cationic polymers are toxic in nature with poor hemocompatibility. In order to circumvent this problem so that compatibility can be improved, modifications with various molecules have been investigated (Fig. 8.3). Among these, polysaccharides are gaining significant importance owing to their highly cytocompatible and hemocompatible nature.

8.5.1 Gellan gum

Gellan gum is an anionic polysaccharide produced by a microorganism, *Sphingomonas elodea*. It is composed of a repeating unit of tetrasaccharide consisting of (1–3)-β-D-glucose, (1–4)-β-D-glucuronic acid, (1–4)-β-D-glucose, and (1–4)-α-L-rhamnose. Gellan gum can be converted to gels by ionic cross-linking and altering the temperature (Crescenzi et al., 1990). In the past two decades, gellan gum has gained a lot of attention in biomedical applications and has been investigated for tissue engineering and drug delivery applications and as a bioadhesive material in drug delivery (Smith et al., 2007; Silva et al., 2012; Milan et al., 2019). Although gellan gum has been extensively explored for the abovementioned applications, only very limited work is reported on its gene delivery applications. Goyal et al. investigated the efficacy of polyethylenimine (PEI)-gellan gum toward gene delivery (Ritu et al., 2011). In this study, they blended gellan gum and PEI at different weight ratios to obtain GP complexes and then prepared nanocomplexes with pDNA. These complexes were cytocompatible and the authors observed better transfection efficiency with these complexes when compared with native PEI and commercial transfection reagents in the cancer cell lines and in primary mouse keratinocytes. The GP-NCs were capable of protecting DNA from nucleases and formed stable nanocomplexes with DNA. The efficacy of these GPs was also evaluated toward siRNA delivery, and a knockdown of 77% in GFP expression was observed. The gellan blended PEI was also tested for in vivo cellular and organismal toxicity in *Drosophila*, which demonstrated

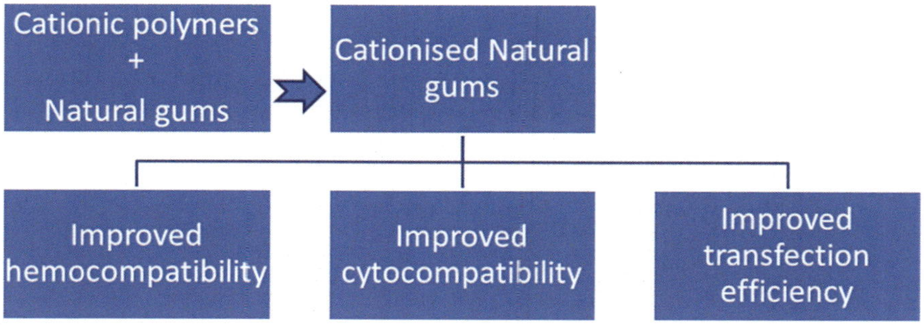

FIG. 8.3 Schematic representing the possible improvement in the properties of the cationic transfecting agents on modification with natural gum polysaccharides.

nonsignificant induction of cellular stress in larvae, and no developmental adversities were observed in larvae or adults. Furthermore, studies in mice also established this blend as an efficient gene delivery system.

8.5.2 Xanthan gum

Xanthan gum is an extracellular anionic heteropolysaccharide secreted by a bacteria *Xanthomonas campestris*. The components of xanthan gum consist of D-glucose units with a trisaccharide side chain which has two mannose units separated by guluronic acid. It is a water-soluble polysaccharide which is nontoxic, biodegradable, and biocompatible, and with Food and Drug Administration (FDA) approval it is being used in the food and pharmaceutical industries (Namita and Jasvirinder, 2018). Xanthan gum is also being widely investigated in biomedical applications as a scaffold and drug delivery matrices due to its abovementioned properties along with its high solubility over a range of pH and its gel forming capacity (Alle et al., 2020; Petri, 2015; Rajesh, 2009; Sethi et al., 2020). There are also reports suggesting that xanthan gum possesses antitumor effects as well. Takeuchi et al. studied the influence of xanthan gum on macrophages and antitumor effect on tumor-bearing mice following oral administration of xanthan gum (Takeuchi et al., 2009). In vitro culture of macrophages in the presence of xanthan gum led to the production of IL-12 and TNF-α. The authors observed that following oral administration of xanthan, the tumor growth was significantly reduced in tumor-bearing mice. Fernandez-Piñeiro et al. developed a xanthan-based gene delivery vector for targeting endothelial cells (Fernandez-Piñeiro et al., 2018). Xanthan was used to functionalize the sorbitan monooleate nanoparticles to create a hydrophilic negatively charged surface coating. This will provide stability to the nanoparticles, and xanthan has an inherent targeting property to endothelial cells. By functionalizing with xanthan, the properties of the sorbitan monooleate particles were improved. In this gene delivery system, the plasmid was encapsulated within the particle rather than forming a complex. These nanoparticles were cytocompatible, stable, protected DNA from enzymatic degradation, and could deliver the gene intracellularly. Owing to its negative charge derived from xanthan, nonspecific interactions could be minimized. In addition, due to the presence of mannose groups, endothelial targeting was possible and according to the report, in vivo plasmid DNA delivery to the vascular endothelium of lung, liver, and kidney could be achieved. The authors performed biodistribution studies with pEGFP loaded xanthan nanoparticles by intravenous injection to mice through the tail vein. It was found that there was no colocalization of GFP expression in Kupffer cells. As per the conclusion, the nanoparticles target the endothelial cells and could escape the reticuloendothelial system (RES). This study highlights the usefulness of xanthan in endothelial targeting and as a transfecting agent.

8.5.3 Guar gum

Guar gum is derived from the seeds of a plant *Cyamopsis tetragonoloba* which belongs to the Leguminosae family (Whistler and Hymowitz, 1979). Guar gum (GG) polysaccharide is a high molecular weight linear chain galactomannan and consists of (1–4)-β-D-mannopyranosyl units with (1–6)-α-D-galactopyranosyl residues as side chains. It is reported that owing to its very high molecular weight and correspondingly high number of hydroxyl groups, it undergo hydrogen bonding in aqueous solutions and is highly soluble in even cold water (Diksha and Sunil, 2021). Because of this hydrogen bonding tendency, intermolecular chain entanglement occurs and consequently its properties such as gelling and thickening properties enhances and also results in high viscosity even at low concentration. Guar gum is approved for oral consumption and is under the Generally Recognized As Safe (GRAS) category. Guar gum is widely used in the pharmaceutical industry as thickening, stabilizing, and binding agents. As for biomedical applications, owing to its mucoadhesive nature, it is considered as an attractive material for developing drug delivery systems (Diksha and Sunil, 2021; Prabaharan, 2011). Reports exist on its use as a controlled drug delivery matrix and also as wound healing material (Vipul et al., 2013; Diksha and Sunil, 2021; Aminabhavi et al., 2014). Although guar gum has been explored widely for controlled drug delivery applications in the forms of hydrogels, nanofibers, nanoparticles, etc., reported gene delivery vectors based on guar gum are insufficient. Jana et al. reported a carboxymethylated guar gum-grafted-polyethyleneimine copolymer (CMGG-*g*-PEI) toward gene delivery application. CMGG-*g*-PEI exhibited good binding ability with plasmid DNA (pDNA) and formed nanocomplexes of size ranging from 150 to 200 nm at a polymer to pDNA weight ratio of 10:1 and above. CMGG-g-PEI was found to be less toxic in nature and hemocompatible. The in vitro gene transfection efficiency of the CMGG-g-PEI/pDNA complex was evaluated using an A549 cell line and was found to be a better transfecting agent than native PEI (Jana et al., 2016). In another study by the same group, they conjugated low molecular weight PEI to the guar gum (Piyali et al., 2020). Here, low molecular weight PEI of 2 kDa was conjugated with aminated guar gum using 4-bromo-1,8-naphthalic anhydride as a coupling reagent. The resultant product GG-g-LPEI (GNP) was cytocompatible with tested cell lines MDA-MB-231 and HeLa. The material formed stable complexes with pDNA even at low weight ratios and exhibited good transfection efficiency. The authors attributed this higher transfection efficiency to high buffering capacity, which was comparable with unmodified low molecular weight PEI. The vector displayed differential transgene expression capabilities in different cell lines. Compared to HeLa, transfection was higher in the MDA-MB 231 cell line, which is due to the overexpression of mannose receptors in these cells.

8.5.4 Alginate

Alginate is a linear anionic polysaccharide produced by seaweed containing blocks of (1,4)-linked β-D-mannuronate (M) and α-L-guluronate (G) residues (Haug, 1959). Alginate is one of the most extensively investigated biopolymers for numerous biomedical applications. It possesses attractive characteristics such as gelling capacity, biocompatibility, low toxicity, etc. Alginates have gained wide attention in drug delivery, wound dressing, and tissue engineering applications. Alginate is a GRAS category material and commercial wound dressings based on alginate are available. Wounds need a moist environment for better healing and alginate dressings provide this; its swelling capacity also makes alginate useful in managing exudating wounds. Alginate matrices, nanoparticles, microparticles, etc., have been widely explored for sustained or controlled delivery of various type of therapeutic molecules (Kuen and David, 2012; Ilekuttige et al., 2020; Alfatama et al., 2018). Limited studies have been carried out on gene delivery applications too (Amiji and Jain, 2012; Zhao et al., 2012). Amiji et al. used alginate as a noncondensation delivery system. Here they developed calcium cross-linked alginate microparticles for gene delivery to macrophages. They observed good cellular uptake and transfection in macrophages in comparison with conventional gene delivery systems. These microparticles were nontoxic and the authors suggest the usage of this system for macrophage targeted antiinflammatory gene delivery (Amiji and Jain, 2012). Zhong et al. used alginate in combination with nano calcium carbonate for gene delivery (Zhao et al., 2012). Alginate aided in the formation of DNA-loaded nanoparticles in the presence of calcium carbonate owing to coprecipitation. It was observed that the presence of alginate molecules retarded the growth of calcium carbonate-based coprecipitates. Due to the presence of alginate, the DNA-loaded alginate/$CaCO_3$ nanoparticles exhibited smaller size and their stability was enhanced in the aqueous solution. The authors explained this phenomenon as follows: in the basic and neutral pH the alginate chains exist in the form of stretching conformation due to the repulsion between carboxyl ions, as the pK_a is pH 3.5. Furthermore, in the presence of Ca^{2+} ions the carboxyl groups bind these ions and become more condensed and those carboxyl groups which are not bound to calcium ions repel each other, leading to affinity to water molecules. The particles formed will have high alginate content in the surface and since the alginate remains ionized at pH above its pK_a, it will have a negative charge which will impart colloidal stability to the nanoparticles in the aqueous medium. The in vitro gene transfection efficacy was evaluated in 293 T and HeLa cells using plasmid pGL3-Luc as a reporter gene, and it was observed that alginate-modified nanoparticles exhibited significantly high transfection efficiency compared with the nanoparticles without alginate modification. It was concluded that alginate modification improved the calcium carbonate coprecipitation technique for the preparation of gene and drug delivery systems. As an extension of the abovementioned study, Zhao et al. reported alginate/$CaCO_3$/DNA/DOX nanoparticles for codelivery of p53 plasmid and doxorubicin hydrochloride. This system was also found to be an efficient carrier for dual delivery of gene and drug (Dong et al., 2012).

8.5.5 Dextran

Dextran is a gum produced by the bacterium *Leuconostoc mesenteroides* from sucrose, and is a linear polysaccharide composed of α-D-glucose linked by α-(1→6) glycosidic bonds. Dextran has a lot of hydroxyl groups, hence it is a very water-soluble polysaccharide (Anthony, 1993). It is nontoxic, biocompatible, biodegradable, and thus is used extensively in medical, food, and agricultural industries as a promising material. Dextran is considered as a GRAS category polymer and hence is in wide use in the biomedical industry. One of its major applications is as a plasma expander and also as a cholesterol-lowering agent and as a substitute to blood plasma (Grönwall and Ingelman, 1945). Dextran acts as a substitute of blood proteins such as albumins and provide colloid osmotic pressure (Klotz and Kroemer, 1987). It is extensively used in various drug delivery applications and it is easily modifiable, resulting in numerous derivatives and conjugates (Qiaobin et al., 2021). Dextran, being highly soluble, with hydroxyl groups that can be modified, biocompatible, nonimmunogenic, and easily available, is extensively used for developing drug delivery systems (Dhaneshwar et al., 2006; Varshosaz, 2012; Lai et al., 2009; Ferreira et al., 2018; Torrieri et al., 2020; Huang and Huang, 2019). Among gums, dextran is probably the most widely explored one for developing gene delivery systems (Tseng and Jong, 2003).

Dextran-PEI was developed to improve the compatibility of PEI (Kuo et al., 2011). PEI is a highly efficient transfecting agent; however, its very dense cationic charges limit its systemic administration. Tseng et al. conjugated dextran to PEI in order to reduce the interaction with blood components. In this study, dextrans of different molecular weight were grafted to nonlinear and branched PEI at varying degrees. The molecular weights of dextrans used were 1500 and 10,000 Da. The dextran-grafted PEI was used to develop nanoplexes with DNA and the stability of these was evaluated in the presence of serum albumin. PEI grafted with dextran of higher molecular weight provided an adequate shielding effect and protected the nanoplexes from displacement in the presence of serum albumin. However, nanoplexes developed using low molecular weight dextran-conjugated PEI resulted in aggregation in the presence of BSA. The authors concluded that the dextran-grafting of branched PEI improved the stability of the nanocomplexes and had potential to conjugate with ligands for in vivo targeted gene delivery.

Kuo et al. have also reported dextran-PEI conjugates for dual delivery of gene and drug (Zink et al., 2009). They developed biodegradable DEX–ADM–PEI nanoparticles where low molecular weight PEI was conjugated to dextran. Dextran–graft–PEI was synthesized by an imine reaction between oxidized dextran and low molecular weight PEI. These nanoplexes were meant for dual delivery of adriamycin and gene. The authors established its high efficiency in delivering the drug intracellularly and also gene transfection efficiency.

Erbacher et al. developed various conjugates of PEI with saccharide, PEG, antibodies, etc., and evaluated the efficacy of these as gene delivery vectors (Erbacher et al., 1999). Their findings were interesting. PEI/DNA nanoplexes, being highly positively charged, will interact with proteins and blood components which are mostly negatively charged. In order to reduce the surface positivity and thereby lower the nonspecific interactions, partial modification of primary amine groups in PEI has been widely explored. However, there are reports suggesting aggregation though the surface charge gets reduced. It is suggested that grafting the primary amino groups with molecules such as polysaccharides prevents aggregation by increasing the solubility of the complexes. They used small molecules as well as high molecular weight molecules for grafting, and found that the small molecules were incapable of preventing aggregation and had to have high N/P ratios to obtain a smaller size. In order to overcome the aggregation tendency shown in the case of low molecular weight saccharides, PEI was conjugated with dextran, which is a polysaccharide. With modification using high chain length saccharides, it was expected that the solubility of the complexes would be increased. By using varying amounts of dextran for conjugation, different degrees of derivatization of 2.7%, 6.0%, and 9.5% were obtained. The physicochemical characterization of these nanoplexes showed stability in DNA complexation. However, transmission electron microscopy (TEM) analysis showed that the shapes of these nanoplexes were different with varying degrees of derivatization, forming shapes from circular donuts to pretzels, and at higher ratios tangles of strings were also observed. In addition, from the surface charge analysis by measuring zeta potential, it was shown that a higher degree of derivatization reduced the surface charge significantly. The transfection efficiency also varied with degree of derivatization of PEI with dextran. They reported low transfection efficacy in highly derivatized conjugates. The transfection efficiency of PEI-dextran 2.7% was one-third that of the transfection efficiency of PEI which further reduced, and for PEI-dextran 9.5%, there was no gene expression.

Due to the poor transfection efficiency of the dextran-modified PEI, Chu et al. (2013) designed its targetable version. Dextran was grafted with low molecular weight branched PEI of 1800 and 800 Da, and was conjugated with a targeting ligand. The developed nanoplexes exhibited low cytotoxicity on different cell lines such as HeLa, 293 T, and HepG2. These polymers were also hemocompatible and had very good transfection efficiency. The targeting ligand used was peptide of the sequence as given, PKKKRKV (PV7). It is a nuclei-localized signal that can recognize the nuclei transport protein on the nuclei membrane and transport the delivery system into the nuclei. The peptide PV7 was noncovalently bound to PEI-modified dextran to enable nuclear targeting. The DNA was labeled with Cy3 and the intracellular uptake was monitored, and the presence of the polyplexes was observed in most cells after 4 h of incubation. The polyplexes were mainly localized in the perinuclear area. Compared to the unmodified nanoplexes, those with targeting ligand had a very high localization in the nucleus.

The authors also reported better transfection efficiency for PV7-conjugated carriers as they mediate nuclear entry via importin by nuclear localization signals (NLS). The transfection efficiency in vivo was evaluated in tumor-bearing mice using pEGFP. Following transfection of pEGFP with ligand-bearing dextran-PEI, the presence of GFP was seen in tumor and organs such as lungs, liver, spleen, and kidneys. Thus by conjugating with a targeting ligand, the transfection efficiency was improved significantly and by in vivo experiments in tumor-bearing mice the authors could show localization of the polyplexes in the tumor site (Erbacher et al., 1999). Thus it was concluded that PV conjugated dextran-*g*-PEI is a potential candidate for effective and biocompatible gene delivery systems.

Dextran-oligoamine conjugates were developed using various oligoamines and were evaluated for their gene transfection efficiency both in vitro and in vivo (Azzam et al., 2003; Hosseinkhani et al., 2004). Azzam et al. initially developed cationic dextrans using four different oligoamines and at different ratios. Dextran of molecular weight 40 kDa was oxidized with periodate and then conjugated using oligoamines such as spermine, spermidine, tetramine, and *N*,*N*-dimethyl propanediamine. The oxidized dextran was then treated with these oligoamines and conjugation occurred by reductive animation to obtain the corresponding imine-conjugates. Furthermore, the resultant conjugates were reduced using excess sodium borohydride to obtain the corresponding amine-conjugates with about 30%–40% yield. The conjugates were then evaluated for their nanoplex-forming ability, and their stability was also evaluated. From this study, the authors could conclude that among the four different oligoamines only the dextran-spermine conjugate had good transfection efficiency; this was attributed to the amino content and molecular weight. They also suggested that the structure of polycation plays a significant role in its transfection efficiency. From the same group, it was reported that of the more than 300 different polycations developed using various natural polysaccharides and oligoamines having two to four amino groups by reductive amination, dextran-spermine was found to be the most efficient transfecting agent. Furthermore, the transfection efficiency of dextran–spermine was assessed in vitro on HEK293 and NIH3T3 cell lines and found to be as high as the commercially available transfecting agents such as DOTAP. Dextran spermine conjugate was further modified with polyethylene glycol, which resulted in a high transfection yield in a serum-rich medium. In vivo evaluations were done by administering the nanoplexes by systemic and intramuscular administration. The mice were injected with dextran–spermine–pSV-LacZ complex intramuscularly, which resulted in high levels of local gene expression compared to that of low expression of the naked DNA. However, intravenous injection of these nanoplexes did not gain any gene expressions in any organ, whereas PEGylated complexes showed high gene expression and were mainly observed in the liver. The authors evaluated the gene expression levels PEGylated dextran–spermine–pSV-LacZ complex in mice following preadministration of arabinogalactan and D-galactosylated bovine albumin. They found that gene expression in the liver was significantly reduced followed by this preadministration, while mannosylated albumin had no influence on the liver level of gene expression. This is because both arabinogalactan and D-galactosylated bovine albumin interact with the asialoglycoprotein receptor of liver parenchymal cells, resulting in inhibition of uptake. From their observations, the authors concluded that targeting of the PEGylated dextran–spermine–plasmid DNA complex to the liver could be via the galactose receptor of the liver parenchymal cells rather than the mannose receptor (Hosseinkhani et al., 2004).

Another group reported the development of bioinspired nonviral vectors by developing dextran esters of various amino acids. Carbonyl diimidazole-activated dextran was reacted with Boc-amino-protected amino acids such as glycine, β-alanine, and L-lysine. The authors developed an array of esters and investigated the relationship between polymer structure and complex formation, and subsequently its stability, toxicity, and transfection. They found that the esters developed from β-alanine and L-lysine could form nanosized complexes of size <110 nm with positive zeta potential. With a higher degree of substitution, the L-lysine derivatives were more efficient in forming nanoplexes and also exhibited good stability in the presence of endonucleases. However, interestingly the β-alanine-bearing dextran esters displayed higher transfection efficiency, which was attributed to more efficient release of DNA from the polyplexes (Zink et al., 2009).

From our own group we developed a dextran-histidine conjugate and attempted to elucidate the capability of dextran, by modification with histidine, to be an efficient, safe, and promising nucleic acid delivery system in gene therapy. The derivative was found to be 6.7-fold more efficient than PEI in its transfecting capability. Mechanisms involved in cellular internalization and vector unpacking were also investigated. The material was found to be a highly efficient and cytocompatible transfecting agent (Jane et al., 2012).

In another work, we grafted the cationized dextran with diethyl aminoethyl methacrylate (DEAEM) for gene delivery in cancer cells (Sherly et al., 2020). PEI-modified dextran was grafted with a DEAEM monomer via the Michael addition reaction to obtain DPD I and DPD II. These polymers were found to be cytocompatible to various cells, had good buffering capacity, and formed nanocomplexes with DNA exhibiting positive zeta potential. The transfection efficiency of these vectors in cell lines C6 and HeLa was analyzed using p53 plasmid, which demonstrated good transfection. Biodistribution studies DPD II in BALB/c mice showed that it mainly accumulated in liver tissue and not in any vital organs including the brain, lungs, and heart; it also demonstrated good renal clearance.

8.6 Conclusions

Though various gums have received wide attention in the fields of drug delivery and tissue engineering, only limited studies are reported on gene delivery systems. Detailed studies are needed to evaluate and understand the exact utility of gums as gene delivery systems. How the presence of this category of polysaccharide can influence or improve transection needs thorough investigation. This will enable the potential use of naturally occurring materials as nonviral vectors.

References

Alfatama, M., Lim, L.Y., Wong, T.W., 2018. Alginate–C18 conjugate nanoparticles loaded in tripolyphosphate-cross-linked chitosan-oleic acid conjugate-coated calcium alginate beads as oral insulin carrier. Mol. Pharm. 15, 3369–3382.

Alle, M., Bhagavanth Reddy, G., Kim, T.H., Park, S.H., Lee, S.-H., Kim, J.-C., 2020. Doxorubicin-carboxymethyl xanthan gum capped gold nanoparticles: microwave synthesis, characterization, and anti-cancer activity. Carbohydr. Polym. 229, 115511.

Alonso-Sande, M., Teijeiro, D., Remuñán-López, C., Alonso, M.J., 2009. Glucomannan, a promising polysaccharide for biopharmaceutical purposes. Eur. J. Pharm. Biopharm. 72, 453–462.

Amiji, M., Jain, S., 2012. Calcium alginate microparticles as a non-condensing DNA delivery and transfection system for macrophages. Pharm. Eng. 32, 1–8.

Aminabhavi, T.M., Nadagouda, M.N., Joshi, S.D., More, U.A., 2014. Guar gum as platform for the oral controlled release of therapeutics. Expert Opin. Drug Deliv. 11, 753–766.

Anthony, N.D.B., 1993. Dextran. In: Roy, L.W., James, N.B. (Eds.), Industrial Gums, third ed. Academic Press, pp. 399–425.

Azzam, T., Eliyahu, H., Makovitzki, A., Domb, A., 2003. Dextran-spermine conjugate: an efficient vector for gene delivery. Macromol. Symp. 195, 247–262. https://doi.org/10.1002/masy.200390130.

Chu, M., Dong, C., Zhu, H., Cai, X., Dong, H., Ren, T., Su, J., Li, Y., 2013. Biocompatible polyethylenimine-graft-dextran catiomer for highly efficient gene delivery assisted by a nuclear targeting ligand. Polym. Chem. 4, 2528.

Crescenzi, V., Dentini, M., Coviello, T., 1990. Solutions and gelling properties of microbial polysaccharides of industrial interest: the case of gellan. In: Dawes, E.A. (Ed.), Novel Biodegradable Microbial Polymers. Kluwer, Dordrecht, pp. 227–284.

Cui, S.W., Wang, Q., 2005. Functional properties of carbohydrates: polysaccharide gums. In: Hui, Y.H., Sherkat, Frank (Eds.), Handbook of Food Science, Technology, and Engineering. 4. CRC Press, Boca Raton, FL, pp. 63–80.

De Smedt, S.C., Demeester, J., Hennink, W.E., 2000. Cationic polymer based gene delivery systems. Pharm. Res. 17, 113–126.

Dhaneshwar, S.S., Mini, K., Gairola, N., Kadam, S., 2006. Dextran: a promising macromolecular drug carrier. Indian J. Pharm. Sci. 68, 705.

Diksha, V., Sunil, K.S., 2021. Recent advances in guar gum based drug delivery systems and their administrative routes. Int. J. Biol. Macromol. 181, 653–671.

Dong, Z., Chuan, J.L., Ren-Xi, Z., Si-Xue, C., 2012. Alginate/$CaCO_3$ hybrid nanoparticles for efficient codelivery of antitumor gene and drug. Mol. Pharm. 9, 2887–2893. https://doi.org/10.1021/mp3002123.

Erbacher, P., Bettinger, T., Belguise-Valladier, P., Zou, S., Coll, J.-L., Behr, J.-P., Remy, J.-S., 1999. Transfection and physical properties of various saccharide, poly(ethylene glycol), and antibody-derivatized polyethylenimines (PEI). J. Gene Med. 1, 210–222. https://doi.org/10.1002/(SICI)1521-.

Fernandez-Piñeiro, I., Alvarez-Trabado, J., Márquez, J., Badiola, I., Sanchez, A., 2018. Xanthan gum-functionalised span nanoparticles for gene targeting to endothelial cells. Colloids Surf. B: Biointerfaces 170, 411–420.

Ferreira, M.P., Talman, V., Torrieri, G., Liu, D., Marques, G., Moslova, K., Liu, Z., Pinto, J.F., Hirvonen, J., Ruskoaho, H., 2018. Dual-drug delivery using dextran-functionalized nanoparticles targeting cardiac fibroblasts for cellular reprogramming. Adv. Funct. Mater. 28, 1705134.

Grönwall, A., Ingelman, B., 1945. Dextran as a substitute for plasma. Nature 155, 3924.

Haug, A., 1959. Fractionation of alginic acid. Acta Chem. Scand. 13, 601–603.
Hosseinkhani, H., Azzam, T., Tabata, Y., et al., 2004. Dextran–spermine polycation: an efficient nonviral vector for in vitro and in vivo gene transfection. Gene Ther. 11, 194–203. https://doi.org/10.1038/sj.gt.3302159.
Huang, S., Huang, G., 2019. The dextrans as vehicles for gene and drug delivery. Future Med. Chem. 11, 1659–1667.
Ilekuttige, P.S.F., Won, W.L., Eui, J.H., Ginnae, A., 2020. Alginate-based nanomaterials: fabrication techniques, properties, and applications. Chem. Eng. J. 391, 123823.
Jana, P., Sarkar, K., Mitra, T., Chatterjee, A., Gnanamani, A., Chakraborti, G., Kundu, P., 2016. Synthesis of a carboxymethylated guar gum grafted polyethyleneimine copolymer as an efficient gene delivery vehicle. RSC Adv. 6, 13730–13741.
Jane, J.T., Rekha, M.R., Sharma, C.P., 2012. Unraveling the intracellular efficacy of dextran-histidine polycation as an efficient nonviral gene delivery system. Mol. Pharm. 9 (1), 121–134.
Klotz, U., Kroemer, H., 1987. Clinical pharmacokinetic considerations in the use of plasma expanders. Clin. Pharmacokinet. 12, 123–135.
Kuen, Y.L., David, J.M., 2012. Alginate: properties and biomedical applications. Prog. Polym. Sci. 37, 106–125.
Kuo, S., Jing, W., Jian, Z., Min, H., Changsheng, L., Tongyi, C., 2011. Dextran–g–PEI nanoparticles as a carrier for co-delivery of adriamycin and plasmid into osteosarcoma cells. Int. J. Biol. Macromol. 49, 173–180.
Lai, K., Wang, Y.-Y., Hanes, J., 2009. Mucus-penetrating nanoparticles for drug and gene delivery to mucosal tissues. Adv. Drug Deliv. Rev. 61, 158–171.
Milan, M., Ivana, P.-L., Branko, B., Amit, K.N., Md Saquib, H., 2019. Gellan gum in drug delivery applications. In: Hasnain, M.S., Amit, K.N. (Eds.), Natural Polysaccharides in Drug Delivery and Biomedical Applications. Academic Press, pp. 145–186.
Namita, J., Jasvirinder, S.K., 2018. Microbial polysaccharides in food industry. In: Grumezescu, A.M., Holban, A.M. (Eds.), Handbook of Food Bioengineering, Biopolymers for Food Design. Academic Press, pp. 95–123.
Nikunj, S., Verma, I.M., 2000. Gene therapy: trials and tribulations. Nat. Rev. Genet. 1, 91–99.
Petri, D.F.S., 2015. Xanthan gum: a versatile biopolymer for biomedical and technological applications. J. Appl. Polym. Sci. 132, 42035.
Piyali, J., Santanu, G., Kishor, S., 2020. Low molecular weight polyethyleneimine conjugated guar gum for targeted gene delivery to triple negative breast cancer. Int. J. Biol. Macromol. 161, 1149–1160.
Prabaharan, M., 2011. Prospective of guar gum and its derivatives as controlled drug delivery systems. Int. J. Biol. Macromol. 49, 117–124.
Qiaobin, H., Yingjian, L., Yangchao, L., 2021. Recent advances in dextran-based drug delivery systems: from fabrication strategies to applications. Carbohydr. Polym. 264, 117999.
Rajesh, N., 2009. Siddaramaiah feasibility of xanthan gum–sodium alginate as a transdermal drug delivery system for domperidone. J. Mater. Sci. Mater. Med. 20, 2085–2089.
Ritu, G., Tripathi, S.K., Shilpa, T., Ravi Ram, K., Ansari, K.M., Shukla, Y., Kar Chowdhuri, D., Kumar, P., Gupta, K.C., 2011. Gellan gum blended PEI nanocomposites as gene delivery agents: evidences from in vitro and in vivo studies. Eur. J. Pharm. Biopharm. 79, 3–14.
Sethi, S., Saruchi, Kaith, B.S., et al., 2020. Cross-linked xanthan gum–starch hydrogels as promising materials for controlled drug delivery. Cellulose 27, 4565–4589.
Sherly, C.D., Rekha, M.R., Harikrishnan, V.S., 2020. Cationised dextran and pullulan modified with diethyl aminoethyl methacrylate for gene delivery in cancer cells. Carbohydr. Polym. 242, 116426. https://doi.org/10.1016/j.carbpol.2020.116426.
Silva, N.A., Cooke, M.J., Tam, R.Y., Sousa, N., Salgado, A.J., Reis, R.L., Shoichet, M.S., 2012. The effects of peptide modified gellan gum and olfactory ensheathing glia cells on neural stem/progenitor cell fate. Biomaterials 33, 6345–6354.
Smith, A.M., Shelton, R., Perrie, Y., Harris, J.J., 2007. An initial evaluation of gellan gum as a material for tissue engineering applications. J. Biomater. Appl. 22, 241–254.
Suhail, A., Mudasir, A., Kaiser, M., Roli, P., Saiqa, I., 2019. A review on latest innovations in natural gums based hydrogels: preparations & applications. Int. J. Biol. Macromol. 136, 870–890.
Takeuchi, A., Kamiryou, Y., Yamada, H., Eto, M., Shibata, K., Haruna, K., Naito, S., Yoshikai, Y., 2009. Oral administration of xanthan gum enhances antitumor activity through toll-like receptor 4. Int. Immunopharmacol. 9, 1562–1567.
Torrieri, G., Fontana, F., Figueiredo, P., Liu, Z., Ferreira, M.P., Talman, V., Martins, J.P., Fusciello, K., Moslova, T., 2020. Teesalu dual-peptide functionalized acetalated dextran-based nanoparticles for sequential targeting of macrophages during myocardial infarction. Nanoscale 12, 2350–2358.

Tseng, W.-C., Jong, C.-M., 2003. Improved stability of polycationic vector by dextran-grafted branched polyethylenimine. Biomacromolecules 4, 1277–1284.

Varshosaz, J., 2012. Dextran conjugates in drug delivery. Expert Opin. Drug Deliv. 9, 509–523.

Vipul, D.P., Girish, K.J., Naresh, G.M., Narayan, P.R., 2013. Pharmaceutical applications of various natural gums, mucilages and their modified forms. Carbohydr. Polym. 92, 1685–1699.

Whistler, R.L., Hymowitz, T., 1979. Guar: Agronomy, Production, Industrial Use, and Nutrition. Purdue University Press, West Lafayette.

Zhao, D., Zhuo, R.-X., Cheng, S.-X., 2012. Alginate modified nanostructured calcium carbonate with enhanced delivery efficiency for gene and drug delivery. Mol. BioSyst. 8, 753–759.

Zink, M., Hotzel, K., Schubert, U.S., Heinze, T., Fischer, D., 2009. Amino acid-substituted dextran-based non-viral vectors for gene delivery. Macromol. Biosci. 19, 1900085.

CHAPTER

9

Locust bean gum-based micro- and nanomaterials for biomedical applications

R.S. Soumya[a], *K.G. Raghu*[a], *and Annie Abraham*[b]

[a]Biochemistry and Molecular Mechanism Laboratory, Agroprocessing and Technology Division, CSIR-National Institute for Interdisciplinary Science and Technology (NIIST), Thiruvananthapuram, Kerala, India [b]Department of Biochemistry, University of Kerala, Kariavattom, Thiruvananthapuram, Kerala, India

9.1 Introduction

In recent decades, polysaccharides have been judged to have many useful applications in the biopharmaceutical field (Dionísio and Grenha, 2012). In drug delivery systems, natural polysaccharides have received more and more consideration (Miao et al., 2018). Polysaccharides are considered as the most favorable materials for the synthesis of micro and nanometric carriers as they have many reactive groups, a varied range of molecular weights, and variable chemical compositions, contributing to their various structural properties (Liu et al., 2008). They can be easily modified chemically, bringing about many polysaccharide derivatives owing to many derivable groups on their molecular chains (e.g., hydroxyl). Various chemical and physical modifications such as cross-linking, grafting, esterification, oxidation, and combination with a variety of other molecules can be performed to produce semisynthetic and composite materials with extended functionality (Chavali et al., 2013). Natural polymers have an essential role in controlling the drug release rate from pharmaceutical formulations (Gopinath et al., 2018). These polysaccharide polymers are biocompatible and interact with living cells, making them compliant and appropriate biomaterials for long-term systemic circulation and for targeted drug delivery systems (Malafaya et al., 2007).

Micro- and Nanoengineered Gum-Based Biomaterials for Drug Delivery and Biomedical Applications
https://doi.org/10.1016/B978-0-323-90986-0.00012-1

In most of the higher plants, polysaccharide hydrocolloids are usually found, and are abundant in nature, including mucilages, gums, and glucans. As a product of their resistance mechanism, higher plants produce gums (Goswami and Naik, 2014). Chemically, natural gums are polysaccharides, with one or more types of monosaccharide units or their derivatives connected by a variety of linkages and structures to form the macromolecular structure. Through hydrolysis, gums will yield simple sugars like mannose, galactose, xylose, arabinose, glucose, or uronic acids (Eggleston et al., 2018). They have heterogeneous conformation. Natural gums are established as traditional excipients in various preparations (Choudhary and Pawar, 2014). The conventional assessment of excipients is that they are inert with no therapeutic effect and have no influence on drug activity, and with the aid of natural gums, the level of absorption of a drug can be efficiently altered (Deep et al., 2019). As these polysaccharide gums are biodegradable, maintainable, and safe, scientists have extensively researched them and used them as raw materials in industries (Jania et al., 2009). In the pharmaceutical field, polysaccharides have a wide range of applications after being isolated from various plant sources like locust bean gum (LBG), xanthan gum, tamarind gum, and guar gum. As these are all galactomannans with mannose and galactose residues. Polysaccharides as a whole are termed for the sugar unit they comprise, glucose-based polysaccharides named glucans, while mannose-based polysaccharides are mannans (d'Ayala et al., 2008). Natural biopolymers demonstrate, as a remarkable example, in what way all the properties exhibited by biological materials and systems are entirely determined by the physicochemical properties of the monomers and their sequence (Malafaya et al., 2007). These polymers have many outstanding merits over synthetic ones, show specifically enhanced cell adhesion capacity, and have mechanical properties like natural tissues (Pollard et al., 2008). Table 9.1 represents the limitation and uses of natural polymer-based nanoparticles for drug delivery application. In this chapter, critical features of LBG are identified, with specific focus on the properties that closely affect LBG's biopharmaceutical applications, and the reported biopharmaceutical applications as microparticles and nanoparticles are explored and discussed.

TABLE 9.1 Advantages and limitations of natural polymer-based nanoparticles in drug delivery.

Advantages	Limitations
High surface/volume ratio	Undefined physical shape
Surface alteration can be done simply	Inadequate capacity to coassociate with other functional molecules
Maximized interaction with mucosae	Unidentified toxicity profile
Increased drug absorption in desired site	Absence of appropriate large-scale production methods
Capability to enter cells	Short stability in some biological fluids
Safeguard encapsulated molecules	Affinity for aggregation
Opportunity to deliver controlled release	Poor loading capacity (unsuitable for less potent drugs)
Provides of targeted delivery	Small size can provide entry to unwanted environments

9.2 Locust bean gum

Locust bean gum (LBG) is a nonionic, neutral, nonstarch polysaccharide made up of mannose and galactose units, which are named galactomannans (Prajapati et al., 2013). LBG is otherwise called carob bean gum, carob seed gum, carob flour, or even ceratonia, and is derived from the seeds of the leguminous plant *C. siliqua* Linn., of the Fabaceae family. This gum is usually cultivated in the Mediterranean region and to a lesser extent in California. By milling the endosperms, the brown pods or beans of the locust bean (Fig. 9.1) are processed to form LBG (Popa, 2011).

LBG consist of β-D-mannose units linked by 1,4-linkages to form long straight chains. To mannose residues, D-galactose units are attached by 1,6-glycosidic linkages (Daas et al., 2000). LBG has an average ratio of 1:4 galactose to mannose units; the galactose units attached in blocks known as substituted or hairy regions. This leaves lengths of mannose chains containing no side groups, known as unsubstituted or smooth regions (Pegg, 2012). Distribution of galactose units in the long main chain influence the galactomannans' physical and chemical properties. This proportion is the foremost characteristic affecting the solubility of galactomannan. If the side chain of galactose is lengthier, the synergic effect is more effective with other polymers and shows greater functionality (Richardson et al., 1998; Parvathy et al., 2005). Depending on the plant origin, the mannose elements form a linear chain linked with galactopyranosyl residues at the side chain at variable distances. Meanwhile, it is a neutral polymer, and its viscosity and solubility are thus little affected by pH variations within the range of 3–11 (Beneke et al., 2009). The reported molecular weight of LBG is between 50 and 1000 kDa and it is incompletely soluble in cold water (Dakia et al., 2008). A temperature of at least 80 °C is required for full hydration. LBG solutions show good stability for most food processing operations and recipes; it is an efficient thickener due to its high viscosity at low concentrations and has good thermal and pH stability except in highly acidic conditions (Dionísio and Grenha, 2012).

LBG is obtained from carob bean seeds by extracting the seeds with aqueous alkali or water (Hirst and Jones, 1948; Smith, 1948), and the galactomannan content in the seeds can reach 85% (Aravamudhan et al., 2014). The ratio of protein, crude fiber, fat, and galactomannan

FIG. 9.1 Locust bean with seed and gum. *From (n.d.). https://www.aepcolloids.com/products/locust-bean-gum.*

in LBG powder was 5.0%:1.0%:0.5%:80–85% (Aravamudhan et al., 2014; Samil Kök, 2007). For commercial applications, LBG powder is being used more and more widely. Commercially available LBG has a moisture content of about 5%–12%, acid-soluble ash 1.7%–5%, ash 0.4%–1.0%, and protein 3%–7%. Refined LBG is composed of moisture of about 3%–10%, acid-soluble matter 0.1%–3%, ash 0.1%–1%, and protein 0.1%–0.7% (Verma et al., 2019). In LBG, approximately 32% albumin and globulin content has been reported, while 68% is glutelin (Smith et al., 2010). Impurities mostly include ash and acid-insoluble matter.

There are reports that LBG is biodegradable, biocompatible, nonteratogenic, and nonmutagenic (Dionísio and Grenha, 2012) with a mucoadhesive property. Its degradation products are excreted readily. In most areas of the world, LBG usage is allowed mainly in the food industry as a thickener, stabilizer, emulsifier, and gelling agent (E410), and it is classified by the Food and Drug Administration (FDA) as a Generally Recognized As Safe (GRAS) material (Mortensen et al., 2017). As an excipient in drug formulations, it is used in the pharmaceutical industry and also has biomedical applications. Several properties make LBG a suitable material for drug delivery (Prajapati et al., 2013). According to the Joint FAO/WHO Expert Committee on Food Additives held in Geneva, April 1975 it is nonteratogenic and nonmutagenic. It has an acceptable shelf life and a biosorbable property. LBG is nondigestible and may be classed as a soluble fiber (Pegg, 2012).

To control systemic or local delivery of biologically active agents, LBG can be used as bioadhesive polymers. In pharmaceutical and biomedical areas, LBG can easily be modified to meet applicants' demands (Dionísio and Grenha, 2012). In this context, for designing drug delivery systems, the use of LBG has been demonstrated, for dose defined delivery, at a chosen rate, to a targeted biological site. As a natural polymer, LBG's use in the biopharmaceutical area had received increased interest, and this chapter focuses on the present use and the varied application of LBG in pharmaceutical areas.

9.3 Biodegradation of LBG

The in vivo biodegradability of polymers is significant for producing polymeric-based materials in biopharmaceutical applications (Song et al., 2018). Subsequently, without the need for additional interventions, the removal by the organism upon administration is essential. By enzymatic action, microorganisms, and pH action, encompassing complex biological, physical, and chemical processes, natural polymers' biodegradation can be employed. These procedures lead to polymer chain breakdown and promote modification of properties like solubility and molecular weight. Many enzymes exist within the organism that might split the LBG macromolecule; its biodegradation is probably driven by enzymatic activity. The effective route that guarantees the degradation of the polysaccharide is oral administration. In the human colonic region, the biodegradation of LBG occurs mainly by the activity of the enzyme β-mannanase (Nakajima and Matsuura, 1997; Alonso-Sande et al., 2009), thus aiding the source for developing many colonic drug delivery approaches that comprise LBG. The two enzymes, β-mannosidase and α-galactosidase, are involved in the entire degradation of LBG. These enzymes act individually, producing D-mannose and D-galactose (Soni et al., 2016; Kurakake and Komaki, 2001). These enzymes were also identified in human fecal contents, supporting this polysaccharide's potential in colonic delivery applications and confirming its degradation upon oral administration (Jain et al., 2007).

9.4 Bioactivity of LBG

Numerous works define the prospective of LBG as a bioactive material. In 1983, LBG was reported for the first time as having a hypolipidemic effect due to its high content of insoluble fiber, which thereby reduced the level of low-density lipoprotein (LDL) cholesterol (Zavoral et al., 1983). Reports validated the useful properties of LBG in regulating hypercholesterolemia (Evans et al., 1992; Zunft et al., 2003). This possible advantage has generally directed some authors to suggest daily consumption of food products augmented with fiber. Additionally, there are reports that the LBG could decrease the amount of hepatic synthesis of cholesterol and it has also been suggested to manage diabetes (Brennan, 2005; Tsai and Peng, 1981). In addition, due to its high gelling ability, LBG is included in dietary products as its ingestion causes a sensation of satiety (Dakia et al., 2007). An LBG-based capsules formulation is available for appetite suppression (Carob gum—Arkopharma Arkocápsulas®).

9.5 Pharmaceutical applications of LBG

The intake of natural polymers has been newly augmented in the field of drug delivery. In recent decades, many attempts have been made to develop suitable delivery systems that elude or reduce side effects while refining the therapeutic ability. For the development of several novel drug delivery systems, LBG demonstrated extensive applications. To make products extra appealing, the natural origin and specific individual characteristics are assets. In the field of drug delivery, many efforts have been devoted to developing an appropriate delivery system that avoids or minimizes side effects while improving therapeutic efficacy. The importance of natural polymers in the pharmaceutical formulation is exceptionally diverse, encompassing stable films, monolithic matrix systems, beads, implants, microparticles, nanoparticles, inhalable and injectable systems, and viscous liquid and gel formulations (Saha et al., 2018). Polymeric materials have diverse purposes within these dosage forms, such as matrix formers, binders, coatings, drug release modifiers, thickeners, or stabilizers, viscosity enhancers, disintegrators, bioadhesive, solubilizers, emulsifiers, suspending agents, and gelling agents (Beneke et al., 2009; Saha et al., 2018). There are reports based on the use of LBG in drug delivery, i.e., allied with formulations intended for colonic delivery, thus promoting colonic microflora's ability to degrade LBG (Hirsch et al., 1999; Raghavan et al., 2002). Additional uses include topical, ocular, and buccal delivery systems. LBG use as a tablet matrix is related to the point that polysaccharides are generally used to play a significant role in drug release mechanisms from matrixes. In most circumstances, it is detected that LBG linked with another polymer gives an enhanced outcome in controlled drug release systems (Hoffman, 2012).

9.6 LBG microparticles

Microparticles have been considered to ensure the drug's safety until they reach the place of action and also have to provide the required aerodynamic features. For carrying drugs and other biomaterials, a microparticulate system is used in the therapeutic field (El-Sherbiny et al., 2015). Its several applications include making reliable entities from oils, changing solubility, controlling taste, odor, carry over volatilization, protecting drugs from getting oxidized, and avoiding incompatibilities (Snehalatha et al., 2013). LBG microparticles were the foremost ones used in polysaccharide-based lung delivery applications. LBG microparticles have been used as matrix material to transport antitubercular drugs, due to macrophages' ability to phagocytose with preference materials containing mannose and galactose as structural units (Sharma et al., 2016a, b; Suzuki and Lim, 1994; Ahsan et al., 2002; East and Isacke, 2002). Alves et al. (2016) reported inhalable antitubercular therapy using spray-dried LBG microparticles (Alves et al., 2016). Their study showed that the LBG microparticles were modified with aerodynamic properties for reaching the alveolar zone to favor phagocytosis. The two first-line antitubercular drugs, isoniazid and rifabutin, can associate with the prepared LBG microparticles. In A549 and macrophage-like THP-1 cells, the cytotoxicity of drug formulations has been checked, and LBG microparticles ability to capture macrophages' was established (Alves et al., 2016).

Rodrigues et al. (2017) reported about inhalable LBG microparticles, i.e., dual antibiotherapy for tuberculosis. This work showed the delivery of two antitubercular drugs' by inhalation and improved internalization by macrophages (Rodrigues et al., 2017). Moreover, the LBG microparticles' respirability was studied to regulate isoniazid's and rifabutin's delivery to the lungs. LBG microparticles have been widely used to formulate oral delivery systems due to the specific recognition of mannose and galactose residues of LBG by macrophage surface receptors (East and Isacke, 2002). Grenha et al. (2020) reported the therapeutic efficiency in the murine model of tuberculosis infection using inhalable LBG microparticles coassociating with antitubercular drugs, isoniazid and rifabutin (Grenha et al., 2020). There are reports on the existence of β-mannosidase in lung tissue (Alkhayat et al., 1998), an enzyme with the reported capability to degrade LBG (Dionísio and Grenha, 2012). LBG microparticles, upon delivery, slowly dissolve in contact with the lung lining fluid, and in that way, phagocytosis by macrophages is expected to occur (Grenha et al., 2020).

9.7 LBG nanoparticles

In the oral delivery of biopharmaceuticals, numerous studies report the active part of polymeric nanoparticles in augmenting oral drug bioavailability by enabling cell internalization (Csaba et al., 2006; Kadiyala et al., 2010). Nanotechnology deals with preparing and applying constituents at different matter levels, which is essential to offer novel materials with a nano size range (Jeevanandam et al., 2018). Nanoencapsulation of drugs exemplifies the production of drug-loaded particles, with a diameter of 1–100 nm. These particles are called nanoparticles and can be used as nanocarriers. These nanocarriers may have several morphologies, e.g., nanospheres, nanofibers, nanocapsules, etc. (Hagens et al., 2007; Sharma et al., 2016a, b). The use of natural gums for the formation of nanostructures is still in its initial stages, and moderately limited studies have been reported about the use of natural gums in nanomedicine development (Deshmukh and Aminabhavi, 2021). In the biomedical field, nanoparticles are widely used to decrease the toxicity and side effects of drugs. Natural polymers or biopolymers are acquired from natural resources like algae, bacteria, green plants, and animals, as they are polymers or polymer matrix composites. Because of several improved individual characteristics like relatively safety, biocompatibility, biodegradability, richness in nature, and low cost in comparison to bioresorbable synthetic polymers, studies that have been initiated to produce nanoparticles from natural resources. The main aim is to make nanoparticles for advanced pharmaceutical applications against different diseases and uplift their drug delivery application. In drug delivery systems, widely used natural polymers comprise collagen, alginate, gelatine, albumin, and chitosan (Bhatia, 2016). The free carboxyl and hydroxyl groups in polysaccharide backbone are mainly used for the chemical functionalization that will produce derivatives with functions that can be molded for desired applications (Mizrahy and Peer, 2012). Polymeric nanoparticles are efficient drug delivery agents that successfully meet the challenge of drugs and pharmaceutical delivery. One of the natural polysaccharides, LBG, is used for nanoparticle synthesis, which may increase nanoparticle properties regarding its oral immunization application facilitated by the nanoparticles. The great potentiality of LBG is due to its chemical structure, being composed of mannose and galactose residues.

Braz et al. (2017) reported the synthesis of LBG and chitosan-based nanoparticles for immunoadjuvant therapy for oral vaccination. By the polyelectrolyte complexation method, the nanoparticles were synthesized and were offered as adjuvants in oral immunization, involving two model antigens (HE antigenic complex and OVA) without conceding their structural integrity (Braz et al., 2017). The model antigen linking efficacies of the nanoparticles for ovalbumin (26%) and a cellular antigen extract of *Salmonella enterica* serovar Enteritidis antigen (32%). The antigenicity did not influence the attachment process. At the end of 24 h, the nanoparticles loaded with 8% salmonella antigen 21.6% released, and in simulated gastric fluid and simulated intestinal fluid, it is 13.2% antigen, respectively. The ovalbumin release has been reported in different behavior. In this case, after 2 h, the release in acidic medium became doubled, though at the end of 4 h the release was 3.0% in the simulated intestinal fluid. This revealed a strong association of antigens with the nanoparticles until they contacted the Peyer's patches' M cells, which facilitated their release. There was a balanced Th1/Th2 immune response when 100 μg ovalbumin-loaded nanoparticles were orally immunized to BALB/c mice. These outcomes were favorable for mucosal antigen delivery.

Aceclofenac-loaded IPN nanocomposites have been synthesized using chitosan (CS) and LBG with glutaraldehyde as a cross-linker. A size of 318 nm with drug entrapment efficiency 78.92% has been obtained for 1:5 mass ratio LBG:CS. However, up to 8 h CS:LBG showed the slowest drug-release profiles in phosphate buffer solution (pH 6.8). In an acidic medium or low pH, the composite systems efficiently inhibited the drug's burst release (Jana and Sen, 2017). The drug-release data was validated well with the swelling properties of the nanocomposites. Moretton and coworkers synthesized hydrolyzed LBG/chitosan nanoparticles of size 263–340 nm by the technique of ionotropic gelation. Compared with uncoated nanoparticles, an improvement in drug encapsulation efficiency of 12.9-fold has been reported for hydrolyzed LBG due to its coating of poly(ε-caprolactone)-b-poly (ethylene-glycol)-b-poly(epsilon-caprolactone) flower-like polymeric micelles. The uptake of the coated systems by RAW 264.7 murine macrophages is due to the successive assembly of rifampicin in the cells (Moretton et al., 2013).

Braz et al. (2017), through controlled polyelectrolyte complexation, reported the preparation of LBG-based nanoparticles between positively charged LBG derivative or chitosan (CS) and negatively charged LBG carboxylates or sulfates. For 1:1 CS/LBG sulfate mass ratio, the minimum size obtained was 364 nm. A mean size of 479 nm was obtained for CS/LBG carboxylate (1:1) nanoparticles, i.e., an increase of 30% compared to the CS/LBG sulfate formulation. Nanoparticles of $\approx$206 nm with spherical morphology have been obtained for amine and sulfate derivatives (1:2). When tested on Caco2 cells, the viability was above 70% (concentration range of 0.1–1.0 mg/mL), except for the amine derivative. Even after 3 h exposure, cell viability of 30% has been shown for ammonium derivative was for all the tested concentrations. This indicated the cytotoxic nature of ammonium derivative of LBG. A minor effect of ammonium derivative was reported in Caco-2 cell viability in the form of nanoparticles.

For producing nanoparticles, Zaritski et al. (2019) reported about LBG chemical functionalization. Firstly, under acidic conditions, LBG was hydrolyzed, and along with ceric ammonium nitrate (initiator) and tetramethylethylenediamine (activator), via free radical polymerization, it was grafted with poly(methyl methacrylate) (PMMA). The graft copolymer produced nanoparticles of size 141 nm with a polydispersity index of 0.1. A hydrophobic tyrosine kinase inhibitor, Imatinib, was incorporated into the nanoparticles, and the nanoparticles revealed low drug encapsulation efficiency (7.5%). The nanoparticles' ability to dynamically target glucose transporter-1 was evaluated by fluorescence confocal microscopy and imaging flow cytometry. Solid pediatric tumors rhabdomyosarcoma (RMS) and Ewing sarcoma (ES) overexpress glucose transporter-1. The results showed the internalization of nanoparticles by the tumor cells and suggested a striking correlation between the nanoparticles' intratumoral accumulation and the glucose transporter-1 gene expression level in each patient-derived tumor cell line (Zaritski et al., 2019).

In our study, we incorporate allicin to LBG nanoparticles (LBGAN) to augment the therapeutic effect and stability. From the garlic intake through food, allicin is not synthesized in the body, as allinase is irreversibly inactivated in acidic pH. In addition, allicin breaks quickly upon heating. For the production of LBG nanoparticles, the nanoprecipitation technique was used. For that, with the enzyme α-galactosidase in a citrate phosphate buffer (pH 5.2), LBG was depolymerized at 30 °C for 24 h . The particle size for LBG was reported below 100 nm, and upon incorporation of allicin, the size of LBGAN increased to 100 nm. Fig. 9.2 represents the distribution of nanoparticles of different sizes by SEM and TEM analysis (Fig. 9.2A and B). From the MTT (3-(4,5-dimethylthiazol-2-yl)-2,5-diphenyltetrazolium bromide) assay, the nanoparticles employed maximum toxicity (10.51%), and the nanoparticles efficiently safeguarded the cells from cell death. In the Sprague-Dawley rat model, it was evident that there is no significant alteration in the marker enzyme activities. Thus, allicin incorporation into the LBG nanoparticles contributed stability to allicin and increased its pharmacological activity. The LBGAN treated group (1 mg/kg bodyweight) showed normal histopathology of the kidney, liver tissues, aorta, and heart, and LBGAN nanoparticles were found to be safe compared to rats treated with control and allicin (Soumya et al., 2018).

Tagad et al. (2014) emphasized that for the synthesis of gold nanoparticles, LBG acts as a reducing agent. Undeniably, the hydroxyl (OH) groups in the LBG polysaccharide and the

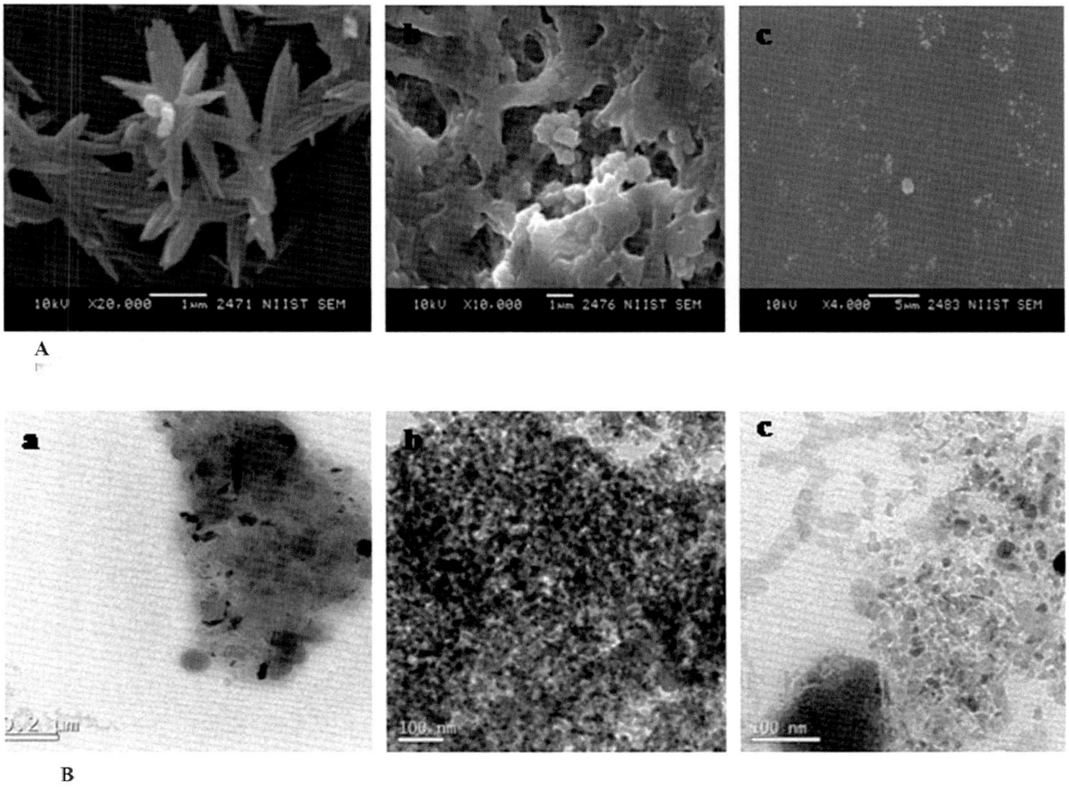

FIG. 9.2 (A) SEM images of (a) allicin (b) LBG nanoparticles, and (c) LBGAN. (B) TEM images of (a) allicin (b) LBG nanoparticles, and (c) LBGAN.

hemiacetal reducing ends performed as reaction cores and assisted the reduction of Au^{3+} to Au^{0}. At 537 nm, the surface plasmon peak indicated the formation of gold nanoparticles (Tagad et al., 2014). A surfactant-assisted homogenization/reticulation technique has been proposed by Maiti et al. for the development of carboxymethyl LBG nanoparticles. Briefly, particles at the nanoscale level were produced (43.82–197.70 nm) from an aqueous gum solution with nonionic surfactant (Span 80), and ionic cross-linking of lamivudine (Al^{3+} ions). A maximum of ~44% drug entrapment efficiency was prominent irrespective of the cross-linker strength, drug:polymer ratio, and no validation between the chemical interaction of polymer and drug. Additionally, in the nano reticulations the drug presumed amorphous dispersion. For a period of >8.5 h, the nanoparticles released the drug gradually in a medium (acidic) and controlled the simulated intestinal fluid release. Due to the controlled drug release property, the particles could avoid the drug dose-related systemic toxicity. The drug targeting ability of the nano reticulations is yet to be investigated (Maiti et al., 2014).

9.8 Conclusions

The emerging field in drug delivery systems is based on polysaccharide-based systems. The main objective for searching for a novel drug delivery system is to overcome the shortcomings of preparing cost, availability, toxicity, and compatibility. LBG is being used in several distinct functions in biopharmaceutical applications, varying from controlled release excipient to tablet disintegrant. Considerable research is required to unveil the real potential that polysaccharides might possess. This chapter focusses on the possibilities of LBG as potential pharmaceutical excipients in micro and nano form. Hence, shortly new LBG-based delivery systems will be developed, which may enrich the usefulness of LBG-based formulations in terms of their category and function.

References

Ahsan, F., Rivas, I.P., Khan, M.A., Torres Suarez, A.I., 2002. Targeting to macrophages: role of physicochemical properties of particulate carriers—liposomes and microspheres—on the phagocytosis by macrophages. J. Control. Release 79, 29–40. https://doi.org/10.1016/s0168-3659(01)00549-1.

Alkhayat, A.H., Kraemer, S.A., Leipprandt, J.R., Macek, M., Kleijer, W.J., Friderici, K.H., 1998. Human beta-mannosidase cDNA characterization and first identification of a mutation associated with human beta-mannosidosis. Hum. Mol. Genet. 7, 75–83. https://doi.org/10.1093/hmg/7.1.75.

Alonso-Sande, M., Teijeiro-Osorio, D., Remuñán-López, C., Alonso, M.J., 2009. Glucomannan, a promising polysaccharide for biopharmaceutical purposes. Eur. J. Pharm. Biopharm. 72, 453–462. https://doi.org/10.1016/j.ejpb.2008.02.005.

Alves, A.D., Cavaco, J.S., Guerreiro, F., Lourenço, J.P., Rosa da Costa, A.M., Grenha, A., 2016. Inhalable antitubercular therapy mediated by locust bean gum microparticles. Molecules 2, 702. https://doi.org/10.3390/molecules21060702.

Aravamudhan, A., Ramos, D.M., Nada, A.A., Kumbar, S.G., 2014. Natural polymers: polysaccharides and their derivatives for biomedical applications. In: Kumbar, S.G., Laurencin, C.T., Deng, M. (Eds.), Natural and Synthetic Biomedical Polymers. Elsevier, Oxford, pp. 67–89, https://doi.org/10.1016/B978-0-12-396983-5.00004-1 (Chapter 4).

Beneke, C.E., Viljoen, A.M., Hamman, J.H., 2009. Polymeric plant-derived excipients in drug delivery. Molecules 14, 2602–2620. https://doi.org/10.3390/molecules14072602.

Bhatia, S., 2016. Natural Polymer Drug Delivery Systems: Nanoparticles, Plants, and Algae. Springer Nature, Switzerland, https://doi.org/10.1007/978-3-319-41129-3.

Braz, L., Grenha, A., Ferreira, D., Rosa da Costa, A.M., Gamazo, C., Sarmento, B., 2017. Chitosan/sulfated locust bean gum nanoparticles: in vitro and in vivo evaluation towards an application in oral immunization. Int. J. Biol. Macromol. 96, 786–797. https://doi.org/10.1016/j.ijbiomac.2016.12.076.

Brennan, C.S., 2005. Dietary fibre, glycaemic response, and diabetes. Mol. Nutr. Food Res. 49, 560–570. https://doi.org/10.1002/mnfr.200500025.

Chavali, M., Parepalli, Y., Chellappa, K., 2013. Polysaccharide nanoparticles based drug delivery—an overview. Int. J. Nanosci. Technol. 2, 1–7.

Choudhary, P.D., Pawar, H.A., 2014. Recently investigated natural gums and mucilages as pharmaceutical excipients: an overview. J. Pharm. (Cairo) 2014, 204849. https://doi.org/10.1155/2014/204849.

Csaba, N., Garcia-Fuentes, M., Alonso, M.J., 2006. The performance of nanocarriers for transmucosal drug delivery. Expert Opin. Drug Deliv. 3, 463–478. https://doi.org/10.1517/17425247.3.4.463.

d'Ayala, G.G., Malinconico, M., Laurienzo, P., 2008. Marine derived polysaccharides for biomedical applications: chemical modification approaches. Molecules 13, 2069–2106. https://doi.org/10.3390/molecules13092069.

Daas, P.J., Schols, H.A., de Jongh, H.H., 2000. On the galactosyl distribution of commercial galactomannans. Carbohydr. Res. 329, 609–619. https://doi.org/10.1016/s0008-6215(00)00209-3.

Dakia, P.A., Wathelet, B., Paquot, M., 2007. Isolation and chemical evaluation of carob (*Ceratonia siliqua* L.) seed germ. Food Chem. 102, 1368–1374.

Dakia, P.A., Blecker, C., Robert, C., Wathelet, B., Paquot, M., 2008. Composition and physicochemical properties of locust bean gum extracted from whole seeds by acid or water dehulling pre-treatment. Food Hydrocoll. 22, 807–818.

Deep, A., Rani, N., Kumar, A., Nandal, R., Sharma, P.C.S., Sharma, A.K., 2019. Prospective of natural gum nanoparticulate against cardiovascular disorders. Curr. Chem. Biol. 13, 197–211. https://www.eurekaselect.com/171088/article. (Accessed 1 April 21).

Deshmukh, A.S., Aminabhavi, T.M., 2021. Pharmaceutical applications of various natural gums natural gums. In: Ramawat, K.G., Mérillon, J.-M. (Eds.), Polysaccharides: Bioactivity and Biotechnology. Springer International Publishing, Cham, pp. 1–30, https://doi.org/10.1007/978-3-319-03751-6_4-1.

Dionísio, M., Grenha, A., 2012. Locust bean gum: exploring its potential for biopharmaceutical applications. J. Pharm. Bioallied Sci. 4, 175–185. https://doi.org/10.4103/0975-7406.99013.

East, L., Isacke, C.M., 2002. The mannose receptor family. Biochim. Biophys. Acta Gen. Subj. 1572, 364–386. https://doi.org/10.1016/s0304-4165(02)00319-7.

Eggleston, G., Finley, J.W., de Man, J.M., 2018. Carbohydrates. In: Principles of Food Chemistry. Food Science Text Series. Springer, Cham, https://doi.org/10.1007/978-3-319-63607-8_4.

El-Sherbiny, I.M., El-Baz, N.M., Yacoub, M.H., 2015. Inhaled nano- and microparticles for drug delivery. Glob Cardiol. Sci. Pract. 2015, 2. https://doi.org/10.5339/gcsp.2015.2.

Evans, A.J., Hood, R.L., Oakenfull, D.G., Sidhu, G.S., 1992. Relationship between structure and function of dietary fibre: a comparative study of the effects of three galactomannans on cholesterol metabolism in the rat. Br. J. Nutr. 68, 217–229. https://doi.org/10.1079/bjn19920079.

Gopinath, V., Saravanan, S., Al-Maleki, A.R., Ramesh, M., Vadivelu, J., 2018. A review of natural polysaccharides for drug delivery applications: special focus on cellulose, starch and glycogen. Biomed. Pharmacother. 107, 96–108. https://doi.org/10.1016/j.biopha.2018.07.136.

Goswami, S., Naik, D.S., 2014. Natural gums and its pharmaceutical application. J. Sci. Innov. Res. 3, 112–121.

Grenha, A., Alves, A.D., Guerreiro, F., Pinho, J., Simões, S., Almeida, A.J., Gaspar, M.M., 2020. Inhalable locust bean gum microparticles co-associating isoniazid and rifabutin: therapeutic assessment in a murine model of tuberculosis infection. Eur. J. Pharm. Biopharm. 147, 38–44. https://doi.org/10.1016/j.ejpb.2019.11.009.

Hagens, W.I., Oomen, A.G., de Jong, W.H., Cassee, F.R., Sips, A.J.A.M., 2007. What do we (need to) know about the kinetic properties of nanoparticles in the body? Regul. Toxicol. Pharmacol. 49, 217–229. https://doi.org/10.1016/j.yrtph.2007.07.006.

Hirsch, S., Binder, V., Schehlmann, V., Kolter, K., Bauer, K.H., 1999. Lauroyldextran and crosslinked galactomannan as coating materials for site-specific drug delivery to the colon. Eur. J. Pharm. Biopharm. 47, 61–71. https://doi.org/10.1016/s0939-6411(98)00089-7.

Hirst, E.L., Jones, J.K.N., 1948. The galactomannan of carob-seed gum (gum gatto). J. Chem. Soc. 10, 1278–1282. https://doi.org/10.1039/jr9480001278.

Hoffman, A.S., 2012. Hydrogels for biomedical applications. Adv. Drug Deliv. Rev. 64, 18–23. https://doi.org/10.1016/j.addr.2012.09.010.

Jain, A., Gupta, Y., Jain, S.K., 2007. Perspectives of biodegradable natural polysaccharides for site-specific drug delivery to the colon. J. Pharm. Pharm. Sci. 10, 86–128.

Jana, S., Sen, K.K., 2017. Chitosan-locust bean gum interpenetrating polymeric network nanocomposites for delivery of aceclofenac. Int. J. Biol. Macromol. 102, 878–884. https://doi.org/10.1016/j.ijbiomac.2017.04.097.

Jania, G.K., Shahb, D.P., Prajapatia, V.D., Jainb, V.C., 2009. Gums and mucilages: versatile excipients for pharmaceutical formulations. Asian J. Pharm. Sci. 4, 309–323.

Jeevanandam, J., Barhoum, A., Chan, Y.S., Dufresne, A., Danquah, M.K., 2018. Review on nanoparticles and nanostructured materials: history, sources, toxicity and regulations. Beilstein J. Nanotechnol. 9, 1050–1074. https://doi.org/10.3762/bjnano.9.98.

Kadiyala, I., Loo, Y., Roy, K., Rice, J., Leong, K.W., 2010. Transport of chitosan-DNA nanoparticles in human intestinal M-cell model versus normal intestinal enterocytes. Eur. J. Pharm. Sci. 39, 103–109. https://doi.org/10.1016/j.ejps.2009.11.002.

Kurakake, M., Komaki, T., 2001. Production of beta-mannanase and beta-mannosidase from aspergillus awamori K4 and their properties. Curr. Microbiol. 42, 377–380. https://doi.org/10.1007/s002840010233.

Liu, Z., Jiao, Y., Wang, Y., Zhou, C., Zhang, Z., 2008. Polysaccharides-based nanoparticles as drug delivery systems. Adv. Drug Deliv. Rev. 60, 1650–1662. https://doi.org/10.1016/j.addr.2008.09.001.

Maiti, S., Mondol, R., Sa, B., 2014. Nanoreticulations of etherified locust bean polysaccharide for controlled oral delivery of lamivudine. Int. J. Biol. Macromol. 65, 193–199. https://doi.org/10.1016/j.ijbiomac.2014.01.036

Malafaya, P.B., Silva, G.A., Reis, R.L., 2007. Natural-origin polymers as carriers and scaffolds for biomolecules and cell delivery in tissue engineering applications. Adv. Drug Deliv. Rev. 59, 207–233. https://doi.org/10.1016/j.addr.2007.03.012.

Miao, T., Wang, J., Zeng, Y., Liu, G., Chen, X., 2018. Polysaccharide-based controlled release systems for therapeutics delivery and tissue engineering: from bench to bedside. Adv. Sci. 5. https://doi.org/10.1002/advs.201700513, 1700513.

Mizrahy, S., Peer, D., 2012. Polysaccharides as building blocks for nanotherapeutics. Chem. Soc. Rev. 41, 2623–2640. https://doi.org/10.1039/c1cs15239d.

Moretton, M.A., Chiappetta, D.A., Andrade, F., das Neves, J., Ferreira, D., Sarmento, B., Sosnik, A., 2013. Hydrolyzed galactomannan-modified nanoparticles and flower-like polymeric micelles for the active targeting of rifampicin to macrophages. J. Biomed. Nanotechnol. 9, 1076–1087. https://doi.org/10.1166/jbn.2013.1600.

Mortensen, A., Aguilar, F., Crebelli, R., Domenico, A.D., Frutos, M.J., Galtier, P., Gott, D., Gundert-Remy, U., Lambré, C., Leblanc, J.-C., Lindtner, O., Moldeus, P., Mosesso, P., Oskarsson, A., Parent-Massin, D., Stankovic, I., Waalkens-Berendsen, I., Woutersen, R.A., Wright, M., Younes, M., Brimer, L., Peters, P., Wiesner, J., Christodoulidou, A., Lodi, F., Tard, A., Dusemund, B., 2017. Re-evaluation of locust bean gum (E 410) as a food additive. EFSA J. 15. https://doi.org/10.2903/j.efsa.2017.4646, e04646.

Nakajima, N., Matsuura, Y., 1997. Purification and characterization of konjac glucomannan degrading enzyme from anaerobic human intestinal bacterium, *Clostridium butyricum-Clostridium beijerinckii* group. Biosci. Biotechnol. Biochem. 61, 1739–1742. https://doi.org/10.1271/bbb.61.1739.

Parvathy, K.S., Susheelamma, N.S., Tharanathan, R.N., Gaonkar, A.K., 2005. Simple non-aqueous method for carboxymethylation of galactomannans. Carbohydr. Polym. 62, 137–141.

Pegg, A.M., 2012. 8—The application of natural hydrocolloids to foods and beverages. In: Baines, D., Seal, R. (Eds.), Natural Food Additives, Ingredients and Flavourings, Woodhead Publishing Series in Food Science, Technology and Nutrition. Woodhead Publishing, pp. 175–196, https://doi.org/10.1533/9780857095725.1.175.

Pollard, M., Kelly, R., Fischer, P., Windhab, E., Eder, B., Amadò, R., 2008. Investigation of molecular weight distribution of LBG galactomannan for flours prepared from individual seeds, mixtures, and commercial samples. Food Hydrocoll. 22, 1596–1606. https://doi.org/10.1016/j.foodhyd.2007.11.004.

Popa, V., 2011. Polysaccharides in Medicinal and Pharmaceutical Applications. Smithers Information Limited, United Kingdom, ISBN: 9781847354389.

Prajapati, V.D., Jani, G.K., Moradiya, N.G., Randeria, N.P., Nagar, B.J., 2013. Locust bean gum: a versatile biopolymer. Carbohydr. Polym. 94, 814–821. https://doi.org/10.1016/j.carbpol.2013.01.086.

Raghavan, C.V., Muthulingam, C., Jenita, J.A., Ravi, T.K., 2002. An in vitro and in vivo investigation into the suitability of bacterially triggered delivery system for colon targeting. Chem. Pharm. Bull. (Tokyo) 50, 892–895. https://doi.org/10.1248/cpb.50.892.

Richardson, P.H., Willmer, J., Foster, T.J., 1998. Dilute solution properties of guar and locust bean gum in sucrose solutions. Food Hydrocoll. 12, 339–348. https://doi.org/10.1016/s0268-005x(98)00025-3.

Rodrigues, S., Alves, A.D., Cavaco, J.S., Pontes, J.F., Guerreiro, F., Rosa da Costa, A.M., Buttini, F., Grenha, A., 2017. Dual antibiotherapy of tuberculosis mediated by inhalable locust bean gum microparticles. Int. J. Pharm. 529, 433–441. https://doi.org/10.1016/j.ijpharm.2017.06.088.

Saha, T., Masum, Z.U., Mondal, S.K., Hossain, M.S., Jobaer, M.A., Shahin, R.I., Fahad, T., 2018. Application of natural polymers as pharmaceutical excipients. Glob. J. Life Sci. Biol. Res. 4, 1–8. https://doi.org/10.35248/gjlsbr.2018.4.1.

Samil Kök, M., 2007. A comparative study on the compositions of crude and refined locust bean gum: in relation to rheological properties. Carbohydr. Polym. 70, 68–76. https://doi.org/10.1016/j.carbpol.2007.03.003.

Sharma, N., Deshpande, R.D., Sharma, D., Sharma, R.K., 2016a. Development of locust bean gum and xanthan gum based biodegradable microparticles of celecoxib using a central composite design and its evaluation. Ind. Crop. Prod. 82, 161–170. https://doi.org/10.1016/j.indcrop.2015.11.046.

Sharma, A.K., Kumar, A., Taneja, G., Nagaich, U., Deep, A., Rajput, S.K., 2016b. Synthesis and preliminary therapeutic evaluation of copper nanoparticles against diabetes mellitus and -induced micro- (renal) and macro-vascular (vascular endothelial and cardiovascular) abnormalities in rats. RSC Adv. 6, 36870–36880. https://doi.org/10.1039/C6RA03890E.

Smith, F., 1948. The constitution of carob gum. J. Am. Chem. Soc. 70, 3249–3253. https://doi.org/10.1021/ja01190a013.

Smith, B.M., Bean, S.R., Schober, T.J., Tilley, M., Herald, T.J., Aramouni, F., 2010. Composition and molecular weight distribution of carob germ protein fractions. J. Agric. Food Chem. 58, 7794–7800. https://doi.org/10.1021/jf101523p.

Snehalatha, M., Kolachina, V., Saha, R.N., Babbar, A.K., Sharma, N., Sharma, R.K., 2013. Enhanced tumor uptake, biodistribution and pharmacokinetics of etoposide loaded nanoparticles in Dalton's lymphoma tumor bearing mice. J. Pharm. Bioallied Sci. 5, 290–297. https://doi.org/10.4103/0975-7406.120081.

Song, R., Murphy, M., Li, C., Ting, K., Soo, C., Zheng, Z., 2018. Current development of biodegradable polymeric materials for biomedical applications. Drug Des. Dev. Ther. 12, 3117–3145. https://doi.org/10.2147/DDDT.S165440.

Soni, H., Rawat, H.K., Pletschke, B.I., Kango, N., 2016. Purification and characterization of β-mannanase from aspergillus terreus and its applicability in depolymerization of mannans and saccharification of lignocellulosic biomass. 3 Biotech 6, 136. https://doi.org/10.1007/s13205-016-0454-2.

Soumya, R.S., Sherin, S., Raghu, K.G., Abraham, A., 2018. Allicin functionalized locust bean gum nanoparticles for improved therapeutic efficacy: an in silico, in vitro and in vivo approach. Int. J. Biol. Macromol. 109, 740–747. https://doi.org/10.1016/j.ijbiomac.2017.11.065.

Suzuki, S., Lim, J., 1994. Microencapsulation with carrageenan-locust bean gum mixture in a multiphase emulsification technique for sustained drug release. J. Microencapsul. 11, 197–203. https://doi.org/10.3109/02652049409040451.

Tagad, C.K., Rajdeo, K.S., Kulkarni, A., More, P., Aiyer, R.C., Sabharwal, S., 2014. Green synthesis of polysaccharide stabilized gold nanoparticles: chemo catalytic and room temperature operable vapor sensing application. RSC Adv. 4, 24014–24019. https://doi.org/10.1039/C4RA02972K.

Tsai, A.C., Peng, B., 1981. Effects of locust bean gum on glucose tolerance, sugar digestion, and gastric motility in rats. J. Nutr. 111, 2152–2156. https://doi.org/10.1093/jn/111.12.2152.

Verma, A., Tiwari, A., Panda, P.K., Saraf, S., Jain, A., Jain, S.K., 2019. 8—Locust bean gum in drug delivery application. In: Hasnain, M.S., Nayak, A.K. (Eds.), Natural Polysaccharides in Drug Delivery and Biomedical Applications. Academic Press, pp. 203–222, https://doi.org/10.1016/B978-0-12-817055-7.00008-X.

Zaritski, A., Castillo-Ecija, H., Kumarasamy, M., Peled, E., Sverdlov Arzi, R., Carcaboso, Á.M., Sosnik, A., 2019. Selective accumulation of galactomannan amphiphilic nanomaterials in pediatric solid tumor xenografts correlates with GLUT1 gene expression. ACS Appl. Mater. Interfaces 11, 38483–38496. https://doi.org/10.1021/acsami.9b12682.

Zavoral, J.H., Hannan, P., Fields, D.J., Hanson, M.N., Frantz, I.D., Kuba, K., Elmer, P., Jacobs, D.R., 1983. The hypolipidemic effect of locust bean gum food products in familial hypercholesterolemic adults and children. Am. J. Clin. Nutr. 38, 285–294. https://doi.org/10.1093/ajcn/38.2.285.

Zunft, H.J.F., Lüder, W., Harde, A., Haber, B., Graubaum, H.J., Koebnick, C., Grünwald, J., 2003. Carob pulp preparation rich in insoluble fibre lowers total and LDL cholesterol in hypercholesterolemic patients. Eur. J. Nutr. 42, 235–242. https://doi.org/10.1007/s00394-003-0438-y.

CHAPTER

10

Alginate microspheres: Synthesis and their biomedical applications

Nguyen Thi Thanh Uyen[a,b], *Syazana Ahmad Zubir*[a,b], *Tuti Katrina Abdullah*[a,b], *and Nurazreena Ahmad*[a,b]

[a]School of Materials and Mineral Resources Engineering, Engineering Campus, Unversiti Sains Malaysia, Nibong Tebal, Pulau Pinang, Malaysia [b]Biomaterials Research Niche Group, School of Materials and Mineral Resources Engineering, Engineering Campus, Universiti Sains Malaysia, Nibong Tebal, Pulau Pinang, Malaysia

10.1 Introduction

Alginate acid, also called algin or alginate, is an anionic polysaccharide found as a structural component in the cell walls of brown algae (*Phaeophyceae*), including *Laminaria hyperborea*, *Laminaria digitata*, *Laminaria japonica*, *Ascophyllum nodusum*, and *Macrocystis pyrifera*, and as a capsular polysaccharide in some bacteria. Marine alginate is typically extracted by treatment in aqueous alkali solutions, while bacterial alginate can be produced by *Azotobacter* and *Pseudomonas* (Smidsrod and Skjak-Bræk, 1990). Although bacterial synthesis of alginate provides superior chemical and physical properties compared to alginate extracted from seaweed, at present, all commercial alginates are extracted from algal sources only (Dhamecha et al., 2019; Sun and Tan, 2013). Alginates, being one of the most abundantly found polysaccharides, were first described by a British chemist, E.C.C. Standford, in 1881. Since then, extensive research has been done to elucidate the chemical structure of alginic acid (Draget, 2009). The interest in alginate research regarding biomedical applications is attributed to the many outstanding properties of alginate such as its biocompatibility, low toxicity, relatively low cost, derivation from natural sources, chemically compatible material, ease of availability, and feasible synthesis method (Dhamecha et al., 2019; Sun and Tan, 2013).

Micro- and Nanoengineered Gum-Based Biomaterials for Drug Delivery and Biomedical Applications
https://doi.org/10.1016/B978-0-323-90986-0.00004-2

Alginates are built from linear copolymers containing blocks of (1 → 4)-linked β-D-mannuronic acid (M) and α-L-guluronic acid (G). These blocks can create three different forms of polymer segments: consecutive G residues, consecutive M residues, and alternating MG residues (Dhamecha et al., 2019; Sun and Tan, 2013). Depending on the source of alginate, variation on the M and G contents as well as the length of each block can be obtained, and this variation enables modifications and synthesis of derivatives (Paques, 2015; Pawar and Edgar, 2012). These variations allow alginate to be tailored to have various structures, properties, functions, and applications. In biomedical applications, alginates can be found in pharmaceutical applications as drug and protein delivery systems, wound dressings for the treatment of acute or chronic wounds, cell culture, tissue regeneration with protein, and cell delivery for regeneration of various tissues and organs in the body (Lee and Mooney, 2012). However, most importantly the ability of alginate to be cross-linked with divalent ions to form hydrogels for cell or drug encapsulation has been proven to be most advantageous for biomedical applications (Pawar and Edgar, 2012).

This chapter focuses on the synthesis of alginate from algal source and provides a brief description on the extraction method of alginic acid from bacteria. The derivative structures and chemical and physical properties of alginate will also be presented. It should be noted that across the literature, the terms alginate microspheres (AMs), alginate microcapsules (AMCs), alginate microbeads, alginate microgels, and alginate microparticles are used interchangeably to discuss the use of alginate to encapsulate various forms of small and large molecules with diverse chemical properties. Thus, in this chapter, all the above formulations will be referred as alginate microspheres (AMs). For biomedical applications, AMs can be prepared by various methods. However, the methods discussed in this chapter are the most common methods used due to ease of preparation and cost-effectiveness; they include the dripping method, emulsification/gelation method, microfluidic method, and spray-drying method. Advantages, disadvantages, and the various parameters that enable the control of AMs size are also discussed. Finally, this chapter describes the applications of AMs for drugs, cells, proteins, vaccine, bacteria encapsulation, and as hemostatic materials.

10.2 Structure and physicochemical properties of alginates

10.2.1 Sources of alginates

Alginates are classified as natural exopolysaccharides which are produced by seaweed and bacteria belonging to the genera *Pseudomonas* and *Azotobacter* (Rehm and Moradali, 2018). There are mainly two sources for the synthesis of alginates: algae and bacteria. An English chemist, E.-C.C. Stanford, first discovered the alginates in the late 19th century (E.C.S., 1883), and isolated the polysaccharide-based alginate from marine macroalgae. In 1964, alginates from a bacterial source were first identified from the mucoid strain of *P. aeruginosa* when the analysis was done on the *Pseudomonas* bacterium isolated from sputum of cystic fibrosis patients (Linker and Jones, 1966). Commercial alginates are produced mainly from algal sources. Alginates isolated from bacteria are normally used on a small scale for research study since they are not economically practical for commercial production (Skjk-Bræk et al., 1986).

Algal source

The major source of alginates can be found in the cell walls and in the intracellular spaces of brown seaweed. The function of the alginate molecules is to prepare a suitable condition for the plant to grow in the sea by providing the flexibility and strength to the plant. Alginates are extracted on a large scale from various brown marine algae of the *Phaeophyceae* family including the species of *Ascophyllum nodosum*, *L. hyperborea*, and *M. pyrifera*. The other sources for the extraction of alginates are *Ecloniamaxima*, *L. digitata*, *Sargassum*, *Lesonia nigrescens*, and *Laminaria japonica* species (Clementi, 1997; Konda et al., 2015; Szekalska et al., 2016b).

Multiple steps are involved in the extraction of alginates from algal sources. The extraction of alginate from algal materials is schematically illustrated in Fig. 10.1. In the first step, the dried harvested algae materials are treated with mineral acid to remove the counter ions by the proton exchanging reaction. This reaction changes the salts of alginic acid (alginates) in the algae into free alginic acid. Next, the process involves the neutralization of alginic acid using sodium carbonate or sodium hydroxide as alkali agents to form water-soluble sodium alginate. The precipitates are then removed by centrifugation or other separation techniques like floatation and shifting followed by filtration. The soluble alginates are recovered by precipitation with either calcium chloride or mineral acid to obtain calcium alginate fiber or alginic acid gel, respectively. Calcium alginate is also treated with mineral acid to form alginic acid. Finally, the alginic acid is mixed with sodium carbonate to produce sodium alginate.

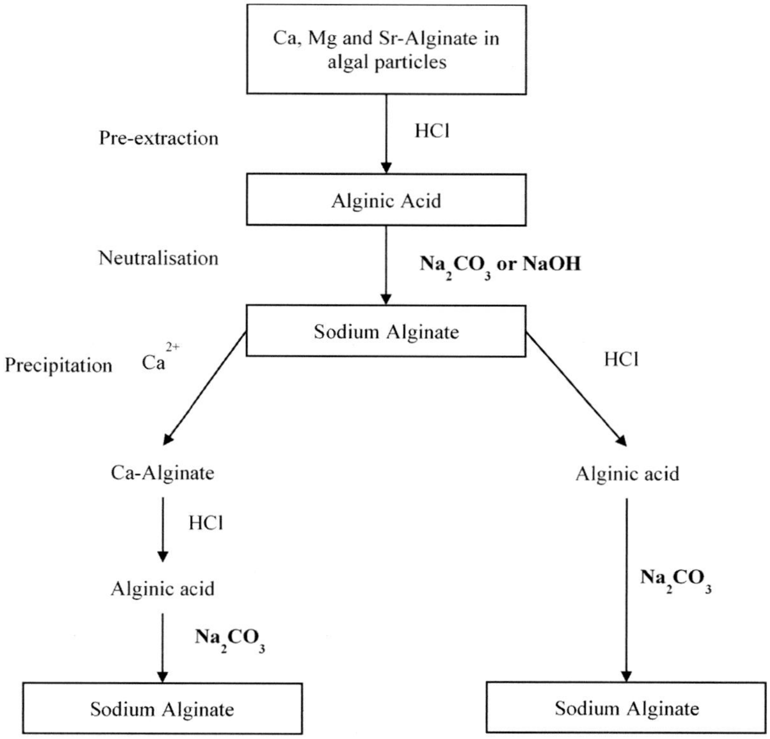

FIG. 10.1 Principle scheme for the isolation of alginate from seaweeds (Draget, 2009).

Bacterial source

Different properties of alginates can be produced from bacterial species belonging to the genera *Pseudomonas* and *Azotobacter* (Sabra and Zeng, 2009). Research into the molecular mechanisms of bacterial alginate biosynthesis has been performed mostly on the opportunistic human pathogen *Pseudomonas aeruginosa* or the soil-dwelling *Azotobacter vinelandii* (Hay et al., 2013). Although these two genera possess similar molecular mechanisms to produce alginate, they secrete alginate with different purposes and with different material properties. Some *P. aeruginosa* strains which are known as mucoid strains are used in the development of thickly natured and highly structured biofilms (Nivens et al., 2001; Remminghorst et al., 2009). *Azotobacter* has ability to produce a stiffer alginate due to the high concentration of guluronate residue (G) in the alginate molecules which remain closely connected with the cells. *A. vinelandii* has been evaluated as a potential source for producing bacterial alginates on a commercial scale. However, at present, algae remains the only source for the large-scale production of alginate.

Synthesis of bacterial alginates

Alginate biosynthesis involves the oxidation of a carbon source to acetyl-CoA, which enters the tricarboxylic acid (TCA) cycle to be converted to fructose-6-phosphate (F6P) via gluconeogenesis (Pawar and Edgar, 2012). Synthesis of bacterial alginates involves four stages: (1) synthesis of precursor; (2) polymerization and cytoplasmic membrane transfer; (3) periplasmic transfer and modification; and (4) export through the outer membrane (Remminghorst et al., 2009).

The synthesis of GDP-mannuronic acid precursor occurs inside the bacterial cell in cytosol. The process requires a series of four enzymatic steps, beginning with the conversion of fructose-6-phosphate (F6P) to mannose-6-phosphate (M6P). The conversion of this central sugar metabolite is catalyzed by bifunctional enzyme phosphomannose isomerase (PMI)/guanosine-diphosphomannose pyrophosphorylase (GMP), designated AlgA (PMI-GMP) (Remminghorst et al., 2009). The M6P is then directly converted to mannose-1-phosphate (M1P) by phosphomannomutase AlgC. The process continues by the conversion to GDP-mannose, which is catalyzed by AlgA. The final stage of the synthesis of the precursor involves the oxidation of GDP-mannose to GDP-mannuronic acid by AlgD (GDP-mannose-dehydrogenase). This is the irreversible step in alginate precursor formation (Urtuvia et al., 2017).

The next step involves the transfer of GDP-mannuronic acid precursor across the cytoplasmic membrane and the polymerization of the monomers to polymannuronate. The polymerization process is carried out by two enzymes: Alg8 (glycosyltransferase/polymerase) and Alg44 (*co*-polymerase) (Remminghorst et al., 2009). Alg8 catalyzes the transfer of sugar molecule from GDP-mannuronic acid to an acceptor molecule which is a growing carbohydrate chain. Alg44 also plays an important role for alginate polymerization, where it acts as part of the periplasmic scaffold. It may provide a bridge between the cytoplasmic membrane protein Alg8 and the export protein AlgE in the outer membrane.

Modification of bacterial alginates occurs almost entirely in the periplasm. This suggests that the alginate is synthesized as polymannuronate and its modification occurs at polymer level. There are three classes of alginate-modifying enzymes: transacetylases, mannuronan C5-epimerases, and lyases. Understanding the reaction patterns of alginate-modifying enzymes might open up the production of tailor-made alginates (Hay et al., 2013).

Transacetylation occurs at the O-2 and/or O-3 position, and only occurs in mannuronic acid residues. Acetylation of alginate indirectly controls epimerization and length of the alginate polymer by preventing the epimerization of guluronic acid residues and degradation of the alginate chain. Guluronic acid residues are introduced into polymannuronate by mannuronan C5-epimerase. Finally, alginate is released from the cell through transmembrane porine (Urtuvia et al., 2017).

10.2.2 Molecular structure of alginates

Alginates are unbranched polysaccharides comprised of two uronic acids residues including α-L-guluronic acid (G block) and β-D-mannuronic acid (M block) (Fig. 10.2) linked by a 1,4-glycosidic bond (Goh et al., 2012; Rehm and Moradali, 2018), organized in homogeneous (poly-G, poly-M) or heterogeneous (MG) block patterns (Fig. 10.3). The proportion and sequence of M and G residues depend on the sources of alginates, thus they influence the physical strength and other physical properties of the alginate structure (BeMiller, 1999; Rhein-Knudsen et al., 2017). Alginates extracted from algae usually show a high content of G blocks while alginates produced by *P. aeruginosa* do not possess G blocks. In bacterial alginates, the presence of acetylated groups has been detected at the O-2 and/or O-3 positions which is not found in algal alginates (Goh et al., 2012).

The physicochemical properties and degree of polymerization depend on the arrangement of the blocks. The frequency of constituting blocks possesses a significant role to the intrinsic viscoelasticity of alginates as flexibility decreases in the order of MG block > MM block > GG block (Rehm and Moradali, 2018). Therefore, alginates with G-rich blocks will generally form hard and brittle gels, while soft and elastic gels can be produced by alginates with M-rich blocks. In the presence of divalent cations, alginates have the ability to form hydrogel and cross-linked polymeric scaffolds by efficiently and selectively binding with the ions. The cations act as cross-linkers between the functional groups of alginate chains. The affinity of alginates toward different divalent ions was found to increase in the order of $Mg^{2+} < Mn^{2+} < Ca^{2+} < Sr^{2+} < Ba^{2+} < Cu^{2+} < Pb^{2+}$ (Cohen and Merzendorfer, 2019).

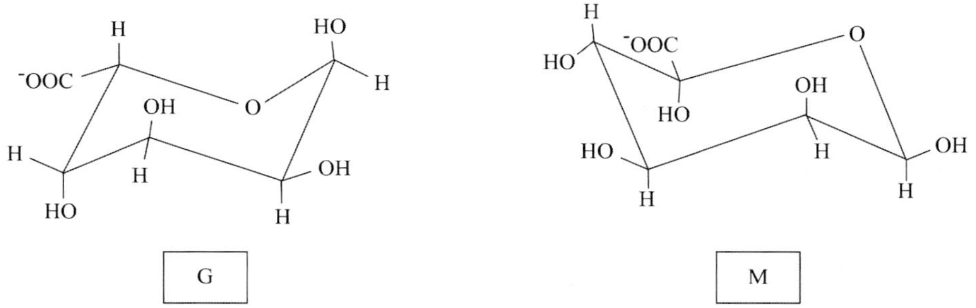

FIG. 10.2 α-L-guluronic acid (G block) and β-D-mannuronic acid (M block).

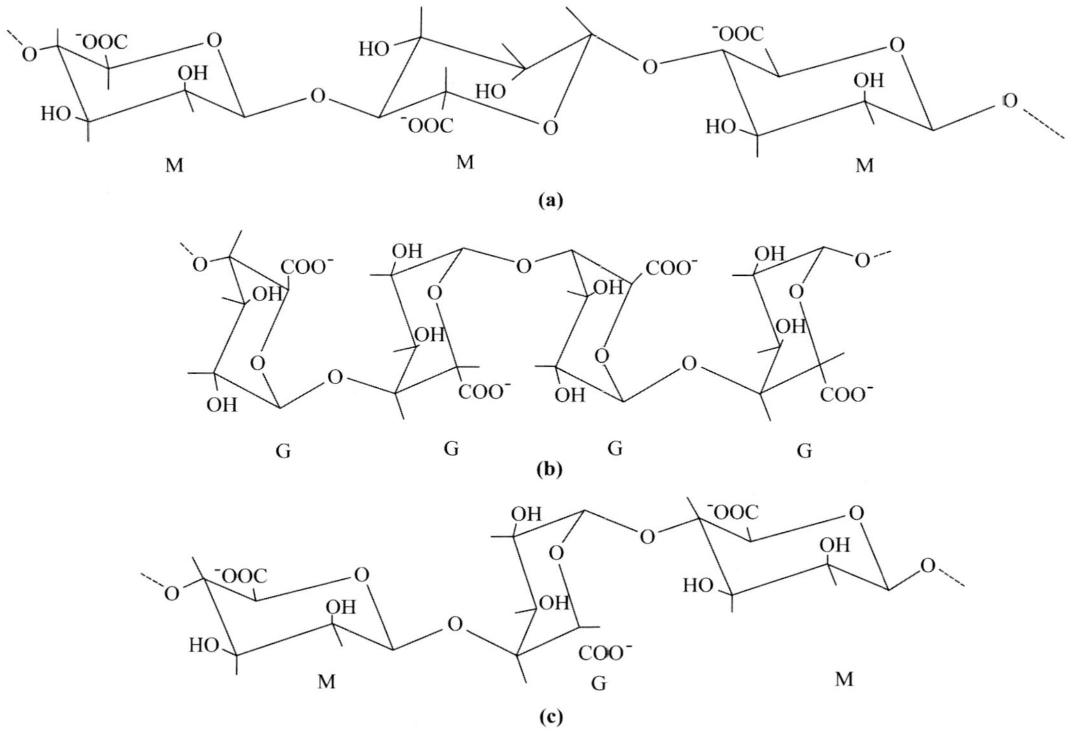

FIG. 10.3 (A) Homopolymeric blocks of MM; (B) homopolymeric blocks of GG; and (C) heteropolymeric blocks of MG.

10.2.3 Properties of alginates

Molecular weight

The viscosity of alginates is highly influenced by their molecular weight, and the rigidity and extension of the polymer chain. Different sources of alginates as well as the extraction factor and conditions will give the variation in the molecular weight (Jiao et al., 2019). A highly viscous gel can be produced by high molecular weight alginates, which is not feasible for industrial applications (Kong et al., 2003).

The two most common methods for averaging molecular weight are the number-average, M_n, and the weight-average, M_w. The number-average and weight-average weigh the polymer molecules according to the number (M_n) and weight (M_w) of molecules in a population having a specific molecular weight (Draget, 2009). The *polydispersity index* (PI), which is used to indicate the distribution of polymer chain molecular weight, can be calculated by M_w/M_n. The molecular weight distribution has a significant influence on the uses of alginates, as the short G-blocks which consist of low molecular weight may not take part in the gel network, and thus do not contribute to the gel strength. Depending on the application, alginates can have a wide range of average molecular weight (50–100,000 residues). Generally, the molecular weight of commercially available sodium alginates ranges between 32,000 and 400,000 g/mol (Rehm and Moradali, 2018).

Ramos et al. (2018) reported that alginate with high G content, even with low molecular weight, is able to form bigger microspheres compared with alginate with low G content. In the study, it was noted that content of G block played a more crucial role in the formation of alginate microspheres than the molecular weight. This is thought to be due to the excess of calcium ions in the high G content alginate that are able to create more interactions between the molecules and thus give stronger and bigger gels. On the other hand, in a study conducted by Jiao et al. (2019) on the use of alginate as an emulsifier, it was found that with decreasing molecular weight, the interfacial tension of the emulsion and the Turbiscan stability index (TSI) increase while the zeta potential of the sodium alginate decreases. From these findings, it was concluded that molecular weight has a significant influence on the application of sodium alginate as an emulsifier.

Solubility

The aqueous solubility of alginates is influenced by the metal cations, solvent pH, and total ionic strength of solution (van den Brink et al., 2009). Potassium, sodium, and ammonium alginates are soluble in cold and hot aqueous media. Due to the presence of carboxylic acid groups along the polymer chain, water-soluble alginate is able to form a gel in the presence of divalent metal cations (Li and Dai, 2006). However, these alginates are not soluble in ethanol, acetone, and ether. The pH of the medium also has a significant influence on the solubility of alginates. In general, most of the alginates show very low solubility in the acidic environment. The solubilities of alginates in various conditions are presented in Table 10.1.

Alginate solubility when encapsulated with a drug is of great interest as the solubility of alginate in various mediums affects the ability of the microspheres to prolong the drug release in the targeted site and to minimize or eliminate burst effect. Soni et al. (2010), in a study to investigate the drug release of theophylline encapsulated alginate microspheres, reported that in acidic condition, alginate microspheres shrink due to the tightening of the gel meshwork, while in basic condition the alginate microspheres eroded. Thus, this condition allows theophylline to be released in a sustained manner by diffusion and slow erosion of the alginate walls.

Biocompatibility

Alginates have received significant attention in pharmaceutical and biomedical applications due to their biocompatibility and nonimmunogenic and nontoxic properties (Sachan

TABLE 10.1 Solubilities of alginates in various conditions.

Alginates	Acidic environment	Alkaline environment	Presence of divalent/trivalent cations
Alginic acid	Insoluble	Soluble	Insoluble
Sodium alginate	Insoluble	Soluble	Insoluble
Potassium alginate	Insoluble	Soluble	Insoluble
Calcium alginate	Insoluble	Insoluble	Insoluble
Ammonium alginate	Soluble	Soluble	Soluble

From Hasnain, M.S., Jameel, E., Mohanta, B., Dhara, A.K., Alkahtani, S., Nayak, A.K., 2020. Alginates in Drug Delivery. A.K. Nayak, M.S. Hasnain (Eds.).

et al., 2009). Although alginate gel is not degradable in the mammalian digestive tract, it has the ability to dissolve as a result of elution of cross-linking calcium ions. Furthermore, only small alginate molecules are excreted by the renal clearance threshold (Clementi, 1997; Konda et al., 2015; Szekalska et al., 2016a).

Many in vitro and in vivo studies have been extensively conducted on the biocompatibility of alginates. However, there is still debate arising concerning the impact of alginate compositions as well as their level of purity. It has been reported that mannuronic acid residues (M blocks) are the active cytokine inducers in alginates (Marit et al., 1991). On the contrary, another study found little or no immunoresponse around alginate implants (Zimmermann et al., 1992). These results suggest an impurity factor which may influence the immunogenic response of the implantation site. Various impurities such as heavy metals, endotoxins, and polyphenols may exist due to the natural sources of the alginate extraction, which may lead to an intense host immune response. This condition could decrease the biocompatibility of the alginates. Importantly, multistep alginate purification did not induce any significant foreign body reaction by reducing the impurity level, thus improving the biocompatibility properties (Torres et al., 2019).

Degradation

Alginates are prone to chain degradation in the presence of reductants such as hydroquinone, sodium sulfite, sodium hydrogen sulfide, cysteine, ascorbic acid, hydrazine, and leucomethylene blue (Pawar and Edgar, 2012). The degradation process occurs not only in the presence of acid and bases but also at neutral pH values. The rate of degradation increases with an increase in phenolics compounds. Alginate degradation was also induced by the sterilization techniques such as heat treatment, autoclaving, ethylene oxide treatment, and γ-irradiation. The mechanisms of alginate degradation involve the formation of peroxide, which leads to the creation of free radicals, which in turn causes the breakdown of alginate chains.

Mucoadhesiveness

The presence of a free carboxyl group in alginate enables the latter to have good mucoadhesive properties. These properties allow prolonged residence time for a drug at the application site. Mucin, which is the main component of mucus, can react with other polymers via their ability to form electrostatic, disulfide, hydrogen bonding, or hydrophobic interactions with the polymers, giving them adhesion properties (Bansil and Turner, 2006). In the case of alginate, the availability of the free carboxyl group in alginate interacts with the mucin by hydrogen and electrostatic bonding, and gives alginate its mucoadhesive properties (Szekalska et al., 2016a). It is worth noting that the polymers' molecular weight, flexibility of the polymer chain, hydrogen bonding capacity, cross-linking density, charge, concentration, and hydration degree of the polymer greatly influence the mucoadhesiveness of the polymer (Shaikh et al., 2011). In the case of alginate, environmental pH also influences its mucoadhesive properties, because only an ionized carboxyl group will interact with the mucosal tissue (Szekalska et al., 2016a; Sosnik, 2014). Alginate mucoadhesive properties have been utilized to encapsulate an antidiabetic drug, glipizide has been employed to increase the bioavailability of the drug in the gastrointestinal tract (Allamneni et al., 2012), alginate has been combined with liposomes as an oral mucoadhesive delivery system for the treatment of oral cancers (Shtenberg et al., 2018), and alginate microspheres have been used for nasal delivery of carvedilol (Patil and Sawant, 2009), and encapsulation of many other drugs for various treatment via buccal (Laffleur, 2014) and ocular routes (Costa et al., 2015).

10.3 Fabrication of alginate microspheres

Various preparation methods have been reported in the literature for alginate microspheres (AMs), including the dripping method, emulsification/gelation method, microfluidic method, and spray-drying method. The characteristics of AMs and particle size control methodology can be monitored via these methods. In addition, the advantages and disadvantages of these methods will be discussed in this section.

10.3.1 Dripping method

The dripping method is the least complex approach to produce AMs. It is also known as the ionic gelation method or extrusion method. The method is based on external gelation, in which the Ca^{2+} ions diffuse from an external source into the alginate solution (Lupo et al., 2014). Briefly, when a drop of sodium alginate solution is added into calcium (Ca^{2+}) ions, gelation quickly happens on its surface, preventing its dissolution. Then Ca^{2+} ions slowly diffuse inside the drop to induce gelation in the entirety of its volume, and the drop turns into a particle. For drug encapsulation, the drug is mixed with sodium alginate solution prior to dropping into calcium solution (Patil and Sawant, 2009). Fig. 10.4 shows a schematic of the dripping method for alginate fabrication (Rajmohan and Bellmer, 2019). This method is very convenient because it does not require complex equipment and skill; a syringe or pipette and stirrer or ultrasonic bath are sufficient (Mokhtare et al., 2017; Priya and Deshmukh, 2017). However, the method is generally limited to the size of beads because of difficulties with producing small alginate droplets. The alginate beads produced by this method have a diameter larger than 200 μm, which depends on the diameter of the nozzle extruder (Pestovsky and Martínez-Antonio, 2020). In addition, dripping method has the drawback regarding large-scale production because beads are formed one by one (Lupo et al., 2014).

For instance, Iswandana et al. (2018) successfully encapsulated tetrandrine into calcium-alginate beads by using the dripping method. In their study, they reported that the obtained beads were almost spherical and had a particle size distribution of 742–780 μm. Similarly,

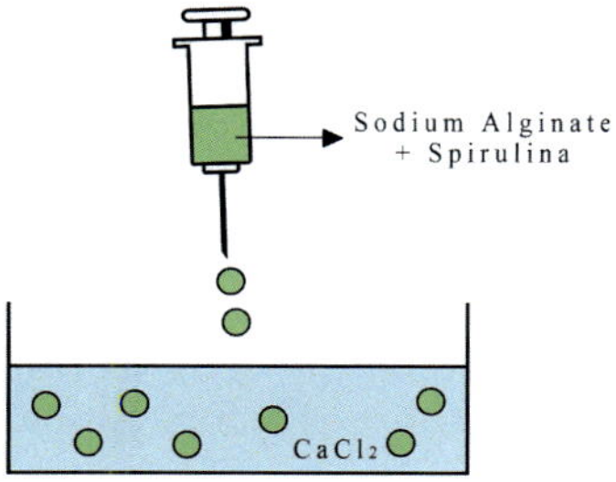

FIG. 10.4 Schematic representation of the formulation of spirulina-loaded AMs by using dripping method (Rajmohan and Bellmer, 2019). *From Rajmohan, D., Bellmer, D., 2019. Characterization of spirulina-alginate beads formed using ionic gelation. Int. J. Food Sci. 2019. http://www.hindawi.com/journals/ijfs/.*

spirulina-encapsulated alginate beads were reported by Rajmohan and Bellmer (2019). They found that the size of alginate beads was in the range of 2.11–4.5 mm by using a syringe with a 26G needle. Lin et al. (2016) reported the encapsulation of astaxanthin in alginate beads using the syringe through a 27G needle. The mean size was in the range of 566–1460 μm. They found that a higher concentration of sodium alginate led to a greater size of alginate beads. This is because the drop size of the alginate solution was expected to increase with increasing the viscosity of alginate solution resulting from higher alginate concentration. In addition, the method produced alginate beads with high loading efficiency ranging from 82% to 100%.

10.3.2 Emulsification/gelation method

The emulsification/gelation method is a useful method to control the particle size (<100 μm) of AMs over the dripping method. However, the AMs have poor dispersity due to the coalescence of droplets during the gelation process. In addition, it is inherently challenging to prepare AMs with narrow size distribution and uniform shape by the emulsification/gelation method. In this method, the preparation of emulsion microspheres may then be carried out by external or internal gelation.

Emulsification/external gelation method

In the emulsification/external gelation method, the emulsion is formed from the dispersed phase of alginate solution in a nonaqueous continuous phase, added later to the medium the Ca^{2+} ions to induce the gelation of droplets and promote separation of the formed microspheres (Lupo et al., 2014; Spadari et al., 2017). Briefly, sodium alginate solution containing a drug that is being encapsulated is emulsified in an oil phase. Surfactant which is added to the oil phase is usually used to stabilize the emulsions. Based on the external gelation, the Ca^{2+} ions are added, hardening the droplets to microspheres. Finally, the AMs are separated by centrifugation, washed, and dried.

The advantages of the method are the characteristics of AMs as well as the particle size being easily controlled from nanometers to millimeters by controlling the viscosity of oil phase, type, and concentration of sodium alginate, surfactant concentration, and emulsification conditions. For instance, the emulsification/external gelation method was used to prepare risedronate sodium-loaded AMs by Gedam et al. (2018). Their study showed that when the alginate concentration was increased, the size of the microspheres also proportionally increased. This could be attributed to an increase in the relative viscosity at a higher concentration of alginate and the formation of larger particles during emulsification. In addition, an increase in the particle size was observed to increase in Ca^{2+} ions concentration as well as cross-linking time. The higher amount of Ca^{2+} will result in relatively more cross-linking of the gluronic acid units of sodium alginate, thereby leading to the formation of larger microspheres. Similarly, increasing the cross-linking time will increase the extent of cross-linking and thereby increase the size of the microspheres.

The drawback of the emulsification/external gelation method is the random disruption of the emulsion droplets which may cause a significant clumping of microspheres before properly hardening (Song et al., 2013).

Emulsification/internal gelation method

In the emulsification/internal gelation method, an emulsion between the aqueous phase and oil phase is formed similarly to the emulsification/external gelation method. However, this method differs from the emulsification/external gelation method mainly in the mechanism of gelation. An insoluble calcium salt is mixed with alginate solution and the mixture is added into an oil phase to form an emulsion. This salt is already present inside the droplets of the emulsion before gelation occurs. The latter is acidified to release Ca^{2+} from the insoluble salt for alginate gelation to occur (Kyzioł et al., 2017; Lupo et al., 2014; Martín et al., 2015; Soni et al., 2010). A scheme of emulsification/internal gelation method is shown in Fig. 10.5.

The method has the same advantages as those for the emulsification/external gelation method. The homogeneity of particles is another advantage over the emulsification/external gelation method. However, the main drawbacks of microspheres obtained by the emulsification/internal gelation method include poor encapsulation efficiency of water-soluble drugs and low production yield (Szekalska et al., 2018). Furthermore, acid concentration must be carefully calculated to avoid overdosing in this method. In addition, this method is a more complex formulation because of the requirement for finely dispersed insoluble calcium salt and the presence of acid (Pestovsky and Martínez-Antonio, 2020).

Song et al. (2013) prepared probiotics/calcium alginate beads to compare the bead characteristics by using the emulsification/external gelation and emulsification/internal gelation methods. Their results showed that beads prepared by emulsification/internal gelation appeared to have much better morphology with spherical shape and smooth surface. Because $CaCO_3$ powder was finely dispersed in an alginate solution of the emulsion, the dissociated Ca^{2+} by adding acetic acid can cause gelation without disrupting the emulsion droplet to

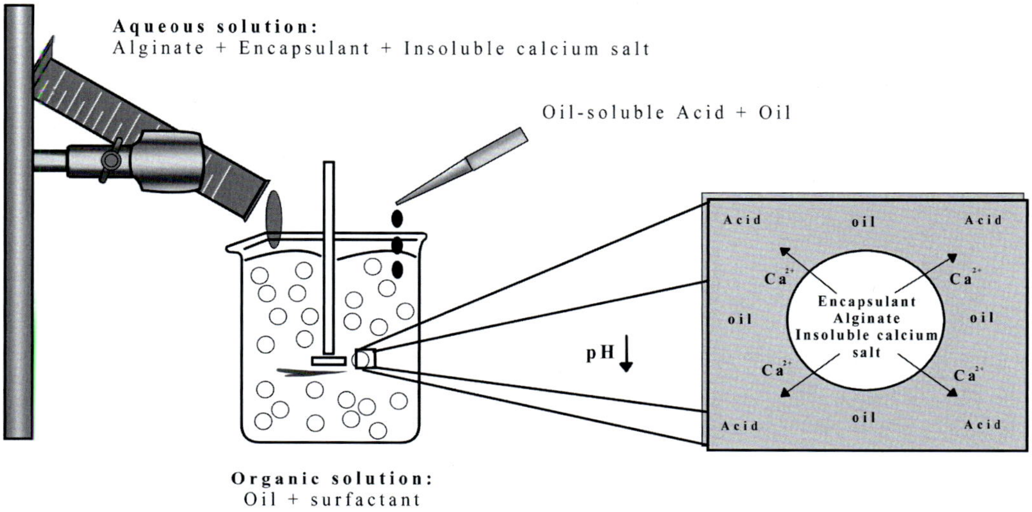

FIG. 10.5 Schematic representation of the fabrication of drug-loaded AMs by using the emulsification/internal gelation method. *From Reis, C.P., Neufeld, R.J., Ribeiro, A.J., Veiga, F., Remminghorst, U., Hay, I.D., Rehm, B.H.A., 2006. Molecular characterization of Alg8, a putative glycosyltransferase, involved in alginate polymerization. Nanomedicine Nanotechnol. Biol. Med. 2 (1), 176–183.*

retain the spherical shape, whereas, for emulsification/external gelation, the random disruption of the emulsion droplets and diffusion of added Ca^{2+} were considered to cause random and nonuniform gelation before properly hardening. In terms of particle size, the beads created by the emulsification/external gelation and emulsification/internal gelation methods had average diameters of 325.4 and 151.1 μm, respectively. The larger mean size was attributed to the collision of two or more droplets of the dispersed alginate and Ca^{2+} ion droplets. Furthermore, the span values of particle size distribution were reported to be 1.59 by using emulsification/external gelation and 1.11 by using emulsification/internal gelation. A narrower size distribution of beads was revealed by using the emulsification/internal gelation method.

10.3.3 Microfluidic method

The microfluidic method has been used for alginate fabrication to overcome the challenge in the agglomeration as well as wide size distribution by the emulsification/gelation method. This method can be used to produce alginate beads with particles smaller than 300 μm. It is a technique used to manipulate fluids in channels of micrometric dimensions. A continuous phase and a dispersed phase are injected separately and then mixed in a junction that connects the channels. When partially miscible phases are mixed in the junction, the droplets can be generated, as shown in Fig. 10.6 (Dhamecha et al., 2019; Sun and Tan, 2013). The principle is similar to that of emulsification, which consists of blending two immiscible phases. The advantage of the microfluidic method permits generation of discrete and uniform microspheres due to precise control over experimental conditions such as channel geometry, flow rates, viscosities of phases, etc. Furthermore, the microspheres can be generated without using a surfactant, which is impossible with conventional emulsification (Zhang et al., 2020).

Chen et al. (2013) prepared AMs with immobilized antibodies using the microfluidic method. Their study showed that the alginate droplet size can be controlled by the flow rates of the two phases or the viscosity ratio of the two phases. The alginate droplet size decreases with increasing flow rate of the oil phase or with decreasing flow rate of the aqueous alginate phase. The alginate gel microspheres showed a uniform size with a mean particle size ranging from 50 to 60 μm.

10.3.4 Spray-drying method

Spray-drying is a method based on atomization of a liquid feed into fine droplets that are sprayed together with hot gas into a drying chamber. Briefly, the alginate solution is fed into the drying chamber through an atomizer and atomization occurs by centrifugal, pressure, or kinetic energy. The small droplets generated are subjected to fast solvent evaporation, leading to the formation of dry microspheres. Then, AMs are separated from the drying gas by a cyclone (Sosnik, 2014). Fig. 10.7 shows a schematic of the spray-drying process.

Spray-drying is widely used as it is rapid, continuous, cost-effective, and has high drug encapsulation efficiency. It is used to produce AMs that have a narrow size distribution and a mean particle size of less than 10 μm. The properties of the final microspheres, i.e., particle size and size distribution, surface characteristics, agglomerations, etc., can be controlled by the type

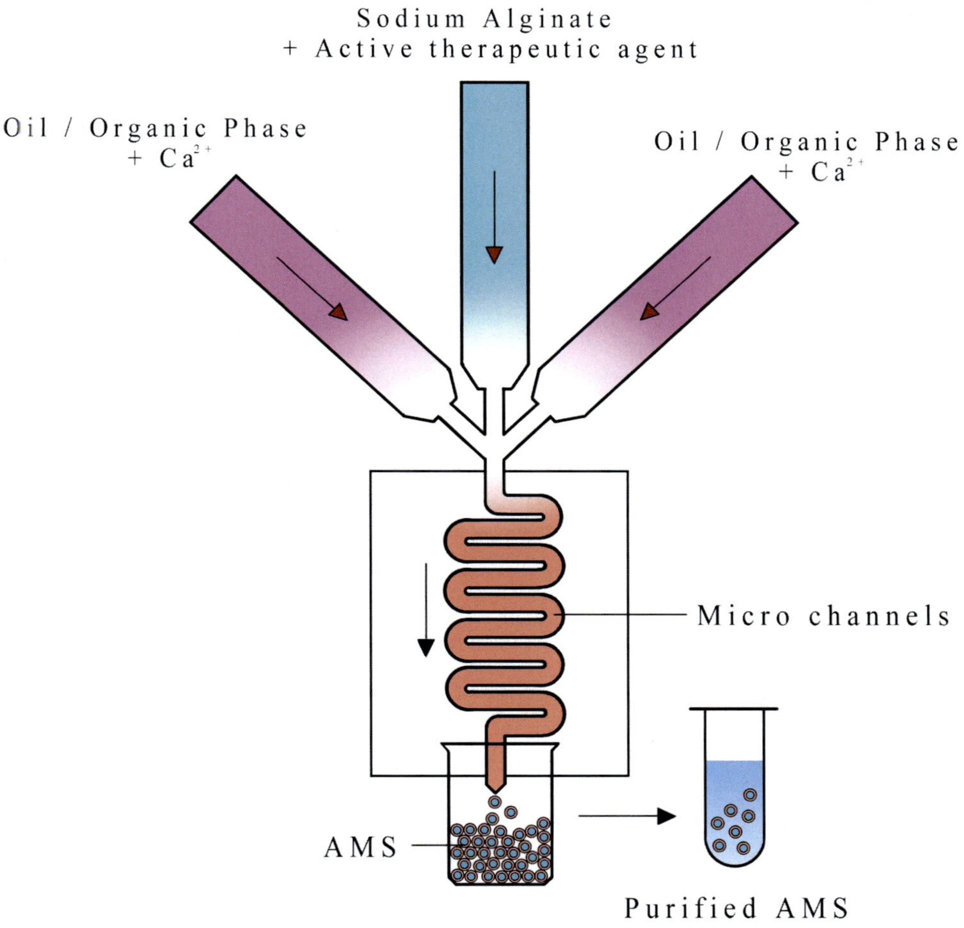

FIG. 10.6 Schematic representation of the formulation of AMs by the microfluidics method. *From Dhamecha, D., Movsas, R., Sano, U., Jyothi, U., Menon, 2019. Applications of alginate microspheres in therapeutics delivery and cell culture: past, present and future. Int. J. Pharm. 569.*

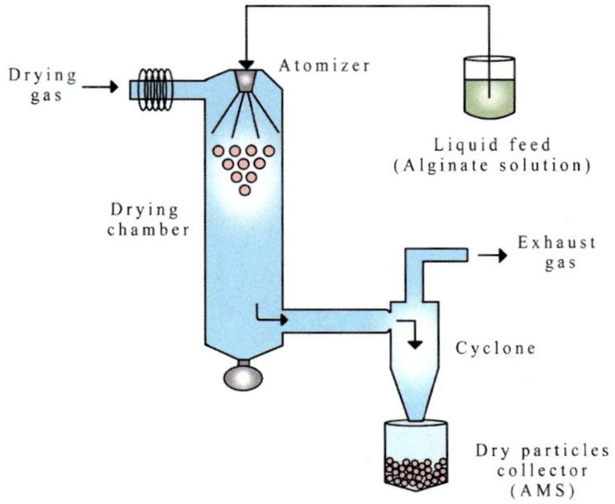

FIG. 10.7 Schematic representation of the formulation of AMs by the spray-drying method (Sosnik and Seremeta, 2015). *From Sosnik, A., Seremeta, K.P., 2015. Advantages and challenges of the spray-drying technology for the production of pure drug particles and drug-loaded polymeric carriers. Adv. Colloid Interface Sci. 223, 40–54.*

of the alginate polymer, drug/polymer ratio, properties of the drug, air flow rate, feed flow rate, and atomizer speed (Hariyadi et al., 2016; Szekalska et al., 2015). In addition, the method can improve process efficiency since it combines drying and matrix encapsulation in a single step (Bowey et al., 2013). However, one of the drawbacks of spray-drying is the low product yield due to the loss of product in the walls of the drying chamber (Sosnik and Seremeta, 2015).

Insulin-loaded AMs were prepared by Bowey et al. (2013) through the spray-drying method. In their study, it was found that the yield of AMs was low at 31% ± 4% due to particle accumulation on the walls of the drying chamber and cyclone, as well as the submicron-sized particles that were lost in the fines and exhaust gas when the stream was passed through the cyclone. It is possible to achieve higher yields when scaled up to industrial dryers, approaching 100% since the fraction lost due to wall adhesion becomes an increasingly smaller component of the overall production volume. The obtained microspheres also showed narrow size distribution with a mean size of 2.1 ± 0.3 μm. In another study, metformin hydrochloride entrapped AMs were conducted by Szekalska et al. (2016a), and it was found that the obtained microspheres were spherical with narrow size distribution and had a mean size of 1.6–5.7 μm. The drug loading was in the range between 37.3 ± 3.8% and 75.6 ± 1.4%. It is worth noting that drug encapsulation efficiency in all formulations was higher than 100%. This can be explained by a partial loss of the alginate polymer during the preparation process, which decreased the theoretical alginate mass and changed the theoretical drug content to values higher than the previous ones.

10.4 Alginate microsphere and its biomedical applications

Alginate which can be derived from marine algae or soil bacteria falls under the classification of natural polymer. Owing to its nontoxicity, excellent biocompatibility, and biodegradability, it has attracted tremendous attention among researchers for the formulation of biopolymers for medical applications (Lee and Mooney, 2012). Alginate may be fabricated in the form of microspheres and microcapsules, and the terms are often used interchangeably with the term microgels, microparticles, or microbeads (Dhamecha et al., 2019; Sun and Tan, 2013). Thus, for this review the term of microspheres will be used and this term covers the term microgels, microparticles, and microbeads.

Among the popular biomedical applications of AMs are encapsulation of small or large molecules for targeted therapeutic delivery and tissue engineering (Puscaselu et al., 2020). Hence, this review focuses on recent reports on the application of AMs in encapsulating drugs, cells, and proteins for targeted delivery.

10.4.1 Drugs-encapsulated AMs

AMs have been used as drug carrier systems with the incorporation of different active agents in their matrix. Various active substances, depending on their function, can be encapsulated for targeted delivery such as pregabalin (Arafa and Ayoub, 2017), metformin (Maestrelli et al., 2017), retinoic acid (Wang et al., 2018), ibuprofen (Raha et al., 2018), curcumin (Uyen et al., 2020), ratinidine (Szekalska et al., 2015), etc. The entrapped active substance is released upon reaching the targeted site in a prolonged and controlled manner, and hence preserves the stability and availability of the drugs.

Currently, much research on drugs-encapsulated AMs is being carried out in attempts to find the best processing conditions and formulations to achieve the desired properties that meet specifications for drug delivery applications. Sadeghi et al. (2021) investigated the effect of processing parameters on the size, shape, and morphology of calcium-cross-linked AMs. The AMs were produced using the water-in-oil (w/o) emulsification/internal gelation method and incorporated with simvastation lactone (SML) and simvastatin β-hydroxyacid (SMA) drugs in a separate system. The optimized AMs were obtained using a 10:1 weight ratio of calcium carbonate/alginate, 2 wt% alginate concentration, 1:20 w/o volume ratio, 5% emulsifier volume concentration, and 1000 rpm emulsification speed. The encapsulation efficiencies of SML- and SMA-loaded AMs were 73% and 69%, respectively, and they exhibited a pH-responsive swelling behavior with controlled release profile. In another work, Martín et al. (2015) have reported on the formulation of nystatin drug-encapsulated AM via the emulsification/internal gelation method for oral candidiasis treatment. The microsphere was spherical in shape and the mean size was between 85 and 135 μm with narrow distribution. An in vivo study has shown that the nystatin was retained in the mucosa and free from systemic absorption or tissue damage, confirming the safety of the treatment.

AMs with the inclusion of gallic acid (GA), a polyphenol commonly found in red fruits, were varied at different ratios of GA/SA and $CaCl_2$ cross-linker times for the treatment of colorectal cancer (Celep et al., 2020). Outcomes of the study showed that the encapsulation efficiency of AMs was between 11.26% and 72.64% and the value decreased on increasing of cross-linking time. The encapsulated AMs demonstrated a controlled release profile and yields of 15.55%–80.27% were obtained. Another study developed a novel method to prepare retinoic acid (RA)-loaded AMs to treat fundus disease (Wang et al., 2018). The RA-encapsulated AMs exhibited a relatively stable and slow release of drug from the microsphere and the results have shown that the AMs are safe and biocompatible in intravitreal injections.

Despite the excellent properties of AMs for drug encapsulation, its microporous structure leads to rapid dissolution of microsphere, and hence causes sudden release of the encapsulated drug together with a low encapsulation efficiency (Goh et al., 2012). The stability of drugs and encapsulation efficiency can be improved by blending AMs with other substance in the formation of an AMs composite as a coating material (Nayak et al., 2013). For example, Ghumman et al. (2019) formulated AMs blended with *taro corm* mucilage and incorporated with pregabalin drug for neuropathic pain treatment. They found out that the drug encapsulation efficiency increased from 86% for AMs alone to 89% for mucilage-blended AMs. The in vitro study showed sustained drug release and better drug bioavailability of pregabalin loaded mucilage-AMs. In another study conducted by the same research group, linseed mucilage was blended with AMs and incorporated with metformin HCl drug; this was shown to exhibit encapsulation efficiency between 77.9% and 92.3% with sustained release of drug (Ghumman et al., 2020).

Recently, Hamed et al. (2020) developed nanoclay-based alginate-chitosan nanocomposite microspheres with incorporation of curcumin as antioxidant and omega-3 rich oil as the core material. The microspheres were spherical in shape with an average size of 139–153 nm and were prepared via a three-step process: oil-in-water emulsification, gelation, and microencapsulation methods. The alginate-chitosan nanocomposite microsphere-encapsulated fish oil exhibited improved encapsulation efficiency (97.5%) compared to AMs alone (91.87%). An antibacterial activity study and release profile have shown that the encapsulated nanocomposite microspheres have potential use in food and pharmaceutical products as an antibacterial agent. Other studies conducted on drug-encapsulated AMs for biomedical applications are summarized in Table 10.2.

TABLE 10.2 Drugs-encapsulated AMs for different biomedical applications.

Drug type	Materials	Methods	Main findings	Reference
Doxorubicin hydrochloride	Hydroxyapatite/sodium alginate/chitosan (HA/SA/CS) composite microspheres	Emulsification cross-linking	High efficiency encapsulation of 93%–98% with 46%–49% drug loading. Possessed good biocompatibility and 90% cell viability for drug delivery and treatment of bone defects	Bi et al. (2019)
Curcumin	Alginate microspheres	Emulsification/gelation	Encapsulation efficiency of 53.6%, 20.2–72.3 μm particle size. Exhibited sustain released up to 672 h and degrade within 42 h. A potential therapeutic delivery via parenteral injection route	Uyen et al. (2020)
Vancomycin	Strontium-substituted hydroxyapatite microspheres (SrHAs)	Emulsification cross-link	Good in vitro bioactivity and osteogenic performance. Showed enhanced MC3T3-E1 cell proliferation. High drug loading efficiency and sustained drug release for bioactive drug carriers and bone filling materials	Li et al. (2018)
Doxorubicin	Alginate microspheres encapsulated with in situ-formed bismuth sulfide nanoparticles (Bi_2S_3@BCA)	Droplet-based microfluidic	Sustained drug release and enhanced release rate under NIR laser irradiation. Combination with chemotherapy acts synergistically in killing tumor cells for tumor/cancer therapy	Zou et al. (2019)
Nitrendipine	Sodium alginate grafted acrylamide	Emulsification	Improved drug bioavailability and sustained slow drug discharge in both stomach and intestinal media. Suitable as controlled drug delivery system	Subramanian et al. (2020)
Metformin	Thiolated sodium alginate (TSA)	Esterification	Non-Fickian drug releases of 2–7 μm size microspheres. Useful excipient with improved gastric residence time and stability used to treat hypertension	Khalid et al. (2020)

10.4.2 Cells-encapsulated AMs

Cells encapsulation technology commences with the aim to counter the problem of graft rejection in tissue engineering application. In cells encapsulation, the allogenic and xenogeneic cells are isolated within the polymer microsphere. The semipermeable polymeric membrane which surrounds the cells allows two-way movement of molecules; however, the inflow of nutrients, oxygen, and growth factor, which are important for cell metabolism and the outflow of waste and protein, prevent the diffusion of host cells. Cells encapsulation permits sustained and prolonged release of therapeutic molecules at the targeted tissue. The cells encapsulation technique allows the incorporation not only of human cells, but also of artificial cells.

Among other polymers, alginate is widely used in incorporating living cells either alone or in composite form (Goh et al., 2012). AMs are semipermeable, noncytotoxic, and have been shown to exhibit immune protection for various cells recipients (Chang, 1998) in treating different diseases. Many studies reported the use of AMs as encapsulating material for cells able to stimulate cell proliferation and differentiation as well as possessing high viability (Lee and Mooney, 2012), characteristics which are essential in stimulating the regeneration of injured or diseased tissues.

AMs have often undergone surface modification using proteins and peptides to enhance their adhesive interactions further with different cell types. In one study, gelatin was added with the aim to enhance the cell adhesion rate of AMs due to the high charge density of alginate (Amini et al., 2020). The gelatin-AMs was incorporated with rat cardiomyoblasts cell for the treatment of cardiovascular disease. The microsphere produced was noncytotoxic and an in vitro study has shown that the rat cardiomyoblasts cell-encapsulated microsphere lowered the oxidative stress and exhibited a cytofunctional effect. In another study, arginylglycylaspartic acid (RGD)-modified alginate/laponite microsphere incorporated with human dental pulp stem cells (hDPSCs) and vascular endothelial growth factor (VEGF) was prepared via the electrostatic microdroplet method for vascularized dental pulp regeneration in an injectable form (Zhang et al., 2020). Laponite acted as an adjuvant for VEGF which helps to boost the angiogenesis process. It was reported that the VEGF bioactivity was maintained while sustained release of VEGF was observed after 28 days and more than 85% of cell viability was obtained.

In another work, Shi et al. (2016) investigated the combination of milk and alginate microspheres in encapsulating probiotics (*Enterococcus faecalis* HZNU P2) to improve its viability in simulated gastric fluid (SGF) and bile salt solutions. The results showed excellent viability of encapsulated cells after 2 h of incubation in SGF and bile salt solutions as well as a complete release of cells in simulated intestine fluid within 1 h. In addition, the encapsulated probiotic cells were well preserved during storage at different temperatures. Recently, Bochenek et al. (2018) reported on clinical trials of their three formulations of chemically modified AMs encapsulated with pancreatic islet cells for type 1 diabetes treatment. The formulations showed reduced foreign body response. On the other hand, one of the formulations was viable and responsive to glucose for 4 months without the need to suppress the immune system.

Regenerative medicine which involves tissue-engineered products often requires long-term storage. The cryopreservation technique is often used to store the products as it can accommodate immediate demand and facilitate transportation. A study conducted by Mohanty et al. (2016) demonstrated that the concentration of alginate as coating material plays a vital

TABLE 10.3 Cells-encapsulated AMs in biomedical applications.

Cell type	Materials	Methods	Main finding	References
Rat cardiomyoblasts	Alginate-gelatin microspheres	Electrostatic droplet method	Viability of rat cardiomyoblasts cell line after 7 days. Efficiency of microencapsulation on cardiomyocytes lineage dynamics for cardiovascular disease	Saberianpour et al. (2019)
Human mesenchymal stem cells (MSCs)	β-Cyclodextrin grafted methacrylated alginate	Gelation/custom microfluidics devices	Size of droplets depends on the dispersion flow. Droplet size of 15 μm with dispersion flow of 0.0004 mL/min to encapsulate MSCs. Maintained cell viability	Etter et al. (2018)
Insulin-producing cells	Alginate-based microbeads and multicomponents microcapsules	Confocal Raman microscopy (CRM)	Multicomponent microcapsules retained structure upon implantation and acts as immunoprotection. CRM was used to develop new cell encapsulation system	Kroneková et al. (2018)
Insulin producing 1.1B4 cell line	Alginate-poly-L-lysine-alginate (APA)	Electrostatic droplet generator	The microfluidic devices as an alternative for in vivo environments. Enriched medium provides constant insulin secretion and secretion booster potentially be used to treat diabetes	Acarregui et al. (2018)

role in protecting fibroblast cells during cryopreservation in the presence of low cryoprotectant (CPA) or a CPA-free environment. At high concentrations of alginate, a comparable metabolic and angiogenic activity of CPA-free frozen cells was observed compared to cells-encapsulated AMs frozen in the presence of low CPA. Other works performed regarding cell-encapsulated AMs for biomedical applications are tabulated in Table 10.3.

10.4.3 Proteins encapsulated AMs

Development of innovative protein for regenerative medicine has encouraged the advancement in biopharmaceuticals with the potential to cure disease and rejuvenate tissue. However, delivery of these proteins to the targeted site remains a challenging task. This is due to the structures of protein that are prone to hydrolytic and enzymatic degradation (Sinha and Trehan, 2003), and thus delivery of these proteins to specific sites in active form is difficult to achieve. One of the methods to enable delivery of proteins in active form is by encapsulation of these proteins in various polymeric matrices microspheres either in synthetic polymers such as polylactide acid (PLA) and poly (lactic-*co*-glycolic-acid (PLGA) or in natural occurring polymers including polysaccharides such as chitosan, alginate, and hyaluronic acid, peptides, and polyesters. However, preparation for protein encapsulation microspheres in synthetic polymers requires a harsh environment whereby the use of organic solvents would cause protein denaturation (Xing et al., 2019; Zhou et al., 2001). Therefore, to retain protein bioactivity, encapsulation of proteins in natural polymers microspheres is suitable as the synthesis often requires mild preparation conditions that do not alter the protein properties (Joye and McClements, 2014).

For the success of protein encapsulation and delivery of microspheres to the targeted site, it is desirable that the microspheres are able to preserve the protein stability, sufficient encapsulation efficiency, minimal burst effect, and controlled release of proteins (Wee and Gombotz, 1998; Wells and Sheardown, 2007; Zhou et al., 2001). Alginate, which is known for its biocompatible, nontoxic, and nonimmunogenic properties, is widely researched as a candidate for protein encapsulation for numerous biomedical applications. However, the unique aspects of alginate that attracts attention as a matrix for protein encapsulation are its relatively inert aqueous environment within the matrix, which requires a mild room temperature encapsulation process, which is free for organic solvent, high gel porosity that allows high diffusion rates of macromolecules, and the ability to control the microspheres' sizes and porosity (Wee and Gombotz, 1998).

In a study done by Zhai et al. (2015), it was found that alginate/PLGA composites showed higher encapsulation efficiency with reduced initial burst release of bovine serum albumin (BSA) compared to PLGA microspheres alone. The use of surfactant during the preparation of alginate/PLGA composite microspheres is thought to increase the encapsulation efficiency. It was also observed that higher protein content was retained and the erosion and degradation rates of the alginate/PLGA composites were slower. These observations suggest a better control of hydrophilic protein release. In an earlier study, Zheng et al. (2004) also reported that an improved in protein entrapment efficiency and a decrease in burst effect were observed in alginate-chitosan-PLGA composite microspheres. In this study, double-walled alginate-chitosan microspheres were fabricated by a modified emulsification method. To further sustain protein released from the alginate-chitosan microspheres, these microspheres were further incorporated in the PLGA to form the microspheres composites. Wells and Sheardown (2007) developed a novel encapsulation technique that enables high encapsulation efficiency to allow prolonged released of the protein. In this technique, alginate microspheres were soaked in a protein-containing sodium chloride (NaCl) solution and followed by recrosslinking with calcium chloride ($CaCl_2$). Using this method, albumin, lysozyme, and chymotrypsin were encapsulated in alginate. The results obtained showed that prolonged released of these proteins was observed depending on the nature of the release media.

Alginate microspheres have also been investigated as a candidate for protein drugs delivery via the oral route. Among the attractive properties of alginate are its bio-adhesiveness and pH sensitivity along with its biocompatibility, nontoxicity, and mild gelation conditions. However, the most challenging task in preparing for oral delivery is the loss of drug during preparation due to leaching out of drugs through the pores of the microspheres (George and Abraham, 2006). Therefore, to overcome this limitation, some modifications have been adopted, such as covalent chemical modifications (Bernknop-Schnurch et al., 2001) or formation polyelectrolytes complexes such as alginate-chitosan complexes (Hari et al., 1996) or complexes with other polycations.

10.4.4 Tissue engineering

Tissue regeneration deals with the process of regrowth of the remaining tissue (of the lost part) to restore the damaged tissue/organ to its original condition. Due to the current problems faced with artificial tissues/organs and tissue/organ transplantation, technology related to tissue regeneration has evolved and attracted massive attention among scientists and clinicians; this technology is tissue engineering. Three major constituents are involved in tissue engineering: cells as the building blocks of tissue, scaffold as the matrix, and signals to stimulate cell regrowth (Kaoud, 2018; Yang et al., 2021). Tissue engineering has displayed encouraging findings in the regeneration of various tissues and organs such as skin, bone, cartilage, the cardiovascular system, and liver (Dzobo et al., 2018; Hellman, 2008).

Alginate microspheres have been explored extensively for potential applications in tissue engineering owing to their ease of fabrication into various forms fitting the natural physiological conditions to ensure cell functions and viability (Venkatesan et al., 2014). Alginate also possesses good mechanical integrity, and appropriate viscoelastic properties make it suitable for tissue reconstruction of bone and cartilage as a supporting material (Szekalska et al., 2016a). In tissue engineering, alginate is commonly used as a template or scaffold for cells to attach, proliferate, and function in promoting tissue regeneration (Bahrami et al., 2019; Farokhi et al., 2020; Sahoo and Biswal, 2021). The alginate scaffold can be developed into different forms such as microbead, injectable hydrogel, and preformed scaffold (Yang et al., 2021; Zhang et al., 2020). Studies have been reported on the potential use of alginate as a scaffold for bone (Venkatesan et al., 2014), cartilage (Farokhi et al., 2020; Xu et al., 2019), cardiac (Liberski et al., 2016), neural (Madhusudanan et al., 2020), and dental tissue engineering (Diniz et al., 2016; Moshaverinia et al., 2012).

Alginate is often combined with other materials to get optimized formulations in fabricating the scaffold in order to meet intended requirements for tissue engineering applications. For bone tissue regeneration, inorganic materials such as bioactive glass and calcium phosphate are commonly added to alginate to improve the bone formation and provide appropriate mechanical integrity (Sahoo and Biswal, 2021). Ho and coworkers evaluated the osteoconductive behavior of TEMPO-oxidized cellulose nanofiber (TOCNF) reinforced β-tricalcium phosphate (β-TCP)/sodium alginate injectable microspheres in healing bone defects (Ho et al., 2020). The in vivo study revealed better bone formation in a rat model compared to control and showed good progress of bone formation after 4 weeks of implantation, suggesting the potential use of the β-TCP/AM for bone regeneration. In another study, bioactive glass was incorporated to improve the mechanical properties of gelatin/alginate hydrogel scaffold for bone tissue regeneration (Sarker et al., 2016). Via in vitro analysis, it was found that the incorporation of 1% bioactive glass in hydrogel scaffold has enhanced the cell attachment and proliferation.

Stem cells are usually mixed with injectable scaffold as a template to regenerate pulp dentin (Wang et al., 2018), periodontal tissue (Li and Dai, 2006), and bioroot (Manivasagam et al., 2019) for regenerative endodontic treatment. Among the popular stem cells used for pulp dentin regeneration are stem cells from apical papilla (soft tissue attached to the end of immature permanent teeth) known as SCAP, stem cells derived from exfoliated deciduous teeth, and dental pulp stem cells (DPSCs) (Chang, 1998). Alginate scaffold applied to the cultured tooth slice (Dobie et al., 2002) and incorporated with human DPSCs (Kumabe et al., 2006) has been developed for dental pulp regeneration and the results have shown that the pulp cells are able to differentiate and form odontoblast-like cells as well as upregulate matrix secretion. The findings suggest that the alginate scaffold is suitable as a matrix for dental pulp regeneration. In another study, Bhoj et al. (2015) fabricated an RGD-alginate scaffold via simple templating for pulp tissue engineering. Two stem cells (DPSCs and human umbilical vein endothelial cells) and two growth factors (VEGF and fibroblast growth factor) were added to the alginate scaffold; the findings showed enhanced proliferation of the combined cells due to the presence of microenvironments as a result of double growth factors.

In endodontic tissue regeneration, alginate as a scaffold is commonly combined with other materials to improve its mechanical strength and product variability, and reduce potential pathogen transmission (Raddall et al., 2019). Devillard et al. (2017) developed a simple method to fabricate a collagen/alginate scaffold incorporated with SCAP for endodontic regeneration upon root canal treatment. Collagen was introduced since the stem cells are easily adhered to and integrated within the collagen scaffold. The study showed that the stem cells are able to attach, proliferate, and differentiate, and hence provide a suitable healing environment for root canal treatment. Devillard et al. (2017) developed a simple method to fabricate collagen/alginate scaffold incorporated with SCAP for endodontic regeneration upon root canal treatment. Collagen was introduced since the stem cells are easily adhered to and integrated within the collagen scaffold.

Recently, a study was conducted by Yang and coworkers on the fabrication of Ginsenoside Rg1 and adipose tissue derived stem cells (ADSCs)-encapsulated laminin/alginate microspheres for breast reconstruction after lumpectomy (Yang et al., 2021). The results demonstrated that the injectable microspheres were able to provide good environment for the ADSCs to proliferate and differentiate with no cytotoxic, genotoxic, or systemic toxic effects. The prolonged release of Rg1 improved the ADSCs' survival rate and encouraged more cells to vascularize, hence promoting tissue growth and repair in breast reconstruction. It is suggested that the microspheres prepared are suitable as a scaffold for adipose tissue engineering.

10.4.5 Other applications

AMs usage in biomedical applications has been explored over the years for its feasibility to be used in vaccine delivery. Interest arises due to the increased demand for vaccination via an oral route. Vaccination via this route eliminates the need for booster doses and the need for repetitive injections. Alginate microspheres have gained interest as the candidate for oral vaccination carrier due to their biocompatibility and mucoadhesive properties. The presence of free carboxyl groups in alginates allows the interaction between alginates and mucin by hydrogen and electrostatic bonding and thus gives alginate good mucoadhesive properties (Szekalska et al., 2016a). AMs have been broadly used in the advancement of controlled release systems for small molecules and proteins delivery, and in the development of a scaffold for tissue or organ regeneration (Martins et al., 2007; Reyes et al., 2006; Seal et al., 2001). However, in the case of vaccine delivery via the oral route, a major issue faced by AMs and other biodegradable polymers is their poor bioavailability. This is because orally administered drugs need to pass through the intestinal wall and liver before reaching the systemic circulation, and thus these biodegradable polymers (natural or synthetic polymers) act as a protective layer, defending the substances from the harsh acidic environment of the stomach or by providing sustained released of drugs (Jana et al., 2021).

In a study by Mittal et al. (2001), alginate microspheres containing plasmid DNA and bovine adenovirus type 3 were used as the vehicle to deliver DNA-based vaccines. The objective of the study was to investigate the effect of the route of administration whereby mice were inoculated orally, intranasally, intramuscularly, subcutaneously, or intraperitoneally. The results obtained demonstrated that alginate microspheres successfully deliver the adenovirus with or without plasmid DNA. Alginate microspheres delivered via mucosal delivery showed a stronger mucosal response while alginate microspheres delivered by systemic delivery showed a stronger systemic response. Another study by Zheng et al. (2004) aiming to develop single-shot formulation for the hepatitis B vaccine developed a novel type of composite microspheres of alginate-chitosan-PLGA. The motivation to develop this composite was to overcome the limitations of PLGA microspheres, which have shown low loading efficiency and high burst effect that may result in an insufficient dose. The results from the study showed that the novel composite of alginate-chitosan-PLGA has improved the low loading entrapment efficiency of the PLGA microspheres and maintained higher antibody levels when compared to double-dose injections.

For the encapsulation of bacteria, specifically probiotics (*Lactobacillus acidophilus and Bifodobacteria*), AMs have gained attention due to their acid-gel character, mild gelling condition, Generally Recognized As Safe (GRAS) status, and nontoxicity (Li et al., 2020). These unique properties, together with AMs' feasibility of fabrication methods such as extrusion, emulsification, and spray-drying, have attracted researchers to encapsulate bacteria in AMs. For the delivery of bacteria, oral delivery is the most convenient method of delivery. However, it was observed that the availability and activity of the bacteria were significantly decreased after oral administration (Puscaselu et al., 2020). In addition, preservation of these microorganisms' viability is also challenging during processing, storage, and transition through the gastrointestinal tract (Dong et al., 2013). Liu et al. (2018) investigated the in vitro gastrointestinal digestion and storage properties of probiotic bacteria encapsulated in sodium alginate and sodium caseinate microspheres or in soy protein isolated microspheres synthesis via spray-drying technology. One of the limitations that was of concern in the study was the high viscosity of alginate, which limits the usage of alginate to encapsulate probiotic bacteria via the spray-drying technique. Therefore, in the study, a low-viscosity alginate was selected to utilize the spray-drying technique. It was concluded that for the survival of the probiotics, encapsulation of the bacteria in sodium caseinate or in soy protein could effectively maintain the survival of the bacteria. However, bacteria encapsulated in sodium alginate and sodium caseinate showed a better storage property. Another study supported the theory that encapsulation in sodium alginate improved the bacteria availability during the spray-drying technique (Chang et al., 2020). Researchers are also focusing on several other bacteria such as *Salmonella*, *Escherichia*, *Clostridium*, *Caulobacter*, *Listeria*, *Proteus*, and *Streptococcus* as potential tumor-targeting agents (Park et al., 2014). However, using these bacteria comes with its complications such as toxicity, sepsis, inflammation, and minimal reachable bacteria to the targeted site. One of the methods suggested to overcome these complications involves encapsulating the bacteria into a biodegradable and biocompatible polymer. Encapsulation will allow the bacteria or active core ingredients to be separated from the environment by providing a protective film or coating. Donthidi et al. (2010) reported that many studies have showed that calcium alginate-encapsulated cultures increase the survival of bacteria under different condition when compared to the nonencapsulated ones. It was also proven in the study that using lecithin and a prebiotic starch with alginate for the encapsulation of a probiotic increases the probiotic survival.

In recent years, alginate has been extensively researched as a candidate for hemostatic treatment. Treating massive bleeding is of importance as uncontrolled bleeding has contributed to deaths in many traumatic injuries. Despite the inherent properties of alginate, which is desirable in the research for various biomaterials and biomedical applications, its nonadhesive properties has limited its applications as a hemostatic material. Therefore, in a recent study done by Jin et al. (2020), alginate-based composite microspheres were synthesized to optimize the feasibility of alginate as a hemostatic material. In this study, a composite of alginate, carboxymethyl chitosan, and collagen microspheres were synthesized and this composite was then coated with berberine, an ammonium salt which exhibits antibacterial activity, to enhance the performance of the composite as a hemostatic material. In treating massive bleeding, it is desirable to reduce the risk of wound infection while at the same time the hemostatic materials must be able to promote blood coagulation to stop blood loss. Therefore, current hemostatic materials are designed to incorporate drug or tissue factor to trigger blood coagulation. Liu et al. (2020) incorporated tissue factor (TF)-liposome with alginate to form a composite paste and ionically cross-linked hydrogel. The result obtained showed that the lapidated TF was evenly distributed in the alginate matrix, and it was noted that the release kinetics of lapidated TF was tunable. Importantly, the study demonstrated that TF-liposomes release from alginate matrix has shown excellent procoagulant properties and exhibits better hemostatic performance.

In another study, thrombin-loaded alginate microsphere were synthesized. Thrombin is a hemostatic drug that could provide a fast hemostatic effect. This is accomplished by accelerating platelets' formation and conversion of fibrinogen to fibrin (Xuan et al., 2017). Thrombin-loaded alginate microspheres were prepared using the electrostatic droplet technique. In vivo results on the embolization ability of thrombin-loaded alginate microspheres showed that blood flow was stopped immediately once these microspheres were delivered to the injured hepatic artery. Another study by Rong et al. (2017), who further investigated the feasibility of thrombin-loaded alginate microspheres, concluded that these have the advantages of rapid and reliable embolic hemostasis and have the potential as a candidate for transcatheter arterial embolization for treating solid visceral organs' rupture and bleeding.

In a review article by Puscaselu et al. (2020), it was stated that the applications of alginates have been expanded for the development of products used in the management of metabolic disorders such as obesity and diabetes. For the purpose of treating metabolic disorders, it is well established that diet plays the most crucial role. Therefore, alginate, which is also a known dietary fiber, has the ability to regulate appetite by reducing gastric emptying, nutrient absorption, and overall feelings of fullness. Therefore, it can be seen not only that alginate has potential in biomedical applications but also that alginate is gaining interest in food technology. In addition to the many possible applications of alginate, development of this biomaterial remains challenging in terms of technology transfer, manufacturing process, consumer acceptance, regulatory safety requirements, and environmental concerns.

10.5 Conclusions

Although alginate contains only two monomer units M and G linked with 1,4-linkages giving it a simpler looking structure, this structure enables alginates to exhibit diverse chemical composition and monomer sequence, resulting in a large variety of physical and biological properties. With these unique characteristics, alginate has become one of the most common natural polymers used in the formation of microspheres for various applications. Researchers are exploring various possible applications and methods to design this natural multifunctional polymer for controlled release, targeted drug delivery, as a matrix for three-dimensional tissue cultures, adjuvants of antibiotics, antiviral agents, or in treatment of various diseases such as diabetics and treating metabolic disorders. Along with the demand to fulfill the needs for various applications, nevertheless advancements in the techniques to synthesize alginate are also being explored to synthesize alginate microspheres with controlled physical and chemical properties. With new developments in extraction methods and synthesis techniques, alginate could fulfill the demanding needs for biomedical applications particularly as a drug delivery system.

References

Acarregui, A., Ciriza, J., del Burgo, L.S., Gurruchaga, H., Yeste, J., Xavier, I., Orive, G., Hernández, R.M., Villa, R., Pedraz, J.L., 2018. Characterization of an encapsulated insulin secreting human pancreatic beta cell line in a modular microfluidic device. J. Drug Target., 36–44.

Allamneni, Y., Reddy, B.V.V.K., Dayananda Chari, P., Venkata Balakrishna Rao, N., Chaitaya Kumar, S., Kalekar, A.-K., 2012. Performance evaluation of mucoadhesive potential of sodium alginate on microspheres containing an anti-diabetic drug: glipizide. Int. J Pharm. Sci. Drug Res. 4 (2), 115–122.

Amini, H., Hashemzadeh, S., Heidarzadeh, M., Mamipour, M., Yousefi, M., Saberianpour, S., Rahbarghazi, R., Nouri, M., Sokullu, E., 2020. Cytoprotective and cytofunctional effect of polyanionic polysaccharide alginate and gelatin microspheres on rat cardiac cells. Int. J. Biol. Macromol. 161, 969–976. http://www.elsevier.com/locate/ijbiomac.

Arafa, M.G., Ayoub, B.M., 2017. DOE optimization of nano-based carrier of pregabalin as hydrogel: new therapeutic & chemometric approaches for controlled drug delivery systems. Sci. Rep. 7, 41503.

Bahrami, N., Farzin, A., Bayat, F., Goodarzi, A., Salehi, M., Karimi, R., Mohamadnia, A., Parhiz, A., Jafar, A., 2019. Optimization of 3D alginate scaffold properties with interconnected porosity using freeze-drying method for cartilage tissue engineering application. Arch. Neurosci. 6 (4).

Bansil, R., Turner, B.S., 2006. Mucin structure, aggregation, physiological functions and biomedical applications. Curr. Opin. Colloid Interface Sci. 11, 164–170.

BeMiller, J.N., 1999. Structure-property correlations of non-starch food polysaccharides. Macromol. Symp. 140, 1–15. http://www3.interscience.wiley.com/journal/60500249/home.

Bernknop-Schnurch, A., Kast, C.E., Richter, M.F., 2001. Improvement in the mucoadhesive properties of alginate by the covalent attachment of cysteine. J. Control. Release 71 (3), 277–285.

Bhoj, M., Zhang, C., Green, D.W., 2015. A first step in de novo synthesis of a living pulp tissue replacement using dental pulp MSCs and tissue growth factors, encapsulated within a bioinspired alginate hydrogel. J. Endod. 41, 1100–1107.

Bi, Y.-g., Lin, Z.-t., Deng, S.-t., 2019. Fabrication and characterization of hydroxyapatite/sodium alginate/chitosan composite microspheres for drug delivery and bone tissue engineering. Mater. Sci Eng. C Mater. Biol. Appl., 576–583.

Bochenek, M.A., Veiseh, O., Vegas, A.J., McGarrigle, J.J., Qi, M., Marchese, E., Omami, M., Doloff, J.C., Mendoza-Elias, J., Nourmohammadzadeh, M., Khan, A., Yeh, C.C., Xing, Y., Isa, D., Ghani, S., Li, J., Landry, C., Bader, A.R., Olejnik, K., Oberholzer, J., 2018. Alginate encapsulation as long-term immune protection of allogeneic pancreatic islet cells transplanted into the omental bursa of macaques. Nat. Biomed. Eng. 2 (11), 810–821. http://www.nature.com/natbiomedeng/.

Bowey, K., Swift, B.E., Flynn, L.E., Neufeld, R.J., 2013. Characterization of biologically active insulin-loaded alginate microparticles prepared by spray drying. Drug Dev. Ind. Pharm. 39 (3), 457–465.

Celep, A.G.S., Demirkaya, A., Solok, E.K., 2020. Antioxidant and anticancer activities of gallic acid loaded sodium alginate microspheres on colon cancer. Curr. Appl. Phys. https://www.sciencedirect.com/science/article/abs/pii/S1567173920301164.

Chang, T.M.S., 1998. Pharmaceutical and therapeutic applications of artificial cells including microencapsulation. Eur. J. Pharm. Biopharm. 45 (1), 3–8. http://www.elsevier.com/locate/ejphabio.

Chang, Y., Yang, Y., Xu, N., Mu, H., Zhang, H., Duan, J., 2020. Improved viability of Akkermansia muciniphila by encapsulation in spray dried succinate-grafted alginate doped with epigallocatechin-3-gallate. Int. J. Biol. Macromol. 159, 373–382.

Chen, W., Kim, J.H., Zhang, D., Lee, K.H., Cangelosi, G.A., Soelberg, S.D., Furlong, C.E., Chung, J.H., Shen, A.Q., 2013. Microfluidic one-step synthesis of alginate microspheres immobilized with antibodies. J. R. Soc. Interface 10 (88). http://rsif.royalsocietypublishing.org/content/10/88/20130566.full.pdf+html.

Clementi, F., 1997. Alginate production by Azotobacter vinelandii. Crit. Rev. Biotechnol. 17 (4), 327–361.

Cohen, E., Merzendorfer, H., 2019. Extracellular Sugar-Based Biopolymers Matrices. Springer.

Costa, J.R., Silva, N.C., Sarmento, B., Pintado, M., 2015. Potential chitosan-coated alginate nanoparticles for ocular delivery of daptomycin. Eur. J. Clin. Microbiol. Infect. Dis. 34 (6), 1255–1262.

Devillard, R., Rémy, M., Kalisky, J., Bourget, J.M., Kérourédan, O., Siadous, R., Bareille, R., Amédée-Vilamitjana, J., Chassande, O., Fricain, J.C., 2017. In vitro assessment of a collagen/alginate composite scaffold for regenerative endodontics. Int. Endod. J. 50 (1), 48–57.

Dhamecha, D., Movsas, R., Sano, U., Jyothi, U., Menon, 2019. Applications of alginate microspheres in therapeutics delivery and cell culture: past, present and future. Int. J. Pharm. 569.

Diniz, I.M., Chen, C., Ansari, S., Zadeh, H.H., Moshaverinia, M., Chee, D., Marques, M.M., Shi, S., Moshaverinia, A., 2016. Gingival mesenchymal stem cell (GMSC) delivery system based on RGD-coupled alginate hydrogel with antimicrobial properties: A novel treatment modality for peri-implantitis. J. Prosthodont. 25 (2), 105–115.

Dobie, K., Smith, G., Sloan, A.J., Smith, A.J., 2002. Effects of alginate hydrogels and TGF-beta 1 on human dental pulp repair in vitro. Connect. Tissue Res. 43 (2–3), 387–390.

Dong, Q.Y., Chen, M.Y., Xin, Y., Qin, X.Y., Cheng, Z., Shi, L.E., Zhen, X.T., 2013. Alginate-based and protein-based materials for probiotics encapsulation: a review. Int. J. Food Sci. Technol. 48, 1339–1351.

Donthidi, A.R., Tester, R.F., Aidoo, K.E., 2010. Effect of lecithin and starch on alginate-encapsulated probiotic bacteria. J. Microencapsul. 27 (1), 67–77.

Draget, K.I., 2009. Alginates. In: Handbook of Hydrocolloids, second ed. Elsevier Inc, pp. 807–828. http://www.sciencedirect.com/science/book/9781845694142.

Dzobo, K., Thomford, N.E., Senthebane, D.A., Shipanga, H., Arielle Rowe, A., Dandara, C., Pillay, M., Motaung, K.-S.C.M., 2018. Advances in regenerative medicine and tissue engineering: innovation and transformation of medicine. Stem Cells Int. 2018, 1–24.

E.C.S, 1883. Scientific American supplement. Am. J. Pharm. 617 (396), 1835–1907.

Etter, J.N., Karasinski, M., Ware, J., Oldinski, R.A., 2018. Dual-crosslinked homogeneous alginate microspheres for mesenchymal stem cell encapsulation. Biomater. Synth. Charact., 1–10.

Farokhi, M., Shariatzadeh, F.J., Solouk, A., Mirzadeh, H., 2020. Alginate based scaffolds for cartilage tissue engineering: a review. Int. J. Polym. Mater. Polym. Biomater. 69 (4), 230–247.

Gedam, S., Jadhav, P., Talele, S., Jadhav, A., 2018. Effect of crosslinking agent on development of gastroretentive mucoadhesive microspheres of risedronate sodium. Int. J. Appl. Pharm. 10 (4), 133–140. https://innovareacademics.in/journals/index.php/ijap/article/download/26071/15190.

George, M., Abraham, T.E., 2006. Polyionic hydrocolloids for the intestinal delivery of protein drugs; alginate and chitosan—a review. J. Control. Release 114 (1), 1–14.

Ghumman, S.A., Bashir, S., Noreen, S., Khan, A.M., Malik, M.Z., 2019. Taro-corms mucilage-alginate microspheres for the sustained release of pregabalin: in vitro & in vivo evaluation. Int. J. Biol. Macromol. 139, 1191–1202. http://www.elsevier.com/locate/ijbiomac.

Ghumman, S.A., Noreen, S., tul Muntaha, S., 2020. Linum usitatissimum seed mucilage-alginate mucoadhesive microspheres of metformin HCl: fabrication, characterization and evaluation. Int. J. Biol. Macromol. 155, 358–368. http://www.elsevier.com/locate/ijbiomac.

Goh, C.H., Heng, P.W.S., Chan, L.W., 2012. Alginates as a useful natural polymer for microencapsulation and therapeutic applications. Carbohydr. Polym. 88 (1), 1–12.

Hamed, S.F., Hashim, A.F., Abdel Hamid, H.A., Abd-Elsalam, K.A., Golonka, I., Musiał, W., El-Sherbiny, I.M., 2020. Edible alginate/chitosan-based nanocomposite microspheres as delivery vehicles of omega-3 rich oils. Carbohydr. Polym. 239. http://www.elsevier.com/wps/find/journaldescription.cws_home/405871/description#description.

Hari, P.R., Chandy, T., Sharma, C.P., 1996. Chitosan/calcium–alginate beads for oral delivery of insulin. J. Appl. Polym. Sci. 59 (11), 1795–1801.

Hariyadi, D.M., Hendradi, E., Irawan, M.B., 2016. Preparation and characterization of Ba-alginate microspheres containing ovalbumin. J. Farmasi Indonesia 8 (1).

Hay, I.D., Rehman, Z.U., Moradali, M.F., Wang, Y., Rehm, B.H.A., 2013. Microbial alginate production, modification and its applications. Microb. Biotechnol. 6 (6), 637–650.

Hellman, K.B., 2008. Tissue engineering: translating science to products. In: Ashammakhi, N., Reis, R., Chiellini, F. (Eds.), Topics in Tissue Engineering. vol. 4. Oulu University.

Ho, H.V., Tripathi, G., Gwon, J., Lee, S.-Y., Lee, B.-T., 2020. Novel TOCNF reinforced injectable alginate/β-tricalcium phosphate microspheres for bone regeneration. Mater. Des. 194, 108892.

Iswandana, R., Putri, K.S.S., Wulandari, F.R., Najuda, G., Sari, S.P., Djajadisastra, J., 2018. Preparation of calcium alginate-tetrandrine beads using ionic gelation method as colon-targeted dosage form. J. Appl. Pharm. Sci. 8 (5), 68–74. http://www.japsonline.com/admin/php/uploads/2627_pdf.pdf.

Jana, P., Shyam, M., Singh, S., Jayaprakash, V., Dev, A., 2021. Biodegradable polymers in drug delivery and oral vaccination. Eur. Polym. J. 142, 110155.

Jiao, W., Chen, W., Mei, Y., Yun, Y., Wang, B., Zhong, Q., Chen, H., Chen, W., 2019. Effect of molecular weight and guluronic acid/mannuronic acid ratio on the rheological behavior and stabilizing property of sodium alginate. Molecules 24 (23), 4347.

Jin, J., Xu, M., Liu, Y., Ji, Z., Dai, K., Zhang, L., Wang, L., Ye, F., Chen, G., Lv, Z., 2020. Alginate-based composite microspheres coated by berberine simultaneously improve hemostatic and antibacterial efficacy. Colloids Surf. B Biointerfaces 194, 111168.

Joye, I.J., McClements, D.J., 2014. Biopolymer-based nanoparticles and microparticles: fabrication, characterization, and application. Curr. Opin. Colloid Interface Sci. 19 (5), 417.

Kaoud, H.A.E.-S., 2018. In: Kaoud, H.A.h.E.S. (Ed.), Introductory Chapter: Concepts of Tissue Regeneration, Tissue Regeneration. Intech Open. https://www.intechopen.com/books/tissue-regeneration/introductory-chapter-concepts-of-tissue-regeneration.

Khalid, S., Abbas, G., Hanif, M., Shah, S., Hussain Shah, S.N., Jalil, A., Yaqoob, M., Ameer, N., Anum, A., 2020. Thiolated sodium alginate conjugates for mucoadhesive and controlled release behavior of metformin microspheres. Int. J. Biol. Macromol., 2691–2700.

Konda, N.V.S.N.M., Singh, S., Simmons, B.A., Klein-Marcuschamer, D., 2015. An investigation on the economic feasibility of macroalgae as a potential feedstock for biorefineries. Bioenergy Res. 8 (3), 1046–1056. http://www.springer.com/life+sci/plant+sciences/journal/12155.

Kong, H.J., Smith, M.K., Mooney, D.J., 2003. Designing alginate hydrogels to maintain viability of immobilized cells. Biomaterials 24 (22), 4023–4029. http://www.journals.elsevier.com/biomaterials/.

Kroneková, Z., Pelach, M., Mazancová, P., Uhelská, L., Treľová, D., Rázga, F., Némethová, V., Szalai, S., Chorvát, D., McGarrigle, J.J., Omami, M., Isa, D., Ghani, S., Majková, E., Oberholzer, J., Raus, V., Šiffalovič, P., Lacík, I., 2018. Structural changes in alginate- based microspheres exposed to in vivo environment as revealed by confocal Raman microscopy. Sci. Rep., 1–12.

Kumabe, S., Nakatsuka, M., Kim, G.S., Jue, S.S., Aikawa, F., Shin, J.W., Iwai, Y., 2006. Human dental pulp cell culture and cell transplantation with an alginate scaffold. Okajimas Folia Anat. Jpn. 82 (4), 147–156.

Kyzioł, A., Mazgała, A., Michna, J., Regiel-Futyra, A., Sebastian, V., 2017. Preparation and characterization of alginate/chitosan formulations for ciprofloxacin-controlled delivery. J. Biomater. Appl. 32 (2), 162–174. http://jba.sagepub.com/.

Laffleur, F., 2014. Mucoadhesive polymers for buccal drug delivery. Drug Dev. Ind. Pharm. 40 (5), 591–598.

Lee, K.Y., Mooney, D.J., 2012. Alginate: properties and biomedical applications. Prog. Polym. Sci. (Oxf.) 37 (1), 106–126. http://www.sciencedirect.com/science/journal/00796700.

Li, Y., Dai, X.Q., 2006. Fabrics. In: Biomechanical Engineering of Textiles and Clothing. Woodhead Publishing Series.

Li, G., Han, N., Zhang, X., Yang, H., Cao, Y., Wang, S., Fan, Z., 2018. Local injection of allogeneic stem cells from apical papilla enhanced periodontal tissue regeneration in Minipig model of periodontitis. Biomed. Res. Int. 2018, 1–8.

Li, S., Jiang, W., Zheng, C., Shao, D., Liu, Y., Huang, S., Han, J., Ding, J., Tao, Y., Li, M., 2020. Oral delivery of bacteria: basic principles and biomedical applications. J. Control. Release 327, 801–833.

Liberski, A., Latif, N., Raynaud, C., Bollensdorff, C., Yacoub, M., 2016. Alginate for cardiac regeneration: from seaweed to clinical trials. Global Cardiol. Sci. Pract. 1.

Lin, S.F., Chen, Y.C., Chen, R.N., Chen, L.C., Ho, H.O., Tsung, Y.H., Sheu, M.T., Liu, D.Z., 2016. Improving the stability of astaxanthin by microencapsulation in calcium alginate beads. PLoS One 11 (4).

Linker, A., Jones, R.S., 1966. A new polysaccharide resembling alginic acid isolated from pseudomonads. J. Biol. Chem. 241 (16), 3845–3851.
Liu, H., Gong, J., Chabot, D., Miller, S.S., Cui, S.W., Zhong, F., Wang, Q., 2018. Improved survival of lactobacillus zeae LB1 in a spray dried alginate-protein matrix. Food Hydrocoll. 78, 100–108.
Liu, C., Shi, Z., Sun, H., Mujuni, C.J., Zhao, L., Wang, X., Huang, F., 2020. Preparation and characterization of tissue-factor-loaded alginate: toward a bioactive hemostatic material. Carbohydr. Polym. 249, 116860.
Lupo, B., Maestro, A., Porras, M., Gutiérrez, J.M., González, C., 2014. Preparation of alginate microspheres by emulsification/internal gelation to encapsulate cocoa polyphenols. Food Hydrocoll. 38, 56–65.
Madhusudanan, P., Raju, G., Shankarappa, S., 2020. Hydrogel systems and their role in neural tissue engineering. J. R. Soc. Interface 17 (162), 1–13.
Maestrelli, F., Mura, P., González-Rodríguez, M.L., Cózar-Bernal, M.J., Rabasco, A.M., Di Cesare Mannelli, L., Ghelardini, C., 2017. Calcium alginate microspheres containing metformin hydrochloride niosomes and chitosomes aimed for oral therapy of type 2 diabetes mellitus. Int. J. Pharm. 530 (1–2), 430–439. http://www.elsevier.com/locate/ijpharm.
Manivasagam, G., Reddy, A., Sen, D., Nayak, S., Mathew, M.T., Rajamanikam, A., 2019. Dentistry: restorative and regenerative approaches. In: Narayan, R. (Ed.), Encyclopedia of Biomedical Engineering. Elsevier, pp. 332–347.
Marit, O., Kjetill, Ø., Gudmund, S.-B., Olav, S., Patrick, S.-S., Terje, E., 1991. Induction of cytokine production from human monocytes stimulated with alginate. J. Immunother., 286–291.
Martín, M.J., Calpena, A.C., Fernández, F., Mallandrich, M., Gálvez, P., Clares, B., 2015. Development of alginate microspheres as nystatin carriers for oral mucosa drug delivery. Carbohydr. Polym. 117, 140–149. http://www.elsevier.com/wps/find/journaldescription.cws_home/405871/description#description.
Martins, S., Sarmento, B., Souto, E.B., Ferreire, D.C., 2007. Insulin-loaded alginate microspheres for oral delivery: effect of polysaccharide reinforcement on physiochemical properties and release profile. Carbohydr. Polym. 69 (4), 725–731.
Mittal, S.K., Aggarwal, N., Sailaja, G., Oplhen, A.V., HogenEsch, H., North, A., Hays, J., Moffatt, S., 2001. Immunization with DNA, adenovirus or both in biodegradable alginate microspheres: effect of route of inoculation on immune response. Vaccine 19, 253–263.
Mohanty, S., Wu, Y., Chakraborty, N., Mohanty, P., Ghosh, G., 2016. Impact of alginate concentration on the viability, cryostorage, and angiogenic activity of encapsulated fibroblasts. Mater. Sci. Eng. C 65, 269–277.
Mokhtare, B., Cetin, M., Ozakar, R.S., Bayrakceken, H., 2017. In vitro and in vivo evaluation of alginate and alginate chitosan beads containing metformin hydrochloride. Trop. J. Pharm. Res. 16 (2), 287–296. http://www.tjpr.org/admin/12389900798187/2017_16_2_5.pdf.
Moshaverinia, A., Chen, C., Akiyama, K., Ansari, S., Xu, X., Chee, W.W., Schricker, S.R., Shi, S., 2012. Alginate hydrogel as a promising scaffold for dental-derived stem cells: an in vitro study. J. Mater. Sci. Mater. Med. 23 (12), 3041–3051.
Nayak, A.K., Pal, D., Hasnain, S.M., 2013. Development, optimization and in vitro-in vivo evaluation of pioglitazone-loaded jackfruit seed starch-alginate beads. Curr. Drug Deliv. 10 (5), 608–619.
Nivens, D.E., Ohman, D.E., Williams, J., Franklin, M.J., 2001. Role of alginate and its O acetylation in formation of Pseudomonas aeruginosa microcolonies and biofilms. J. Bacteriol. 183 (3), 1047–1057.
Paques, J.P., 2015. In: Sagis, L.M.C. (Ed.), Microencapsulation and Microspheres for Food Applications. Elsevier.
Park, S.J., Lee, Y.K., Cho, S., Uthaman, S., Park, I.-K., Min, J.-J., Seong, Y.K., Park, J.-O., Park, S., 2014. Effect of chitosan coating on a bacteria-based alginate microbot. Biotechnol. Bioeng. 112, 769–776.
Patil, S.B., Sawant, K.K., 2009. Development, optimization and in vitro evaluation of alginate mucoadhesive microspheres of carvedilol for nasal delivery. J. Microencapsul. 26 (5), 432–443.
Pawar, S.N., Edgar, K.J., 2012. Alginate derivatization: a review of chemistry, properties and applications. Biomaterials 33 (11), 3279–3305.
Pestovsky, Y.S., Martínez-Antonio, A., 2020. The synthesis of alginate microparticles and nanoparticles. DDIPI J. 3 (1).
Priya, P., Deshmukh, 2017. Formulation and evaluation of microspheres of glibenclamide by ionotropic gelation method. Indo Am. J. Pharm. Res. 7 (09).
Puscaselu, G., Lobiuc, A., Dimian, M., Covasa, M., 2020. Alginate: from food industry to biomedical applications and management of metabolic disorders. Polymers 12 (10).
Raddall, G., Mello, I., Leung, B.M., 2019. Biomaterials and scaffold design strategies for regenerative endodontic therapy. Front. Bioeng. Biotechnol. 7 (317), 1–13.
Raha, A., Bhattacharjee, S., Mukherjee, P., Paul, M., Bagchi, A., 2018. Design and characterization of ibuprofen loaded alginate microspheres prepared by ionic gelation method. Int. J. Pharm. Res. Health Sci. 6, 2713–2716.

Rajmohan, D., Bellmer, D., 2019. Characterization of spirulina-alginate beads formed using ionic gelation. Int. J. Food Sci. 2019. http://www.hindawi.com/journals/ijfs/.

Ramos, P.E., Silva, P., Alario, M.M., Pastrana, L.M., Teixeira, J.A , Cerqueira, M.A., Vicente, A.A., 2018. Effect of alginate molecular weight and M/G ratio in beads properties foreseeing the protection of probiotics. Food Hydrocoll. 77, 8–16.

Rehm, B.H.A., Moradali, F., 2018. Alginates and Their Biomedical Applications. Springer.

Remminghorst, U., Hay, I.D., Rehm, B.H.A., 2009. Molecular characterization of Alg8, a putative glycosyltransferase, involved in alginate polymerisation. J. Biotechnol. 140 (3–4), 176–183.

Reyes, N., Rivas-Ruiz, I., Dominguez-Espinosa, R., Solis, S., 2006. Influence of immobilization parameters on endopolygalacturonase productivity by hybrid aspergillus sp. HL entrapped in calcium alginate. Biochem. Eng. J. 32 (1), 43–48.

Rhein-Knudsen, N., Ale, M.T., Ajalloueian, F., Meyer, A.S., 2017. Characterization of alginates from Ghanaian brown seaweeds: Sargassum spp. and Padina spp. Food Hydrocoll. 71, 236–244.

Rong, J., Liang, M., Xuan, F., Sun, J., Zhao, L., Zheng, H., Tian, X., Liu, D., Zhang, Q., Peng, C., Li, F., Wang, X., Han, Y., Yu, W., 2017. Thrombin-loaded alginate-calcium microspheres: a novel hemostatic embolic material for transcatheter arterial embolization. Int. J. Biol. Macromol. 104, 1302–1312.

Saberianpour, S., Karimi, A., Nemati, S., Amini, H., Sardroud, H.A., Khaksar, M., Mamipour, M., Nouri, M., Rahbarghazi, R., 2019. Encapsulation of rat cardiomyoblasts with alginate-gelatin microspherespreserves stemness featurein vitro. Biomed. Pharmacother., 402–407.

Sabra, W., Zeng, A.P., 2009. In: Rehm, B.H.A. (Ed.), Microbial Production of Alginates: Physiology and Process Aspects. Springer.

Sachan, N.K., Pushkar, S., Jha, A., Bhattcharya, A., 2009. Sodium alginate: the wonder polymer for controlled drug delivery. J. Pharm. Res. 2 (8), 1191–1199. http://jprsolutions.info/files/final-file-56b22486ed6434.87804114.pdf.

Sadeghi, D., Solouk, A., Samadikuchaksaraei, A., Seifalian, A.M., 2021. Preparation of internally-crosslinked alginate microspheres: optimization of process parameters and study of pH-responsive behaviors. Carbohydr. Polym. 255. http://www.elsevier.com/wps/find/journaldescription.cws_home/405871/description#description.

Sahoo, D.R., Biswal, T., 2021. Alginate and its application to tissue engineering. SN Appl. Sci. 3.

Sarker, B., Li, W., Zheng, K., Detsch, R., Boccaccini, A.R., 2016. Designing porous bone tissue engineering scaffolds with enhanced mechanical properties from composite hydrogels composed of modified alginate, gelatin, and bioactive glass. ACS Biomater. Sci. Eng. 2 (12), 2240–2254.

Seal, B., Otero, T., Panitch, A., 2001. Polymeric biomaterials for tissue and organ regeneration. Mater. Sci. Eng. R. Rep. 34 (4–5), 147–230.

Shaikh, R., Raj Singh, T.R., Garland, M.J., Woolfson, D.A., Donelly, R.F., 2011. Mucoadhesive drug delivery system. J. Pharm. BioAllied Sci. 3 (1), 89–100.

Shi, L.E., Zheng, W., Zhang, Y., Tang, Z.X., 2016. Milk-alginate microspheres: protection and delivery of enterococcus faecalis HZNU P2. LWT Food Sci. Technol. 65, 840–844. http://www.elsevier.com/inca/publications/store/6/2/2/9/1/0/index.htt.

Shtenberg, Y., Goldfeder, M., Prinz, H., Shainsky, J., Ghantous, Y., El-Naaj, I.A., Schroeder, A., Bianco-Peled, H., 2018. Mucoadhsive alginate pastes with embedded liposomes for local oral drug delivery. Int. J. Biol. Macromol. 11, 62–69.

Sindhu, S., Kumar, A., Aruchamy, B., Prasanth, A.G., Kumar, A.S., Shruthi, B.S, 2020. Preparation, characterization, and drug release behavior of nitrendipine encapsulated grafted copolymer microspheres. Mater. Today: Proc., 2296–2300.

Sinha, V.R., Trehan, A., 2003. Biodegradable microspheres for protein delivery. J. Control. Release 90 (3), 261–280.

Skjk-Bræk, G., Grasdalen, H., Larsen, B., 1986. Monomer sequence and acetylation pattern in some bacterial alginates. Carbohydr. Res. 154 (1), 239–250.

Smidsrod, O., Skjak-Bræk, G., 1990. Alginate as immobilization matrix for cells. Trends Biotechnol. 8, 71–78.

Song, H., Yu, W., Gao, M., Liu, X., Ma, X., 2013. Microencapsulated probiotics using emulsification technique coupled with internal or external gelation process. Carbohydr. Polym. 96 (1), 181–189.

Soni, M.L., Kumar, M., Namdeo, K.P., 2010. Sodium alginate microspheres for extending drug release: formulation and in vitro evaluation. Int. J. Drug Deliv. 2, 64–68.

Sosnik, A., 2014. Alginate particles as platform for drug delivery by the oral route: state-of-the-art. ISRN Pharm., 1–17.

Sosnik, A., Seremeta, K.P., 2015. Advantages and challenges of the spray-drying technology for the production of pure drug particles and drug-loaded polymeric carriers. Adv. Colloid Interf. Sci. 223, 40–54.

Spadari, C.d.C., Lopes, L.B., Ishida, K., 2017. Potential use of alginate-based carriers as antifungal delivery system. Front. Microbiol. 8. http://journal.frontiersin.org/article/10.3389/fmicb.2017.00097/full.

Sun, J., Tan, H., 2013. Alginate-based biomaterials for regenerative medicine applications. Materials 6 (4), 1285–1309. http://www.mdpi.com/1996-1944/6/4/1285/pdf.
Szekalska, M., Amelian, A., Winnicka, K., 2015. Alginate microspheres obtained by the spray drying technique as mucoadhesive carriers of ranitidine. Acta Pharma. 65 (1), 15–27. http://www.degruyter.com/view/j/acph.
Szekalska, M., Puciłowska, A., Szymańska, E., Ciosek, P., Winnicka, K., 2016a. Alginate: current use and future perspectives in pharmaceutical and biomedical applications. Int. J. Polym. Sci. 2016, 1–17.
Szekalska, M., Wroblewska, M., Sosnowska, K., Winnicka, K., 2016b. Influence of sodium alginate on hypoglycemic activity of metformin hydrochloride in the microspheres obtained by the spray drying. Int. J. Polym. Sci. 2016. http //www.hindawi.com/journals/ijps/.
Szekalska, M., Sosnowska, K., Czajkowska-Kósnik, A., Winnicka, K., 2018. Calcium chloride modified alginate microparticles formulated by the spray drying process: a strategy to prolong the release of freely soluble drugs. Materials 11 (9). http://www.mdpi.com/1996-1944/11/9/1522/pdf.
Torres, M.L., Fernandez, J.M., Dellatorre, F.G., Cortizo, A.M., Oberti, T.G., 2019. Purification of alginate improves its biocompatibility and eliminates cytotoxicity in matrix for bone tissue engineering. Algal Res. 40, 101499.
Urtuvia, V., Maturana, N., Acevedo, F., Pena, C., Diaz-Barrera, A., 2017. Bacterial alginate production: an overview of its biosynthesis and potential industrial production. World J. Microbiol. Biotechnol. 33 (11).
Uyen, T., Hamid, N.T., Thi, Z.A.A., Ahmad, L.A., 2020. Synthesis and characterization of curcumin loaded alginate microspheres for drug delivery. J. Drug Deliv. Sci. Technol. 58.
van den Brink, P., Zwijnenburg, A., Smith, G., Temmink, H., van Loosdrecht, M., 2009. Effect of free calcium concentration and ionic strength on alginate fouling in cross-flow membrane filtration. J. Membr. Sci. 345 (1–2), 207–216.
Venkatesan, J., Nithya, R., Sudha, P.N., Kim, S.K., 2014. Role of alginate in bone tissue engineering. Adv. Food Nutr. Res. 73, 45–57.
Wang, F., He, S., Chen, B., 2018. Retinoic acid-loaded alginate microspheres as a slow release drug delivery carrier for intravitreal treatment. Biomed. Pharmacother. 97, 722–728. http://www.elsevier.com/locate/biomedpharm.
Wee, S., Gombotz, W.R., 1998. Protein release from alginate matrices. Adv. Drug Deliv. Rev. 31 (3), 267–285.
Wells, L.A., Sheardown, H., 2007. Extended release of pI proteins from alginate microspheres via a novel encapsulation technique. Eur. J. Pharm. Biopharm. 65 (3), 329–335.
Xing, L., Sun, J., Tan, H., Yuan, G., Li, J., Jia, Y., Xiong, D., Chen, G., Lai, J., Lin, Z., Chen, Y., Niu, X., 2019. Covalently polysaccharide-based alginate/chitosan hydrogel embedded alginate microspheres for BSA encapsulation and soft tissue engineering. Int. J. Biol. Macromol. 127, 340–348.
Xu, Y., Peng, J., Richards, G., Lu, S., Eglin, D., 2019. Optimization of electrospray fabrication of stem cell-embedded alginate-gelatin microspheres and their assembly in 3D-printed poly(ε-caprolactone) scaffold for cartilage tissue engineering. J. Orthop. Transl. 18, 128–141.
Xuan, F., Rong, J., Liang, M., Zhang, X., Sun, J., Zhao, L., Li, Y., Liu, D., Li, F., Wang, X., Han, Y., 2017. Biocompatibility and effectiveness of new hemostatic embolization agent: thrombin loaded alginate calcium microsphere. BioMed Res Int., 1–10.
Yang, I.H., Chen, Y.S., Li, J.J., Liang, Y.J., Lin, T.C., Jakfar, S., Thacker, M., Wu, S.C., Lin, F.H., 2021. The development of laminin-alginate microspheres encapsulated with Ginsenoside Rg1 and ADSCs for breast reconstruction after lumpectomy. Bioact. Mater. 6 (6), 1699–1710.
Zhai, P., Chen, X.B., Schreyer, D.J., 2015. PLGA/alginate composites microspheres for hydrophilic protein delivery. Mater. Sci. Eng. C 56, 251–259.
Zhang, C., Grossier, R., Candoni, N., Veesler, S., 2020. Preparation of Alginate Hydrogel Microparticles Using Droplet-Based Microfluidics: A Review of Methods. ArXiv https://arxiv.org.
Zheng, C.-H., Gao, J.-Q., Zhang, Y.-P., Liang, W.-Q., 2004. A protein delivery system: biodegradable alginate-chitosan-poly(lactic-co-glycolic acid) composite microspheres. Biochem. Biophys. Commun. 323 (4), 1321–1327.
Zhou, S., Deng, X., Li, X., 2001. Investigation on a novel core-coated microspheres protein delivery system. J. Control. Release 75 (1–2), 27–36.
Zimmermann, U., Klöck, G., Federlin, K., Hannig, K., Kowalski, M., Bretzel, R.G., Horcher, A., Entenmann, H., Sieber, U., Zekorn, T., 1992. Production of mitogen-contamination free alginates with variable ratios of mannuronic acid to guluronic acid by free flow electrophoresis. Electrophoresis 13 (1), 269–274.
Zou, Q., Liao, Y., Wang, Q., Yang, Y., 2019. Microfluidic one-step preparation of alginate microspheres encapsulated with in situ-formed bismuth sulfide nanoparticles and their photothermal effect. Eur. Polym. J. 115, 282–289.

CHAPTER

11

Biomedical applications of cashew gum-based micro- and nanostructures

Gouranga Nandi[a] *and Subhankar Mukhopadhyay*[b]

[a]**Division of Pharmaceutics, Department of Pharmaceutical Technology, University of North Bengal, Raja Rammohunpur, West Bengal, India** [b]**Key Laboratory of Modern Chinese Medicines, China Pharmaceutical University, Nanjing, P.R. China**

11.1 Introduction

Gums obtained from natural sources are considered as potential biomaterials which have tremendous applications in the biomedical and pharmaceutical fields (Kumar et al., 2012; Ribeiro et al., 2016). High drug loading capacity, low toxicity, biodegradability, biocompatibility, target ability, ease of surface modification, ease of synthesis, and low cost in production make them promising candidates for controlled or sustained drug delivery systems as nanocarriers and bioimaging, biosensing, and theranostic agents (Pinto et al., 2002; Sosnik et al., 2014; Yang et al., 2015). Algae, exudates, and seeds are among various sources of natural gums (Pinto et al., 2002; Hosseini et al., 2015; Lee et al., 2016; Mudgil et al., 2016).

From ancient times, people have relied on plants to recover from certain diseases (Kosalge and Fursule, 2009). Nowadays, much research has ensured the biomedical uses of medicinal herbs containing phytochemicals such as polysaccharides, alkaloids, triterpenes, flavonoids, etc. (Dutra et al., 2016). Comparatively greater molecular weight of polysaccharides has significant influence over the control release of payloads and its biodegradation before excretion by the kidneys (Basu et al., 2015). In addition to their aforementioned benefits, they also show some limitations like reduced viscosity while storing, hydration, lower chemical stability, and mechanical characteristics (Kulkarni et al., 2012). These pitfalls could be overcome by adopting chemical cross-linking (Chaurasia et al., 2006; Maiti et al., 2011), carboxymethylation (Ray et al., 2010; Kaur et al., 2012a; Dey et al., 2013), polymer grafting (Malik and Ahuja, 2011; Malik et al., 2012), thiolation (Kaur et al., 2012b), polymer blending (Ahuja et al., 2010; Nayak et al., 2012, 2013; Nayak and Pal, 2013), etc.

Micro- and Nanoengineered Gum-Based Biomaterials for Drug Delivery and Biomedical Applications
https://doi.org/10.1016/B978-0-323-90986-0.00001-7

The cell walls of the plants are mainly composed of bio-macromolecules known as polysaccharides. These polysaccharides are responsible for providing structural stability to the plant. The exudate from *Anacardium occidentale* Linn, commonly known as the cashew tree, is the source of cashew gum (CG) polysaccharide (da Silva et al., 2018; Cardial et al., 2019). The cashew tree is widely grown in tropical regions like Northeast Brazil and the hydrophilic branched polysaccharide known as cashew gum obtained from the exudates of this plant has been found to possess ethnopharmacological potential, low toxicity, and reduced viscosity (Kulkarni et al., 2012; Da Silveira Nogueira Lima et al., 2002; Agra et al., 2007). Cashew gum has been used as an excellent excipient in pharmaceutical dosage forms and in the food industry as a food additive and viscosity builder due to its chemical inertness and physiological compatibility (Pandey and Khuller, 2004; Chamarthy and Pinal, 2008; Alonso-Sande et al., 2009). This chapter reviews the collection of crude cashew gum, and its extraction, purification, chemical structure, and applications in fabrication of various micro- and nanostructured drug delivery and biomedical devices.

11.2 Isolation and purification of cashew gum

Cashew gum is extracted from the bark exudate of *A. occidentale Linn*, which belongs to the Anacardiaceae family. This tree is native to Northeast Brazil but is also found in other tropical and subtropical countries like India, Tanzania, Kenya, and Mozambique (Menestrina et al., 1998; Cunha et al., 2009). However, various infectious diseases may hinder the growth of the cashew tree. *Analeptestrifascita* is a monotypic beetle genus which is very harmful for the cashew tree cultivated in Nigeria (Freire et al., 2002; Cardoso et al., 2004; Asogwa et al., 2011).

Beside its gum, cashew nuts are also used to prepare many foods. Similarly to the latex extraction process from rubber plants, here also resin is obtained naturally by the bark's epithelial cells lining canals. Physical, chemical, or natural stimuli play important roles in isolating gum from the tree. Incisions into the bark or administrating chemical compounds such as benzoic acid derivatives, 2-chloroethylphosphonic acid, ethylene oxide, etc., inside the bark help to extract the CG. The quality and quantity of the obtained gum depend upon the age of the tree and the environmental conditions. The isolation could be done from any part of the plant but differs in quantity and quality (Kumar et al., 2012).

Research by Torquato et al. demonstrated the process of polysaccharide isolation from the gum previously obtained from the cashew tree. Later, scientists made some modifications to the proposed method. The general scheme of the extraction and purification process of cashew gum is presented in Fig. 11.1. At first, trituration of exudate was done in a knives mill followed by solubilization of triturates in water at 300:1 (g/L) ratio. Then filtration and centrifugation at 10,000 rpm for 10 min (at room temperature) were carried out to discard the residues. Ninety-six percent ethanol was taken to precipitate the supernatant in the proportion of 1:3 (v/v). Then it was left for 1 day in a refrigerator. After extraction of ethanol, the proper drying of the precipitate was ensured by drying at 60 °C in an air-flow oven. Finally, milling was done to the sample to get CG, which was subjected to lyophilization and kept in proper sealed packets for use in future (Torquato et al., 2004).

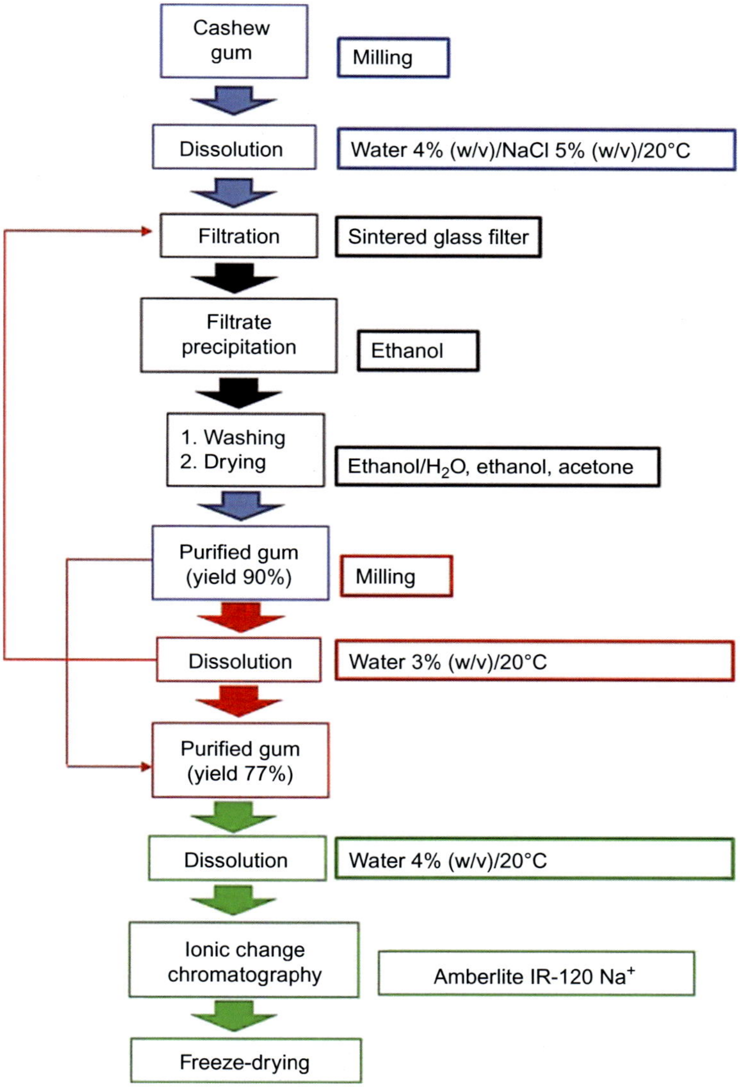

FIG. 11.1 Flow chart of purification of cashew gum from its crude form (Ribeiro et al., 2016). *From Ribeiro, A.J., De Souza, F.R.L., Bezerra, J.M.N.A., Oliveira, C., Nadvorny, D., De La Roca Soares, M.F., Nunes, L.C.C., Silva-Filho, E.C., Veiga, F., Soares Sobrinho, J.L., 2016. Gums' based delivery systems: review on cashew gum and its derivatives. Carbohydr. Polym. 147, 188–200. https://doi.org/10.1016/j.carbpol.2016.02.042.*

Proteins, nucleic acids, lignins, and inorganic salts are very common types of contaminants associated with polysaccharides obtained from natural sources like gums. The removal of such contaminants is essential to obtain purified gums. Various purification techniques such as complexation of metallic ions or quaternary ammonium salts, precipitation with alcohol or acetone, chromatographic separation, and drying were employed to achieve this goal.

Another method of purification of cashew gum was described where sodium chloride was added to put back cations in CG. At first, CG was solubilized in water at a concentration of 5% (w/v) at room temperature (25 ± 2 °C). Then it was stirred for 12 h in a magnetic stirrer. Filtration was carried out to discard the impurities. To get sodium salt polymer of CG, the pH of the system was regulated to 7; this was followed by inclusion of sodium chloride. Ethanol was employed to precipitate the CG and therefore vacuum filtration and drying were carried out to get the final product. Gel permeation chromatography is a popular choice to evaluate the molar mass of the CG by taking an ultrahydrogel linear column at a temperature of 28 °C (Araruna et al., 2020).

Britto et al. demonstrated an easier method of CG purification where CG was dissolved in ultrapure water (20 °C) and allowed to stand for 1 h to precipitate the CG. Then the supernatant was removed and the percentage yield of CG was approximately 90%, which was recovered by vacuum filtration (de Britto et al., 2012). Though the process looks simpler than the previous one, this experiment lacks some critical information such as concentration of CG solution in water, pH regulation, type of solvent used to precipitate the CG, etc. However, the same researchers developed their newest purification process where all the significant parameters were well elaborated (Forato et al., 2015). Thus, in a nutshell the dried and purified CG was synthesized from the bark-free exudate of the cashew tree followed by grinding, solubilization in water, centrifugation, alcohol precipitation, and vacuum drying.

11.3 Chemical composition and molecular structure of cashew gum

CG is a complex highly branched acidic polysaccharides extracted from the stem exudate of the cashew tree (Da Silveira Nogueira Lima et al., 2002). Abundant availability of the exudates and low extraction cost make this gum useful carrier in the micro- and nano-drug delivery field. The structure of CG is comprised of galactose (1 → 3) bonds in the major chain with the galactose (1 → 6) bonds, which is confirmed by the methylation analyses, degradation products, and sequential Smith degradations (Kumar et al., 2012; Souza Filho et al., 2018). Seventy-two percent galactose, 14% glucose, 4.6% arabinose, 4.7% glucuronic acid, and 3.2% rhamnose are the crucial monosaccharide moieties responsible for constituting the structure of CG (de Paula et al., 1998). The chemical structure of the repeating units of CG-polymeric backbone is presented in Fig. 11.2.

The CG polysaccharide is a low-viscosity compound that is prone to contamination by microbes, and its stability gets affected after long-term storage due to a fall in viscosity (De Paula and Rodrigues, 1995). These drawbacks could be overcome by chemical alteration to the CG that further may improve the overall quality of the CG formulation (Zhang et al., 2009; Rana et al., 2011). CG is a hydrophilic polysaccharide which can be modified by certain functionalizing agents or groups to make it an amphiphilic polymer. That modification of the CG will help to incorporate and entrap hydrophobic therapeutic candidates into it (Pitombeira et al., 2015). In addition, the hydrophobicity to the CG can be induced by modification with acetic anhydride. Pyridine and sulfuric acid were used as catalyzing agents in that reaction. Thus synthesized CG was utilized as a drug delivery vehicle. The introduction of an acetyl group in the structure of CG was executed to make it a biodegradable nanocarrier

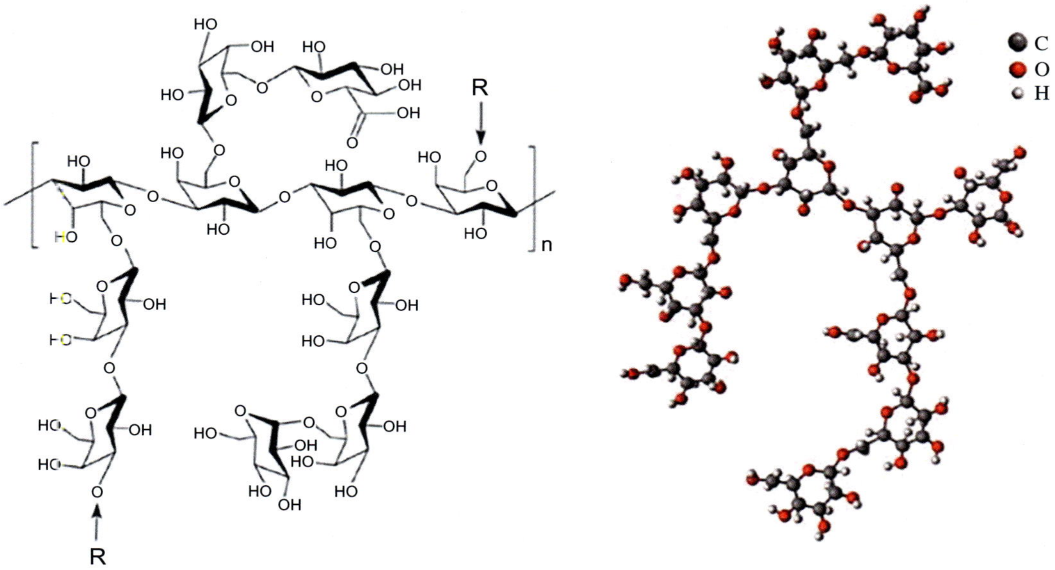

FIG. 11.2 Chemical structure of cashew gum (Barros et al., 2020). *From Barros, A.B., Moura, A.F., Silva, D.A., Oliveira, T.M., Barreto, F.S., Ribeiro, W.L.C., Alves, A.P.N.N., Araújo, A.J., Moraes Filho, M.O., Iles, B., Medeiros, J.V.R., Marinho-Filho, J.D.B., 2020. Evaluation of antitumor potential of cashew gum extracted from* Anacardium occidentale *Linn. Int. J. Biol. Macromol. 154, 319–328. https://doi.org/10.1016/j.ijbiomac.2020.03.096.*

in novel drug delivery field. This alteration ensures the release of the active pharmaceutical ingredient in a controllable manner and also protects it from the gastrointestinal environment (Dias et al., 2016). The acetylated CG possessed a greater anionic property than the normal CG. This anionic CG showed more interaction with cationic biopolymers to synthesize nanoparticles (Silva et al., 2019a).

11.4 Physiochemical characteristics of cashew gum

The origin and age of the cashew tree have a significant influence on its physiological properties. Viscosity is one of the key physical properties which ensure the quality of the cashew gum. Higher viscosity determines good-quality gum. Some researchers have done an experiment by collecting CG from below and above cashew trees 10 years of age that are from four different sources. Here, the viscosity of CG was compared with gum arabic which shows that CG possesses more viscosity than its other counterpart. CG with 1% concentration exhibited 10.03 cPs viscosity whereas gum arabic only showed 5 cPs. This confirmed that CG is a more competent stabilizer or thickening candidate. CG with good viscosity can be obtained from aged trees. Differences in pH and molecular weight may be the reasons behind the variation in the CG viscosity (Kumar et al., 2012).

CG is a water-soluble polysaccharide that forms an opaque solution (Ribeiro et al., 2016). Other derivatives and adulterants associated with CG polysaccharide may be responsible for this. Temperature and pH play critical roles in the solubility of CG. With an increase of temperature, the solubility also increases. However, a lower pH may hinder solubility (Zakaria and Rahman, 1996; Owusu et al., 2005). It is also believed that CG shows good solubility because of the presence of galacturonic and glucuronic acids (da Silva et al., 2007).

11.5 Chemical modifications of cashew gum

Cashew gum has been chemically modified or tailored in order to make it suitable for specific applications. Different kinds of chemical modifications of cashew gum polysaccharide moiety have been reported, and these include acetylation, carboxymethylation, phthalation, and graft copolymerization (Silva et al., 2006; Rana et al., 2011; Guilherme et al., 2005).

11.5.1 Acetylation of cashew gum

Cashew gum has been derivatized through acetylation to impart hydrophobicity to its moiety, which finally results in a hydrophilic-lipophilic balance (HLB) to cashew gum polysaccharide. Acetylated cashew gum has been fabricated into self-assembled polymeric micelles or nanoparticles for encapsulation of hydrophobic drugs. Lima et al. recently described self-assembled nanoparticles of acetylated cashew gum for encapsulation of hydrophobic antifungal drug amphotericin B (Lima et al., 2018). Acetylated derivative of the gum was obtained by mixing the gum with acetic anhydride and acetic acid in a particular ratio and subsequent freezing at 5 °C and heating at temperatures of 50–90 °C. The derivative was finally collected by precipitation using ethanol. The natural alkaloid epiisopiloturine has also been encapsulated by acetylated cashew gum nanoparticles in order to increase its solubility as well as sustain its release (Rodrigues et al., 2019).

11.5.2 Carboxymethylation of cashew gum

Carboxymethyl derivative of cashew gum has been reported for fabrication of various drug delivery systems and biomedical applications such as antibody immobilizations for biosensors. The Williamson etherification method has been employed for carboxymethylation of natural polysaccharides where the gum is first kneaded with 10 M sodium hydroxide solution to obtain gum alkoxide; this is followed by addition of monochloroacetic acid to impart carboxymethyl groups to the –OH groups of the polysaccharide moiety. Finally, the carboxymethyl gum is precipitated with acetone or another suitable organic solvent (Meloa et al., 2020). Carboxymethylation introduces –COOH groups to polysaccharide moiety, which results in an increase in hydration capacity of the gum. Carboxymethylation of gum has also been found to be suitable for the preparation of antibacterial silver nanoparticles (Araruna et al., 2020).

11.5.3 Phthalation of cashew gum

Like acetylation of natural polysaccharides, phthalation has also been reported as another approach to impart hydrophobicity to polysaccharide moiety. Phthalation of –OH groups of gum polysaccharides has been described to increase drug entrapment efficiency and sustained release capacity. Oliveira and coresearchers described a method of introduction of phthalic groups in cashew gum polysaccharide backbone, where the gum was mixed with a solution of phthalic anhydride followed by microwave irradiation for a specified time (Oliveira et al., 2021). The reaction parameters such as reaction time and microwave irradiation magnitude were optimized in the study. Finally, phthalated cashew gum was fabricated into benznidazole-loaded nanoparticles.

11.5.4 Graft copolymerization of cashew gum

Graft copolymer of cashew gum and poly (L-lactide) has been described in order to make a cashew gum amphiphilic polymer for encapsulation of hydrophobic drugs through micellar solubilization and therefore improve aqueous solubility and subsequent bioavailability (Richter et al., 2020). For the grafting of L-lactide onto cashew gum backbone, first the gum was dissolved in dimethylsulfoxide and subsequently a monomer and triethylamine were mixed with it at temperatures of 70–75 °C under magnetic stirring in a nitrogen atmosphere. After 2 h, the nitrogen flow was interrupted and the reaction mixture was left for 10 h. Finally, the mixture was filtered, dialyzed against water, and freeze-dried. The unreacted monomer and homopolymer were extracted from the copolymer using hexane. Cashew gum-grafted-poly(L-lactide) copolymer was described to prepare self-assembled nanoparticles of hydrophobic antifungal drug, amphotericin B, in order to increase its solubility as well as its antifungal activity (Richter et al., 2020). Another graft copolymer of cashew gum was reported for fabrication of nanoparticles, where cashew gum was grafted with poly(N-isopropylacrylamide) using a free radical initiator. The graft copolymer was fabricated into self-assembled epirubicin-loaded nanoparticles for its controlled release (Abreu et al., 2016).

11.6 Cashew gum-based microstructures

A wide range of microstructures based on native cashew gum or its various derivatives have been reported, which includes microspheres, microcapsules, and microbeads. The preparations of such microstructures are based on different green approaches such as polyelectrolyte complexation, ionotropic gelation, and spray-drying. Some examples of cashew gum-based microstructures with drug delivery applications are presented in Table 11.1.

11.6.1 Microspheres based on polyelectrolyte complex

Microspheres of carboxymethyl cashew gum (CCG) with chitosan were fabricated by needle extrusion of CCG solution into a chitosan solution and subsequent gelation via polyelectrolyte complexation for delivery of model protein bovine serum albumin (BSA) (Magalhães Jr. et al., 2009). Two different grades of carboxymethyl cashew gum having different degrees

TABLE 11.1 Some examples of cashew gum-based microstructures with biomedical applications.

Types of microstructures	Fabrication techniques	Applications	References
Carboxymethyl cashew gum-chitosan microspheres	Needle extrusion; polyelectrolyte complexation	Protein delivery via oral route	Magalhães Jr. et al. (2009)
Microbeads	Needle extrusion; ionotropic gelation	Controlled delivery of isoxuprine	Das et al. (2014)
Microcapsules	Emulsification followed by spray-drying	Encapsulation of essential oil	Fernandes et al. (2016)
Electrosprayed microparticles	Emulsification followed by electrospraying	Encapsulation of β-carotene	Vázquez-González et al. (2021)

of carboxymethyl substitutions were used in the preparation. The study showed higher swelling of the batch with higher degree of substitution, which might be due to the higher number of polar carboxylic groups in the polymeric backbone with higher degrees of substitution. Faster BSA release rates were found for microparticles prepared with the higher degree of substitution, whereas those prepared with a degree of substitution of 0.16 were found to take twice the time for similar release.

11.6.2 Microspheres based on ionic cross-linking

Microbeads of isoxsuprine hydrochloride composed of Zn^{++}-cross-linked composite matrix of sodium alginate and carboxymethyl cashew gum have been described (Das et al., 2014). The ionotropic gelation technique was employed for the preparation of microbeads using zinc sulfate as a cross-linker. The influences of polymer ratio and concentration of $ZnSO_4$ on drug entrapment efficiency and drug release at 7 h were studied and optimized using 2^3 full factorial design. Optimized microbead formulation showed nearly 80% drug entrapment efficiency and only around 60% drug release at 7 h. An in vitro drug release study also exhibited excellent sustained-release capacity with the Korsmeyer-Peppas model release-mechanism in a phosphate buffer with pH 6.8.

11.6.3 Cashew gum-based microcapsules

Microcapsules of ginger essential oil have been described, which were prepared with cashew gum along with inulin as wall forming material by ultrasound-assisted emulsification and subsequent spray-drying. In the study, the dissolution rate of the wall of the microcapsules was found to be proportional to the concentration of inulin in wall composition and it reached a maximum value of about 90% at the highest inulin concentration. Therefore, encapsulation efficiency was found to be as low as 15.8% at the highest inulin concentration. The treatment of the microcapsules with cashew gum as an encapsulant showed the highest water absorption capacity. The matrix of cashew gum and inulin in a 3:1 (w/w) ratio was found to exhibit the highest encapsulation efficiency and best morphology, having no cracks on the surface (Fernandes et al., 2016).

11.6.4 Electrosprayed microparticles

β-Carotene-encapsulated microparticles of cashew gum have been prepared by a novel electrospraying method employing a green approach (Vázquez-González et al., 2021). β-Carotene was first dissolved in castor oil and then emulsified as an internal dispersed phase in an aqueous solution of cashew gum as an external continuous phase. Two different phase-volume ratios (internal phase:external phase) of 20:80 and 10:90 were used in the study. The oil-in-water (o/w) emulsion was then sprayed using the electrospraying method and dried in order to obtain spherical microparticles. Spherical microparticles with smooth surface and average diameters of 3–6 μm were obtained from the study. The particles obtained from the emulsion with phase volume ratio of 10:90 (v/v) exhibited a loading capacity of $0.075 \pm 0.006\%$ and a minor amount of extractable β-carotene of $10.75 \pm 2.42\%$. No drug-excipients interaction was found in the ATR-FTIR study. This study demonstrated cashew gum as a viable polymer candidate for encapsulation of different bioactive compounds through the electrospraying method for drug delivery, biomedical, agricultural, and food applications.

11.6.5 Cashew gum-based chemically cross-linked microparticles

Pequi oil-loaded cashew gum-based microparticles have been developed by the coacervation and phase separation process followed by chemical cross-linking to increase the rigidity of the microparticles. The purposes of the encapsulation of pequi oil in the proposed microparticles were to increase the stability of the oil and mask unpleasant flavors and aromas. Complex coacervation is a technique that is based on the electrostatic interaction between two oppositely charged biopolymers and subsequent formation of polyelectrolyte complex. The microparticles formed due to coacervation were further cross-linked with tannic acid to increase their rigidity. Cashew gum is an anionic biopolysaccharide and gelatin was used as a cationic polymer for the formation of polyelectrolyte complex. The ratio of cashew gum, gelatin, and oil was 2:1:1 (m/m/m) in the study. The cross-linking process was carried out with tannic acid for 30 min at 40 °C. The optimized formulation of the microparticles by the rotational central compound design for microparticle formation was found to be 0.65% biopolysaccharides (cashew gum and gelatin) and 6.9% tannic acid. The study showed that an increase in the percentage of tannic acid in the cross-linking of the microparticles increased the yield and encapsulation efficiency. The degradation temperature was found to increase due to cross-linking, which indicatedimproved the thermal stability of the particles due to cross-linking (Alexandre et al., 2019).

Another work has been reported where rosemary essential oil was encapsulated in microparticles composed of cashew gum and galactomannan composite cross-linked by sodium trimetaphosphate in order to increase stability as well as to ensure controlled release. The oil was emulsified in an aqueous solution of cashew gum and galactomannan, and during emulsification the polysaccharides were cross-linked. The emulsion was finally atomized to obtain oil-loaded microparticles, which were evaluated for moisture, solubility, particle size, morphology, encapsulation efficiency, antimicrobial activity, and chemical composition of the essential oil. The galactomannan-cashew gum composite exhibited higher encapsulation efficiency and sustained oil release compared to the single cashew gum matrix, which substantiates the theory that incorporation of galactomannan in the cashew gum matrix improved the characteristics of microparticles due to the galactomannan's high emulsifying capacity even in low concentrations. The antimicrobial and other chemical properties of the essential oil were found to be retained after the encapsulation procedure (Mendes et al., 2020).

11.7 Cashew gum-based nanostructures

Different types of nanostructures based on cashew gum and its different derivatives such as nanoparticles, nanosuspension, nanoemulsion, self-assembled nanomicelles, nanogels, etc., have been reported. Various drug delivery applications have been exhibited by these nanostructures. Some examples of such nanostructures are presented in Table 11.2.

TABLE 11.2 Examples of cashew gum-based nanostructures with biomedical applications.

Types of nanostructures	Fabrication techniques	Applications	References
Self-assembled nanoparticles	Acetylation of cashew gum; self-assembling; water dialysis using dialysis membrane	Encapsulation and sustained release of hydrophobic amphotericin B, epiisopiloturine	Lima et al. (2018) and Rodrigues et al. (2019)
Cashew gum-stabilized silver nanoparticles	Reduction of silver nitrate by cashew gum and/or carboxymethyl cashew gum	Antibacterial activity of silver nanoparticles embedded in cashew gum	Araruna et al. (2020)
Phthalated cashew gum nanoparticles	Phthalation of cashew gum; nanoprecipitation and subsequent solvent evaporation	Protection of benznidazole from gastric fluid; useful in Chagas disease	Oliveira et al. (2021)
Self-assembled nanoparticles	Cashew-gum-grafted-poly(L-lactide) copolymer; self-assembling	Micellar solubilization of hydrophobic drug amphotericin B	Richter et al. (2020)
Chitosan-cashew gum nanogel	Polyelectrolyte complexation (PEC); emulsification of oil in PEC; spray-drying	Encapsulation of *Lippia sidoides* oil; improved larvicide activity	Abreu et al. (2012)
Self-assembling nanoparticles	Cashew-gum-grafted-poly(N-isopropylacrylamide) copolymer; dialysis against water; self-assembling	Encapsulation of epirubicin; controlled release	Abreu et al. (2016)
Transdermal nanoparticles	Nanoprecipitation and dialysis against water	Transdermal delivery of diclofenac diethylamine with improved permeation	Dias et al. (2016)
Cashew gum nanoparticles as Pickering emulsion stabilizer	Acetylation of cashew gum; nanoparticles formation by nanoprecipitation; emulsification with nanoparticles	Stabilization of Pickering emulsion as indomethacin carrier	Cardial et al. (2019)
Silver nanoparticles stabilized by phthalated cashew gum	Phthalation of cashew gum; reduction of silver nitrate solution by phthalated cashew gum	Green approach for preparation of antibacterial silver nanoparticles	Oliveira et al. (2019)

TABLE 11.2 Examples of cashew gum-based nanostructures with biomedical applications—cont'd

Types of nanostructures	Fabrication techniques	Applications	References
Self-assembled nanoparticles of acetylated cashew gum	Acetylation of cashew gum; dialysis against water	Controlled release of indomethacin	Pitombeira et al. (2015)
Nanocomposite	Dispersion of kaolinite in aqueous DMSO followed by mixing of cashew gum	Controlled release of doxazosin mesylate	Silva et al. (2020)
Nanoparticles based on polyelectrolyte complex	Solvent-free acetylation of cashew gum; polyelectrolyte complexation with chitosan	Controlled-release oral delivery of insulin	Silva et al. (2019a)

11.7.1 Self-assembled nanoparticles

Synthesis and characterizations of a stimuli-responsive copolymer of cashew gum (CG) have been described, where N-isopropylacrylamide (NIPA) has been grafted onto the cashew gum polysaccharide backbone by free radical grafting. The study exhibited the thermo-responsiveness of the copolymer, which might be useful in the fabrication of various stimuli-responsive drug delivery systems. The critical aggregation concentration of the copolymer was found to be higher at 25 °C compared to that at 50 °C. Smaller nanoparticles ranging from 12 to 21 nm were found to form at temperatures lower than critical solution temperature (CST) depending upon the ratio of CG to NIPA, whereas larger particles were formed at temperatures greater than CST. No cytotoxicity of the proposed nanoparticles was found in the cytotoxic study with the Caco-2 and HT29-MTX intestine cell lines. The nanoparticles with a CG:NIPA ratio of 1:1 showed 64% aggregation efficiency, 22% epirubicin loading, and sustained drug release, which indicates the potential of the copolymer as a controlled release drug delivery smart biopolymer (Abreu et al., 2016).

Cashew gum has been chemically modified by acetylation and subsequently fabricated into self-assembled nanoparticles for encapsulation of hydrophobic drugs such as amphotericin B (Lima et al., 2018). The effects of various reaction parameters on the yield, degree of acetylation, and physicochemical properties of acetylated gum have been analyzed by 2^3 factorial design. The acetylated form of CG was found to form self-assembled nanoparticles having particle sizes ranging from 190 to 300 nm when added in water. However, the acetylated derivatives of CG having a degree of acetylation below 1.5 were found to form relatively finer particles. The study also revealed that the nanoparticles were capable of encapsulating about 70% of the drug and releasing the drug at a sustained rate.

An acetyl derivative of cashew gum has been described as a controlled release polymer for encapsulation and sustained release of the natural alkaloid epiisopiloturine having antiinflammatory and antischistosomal effects. Acetyl groups are introduced into functional groups of polymeric moiety of natural polysaccharides in order to impart partial hydrophobicity to them, and this approach makes the polysaccharides suitable for fabrication of nanoparticles and drug encapsulation. In this study, the drug-loaded nanoparticles were found to increase the solubility of epiisopiloturine and sustain its release. The nanoparticles were also found to be spherical in shape with significant drug encapsulation efficiency (about 55%) and stable in physiological fluids. An in vitro release study showed the sustained-release capacity of the nanoparticles over a period of 6 h with Fickian diffusion mechanism (Rodrigues et al., 2019).

Another work also described the application of acetyl derivative of cashew gum in the fabrication of self-assembled nanoparticles. The self-assembled nanoparticles were obtained from the dialysis of an organic solution of acetylated cashew gum in dimethylsulfoxide against water. The average diameter and the critical aggregation concentration in water were found to be 179 nm and 2.1×10^{-3} g/L, respectively. A model hydrophobic drug indomethacin was successfully encapsulated in the nanoparticles and the in vitro study revealed the sustained-release capacity over a period of 72 h (Pitombeira et al., 2015).

CG has also been modified through graft copolymerization with poly(L-lactide) in order to impart hydrophobicity to it and therefore make it an amphiphilic biopolymer (Richter et al., 2020). Amphiphilic polymers are capable of forming self-assembled micelles or nanoparticles which entrap hydrophobic drug molecules and increase their aqueous solubility through micellar solubilization. This approach has been applied to enhance the bioavailability of hydrophobic drugs as well as their systemic toxicity. The study showed that drug entrapment resulted in an increase in particle size compared to placebo nanoparticles. Nanoparticles with a drug-polymer ratio of 1:10 showed higher stability than nanoparticles with a 1:1 ratio. The approach has shown higher drug entrapment efficiency, low hemolytic effect, and optimum antifungal activity.

11.7.2 Silver nanoparticles stabilized with cashew gum

Green preparation of silver nanoparticles using various natural polysaccharides has been drawing attention due to its simplicity, environmental friendliness, and cost effectiveness. A green synthetic route for the preparation of silver nanoparticles stabilized by cashew gum and its carboxymethyl derivative has been described (Araruna et al., 2020). The study utilized microwave energy for rapid synthesis and showed significant yield of stabilized silver nanoparticles with significant antibacterial efficacy against *Staphylococcus aureus* and *Escherichia coli*. The study shows that higher pH results in formation of well-dispersed, spherical, and stable nanoparticles having significant antibacterial activity in the case of both native CG and carboxymethyl CG.

A solvent-free environmentally benign method for the preparation of silver nanoparticles using phthalated cashew gum has been described. Phthalated cashew gum was used in the green route preparation of silver nanoparticles, whereas the conventional route used sodium borohydride as a reducing agent. The average diameter and the zeta potential of the silver nanoparticles were found to be 51.9 nm and −55.8 mV, respectively. The silver nanoparticles prepared by the conventional method were found to be larger in size compared to those of the green route. Silver nanoparticles synthesized using phthalated cashew gum exhibited significant antimicrobial activity against *S. aureus* and *E. coli* (Oliveira et al., 2019).

11.7.3 Cashew gum-based nanoparticles by nanoprecipitation techniques

Nanoprecipitation has been considered as a very simple, semigreen, and effective route of nanoparticle preparation. This method is useful for synthetic polymers as they are hydrophobic and soluble in water-miscible organic solvents. On the other hand, the synthetic polymers are losing their importance due to their toxicity, poor biocompatibility, nonbiodegradability, and higher cost. Natural polysaccharides are biocompatible, less toxic, biodegradable, and cost-effective, but being hydrophilic they are not suitable for nanoprecipitaion. To overcome this problem, natural polysaccharides have been chemically modified to impart hydrophobicity to their chemical moiety and make them suitable for preparation of nanoparticles by nanoprecipitation. Cashew gum has recently been phthalated in order to make it hydrophobic and fabricated into nanoparticles for encapsulation of benznidazole (Oliveira et al., 2021). The study showed that phthalated cashew gum nanoparticles of benznidazole have potential as a new alternative for improving the treatment of Chagas disease.

11.7.4 Cashew gum-based nanogel

Nanogels based on polyelectrolyte complex of cashew gum with chitosan were described for encapsulation of *Lippia sidoides* oil and subsequent improvement of its larvicidal activity. The nanogel particles were prepared by formation of polyelecrolyte complex (PEC) of cashew gum and chitosan, subsequent emulsification of the oil in the PEC coacervate, and spraying of the emulsion in a spray-dryer chamber. The optimized nanogel having a composition of total polymer concentration 5%, polymer ratio 1:1, and matrix-oil ratio 5:1 was found to exhibit 70% drug entrapment efficiency, 11.8% drug loading, sustained drug release, and improved larvicidal activity (Abreu et al., 2012).

11.7.5 Cashew gum-based nanoparticles for transdermal applications

The presentation of a drug in the nanoencapsulated form at skin surface has gained attention to increase the transdermal permeation of the drugs. Chemically modified biopolysaccharides have been explored for the fabrication of drug-loaded nanoparticles due to their lower toxicity and their environmentally friendly preparation. Dias and coresearchers described a work where cashew gum was chemically modified by acetylation and subsequently used in the fabrication of drug-loaded nanoparticles to improve the transdermal permeation of an antiinflammatory drug, diclofenac diethylamine (Dias et al., 2016). In the study, nanoparticles were prepared by both nanoprecipitation and dialysis methods. The particles prepared by the dialysis method were found to be larger in size than those by nanoprecipitation, but the particles obtained from nanoprecipitation showed better yield and stability. The encapsulation efficiency was found to be significant (>60%) for both the methods. The biocompatibility was checked by the cytotoxicity assay, which demonstrated no significant effect on the cell viability. The in vitro drug-release and permeation study revealed the controlled pattern of drug release and 90% transdermal drug penetration.

11.7.6 Cashew gum-based nanoparticles as Pickering emulsion stabilizer

An interesting application of chemically modified cashew gum in the stabilization of Pickering emulsion has recently been described. In this study, cashew gum was first acetylated and various degrees of acetylation were then used to stabilize Pickering emulsions. The study revealed the improved stability of the emulsion formulated with acetyl derivative of cashew gum with globule size from 269 to 312 nm, unimodal size distribution, encapsulation efficiency up to 52%, and a controlled-release profile of encapsulated indomethacin (Cardial et al., 2019).

11.7.7 Cashew gum-based bionanocomposite

A cashew gum and kaolinite based nanocomposite has been demonstrated for controlled release of doxazosin mesylate. The formation of nanocomposite and successful drug incorporation were ratified by X-ray diffraction, differential scanning calorimetry, and Fourier transform infra-red spectroscopy. The effects of solution pH, adsorbent dose, initial drug concentration, contact time, and temperature were systematically investigated in the study. The results showed the achievement of equilibrium at around 60 min with a maximum adsorption capacity of 31.5 ± 2.0 mg g^{-1} at a pH of 3.0 and 25 °C. A hydrogen bonding contribution was found in doxazocin mesylate incorporation. The in vitro drug release study showed around 16% drug release at pH 1.2 and 77% at pH 7.4 (Silva et al., 2020).

11.7.8 Cashew gum nanoparticles based on polyelectrolyte complexation

Polyelectrolyte complexation has been considered as another attractive approach for the preparation of nanoparticles because of its simplicity, environmental friendliness, and biocompatibility. Cashew gum is an excellent anionic polyelectrolyte member for polyelectrolyte complex formation. Several studies have been reported demonstrating the preparation of nanoparticles with cashew gum and its chemically modified forms. An acetyl derivative of cashew gum has been described with a fast, simple, solvent-free, and low-cost methodology. The derivatized gum was used as a platform for protein delivery systems using insulin as a model drug. The study demonstrated successful formation of drug-loaded nanoparticles with an average size of 460 nm, drug entrapment efficiency of 52.5%, zeta potential of +30.6 mV, and a sustained-release profile of insulin over 24 h (Silva et al., 2019a).

11.7.9 Nanoparticles prepared by ionotropic gelation

Insulin is administered through the subcutaneous route; this is invasive and sometimes not suitable for self-administration, and requires assistance from a healthcare professional. This demands peroral delivery systems for insulin, but the main constraint of peroral delivery of insulin is its degradation in gastric fluid in the presence of proteolytic enzymes as insulin is a protein hormone. The oral route would be the most physiological and convenient option if this constraint could be overcome. A nanostructured peroral system was developed for oral insulin delivery based on cashew gum. Due to the presence of –COOH groups in the glucuronic acid moieties in its polymeric backbone, cashew gum is capable of undergoing ionotropic gelation and nanoparticles can be fabricated using this property employing a green approach. Insulin-loaded nanoparticles were prepared through the ionotropic gelation method integrating cashew gum, dextran sulfate, and poloxamer. The nanoparticles were further stabilized with chitosan and poly(ethylene glycol) and finally coated with albumin. The particles were characterized for particle size, size distribution, zeta potential, and drug entrapment efficiency. Cashew gum was also characterized prior to formulation by determining the molar mass (2.35×104 g/mol). The bare particles were found to have an average size of 156 nm, whereas coated particles had an average size of 5387 nm with 92% drug entrapment efficiency and zeta potential of −51 mV. These parameters indicate the electrostatic stabilization of the nanoparticles and suggest an innovative cashew gum-based system for peroral administration of insulin (Silva et al., 2019b).

11.8 Conclusions

Cashew gum has been gaining importance as a potential drug delivery polymeric excipient due to its easy availability, low cost, low toxicity, environmentally friendly fabricability, biodegradability, and broad-spectrum physicochemical compatibility with various drugs. Recently, a wide range of drug delivery systems and biomedical soft devices such as wound healing dressings, tissue engineering scaffolds, along with various micro- and nanostructured-based devices have been reported with promising results. Different studies showed the potential of native as well as chemically modified forms of cashew gum as a polymeric platform for various biomedical applications. Cashew gum has also been used in food industries as a thickener, diluant, emulsifier, and sweetener. The presence of carboxylic groups in the glucuronic moiety makes it an important member of anionic biopolysaccharides such as alginate, gum acacia, gellan, pectin, etc., and suitable for polyelectrolyte complexation. In this chapter, the chemical structure of cashew gum, its various chemical modifications, and their applications in the design and fabrication of various micro- and nanostructured drug delivery devices have been presented. However, more substantive studies are required to customize its physicochemical properties such as solubility, rheology, viscoelasticity, etc. to meet the intentions of a particular drug delivery strategy and establish its potential as an industrially applicable polymer for manufacturing of commercially available pharmaceutical dosage forms and cosmetics.

References

Abreu, F.O.M.S., Oliveira, E.F., Paula, H.C.B., de Paula, R.C.M., 2012. Chitosan/cashew gum nanogels for essential oil encapsulation. Carbohydr. Polym. 89, 1277–1282.

Abreu, C.M.W.S., Paula, H.C.B., Seabra, V., Feitosa, J.P.A., Sarmento, B., de Paula, R.C.M., 2016. Synthesis and characterization of non-toxic and thermo-sensitive poly(N-isopropylacrylamide)-grafted cashew gum nanoparticles as apotential epirubicin delivery matrix. Carbohydr. Polym. 154, 77–85.

Agra, M.D.F., Freitas, P.F.D., Barbosa-Filho, J.M., 2007. Synopsis of the plants known as medicinal and poisonous in northeast of Brazil. Rev. Bras 17 (1), 114–140.

Ahuja, M., Yadav, M., Kumar, S., 2010. Application of response surface methodology to formulation of ionotropically gelled gum cordia/gellan beads. Carbohydr. Polym. 80 (1), 161–167.

Alexandre, J.D.B., Barroso, T.L.C.T., Oliveira, M.D.A., Mendes, F.R.D.S., Costa, J.M.C.D., Moreira, R.D.A., et al., 2019. Cross-linked coacervates of cashew gum and gelatin in the encapsulation of Pequi oil. Cienc. Rural 49 (12), e20190079.

Alonso-Sande, M., Teijeiro-Osorio, D., Remuñán-López, C., Alonso, M., 2009. Glucomannan, a promising polysaccharide for biopharmaceutical purposes. Eur. J. Pharm. Biopharm. 72 (2), 453–462.

Araruna, F.B., Oliveira, T.M., Quelemes, P.V., Nobre, A.R., Pl'acido, A., Vasconcelos, A.G., Paula, R.C.M., Mafud, A.-C., Almeida, M.P., Delerue-Matos, C., Mascarenhas, Y.P., Eaton, P., Leite, J.R.S.A., Silva, D.A., 2020. Antibacterial application of natural and carboxymethylated cashew gum-based silver nanoparticles produced by microwave-assisted synthesis. Carbohydr. Polym. 241, 115260.

Asogwa, E., Ndubuaku, T., Hassan, A., 2011. Distribution and damage characteristics of Analeptes trifasciata Fabricius 1775 (Coleoptera: Cerambycidae) on cashew (*Anacardium occidentale* Linnaeus 1753) in Nigeria. Agric. Biol. J. N. Am. 2 (3), 421–431.

Barros, A.B., Moura, A.F., Silva, D.A., Oliveira, T.M., Barreto, F.S., Ribeiro, W.L.C., Alves, A.P.N.N., Araújob, A.J., Filho, M.O.M., Iles, B., Medeiros, J.V.R., Marinho-Filho, J.D.B., 2020. Evaluation of antitumor potential of cashew gum extracted from *Anacardium occidentale* Linn. Int. J. Biol. Macromol. 154, 319–328.

Basu, A., Kunduru, K.R., Abtew, E., Domb, A.J., 2015. Polysaccharide-based conjugates for biomedical applications. Bioconjug. Chem. 26 (8), 1396–1412.

Cardial, M.R.L., Paula, H.C., da Silva, R.B.C., da Silva Barros, J.F., Richter, A.R., Sombra, F.M., de Paula, R.C., 2019. Pickering emulsions stabilized with cashew gum nanoparticles as indomethacin carrier. Int. J. Biol. Macromol. 132, 534–540.

Cardoso, J., Santos, A., Rossetti, A., Vidal, J., 2004. Relationship between incidence and severity of cashew gummosis in semiarid North-Eastern Brazil. Plant Pathol. 53 (3), 363–367.

Chamarthy, S.P., Pinal, R., 2008. Plasticizer concentration and the performance of a diffusion-controlled polymeric drug delivery system. Colloids Surf. A Physicochem. Eng. Asp. 331 (1–2), 25–30.

Chaurasia, M., Chourasia, M.K., Jain, N.K., Jain, A., Soni, V., Gupta, Y., Jain, S.K., 2006. Cross-linked guar gum microspheres: a viable approach for improved delivery of anticancer drugs for the treatment of colorectal cancer. AAPS PharmSciTech 7 (3), E143.

Cunha, P.L.R.D., Paula, R.C.M.D., Feitosa, J., 2009. Polysaccharides from Brazilian biodiversity: an opportunity to change knowledge into economic value. Quim. Nova 32 (3), 649–660.

da Silva, D.A., de Paula, R.C., Feitosa, J.P., 2007. Graft copolymerisation of acrylamide onto cashew gum. Eur. Polym. J. 43 (6), 2620–2629.

da Silva, D.P.B., Florentino, I.F., da Silva Moreira, L.K., Brito, A.F., Carvalho, V.V., Rodrigues, M.F., Fernandes, K.F., 2018. Chemical characterization and pharmacological assessment of polysaccharide free, standardized cashew gum extract (*Anacardium occidentale* L.). J. Ethnopharmacol. 213, 395–402.

Da Silveira Nogueira Lima, R., Rabelo Lima, J., Ribeiro de Salis, C., de Azevedo Moreira, R., 2002. Cashew-tree (*Anacardium occidentale* L.) exudate gum: a novel bioligand tool. Biotechnol. Appl. Biochem. 35 (1), 45–53.

Das, B., Dutta, S., Nayak, A.K., Nanda, U., 2014. Zinc alginate-carboxymethyl cashew gum microbeads for prolongeddrug release: development and optimization. Int. J. Biol. Macromol. 70, 506–515.

de Britto, D., de Rizzo, J.S., Assis, O.B., 2012. Effect of carboxymethylcellulose and plasticizer concentration on wetting and mechanical properties of cashew tree gum-based films. Int. J. Polym. Anal. Charact. 17 (4), 302–311.

De Paula, R., Rodrigues, J., 1995. Composition and rheological properties of cashew tree gum, the exudate polysaccharide from *Anacardium occidentale* L. Carbohydr. Polym. 26 (3), 177–181.

de Paula, R.C., Heatley, F., Budd, P.M., 1998. Characterization of Anacardium occidentale exudate polysaccharide. Polym. Int. 45 (1), 27–35.

Dey, P., Maiti, S., Sa, B., 2013. Gastrointestinal delivery of glipizide from carboxymethyl locust bean gum–Al^{3+}—-alginate hydrogel network: in vitro and in vivo performance. J. Appl. Polym. Sci. 128 (3), 2063–2072.

Dias, S.F.L., Nogueira, S.S., Dourado, F.F., Guimarães, M.A., Pitombeira, N.A.O., Gobbo, G.G., Primo, F.L., Paula, R.-C.M., Feitosa, J.P.A., Tedesco, A.C., Nunes, L.C.C., Leite, J.R.S.A., da Silva, D.A., 2016. Acetylated cashew gum-based nanoparticles for transdermal delivery of diclofenac diethyl amine. Carbohydr. Polym. 143, 254–261.

Dutra, R.C., Campos, M.M., Santos, A.R., Calixto, J.B., 2016. Medicinal plants in Brazil: pharmacological studies, drug discovery, challenges and perspectives. Pharmacol. Res. 112, 4–29.

Fernandes, R.V.D.B., Botrel, D.A., Silva, E.K., Borges, S.V., Oliveira, C.R.D., Yoshida, M.I., Feitosa, J.P.D.A., Paula, R.-C.M., 2016. Cashew gum and inulin: new alternative for ginger essential oil microencapsulation. Carbohydr. Polym. 153, 133–142.

Forato, L.A., de Britto, D., de Rizzo, J.S., Gastaldi, T.A., Assis, O.B., 2015. Effect of cashew gum-carboxymethylcellulose edible coatings in extending the shelf-life of fresh and cut guavas. Food Packag. Shelf Life 5, 68–74.

Freire, F., Cardoso, J., Dos Santos, A., Viana, F., 2002. Diseases of cashew nut plants (*Anacardium occidentale* L.) in Brazil. Crop Prot. 21 (6), 489–494.

Guilherme, M.R., Reis, A.V., Takahashi, S.H., Rubira, A.F., Feitosa, J.P., Muniz, E.C., 2005. Synthesis of a novel superabsorbent hydrogel by copolymerization of acrylamide and cashew gum modified with glycidyl methacrylate. Carbohydr. Polym. 61 (4), 464–471.

Hosseini, A., Jafari, S.M., Mirzaei, H., Asghari, A., Akhavan, S., 2015. Application of image processing to assess emulsion stability and emulsification properties of Arabic gum. Carbohydr. Polym. 126, 1–8.

Kaur, H., Ahuja, M., Kumar, S., Dilbaghi, N., 2012a. Carboxymethyl tamarind kernel polysaccharide nanoparticles for ophthalmic drug delivery. Int. J. Biol. Macromol. 50 (3), 833–839.

Kaur, H., Yadav, S., Ahuja, M., Dilbaghi, N., 2012b. Synthesis, characterization and evaluation of thiolated tamarind seed polysaccharide as a mucoadhesive polymer. Carbohydr. Polym. 90 (4), 1543–1549.

Kosalge, S., Fursule, R., 2009. Investigation of ethnomedicinal claims of some plants used by tribals of Satpuda Hills in India. J. Ethnopharmacol. 121 (3), 456–461.

Kulkarni, R.V., Mutalik, S., Mangond, B.S., Nayak, U.Y., 2012. Novel interpenetrated polymer network microbeads of natural polysaccharides for modified release of water soluble drug: in-vitro and in-vivo evaluation. J. Pharm. Pharmacol. 64 (4), 530–540.

Kumar, A., Moin, A., Ahmed, A., Shivakumar, G., H., 2012. Cashew gum a versatile hydrophyllic polymer: a review. Curr. Drug Ther. 7 (1), 2–12.

Lee, W.-K., Lim, P.-E., Phang, S.-M., Namasivayam, P., Ho, C.-L., 2016. Agar properties of Gracilaria species (Gracilariaceae, Rhodophyta) collected from different natural habitats in Malaysia. Reg. Stud. Mar. Sci. 7, 123–128.

Lima, M.R., Paula, H.C.B., Abreu, F.O.M.S., Silva, R.B.C., Sombra, F.M., Paula, R.C.M., 2018. Hydrophobization of cashew gum by acetylation mechanism and amphotericin B encapsulation. Int. J. Biol. Macromol. 108, 523–530.

Magalhães Jr., G.A., Santos, C.M.W., Silva, D.A., Maciel, J.S., Feitosa, J.P.A., Paula, H.C.B., de Paula, R.C.M., 2009. Microspheres of chitosan/carboxymethyl cashew gum (CH/CMCG): effect of chitosan molar mass and CMCG degree of substitution on the swelling and BSA release. Carbohydr. Polym. 77, 217–222.

Maiti, S., Ranjit, S., Mondol, R., Ray, S., Sa, B., 2011. Al+ 3 ion cross-linked and acetalated gellan hydrogel network beads for prolonged release of glipizide. Carbohydr. Polym. 85 (1), 164–172.

Malik, S., Ahuja, M., 2011. Gum kondagogu-g-poly (acrylamide): microwave-assisted synthesis, characterisation and release behaviour. Carbohydr. Polym. 86 (1), 177–184.

Malik, S., Kumar, A., Ahuja, M., 2012. Synthesis of gum kondagogu-g-poly (N-vinyl-2-pyrrolidone) and its evaluation as a mucoadhesive polymer. Int. J. Biol. Macromol. 51 (5), 756–762.

Meloa, A.M.A., Oliveira, M.R.F., Furtado, R.F., Borges, M.F., Biswas, A., Cheng, H.N., Alvesa, C.R., 2020. Preparation and characterization of carboxymethyl cashew gum grafted with immobilized antibody for potential biosensor application. Carbohydr. Polym. 228, 115408.

Mendes, L.G., da Silva Mendes, F.R., Furtado, R.F., Freire, G.A., Sales, G.W.P., da Costa, J.M.C., et al., 2020. Use of cashew gum combined with galactomannan for encapsulation of Rosmarinus officinalis essential oil. J. Environ. Anal. Prog. 5 (4), 369–380.

Menestrina, J.M., Iacomini, M., Jones, C., Gorin, P.A., 1998. Similarity of monosaccharide, oligosaccharide and polysaccharide structures in gum exudate of *Anacardium occidentale*. Phytochemistry 47 (5), 715–721.

Mudgil, D., Barak, S., Khatkar, B., 2016. Effect of partially hydrolyzed guar gum on pasting, thermo-mechanical and rheological properties of wheat dough. Int. J. Biol. Macromol. 93, 131–135.

Nayak, A.K., Pal, D., 2013. Blends of jackfruit seed starch–pectin in the development of mucoadhesive beads containing metformin HCl. Int. J. Biol. Macromol. 62, 137–145.

Nayak, A.K., Das, B., Maji, R., 2012. Calcium alginate/gum Arabic beads containing glibenclamide: development and in vitro characterization. Int. J. Biol. Macromol. 51 (5), 1070–1078.

Nayak, A.K., Pal, D., Das, S., 2013. Calcium pectinate-fenugreek seed mucilage mucoadhesive beads for controlled delivery of metformin HCl. Carbohydr. Polym. 96 (1), 349–357.

Oliveira, A.C.D.J., Araújo, A.R.D., Quelemes, P.V., Nadvorny, D., Soares-Sobrinho, J.L., Leite, J.R.S.A., Silva-Filho, E.C.D., Silva, D.A.D., 2019. Solvent-free production of phthalated cashew gum for green synthesis of antimicrobial silver nanoparticles. Carbohydr. Polym. 213, 176–183.

Oliveira, O.C.J., Chaves, L.L., Ribeiro, F.O.S., Lima, L.R.M., Oliveira, T.C., García-Villén, F., Viseras, C., Paula, R.C.M., Rolim-Neto, P.J., Hallwass, F., Silva-Filho, E.C., Silva, D.A., Soares-Sobrinho, J.L., Soares, M.F.L.R., 2021. Microwave-initiated rapid synthesis of phthalated cashew gum for drug delivery systems. Carbohydr. Polym. 254, 117226.

Owusu, J., Oldham, J., Oduro, I., Ellis, W., Barimah, J., 2005. Viscosity studies of cashew gum. Trop. Sci. 45 (2), 86–89.

Pandey, R., Khuller, G., 2004. Polymer based drug delivery systems for mycobacterial infections. Curr. Drug Deliv. 1 (3), 195–201.

Pinto, A.C., Silva, D.H.S., Bolzani, V.D.S., Lopes, N.P., Epifanio, R D.A., 2002. Produtos naturais: atualidade, desafios e perspectivas. Quim. Nova 25, 45–61.

Pitombeira, N.A., Neto, J.G.V., Silva, D.A., Feitosa, J.P., Paula, H.C., de Paula, R.C., 2015. Self-assembled nanoparticles of acetylated cashew gum: characterization and evaluation as potential drug carrier. Carbohydr. Polym. 117, 610–615.

Rana, V., Rai, P., Tiwary, A.K., Singh, R.S., Kennedy, J.F., Knill, C.J., 2011. Modified gums: approaches and applications in drug delivery. Carbohydr. Polym. 83 (3), 1031–1047.

Ray, R., Maity, S., Mandal, S., Chatterjee, T.K., Sa, B., 2010. Development and evaluation of a new interpenetrating network bead of sodium carboxymethyl xanthan and sodium alginate. Pharmacol. Pharm. 1 (01), 9.

Ribeiro, A.J., de Souza, F.R.L., Bezerra, J.M., Oliveira, C., Nadvorny, D., Monica, F., Sobrinho, J.L.S., 2016. Gums' based delivery systems: review on cashew gum and its derivatives. Carbohydr. Polym. 147, 188–200.

Richter, A.R., Carneiro, M.J., Sousa, N.A., Pinto, V.P.T., Freire, R.S., Sousa, J.S., Mendes, J.F.S., Fontenelle, R.O.S., Feitosa, J.P.A., Paula, H.C.B., Goycoolea, F.M., Paula, R.C.M., 2020. Self-assembling cashew gum-graft-polylactide copolymer nanoparticles as a potential amphotericin B delivery matrix. Int. J. Biol. Macromol. 152, 492–502.

Rodrigues, J.A., Araújo, A.R., Pitombeira, N.A., Plácido, A., Almeida, M.P., Veras, L.M.C., Delerue-Matos, C., Lima, F.C.D.A., Batagin-Neto, A., Paula, R.C.M., Feitosa, J.P.A., Eaton, P., Leite, J.R.S.A., Silva, D.A., 2019. Acetylated cashew gum-based nanoparticles for the incorporation of alkaloid epiisopiloturine. Int. J. Biol. Macromol. 128, 965–972.

Silva, D.A., Feitosa, J.P., Maciel, J.S., Paula, H.C., de Paula, R.C., 2006. Characterization of crosslinked cashew gum derivatives. Carbohydr. Polym. 66 (1), 16–26.

Silva, E.D.L.V., de Jesus Oliveira, A.C., Patriota, Y.B.G., Ribeiro, A.J., Veiga, F., Hallwass, F., Wanderley, A.G., 2019a. Solvent-free synthesis of acetylated cashew gum for oral delivery system of insulin. Carbohydr. Polym. 207, 601–608.

Silva, E.D.L.V., Oliveira, A.C.D.J., Silva-Filho, E.C., Ribeiro, A.J., Veiga, F., Soares, M.F.D.L.R., et al., 2019b. Nanostructured polymeric system based of cashew gum for oral admnistration of insulin. Matéria (Rio J.) 24 (3), e-12399.

Silva, M.C.C., Santos, M.S.F., Bezerra, R.D.S., Araújo-Júnior, E.A., Osajima, J.A., Santos, M.R.M.C., Fonseca, M.G., Silva-Filho, E.C., 2020. Kaolinite/cashew gum bionanocomposite for doxazosin incorporation and its release. Int. J. Biol. Macromol. 161, 927–935.

Sosnik, A., das Neves, J., Sarmento, B., 2014. Mucoadhesive polymers in the design of nano-drug delivery systems for administration by non-parenteral routes: a review. Prog. Polym. Sci. 39 (12), 2030–2075.

Souza Filho, M.D., Medeiros, J.V., Vasconcelos, D.F., Silva, D.A., Leódido, A.C., Fernandes, H.F., Pinto, G.R., 2018. Orabase formulation with cashew gum polysaccharide decreases inflammatory and bone loss hallmarks in experimental periodontitis. Int. J. Biol. Macromol. 107, 1093–1101.

Torquato, D., Ferreira, M., Sá, G., Brito, E., Pinto, G., Azevedo, E., 2004. Evaluation of antimicrobial activity of cashew tree gum. World J. Microbiol. Biotechnol. 20 (5), 505–507.

Vázquez-González, Y., Prieto, C., Filizoglu, M., Ragazzo-Sánchez, J.A., Calderón-Santoyo, M., Furtado, R., et al., 2021. Electrosprayed cashew gum microparticles for the encapsulation of highly sensitive bioactive materials. Carbohydr. Polym. 264, 118060.

Yang, J., Han, S., Zheng, H., Dong, H., Liu, J., 2015. Preparation and application of micro/nanoparticles based on natural polysaccharides. Carbohydr. Polym. 123, 53–66.

Zakaria, M.B., Rahman, Z.A., 1996. Rheological properties of cashew gum. Carbohydr. Polym. 29 (1), 25–27.

Zhang, H.-Z., Gao, F.-P., Liu, L.-R., Li, X.-M., Zhou, Z.-M., Zhang, Q.-Q., 2009. Pullulan acetate nanoparticles prepared by solvent diffusion method for epirubicin chemotherapy. Colloids Surf. B: Biointerfaces 71 (1), 19–26.

CHAPTER

12

Dextran-based micro- and nanobiomaterials for drug delivery and biomedical applications

Yeliz Basaran Elalmis[a], *Ecem Tiryaki*[a,b], *Burcu Karakuzu Ikizler*[a], and *Sevil Yucel*[a]

[a]Department of Bioengineering, Yildiz Technical University, Istanbul, Turkey
[b]Department of Physical Chemistry, Biomedical Research Center, Southern Galicia Institute of Health Research and Biomedical Research Networking Center for Mental Health, Universidade de Vigo, Vigo, Spain

12.1 Introduction

Dextran is a significant polysaccharide which is used in biomedical and pharmaceutical applications. It has been known for many years, but its value was exposed in the 1940s due to its unique properties like biocompatibility, hydrophilicity, biodegradability, water solubility, and stability. In particular, the clinical applications of dextran and its derivatives have expanded remarkably throughout the years (de Belder, 1996; Wang et al., 2016).

Dextran, a prevalent biological macromolecule, is a polymer formed of α-D-glucose monomers (Huang and Huang, 2018). Dextran can be easily modified chemically with various functional groups via reactions such as oxidation due to its functional hydroxyl groups to build different structures (e.g., spherical, three-dimensional networks) (Huang and Huang, 2018; Sun and Mao, 2012). Dextran, in addition to its features mentioned above, improves the drug delivery system (DDS) stability and also eliminates DDS accumulation in blood circulation. Due to the ligand activity of the polysaccharides (e.g., dextran, heparin sulfate, chitosan), the target cell receptor interacts specifically with these polysaccharide molecules and phagocytosis is triggered when DDS is modified with these molecules. Several drugs/active molecules can be easily encapsulated thanks to its hydrophilic structure, and drug loading and

Micro- and Nanoengineered Gum-Based Biomaterials for Drug Delivery and Biomedical Applications
https://doi.org/10.1016/B978-0-323-90986-0.00011-X

release behaviors can be adjusted in a controlled manner with their surface modifications (Huang and Huang, 2018). In addition, the use of dextran as a coating material with metallic nanoparticles greatly improves the biocompatibility and stability of these particles during their clinical applications (Predescu et al., 2018; Tassa et al., 2011).

This chapter focuses on micro- and nanostructured dextran-based biomaterials production and applications in drug delivery and biomedical areas. Dextran, its derivatives, and hybrid dextran systems have been extensively reviewed as drug delivery, wound dressing material, imaging probes, and tissue engineering materials.

12.1.1 Structure and origin of dextran

Dextran is a neutral polysaccharide that consists of α-1,6 linked D-glucose units varying from 50% to 97% of the total glycosidic bonds and with a few percent of α-1,2, α-1,3, or α-1,4 linked side chains depending on the production process. Molecular weights can change from 5 to 500 kDa (Maia et al., 2014; Wang et al., 2016). Dextrans of diverse chemical composition are synthesized by the family of *Lactobacillaceae* and mostly from *Leuconostoc mesenteroides*, *Leuconostoc dextranicum*, and *Streptobacterium dextranicum*. In 1861, Pasteur discovered slime-producing bacteria, and in 1878 van Tieghem named that bacteria as *L. mesenteroides*, used for general pharmaceutical purposes. The name "dextran" was given to this carbohydrate by Dhaneshwar et al. (2006), Gyles et al. (2017), Heinze et al. (2006), and Varghese et al. (2019).

Branching degrees of dextran vary from 0.5% to 60%, depending on the origin of dextrans. Furthermore, the origin may affect the molecular weight distribution of polymers. Branching degree and molecular weight distribution of dextrans determine the physicochemical properties. For instance, the water solubility of dextrans decreases by an increase in the branching degree of dextran. If the α-1,3 linkages are branching more than 43%, dextran has been considered insoluble in water while low degree branching dextrans (0.5%) are highly soluble in water. Dextran polymers are relatively stable under a soft acidic and basic environment while being highly soluble in water (Mehvar, 2000; Walker, 1978).

Dextran can be simply modified chemically owing to its countless reactive hydroxyl groups and can be used in several fields (Gyles et al., 2017). In addition, these hydroxyl groups (–OH) allow connection points to create conjugates with various materials. For example, dextran-based amphiphiles can be created with the modification of hydroxyl groups with hydrophobic groups. Thus, chemotherapeutic materials that are weakly soluble in water can be encapsulated with dextran-based amphiphiles and self-assemble into nanocarriers (Gaspar et al., 2016).

12.1.2 Derivatives of dextran

Chemical modification of the surface groups on dextrans extends their application field. Some of the frequently used derivatives are presented in Fig. 12.1, and the studies on their usage are mentioned in the following sections.

Dextran aldehyde is produced by periodate oxidation of dextran. Dextran aldehyde has the greatest amount of aldehyde groups due to the attachment of two vicinal aldehyde groups per glucose unit when the oxidation is fully complete. Generally, dextran aldehyde has been

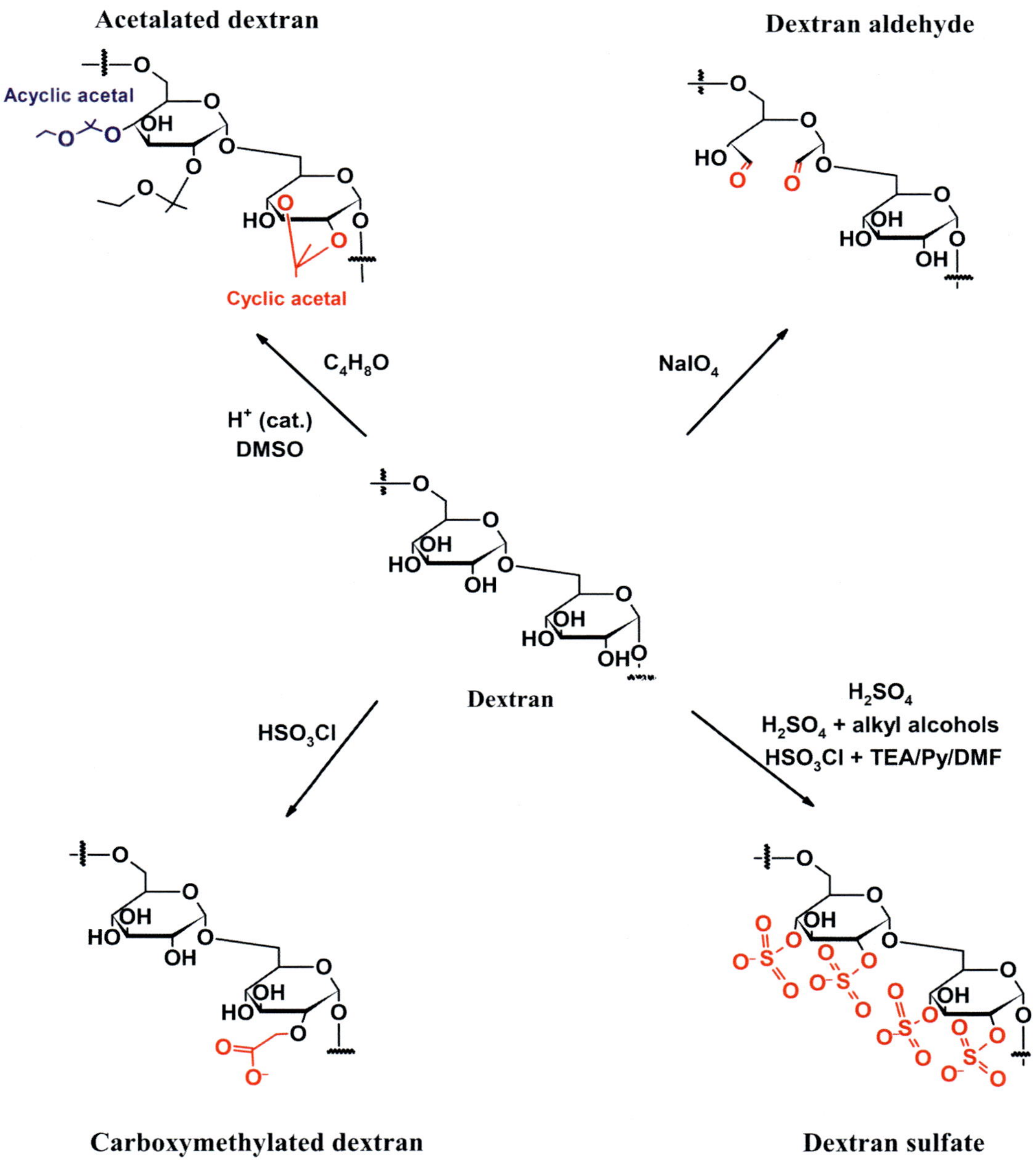

FIG. 12.1 Schematic representation of frequently used derivatives of dextran.

used as a superior cross-linking agent, especially for proteins (Cejudo-Sanches et al., 2020). Generally, oxidation is performed in aqueous solutions; nevertheless, to increase the solubility of bulky organic substrates, certain cosolvents may be used, such as ethanol, methanol, or acetic acid (Tacias-Pascacio et al., 2019).

Dextran sulfate has been examined as a possible replacement for heparin in anticoagulant treatment and research has shown that the highest anticoagulant properties are obtained with the lowest molecular weight samples (Lee, 2017). Dextran sulfate synthesis is performed using various methods. In the beginning, concentrated or diluted sulfuric acid solution was used to treat dextran, and depolymerization occurred. In addition, alkyl alcohols combined with sulfuric acids and generated the alkyl sulfates as reactive species. Chlorosulfonic acid and sulfur trioxide are the other sulfating sources that can be complexed with triethylamine (TEA), pyridine (Py), or N,N-dimethylformamide (DMF) (Heinze et al., 2006).

Carboxymethylated dextran has features with possible applications in the food, chemical, pharmaceutical, and cosmetic industries (Li et al., 2021). Carboxymethylation of dextran is performed with monochloroacetic acid in a water/organic solvent mixture (Heinze et al., 2006).

Iron dextran solutions are used extensively for treating anemia in newborn piglets. The solution caused an increase after intravenous infusion in hemoglobin when it was examined in humans. The best results are obtained when the solution is delivered together with glucose. The solution is prepared with alkali heating and then neutralized in ferric acid solutions. The mixture of 5% iron and 20% dextran solution is suitable for intramuscular and intravenous injection for the treatment of iron deficiency anemia (Lee, 2017).

12.1.3 Properties of dextran

Molecular weight

Dextran polymers have considerable potential in terms of clinical and pharmacological purposes. Hence, it is necessary to understand the physicochemical properties of dextran entirely to achieve successful applications. The dextran fraction's molecular weight distribution is related to some of the physical properties and its numerous pharmacological properties. The ratio of weight average molecular weight (Mw) to number average molecular weight (Mn) of dextran with lower values like 1.1 points out sharp fractions and higher values like 2.0 showing broader fractions (de Belder, 1996).

The molecular weight of dextran increases in consequence of increasing branch density and so polydispersity increases too. Native dextran has a high polydispersity with a high average molecular weight that varies from 9×10^6 to 5×10^8 g/mol. Different techniques can be applied to determine the weight average molecular weight and number average molecular weight. Ultracentrifugation, small-angle neutron scattering, light scattering, and viscometry techniques are used for Mw determination, and membrane osmometry and end group analysis are used for Mn determination. In addition, size exclusion chromatography (SEC) is a beneficial instrument to examine the molecular weight distribution (Heinze et al., 2006).

Physicochemical properties

Dextran dissolves easily in water and electrolyte; concentrated aqueous solutions can also be prepared. Dextran also dissolves in ethylene glycol, glycerol, dimethyl sulfoxide, and formamide, unlike monohydric alcohols. Dextran fractions of molecular weight $< 20{,}000$ may need powerful heating to convert into the solution because of the exact crystallization degree. Dextran solutions can be autoclaved at pH 4–6. Color change may occur above these pH values and especially with low molecular weight. The viscosity of dextran solutions is relevant to molecular weight; there is also a relationship between viscosity and temperature, concentration, and molecular weight parameters (de Belder, 1996).

The colloid osmotic pressure, oncotic pressure, is essential for several clinical applications. The diffusion of small dextran molecules through the membranes will decrease the oncotic pressures; therefore, the membranes' permeability must be considered. It should be noted that dextran solutions may interact with other macromolecules, increasing the oncotic pressure. Dextran molecules penetrate the glomerular membrane under 15.000 molecular weight without limitation, and the diffusion of the molecules is limited with increasing molecular weight, and at approximately 50,000 the diffusion is nearly zero (Arturson and Wallenius, 1964; de Belder, 1996; Rowe, 1955).

Dextrans have been utilized clinically in different ways such as plasma volume expander, peripheral flow promotion, and antithrombotic agents for more than 70 years. Dextrans have also been examined as macromolecular carriers for drug and protein delivery systems to provide and enhance the durability of therapeutic agents in blood circulation. This durability provides a more extended half-life in blood for high molecular weight dextran conjugates compared to naked drug or protein. Additively, dextrans may be targeted specifically to the region of interest via active or passive targeting. Lastly, dextrans have been used to reduce the in vivo immunogenicity and improve the in vitro stability of protein or enzymes (Mehvar, 2000; Thoren, 1980).

Metabolism of dextrans

Depolymerization of dextrans occurs via dextranase (α-1,6-glucosidases), which is present in different parts of the organism like the liver, kidney, and spleen, and close to the end of the gastrointestinal system. Higher amounts of dextranase are located in the spleen and liver (Larsen, 1989).

Dextran can be eliminated in two ways in the liver: depolymerization by dextranase enzymes, and excretion into the bile. Although a high amount of dextranase is present in the liver, depolymerization seems to be slow in some research. In vitro studies showed that chemical modification of dextrans would reduce the metabolism by the dextranase enzymes. For this reason, it should be noted that dextran conjugates with drugs or proteins may exhibit different in vivo depolymerization behavior (Larsen, 1989; Mehvar, 2000).

When dextran infuses in the body, the ratio of 50% is practically excreted by the kidney in the first 24 h, and in the following 48 h, this rate rises to 70%. The rest of it has the highest molecular weight, partly eliminated by the gastrointestinal tract and partly metabolized by dextranase in the spleen and liver (Aronson, 2016).

As additional information on this subject, natural dextran may be immunogenic for humans and some animals; contrary to this, low molecular weight dextrans have no immunogenic effect (de Belder, 1996).

12.2 Application of dextrans

Dextrans have unique properties for medical usages, such as antithrombotic activity, reducing erythrocyte aggregation, and platelet adhesiveness. They are also used as a liquid plasma expander after acute blood loss (Karewicz, 2014). Low molecular weight dextran is commonly used as an aid to increase blood flow in peripheral vascular disease patients. In some research, peripheral resistance reduces after the infusion of low molecular weight dextran (Gelin and Thoren, 1961) and a relationship has been found between increased blood flow and decreased blood viscosity (Dormandy, 1971; Humphreys et al., 1976). Dextran 40 enhances the blood flow, presumably generated by a decrease in blood viscosity and erythrocyte aggregation inhibition.

6% or 10% aqueous solutions of clinical dextrans (40,000, 60,000, and 70,000 g/mol molecular weights) are used to substitute medium blood losses. Biological stability in the blood circulation is provided by high percentage α-1,6 linkages. Dextran basically replaces blood proteins like albumins and draws fluid from the interstitial area within the plasma to produce colloid osmotic pressure. Clinical dextran used as an antithrombotic material presents preventative treatment for deep venous thrombosis and postoperative pulmonary emboli (Heinze et al., 2006).

Dextran has some unique properties like biocompatibility, biodegradability, ease in accessibility, and low cost. It is also used in vaccines for antigen delivery according to these features and further self-adjuvant property (Karandikar et al., 2017). In addition, dextran has remarkable effects on organ perfusion solutions and protection and is used in the protection of viable organs like the kidneys, liver, lungs, pancreas, and corneas (de Belder, 1996).

Recent studies of medical uses and specific assignments of dextran are investigated in the following sections in detail.

12.3 Dextran-based micro- and nanogels

Hydrogels are hydrophilic gel networks obtained via chemical or physical cross-linking. They are able to swell quickly and retain a substantial amount of water while preserving their three-dimensional (3D) structure or remaining nonsoluble in aqueous media (Raemdonck et al., 2008; Salimi-Kenari et al., 2018; Su et al., 2016). Formulation of hydrogels with various physical forms, such as membranes, microgels, and nanogels, is generally possible. Microgels and nanogels are hydrogel particles with micro and nano sizes, respectively. These particulate networks formed by polymers (hydrophilic or amphiphilic) express unique features due to the combined hydrogel and micro – /nano-sized particle characteristics, such as high surface area, small particle size, and high amount of water uptake. Hydrogel particles are biocompatible, stimuli-responsive, and high loading capacity materials, which make them promising delivery materials (Su et al., 2016).

12.3.1 Dextran-based microgels

Microgel particles can be described as three-dimensional networks (microscopic) formed of cross-linked polymeric molecules dispersed in a suitable solvent (Farjami and Madadlou, 2017; Shewan and Stokes, 2013; Thorne et al., 2011). Microgel stability depends on the covalent bonds present and strong interactions (noncovalent). Colloidal microgels are generally categorized as nanogel (diameter < 0.5 μm) and microgel (diameter between 0.5 and 5 μm). Tendency of high water sorption of most particulate microgels depend on the hydrophilic character of polymer chains forming the microgel structures. Microgel particles demonstrate swelling behavior rather than dissolution in aqueous media, which is due to the cross-linked structure of the microgel (Farjami and Madadlou, 2017; Hamidi et al., 2008). They demonstrate swelling-deswelling properties reversibly due to the exterior stimuli (e.g., change in temperature and pH, solvent content, and ionic strength) allowing engineered particle design via interpolymer and polymer-water interaction modulation (Farjami and Madadlou, 2017). Simplicity in fabrication, stimuli-responsive degradation behavior, or volume variations of microgels lead to a variety of applications. One of these applications is controlled drug and protein delivery in the body by microgel materials; drug or protein release from microgel materials can be triggered by media stimuli such as pH, temperature, or enzyme presence. Furthermore, functionalization of microgels with targeting molecules is also possible in order to deliver drug or protein molecules to the required site (Zhang et al., 2018). Polysaccharide microgels design and fabrication is one of the progressive areas of microencapsulation scientific research and technology because of their high biomedical application potential. Formulation of polysaccharides such as dextran into cross-linked spherical microgels, which have a reactive high surface area, presents a remarkable applicability potential as bioactive molecule carriers. Dextran-based microgels can be used for various purposes such as drug delivery, microbial growth supports, and hemostatic agents. Dextran microgels, when applied in medical areas, particularly as hemostatic agents, have two key parameters, mean particle size and distribution, which significantly affect crucial properties (water uptake capacity and kinetics) of the ultimate microgels (Salimi-Kenari et al., 2016). Fabrication of microgel particles with a variety of complex morphologies (e.g., core-shell particles, microgel-enclosed nanoparticles, Janus microspheres) is possible for various application purposes including sensing and delivery. In applications of microgel particles (e.g., tissue and biomedical engineering), biocompatibility and biodegradability are crucial subjects that need to be considered (Farjami and Madadlou, 2017).

Fabrication

Natural biopolymer-based microgels can be fabricated using physicochemical (molecular association procedures) and mechanical (emulsion-based procedures) approaches. Structure, physicochemical features, and electrical charge of fabricated microgels are determined by the microgel fabrication technique used and the conditions of the microgel formation media. pH, temperature, concentration of the biopolymer, type of solvent, cross-linking agent and degree of the cross-linking, and ionic strength are among these microgel formation media conditions. Microgel particle fabrication methods can be classified as (i) molecular association-based methods and (ii) mechanical applications (Farjami and Madadlou, 2017).

Su et al. (2016) developed dextran-based microgels via the formation of Schiff base in a water-in-oil (W/O) microemulsion between the dextran aldehyde and ethylenediamine. Cross-linking process in W/O microemulsion (inverse) includes the heterogeneous gelation of polymers used. Generally, a water-soluble polymer is dispersed into aqueous micro-sized droplets that present in a continuous phase (organic), then with the use of a cross-linker (water soluble), the polymer is cross-linked in these surfactant stabilized droplets in order to form the particulate microgels. Su et al. (2016) also labeled the fabricated microgel particles using aminofluorescein by linking the free amino groups of fluorescein to free aldehyde groups left on dextran-based hydrogel network by means of a Schiff base reaction. Linkages formed via

Schiff base reactions are susceptible to hydrolysis and decreases in media pH lead to decreased stability. Thus the obtained microgel demonstrated pH-responsive degradation (Su et al., 2016).

He et al. (2021) prepared Fe_3O_4 nanoparticles coated with dextran at various particle sizes and physically embedded them into dextran microgels which were formed through a Schiff base reaction in a W/O inverse microemulsion between the dextran aldehyde and diamine; as a result, magnetic dextran microgels (MDMGs) containing magnetic nanoparticles (MNPs) were obtained. Furthermore, He et al. (2021) encapsulated doxorubicin (DOX) into MDMGs by dissolving doxorubicin hydrochloride in an aqueous phase, containing a dextran aldehyde and MNP mixture, to form a W/O inverse microemulsion. Consequently, dual-responsive (pH and magnetic field) DOX-releasing MDMGs were obtained that can find use as a stimuli-sensitive drug delivery material (He et al., 2021).

Drug delivery applications

Drug delivery techniques can significantly influence the drug therapeutic efficacy. Conventional drug delivery methods (oral and injection) lead to initial instantaneous increase in drug concentration (over the therapeutic range) followed by an instantaneous decrease in drug concentration (beneath the effective therapeutic range). Serious toxicity risk and associated complications can be caused by initial high concentrations of effectual drugs. Thus, constant drug release for a long period is the focus of the controlled drug release methods (Zhang et al., 2011). Furthermore, traditional cancer drugs are not able to discriminate between cancer and healthy cells. Thus, a variety of drug delivery systems (DDSs) that can selectively release the drug at cancer tissue have been studied. Introducing groups to target the carriers to perform targeted delivery has been the focus of the most DDSs. These types of DDSs are capable of reaching the site of target but unexpected release of drugs from such DDSs before reaching the site of target in the blood circulation limits their applications. Thus, intelligent DDS designs that respond to particular physiopathological signals are of great importance (Zhang et al., 2012).

Polymeric microgels with pH-responsive degradation and volume transition properties have possible applications for cancer treatment as drug delivery materials. Around cancer tissues, pH value is lower (pH 5–6) with regard to blood and normal tissues (pH 7.4), thus drug loaded to the pH-responsive microgels can be released in acidic media from these pH-responsive microgels. pH-responsive behavior can be obtained in two ways: (i) -pH-sensitive polymer usage and (ii) microgels designed with pH-degradable cross-links. For in vivo applications, degradable microgel carriers are preferred, thus pH-responsive cross-links (e.g., Schiff base) are adequate for pH-responsive microgel fabrication (Zhang et al., 2018).

Dextran-based microgels show good biocompatibility, and have therefore been investigated as potential carriers for protein/peptide drugs. Predominant linkages of dextran structure are α-1,6 glucosidic linkages, which are susceptible to hydrolysis by dextranase enzyme. Several organs (colon, liver, kidney, spleen) in the body contain dextranases. According to semiquantitative analyses results, high-level dextranase activity is demonstrated by the liver, midlevel activity by the spleen, and low-level activity by the kidneys in humans. The jejunum's distant part showed the highest dextranase activity in the gastrointestinal tract; however, dextranase is not present in blood. Hydrolysis of dextran by dextranase attracted considerable interest for drug release from dextran hydrogels in a controlled and targeted manner, as well as conjugates of dextran with therapeutic materials (Widenbring et al., 2014).

Zhang et al. (2018) demonstrated a facile fabrication method for pH-sensitive and fluorescence inorganic/organic microgels. They used amino-modified ZnO quantum dots (ZnO QDs) to cross-link carboxymethyl dextran (CMD) and obtain ZnO@dextran microgels. Obtained microgels had a mean diameter of around 5 μm and showed strong fluorescence under ultraviolet irradiation (365 nm). Cross-linkage by ZnO QDs in the structure of hybrid microgels enabled microgel degradation in mild acidic media because of ZnO QD pH sensitivity. Furthermore, DOX was loaded to microgels and used as pH-responsive DOX-releasing carriers of drugs. They could monitor microgel degradation and DOX release via microgel fluorescence intensity detection. It was stated by the authors that synergistic therapy is also possible with ZnO@dextran microgels due to ZnO QD cytotoxicity (Zhang et al., 2018).

12.3.2 Dextran-based nanogels

Developments in the nanotechnology field has revealed the need to develop nanogel systems that prove their potential to deliver drugs in a controlled, sustainable, and targeted manner. With the developments in the polymer sciences field, preparation of intelligent nano systems that can be effective for the advancement of treatment and clinical trials has become unavoidable (Sultana et al., 2013). Nanogels that are 3D nano-sized hydrogel particles have several attractive features such as stimuli-responsive structure, controlled drug release, great capacity of bioactive material loading, high water sorption and mechanical stability, and small particle size. These features of nanogels make them potential delivery systems for bioactive materials. Nanogels based on natural biopolymers have attracted increasing interest as delivery systems since they are biodegradable and biocompatible, and show similarity to the extracellular matrix (ECM) macromolecular constituents. Furthermore, various biopolymers (e.g., polysaccharide, protein) contain a considerable amount of functional groups utilizable for more bio-conjugations that can allow adaptation of hydrogel performance to specific applications. Studies to fabricate stable nanogels using polysaccharides and proteins demonstrated that formed biopolymer particle size, stability, and charge properties depend on factors including pH, protein/polysaccharide ratio, type of biopolymer, ionic strength, heating temperature, and time (Jin et al., 2016).

Core-shell nanogel structures mostly became the focus of drug delivery research. In general, a nanogel is stabilized in water (or biological fluids) by a hydrophilic shell, which at the same time provides a "stealth" featured nanogel that can avoid from the recognition of the mononuclear phagocytic system and opsonization. Drug loading via hydrophobic and electrostatic interactions is allowed by the hydrophobic core. Polyethylene glycol-poly (ethylene imine) and natural polymers are among the commonly used polymers in nanogel preparation. Polysaccharide usage (e.g., dextran) to prepare the hydrophilic shell is especially attractive in biomedical applications, due to well-demonstrated biodegradability and biocompatibility, and features including providing multiple sites for the specific ligand attachments, and eliminating protein adsorption by providing steric protection (Ferrer et al., 2013).

Fabrication

Nanogels can be fabricated using natural (e.g., dextran, chitosan, albumin) as well as synthetic polymers. W/O heterogeneous emulsion, homogenous aqueous gelation, chemical cross-linking, and spray-drying techniques are among the methods that can be used in nanogel design (Kaur et al., 2019).

Yu et al. (2020) developed dextran-based (Dex-SS) nanogels in W/O inverse microemulsion using a facile technique based on the formation of a Schiff base involving disulfide (–S–S–) between dextran aldehyde and cystamine (Cys). They synthesized aldehyde derivative of dextran (Dex-CHO) by periodate oxidation and used it in the hydrogel synthesis. To prepare Dex-SS nanogels, an aqueous solution containing Cys and $NaHCO_3$ (molar ratio $NaHCO_3$/Cys:2/1) was prepared; subsequently an aqueous solution of Dex-CHO was included under ultrasonication into cyclohexane solution (containing Span 80 and Tween 80). Following W/O inverse microemulsion formation, Dex-SS nanogels were obtained by the addition of solution of Cys (Yu et al., 2020).

Su et al. (2018) used the inverse micromulsion method to prepare dextran-based nanogel to which DOX was covalently conjugated via a Schiff base formation. Dextran-based nanogels were loaded with DOX covalently by the formation of Schiff base between the DOX $-NH_2$ groups and Dex-CHO aldehyde groups, then the formed conjugate (DOX/Dex-CHO) was cross-linked using ethylenediamine in order to obtain DOX-loaded dextran nanogel (DOX@Dex) via a Schiff base reaction. The calculated DOX-HCl amount was included in a Dex-CHO aqueous solution to synthesize a DOX/Dex-CHO conjugate. Su et al. used the W/O inverse microemulsion cross-linking method to synthesize a DOX@Dex nanogel. This nanogel was synthesized by the addition of an aqueous DOX/Dex-CHO conjugate solution under ultrasonication into a cyclohexane solution (containing Span 80 and Tween 80), and subsequent addition of ethylenediamine to the formed emulsion (Su et al., 2018).

Inverse emulsion polymerization and photopolymerization methods are among the common methods to synthesize core-shell nanogels. Maillard reaction and subsequent heat gelation is an alternative two-step core-shell nanogel synthesis method. Protein and polysaccharide is linked by the Maillard reaction, which is a nontoxic natural reaction. Subsequently applied heat gelation provides partial protein denaturation and stable nanogel formation. It is a low-cost and green synthesis method which does not require initiator (Ferrer et al., 2013).

Li et al. (2021) developed a novel technique for nanogel preparation with a dextran shell and lysozyme core using Maillard dry-heat and heat gelation processes. Firstly, conjugates of lysozyme and dextran were produced via the Maillard reaction; subsequently, to produce the nanogels, heating was applied to a conjugate solution above the lysozyme denaturation temperature. Obtained nanogels were spherical in shape with a hydrodynamic diameter of about 200 nm (Li et al., 2008).

Drug delivery applications

Nanogels have favorable features over other systems. Drug release can be attained by active and passive release owing to particle size and surface features of nanogels. Furthermore, sustained/targeted drug delivery can be achieved owing to their elevated loading capacity (Kaur et al., 2019). Nanogel versatility makes these materials ideal bioimaging and drug delivery candidates. A considerable amount of water can be absorbed by nanogels (3D hydrophilic networks), during which they maintain their structure. Surface tension between the nanogel and fluid is low when in swollen state, which results in reduced nonspecific interactions (e.g., protein-cell adhesion) and enhanced biocompatibility. Nanogel submicrometer size provides the intake of most of the nanogels by cells. Furthermore, the large surface area of nanogels enables tailoring for a variety of different in vitro and in vivo applications and rapid stimuli responses (Ferrer et al., 2013).

In recent years, polymeric nano-sized carriers with dual and/or multistimuli-responsive features that respond to stimuli (two or more) simultaneously or in sequential mode would lead to an elevated anticancer effect in vitro and/or in vivo due to reported unique drug delivery and release control. Nano-sized carriers responding to unique pathological triggers and allowing release of the drug at the site of a tumor with appropriate therapeutic levels will be significantly beneficial. Tumor tissues have acidic microenvironments between pH 6.5 and 7.2, endosomes between pH 5.0 and 6.5, and lysosomes between 4.5 and 5.0, while the pH of normal tissue, extracellular media, and blood is 7.4. Furthermore, cytosol of cancer cells has a higher reducing glutathione (GSH) level between 2 and 10 mM (high redox potential) compared to normal tissue GSH level (between 2 and 10 μM) (Yu et al., 2020). Yu et al. (2020) obtained DOX conjugated dextran-based (Dex-SS) nanogels with the formation of disulfide containing a Schiff base as both acidic and reductive (GSH) media-sensitive DOX release systems. They stated that resulting dextran nanogels demonstrated dual sensitive (low pH/high GSH) release behavior. Uptake by cancer cells (H1299 and Hela cells) of DOX-loaded nanogel was demonstrated using in vitro cell labeling studies, and effective DOX release in endocytic vesicles was observed (Yu et al., 2020).

Gene delivery applications

Recently, dextran has been largely preferred in gene delivery applications due to biodegradability, biocompatibility, cell endocytosis promotion ability, and prolonging polyplex circulation time in vivo. Dextran is also utilized widely in biomedical areas including immunotherapy due to its specific binding to C-type lectin dextran-binding receptors. Thus, dextran is an important constituent of nano-sized carriers. Studies showed that gene carrier cationic polymers with dextran modification can lead to effectively reduced biological toxicity, and enhanced stability for the formed polyplex in blood serum. Positively charged polymers such as poly(ethyleneimine) (PEI) and poly-L-lysine (PLL) and their complexes are used as nonviral vectors with higher transfection efficiency of genetic materials (Uddin and Islam, 2006).

de Belder (1996) and Wang et al. (2016) designed a type of core/shell, which consists of a "core" formed by positively charged peptide/gene coacervate encapsulated using a dextran nanogel "shell" (DNSC). They introduced disulfide bonds at connection points between dextran and positively charged peptide in order to release genes effectively in target cells. Thus, dextran-polyacrylic acid (Dex-PAA) redox-sensitive nanogels were prepared via disulfide cross-linking, and in order to form the paptide/gene coacervate "core," a cationic peptide was selected that also could be cross-linked by disulfide bonds. Wang et al. (2020) showed that the DNSC could condense genes effectively and release genes in reducing media. Cationic peptide/gene coacervate "core" could be effectively shielded by the dextran-based nanogel "shell," reducing positively charged gene carrier side effects. The DNSC demonstrated high transfection effect and low cytotoxicity. Thus, it was concluded that the DNSC provided an efficient stimuli-responsive gene carrier (Wang et al., 2020).

Small interfering RNA (siRNA) inhalation therapy is an encouraging approach for pulmonary disorder treatment. However, clinical applications are critically limited due to deficiency of convenient delivery systems. De Backer et al. (2015) designed novel hybrid bioinspired nanoparticles with a core-shell structure containing a core from siRNA-loaded dextran nanogel (siNG) and an outer shell from a pulmonary surfactant (Curosurf®). The surfactant shell leads to enhanced colloidal stability and eliminates siRNA release when competing polyanions are available, and these competing polyanions are abundantly available in biofluids. De Backer et al. (2015) determined the effect of the surfactant shell on the siNG biological effectivity in lung cancer cells. Cellular siNG uptake is considerably reduced in the the presence of a surfactant. They included folate as a targeting ligand in order to overcome the observed cellular dose reduction and promote receptor-mediated endocytosis. Consequently, the obtained system considerably enhanced both gene silencing and cellular uptake, and achieved effective knockdown at low siRNA concentrations (nM range) (De Backer et al., 2015).

Imaging applications

Electrical, magnetic, and optical features can be included in nanogels by the addition of inorganic materials for various medical applications such as diagnostic and imaging (Kaur et al., 2019).

Ferrogels, which are hybrid magnetic hydrogels, are formed of magnetic nanoparticle-inserted polymeric hydrogel network structures. These materials were extensively studied due to their superparamagnetic features and fast external magnetic field response. Magnetic resonance imaging (MRI), cancer treatment via hyperthermia, and remote operated drug delivery applications are among the ferrogel studies. Ferrogels containing superparamagnetic iron oxide nanoparticles (SPIONPs) can be prepared via several developed techniques such as in situ precipitation, physical blending, and grafting-onto techniques. Presynthesized SPIONPs are generally mixed with a hydrogel precursor aqueous solution in the blending technique, and hydrogel matrix-encapsulated SPIONPs would be obtained via subsequent cross-linking of the mixture. The blending technique, which is the most widely applied technique for the synthesis of ferrogels as biomaterials, allows Fe_3O_4 nanoparticle size control and mild reaction conditions, and thus is advantageous (Su et al., 2019).

Su et al. (2019) synthesized dextran nanogel using the W/O inverse microemulsion method via the Schiff base formation technique and encapsulated Fe_3O_4 nanoparticles (NPs) into the dextran gel by the physical blending method. The magnetic dextran NGs developed have capabilities as pH-sensitive drug carriers and MRI probes. For this, first Fe_3O_4 NPs, using the coprecipitation technique, were synthesized and blended directly with an aqueous solution of Dex-CHO, subsequently addition of ethylenediamine cross-linker to the resulting mixture in W/O inverse microemulsion-based nanoreactors resulted in desired ferrogels. Schiff bases, which are degraded at low pH values, would lead to cargo release at low pH (acidic) media from the Schiff base containing carrier materials in a pH-responsive manner. Su et al. (2019) showed via in vitro MRI studies that dextran ferrogel-encapsulated iron oxide NPs demonstrated 5.6-fold increased T2 relaxivity value compared to iron oxide NPs alone. Multi-Schiff base linkages and aldehyde groups present on the hydrogel network enable multifunctional nanosystem designs such as MRI-guided pH-responsive DDSs (Su et al., 2019).

Li et al. (2014) reported the development of polymeric nanogel emitting near-infrared (NIR) for fluorescence mapping (noninvasive) of sentinel lymph node (SLN) in a mouse. Developed nanogels had about 28 nm diameter, which is optimal for uptake in SLN. First, self-assembled nanogel was obtained from the conjugate of dextran-deoxycholic acid, which was formed by disulfide linkages. Next, Cy7 and NIR dye were coupled to the nanogel formed in the first step and Dex–Cy7 nanogel was obtained. Analysis of fluorescence imaging demonstrated that photostability of Dex–Cy7 was enhanced in comparison to Cy7 alone. After Dex–Cy7 nanogel intradermal injection into the front paw of a mouse, nanogels could migrate into the axillary lymph node of the mouse. The fluorescence intensity shown was higher and the retention time was longer for Dex-Cy7 nanogel compared to Cy7, in the node. Localization of nanogels was revealed to be in the lymph node central region using an immunohisto-fluorescence assay, and the uptake of nanogels was mostly by macrophages. Low cytotoxicity (up to 1000 μg/mL polymer concentration) and a harmless nature to mice's normal livers and kidneys (at 1.25 mg/kg intravenous dose) was shown via in vitro/in vivo toxicity results. The authors suggested that biodegradable dextran-deoxycholic acid conjugate-based NIR-emitting nanogels demonstrated great potential as safe fluorescence nanoprobes with the advantage of noninvasive SLN mapping (Li et al., 2014).

12.4 Dextran-based electrospun nanofibers

The most common technique for nanofiber (NF) production is electrospinning, and the bioengineering area is the one of the potential areas for electrospun fiber usage. Natural polymers are commonly preferred materials in the biomedical field due to their high biocompatibility and biofunctional features. Natural polymers (e.g., silk, gelatin, dextran) were used to obtain NFs via the electrospinning technique in order to develop materials suitable for biomedical applications. Polysaccharides and proteins are the natural polymers mostly used in electrospun fiber fabrication (Cengiz-Çallıoğlu, 2014). Continuous nano- and micro-sized polymer fibers can be fabricated using the electrospinning technique. Obtained NFs are great candidates for DDSs, wound dressing materials, and tissue engineering scaffolds. NFs with nanometer diameters are possible with electrospinning technique using polymers (synthetic and natural), leading to very high specific surface area values compared to microfibers fabricated via conventional dry/wet and melt spinning techniques. Thus, large surface area products can be developed using electrospun NFs (Maslakci et al., 2017).

12.4.1 Fabrication

Electrospinning, which is a cost-efficient, adaptable, simple, and scalable system, uses an electrical field (high voltage) to fabricate aligned or random NFs using many natural and synthetic polymers (Unnithan et al., 2012). Direct blend, coaxial, and emulsion electrospinning techniques are generally applied techniques for fibrous polymer/DDSs fabrication. Core-shell structured NFs can be fabricated via emulsion electrospinning, which is a simple technique (Moydeen et al., 2018).

Unnithan et al. (2012) prepared polyurethane (PU) dextran (Dex) electrospun nanofiber mats with ciprofloxacin HCl (CipHCl). PU and Dex were dissolved in DMF:THF (1:1) solvent and CipHCl was added to the polymer solution. Nanofibers were electrospun at 22 kV and 15 cm tip-to-collector distance; the obtained PU-Dex-Cip nanofibers were dried under vacuum in an oven at 30 °C (Unnithan et al., 2012). Moydeen et al. (2018) fabricated core-shell nanofibers via emulsion electrospinning using poly(vinyl alcohol) (PVA) and dextran sulfate (Dex), and ciprofloxacin (Cipro) as a model drug. They prepared PVA/Dex blend solutions with different PVA/Dex ratios, and the model drug Cipro was dissolved in plant oil with heating and allowed to cool. Cipro solution was added to the PVA/Dex mixture under constant stirring in order to obtain the emulsion. The emulsion was drawn into a syringe and electrospun at 15 cm tip-to collector distance, 0.5 mL/h feed rate, and 15 kV voltage conditions. NF mat was collected during 30 h onto an aluminum foil (Moydeen et al., 2018).

12.4.2 Wound dressing and tissue engineering applications

Skin, which is the largest organ, has an important role in the protection of humans against external factors including chemical, thermal, osmotic, pathological, mechanical, and photo-induced damage. Protection against microorganisms during wound healing, which is a complex and painful period, is crucial after an injury (Nematpour et al., 2020).

Wound dressing has an important role in the treatment of certain open wounds such as thermal and chronic wounds. Wound beds with their moist, nutritious, and warm media provide conditions ideal for the growth of microorganisms. Antimicrobial wound dressing materials ideally should have properties such as enabling moist media for enhanced healing, and broad-spectrum antimicrobial effect, including antibiotic-resistant bacteria. Skin wounds need urgent care to prevent microbial infection and trans-epidermal water loss (Moydeen et al., 2018). Because of their superior architecture, electrospun NFs were stated to be very effective as wound dressing materials. These NFs can simulate ECM media, and thus help host cell growth and the formation of a new natural cellular matrix. Above all, selected therapeutic materials, which can significantly influence the healing of a wound, can be included in the NF with the use of the electrospinning technique (Unnithan et al., 2015).

Unnithan et al. (2012) prepared CipHCl containing PU-Dex-based NF as wound-dressing materials. They obtained uniform and continuous NFs of PU-Dex and CipHCl-loaded PU-Dex blend. Drug addition to the PU-Dex blend led to the size reduction and narrowed distribution of NF diameters. Including dextran in the PU resulted in improved cell attachment and viability. Moreover, these composite NF mats demonstrated elevated antibacterial activity on gram-positive and gram-negative bacteria (Unnithan et al., 2012). Unnithan et al. (2015), in another study, obtained β-estradiol-loaded PU-Dex composite nanofibrous material as a wound dressing by electrospinning. They showed β-estradiol's importance in the healing of cutaneous wounds. Uniform continuous NFs of estradiol-loaded PU-Dex were obtained and implemented as wound dressings successfully (Unnithan et al., 2015).

Tissue engineering of blood vessels with small diameter is a challenge with little progress made in recent decades, despite fair efforts. A variety of synthetic polymers have been investigated in the search for the ideal tissue scaffolding material. These synthetic materials often necessitate surface modification in order to enhance cellular binding due to the absence of domains for cell binding. However, biopolymers such as proteins and polysaccharides mimic the ECM's biochemical nature within the blood vessels and frequently provide enhanced cell-substrate interaction, and thus are attractive materials.

NFs mimic the fibrous protein architecture within blood vessels, and thus have excellent potential as vascular grafts. The orientation and size of the fibers may be tuned using the electrospinning technique, which is a versatile one, to enable contact direction for seeded cells. Moreover, electrospun NFs are easily biofunctionalized by drug inclusion to deliver the biochemical signals sustainably in order to direct cell fate (Shi et al., 2012).

Shi et al. (2012) evaluated the potential use of pullulan/dextran (P/D) electrospun structures for applications in vascular tissue engineering. The P/D NFs obtained had a mean diameter of 323 nm, and similar mechanical features to human arteries. The authors showed that the topography of the NFs provided cell adhesion, and that the endothelial phenotype was maintained on NFs. These NFs supported, over 14 days, the stable confluent monolayer of the endothelial cells. Similar smooth muscle alpha-actin levels and a lower rate of proliferation were demonstrated by the smooth muscle cells (SMCs) that were seeded on P/D NFs compared to the cells on 2D cultures. Observations obtained by Shi et al. (2012) suggested that these NFs promoted in SMCs the generation of quiescent contractile phenotype.

12.4.3 Drug delivery applications

Development of controlled drug delivery systems (CDDSs) aimed to enhance the therapeutic features of the drug and to make it safer, efficient, and reliable. Reduced drug toxicity, increased efficiency and selectivity, and prolonged lifetime are advantages offered by CDDSs. Liposomes, microspheres, hydrogels, nanoparticles, and nanofibers are among the drug delivery carriers. Electrospun NFs, due to their advantageous features such as high surface area and porosity, and ECM mimicking structure, were investigated as new drug delivery structures. Electrospun NFs, in order to develop DDSs, can be fabricated via the electrospinning method in which a polymer solution and drug are generally combined and subsequently electrospun. The key factor for improved sustainable release is the successful drug loading into the NF (Moydeen et al., 2018).

Moydeen et al. (2018) prepared PVA/Dex core-shell NFs using emulsion electrospinning (PDC-nE) and used them as DDS. Furthermore, they prepared a blend and coaxial PVA/Dex/Cipro-HCl NFs (PDC-10B and PDC-10C, respectively) for comparison. Scanning electron microscope micrographs revealed that increased Dex content resulted with increased fiber diameters. PDC-nE demonstrated slow release kinetics when compared to PDC-10B (Moydeen et al., 2018).

Maslakci et al. (2017) fabricated uniform NFs using polyvinylpyrrolidone and dextran polymers and ibuprofen and acetylsalicylic acid drugs via the electrospinning technique as a DDS. Reduced size and narrowed distribution of NF diameters were observed with the Dext T10 (Dextran, Mw:10,000) and Dext T40 (Dextran, Mw:40,000) addition. Drug solubility in a phosphate buffer increased in the presence of dextran. Antibacterial activity of the drug-included NFs was stated to be very good against gram-positive and gram-negative bacteria (Maslakci et al., 2017).

12.5 Dextran-based micro- and nanoparticles

12.5.1 Dextran-based microparticles

Microparticles (MPs) are the materials ranging in size from 1 to 1000 μm, that can be synthesized from natural or synthetic materials. Generally, methods such as spray-drying, phase separation, micro grinding, suspension/emulsion polymerization, solvent evaporation, microfluidics, and electrospray have been used to produce MPs (Morais et al., 2020). Dextran MPs (Dex-MPs) with tunable properties serve as a wide range of application materials. They are commonly used for drug (Chen et al., 2018; Kauffman et al., 2012; Meenach et al., 2012), biomolecules (Bachelder et al., 2010; Kanthamneni et al., 2012), or gene delivery applications (Cohen et al., 2011).

Fabrication

Depending on their application, Dex-MPs can be synthesized in aqueous or organic solvents. In general, the aqueous two-phase system with two different water-soluble polymers is preferred for fabrication of dextran-based MPs supported by the self-assembly phase separation method (Stenekes et al., 1998). The poly(ethylene glycol) (PEG)/dextran system is the most studied one to produce Dex-MPs via this method. The principle of the synthesis that produces cross-linked dextran is based on the polymerization of emulsified methacrylate modified dextran (Dex-MA) in a PEG aqueous solution (Stenekes et al., 1998; Yeredla et al., 2016). In addition, thermo-sensitive polymers such as ethylene oxide-propylene oxide random copolymers and Pluronic F127 are also used instead of PEG in the synthesis of dextran MPs via this approach (Forciniti, 2000; Yeredla et al., 2016).

Acetalated dextran (Ac-Dex) with its organic-soluble, pH-sensitive, and biodegradable nature is the widely preferred dextran derivate to synthesize Dex-MPs. Diols in the dextran backbone can be acetalated to transform the hydrophilic polymer into a hydrophobic structure to give Ac-Dex. The size, degradation rate, and encapsulation properties of the MPs can be adjusted by varying the acetalation degree (cyclic/acyclic acetal ratio). Ac-Dex MPs, which are acid-sensitive, degrade faster under lysosomal conditions (<pH 5) compared to the extracellular media (pH 7.4), due to the pH-dependent hydrolysis of acetal. There are various studies that manufacture Ac-Dex-based MPs by the electrospray method (Collier et al., 2018; Duong et al., 2013), or single/double emulsion polymerization (Kanthamneni et al., 2012; Suarez et al., 2013). Electrospray technology facilitates the manufacture of MPs with highly controllable sizes, dispersions, and unique surface morphology. Moreover, it is a commonly preferred approach in the biomedical field as it enables straightforward encapsulation of active molecules/drugs during the synthesis and controlled functionalization of biodegradable polymeric MPs (Morais et al., 2020).

Single and double emulsion polymerization techniques are other main approaches used to synthesize Dex-MPs. In the single emulsion systems, dextran is first dissolved/dispersed in an aqueous medium, then transferred to a nonaqueous medium. In the second stage, MPs are obtained via cross-linking by heating or chemical cross-linkers such as glutaraldehyde. Differently, in the double emulsion technique, there are three different phases as water/oil/water (W/O/W). In addition to allowing the synthesis of MPs, it also enables the active ingredients to be dissolved in different phases to encapsulate these components into MPs during synthesis. After the aqueous and organic phases are mixed, the solvent from the emulsion is removed by being subjected to evaporation or extraction steps. The emulsion is then placed in the aqueous environment where the organic phase is removed, and solid MPs are obtained by filtration (Ramteke, 2012). Suarez et al. (2013) used the W/O/W double emulsion to synthesize Ac-Dex MPs for the delivery of protein therapeutics to the heart after myocardial infarction. For this, hydrophobic Ac-Dex was dissolved in an organic phase (dichloromethane) including a surfactant and mixed with an aqueous phase to obtain a primary emulsion. The base emulsion was injected to a second aqueous solution including PVA, which initiated polymerization. Finally, Dex-MPs were obtained by evaporating the solvent and washing the particles (Suarez et al., 2013).

Drug delivery applications

Dex-MPs, which are able to encapsulate a variety of hydrophilic or hydrophobic drugs, provide targeted and controlled/prolonged release of these drugs thanks to their simple surface functionalization and biostability (Suarez et al., 2013).

Ac-Dex MPs allow adjustable release with the potential to deliver differently featured molecules (e.g., charged/uncharged, hydrophilic/hydrophobic), and release encapsulated therapeutics in a stimuli-response manner at acidic pH. The release rate of active molecules from Ac-Dex MPs can be adjusted easily by changing the acetalization degree of Ac-Dex. For this, the ratio of cyclic/acyclic acetals is adjusted by varying the reaction time of synthesis. Since cyclic acetals degrade more slowly than acyclic acetals, the ratio between them directly affects the rate of degradation of the MPs (Lee, 2017). Ac-Dex demonstrated adjustable release of several loads, such as imiquimod, ovalbumin, rapamycin, horseradish peroxidase, plasmid DNA, siRNA, and cathomycin. The particle size can be adjusted according to the purpose of application. For instance, design of particles in the range of 100 nm to 10 μm size is essential for cellular uptake and lysosomal degradation (e.g., immunotherapy, gene delivery, and pulmonary drug delivery applications); on the other hand, particle diameters between 20 and 100 μm are favorable in order to prevent cell uptake and enhance interstitial retention time to use Ac-Dex for intramuscular administration in ischemic conditions (Suarez et al., 2013). Degradation velocity of the particles can be designated from minutes to days under endosomal conditions (pH 5) (Cohen et al., 2010). Although it has not yet been tested in vivo, Ac-Dex degradation is estimated to have dextran, trace quantities of acetone and methanol as products used in several FDA-approved products (Suarez et al., 2013). Kauffman et al. (2012) synthesized rapamycin-loaded Ac-Dex MPs using a single emulsion (W/O) method. Rapamycin is an immunosuppressant with serious side effects in its clinical delivery which is used for autoimmune disorders. The authors aimed to reduce its side effect via encapsulation in Ac-Dex MPs for passive phagocyte targeting. Ac-Dex experienced controlled burst degradation in the phagosome (pH 5) at low pH conditions and slower degradation rate in extracellular media (pH 7.4). This degradation behavior makes Ac-Dex an ideal immune application candidate. In vivo studies with RAW macrophages have shown that rapamycin-loaded particles cause low toxicity and reduced nitric oxide production (Kauffman et al., 2012).

Dex-MPs are frequently used in protein-based (subunit) vaccine studies to deliver protein-based molecules and to maintain their stability (Kanthamneni et al., 2012). Poly(lactic-*co*-glycolic acid) (PLGA) is widely preferred as a DDS due to its degradation ability and biocompatibility. Nevertheless, the necessity of particles sensitive to acid environment in vaccine studies restricts the use of PLGA. In vaccine studies, protein delivery is achieved by the degradation of acid-sensitive materials in the phagolysosomes of immune effector cells. In the use of acid-insensitive particles such as PLGA, the mechanism cannot be easily controlled; in addition, the accumulation of lactic acid and glycolic acid resulting from PLGA degradation is disadvantageous in terms of the stability of protein-containing vaccines. Therefore, the use of acid-sensitive Ac-Dex MPs in this field provides a great advantage (Bachelder et al., 2010).

Gene delivery applications

Biopolymer-based MPs are effective genetic material carriers that have different properties from nanoparticle systems. Thanks to their size, they provide passive targeting in vaccine studies, while also eliminating toxicity and stability problems encountered in nanoparticle-based carriers (Cohen et al., 2010).

Cohen et al. (2010) synthesized acetalated dextran-based MPs to optimize the particle properties on their gene delivery applications. For this, the authors evaluated the effect of the acetylation degree on the degradation behavior of polymer. Further, Ac-Dex MPs were blended with a cationic polymer, loaded with luciferase reporter plasmid via the double emulsion technique, and modified with cell-penetrating peptide (CPP) to enhance the transfection ability (Cohen et al., 2010). The same authors later fabricated spermine-modified Ac-Dex and encapsulated siRNA into the particles. Spermine is an oligoamine with broad actions on cellular metabolism and has a cationic nature, which is advantageous for gene delivery. The authors hypothesized that siRNA can be loaded effectively to cationic spermine-Ac-Dex particles, which also enhances cellular uptake by the facilitated bindings to the negatively charged cell membranes. As a result of cell culture experiments, it was observed that the particles suppressed the expression of luciferin in HeLa-luc cells with low dose-dependent toxicity (Cohen et al., 2011).

12.5.2 Dextran-based nanoparticles

Polymeric NPs have attracted attention in recent years with their properties thanks to their small sizes that vary from 1 to 1000 nm. Polymeric NPs have many advantages for utilization as drug carriers. These advantages are the potential use for controlled release, protection capability of drug and the biologically active molecules from the surroundings, improvement of therapeutic index, and bioavailability.

Dex-NPs have become a focus for researchers with their wide variety of applications like nanomedicine, tissue engineering, wound healing, and gene and drug delivery.

Fabrication

Dextran is not able to self-assemble into NPs in its natural state. Thus, different methodologies have been developed for manufacturing dextran-based NPs (Abid et al., 2020; Coombes et al., 1997). The modification of dextran surface with reactive groups is a widely used strategy for this purpose (Abid et al., 2020). Dextran NPs can be synthesized via covalently linking the modified/unmodified dextran surface with various cross-linkers (Chalasani et al., 2007; Li et al., 2009; Tripathi et al., 2011). Following the cross-linking reactions of dextran NPs, grafting of the surface with various biocompatible polymers can enhance their usage in biomedical applications. For instance, grafting of Dex-NPs with PEI has improved the biocompatibility and biostability, which is commonly preferred in gene transfer applications (Tripathi et al., 2011). Similarly, PEG-grafted Dex-NPs were used as reduction and pH dual responsive DDS for intracellular targeting (Lian et al., 2017).

The surface modifications of dextran structure give an advantage to obtain conjugated NP systems. Various studies have performed conjugation reactions through modified dextran to synthesize Dex-NPs. For instance, Thambi et al. (2014) synthesized CMD NPs via conjugation between amine-functionalized lithocholic acid (hydrophobic) and CMD backbone (hydrophilic). The amphiphilic conjugates obtained had disulfide bonds between the polymers that are particularly breakable by glutathione in the cytoplasm. As a result, designed NPs provided the targeted delivery for the tumor cells by this approach (Thambi et al., 2014). Similarly, Schiff base reactions are the approach mostly applied to synthesize Dex-NPs. Dex-CHO is a useful dextran derivative used to synthesize Dex-NPs through Schiff base reactions, which are specific reactions for aldehyde groups under basic conditions. Wasiak et al. (2016) synthesized Dex-CHO based NPs via this process for DOX drug delivery. During the Schiff's base process, pH-sensitive carbon-nitrogen (C=N) bonds were obtained between the aldehyde groups of Dex-CHO and the amino groups of cross-linker (aliphatic amine). The molecular weight and oxidation degree of dextran, type of cross-linker, and degree of aldehyde group substitution are effective on the final properties of the Dex-NPs (Wasiak et al., 2016).

Dextran has been also used as covering material to improve the properties of metallic, inorganic, or organic NPs. Among them, dextran-coated magnetic iron oxide (Fe_3O_4) NPs are commonly synthesized and evaluated for different applications. Generally, Fe_3O_4 NPs are produced via various approaches. However, the difficulties in maintaining their stability cause limitations in their applications. In order to overcome these limitations, the NPs can be coated with an oxygen-impermeable layer to prevent oxygen from reaching the NPs surface and changing its magnetic properties (Kamalzare et al., 2019; Predescu et al., 2018). One of the reasons that dextran is preferred for surface coatings is the presence of appropriately sized dextran chains that allow efficient interactions with the surface of Fe_3O_4 NPs (Laurent et al., 2008). Synthesis of dextran-coated metallic NPs are generally performed via conjugation of functional groups between the metal surface and dextran. There are different routes to obtain dextran-coated Fe_3O_4 NPs, such as coating the surfaces with dextran after the synthesis of Fe_3O_4 NPs or synthesizing them in the dextran aqueous solution to obtain directly coated particles (Naha et al., 2019; Tassa et al., 2011).

Drug delivery applications

As DDSs, dextran NPs not only increase the effectiveness and stability of drugs but also improve their transport and distribution in the body and allow them to pass through many biological barriers intact. Thanks to the hydrophilic nature of dextran, various hydrophobic bioactive molecules can be easily encapsulated and these molecules are prevented from being removed by the mononuclear phagocyte system by ensuring in vivo stability during transport

(Huo et al., 2020). Drug-loaded Dex-NPs can be prepared by various methods according to the structure and property of dextran. Generally, hydroxyl groups in the dextran structure are involved in the incorporation of drug molecules into the structure. In addition, surface modifications of Dex-NPs performed by these hydroxyl groups often ensure the drug encapsulation and also selectivity for target organs, tissues, and cells (Huang and Huang, 2018). Functionalization of dextran with several reactive agents such as deoxycholic acid (Huo et al., 2020), carboxymethyl (Thambi et al., 2014), and methacrylate (Gracia et al., 2017) provide incorporation of several drugs and biomolecules via conjugation with these reactive groups.

The drug release behavior of dextran NPs depends on the particle properties such as the nature and molecular weight of dextran, size of the NPs, or particle-drug interactions. In general, the drug release is triggered by external stimuli. For instance, dextran has a naturally enzyme-triggered degradation ability since dextranase enzyme can degrade dextran backbone. Thus, dextran provide targeting delivery for various molecules encapsulated in dextran NPs via their controlled release throughout dextran degradation in the presence of enzyme (Abid et al., 2020; Tiryaki et al., 2020). Also, as mentioned above, Ac-Dex structure is a pH-sensitive material which can be triggered in acidic pH while being stable in physiological pH 7.4. Thus, Ac-Dex provide pH-triggered drug release behavior in their drug delivery applications. Torrieri et al. (2020), designed a putrescine modified Ac-Dex (Putre-Ac-Dex) based nano system by the single emulsion approach and functionalized their surfaces with atrial natriuretic peptide and TT1 peptides. Putrescine modification provided reactive groups on the surface of the particles to achieve effective interactions with peptides which have targeting capability for the infarcted cardiac cells and enhanced the biocompatibility during treatment (Torrieri et al., 2020). In another study, Ac-Dex structures were used for cancer-chemo immunotherapy by modification with spermine. For this purpose, spermine-Ac-Dex particles were synthesized and loaded with two molecules which are effective for the apoptosis of tumor cells by enhancing the macrophage response. The developed NPs suppressed the growth of tumor cells through pH-controlled release behavior and induced endosomal escape thanks to the presence of dextran (Bauleth-Ramos et al., 2017).

In addition, dextran is a preferred biopolymer for synthesis of protein-based NPs via a facile, green, and convenient self-assembly approach. Protein-dextran conjugates are known to improve the encapsulation properties of hydrophobic drugs/active ingredients and provide better stability than protein alone. The use of biocompatible polymers such as dextran provides an alternative to cross-linkers such as glutaraldehyde, which may have toxic effects, while at the same time avoiding protein-induced agglomerations. Accordingly, various protein-dextran conjugates such as whey protein-dextran (Fan et al., 2017), lysozyme-dextran (Li et al., 2008), bovine serum albumin (BSA)-dextran (Fan et al., 2018; Xia et al., 2015), and ovalbumin-dextran (Feng et al., 2016; Li and Gu, 2014) have been developed and used as effective drug or biomolecule delivery systems.

Dextran-coated metal NPs for drug delivery

The coating of metallic particles with dextran is of great importance in ensuring the stability, biocompatibility, and efficiency of the particles. Dex-coated metal NPs have high selectivity due to the presence of dextran, minimizing the damage of drug molecules to healthy cells, reducing side effects, improving the therapeutic effect of the drug, and increasing cellular uptake and the inhibition of tumor cells (Huang and Huang, 2018). Dex-coated Fe_3O_4 NPs are a widely used example of polymer-metal composite NPs in biomedical applications (Abdollah et al., 2018; Remya et al., 2016). The use of Fe_3O_4 NPs with magnetic properties is an important approach for targeted DDSs, since they are triggered by external stimuli and thus provide targeting to diseased tissues (Huang and Huang, 2019). Additionally, dextran coating provides biocompatibility for Fe_3O_4 NPs which have several cytotoxic effects in their clinical usage related to iron ion release with their excessive amounts. Iron releases often cause the formation of reactive oxygen species (ROS) that damage cellular organelles through lipid peroxidation (Remya et al., 2016). In this manner, Feridex, which is the United States Food and Drug Administration (FDA) approved dextran coated Fe_3O_4 NP, was developed as MRI contrast agent with high biocompatibility and effectiveness (Naha et al., 2019). The studies confirmed that coating with dextran significantly reduces the Fe_3O_4 NPs-mediated cytotoxic effects and has better biocompatibility and stability compared with the uncoated NPs (Badman et al., 2020; Bolandparvaz et al., 2020; Remya et al., 2016; Tingirikari et al., 2017). Unfortunately, due to the biodegradable nature of dextran, degradation by dextranase enzymes can reduce its ability for inhibition of Fe_3O_4 NPs toxicity (Badman et al., 2020; Mohammadi et al., 2018). In order to prevent this, Mohammadi et al. (2018), aimed to prevent iron ion leakage by coating NPs with a double-layer polymer containing dextran and PEG. The authors observed that double-layer coating of Fe_3O_4 NPs not only improved colloidal stability, but also reduced the formation of ROS compared to dextran-coated Fe_3O_4 NPs. In addition, the effect of the PEG/dextran bilayer on the biodistribution and pharmacokinetics of the NPs was evaluated, and the results showed that the proper surface coating of Fe_3O_4 NPs allowed them to remain in circulation in the body for a longer time (Mohammadi et al., 2018).

Besides drug delivery applications, Dex-coated NPs have been evaluated for hyperthermia, phototherapy, photodynamic therapy, and sensor and bioimaging applications. Sadaphal et al. (2018) synthesized magnetite/gold-NPs deposited Dex-coated carbon nanotubes for hyperthermia applications. In this study, dextran was preferred because it enhances the stability of the particles in aqueous solutions and prevents their agglomeration to a great extent (Sadaphal et al., 2018). Naha et al. (2020) synthesized Dex-coated cerium oxide (Ce) NPs as a contrast agent to use in clinical imaging by computed tomography. Ce NPs can produce strong X-ray attenuation while dextran would encourage accumulation in inflammation sites during the treatment of inflammatory bowel disease. According to the authors' results, Dex-Ce NPs were found to be protective against oxidative damage and more than 97% of oral doses were cleared from the body within 24 h (Naha et al., 2020).

Dextran-drug conjugates

Conjugation of the drug molecules with biopolymers is a widely used approach in drug delivery applications in order to improve the encapsulation efficiency and stability of drugs and to prevent uncontrollable releases during their delivery (Lee et al., 2018). In this approach, the therapeutic properties of drug molecules are combined with the biological properties of dextran. The common synthesis methods to fabricate dextran-drug conjugates include the covalent binding with functional groups of drugs following by conjugation of dextran backbone by several linkers. Direct esterification, carbonyldiimidazole activation, carbonate or carbamate esters, periodate oxidation, cyanogens bromide activation, and etherification are the main methods to obtain conjugated dextran before binding with drug molecules (Varshosaz, 2012).

Qi et al. (2017), synthesized ibuprofen (IBU) dextran conjugates to encapsulate methylprednisolone (MetP) molecules for treatment of spinal cord injury. Applying DDSs is an important strategy to reduce side effects caused by excess amounts of MetP usages during the treatment of spinal cord injuries. The authors claimed that the undesired side effects of the MetP drug during the conventional treatments can be avoided by the incorporation of the MetP drug inside the dextran-IBU conjugate. Dextran-drug conjugation was achieved as a result of N,N-carbonyldiimidazole-mediated esterification reaction of hydroxyl groups in dextran and carboxyl groups in IBU. In vivo studies showed that the MetP-loaded NPs had prolonged release with enhanced efficiency on neural activity (Qi et al., 2017).

It is the preferred approach to include stimuli-sensitive structures in polymer-drug conjugates to improve the controlled release of drug molecules. Controlled drug releases can be achieved by activating these components by external trigger such as pH and ROS. Therefore, it is an effective approach to target molecules with these external stimulants to tumor or inflammatory cells with excess ROS and low pH, unlike healthy cells. Lee et al. (2018) fabricated a pH- and ROS-sensitive dextran-drug conjugate for targeting inflammation. For this purpose, the antiinflammatory drug of Naproxen (NAP) was conjugated onto dextran followed by its modification with phenylboronic acid, which has a sensitivity to ROS. Afterward, dual responsive NPs were achieved via blending the Dex-NAP conjugate with the Ac-Dex structure. The complete release of the drug from the conjugate occurred within 20 min, and reduced the pro-inflammatory cytokine levels more efficiently than NAP alone (Lee et al., 2018).

Gene delivery

Polycationic macromolecules with high security and low cost are preferred in gene delivery applications, thus providing resistance against early serum degradation and easy penetration of cell membrane and binding to DNA. However, the presence of these macromolecules alone is not enough to overcome the difficulties in the release of nucleic acids and to increase the efficiency for clinical use. For this purpose, many polycationic conjugates, in which dextran is also used, have been synthesized. Cationic dextran can improve the stability of genetic material and penetrate cell membrane through the action of charge, thus avoiding the degradation by endosomes (Huang and Huang, 2018; Sizovs et al., 2010).

The first approach was carried out to obtain positively charged dextran structures by modification with 2-diethyl-aminoethyl. The encapsulation efficiency of several nucleic acids into the obtained dextran complex was improved thanks to the cationic nature of dextran (Vaheri, 1965; Mehvar, 2000). After this report, Mack et al. (1998) used this complex for transfection of the genetic material to macrophages, which have importance in the immune response and also limitations during their transfection. Later, this approach was improved to prevent the degradation of cationic dextran structures by dextranase enzyme during application. Thus, while the stability of the particles obtained during application was increased, transfection performance and biocompatibility were also improved (Onishi et al., 2005).

Cationic dextran structures can be synthesized via modifications with oligoamines, such as spermine or spermidine, which can improve the interactions with DNA. However, during studies conducted with this approach, a decrease in gene expression may be encountered in environments containing serum (Azzam et al., 2003; Azzam and Domb, 2004). PEGylation of the carrier material is a good strategy for enhancing the transfer of the genetic material in the biological serum. As an example for this approach, Hosseinkhani et al. (2004) fabricated a PEG-modified dextran-spermine conjugate. They boosted gene expression as a result of improved transfection efficiency via delivery of PEGylated dextran-spermine NPs to the cell line (Hosseinkhani et al., 2004). Foerster et al. (2016) synthesized spermine-modified Dex-NPs (160–250 nm) via the double emulsion method to use them for SiRNA targeting. Cy5-tagged siRNA was encapsulated in Dex-NPs during synthesis and the obtained particles were coated with the methoxypolyethylene glycol-N-hydroxysuccinimide ester. PEGylation of Dex-NPs improves their biodistribution and cellular uptake behavior, as well as providing good dispersibility for particles. Designed Dex-NPs did not show toxic effects in their applications (Foerster et al., 2016). In another study, Ac-Dex NPs were modified with poly(arginine) which has alkoxyamine moieties, to improve the penetration efficiency of the particles into cell membranes. The bioluminescence oxidative enzyme expressions were highly enhanced in HeLa cell lines as a result of interactions with Ac-Dex@poly(arginine) NPs (Beaudette et al., 2009).

MicroRNA (miRNA) delivery is an effective approach for the treatment of cancer diseases by means of its ability of regulating tumor cell pathways. However, a disadvantage is their limited efficiency due to their degradation by enzymes during delivery. Yalcin (2019) synthesized dextran-coated Fe_3O_4 NPs loaded with miR-29a and investigated their efficiency on breast cancer cells. miRNA was targeted to tumor cells without losing its effectiveness, and the results demonstrated that antiapoptotic proteins of the tumor cells were suppressed by delivering miRNA (Yalcin, 2019). Kamalzare et al. (2019) fabricated carboxymethyl dextran-coated iron oxide/trimethyl chitosan NPs to deliver siRNA, which is effective in the treatment of HIV-1 infections. Carboxymethyl dextran was used to enhance the biocompatibility and biostability of the NPs. According to the results, the cellular uptake and gene transfection capabilities were improved thanks to the presence of carboxymethyl dextran and hybrid particles suppressed the expression of HIV-1 nef protein by delivering siRNA successfully (Kamalzare et al., 2019).

12.6 Concluding remarks

This chapter focused on dextran-based micro- and nanostructured biomaterials, fabrication techniques, and applications in drug delivery and biomedical areas. Dextran-based biomaterials are advantageous due to their biodegradability, biocompatibility features, and proven clinical applications. A high number of hydroxyl groups allowing chemical modification is another important feature of dextran since it enables the design of novel materials with various structures and controlled properties. In addition, solubility in various solvents and high stability properties of dextran gain high workability features.

Dextrans have unique properties for medical purposes, especially blood-related applications, such as antithrombotic activity, reducing erythrocyte aggregation, liquid plasma expander, and blood flow improver. The biomedical application limits of a material are dependent on the cellular responses in which the micro- and nanoscale structure of a biomaterial is known to play a crucial role. Thus, techniques for the fabrication of materials with controlled micro- and nanoscale properties have gained high interest. In general, water-in-oil microemulsion, heterogeneous emulsion, and chemical cross-linking are among the most abundant dextran-based micro and nanogel fabrication techniques. The aqueous two-phase system is preferred for the fabrication of dextran-based microparticles, while covalently linking the modified/unmodified dextran surface with various cross-linkers is the general dextran-based nanoparticle fabrication strategy. Blend, coaxial, and emulsion electrospinning techniques are applied in the fabrication of dextran-based nanofibers and core-shell nanofibers.

Dextran-based micro- and nanostructured materials are widely investigated for drug delivery applications and are also preferred in gene delivery and imaging application studies. Nanofibrous dextran-based biomaterials are investigated for wound dressing and tissue scaffold applications. The main drawback for the fabrication of dextran-based biomaterials may be the effect of the dextran source changes, thus improved fabrication techniques for stable and controlled structured biomaterials are needed. The clinical applicability of dextran is already proven, which provides a basis for future drug delivery and biomedical research on dextran-based micro- and nanobiomaterials.

References

Abdollah, M.R.A., Carter, T.J., Jones, C., Kalber, T.L., Rajkumar, V., Tolner, B., Gruettner, C., Zaw-Thin, M., Baguña Torres, J., Ellis, M., Robson, M., Pedley, R.B., Mulholland, P., Rafael, T.M.D.R., Chester, K.A., 2018. Fucoidan prolongs the circulation time of dextran-coated iron oxide nanoparticles. ACS Nano 12 (2), 1156–1169.

Abid, M., Naveed, M., Azeem, I., Faisal, A., Faizan Nazar, M., Yameen, B., 2020. Colon specific enzyme responsive oligoester crosslinked dextran nanoparticles for controlled release of 5-fluorouracil. Int. J. Pharm. *586*.

Aronson, J.K., 2016. Dextrans. In: Meyler's Side Effects of Drugs, sixteenth ed. Elsevier, pp. 893–898.

Arturson, G., Wallenius, G., 1964. The renal clearance of dextran of different molecular sizes in normal humans. Scand. J. Clin. Lab. Invest. 16 (1), 81–86.

Azzam, T., Domb, A.J., 2004. Cationic polysaccharides for gene delivery. In: Polymeric Gene Delivery: Principles and Applications. CRC Press, pp. 279–299.

Azzam, T., Eliyahu, H., Makovitzki, A., Domb, A.J., 2003. Dextran-spermine conjugate: an efficient vector for gene delivery. Macromol. Symp. 195 (1), 247–262.

Bachelder, E.M., Beaudette, T.T., Broaders, K.E., Fréchet, J.M.J., Albrecht, M.T., Mateczun, A.J., Ainslie, K.M., Pesce, J.T., Keane-Myers, A.M., 2010. In vitro analysis of acetalated dextran microparticles as a potent delivery platform for vaccine adjuvants. Mol. Pharm. 7 (3), 826–835.

Badman, R.P., Moore, S.L., Killian, J.L., Feng, T., Cleland, T.A., Hu, F., Wang, M.D., 2020. Dextran-coated iron oxide nanoparticle-induced nanotoxicity in neuron cultures. Sci. Rep. 10 (1).

Bauleth-Ramos, T., Shahbazi, M.A., Liu, D., Fontana, F., Correia, A., Figueiredo, P., Zhang, H., Martins, J.P., Hirvonen, J.T., Granja, P., Sarmento, B., Santos, H.A., 2017. Nutlin-3a and cytokine co-loaded spermine-modified acetalated dextran nanoparticles for cancer chemo-immunotherapy. Adv. Funct. Mater. 27 (42).

Beaudette, T.T., Cohen, J.A., Bachelder, E.M., Broaders, K.E., Cohen, J.L., Engleman, E.G., Fréchet, J.M.J., 2009. Chemoselective ligation in the functionalization of polysaccharide-based particles. J. Am. Chem. Soc. 131 (30), 10360–10361.

Bolandparvaz, A., Vapniarsky, N., Harriman, R., Alvarez, K., Saini, J., Zang, Z., Van De Water, J., Lewis, J.S., 2020. Biodistribution and toxicity of epitope-functionalized dextran iron oxide nanoparticles in a pregnant murine model. J. Biomed. Mater. Res. A 108 (5), 1186–1202.

Cejudo-Sanches, J., Orrego, A.H., Jaime-Mendoza, A., Ghobadi, R., Moreno-Perez, S., Fernandez-Lorente, G., Rocha-Martin, J., Guisan, J.M., 2020. High stabilization of immobilized Rhizomucor miehei lipase by additional coating with hydrophilic crosslinked polymers: poly-allylamine/aldehyde–dextran. Process Biochem. 92, 156–163.

Cengiz-Çallıoğlu, F., 2014. Dextran nanofiber production by needleless electrospinning process. e-Polymers 14 (1), 5–13.

Chalasani, K.B., Russell-Jones, G.J., Jain, A.K., Diwan, P.V., Jain, S.K., 2007. Effective oral delivery of insulin in animal models using vitamin B12-coated dextran nanoparticles. J. Control. Release 122 (2), 141–150.

Chen, N., Kroger, C.J., Tisch, R.M., Bachelder, E.M., Ainslie, K.M., 2018. Prevention of type 1 diabetes with acetalated dextran microparticles containing rapamycin and pancreatic peptide P31. Adv. Healthc. Mater. 7 (18).

Cohen, J.A., Beaudette, T.T., Cohen, J.L., Broaders, K.E., Bachelder, E.M., Fréchet, J.M.J., 2010. Acetal-modified dextran microparticles with controlled degradation kinetics and surface functionality for gene delivery in phagocytic and non-phagocytic cells. Adv. Mater. 22 (32), 3593–3597.

Cohen, J.L., Schubert, S., Wich, P.R., Cui, L., Cohen, J.A., Mynar, J.L., Fréchet, J.M.J., 2011. Acid-degradable cationic dextran particles for the delivery of siRNA therapeutics. Bioconjug. Chem. 22 (6), 1056–1065.

Collier, M.A., Junkins, R.D., Gallovic, M.D., Johnson, B.M., Johnson, M.M., MacIntyre, A.N., Sempowski, G.D., Bachelder, E.M., Ting, J.P.Y., Ainslie, K.M., 2018. Acetalated dextran microparticles for codelivery of STING and TLR7/8 agonists. Mol. Pharm. 15 (11), 4933–4946.

Coombes, A.G.A., Tasker, S., Lindblad, M., Holmgren, J., Hoste, K., Toncheva, V., Schacht, E., Davies, M.C., Illum, L., Davis, S.S., 1997. Biodegradable polymeric microparticles for drug delivery and vaccine formulation: the surface attachment of hydrophilic species using the concept of poly(ethylene glycol) anchoring segments. Biomaterials 18 (17), 1153–1161.

De Backer, L., Braeckmans, K., Stuart, M.C.A., Demeester, J., De Smedt, S.C., Raemdonck, K., 2015. Bio-inspired pulmonary surfactant-modified nanogels: a promising siRNA delivery system. J. Control. Release 206, 177–186.

de Belder, A.N., 1996. Medical applications of dextran and its derivatives. In: Polysaccharides in Medicinal Applications. Marcel Dekker, Inc, pp. 505–523.

Dhaneshwar, S.S., Kandpal, M., Gairola, N., Kadam, S.S., 2006. Dextran: a promising macromolecular drug carrier. Indian J. Pharm. Sci. 68 (6), 705–714.

Dormandy, J.A., 1971. Influence of blood viscosity on blood flow and the effect of low molecular weight dextran. Br. Med. J. 4 (5789), 716–719.

Duong, A.D., Sharma, S., Peine, K.J., Gupta, G., Satoskar, A.R., Bachelder, E.M., Wyslouzil, B.E., Ainslie, K.M., 2013. Electrospray encapsulation of toll-like receptor agonist resiquimod in polymer microparticles for the treatment of visceral leishmaniasis. Mol. Pharm. 10 (3), 1045–1055.

Fan, Y., Yi, J., Zhang, Y., Wen, Z., Zhao, L., 2017. Physicochemical stability and in vitro bioaccessibility of β-carotene nanoemulsions stabilized with whey protein-dextran conjugates. Food Hydrocoll. 63, 256–264.

Fan, Y., Yi, J., Zhang, Y., Yokoyama, W., 2018. Fabrication of curcumin-loaded bovine serum albumin (BSA)-dextran nanoparticles and the cellular antioxidant activity. Food Chem. 239, 1210–1218.

Farjami, T., Madadlou, A., 2017. Fabrication methods of biopolymeric microgels and microgel-based hydrogels. Food Hydrocoll. 62, 262–272.

Feng, J., Wu, S., Wang, H., Liu, S., 2016. Improved bioavailability of curcumin in ovalbumin-dextran nanogels prepared by Maillard reaction. J. Funct. Foods 27, 55–68.

Ferrer, M.C.C., Sobolewski, P., Composto, R.J., Eckmann, D.M., 2013. Cellular uptake and intracellular cargo release from dextran based nanogel drug carriers. J. Nanosci. Nanotechnol. 4 (1).

Foerster, F., Bamberger, D., Schupp, J., Weilbächer, M., Kaps, L., Strobl, S., Radi, L., Diken, M., Strand, D., Tuettenberg, A., Wich, P.R., Schuppan, D., 2016. Dextran-based therapeutic nanoparticles for hepatic drug delivery. Nanomedicine 11 (20), 2663–2677.

Forciniti, D., 2000. Preparation of aqueous two-phase systems. In: Hatti-Kaul, R. (Ed.), Aqueous Two-Phase Systems: Methods and Protocols. Humana Press, pp. 23–33.

Gaspar, V.M., Moreira, A.F., de Melo-Diogo, D., Costa, E.C., Queiroz, J.A., Sousa, F., Pichon, C., Correia, I.J., 2016. Multifunctional nanocarriers for codelivery of nucleic acids and chemotherapeutics to cancer cells. In: Nanobiomaterials in Medical Imaging: Applications of Nanobiomaterials. Elsevier Inc, pp. 163–207.

Gelin, L.E., Thoren, O.K., 1961. Influence of low viscous dextran on peripheral circulation in man. A plethysmographic study. Acta Chir. Scand. 122, 303–308.

Gracia, R., Marradi, M., Cossío, U., Benito, A., Pérez-San Vicente, A., Gómez-Vallejo, V., Grande, H.J., Llop, J., Loinaz, I., 2017. Synthesis and functionalization of dextran-based single-chain nanoparticles in aqueous media. J. Mater. Chem. B 5 (6), 1143–1147.

Gyles, D.A., Castro, L.D., Silva, J.O.C., Ribeiro-Costa, R.M., 2017. A review of the designs and prominent biomedical advances of natural and synthetic hydrogel formulations. Eur. Polym. J. 88, 373–392.

Hamidi, M., Azadi, A., Rafiei, P., 2008. Hydrogel nanoparticles in drug delivery. Adv. Drug Deliv. Rev. 60 (15), 1638–1649.

He, L., Zheng, R., Min, J., Lu, F., Wu, C., Zhi, Y., Shan, S., Su, H., 2021. Preparation of magnetic microgels based on dextran for stimuli-responsive release of doxorubicin. J. Magn. Magn. Mater. 517.

Heinze, T., Liebert, T., Heublein, B., Hornig, S., 2006. Functional polymers based on dextran. Adv. Polym. Sci. 205 (1), 199–291.

Hosseinkhani, H., Azzam, T., Tabata, Y., Domb, A.J., 2004. Dextran-spermine polycation: an efficient nonviral vector for in vitro and in vivo gene transfection. Gene Ther. 11 (2), 194–203.

Huang, G., Huang, H., 2018. Application of dextran as nanoscale drug carriers. Nanomedicine 13 (24), 3149–3158.

Huang, S., Huang, G., 2019. Preparation and drug delivery of dextran-drug complex. Drug Deliv. 26 (1), 252–261.

Humphreys, W.V., Walker, A., Cave, F.D., Charlesworth, D., 1976. The effect of an infusion of low molecular weight dextran on peripheral resistance in patients with arteriosclerosis. Br. J. Surg. 63 (9), 691–693.

Huo, M., Wang, H., Zhang, Y., Cai, H., Zhang, P., Li, L., Zhou, J., Yin, T., 2020. Co-delivery of silybin and paclitaxel by dextran-based nanoparticles for effective anti-tumor treatment through chemotherapy sensitization and microenvironment modulation. J. Control. Release 321, 198–210.

Jin, B., Zhou, X., Li, X., Lin, W., Chen, G., Qiu, R., 2016. Self-assembled modified soy protein/dextran nanogel induced by ultrasonication as a delivery vehicle for riboflavin. Molecules 21 (3).

Kamalzare, S., Noormohammadi, Z., Rahimi, P., Atyabi, F., Irani, S., Tekie, F.S.M., Mottaghitalab, F., 2019. Carboxymethyl dextran-trimethyl chitosan coated superparamagnetic iron oxide nanoparticles: an effective siRNA delivery system for HIV-1 Nef. J. Cell. Physiol. 234 (11), 20554–20565.

Kanthamneni, N., Sharma, S., Meenach, S.A., Billet, B., Zhao, J.C., Bachelder, E.M., Ainslie, K.M., 2012. Enhanced stability of horseradish peroxidase encapsulated in acetalated dextran microparticles stored outside cold chain conditions. Int. J. Pharm. 431 (1–2), 101–110.

Karandikar, S., Mirani, A., Waybhase, V., Patravale, V.B., Patankar, S., 2017. Nanovaccines for oral delivery-formulation strategies and challenges. In: Nanostructures for Oral Medicine. Elsevier Inc, pp. 263–293.

Karewicz, A., 2014. Polymeric and liposomal nanocarriers for controlled drug delivery. In: Biomaterials for Bone Regeneration: Novel Techniques and Applications. Elsevier Ltd, pp. 351–373.

Kauffman, K.J., Kanthamneni, N., Meenach, S.A., Pierson, B.C., Bachelder, E.M., Ainslie, K.M., 2012. Optimization of rapamycin-loaded acetalated dextran microparticles for immunosuppression. Int. J. Pharm. 422 (1–2), 356–363.

Kaur, M., Sudhakar, K., Mishra, V., 2019. Fabrication and biomedical potential of nanogels: an overview. Int. J. Polym. Mater. Polym. Biomater. 68 (6), 287–296.

Larsen, C., 1989. Dextran prodrugs—structure and stability in relation to therapeutic activity. Adv. Drug Deliv. Rev. 3 (1), 103–154.

Laurent, S., Forge, D., Port, M., Roch, A., Robic, C., Vander Elst, L., Muller, R.N., 2008. Magnetic iron oxide nanoparticles: synthesis, stabilization, vectorization, physicochemical characterizations and biological applications. Chem. Rev. 108 (6), 2064–2110.

Lee, C.J., 2017. Bacterial capsular polysaccharides: immunogenicity and vaccines. In: Polysaccharides in Medicinal Applications. CRC Press, pp. 411–442.

Lee, S., Stubelius, A., Hamelmann, N., Tran, V., Almutairi, A., 2018. Inflammation-responsive drug-conjugated dextran nanoparticles enhance anti-inflammatory drug efficacy. ACS Appl. Mater. Interfaces 10 (47), 40378–40387.

Li, Z., Gu, L., 2014. Fabrication of self-assembled (-)-epigallocatechin gallate (EGCG) ovalbumin-dextran conjugate nanoparticles and their transport across monolayers of human intestinal epithelial caco-2 cells. J. Agric. Food Chem. 62 (6), 1301–1309.

Li, J., Yu, S., Yao, P., Jiang, M., 2008. lysozyme—dextran core—Shell nanogels prepared via a green process. Langmuir 24 (7), 3486–3492.

Li, Y.L., Zhu, L., Liu, Z., Cheng, R., Meng, F., Cui, J.H., Ji, S.J., Zhong, Z., 2009. Reversibly stabilized multifunctional dextran nanoparticles efficiently deliver doxorubicin into the nuclei of cancer cells. Angew. Chem. Int. Ed. 48 (52), 9914–9918.

Li, J., Jiang, B., Lin, C., Zhuang, Z., 2014. Fluorescence tomographic imaging of sentinel lymph node using near-infrared emitting bioreducible dextran nanogels. Int. J. Nanomedicine 9, 5667–5682.

Li, W., Yun, L., Rifky, M., Liu, R., Wu, T., Sui, W., Zhang, M., 2021. Carboxymethylation of (1 → 6)-α-dextran from Leuconostoc spp.: effects on microstructural, thermal and antioxidant properties. Int. J. Biol. Macromol. 166, 1–8.

Lian, H., Du, Y., Chen, X., Duan, L., Gao, G., Xiao, C., Zhuang, X., 2017. Core cross-linked poly(ethylene glycol)-graft-Dextran nanoparticles for reduction and pH dual responsive intracellular drug delivery. J. Colloid Interface Sci. 496, 201–210.

Mack, K.D., Wei, R., Elbagarri, A., Abbey, N., McGrath, M.S., 1998. A novel method for DEAE-dextran mediated transfection of adherent primary cultured human macrophages. J. Immunol. Methods 211 (1–2), 79–86.

Maia, J., Evangelista, M.B., Gil, H., Ferreira, L., 2014. Dextran-based materials for biomedical applications. In: Gil, M.H. (Ed.), Carbohydrates Applications in Medicine. Research Signpost, pp. 31–53.

Maslakci, N.N., Ulusoy, S., Uygun, E., Çevikbaş, H., Oksuz, L., Can, H.K., Uygun Oksuz, A., 2017. Ibuprofen and acetylsalicylic acid loaded electrospun PVP-dextran nanofiber mats for biomedical applications. Polym. Bull. 74 (8), 3283–3299.

Meenach, S.A., Kim, Y.J., Kauffman, K.J., Kanthamneni, N., Bachelder, E.M., Ainslie, K.M., 2012. Synthesis, optimization, and characterization of camptothecin-loaded acetalated dextran porous microparticles for pulmonary delivery. Mol. Pharm. 9 (2), 290–298.

Mehvar, R., 2000. Dextrans for targeted and sustained delivery of therapeutic and imaging agents. J. Control. Release 69 (1), 1–25.

Mohammadi, M.R., Malkovskiy, A.V., Jothimuthu, P., Kim, K.M., Parekh, M., Inayathullah, M., Zhuge, Y., Rajadas, J., 2018. PEG/dextran double layer influences fe ion release and colloidal stability of iron oxide nanoparticles. Sci. Rep. 8 (1), 1–11.

Morais, A.Í.S., Vieira, E.G., Afewerki, S., Sousa, R.B., Honorio, L.M.C., Cambrussi, A.N.C.O., Santos, J.A., Bezerra, R.D.S., Furtini, J.A.O., Silva-Filho, E.C., Webster, T.J., Lobo, A.O., 2020. Fabrication of polymeric microparticles by electrospray: the impact of experimental parameters. J. Funct. Biomater. 11 (1).

Moydeen, A.M., Ali Padusha, M.S., Aboelfetoh, E.F., Al-Deyab, S.S., El-Newehy, M.H., 2018. Fabrication of electrospun poly(vinyl alcohol)/dextran nanofibers via emulsion process as drug delivery system: kinetics and in vitro release study. Int. J. Biol. Macromol. 116, 1250–1259.

Naha, P.C., Liu, Y., Hwang, G., Huang, Y., Gubara, S., Jonnakuti, V., Simon-Soro, A., Kim, D., Gao, L., Koo, H., Cormode, D.P., 2019. Dextran-coated iron oxide nanoparticles as biomimetic catalysts for localized and pH-activated biofilm disruption. ACS Nano 13 (5), 4960–4971.

Naha, P.C., Hsu, J.C., Kim, J., Shah, S., Bouché, M., Si-Mohamed, S., Rosario-Berrios, D.N., Douek, P., Hajfathalian, M., Yasini, P., Singh, S., Rosen, M.A., Morgan, M.A., Cormode, D.P., 2020. Dextran-coated cerium oxide nanoparticles: a computed tomography contrast agent for imaging the gastrointestinal tract and inflammatory bowel disease. ACS Nano 14 (8), 10187–10197.

Nematpour, N., Farhadian, N., Ebrahimi, K.S., Arkan, E., Seyedi, F., Khaledian, S., Shahlaei, M., Moradi, S., 2020. Sustained release nanofibrous composite patch for transdermal antibiotic delivery. Colloids Surf. A Physicochem. Eng. Asp. 586.

Onishi, Y., Eshita, Y., Murashita, A., Mizuno, M., Yoshida, J., 2005. Synthesis and characterization of 2-diethyl-aminoethyl-dextran-methyl methacrylate graft copolymer for nonviral gene delivery vector. J. Appl. Polym. Sci. 98 (1), 9–14.

Predescu, A.M., Matei, E., Berbecaru, A.C., Pantilimon, C., Drăgan, C., Vidu, R., Predescu, C., Kuncser, V., 2018. Synthesis and characterization of dextran-coated iron oxide nanoparticles. R. Soc. Open Sci. 5 (3).

Qi, L., Jiang, H., Cui, X., Liang, G., Gao, M., Huang, Z., Xi, Q., 2017. Synthesis of methylprednisolone loaded ibuprofen modified dextran based nanoparticles and their application for drug delivery in acute spinal cord injury. Oncotarget 8 (59), 99666–99680.

Raemdonck, K., Van Thienen, T.G., Vandenbroucke, R.E., Sanders, N.N., Demeester, J., De Smedt, S.C., 2008. Dextran microgels for time-controlled delivery of siRNA. Adv. Funct. Mater. 18 (7), 993–1001.

Ramteke, K.H., 2012. Microspheres: as carrieres used for novel drug delivery system. IOSR J. Pharm. 2 (4), 44–48.

Remya, N.S., Syama, S., Sabareeswaran, A., Mohanan, P.V., 2016. Toxicity, toxicokinetics and biodistribution of dextran stabilized iron oxide nanoparticles for biomedical applications. Int. J. Pharm. 511 (1), 586–598.

Rowe, D.S., 1955. Colloid osmotic pressures of dextran, serum and dextran-serum mixtures [9]. Nature 175 (4456), 554–555.

Sadaphal, V., Mukherjee, S., Ghosh, S., 2018. Hybrid photomagnetic modulation of magnetite/gold-nanoparticle-deposited dextran-covered carbon nanotubes for hyperthermia applications. Appl. Phys. Express 11 (9).

Salimi-Kenari, H., Imani, M., Nodehi, A., Abedini, H., 2016. An engineering approach to design of dextran microgels size fabricated by water/oil emulsification. J. Microencapsul. 33 (6), 511–523.

Salimi-Kenari, H., Mollaie, F., Dashtimoghadam, E., Imani, M., Nyström, B., 2018. Effects of chain length of the cross-linking agent on rheological and swelling characteristics of dextran hydrogels. Carbohydr. Polym. 181, 141–149.

Shewan, H.M., Stokes, J.R., 2013. Review of techniques to manufacture micro-hydrogel particles for the food industry and their applications. J. Food Eng. 119 (4), 781–792.

Shi, L., Aid, R., Le Visage, C., Chew, S.Y., 2012. Biomimicking polysaccharide nanofibers promote vascular phenotypes: a potential application for vascular tissue engineering. Macromol. Biosci. 12 (3), 395–401.

Sizovs, A., McLendon, P.M., Srinivasachari, S., Reineke, T.M., 2010. Carbohydrate polymers for nonviral nucleic acid delivery. Top. Curr. Chem. 296, 131–190.

Stenekes, R.J.H., Franssen, O., van Bommel, E.M.G., Crommelin, D.J.A., Hennink, W.E., 1998. The preparation of dextran microspheres in an all-aqueous system: effect of the formulation parameters on particle characteristics. Pharm. Res. 15, 557–561.

Su, H., Jia, Q., Shan, S., 2016. Synthesis and characterization of Schiff base contained dextran microgels in water-in-oil inverse microemulsion. Carbohydr. Polym. 152, 156–162.

Su, H., Zhang, W., Wu, Y., Han, X., Liu, G., Jia, Q., Shan, S., 2018. Schiff base-containing dextran nanogel as pH-sensitive drug delivery system of doxorubicin: synthesis and characterization. J. Biomater. Appl. 33 (2), 170–181.

Su, H., Han, X., He, L., Deng, L., Yu, K., Jiang, H., Wu, C., Jia, Q., Shan, S., 2019. Synthesis and characterization of magnetic dextran nanogel doped with iron oxide nanoparticles as magnetic resonance imaging probe. Int. J. Biol. Macromol. 128, 768–774.

Suarez, S., Grover, G.N., Braden, R.L., Christman, K.L., Almutairi, A., 2013. Tunable protein release from acetalated dextran microparticles: a platform for delivery of protein therapeutics to the heart post-MI. Biomacromolecules 14 (11), 3927–3935.

Sultana, F., Manirujjaman, I.-U.-H., Arafat, M., Sharmin, S., 2013. An overview of nanogel drug delivery system. J. Appl. Pharm. Sci. 3 (8), S95–S105.

Sun, G., Mao, J.J., 2012. Engineering dextran-based scaffolds for drug delivery and tissue repair. Nanomedicine 7 (11), 1771–1784.

Tacias-Pascacio, V.G., Ortiz, C., Rueda, N., Berenguer-Murcia, Á., Acosta, N., Aranaz, I., Civera, C., Fernandez-Lafuente, R., Alcántara, A.R., 2019. Dextran aldehyde in biocatalysis: more than a mere immobilization system. Catalysts 9 (7).

Tassa, C., Shaw, S.Y., Weissleder, R., 2011. Dextran-coated iron oxide nanoparticles: a versatile platform for targeted molecular imaging, molecular diagnostics, and therapy. Acc. Chem. Res. 44 (10), 842–852.

Thambi, T., You, D.G., Han, H.S., Deepagan, V.G., Jeon, S.M., Suh, Y.D., Choi, K.Y., Kim, K., Kwon, I.C., Yi, G.R., Lee, J.Y., Lee, D.S., Park, J.H., 2014. Bioreducible carboxymethyl dextran nanoparticles for tumor-targeted drug delivery. Adv. Healthc. Mater. 3 (11), 1829–1838.

Thoren, L., 1980. The dextrans—clinical data. Dev. Biol. Stand. 48, 157–167.

Thorne, J.B., Vine, G.J., Snowden, M.J., 2011. Microgel applications and commercial considerations. Colloid Polym. Sci. 289 (5–6), 625–646.

Tingirikari, J.M.R., Rani, A., Goyal, A., 2017. Characterization of super paramagnetic nanoparticles coated with a biocompatible polymer produced by dextransucrase from Weissella cibaria JAG8. J. Polym. Environ. 25 (3), 569–577.

Tiryaki, E., Başaran Elalmış, Y., Karakuzu İkizler, B., Yücel, S., 2020. Novel organic/inorganic hybrid nanoparticles as enzyme-triggered drug delivery systems: dextran and dextran aldehyde coated silica aerogels. J. Drug Deliv. Sci. Technol. 56.

Torrieri, G., Fontana, F., Figueiredo, P., Liu, Z., Ferreira, M.P.A., Talman, V., Martins, J.P., Fusciello, M., Moslova, K., Teesalu, T., Cerullo, V., Hirvonen, J., Ruskoaho, H., Balasubramanian, V., Santos, H.A., 2020. Dual-peptide functionalized acetalated dextran-based nanoparticles for sequential targeting of macrophages during myocardial infarction. Nanoscale 12 (4), 2350–2358.

Tripathi, S.K., Goyal, R., Gupta, K.C., 2011. Surface modification of crosslinked dextran nanoparticles influences transfection efficiency of dextran-polyethylenimine nanocomposites. Soft Matter 7 (24), 11360–11371.

Uddin, S.N., Islam, K.K., 2006. Cationic polymers and its uses in non-viral gene delivery systems: a conceptual research. Trends Med. Res., 86–99.

Unnithan, A.R., Barakat, N.A.M., Tirupathi Pichiah, P.B., Gnanasekaran, G., Nirmala, R., Cha, Y.S., Jung, C.H., El-Newehy, M., Kim, H.Y., 2012. Wound-dressing materials with antibacterial activity from electrospun polyurethane-dextran nanofiber mats containing ciprofloxacin HCl. Carbohydr. Polym. 90 (4), 1786–1793.

Unnithan, A.R., Sasikala, A.R.K., Murugesan, P., Gurusamy, M., Wu, D., Park, C.H., Kim, C.S., 2015. Electrospun polyurethane-dextran nanofiber mats loaded with estradiol for post-menopausal wound dressing. Int. J. Biol. Macromol. 77, 1–8.

Vaheri, A., Pagano, J.S., 1965. Infectious poliovirus RNA: a sensitive method of assay. Virology 27 (3), 434–436.

Varghese, S.A., Rangappa, S.M., Siengchin, S., Parameswaranpillai, J., 2019. Natural polymers and the hydrogels prepared from them. In: Hydrogels Based on Natural Polymers. Elsevier, pp. 17–47.
Varshosaz, J., 2012. Dextran conjugates in drug delivery. Expert Opin. Drug Deliv. 9 (5), 509–523.
Walker, G.J., 1978. Dextrans. In: Manners, D.J. (Ed.), Biochem. Carbohydrates II. University Park, Baltimore, MD, USA, pp. 75–126.
Wang, R., Dijkstra, P.J., Karperien, M., 2016. Dextran. In: Biomaterials from Nature for Advanced Devices and Therapies. Wiley Blackwell, pp. 307–319.
Wang, C., You, J., Gao, M., Zhang, P., Xu, G., Dou, H., 2020. Bio-inspired gene carriers with low cytotoxicity constructed via the assembly of dextran nanogels and nano-coacervates. Nanomedicine 15 (13), 1285–1296.
Wasiak, I., Kulikowska, A., Janczewska, M., Michalak, M., Cymerman, I.A., Nagalski, A., Kallinger, P., Szymanski, W.W., Ciach, T., 2016. Dextran nanoparticle synthesis and properties. PLoS One 11 (1).
Widenbring, R., Frenning, G., Malmsten, M., 2014. Chain and pore-blocking effects on matrix degradation in protein-loaded microgels. Biomacromolecules 15 (10), 3671–3678.
Xia, S., Li, Y., Zhao, Q., Li, J., Xia, Q., Zhang, X., Huang, Q., 2015. Probing conformational change of bovine serum albumin-dextran conjugates under controlled dry heating. J. Agric. Food Chem. 63 (16), 4080–4086.
Yalcin, S., 2019. Dextran-coated iron oxide nanoparticle for delivery of miR-29a to breast cancer cell line. Pharm. Dev. Technol. 24 (8), 1032–1037.
Yeredla, N., Kojima, T., Yang, Y., Takayama, S., Kanapathipillai, M., 2016. Aqueous two phase system assisted self-assembled PLGA microparticles. Sci. Rep. 6 (1), 1–8.
Yu, K., Yang, X., He, L., Zheng, R., Min, J., Su, H., Shan, S., Jia, Q., 2020. Facile preparation of pH/reduction dual-stimuli responsive dextran nanogel as environment-sensitive carrier of doxorubicin. Polymer 200.
Zhang, J., Li, C., Wang, Y., Zhuo, R.X., Zhang, X.Z., 2011. Controllable exploding microcapsules as drug carriers. Chem. Commun. 47 (15), 4457–4459.
Zhang, J., Xu, X.D., Liu, Y., Liu, C.W., Chen, X.H., Li, C., Zhuo, R.X., Zhang, X.Z., 2012. Design of an \active defense\ system as drug carriers for cancer therapy. Adv. Funct. Mater. 22 (8), 1704–1710.
Zhang, J., Chen, L., Chen, J., Wu, D., Feng, J., 2018. Dextran microgels loaded with ZnO QDs: pH-triggered degradation under acidic conditions. J. Appl. Polym. Sci. 135 (6), 45831.

CHAPTER

13

Gum arabic-based nanocarriers for drug and bioactive compounds delivery

Neda Aliabbasi[a], *Morteza Fathi*[b], *and Zahra Emam-Djomeh*[a]

[a]**Functional Food Research Core (FFRC), Transfer Phenomena Laboratory (TPL), College of Agriculture & Natural Resources, University of Tehran, Karaj, Iran** [b]**Function Health Research Center, Lifestyle Institute, Baqiyatallah University of Medical Sciences, Tehran, Iran**

13.1 Introduction

Gum arabic (GA) is a dried polysaccharide exudate from the stems and branches of *Acacia senegal* or *Acacia seyal*. GA is naturally produced as a response to tissue damage or environmental stress. It has been documented that the gums obtained from both species are acceptable for application in food and pharmaceutical formulations (Phillips et al., 2008). Various research has focused on the origin, structure, and potential applications of GA (Akiyama et al., 1984; Dror et al., 2006; Islam et al., 1997). GA is commonly applied as a food additive in food products, textiles, ceramics, cosmetics, and pharmaceuticals. Regarding food systems, this gum is employed as a stabilizing and gelling agent, drug carrier, etc. (Verbeken et al., 2003). Several studies also reported that GA has biological activities. For instance, GA's ingestion decreased plasma cholesterol concentrations in rats (Kelley and Tsai, 1978). It has also been reported that this polysaccharide could be used in dentistry due to its antimicrobial activity (Ali et al., 2009). This chapter evaluates the proposed GA-based nanocarriers and their fabrication methods. Additionally, their potential application in food and pharmaceutical systems are examined. Overall, despite the safety of GA and excellent physicochemical properties and biological activity, GA-based nanocarriers are promising vehicles in functional food development.

Micro- and Nanoengineered Gum-Based Biomaterials for Drug Delivery and Biomedical Applications
https://doi.org/10.1016/B978-0-323-90986-0.00017-0

13.2 Safety of gum arabic

According to the broad data available from research in animals and humans, the Food and Drug Administration (FDA) approved the Generally Recognized As Safe (GRAS) status for gum arabic (Anderson, 1986). Different studies also investigated the safety of GA. For instance, Anderson et al. (1984) investigated rat hearts' and livers' microstructure after diet supplementation with 0%, 1%, 4%, and 8% (w/w) GA for 28 days. They observed that there were no detectable abnormalities in the tested organelles. Additionally, no inclusions or any other pathological alteration were observed. In another study, Strobel et al. (1982) showed that GA's administration could induce specific systemic immunologic hyporesponsiveness to the antigens. Strobel et al. (1986) also indicated that GA is tolerogenic when encountered by the natural route via the gut.

13.3 Chemical composition and structure

The chemical composition and structure of polysaccharides affect their functional characteristics, including rheological behavior, gelling and emulsifying capacities, water solubility, and potential application as carriers (Cui, 2005; Fathi et al., 2018b). Different techniques such as gel permeation chromatography, gas chromatography, high-pressure anion exchange chromatography, 1D (^{13}C-^{1}H NMR), and 2D NMR (HMBC, HMQC, COSY, NOESY) spectroscopy are commonly used to elucidate the composition and chemical structure of polysaccharides. As a complex heterogeneous anionic polysaccharide, gum arabic has excellent emulsifying capacity and low dispersion viscosity. These unique properties make it ideal for application in different food formulations. Gum arabic is a complex polysaccharide containing arabinogalactan proteins that cannot be removed by purification (Al Assaf et al., 2005; Glicksman, 1983; Imeson, 1992; Whistler, 1993). GA also contains a lipid portion composed of predominantly tetracosanoic acid and phytosphingosine (Svetek et al., 1999; Yadav et al., 2007). It has been reported that the lipid portion of the GA structure also has an important role in the emulsifying capacity of this complex polysaccharide. Compositional properties and some GA properties obtained from *A. senegal* or *A. seyal* are presented in Table 13.1 (Osman et al., 1993b). Different parameters such as source, time of exudation, the plant's age, growing conditions, and polysaccharide contamination alter the composition of hydrocolloids (Fathi et al., 2016; Karamalla et al., 1998). GA is composed of D-galactose, L-arabinose, L-rhamnose, D-glucuronic acid, and 4-*O*-methyl glucuronic acid. Based on gel permeation chromatography analysis, it has been proved that GA consists of three fractions: (i) the major fraction with a molecular weight of <250 kDa (<89% of the total) is composed of galactopyranose backbone, highly branched with units of rhamnose, arabinose and glucuronic acid; (ii) a fraction consists of arabinogalactan-protein complexes (<10% of the total) with a molecular weight of 1400 kDa; and (iii) the third fraction is also composed of arabinogalactan-protein complexes, but it has greater protein content and different amino acid profile when compared to the second fraction (Liu et al., 2009; Osman et al., 1993a; Renard et al., 2006). The excellent emulsifying capacity of GA has been attributed to the high molecular weight fraction presence (Randall et al., 1988, 1989; Ray et al., 1995). The interaction between this fraction and the charged polysaccharide provides a barrier against flocculation and coalescence.

TABLE 13.1 Compositional properties and some GA properties obtained from Acacia Senegal or Acacia Seya (Osman et al., 1993b).

Parameter	Monosaccharide (%)	Acacia Senegal	Acacia Seyal
Monosaccharide composition	Rhamnose	14	3
	Arabinose	29	41
	Galactose	36	32
	Glucuronic acid	14.5	6.5
Nitrogen		0.365	0.147
Protein		2.41	0.97
Weight average molecular weight (kDa)		380	850

13.4 Gum arabic: An excellent polysaccharide for encapsulation of bioactive agents

Micro-/nano-encapsulation of bioactive agents is commonly used to improve their chemical stability, provide a sustained release of entrapped compounds, and manufacture the powdered capsules with improved handling properties (Anandaraman and Reineccius, 1986; Montenegro et al., 2012). The carriers should protect the entrapped ingredients against environmental and physiological conditions during fabrication, storage, and handling to release them into food and pharmaceutical products. As mentioned above, GA consists of three portions: an arabinogalactan molecule responsible for the film-forming characteristics, while the glycoprotein-containing fraction imparts emulsification abilities to the gum. GA also has excellent water solubility (more than 50%), and it forms low-viscosity dispersions at high solids concentrations (Kaushik and Roos, 2007; Thevenet, 1995). The high emulsifying capacity and low viscosity of GA make it ideal for two different encapsulation techniques: spray-drying and complex coacervation. GA is preferred for encapsulation of bioactive agents by drying method due to the following reasons: (i) GA produces emulsions with high stability in relatively high concentrations; (ii) it has low viscosity even at a high concentration that facilitates its pumping during the spray-drying process; and (iii) it has good film-forming properties and can form films after drying. GA is also a suitable candidate for application as a carrier for retaining and protecting bioactive compounds in the coacervation method. The reason for that is the negative surface charge of this polysaccharide above pH 2.2. Positively charged polysaccharides or proteins can interact with GA with negative surface charge (Burgess and Carless, 1984; Fathi et al., 2018a, b). To date, different works have been carried out to replace GA with other hydrocolloids, but so far, no gum has been found that can mimic all of the unique characteristics of GA.

13.5 Different nanocarriers prepared with GA

Different encapsulation techniques produce encapsulated products with various particles sizes and morphologies.

13.5.1 Coacervation

Based on the definition of the International Union of Pure and Applied Chemistry (McNaught and Wilkinson, 1997), coacervation is the separation of a colloidal system into two liquid phases. Two coacervation techniques have some advantages and disadvantages. The driving force for encapsulation of bioactive compounds by coacervation may be either electrostatic interaction between a negatively charged polysaccharide like GA and positively charged bioactive compounds (simple coacervation) or the attraction of negatively charged biopolymers with positively charged polymers (complex coacervation) (Fathi et al., 2014; Gaonkar et al., 2014). The use of complex coacervation methods leads to producing two immiscible liquid phases: (i) the polymer-poor continuous phase and (ii) the polymer-rich dense phase. The latter, also named the coacervate phase, can be used to entrap a wide range of bioactive compounds (Fathi et al., 2018a; Schmitt and Turgeon, 2011). Coacervation is a time-consuming and expensive technique compared to the drying method, which is the current dominant encapsulation method in food systems (De Kruif et al., 2004; Turgeon et al., 2003). However, the coacervation method has different advantages than spray-drying for preserving high-value compounds, such as mild preparation conditions, high encapsulation efficiency (up to 99%), and good controlled-release characteristics (Beindorff and Zuidam, 2010; Burgess, 1994). Various studies have been conducted on applying coacervation techniques to incorporate reactive, sensitive, or volatile additives or nutrients into GA-based nanoparticles, and these will be discussed in the next sections.

13.5.2 Nanoemulsions

Nanoemulsions are very fine oil-in-water dispersions, having droplet diameters smaller than 100 nm. These stable liquid-in-liquid dispersions require high energy input to obtain nanoemulsion (Gupta et al., 2016). Nanoemulsions have attracted the attention of scientists due to their benefits such as optical clarity, ease of preparation, thermodynamic stability, high surface area, and improvement of bioavailability of bioactive compounds (Mason et al., 2006; Silva et al., 2012). Various works have been conducted to use nanoemulsions to encapsulate bioactive compounds and drugs, which will be reviewed in the following sections.

13.5.3 Spray-drying

Spray-drying is the most common method for encapsulation of bioactive compounds. It has numerous advantages, including being fast, cheap, and reproducible (Yeo et al., 2001). In this method, bioactive compounds are first dispersed in a biopolymer solution, and then the resulting mixture is atomized in a heated air chamber to evaporate the solvent rapidly, and forms a dried bioactive agent-loaded particle (Desobry et al., 1997). This method also has some disadvantages, such as high temperature, limiting its application in the food industry.

13.5.4 Other methods

Several encapsulation processes exist in addition to the methods described for the encapsulation of bioactive compounds and drugs. Some of them are including extrusion, molecular complexation, and absorption/desorption systems (Vemmer and Patel, 2013).

13.6 Application of GA nanocarriers for various food bioactive agents

Bioactive agents are usually highly susceptible to environmental and gastrointestinal conditions. Therefore, these compounds' encapsulation is commonly used to increase their stability and present a target delivery system. As mentioned above, natural polymers have desirable structural, physicochemical, and biological activities.

13.6.1 Vitamins

Vitamin D3 and vitamin E are two types of fat-soluble vitamins responsible for improving intestinal absorption of calcium and phosphorus and preventing many disorders, respectively. However, their high sensitivity against freezing, heating, and oxidation has limited their applications in food and pharmaceutical systems. The incorporation of these sensitive compounds into biopolymers-based NPs is a promising approach to overcome this challenge. In a study conducted by Athanasopulos and Carolina (2020), GA and sodium alginate were dried using the pressurized gas expanded liquid technology. This drying technique is commonly used to fabricate micro- or nano-sized particles from high molecular weight polysaccharides. The adsorptive precipitation method was also employed to load the vitamins into particles. The crystalline structure of the samples revealed that vitamin D3 retained some crystalline form on the loaded GA. The release profile of produced nanoparticles showed a sustained release in the simulated intestinal medium. In another work, Moradi and Anarjan (2019) used GA as a natural stabilizing and emulsifying polysaccharide for encapsulation of α-tocopherol. Zeta potential and the mean size of nanoparticles were in the range of zeta −13.5 to −47.8 mv and 10.01–171.2 nm, respectively. The cellular uptake of α-tocopherol was improved significantly by encapsulation. In conclusion, α-tocopherol-loaded GA can be utilized to enhance the nutrition value of water-based food and beverage formulations. The potential of nano-liposomal encapsulated vitamin D for fortifying yogurt powders was investigated by Jafari et al. (2019). They reported that the samples prepared in milk protein concentration of 0%, GA concentration of 1%, the modified starch concentration of 3%, maltodextrin concentration of 15%, and at 180 °C showed acceptable physicochemical properties.

13.6.2 Essential oils

The use of natural preservatives in food formulations has attracted the attention of scientists. Among these natural compounds, essential oils have been extensively investigated. The application of essential oils as an appropriate substitute for synthetic preservatives in food systems is often restricted due to their high sensitivity against heat, light, pressure, and oxygen (Karami-Osboo et al., 2010). Nanoencapsulation is a promising strategy to improve bio-based preservatives' stability and functionality in the food matrix (Solghi et al., 2020). Li et al. (2018) proposed nano-encapsulation of thymol essential oil using Zein/gum arabic. They proved that thymol-loaded nanoparticles could effectively inhibit the growth of *Escherichia coli* (*E. coli*). The high antimicrobial efficacy of thymol-loaded nanoparticles has been attributed to the controlled release of the essential oil and the protective effect of the stable interfacial layer of the nanoparticles. In another study, Niu et al. (2016) formulated a nanoemulsion based on ovalbumin and GA to load thymol oil and then investigated the physical characteristics and antimicrobial capacity of thyme oil emulsion against *E. coli*. It was found that thymol-loaded nanoparticles had greater antimicrobial activity than the pure thyme oil. In another study, Hassani and Hasani (2018) incorporated thyme essential oil into chitosan/GA nanoparticles using the emulsion method. The encapsulation efficiency, particle diameter, and surface charge of the nanoparticles at optimal conditions were 77.67%, 385.2 nm, and +43.17 mV, respectively. Antioxidant and antibacterial activities of encapsulated thymol were greater than those observed for pure thymol.

Cymbopogon citratus essential oil has been nanoencapsulated via cross-linking of GA (Ribeiro et al., 2015). The authors used different concentrations of sodium trimetaphosphate as an effective cross-linking agent for the modification of the physical and chemical properties of GA-based nanoparticles. The particle diameter of the nanoparticles decreased when the cross-linking degree increased. Furthermore, the authors found that the particles' encapsulation efficiency obtained by cross-linked GA was more than that of the unmodified GA. Lv et al. (2014) applied a complex coacervation technique to fabricate jasmine essential oil-loaded gelatin/GA nanoparticles. For this purpose, they used transglutaminase as a cross-linker to entrap jasmine essential oil into the developed nanoparticles. The thermal stability of jasmine essential oil-loaded gelatin/GA nanoparticles was tested, and it was found that the thermal stability of jasmine essential oil heated at 80 °C for 7 h was increased considerably by encapsulation.

13.6.3 Phenolic compounds

Plants are rich in phenolic compounds that show extensive biological activity in health promotion and disease prevention. For instance, these bioactive compounds' antioxidant capacity and antimicrobial activity have been documented in the literature (Balasundram et al., 2006; Velderrain-Rodríguez et al., 2014). Phenolic compounds are not stable over food processing and storage. Furthermore, the protection of phenolic compounds in the gastrointestinal tract could be enhanced by encapsulation. Therefore, various studies have focused on the entrapment of phenolic compounds into polymeric matrices. Epigallocatechin gallate, as a catechin type, has various biological activities, including antimicrobial, anticarcinogenic, and antioxidative activities. However, this bioactive compound has a high sensitivity to light and high temperature (Fujiki et al., 1992). Epigallocatechin-3-gallate has been nano encapsulated via gum arabic and maltodextrin by coacervation technique (Rocha et al., 2011). The encapsulation efficiency of the developed nanostructured vehicle was found to be approximately 85%. The biological activity of free and encapsulated epigallocatechin-3-gallate was evaluated by reducing the cell viability and inducing apoptosis of Du145 prostate cancer cells. The biological activity of epigallocatechin-3-gallate-loaded GA nanoparticles was greater than observed for free epigallocatechin-3-gallate. Rajabi et al. (2019) produced nanoparticles using chitosan/GA by the ionic gelation technique. The particle diameter, PDI, and surface charge of fabricated nanoparticles were in the ranges of 183–295 nm, 0.272–0.612, and 20.5–50.5 mV, respectively. TEM images revealed a spherical-shape with a homogenous particle size distribution for chitosan/GA nanoparticles. The authors showed that the in vitro release profiles of saffron from chitosan/GA nanoparticles in acidic and neutral mediums were considerably different because of alterations in chitosan solubility. According to the results reported by the authors of this paper, the use of developed nanoparticles can be introduced as a promising strategy to produce a stable form of saffron.

In another study, Hassani et al. (2020a) improved the bioavailability of gallic acid, as a natural phenolic compound with therapeutic activity, by encapsulation. For this purpose, they used the ionotropic gelation technique to develop gallic acid-loaded GA nanoparticles. The antioxidant and biological activities of the gallic acid-loaded GA nanoparticles were examined by different assays, including DPPH, nitric oxide scavenging (NO), β-carotene bleaching, and angiotensin-converting enzyme (ACE) inhibitory, cell uptake, and cytotoxicity assays. The results of this work exhibited that gallic acid-loaded GA nanoparticles have very strong antioxidant and anticancer activities. Interestingly, a synergistic effect of increasing the selectivity toward cancer cells and improving the antioxidant capacity was observed with gallic acid-loaded GA nanoparticles than pure gallic acid. Curcumin is a phenolic compound that its antibacterial and antioxidant activities have been deeply investigated in the literature. Ak and Gülçin (2008) examined the antioxidant activity of curcumin using different techniques. They found out that curcumin's antioxidant capacity was comparable to synthetic antioxidant compounds such as BHA, BHT, and Trolox. Additionally, curcumin's antimicrobial activity against various bacteria, viruses, fungi, and parasites has been reported previously (Teow et al., 2016; Zorofchian Moghadamtousi et al., 2014). In a study by Sarika et al. (2015), curcumin's conjugation to GA was carried out to enhance curcumin's solubility and stability. Spherical nano-micelles with a size range of 270 nm were observed. The water solubility and stability of curcumin improved significantly when it was conjugated to GA. From a biological perspective, the conjugate showed more accumulation and toxicity in HepG2 cells. It has been reported that the cytotoxicity influence of curcumin toward carcinoma cells is associated with the induction of apoptosis mediated by the direct release of cytochrome C and the subsequent activation of caspases (Woo et al., 2003). More cytotoxicity effect of conjugate than pure curcumin may be because of the targeting efficiency of the galactose groups present GA (Rigopoulou et al., 2012). Moreover, the improved cytotoxicity capacity of curcumin after encapsulation may be due to its improved water solubility of curcumin confirmed by the authors. Tan et al. (2016) developed a new encapsulation system based on complex coacervation of chitosan and GA. They optimized the formulation of nanocapsules and then characterized their physicochemical properties. It was found that the particle diameter of curcumin-loaded nanoparticles prepared at optimum condition was in the range of 250–290 nm. Encapsulation efficiency and loading capacity of the developed carrier were 90% and 3.8%, respectively. The antioxidant capacity of curcumin-loaded GA/chitosan nanoparticles was greater than that of pure curcumin. The release rate and stability of curcumin in both pure and encapsulated forms in the simulated gastrointestinal medium were evaluated, which proved the efficiency of nanoencapsulation by GA/chitosan. According to these results, chitosan/GA nanoparticles can be introduced as useful vehicles to deliver hydrophobic bioactive ingredients like curcumin in the gastrointestinal tract.

Shahgholian and Rajabzadeh (2016) also used the emulsification-coacervation method to fabricate a nanostructured vehicle based on GA and bovine serum albumin. They used surface response methodology to determine the optimum condition for the fabrication of nanocapsules with the highest encapsulation efficiency. The fabricated nanoparticles showed a sponge-like structure with an encapsulation efficiency of 92%. Considering the high encapsulation efficiency and facility of emulsification-coacervation method, it can be introduced as a useful technique for curcumin encapsulation. Mohammadian et al. (2019) also used a complex coacervation method to prepare coacervates based on GA and whey protein nanofibrils. The curcumin-loaded nanoparticles demonstrated excellent surface activity, and hence, they can be used as valuable candidates to produce new functional food emulsions and beverages. As expected, the authors reported that antioxidant capacity and photo-stability of curcumin were enhanced by encapsulation into whey protein nanofibrils/GA nanoparticles. Curcumin was efficiently loaded in whey protein nanofibrils/GA nanoparticles and sustained release was observed in a simulated gastrointestinal medium. The potential application of GA/sodium alginate as a carrier for the encapsulation of curcumin was examined by Hassani et al. (2020b). They fabricated GA/sodium alginate nanoparticles via the ionotropic gelation technique and then investigated their physicochemical, biological, and in vitro release properties. The particles showed a particle size range of 10 ± 0.3 nm to 190 ± 0.1 nm with a surface charge of -15 ± 0.2 mV. The antioxidant activity of curcumin-loaded GA/sodium alginate nanoparticles was investigated in vitro through a free radical scavenging activity test. This study revealed that the developed encapsulation system could effectively inhibit DPPH radicals. Furthermore, an in vitro cytotoxicity test was done using MTT assay, and the results demonstrated significant anticancer activity of curcumin-loaded GA/sodium alginate nanoparticles against human liver cancer cells (HepG2), colon cancer (HT29), lung cancer (A549), and breast cancer (MCF7) cells. In another work, curcumin was entrapped in GA aldehyde-gelatin nanogels to enhance its bioaccessibility and therapeutic efficacy toward cancer cells. Encapsulation efficiency, surface charge, and particle diameter of the nanogels were 452 ± 8 nm, -27 mV, and $65\% \pm 3\%$, respectively. Cytotoxicity evaluation of both bare and curcumin-loaded nanogels revealed that curcumin-loaded nanogels induce toxicity in MCF-7 cells (Sarika and Nirmala, 2016).

13.6.4 Carotenoids

B-carotene, as a lipophilic bioactive compound, has various beneficial health effects that have been confirmed by a number of works (Álvarez-Henao et al., 2018; Pham-Hoang et al., 2018; Salehi et al., 2019). However, its poor water solubility and high sensitivity against light and pH have limited its application in the food and pharmaceutical industries. In a study by Sheng et al. (2018), a layer-by-layer electrostatic deposition technique was applied to develop a new type of β-carotene bilayer emulsion delivery system. The authors used bovine serum albumin as the inner emulsifier and GA as the outer emulsifier. They first characterized the bilayer emulsion and then investigated the influences of processing conditions, including pH, heating treatment, UV radiation, a strong oxidant, and storage time on the developed nano-encapsulation system's chemical stability. A small particle size (221.27 ± 5.17 nm) with high surface charge (-30.37 ± 0.71 mV), physical stability, and encapsulation efficiency were reported by the authors. Furthermore, the bilayer emulsion showed higher chemical stability under tested environmental stresses than monolayer emulsion. This observation may be due to the existence of a dense and thick layer structure in the developed bilayer emulsion. An oil-in-water nanoemulsion stabilized with WPC and GA was also fabricated by high-pressure homogenization. The particle diameter of the nanoemulsions decreased when the interfacial tension increased. Based on enthalpy–entropy compensation, the authors found that GA was enthalpy driven (Flores-Andrade et al., 2021).

13.6.5 Other bioactive agents

Yüksel and Şahin-Yeşilçubuk (2018) entrapped long-chain fatty acids into gelatin/GA nanoparticles by the complex coacervation method. The impact of different parameters, including biopolymers concentration, bioactive compound: carrier ratio, and speed of homogenization rate on encapsulation process, were examined by the authors. It was found that when the biopolymer concentration, core: wall ratio, and homogenization speed were 2%, 1:1, and 15,000 rpm, respectively, the maximum encapsulation efficiency (84.11 ± 0.77%) was observed. The particle diameter of the samples was in the range 19–263 nm. Furthermore, the nanoparticles showed a low polydispersity index (PDI), which indicates they had a homogeneous size distribution pattern. Sharifi et al. (2021) also employed complex coacervation techniques to encapsulate probiotic *Lactobacillus plantarum* and phytosterols. They used a whey protein isolate and GA alone or in combination with phytosterols by the complex coacervation method. The coacervates were dried using spray-drying and freeze-drying methods to obtain solid nanoparticles. Comparatively, the survivability of *L. plantarum* ATCC 8014 was enhanced when phytosterols were added to the formulation of coacervates. Moreover, the tested bacteria's cell viability in lyophilized nanoparticles was more than those dried using the spray-drying technique. The particle diameter of the whey protein isolate/GA particles whey protein isolate/GA/phytosterols were 196.2 ± 13.72 and 366.5 ± 0.56 nm, respectively. According to SEM images, the nanoparticles dried by freeze-dryer and spray-dryer exhibited a porous morphology and quasispherical configuration, respectively.

The potential application of complex coacervates obtained from albumin and GA was also evaluated by Hedayati et al. (2012). The influences of different factors, including pH, albumin to GA ratio, ionic strength, biopolymer concentration, temperature, agitation rate, and glutaraldehyde concentration on the encapsulation efficiency and particle diameter of the samples, were examined. Furthermore, the effect of the stabilizing process on particle diameter and PDI was tested. It was found that pH, temperature, and ionic had a significant impact on the fabricated systems, but the protein/polysaccharide concentration showed no significant effect on the nanoparticles' properties. At optimum conditions (temperature of 4 °C, pH 4.9, and $Is = 0$), the smallest nanoparticles (108 nm) were produced. In another study, GA and maltodextrin as wall materials were used to encapsulate propolis extract (Šturm et al., 2019). Either freeze-drying or spray-drying techniques were employed to fabricate nanoparticles. It was observed that the propolis extract-loaded nanoparticles contained high phenol contents, as well as having high water dispersibility and antioxidant capacity. Interestingly, the nanoparticle powders showed high water stability for up to 24 h, and a sustained release of entrapped phenols was observed that did not alter across various environmental conditions.

13.7 Conclusions and further remarks

GA is an arabinogalactan polysaccharide with extensive industrial applications as a stabilizer, a thickener, an emulsifier, and in the encapsulation of bioactive agents. Regarding the targeting efficiency of the galactose groups present in GA, the GA-based nanocarriers can be introduced as promising carriers to improve the anticancer activity of bioactive compounds.

References

Ak, T., Gülçin, I., 2008. Antioxidant and radical scavenging properties of curcumin. Chem. Biol. Interact. 174 (1), 27–37. https://doi.org/10.1016/j.cbi.2008.05.003.

Akiyama, Y., Eda, S., Kato, K., 1984. Gum arabic is a kind of arabinogalactan-protein. Agric. Biol. Chem. 48 (1), 235–237. https://doi.org/10.1080/00021369.1984.10866126.

Al Assaf, S., Phillips, G.O., Williams, P.A., 2005. Studies on acacia exudate gums. Part I: the molecular weight of Acacia Senegal gum exudate. Food Hydrocoll. 19 (4), 647–660. https://doi.org/10.1016/j.foodhyd.2004.09.002.

Ali, B.H., Ziada, A., Blunden, G., 2009. Biological effects of gum arabic: a review of some recent research. Food Chem. Toxicol. 47 (1), 1–8. https://doi.org/10.1016/j.fct.2008.07.001.

Álvarez-Henao, M.V., Saavedra, N., Medina, S., Jiménez Cartagena, C., Alzate, L.M., Londoño-Londoño, J., 2018. Microencapsulation of lutein by spray-drying: characterization and stability analyses to promote its use as a functional ingredient. Food Chem. 256, 181–187. https://doi.org/10.1016/j.foodchem.2018.02.059.

Anandaraman, S., Reineccius, G., 1986. Stability of encapsulated orange peel oil. Food Technol. https://doi.org/10.1019/j.fct.2008.07.001.

Anderson, D.M.W., 1986. Evidence for the safety of gum Arabic (Acacia Senegal (L.) Willd.) as a food additive—a brief review. Food Addit. Contam. 3 (3), 225–230. https://doi.org/10.1080/02652038609373584.

Anderson, D.M.W., Ashby, P., Busuttil, A., Kempson, S.A., Lawson, M.E., 1984. Transmission electron microscopy of heart and liver tissues from rats fed with gums arabic and tragacanth. Toxicol. Lett. 21 (1), 83–89. https://doi.org/10.1016/0378-4274(84)90227-3.

Athanasopulos, V., Carolina, A., 2020. Adsorptive Precipitation of Vitamin D3 and Vitamin E on Gum Arabic and Sodium Alginate Using Supercritical Carbon Dioxide., https://doi.org/10.7939/r3-w8tt-q150.

Balasundram, N., Sundram, K., Samman, S., 2006. Phenolic compounds in plants and agri-industrial by-products: antioxidant activity, occurrence, and potential uses. Food Chem. 99 (1), 191–203. https://doi.org/10.1016/j.foodchem.2005.07.042.

Beindorff, C.M., Zuidam, N.J., 2010. Microencapsulation of fish oil. In: Encapsulation Technologies for Active Food Ingredients and Food Processing. Springer, New York, pp. 161–185, https://doi.org/10.1007/978-1-4419-1008-0_6.

Burgess, D.J., 1994. Complex Coacervation: Microcapsule Formation. Springer Science and Business Media LLC, pp. 285–300, https://doi.org/10.1007/978-3-642-78469-9_17.

Burgess, D.J., Carless, J.E., 1984. Microelectrophoretic studies of gelatin and acacia for the prediction of complex coacervation. J. Colloid Interface Sci. 98 (1), 1–8. https://doi.org/10.1016/0021-9797(84)90472-7.

Cui, S.W., 2005. Structural Analysis of Polysaccharides. Food Carbohydrates: Chemistry, Physical Properties, and Applications. CRC Press.

De Kruif, C.G., Weinbreck, F., De Vries, R., 2004. Complex coacervation of proteins and anionic polysaccharides. Curr. Opin. Colloid Interface Sci. 9 (5), 340–349. https://doi.org/10.1016/j.cocis.2004.09.006.

Desobry, S.A., Netto, F.M., Labuza, T.P., 1997. Comparison of spray-drying, drum-drying and freeze-drying for β-carotene encapsulation and preservation. J. Food Sci. 62 (6), 1158–1162. https://doi.org/10.1111/j.1365-2621.1997.tb12235.x.

Dror, Y., Cohen, Y., Yerushalmi-Rozen, R., 2006. Structure of gum arabic in aqueous solution. J. Polym. Sci. B 44 (22), 3265–3271. https://doi.org/10.1002/polb.20970.

Fathi, M., Martín, Á., McClements, D.J., 2014. Nanoencapsulation of food ingredients using carbohydrate based delivery systems. Trends Food Sci. Technol. 39 (1), 18–39. https://doi.org/10.1016/j.tifs.2014.06.007.

Fathi, M., Mohebbi, M., Koocheki, A., 2016. Introducing Prunus cerasus gum exudates: chemical structure, molecular weight, and rheological properties. Food Hydrocoll. 61, 946–955. https://doi.org/10.1016/j.foodhyd.2016.07.004.

Fathi, M., Donsi, F., McClements, D.J., 2018a. Protein-based delivery systems for the nanoencapsulation of food ingredients. Compr. Rev. Food Sci. Food Saf. 17 (4), 920–936. https://doi.org/10.1111/1541-4337.12360.

Fathi, M., Emam-Djomeh, Z., Sadeghi-Varkani, A., 2018b. Extraction, characterization and rheological study of the purified polysaccharide from Lallemantia ibrica seeds. Int. J. Biol. Macromol. 120, 1265–1274. https://doi.org/10.1016/j.ijbiomac.2018.08.159.

Flores-Andrade, E., Allende-Baltazar, Z., Sandoval-González, P.E., Jiménez-Fernández, M., Beristain, C.I., Pascual-Pineda, L.A., 2021. Carotenoid nanoemulsions stabilized by natural emulsifiers: whey protein, gum Arabic, and soy lecithin. J. Food Eng. *290*. https://doi.org/10.1016/j.jfoodeng.2020.110208.

Fujiki, H., Yoshizawa, S., Horiuchi, T., Suganuma, M., Yatsunami, J., Nishiwaki, S., Okabe, S., Nishiwaki-Matsushima, R., Okuda, T., Sugimura, T., 1992. Anticarcinogenic effects of (-)-epigallocatechin gallate. Prev. Med. 21 (4), 503–509. https://doi.org/10.1016/0091-7435(92)90057-O.

Gaonkar, A.G., Vasisht, N., Khare, A., Sobel, R., 2014. Microencapsulation in the Food Industry. Elsevier Science Publishing Company, p. 590.

Glicksman, M., 1983. Food hydrocolloids. In: Natural Plant Exudates–Seaweed Extracts., https://doi.org/10.1039/C5SM02957A.

Gupta, A., Eral, H.B., Hatton, T.A., Doyle, P.S., 2016. Nanoemulsions: formation, properties and applications. Soft Matter, 2826–2841. https://doi.org/10.1039/C5SM02958A.

Hassani, A., Azarian, M.M.S., Ibrahim, W.N., Hussain, S.A., 2020a. Preparation, characterization and therapeutic properties of gum arabic-stabilized gallic acid nanoparticles. Sci. Rep. *10* (1). https://doi.org/10.1038/s41598-020-71175-8.

Hassani, M., Hasani, S., 2018. Nano-encapsulation of thyme essential oil in chitosan-Arabic gum system: evaluation of its antioxidant and antimicrobial properties. Trends Phytochem. Res. 2 (2), 75–82.

Hassani, A., Mahmood, S., Enezei, H.H., Hussain, S.A., Hamad, H.A., Aldoghachi, A.F., Hagar, A., Doolaanea, A.A., Ibrahim, W.N., 2020b. Formulation, characterization and biological activity screening of sodium alginate-gum Arabic nanoparticles loaded with curcumin. Molecules 25 (9). https://doi.org/10.3390/molecules25092244.

Hedayati, R., Jahanshahi, M., Attar, H., 2012. Fabrication and characterization of albumin-acacia nanoparticles based on complex coacervation as potent nanocarrier. J. Chem. Technol. Biotechnol. 87 (10), 1401–1408. https://doi.org/10.1002/jctb.3758.

Imeson, A., 1992. Exudate gums. In: Thickening and Gelling Agents for Food. Springer Nature, pp. 66–97, https://doi.org/10.1007/978-1-4615-3552-2_4.

Islam, A.M., Phillips, G.O., Sljivo, A., Snowden, M.J., Williams, P.A., 1997. A review of recent developments on the regulatory, structural and functional aspects of gum Arabic. Food Hydrocoll. 11 (4), 493–505. https://doi.org/10.1016/S0268-005X(97)80048-3.

Jafari, S.M., Vakili, S., Dehnad, D., 2019. Production of a functional yogurt powder fortified with nanoliposomal vitamin D through spray drying. Food Bioproc. Tech. 12 (7), 1220–1231. https://doi.org/10.1007/s11947-019-02289-9.

Karamalla, K., Siddig, N., Osman, M., 1998. Senegal gum samples were collected between 1993 and 1995 from Sudan. Food Hydrocoll. 12 (4), 5–8. https://doi.org/10.1016/S0268-005X(98.

Karami-Osboo, R., Khodaverdi, M., Ali-Akbari, F., 2010. Antibacterial effect of effective compounds of Satureja hortensis and Thymus vulgaris essential oils against Erwinia amylovora. J. Agric. Sci. Technol. 12 (1), 35–45. http://jast.modares.ac.ir/browse.php?a_code=A-10-380-1&slc_lang=en&sid=1&ftxt=1.

Kaushik, V., Roos, Y.H., 2007. Limonene encapsulation in freeze-drying of gum Arabic-sucrose-gelatin systems. LWT Food Sci. Technol. 40 (8), 1381–1391. https://doi.org/10.1016/j.lwt.2006.10.008.

Kelley, J.J., Tsai, A.C., 1978. Effect of pectin, gum arabic and agar on cholesterol absorption, synthesis, and turnover in rats. J. Nutr. 108 (4), 630–639. https://doi.org/10.1093/jn/108.4.630.

Li, J., Xu, X., Chen, Z., Wang, T., Lu, Z., Hu, W., Wang, L., 2018. Zein/gum Arabic nanoparticle-stabilized Pickering emulsion with thymol as an antibacterial delivery system. Carbohydr. Polym. 200, 416–426. https://doi.org/10.1016/j.carbpol.2018.08.025.

Liu, S., Low, N.H., Nickerson, M.T., 2009. Effect of pH, salt, and biopolymer ratio on the formation of pea protein isolate gum arabic complexes. J. Agric. Food Chem. 57 (4), 1521–1526. https://doi.org/10.1021/jf802643n.

Lv, Y., Yang, F., Li, X., Zhang, X., Abbas, S., 2014. Formation of heat-resistant nanocapsules of jasmine essential oil via gelatin/gum arabic based complex coacervation. Food Hydrocoll. 35, 305–314. https://doi.org/10.1016/j.foodhyd.2013.06.003.

Mason, T.G., Wilking, J.N., Meleson, K., Chang, C.B., Graves, S.M., 2006. Nanoemulsions: formation, structure, and physical properties. J. Phys. Condens. Matter, R635–R666. https://doi.org/10.1088/0953-8984/18/41/r01.

McNaught, A.D., Wilkinson, A., 1997. Compendium of Chemical Terminology. Blackwell Science Oxford.

Mohammadian, M., Salami, M., Alavi, F., Momen, S., Emam-Djomeh, Z., Moosavi-Movahedi, A.A., 2019. Fabrication and characterization of curcumin-loaded complex coacervates made of gum arabic and whey protein nanofibrils. Food Biophys. 14 (4), 425–436. https://doi.org/10.1007/s11483-019-09591-1.

Montenegro, M.A., Boiero, M.L., Valle, L., Borsarelli, C.D., 2012. Gum Arabic: More Than an Edible Emulsifier. Products and Applications of Biopolymers. Vol. 51 InTech.

Moradi, S., Anarjan, N., 2019. Preparation and characterization of α-tocopherol nanocapsules based on gum Arabic-stabilized nanoemulsions. Food Sci. Biotechnol. 28 (2), 413–421. https://doi.org/10.1007/s10068-018-0478-y.

Niu, F., Pan, W., Su, Y., Yang, Y., 2016. Physical and antimicrobial properties of thyme oil emulsions stabilized by ovalbumin and gum Arabic. Food Chem. 212, 138–145. https://doi.org/10.1016/j.foodchem.2016.05.172.

Osman, M.E., Menzies, A.R., Williams, P.A., Phillips, G.O., Baldwin, T.C., 1993a. The molecular characterisation of the polysaccharide gum from Acacia senegal. Carbohydr. Res. 246 (1), 303–318. https://doi.org/10.1016/0008-6215(93)84042-5.

Osman, M.E., Williams, P.A., Menzies, A.R., Phillips, G.O., 1993b. Characterization of commercial samples of gum Arabic. J. Agric. Food Chem. 41 (1), 71–77. https://doi.org/10.1021/jf00025a016.

Pham-Hoang, B.N., Romero-Guido, C., Phan-Thi, H., Waché, Y., 2018. Strategies to improve carotene entry into cells of Yarrowia lipolytica in a goal of encapsulation. J. Food Eng. 224, 88–94. https://doi.org/10.1016/j.jfoodeng.2017.12.029.

Phillips, G.O., Ogasawara, T., Ushida, K., 2008. The regulatory and scientific approach to defining gum arabic (Acacia senegal and Acacia seyal) as a dietary fibre. Food Hydrocoll. 22 (1), 24–35. https://doi.org/10.1016/j.foodhyd.2006.12.016.

Rajabi, H., Jafari, S.M., Rajabzadeh, G., Sarfarazi, M., Sedaghati, S., 2019. Chitosan-gum Arabic complex nanocarriers for encapsulation of saffron bioactive components. Colloids Surf. A Physicochem. Eng. Asp. 578, 123644. https://doi.org/10.1016/j.colsurfa.2019.123644.

Randall, R.C., Phillips, G.O., Williams, P.A., 1988. The role of the proteinaceous component on the emulsifying properties of gum arabic. Food hydrocoll. 2 (2), 131–140.

Randall, R.C., Phillips, G.O., Williams, P.A., 1989. Fractionation and characterization of gum from *Acacia senegal*. Food hydrocoll. 3 (1), 65–75.

Ray, A.K., et al., 1995. Functionality of gum arabic. Fractionation, characterization and evaluation of gum fractions in citrus oil emulsions and model beverages. Food Hydrocoll. 9 (2), 123–131.

Renard, D., Lavenant-Gourgeon, L., Ralet, M.C., Sanchez, C., 2006. Acacia senegal gum: continuum of molecular species differing by their protein to sugar ratio, molecular weight, and charges. Biomacromolecules 7 (9), 2637–2649. https://doi.org/10.1021/bm060145j.

Ribeiro, F.W.M., Laurentino, L., Alves, C.R., Bastos, M., da Costa, J.M.C., Canuto, K.M., Furtado, R.F., 2015. Chemical modification of gum arabic and its application in the encapsulation of Cymbopogon citratus essential oil. J. Appl. Polym. Sci. *132* (8). https://doi.org/10.1002/app.41519.

Rigopoulou, E.I., Roggenbuck, D., Smyk, D.S., Liaskos, C., Mytilinaiou, M.G., Feist, E., Conrad, K., Bogdanos, D.P., 2012. Asialoglycoprotein receptor (ASGPR) as target autoantigen in liver autoimmunity: lost and found. Autoimmun. Rev. 12 (2), 260–269. https://doi.org/10.1016/j.autrev.2012.04.005.

Rocha, S., Generalov, R., Pereira, M.D.C., Peres, I., Juzenas, P., Coelho, M.A.N., 2011. Epigallocatechin gallate-loaded polysaccharide nanoparticles for prostate cancer chemoprevention. Nanomedicine 6 (1), 79–87. https://doi.org/10.2217/nnm.10.101.

Salehi, B., Sharifi-Rad, R., Sharopov, F., Namiesnik, J., Roointan, A., Kamle, M., Kumar, P., Martins, N., Sharifi-Rad, J., 2019. Beneficial effects and potential risks of tomato consumption for human health: an overview. Nutrition 62, 201–208. https://doi.org/10.1016/j.nut.2019.01.012.

Sarika, P.R., Nirmala, R.J., 2016. Curcumin loaded gum Arabic aldehyde-gelatin nanogels for breast cancer therapy. Mater. Sci. Eng. C 65, 331–337. https://doi.org/10.1016/j.msec.2016.04.044.

Sarika, P.R., James, N.R., Kumar, P.R.A., Raj, D.K., Kumary, T.V., 2015. Gum arabic-curcumin conjugate micelles with enhanced loading for curcumin delivery to hepatocarcinoma cells. Carbohydr. Polym. 134, 167–174. https://doi.org/10.1016/j.carbpol.2015.07.068.

Schmitt, C., Turgeon, S.L., 2011. Protein/polysaccharide complexes and coacervates in food systems. Adv. Colloid Interface Sci. 167 (1–2), 63–70. https://doi.org/10.1016/j.cis.2010.10.001.

Shahgholian, N., Rajabzadeh, G., 2016. Fabrication and characterization of curcumin-loaded albumin/gum arabic coacervate. Food Hydrocoll. 59, 17–25. https://doi.org/10.1016/j.foodhyd.2015.11.031.

Sharifi, S., Rezazad-Bari, M., Alizadeh, M., Almasi, H., Amiri, S., 2021. Use of whey protein isolate and gum Arabic for the co-encapsulation of probiotic Lactobacillus plantarum and phytosterols by complex coacervation: enhanced viability of probiotic in Iranian white cheese. Food Hydrocoll. 113, 106496. https://doi.org/10.1016/j.foodhyd.2020.106496.

Sheng, B., Li, L., Zhang, X., Jiao, W., Zhao, D., Wang, X., Wan, L., Li, B., Rong, H., 2018. Physicochemical properties and chemical stability of β-carotene bilayer emulsion coated with bovine serum albumin and arabic gum compared to monolayer emulsions. Molecules 23 (2). https://doi.org/10.3390/molecules23020495.

Silva, H.D., Cerqueira, M.A., Vicente, A.A., 2012. Nanoemulsions for food applications: development and characterization. Food Bioproc. Tech. 5 (3), 854–867. https://doi.org/10.1007/s11947-011-0683-7.

Solghi, S., Emam-Djomeh, Z., Fathi, M., Farahani, F., 2020. The encapsulation of curcumin by whey protein: assessment of the stability and bioactivity. J. Food Process Eng. 43 (6). https://doi.org/10.1111/jfpe.13403.

Strobel, S., Ferguson, A., Anderson, D.M.W., 1982. Immunogenicity of foods and food additives—in vivo testing of gums arabic, karaya and tragacanth. Toxicol. Lett. 14 (3–4), 247–252. https://doi.org/10.1016/0378-4274(82)90059-5.

Strobel, S., Ferguson, A., Anderson, D.M.W., 1986. Immunogenicity, immunological cross reactivity and non-specific irritant properties of the exudate gums, arabic, karaya and tragacanth. Food Addit. Contam. 3 (1), 47–56. https://doi.org/10.1080/02652038609373564.

Šturm, L., Osojnik Črnivec, I.G., Istenič, K., Ota, A., Megušar, P., Slukan, A., Humar, M., Levic, S., Nedović, V., Kopinč, R., Deželak, M., Pereyra Gonzales, A., Poklar Ulrih, N., 2019. Encapsulation of non-dewaxed propolis by freeze-drying and spray-drying using gum Arabic, maltodextrin and inulin as coating materials. Food Bioprod. Process. 116, 196–211. https://doi.org/10.1016/j.fbp.2019.05.008.

Svetek, J., Yadav, M.P., Nothnagel, E.A., 1999. Presence of a glycosylphosphatidylinositol lipid anchor on rose arabinogalactan proteins. J. Biol. Chem. 274 (21), 14724–14733. https://doi.org/10.1074/jbc.274.21.14724.

Tan, C., Xie, J., Zhang, X., Cai, J., Xia, S., 2016. Polysaccharide-based nanoparticles by chitosan and gum arabic polyelectrolyte complexation as carriers for curcumin. Food Hydrocoll. 57, 236–245. https://doi.org/10.1016/j.foodhyd.2016.01.021.

Teow, S.Y., Liew, K., Ali, S.A., Khoo, A.S.B., Peh, S.C., 2016. Antibacterial action of curcumin against staphylococcus aureus: a brief review. J. Trop. Med. 2016. https://doi.org/10.1155/2016/2853045.

Thevenet, F., 1995. Acacia Gums. American Chemical Society (ACS), pp. 51–59, https://doi.org/10.1021/bk-1995-0590.ch005.

Turgeon, S.L., Beaulieu, M., Schmitt, C., Sanchez, C., 2003. Protein-polysaccharide interactions: phase-ordering kinetics, thermodynamic and structural aspects. Curr. Opin. Colloid Interface Sci. 8 (4–5), 401–414. https://doi.org/10.1016/S1359-0294(03)00093-1.

Velderrain-Rodríguez, G.R., Palafox-Carlos, H., Wall-Medrano, A., Ayala-Zavala, J.F., Chen, C.Y.O., Robles-Sánchez, M., Astiazaran-García, H., Alvarez-Parrilla, E., González-Aguilar, G.A., 2014. Phenolic compounds: their journey after intake. Food Funct. 5 (2), 189–197. https://doi.org/10.1039/c3fo60361j.

Vemmer, M., Patel, A.V., 2013. Review of encapsulation methods suitable for microbial biological control agents. Biol. Control 67 (3), 380–389. https://doi.org/10.1016/j.biocontrol.2013.09.003.

Verbeken, D., Dierckx, S., Dewettinck, K., 2003. Exudate gums: occurrence, production, and applications. Appl. Microbiol. Biotechnol. 63 (1), 10–21. https://doi.org/10.1007/s00253-003-1354-z.

Whistler, R.L., 1993. Exudate gums. In: Industrial Gums: Polysaccharides and Their Derivatives. Elsevier BV, pp. 309–339, https://doi.org/10.1016/b978-0-08-092654-4.50016-4.

Woo, J.H., Kim, Y.H., Choi, Y.J., Kim, D.G., Lee, K.S., Bae, J.H., Min, D.S., Chang, J.S., Jeong, Y.J., Lee, Y.H., Park, J.W., Kwon, T.K., 2003. Molecular mechanisms of curcumin-induced cytotoxicity: induction of apoptosis through generation of reactive oxygen species, down-regulation of Bcl-XL and IAP, the release of cytochrome c and inhibition of Akt. Carcinogenesis *24* (7), 1199–1208. https://doi.org/10.1093/carcin/bgg082.

Yadav, M.P., Manuel Igartuburu, J., Yan, Y., Nothnagel, E.A., 2007. Chemical investigation of the structural basis of the emulsifying activity of gum arabic. Food Hydrocoll. 21 (2), 297–308. https://doi.org/10.1016/j.foodhyd.2006.05.001.

Yeo, Y., Baek, N., Park, K., 2001. Microencapsulation methods for delivery of protein drugs. Biotechnol. Bioprocess Eng. 6 (4), 213–230. https://doi.org/10.1007/BF02931982.

Yüksel, A., Şahin-Yeşilçubuk, N., 2018. Encapsulation of structured lipids containing medium- and long chain fatty acids by complex coacervation of gelatin and gum arabic. J. Food Process Eng. 41 (8). https://doi.org/10.1111/jfpe.12907, e12907.

Zorofchian Moghadamtousi, S., Abdul Kadir, H., Hassandarvish, P., Tajik, H., Abubakar, S., Zandi, K., 2014. A review on antibacterial, antiviral, and antifungal activity of curcumin. Biomed. Res. Int., 1–12. https://doi.org/10.1155/2014/186864.

CHAPTER

14

Tamarind gum as a wall material in the microencapsulation of drugs and natural products

Erik Alpizar-Reyes[a,d], *Stefani Cortés-Camargo*[b], *Angélica Román-Guerrero*[c], *and César Pérez-Alonso*[a]

[a]Chemical Engineering Department, Faculty of Chemistry, Autonomous Mexico State University, Toluca, Estado de México, Mexico [b]Nanotechnology Department, Technological University of Zinacantepec, Zinacantepec, Estado de México, Mexico [c]Biotechnology Department, Autonomous Metropolitan University-Iztapalapa, México City, Mexico [d]LabMAT, Department of Civil and Environmental Engineering, University of Bío-Bío, Concepción, Chile

14.1 Introduction

14.1.1 Components of tamarind tree, tamarind pulp, and seed

Tamarind (*Tamarindus indica* L.) is native to the dry savannas of eastern tropical Africa, probably from the island of Madagascar. It was introduced to India by Arab traders, where it adapted perfectly to the agro-ecological conditions of the country's coastal plains. The tamarind was domesticated in the East Indies (Southeast Asia) and the islands of the Pacific. It was introduced to the New World by the Spanish and Portuguese in the 18th century (Rao and Mathew, 2012; Sharma and Bhardwaj, 1997). The main tamarind-producing areas in Mexico are located on the coastal plain of the Pacific Ocean coast and to a lesser extent on the coasts of the Gulf of Mexico. It is exploited commercially for the production of fruit in Colima, Guerrero, Oaxaca, Michoacán, Jalisco, and Veracruz.

Mexican tamarind is a slow-growing and large tree; in optimal conditions of development, the trees reach 24–30 m in height. The tamarind is a tree with evergreen foliage (perennifolio); however, in very dry climates it behaves as a subdeciduous, as it can lose its foliage for a short period of time during hot months. It is a long-lived tree, since it can live more than 200 years.

Micro- and Nanoengineered Gum-Based Biomaterials for Drug Delivery and Biomedical Applications
https://doi.org/10.1016/B978-0-323-90986-0.00016-9

The trunk is short, thick, and straight; the bark has cracks along the trunk and main branches, and presents different shades ranging from ash gray to dark brown (Orozco, 2001).

The leaves are pale to dark green, alternate and parapinnate; they are 7.5–15 cm long and have 10–20 pairs of leaflets. The flowers are hermaphroditic, pale yellow with red or orange veins and measure 2–2.5 cm in diameter, formed in small clusters 5–10 cm long and with 8–14 flowers per cluster (Parrotta, 1990; Viveros-García et al., 2012). The fruit is an indehiscent pod (remains closed when ripe), protruding and oblong, cinnamon brown or greyish brown, slightly curved and flattened. It is 7–20 cm long and 1–3 cm wide. The fruits are formed in abundance along the branches or in the terminal parts of them. When the fruits ripen, their skin is brittle and can break easily. The pulp is the edible part of the fruit, brown in color, firm, viscous, granular, with a pleasant bittersweet flavor and high content of sugar and acid. It has 1–12 seeds of a bright brown color, flat, oval and united by fibers that are located in the pulp of the fruit. In a kilogram there are from 2000 to 2500 seeds (Bhattacharya et al., 1993; Viveros-García et al., 2012).

The pulp of the tamarind fruit has been a very important culinary ingredient in Mexico for a long time. On average, 65.95% moisture, 2.35% protein, 0.48% fat, 26.2% sugars, 0.45% sucrose, 2.7% cellulose, and 1.87% other compounds can be found on average in ripe tamarind pulp. About 60 volatile compounds have also been found in tamarind pulp. There is also a variety of tamarind with a sweeter flavor with a reddish hue. This sweet taste can be caused due to a lower presence in the amount of polyprotonated acids (Shankaracharya, 1998; Viveros-García et al., 2012).

On the other hand, tamarind seed represents on average 35% of all fruit; this is made up of two main parts: the testa, representing 30% of the seed, and the endosperm or kernel, representing the remaining 70%. Analyzing the composition of the seed, it was possible to find proteins in 18%, fats in 7%, carbohydrates in 69%, fiber in 3%, and other compounds with 3% (Bhattacharya et al., 1993; Kumar and Bhattacharya, 2008; Tsuda et al., 1994). The protein contained in the endosperm is abundant in lysine, glutamic acid, aspartic acid, glycine, leucine, and potassium, but deficient in amino acids that contain sulfur groups (Bhattacharya et al., 1994; Kumar and Bhattacharya, 2008).

14.1.2 Technologies of tamarind seed gum extraction

Tamarind seed gum is a polysaccharide which main compound is xyloglucan. It was discovered during World War II when the Forest Research Institute, Dehra Dun, was searching for new sizing materials; nowadays it has different applications including in the food, pharmaceutical, confectionary, cosmetic, and textile industries. This compound was commercially isolated in 1941 by Daurala Sugar Works, India, when it was proposed as a pectin substitute. Tamarind seed gum was commercially produced in 1943 and the extraction method in laboratories was first devised in 1945 (Rao and Mathew, 2012; Sharma and Bhardwaj, 1997).

Most common extraction technologies involve some of the following steps: the first step is washing the crude tamarind seeds with water and subsequent heating to make the testa (seed coating) brittle and friable. The seeds are then decorticated to leave the heavier crushed endosperm, which is finally ground to yield tamarind kernel powder. After this, it is boiled with about 30–40 times its weight of water for about 30–40 min under agitation and allowed to sit

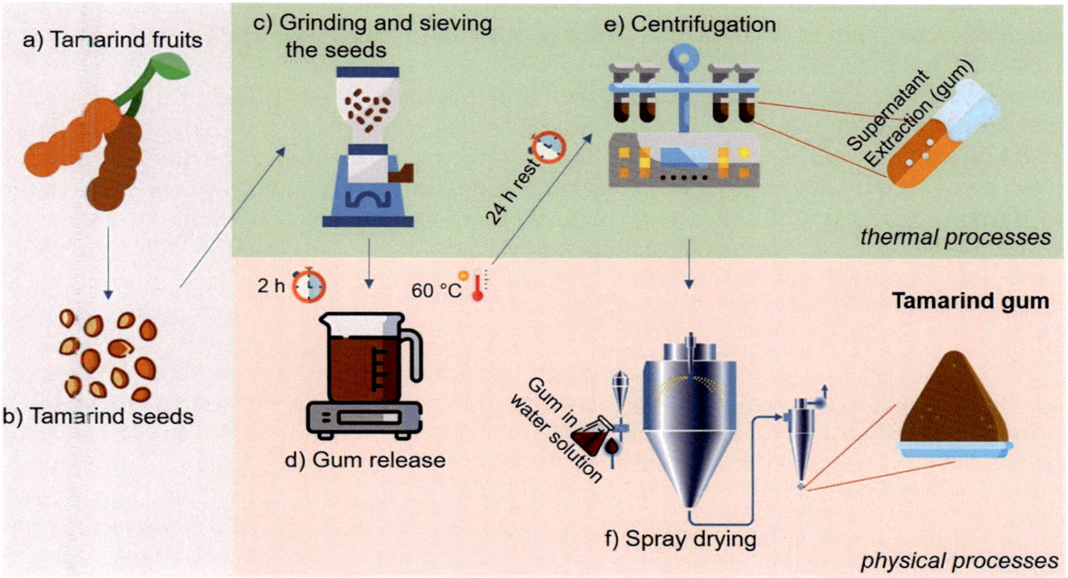

FIG. 14.1 Eco-green-based technology of tamarind gum extraction. Method based on Alpizar-Reyes et al. (2017a): (A and B) The seeds are extracted manually from mature pods of tamarind, then (C) milled and grounded through a 355 μm mesh. (D) 20 g of milled tamarind seeds are dispersed in 1 L of bidistilled water and stirred for 2 h at 60 °C, then the dispersion is kept at 20 °C for 24 h for the gum release, (E) and centrifuged for 8 min at 524 × *g*. The supernatant represents the gum, (F) which is fed at 40 mL/min to a spray-drier with an inlet temperature of 135 ± 5 °C, outlet temperature at 80 ± 5 °C and injecting compressed air at 4 bar. Tamarind gum extracted by an eco-green-based technology uses thermal and physical processes, avoiding chemical pollution.

overnight in a settling tank in order to precipitate and settle out fibers and for the most part of the proteins. The next step is crucial: the noneco-green gum is precipitated with solvents as ethanol and some salts and dried to a constant weight using some of the drying technologies available (Fig. 14.1) (Nishinari et al., 2009; White and Rao, 1953; Yamatoya and Shirakawa, 2003).

Through the methodologies used for tamarind gum extraction, different chemical reagents have been used at commercial and laboratory scale. The original method proposed by Ghose and Krishna (1942) was pouring tamarind kernel powder into cold water at a ratio of 1:10, after that, it was added to boiling water at a ratio of 1:30 or 1:40 and maintained at boiling temperature for 20–30 min; it was then strained through cloth and sulfur dioxide was added to bleach the gum. The mix was left overnight before centrifugation and drum drying. It was observed that sulfur dioxide caused the original gelling properties of the polysaccharide to be lost; very tough gels were formed instead if this treatment was not carried out. Later, it was suggested to add an organic acid to the boiling water in order to obtain the desired degradation degree of the gum that enabled soft gels. The yield obtained with this process was higher than 50% of the tamarind kernel powder. Precipitation with ethanol is an alternative to drum drying; using this technique, the polysaccharide is usually filtered mechanically, dried in an air oven, and finally milled to homogenize particle size. Another method utilized for extraction and purification of the tamarind seed gum consists of dilution, dialysis, and precipitation. From this process a fine white powder with less than 1% protein content is obtained (Rao et al., 1973).

A method patented by Jones et al. in the United States consists of defatting of the tamarind kernel powder using C-6 or C-8 aromatic hydrocarbons or C-1, C-2, or above halogenated lower hydrocarbons or C-1 or C-5 mono or dihydroxy alcohols. The defatted powder is then recovered by filtration or centrifugation and dried. After that, flow properties of the powder are enhanced with salicaceous materials, and the powder is grounded to reduce its size below 100 μm and further classified into three groups: fine fraction rich in protein, moderately fine fraction rich in polysaccharides, and coarser fraction rich in mechanical properties. Finally, the tamarind seed gum can be isolated from the moderately fine fraction (Gupta et al., 2010).

Solvent-induced precipitation was proposed to extract tamarind gum with the next steps: first, the seeds were heated in sand (seed: sand ratio of 1:4) to remove the testa, then the seeds were crushed and soaked for 24 h, after that the dispersion was boiled for 1 h and left to sit for 2 h. The seeds were squeezed and the gum separated was mixed with equal quantity of acetone to precipitate it. Finally, the mucilage was dried at 50 °C, powdered, and passed through sieve number 80. The yield obtained with this procedure was 78% (w/w) from tamarind seeds (Phani Kumar et al., 2011).

Another alternative to solvent employment is to isolate the tamarind seed polysaccharide using a common method. Firstly, a reflux system was used with water as a solvent at 70 °C for about 6 h. The extract was then pressed in cheese cloth bag, cooled to 4 °C, and mixed with alcohol 2:1 (v/v) to achieve tamarind seed polysaccharide precipitation. It was stirred continuously for 15 min and then left to stand for 2 h to allow the formation of coagulates which were further filtered, washed with alcohol, and pressed. Finally, the pressed product was dried at 35–45 °C in a hot air oven, grounded, and sieved through sieve number 20 (Bansal et al., 2013).

There has been considerable progress in the development and usage of various extraction and purification techniques for tamarind gum; their selection is crucial for this polysaccharide's physicochemical and techno-functional properties. Additionally, the environmental impact of complex extraction techniques opened a new window on implementing new eco-green extraction technologies with more minor ecological effects.

Some environmentally friendly methodologies reported for the extraction of tamarind gum are described as follows. Alpizar-Reyes et al. (2017a, b) obtained tamarind gum by manually extracting the tamarind seeds, and milling and grinding them through a mesh. The powdered seeds were added to bidistilled water to achieve a weight ratio of 1:10, and the aqueous dispersion was heated and stirred for 10 min. Additional bidistilled water was then added to reach a weight ratio 1:40 and heated at 80 °C for 60 min. The dispersion was left to stand, centrifuged, and finally, the supernatant was spray-dried (Fig. 14.2) yielding ~29% (w/w) of tamarind gum.

Other solvent-free methods have been also tested, i.e., methods that obtained yields for tamarind gum of ~27%–32% using two distinct methods. The first method consisted of tamarind seeds that were washed and dried at 100 °C, and grinded using a blender. The resultant powder was mixed with water and poured into boiling water, stored overnight, and centrifuged. The supernatant phase was precipitated with aqueous ethanol (95% *v*/v) and then, dried at 50 °C for 4 h in an oven. The second method consisted of the use of

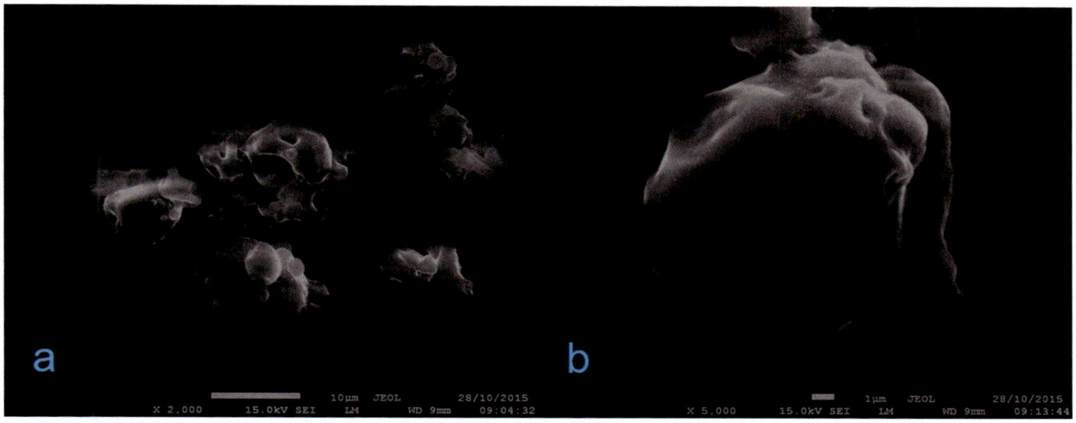

FIG. 14.2 Scanning electron microscopy for tamarind gum (A) at 2000 magnifications and (B) at 5000 magnifications.

defatted tamarind seed powder followed by the same procedure previously described (Chawananorasest et al., 2016).

Different techniques for tamarind gum extraction include that reported by Limsangouan et al. (2019), who applied subcritical water and compared some characteristics from the obtained tamarind gum with that extracted by the "conventional method." The latter consisted of defatting the tamarind kernel powder with hexane at the ratio of 1:5 (w/v) and extracting tamarind gum with boiling distilled water for 30 min. Once the hydrocolloid dispersion was cooled to ambient temperature, it was treated with protease enzyme. The mixture was then centrifuged at 5000 rpm for 15 min at 25 °C, and the supernatant was mixed at the ratio 1:2 (v/v) with aqueous ethanol (95% v/v) for 30 min to induce the formation of a gel-like phase, which was filtered through muslin cloth and dried in a hot air oven at 60 °C for 8 h. The resultant material was grinded and sieved with a common 50-mesh. On the other hand, extraction with subcritical water consisted of mixing the filtered defatted powder with distilled water (5%, w/v), pouring into a high-pressure resistant vessel, and heated until reaching temperatures of 100, 125, 150, 175, and 200 °C. The extracted solution was treated with protease enzyme and precipitated as the conventional method. The extracted gum exhibited higher yields (52%–62%), greater color intensity and water solubility index, but lower molecular weight, holding strength, viscosity, water absorption index, antioxidant capacity, and total phenolic content than that obtained with the conventional extraction method.

14.2 Tamarind gum characterization

14.2.1 Chemical structure and composition

The chemical structure and composition from different extraction techniques of tamarind gum have been studied by many researchers. The chemical structure of tamarind gum (Fig. 14.3) consists of a cellulose-like main chain of a branched polysaccharide with molecular weight of 700–880 kDa (Khounvilay and Sittikijyothin, 2012), composed of a β-(1,4)-D-glucan

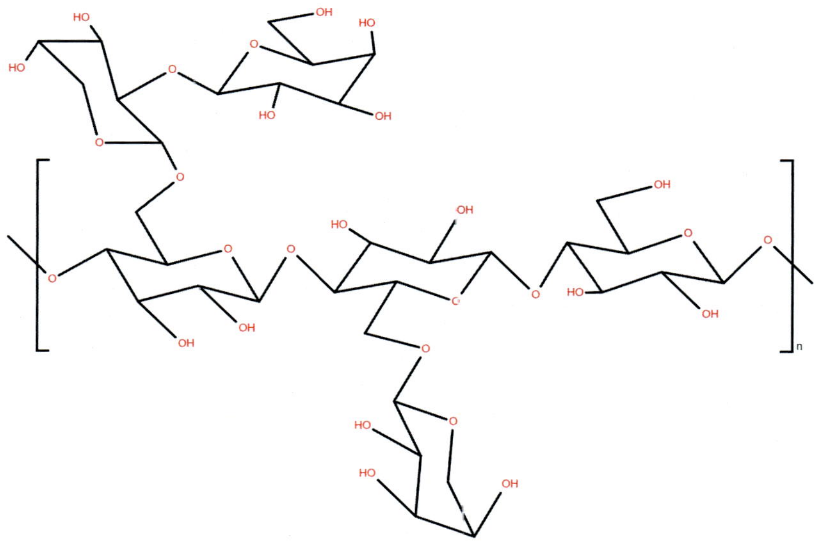

FIG. 14.3 Chemical structure of tamarind gum β-(1,4)-D-glucan backbone chain, partially substituted with side chains of α-(1,4)-D-xylopyranose and (1,6) linked [β-D-galactopyranosyl-(1,2)-α-D-xylopyranosyl] to glucose residues.

backbone chain, partially substituted with side chains of α-(1,4)-D-xylopyranose and (1,6) linked [β-D-galactopyranosyl-(1,2)-α-D-xylopyranosyl] to glucose residues, where tamarind gum is conformed from monomer units that essentially contain three types of sugars, glucose, galactose, and xylose, with a molar ratio of 2.8:2.25:1.0. A chain length of tamarind gum from 300 to 3000 glucose units was reported (Fry, 1989). Moreover, xylose units are shown to be more hydrophilic than glucose units. Due to the presence of hydrophilic and hydrophobic units, tamarind gum shows good solubility in water, even though individual macromolecules do not fully hydrate, resulting in the formation of aggregated species in water even at very dilute solutions. Tamarind gum can be chemically or enzymatically modified to assess improved rheological properties, i.e., gel formation or extra-high viscosity, where this modification has a direct correlation between the structure of a polysaccharide and its surface and functional properties (Kulkarni et al., 2017; Pardeshi et al., 2018).

The yield of the extraction of tamarind gum prior to be used as a wall material in the microencapsulation of drugs and natural products exerts a great influence on the economic aspects, where a higher performance with a lower environmental impact is always desired. In this sense, in Table 14.1, it can be seen that eco-green-based technology (solvent free) has the greatest economic advantages compared to solvent application technologies. Solvent-free processes offer greater competitiveness and a window of economic opportunity for future applications of tamarind gum as a micro-engineered gum-based biomaterial for drug delivery and for biomedical applications.

The chemical composition of tamarind gum (Table 14.2) is dominated by the carbohydrate content (70%–90%) because gums are polysaccharides by nature (Prajapati et al., 2013), in

TABLE 14.1 Yield for tamarind gum with different extraction methods.

Extraction method	(%) reported	References
Eco-green-based technology (solvent free)	29.83 ± 1.32	Alpizar-Reyes et al. (2017a)
Low solvent application (ethanol precipitation)	27.93 ± 0.23	Crispín-Isidro et al. (2019)
Medium solvent application (ethanol precipitation + hexane distillation)	22.12 ± 0.11	
High solvent application (ethanol precipitation + hexane distillation + NaOH)	18.93 ± 0.35	

Summary of the results supported by the references cited.

TABLE 14.2 Summary of the chemical composition in dry basis quantified for tamarind gum with different extraction methods used.

	Extraction method	(%) reported	References
Carbohydrates	Eco-green-based technology (solvent free)	80.25	Alpizar-Reyes et al. (2017b)
		79.76 ± 0.72	Alpizar-Reyes et al. (2017a)
		82.17 ± 0.21	Alpizar-Reyes et al. (2018)
	Milled and grounded seeds (analysis not in dry basis)	80.66	Khounvilay and Sittikijyothin (2012)
	Strong employment of chemicals (ethanol precipitation + hexane distillation + NaOH)	85.74 ± 0.75	Chandra Mohan et al. (2018)
	Low solvent application (ethanol precipitation)	79.24 ± 0.68	Crispín-Isidro et al. (2019)
	Medium solvent application (ethanol precipitation + hexane distillation)	81.29 ± 0.09	
	High solvent application (ethanol precipitation + hexane distillation + NaOH)	89.56 ± 1.48	
Glucose:xylose: galactose ratio	Milled and grounded seeds (analysis not in dry basis)	2.61:1.43:1.00	Khounvilay and Sittikijyothin (2012)
Proteins	Eco-green-based technology (solvent free)	14.24	Alpizar-Reyes et al. (2017b)
		14.78 ± 0.45	Alpizar-Reyes et al. (2017a)
		12.96 ± 0.09	Alpizar-Reyes et al. (2018)
	Milled and grounded seeds (analysis not in dry basis)	13.51	Khounvilay and Sittikijyothin (2012)

Continued

TABLE 14.2 Summary of the chemical composition in dry basis quantified for tamarind gum with different extraction methods used—cont'd

	Extraction method	(%) reported	References
	Strong employment of chemicals (ethanol precipitation + hexane distillation + NaOH)	0.58 ± 0.08	Chandra Mohan et al. (2018)
	Low solvent application (ethanol precipitation)	12.11 ± 0.14	Crispín-Isidro et al. (2019)
	Medium solvent application (ethanol precipitation + hexane distillation)	16.53 ± 0.19	
	High solvent application (ethanol precipitation + hexane distillation + NaOH)	9.00 ± 0.12	
Fats	Eco-green-based technology (solvent free)	4.96	Alpizar-Reyes et al. (2017b)
		4.76 ± 0.36	Alpizar-Reyes et al. (2017a)
		4.28 ± 0.17	Alpizar-Reyes et al. (2018)
	Milled and grounded seeds (analysis not in dry basis)	5.76	Khounvilay and Sittikijyothin (2012)
	Low solvent application (ethanol precipitation)	0.10 ± 0.01	Crispín-Isidro et al. (2019)
	Medium solvent application (ethanol precipitation + hexane distillation)	0.08 ± 0.01	
	High solvent application (ethanol precipitation + hexane distillation + NaOH)	7.20 ± 0.31	
Ashes	Eco-green-based technology (solvent free)	0.55	Alpizar-Reyes et al. (2017b)
		0.70 ± 0.12	Alpizar-Reyes et al. (2017a)
		0.59 ± 0.07	Alpizar-Reyes et al. (2018)
	Milled and grounded seeds (analysis not in dry basis)	0.07	Khounvilay and Sittikijyothin (2012)
	Strong employment of chemicals (ethanol precipitation + hexane distillation + NaOH)	3.68 ± 0.11	Chandra Mohan et al. (2018)
	Low solvent application (ethanol precipitation)	0.72 ± 0.00	Crispín-Isidro et al. (2019)
	Medium solvent application (ethanol precipitation + hexane distillation)	0.94 ± 0.01	
	High solvent application (ethanol precipitation + hexane distillation + NaOH)	0.38 ± 0.07	

Summary of the results supported by the references cited.

agreement to (Crispín-Isidro et al., 2019), the purity of tamarind gum polysaccharide is increased when the carbohydrate content is bigger. In this regard, the amount and type of solvents used for the extraction and purification of tamarind gum have been shown to influence the amount of carbohydrate content in the polysaccharide, especially those employed for removing fats and proteins associated to the polysaccharide backbone in tamarind gum. According to the monosaccharide content, tamarind gum exhibits a ratio of glucose, xylose, and galactose of 2.61:1.43:1.00, allowing it to hold water molecules in relation with monosaccharide composition and structure of gums. The variation in the availability of hydrophilic sites that hold water molecules in the polysaccharide chain may be the cause of their different moisture contents (Hamdani et al., 2019).

Researchers have reported a wide range of protein content of tamarind gum with different extraction techniques between 0.58 and 16.53% (Table 14.2). Hydrophobic proteins in tamarind gum are often responsible for the emulsifying ability main factor that allows tamarind gum to act as a wall material in the microencapsulation of drugs and natural products. Hydrophobic proteins adsorb onto the surface of oil droplets while the hydrophilic carbohydrate moiety inhibits the flocculation and coalescence of molecules through electrostatic and stearic repulsions (Dickinson, 1994).

Solvent-free technologies reported (Table 14.2) higher protein content of 12.96%–14.78% (Alpizar-Reyes et al., 2017a, b, 2018, 2020) and 13.51% (Khounvilay and Sittikijyothin, 2012) for tamarind gum extracted with hot water without any further solvent. Solvent application lowered the protein content, to 12.11% for low solvent usage and 0.58%–9.00% for the strong employment of chemicals (Chandra Mohan et al., 2018; Crispín-Isidro et al., 2019). The poor protein content exhibited is attributed to the extent purification process to which it was subjected (defatted + soluble protein alkaline extraction). It is known that gums precipitation with ethanol can induce the coprecipitation of other materials, such as protein, organic acids, certain salts, and other similar substances (Naod and Tsige, 2012).

Significant differences have been reported in the fat content in tamarind gum, from traces up to 7.20% (Table 14.2), which may be due to the differences in geographical origin, variety, and growing conditions, and in the extraction and purification processes. Fat content of tamarind gum may include saturated as well as unsaturated fatty acids. The fat content present in free solvent (nondefatted) technologies for tamarind gum has been reported as 4.76% (Alpizar-Reyes et al., 2017a) and 5.76% (Khounvilay and Sittikijyothin, 2012). The application of defatting technologies to tamarind gum has significantly reduced the content of fats to values near to traces (0.08%) (Chandra Mohan et al., 2018; Crispín-Isidro et al., 2019).

Ash content of gums has been found varying between 0.07% and 0.94% (Table 14.2). The presence of minerals like calcium, magnesium, manganese, zinc, and lead have been reported in plant gums, so it can be said that plant gums contain various essential nutrients (Fathi et al., 2016; Hamdani et al., 2019; Rezaei et al., 2016).

14.2.2 Infrared (FTIR) evaluation

One of the most successful approaches for studying and understanding the functional groups of various gums, such as tamarind gum, is to employ Fourier Transform Infrared (FTIR) spectroscopy. When new processes to extract tamarind gum are involved, the role of FTIR techniques to analyze the composition of these developments is useful to monitor the specific functional groups. In general, the use of FTIR analysis in the field of gums application is crucial to support the justification of the changes in their properties and performance in various drugs applications.

FTIR spectrums for tamarind gum commonly show two notable regions corresponding to wavenumbers of 3600–2700 cm^{-1} (mainly corresponding to lipids section) and 1800–800 cm^{-1} (consistent to proteins and carbohydrates wavenumber). Independently of the purification process involved on tamarind gum extraction, FTIR spectrums shows a wide band centered at 3310 cm^{-1} attributed to O—H stretching vibration and a sharp band at 2919 cm^{-1} due to aliphatic C—H stretching vibration of the lipid fraction (Alpizar-Reyes et al., 2017a; Chandra Mohan et al., 2018; Crispín-Isidro et al., 2019). Thus, it is evident that the processing conditions are not able to remove the small fat fractions present on tamarind gum.

For the second region, the band for —C=O stretching characteristic amide I band from proteins was observed at 1610 cm^{-1} of the acetylated units (—$CONH_2$ groups); and at 1525 cm^{-1} a strong band is associated to —$NH_3{}^+$ groups for the amide III region (Rezaei et al., 2016) and band at 1280 cm^{-1} is attributed the stretching of the C—O bond (Alpizar-Reyes et al., 2017a). Peaks at 1370, 1150, 1037, 1071, 944, and 987 cm^{-1} are characteristics of xyloglucan (Alpizar-Reyes et al., 2017a), which is an important component of tamarind gum. A band at 1370 cm^{-1} denotes the CH_2 bending of xyloglucan, the soft peak at 1150 cm^{-1} corresponds to O—C—O asymmetric stretching, while peaks at 1037 and 1071 cm^{-1} are the result of C—O and C—C stretching of the xyloglucan ring. A band at 944 cm^{-1} corresponds to the ring vibration of xyloglucan and finally, an 897 cm^{-1} band is due to C—H stretching characteristic of glucose and xylose β-anomeric links (Munir et al., 2016).

For protein structures pertaining to the exploration of eco-green-based technology (solvent free), low solvent application (ethanol precipitation), medium solvent application (ethanol precipitation + hexane distillation), and high solvent application (ethanol precipitation + hexane distillation + NaOH), Gaussian peaks were assigned to the corresponding structure based on their center (Crispín-Isidro et al., 2019). Changes in the secondary structures of the proteins for medium and high solvent application indicated that they were denatured probably due to the thermal and alkaline conditions used to purify tamarind gum. Crispín-Isidro et al. (2019) studied different extraction techniques to purify tamarind gum to a higher degree, but they concluded that the chemical structure of tamarind gum was not altered by the purification methods involved.

When strong chemicals were employed (ethanol precipitation + hexane distillation + NaOH), meaning the employment of multistage alkali wash of tamarind gum, the protein molecules in tamarind seed gum were completely removed (Chandra Mohan et al., 2018) (confirmed by absence of characteristic bands corresponding to primary and secondary amine). Finally, it is possible to establish that eco-green extraction techniques seem to fit better to the actual market requirements.

14.2.3 NMR spectrum

NMR spectroscopy has been used to determine molecular identity in polysaccharide materials, with a particular focus on the monosaccharide ratio that compounds these biopolymers. The ratios of the main monosaccharides were studied by Rodrigues et al. (2018) at 600 MHz spectra by the ^{1}H and ^{13}C NMR and xylose, glucose, and galactose were reported as the main carbohydrate residues of the tamarind gum. The authors reported that an area under glucose (at δ 3.44), xylose (sum of the signals at δ 4.95 and δ 5.56), and galactose (difference between the signal at δ 4.56 and at δ 3.44) resulted in a ratio of glucose: xylose: galactose of 3:2.58:1.63. The ^{13}C NMR was also used to evaluate the glucose: xylose: galactose ratio, which resulted in a glucose: xylose: galactose of 3:2.77:1.46. These studies are in concordance with the results generally reported to be in a molar ratio of glucose:xylose:galactose xyloglucan as 3:2:1 (Gidley et al., 1991; Hamdani et al., 2019; Rodrigues et al., 2018).

14.2.4 Thermal stability

The thermal stability of tamarind gum was investigated for different extraction methods by thermo gravimetric differential scanning calorimetry (TGA-DSC) analysis. Thermo gravimetric analysis (TGA) and differential scanning calorimetry (DSC) curves of tamarind seed remain the same for all tamarind gums despite the extraction technologies applied. In general, the TGA plot for tamarind gum shows two mass loss events. The first mass change from 1.0% to 5.5% was observable at temperatures higher than 75 °C, which can be attributed to the loss of moisture correspondent to free water in the gum particles (Alpizar-Reyes et al., 2017a; Bergstřm et al., 2012; Chandra Mohan et al., 2018; Crispín-Isidro et al., 2019). This transition has been associated to the hydrophilic nature of the functional groups of each polysaccharide. Major weight loss from 50% to 65% for tamarind gum occurred between 175 and 480 °C, which was confirmed by TGA with peak decomposition temperature at 322 °C, and was commonly attributed to the polysaccharide thermal decomposition (Chandra Mohan et al., 2018). TGA results of tamarind gum proved its thermal stability until 170 °C, thus it is useful for any process for drug delivery under this decomposition temperature.

On the other hand, the DSC technique is commonly used for studying thermal transitions that occurred during heating in the presence of an inert atmosphere. Alpizar-Reyes et al. (2017a) reported DSC plots that showed two main regions; the first region was positioned from 65 to 175 °C and, with a peak of 98 °C, indicating the evaporation of free water desorbed from the polysaccharide matrix, thus all the gums were thermally stable independently of the purification process to which they were subjected. The second peak, located at 310 °C, related to an exothermic event due to the polysaccharide decomposition.

14.2.5 X-ray diffraction (XRD) studies

XRD is a technique used to identify whether the nature of the materials is crystalline or amorphous. It will define the quantification of gum materials; therefore, XRD is a versatile tool to examine the nature of tamarind gum. The diffraction curve of tamarind gum belongs to that typically exhibited by an amorphous material indicated by the absence of sharp peaks, with an amorphous halo with a broad band centered at $2\theta = 20$ degree (Alpizar-Reyes et al., 2017a; Kaur et al., 2012a, b; Madgulkar et al., 2016; Premalatha et al., 2017). As a result, it is easy to establish that the extraction technique does not change the general conformation and structure of the gum, and only exerts influence on protein and fat content.

14.3 Functional properties of tamarind gum

Some of the functional properties of tamarind gum, such as solubility, water-holding capacity, oil-holding capacity, emulsion ability, emulsion stability, surface tension, and electrophoretic mobility and rheology, at specific conditions, will be described in the following sections. The study of the functional properties of tamarind gum is important to evaluate the gum's potential technology applications as a drug carrier and control release system, to design and to operate processing equipment, and to control its storage (Kumar and Bhattacharya, 2008).

14.3.1 Water solubility, water-holding capacity, swelling ability, and oil-holding capacity

Tamarind gum is water soluble, and the steric hindrance of its side chain inhibits the aggregation to cellulose-like chains. It forms a homogeneous solution on heating with water while stirring. The chemical modification of this gum, such as carboxymethylation, tends to increase its solubility in cold water (Goyal et al., 2007; Mali et al., 2019). In another study, Alpizar-Reyes et al. (2017a) evaluated the water solubility of tamarind seed gum depending on the temperature, and they found that water solubility of this gum increased from 8% to 21.8% on increasing the temperature from 25 to 65 °C.

Water-holding capacity (WHC) represents the amount of water held and absorbed by the hydrated sample after an external force is applied, and provides information about the stability, yield, and sensory characteristics of a gum. Alpizar-Reyes et al. (2017a) evaluated the WHC of tamarind gum as a function of temperature and found that WHC increased as the temperature increased, going from 0.18 g/g at 25 °C to 1.07 g/g at 65 °C.

Furthermore, tamarind gum swelling studies using water found that tamarind gum swelled 1.6 times its weight, while carboxymethylated tamarind gum (CTG) swelled 2 times with respect to the *volume* of the dry gum (Mali et al., 2019). In addition, the enhancement of substitution degree in the carboxymethylation of gum had a favorable effect on swelling of tamarind gum (Goyal et al., 2007). The rapid swelling ability of tamarind gum suggests its use in hydrophilic matrix tablets and its application in controlled drug delivery (Phani Kumar et al., 2011).

The oil-holding capacity (OHC) of a gum is the absorption of oil through the lateral nonpolar sites within protein molecules. A gum with high OHC may be able to retain oil-based flavoring and enhance mouthfeel in food applications; likewise, it may retain oil-based drugs. Alpizar-Reyes et al. (2017a) evaluated the OHC of tamarind gum as a function of temperature and found that OHC increased as the temperature increased, ranging from 0.068 g/g at 25 °C to 0.133 g/g at 65 °C.

14.3.2 Emulsifying ability and emulsifying stability

Emulsifying ability (EA) measures the ability of an emulsifying agent to form emulsions while the emulsion stability (ES) to heating measures the breakdown of the emulsion when it is heated, where the proteins adsorbed to the surface of the oil droplets unfold and expose nonpolar amino acid, which leads to hydrophobic attraction between droplets and flocculation occurs.

EA measured considering the fraction in weight of tamarind gum and volume of oil of 0.2:1 tamarind gum:oil ratio, the EA obtained was 78.3%, while using 1:1 tamarind gum/oil ratio the EA was 90% (Alpizar-Reyes et al., 2017b). This behavior was attributed to the increment of the gum, and thus to the increment of the branched structure among the surface active to absorb oil molecules that reduced surface tension. Furthermore, the authors evaluated the ES based on mass of tamarind gum/volume of oil ratio and found that decreasing the gum fraction with respect to the oil increased the ES—that is, using 1:1 tamarind gum/oil ratio, the ES was 82.2%, while using 0.2:1 tamarind gum/oil ratio, the ES was 91.1%.

Moreover, the EA of two different formulations of oil/water emulsions to prepare microcapsules was evaluated (Alpizar-Reyes et al., 2020). Emulsion 1 was composed of tamarind gum/sesame oil ratio 1:1 in dry basis, 10% (w/w) of total solid content, and dispersed phase volume fraction $\varphi_{O/W} = 0.05$. Emulsion 2 was composed of tamarind gum/sesame oil ratio 1:2 in dry basis, 15% (w/w) of total solid content, and $\varphi_{O/W} = 0.1$. EA was greater for emulsion 2 (90.28%) than emulsion 1 (86.26%); conversely, ES was greater for emulsion 1 (82.31%) than emulsion 2 (80.33%). Both emulsions showed good stability to heating; however, the reduction on the surface tension for tamarind seed gum used for stabilizing emulsion 2 was attributed to the reduction in its long-term stability.

Tamarind gum has demonstrated good emulsifying properties and was applied in various emulsions studies; for example, Bhattacharya et al. (1993), Kumar and Bhattacharya (2008), and Tsuda et al. (1994) described the formation of stable castor oil-in-water emulsions using tamarind seed gum at 2% *w/v* as stabilizer agent, exhibiting droplet sizes from 1 to 10 μm.

On the other hand, the fabrication of oil-in-water emulsions stabilized with tamarind gum aqueous dispersions (2%, w/w) at different levels of purification was reported (Crispín-Isidro et al., 2019). For emulsions with low purity in tamarind gum, the initial area-volume mean diameter ($d_{3,2}$) was 1.35 ± 0.15 μm; meanwhile, emulsions where medium purity tamarind gum was used displayed droplet sizes of 4.82 ± 0.10 μm, and for emulsions of high purity, tamarind gum of 9.45 ± 0.10 μm. Thus, low-purity gum achieved the smallest droplet size and the highest emulsion stability. Then, the purification of tamarind gum had a significant effect not only on the final chemical composition but also on the emulsifying properties of this polysaccharide.

14.3.3 Surface and interfacial tension

Tamarind gum has surfactant properties since this produces changes in the surface tension of the water. Phani Kumar et al. (2011) determined the surface tension of tamarind gum (0.1%, w/v) by the drop count method, using a stalagmometer, and the obtained result was 83.26 dynes/cm.

The use of tamarind gum with three different levels of purification (low, medium, and high purity), in agreement to its carbohydrate content, was studied for stabilizing canola oil/water emulsions, and the time evolution of the dynamic interfacial tension (σ) was also reported (Crispin-Isidro et al., 2019). All tamarind gum aqueous dispersions exhibited the same trend with an initial sharp decrease of σ during the first 200 s, followed by a slow progressive drop in σ at longer times, until reaching asymptotic values, where σ did not change by more than

0.5 mN/m in 30 min. The main results showed that tamarind gum with the lowest purity decreased the interfacial tension faster than gums with higher purity, due to its relatively high content of surface-active compounds, such as protein and polyphenols, which may contribute to its adsorption at the interface.

14.3.4 Electrophoretic mobility (ζ-potential)

ζ-potential values are related to the stability of the colloidal systems. If the particles in a suspension have a large negative or positive ζ-potential, they tend to repel each other and no tendency for the aggregation of particles can be observed. In contrast, if the particles have a low ζ-potential value, aggregation and flocculation occur (Crispín-Isidro et al., 2019). González-Martínez et al. (2017) reported a ζ-potential value of −12 mV at pH 10.0 in a solution of crude tamarind seed gum.

In addition, tamarind gum at different purity levels (low, medium, and high purity), shows negative ζ-potential values ranging from -11.00 ± 0.19 mV for low-purity tamarind gum to -5.72 ± 0.20 mV for the highest-purity tamarind gum (Crispín-Isidro et al., 2019). The decrease in the ζ-potential values as the purity in the tamarind gum increased was related to the elimination of charged impurities throughout the purification, that contribute to the electrophoretic mobility in the aqueous dispersion.

14.3.5 Rheology

There is a great variety of studies that evaluate the rheological behavior of tamarind gum. Kumar and Bhattacharya (2008) evaluated the rheology of tamarind kernel powder at different concentrations (2%, 4%, 6%, 8%, and 10%, w/w) and these behave like non-Newtonian, shear-thinning fluids with low yield stress values at low concentration and vice versa (Kumar and Bhattacharya, 2008). The Herschel–Burkley model adequately fits the shear stress–shear rate data.

In another study, tamarind seed gum solutions at 20 °C, in a range of concentrations from 0.67% to 5.70% exhibited shear-thinning flow behavior at high shear rate and Newtonian region at low shear rate. At higher concentrations, pronounced shear-thinning was shown and at lower concentrations, the viscosity did not show dependence on shear rate. Tamarind seed gum showed a typical random-coil polymer behavior. Dilute and semidilute regions were observed with slopes of 2.2 and 4.3, respectively. When the specific viscosity at zero shear rate (η_{sp0}) was plotted against the coil overlap parameter (C[η]), it was found the critical concentration $C^*[\eta] = 4.23$, that is, about 0.90% (w/w), which agrees with the results of the majority of coiled hydrocolloids. In addition, a viscoelastic study was carried out and the Cox-Merz rule was applied and well-adjusted at 2.30 and 2.75% (w/w) of tamarind gum (Khounvilay and Sittikijyothin, 2012).

Tamarind gum dispersions reported by Alpizar-Reyes et al. (2018) had non-Newtonian shear-thinning behavior described by the Power law model. In this study, the effects of tamarind gum concentration, temperature, pH, and salt addition were evaluated on apparent viscosity. It was determined that as tamarind gum concentration increased (from 0.5% to 2.0%, w/w), the viscosity and pseudoplasticity of gum dispersions increased; as temperature increased (from 25 to 60 °C), the viscosity and pseudoplasticity of gum dispersions decreased;

as pH increased (from 4 to 10), the apparent viscosity of the gum dispersions increased; and finally, it was observed that the addition of salts (NaCl, KCl, and $CaCl_2$) modified the apparent viscosity differently.The rheological behavior of the tamarind seed gum showed that this fluid is resilient against pH, temperature, salt, and sugar concentrations, therefore, it can be applied as a food additive (Alpizar-Reyes et al., 2018).

A similar rheology study was carried out by Shao et al. (2019), who found that tamarind seed gum aqueous solutions from 0.5% to 4% (w/v) exhibited non-Newtonian shear-thinning behavior described by the Williamson model. Apparent viscosities of tamarind seed gum solutions decreased drastically in an alkaline solution of pH > 10; however, these were only slightly affected by pH < 10. The apparent viscosity of the tamarind seed gum solution at 2% (w/v) decreased slightly with increasing temperature (5–85 °C) at three different shear rates. On the other hand, dynamic oscillatory analysis of tamarind seed gum was evaluated at concentrations from 2% to 10% (w/v). For 2% (w/v) of tamarind seed gum solution, the storage modulus was lower than the loss modulus ($G' < G''$) throughout the frequency range, indicating a viscous-like behavior. As the concentration increased 4%, 8%, and 10% (w/v), the G' and G'' increased, and a crossover occurred at a frequency of 80, 25, and 7 rad/s, respectively. Thus, by increasing the concentration of tamarind seed gum from 2% to 10% (w/v), there was a change in the behavior of the gel from viscous to elastic, and at 10% a weak-gel was formed (Shao et al., 2019).

Crispín-Isidro et al. (2019) studied the flow behavior of tamarind gum solutions with different purity levels at 1.5% and 2.0% (w/w). The apparent viscosity was described by a plateau region (Newtonian behavior) at low shear rate values and a shear-thinning behavior (non-Newtonian) at higher shear rates, and these curves had a better fit to the Ellis model. The low shear viscosity values of the tamarind gum solutions tended to be higher when the concentration and the purification of the gum were higher (Crispín-Isidro et al., 2019).

The rheology of chemically modified tamarind gum has also been evaluated. Carboxymethylated tamarind gum (CMTG) increased the viscosity of the gum since CMTG disrupts the organization and exposes the polysaccharide network to hydration, which results in higher viscosity (Goyal et al., 2007). CMTG can be used as a matrix former and release retardant in the development of novel drug delivery systems (Mali et al., 2019).

In addition, tamarind seed gum has been thiolated by esterification of its hydroxyl groups with thioglycolic acid in order to improve its mucoadhesivity and cohesive properties. Kaur et al. (2012a, b) compared carbopol-based metronidazole gels using thiolated and nonthiolated tamarind seed gum gels, and found higher mucoadhesion to chicken ileum when the thiolated tamarind gum gel was used. The gels containing thiolated tamarind seed gum had the lowest hardness and adhesiveness but the highest cohesiveness.

Tamarind xyloglucans are neutral polysaccharides of low viscosity and low molecular weight compared to other polysaccharides. Tamarind xyloglucans have been used as a gelling agent since these can function as pectin; therefore, they represent an alternative of gelling raw material for the pharmaceutical and food industries (Mishra and Malhotra, 2009).

14.3.6 Functional properties of binary mixtures using tamarind gum

The functional properties of tamarind gum change when it is mixed with other compounds such as starches, gums, proteins, etc., with which it forms binary mixtures.

The addition of tamarind seed gum to starch causes an increase in its viscosity. Pongsawatmanit et al. (2006) evaluated the viscosity of 5% tapioca starch/tamarind xyloglucan mixtures, at different mixing ratios, and found that peak and final viscosities increased with increasing tamarind xyloglucan content. Furthermore, a mechanical system of 5% (w/w) tapioca starch/tamarind xyloglucan mixtures changed from gel behavior to concentrated solution and showed higher loss tangent (G''/G') with increasing tamarind xyloglucan concentration.

Mung bean starch gel increased its peak of viscosity (Liu and Xu, 2019) from 4372 to 10,285 cP adding 10% (w/w) of tamarind gum. The addition of tamarind gum at high concentration to starch generated firm and spring gels, and reduced the gel syneresis. Xie et al. (2020) mixed tamarind gum with three types of corn starches (normal, waxy, and high amylose corn starch), using different amylose-amylopectin ratios, and demonstrated that tamarind gum retards the gelatinization of starch granules mainly affecting amylopectin. Binary mixtures of tamarind gum with normal, and with waxy corn starch formed weak gels, while tamarind gum with high amylose corn starch increased their elastic properties.

On the other hand, tamarind gum has been mixed with other gums. Zhang et al. (2008) mixed tamarind gum (2%, w/w) and sodium alginate (0.6%, w/w), and formed beads using calcium chloride; they evaluated the swelling behavior of this mixture varying pH ($3-11$) and found that in the pH range 3–7 there was no destruction of hydrogel; however, increasing the pH above 7 produced the disintegration of alginate chains and an increase in the swelling degree. In later studies, Nayak and Pal (2011) applied tamarind gum–alginate composite beads for controlled delivery of diclofenac sodium for prolonged period with good results, and they found that swelling and degradation of the beads were affected by the change in pH.

In the optimization of spray-drying tamarind gum, soya protein isolate was added, which produced an increase in yield, a reduction of the gum hygroscopicity, and an increase in the solubility of the powder, since the protein is an efficient carrier agent (Muzaffar and Kumar, 2015). Also using proteins, Jana et al. (2016) used gelatin/carboxymethylated tamarind gum (CMTG) mixtures to control the delivery of aceclofenac and they found that gelatin alone had difficulty to retain a significant amount of the drug, while mixing gelatin with CMTG improved the drug entrapment efficiency above 90%.

Tamarind gum has also been mixed with other substances to modify its functional properties in certain applications; for example, Yadav et al. (2017) studied composite films using polyvinyl alcohol (PVA) and carboxymethyl tamarind gum, and found an increase in firmness of PVA films when CMTG was added, and this was dependent on the gum concentration. Increasing CMTG content in the film also increased the elastic component due to the improvement in the reinforcement effect. The mechanical characteristics of the films improved on adding CMTG, due to an increase of intermolecular hydrogen bonding, an enhanced crystallinity of polymers, and low molecular rearrangement under stress. The mechanical stability of films is important during handling and storage.

Likewise, Sharma et al. (2014) prepared ion gels using tamarind gum and synthetic and bio-based ionic liquids by a heating/cooling process and these gels exhibited thixotropic behavior, which involved the recovery of gel structures after 10 consecutive cycles. The ion gels had superior quality in viscosity, viscoelasticity, and thixotropic behavior compared with the hydrated tamarind gum (Rao and Mathew, 2012; Sharma and Bhardwaj, 1997).

14.4 Tamarind gum-based colloidal systems in food and pharmaceutical applications

As previously mentioned, tamarind gum is extracted from the kernel of tamarind seed, which is a residue from the tamarind pulp industry, mainly composed of a galactoxyloglucan polysaccharide. Tamarind gum is considered as a promising biopolymer due to its applications as a stabilizer, thickener, viscosity enhancer, emulsifier, gelling, drug carrier, release retardant, and binder agent in the food and pharmaceutical industries. When native tamarind gum is dispersed in aqueous media, it displays excellent ability to swell in the aqueous medium and form a mucilaginous solution, which exhibits rheological properties like a pseudoplastic fluid, besides possessing a hydrophilic character that allows it to display gel-forming, biomucoadhesive characteristics, and a stiffer conformation with the large volume of occupancy, making it suitable as a potential excipient in the preparation of hydrogels, oleogels, emulsions, and controlled released systems. Other distinguishing properties of tamarind gum are related to its high swelling index and high thermal stability, making it a suitable excipient for drug delivery systems.

In the food industry, tamarind gum applications include being used as a thickener agent in sauces, fruit pulp beverages, low-fat milk, and cocoa because of its smooth flow and lack of sticky texture, as gelling and water retaining in jellies and pudding by providing an elastic gel behavior to concentrated sugar solutions, and as a suppressor of the aging of starch by conferring heat stability and mechanical strength, improving the texture of starch in bakery products, custard cream, flour paste, stew, and noodles (Ferrero, 2017; Yamatoya et al., 2020). As tamarind gum exhibits good emulsion stability in acidic conditions, Kim et al. (2006) reported that tamarind gum provides benefits in dressing applications by adjusting the texture of dressings in combination with xanthan gum due to its thermoreversible physical gel-like properties. Tamarind gum was also used in frozen desserts, displaying an overrun and suppressing the ice crystal growth and sugar crystallization after storage, associated to its water-retention capabilities, indicating that tamarind gum holds more free water in freezing mixes around the ice crystals and prevents their growth; in addition, tamarind gum-sugar gel becomes harder and more elastic after freeze-thaw, suggesting that freeze-thaw processes make the gel-like network stronger (Yamatoya et al., 2020; Yamatoya and Shirakawa, 2003).

Therefore, in this section, the relationship between physicochemical and functional properties for colloidal applications of tamarind gum and its functionalized derivatives will be described.

14.4.1 Tamarind gum hydrogels

Hydrogels are defined as three-dimensional (3D) cross-linked polymer networks with high capability to uptake a large amount of water and even biological fluids resembling biological tissues (Ahmad et al., 2019; Ahmed, 2015). These 3D networks are connected to each other through cross-linking polymers (either physically or chemically) to render the network insoluble and immersed in an aqueous solution (De et al., 2002; Maharana et al., 2017). The spaces available within the formed 3D network allow the immobilization of aqueous or organic solvents. The cross-linking reaction can be achieved by using cross-linker compounds, chemical modification, grafting, or high energy radiation (gamma or UV rays), being more stable systems than those carried out by chemical processes (Ali and Ahmed, 2018). In addition, the biopolymer network may result in hydrophilic or hydrophobic materials, where hydrophilic components are related to the induction of swelling in the hydrogel structure, whereas hydrophobic components tend to control the swelling rate of the gel as well as the mechanical properties (Zhang et al., 2008).

Tamarind gum functionalization

Tamarind gum does not form gel-like structures when it is placed alone in an aqueous solution; instead, it dissolves in water, yielding high-viscous dispersions, where gelation is hindered by steric hindrance of (1→2)-β-galacto-xylose branches, remaining preferably in sol state. However, when tamarind gum is in the presence of sugars like sucrose, ethanol, polyphenolic compounds, and iodine, their interactions induce the gelled-phase formation (Yamatoya et al., 2020; Yuguchi et al., 2001). Gelation of tamarind gum solution mixed with sugar or alcohol is considered to involve cross-linking of tamarind gum molecule aggregation domains due to a dehydrating action by these additives. It is also reported that these tamarind gum gels can become sols by heating and revert to gel again by cooling, indicating that the transition between sol and gel is thermoreversible, exhibiting high elasticity and low water release, even when single tamarind gum does not form gelled systems by heat treatment (Yamatoya et al., 2020).

Functionalization of tamarind gum by different chemical modifications has been demonstrated to improve the physicochemical properties of native gum. The first modification is tamarind gum degalactosylation, which induces the formation of gelled structures by cross-linking occurred between the degalactosylated tamarind gum molecules, forming arrangements in lateral aggregates that lead to flat plate shapes, which increase as the loss of (1→2)-β-galacto-xylose occurs, resulting in an opaque thermoreversible gelation process (Shigenobu et al., 1999).

Carboxymethylation of tamarind gum (CMTG) consists in the attachment of pendant carboxylic acid groups (–COOH) to the native tamarind gum structure via Williamson' etherification, using monochloroacetic acid and sodium hydroxide reaction at high temperature (Khalil et al., 1990). This reaction leads to a nonspecific degradation via β elimination and/or peeling reaction initiated at decreasing sugar units in the native gum structure due to the highly alkaline pH environment; this derivatization process provokes the disruption in the organization of the macromolecule, which exposes the polysaccharide network to hydration, causing the decrease in the molecular weight of the derivatized gum but increasing the viscosity of the biopolymer aqueous dispersion (Manchanda et al., 2014; Nayak and Pal, 2018; Olusola et al., 2014). This type of modification is one of the most common due to its simplicity, lower costs, and great variety of applications for the resultant CMTG. This CMTG displays higher hydrophilicity and solubility in an aqueous medium than its respective native gum, due to the presence of carboxymethyl groups in the tamarind gum structure; it displays higher viscosity due to its higher swelling capability, as well as higher resistant toward enzymatic attack and therefore lower degradability in aqueous environments, making it a good excipient material in hydrophilic drug delivery systems (Manchanda et al., 2014; Nayak and Pal, 2018).

CMTG is an anionic polysaccharide able to interact with cationic moieties to form a gel structure. Alpizar-Reyes et al. (2017a), Kaur et al. (2012b), Madgulkar et al. (2016), and Premalatha et al. (2017) demonstrated that CMTG can interact with divalent cations like Ca^{2+} and form ionically gelled nanoparticles, such as those formed with alginate and calcium chloride, where the gelled particle sizes were directly dependent on the CMTG and Ca^{2+} cross-linker concentrations. Alpizar-Reyes et al. (2017a), Kaur et al. (2012b), Madgulkar et al. (2016), and Premalatha et al. (2017) showed that CMTG is able to interact with macromolecules like chitosan by the formation of polyelectrolyte complex linkage between free -NH_3^+ of chitosan and -COO^- of CMTG, respectively. In this work, functional properties of hydrogels, like the swelling index and drug carrier, were dependent on the polymer charge and concentration, degree of ionization, cross-linking density, hydrophilicity and hydrophobicity, as well as the pH of the aqueous medium, where higher entrapment efficiencies and better swelling equilibrium were achieved at higher CMTG contents, and closely related to higher viscosity systems. Similar results were observed by Jana et al. (2016), who synthesized gelatin-CMTG hydrogel composites using glutaraldehyde as a cross-linker agent. This gelatin-CMTG hydrogel achieved higher drug entrapment efficiency than that reached by the cross-linked gelatin alone, and the release profile was dependent upon the extended cross-linking and the amount of CMTG. In a similar biopolymer system, Shaw et al. (2015) reported that CMTG exhibited high compatibility to the gelatin phase, resulting in better mechanical and mucoadhesive properties of the hydrogels, which was attributed to the presence of the free carboxylic groups in CMTG, inducing the formation of pH-sensitive swelling and drug release hydrogels, demonstrating the potential applications of this CMTG. The same trend was observed by Meenakshi and Ahuja (2015) when CMTG-polyvinyl alcohol (PVA) cryogels were synthesized using the freeze-thaw method as a physical cross-linking strategy. In this work, lower CMTG concentrations and high PVA contents led to lower release of metronidazole as a core material, achieving an increase in metronidazole release as the CMTG content increased.

Another modification applied to tamarind gum is thio-functionalization (TTG), where tamarind gum polysaccharide is modified by its esterification with thioglycolic acid and confirmed by the S–H stretch in Fourier-transformed infra-red spectra (Manchanda et al., 2014), aiming the improvement of the mucoadhesion of natural polysaccharides. In this sense, Kaur et al. (2012a, b) reported that thio-functionalization in tamarind gum is achieved by the esterification of the hydroxyl groups of galactoxylan moieties with the carboxyl groups of the thioglycolic acid, resulting in a white and water-soluble powder with a higher degree of crystallinity. TTG hydrogel showed an increase of 6.85-fold greater mucoadhesive strength and better surface roughness than tamarind gum hydrogel and commercial formulations.

On the other hand, grafting of tamarind gum (GTG) was reported by Shailaja et al. (2012); this modification overcame some disadvantages, such as uncontrolled rate of hydration, drop viscosity on storage, and susceptibility to microbial degradation, that tamarind gum exhibits. GTG is formed by bonding synthetic monomers molecules onto the polymer chain, combining the functional properties of copolymers and polymers molecules; they are biodegradable to some extent and stable to shearing due to the attachment of flexible synthetic polymers onto the more rigid polysaccharide backbone. Generally, this type of modification is carried out by conventional methods like redox, or nonconventional ones like microwave irradiation, γ-ray irradiation, or electron beams, where the resultant grafted biopolymers exhibit excellent capabilities for controlled drugs release (Ghosh et al., 2010; Rani et al., 2012). The main molecules used for grafting of tamarind gum are polyacrylamide (Sen and Pat, 2009), methyl methacrylate (Shailaja et al., 2012), ethyl acrylate (Del Real et al., 2015), and acrylonitrile (Singh et al., 2009), where the functional properties of tamarind gum and its hydrogels, such as water retention capacities and gelling tendency, were modified after grafting, producing better natural products with less side effects and minimum loss of the initial properties of the substrate used.

Sulfonation in tamarind gum was assessed by swelling the tamarind gum polysaccharide in a dimethylformamide and sulfur-trioxide-pyridine complex; alkylamination, an oxidation of galactosyl hydroxyl methyl groups to formyl groups, by using galactose oxidase catalase; and its cross-linking with epichlorohydrin, which exhibited superior wicking and swelling behavior, as well as better retarding effect in the drug release than native tamarind gum (Kumar et al., 2018; Nayak and Pal, 2018).

Applications of hydrogels from tamarind gum, native or functionalized, are focused mainly on pharmaceutical approaches such as: the production of thickened ophthalmic solutions used as a vehicle for sustained release of drugs due to the mucoadhesive properties that extend the retention time onto the surface of the eye; acyclovir drug loaded in nanoparticles with greater in vivo bioavailability than commercial products where acyclovir is used in a suspension product; and the sustained release of hydrophilic drugs like acetaminophen, caffeine, theophylline, and salicylic acid, or nonpolar compounds like indomethacin, exhibiting a zero-order release rate with the capability to control their release by modifying the diluent or the type of binders (Kumar et al., 2018).

Tamarind gum has also been functionalized by the incorporation of carbon nanotubes (CNTs). Choudhary et al. (2018) incorporated CNTs, hydroxyl functionalized CNTs (OH-CNTs), and carboxyl functionalized CNTs (COOH-CNTs) in tamarind gum hydrogels. Their results showed that CNTs induced modifications in the microstructure by altering the intra- and intermolecular interactions in the hydrogels. These changes provoked modifications in the physicochemical properties of the tamarind gum hydrogels and also in the differential drug release patterns of tigecycline as a model drug. Even when all the tamarind gum-CNTs hydrogels were easily spreadable, differences in the mechanical properties, microarchitecture, topography, and electric impedance were observed, and were dependent on the type of CNTs used.

Hybrid tamarind gum composites

Biopolymer hydrogels are materials that display technological features like environment-responsive, self-healing, self-assembled conductive, and shape memory, and are considered as supramolecular materials (Mahinroosta et al., 2018). In this sense, crude and CMTG functionalized tamarind gums have been used in combination with other macromolecules like proteins and polysaccharides in order not only to improve the controlled and sustained release of drug, but also to release it in specific target sites.

Hybrid functionalized tamarind gum composites formed by a combined process of chemical cross-linking and freeze-drying were reported by Jana et al. (2016), where gelatin and CMTG hybrid composites were formed and tested for controlled delivery of aceclofenac, a nonsteroidal antiinflammatory drug widely used for symptomatic relief of rheumatoid arthritis, osteoarthritis, and ankylosing spondylitis. It was found that the structural arrangement obtained from the composites led to a suppression in the drug core release when the biopolymer particles were subjected to in vitro acidic media, extending the release at neutral pH values in phosphate buffer solution (pH 6.8). This behavior was attributed to the performance of polymer chain relaxation/swelling and diffusion mass transfer mechanisms, as well as the functional properties of CMTG for modulating the release of drugs like aceclofenac at specific triggering agents and aiding to the reduction of gastrointestinal side effects, costs, and patient suffering associated with the frequent dosing of this drug (Jana et al., 2016).

Synthesis of polyvinyl alcohol (PVA) and CMTG hybrid composites was reported (Yadav et al., 2017); these composites were applied for active films loading ciprofloxacin hydrochloride as a drug model. It is noteworthy that despite PVA being a water-soluble biopolymer used in drug delivery systems and wound dressing applications due to its good physicochemical properties, this polymer is lacking in terms of cell-specific bioactivities. Therefore, the addition of CMTG to PVA promoted the formation of films with better mechanical, thermal, and biological properties than the forming biopolymers, where the improvement in these characteristics was dependent on the concentration of CMTG and was associated with the presence of intermolecular hydrogen bonding, with low molecular rearrangement under stress conditions, and enhanced crystallinity. The evaluation of antibacterial activity showed that films based on PVA-CMTG composites loaded with ciprofloxacin hydrochloride did not exhibit significant differences when CMTG content varied in the formulation. However, the evaluation of human epidermal keratinocyte cell (HaCaT) proliferation, as an accepted cellular candidate for probing epidermal biology in vitro, showed that low contents of CMTG in the composite film supported to a greater extent the cell proliferation than films where higher CMTG contents were tested. The explanation for this behavior was related to the use of CMTG as a "chemical cue" where the cell proliferation was dependent on its concentration, joined to the combined effect of surface properties of the biopolymer film and the ligand distribution, showing the good cytocompatibility of the PVA-CMTG film.

Tamarind gum grafting copolymerization

Graft copolymerization in natural polysaccharides is an important resource for developing advanced materials with improved functional properties, allowing them to be used in agricultural materials and their by-products to be used as substitutes for unsustainable synthetic polymeric materials.

Graft copolymers are defined as a long sequence of one polymer (backbone polymer) with one or more branches (grafts) of another (chemically different) polymer. The process of graft copolymer synthesis in natural biopolymers starts with a preformed polymer, as the polysaccharide, then an external agent is used to create free radical sites on the polymer backbone; after this, the monomer is added up through the chain propagation step, leading to the formation of grafted chains. The various methods of graft copolymer synthesis mainly differ in the types of generation of the free radical sites on this preformed polymer (Ghosh et al., 2010; Rani et al., 2012).

An example of this type of tamarind gum (TG) grafting is reported by Singh et al. (2009), where acrylonitrile was grafted on to TG polysaccharide using persulfate/ascorbic acid redox initiator to synthesize the poly(acrylonitrile)-grafted-TG (PAN-g-TG). The main results indicated that grafted materials displayed different water/saline retention, gel forming ability, and enhanced shelf life of the grafted gum solutions with great potential for its commercial utilization in pharmaceutical and industrial approaches.

Polyacrylamide-grafted-tamarind gum (PAM-g-TG) synthesized by conventional redox grafting, microwave-initiated grafting, and microwave-assisted grafting showed that grafted copolymers where a maximum percentage of grafting was achieved exhibited enhanced intrinsic properties, associated to the longer chains of PAM grafted onto the backbone of TG, leading to the increase in the hydrodynamic volume of the biopolymer molecule in a solution and therefore an increase in the intrinsic viscosity and molecular weight (Ghosh et al., 2010).

Ghosh et al. (2010) reported the functional properties of PAM-g-TG in terms of its flocculation characteristics in a kaolin suspension system, where this performance was improved when graft TG copolymers were obtained.

Nandi et al. (2019) synthesized a similar PAM-g-TG system, by a free radical method assisted with microwave with ceric (IV) ammonium nitrate (CAN) as a free radical initiator, where the flocculating potential in a peroral paracetamol suspension was evaluated. It is noteworthy that pharmaceutical dosage forms include flocculated suspensions, which are characterized by the formation of loosely packed cake which can easily be redispersed by gentle shaking. The use of natural hydrophilic polysaccharides, like TG, has demonstrated their capability for acting as suspending agents, joined to their protective and coating roles, which may prevent or induce the cake formation. In this study, the PAM-g-TG exhibited improved thermal stability not only for the biopolymer material, but also for the loaded drug, and showed that flocculation efficiency in paracetamol suspension was enhanced in the grafting as the TG content increased. This behavior was associated to the PAM functionalization by the TG grafting and the incorporation of numerous side branches in the main polymeric backbone, resulting in a comb-like structure (Nandi et al., 2019).

Similar results were reported by Sen and Pat (2009), where PAM-g-CMTG was used as a flocculant agent in a kaolin suspension with better functional properties than those found when CMTG was used for the same purpose. Furthermore, Pal et al. (2012) stated that PAM grafted polysaccharides usually find applications as efficient flocculants agents in environmental or pharmaceutical approaches when are used at low doses, as well as their controlled biodegradable, shear-resistant, inexpensive, and ecofriendly characteristics. Nonetheless, these materials display some drawbacks and limitations such as their low surface area, small hydrodynamic radius, and complicated diffusion processes. Pal et al. (2012) overcame these drawbacks by developing a high-performance nanocomposite based on silica nanoparticle-incorporated PAM-g-CMTG. The functional properties that were evaluated indicated that these nanocomposites exhibited an enhanced adsorption of methylene blue dye and better properties as a flocculant agent in comparison with PAM-g-CMTG. This was attributed to the enhancement of hydrodynamic volume and hydrodynamic radius and their direct correlation to this type of functional property, where surface-active properties determine the efficacy of the materials.

Del Real et al. (2015) reported the graft copolymerization of ethyl acrylate (EA) onto TG (EA-g-TG). Their results showed that free radical polymerization mechanism allowed the grating reaction, which was confirmed by FTIR and NMR ^{1}H spectroscopies. Techno-functional properties of EA-g-TG showed higher thermal stability when compared with TG polysaccharide. Fresh grafted copolymer was only soluble in water, and became insoluble in water and organic solvents after drying. Mechanical properties were increased for tensile strain and also exhibited high biodegradability under anerosion conditions in the presence of the bacterium strain *Alicyliphilus* sp. *BQ1.*, making this copolymer adequate for usage in disposable products (Del Real et al., 2015).

A tamarind gum-chitosan (TG-chitosan) copolymer was proposed for use in producing magnetic microspheres. According to Zhang et al. (2007), the composite magnetic microspheres were synthesized utilizing the suspension cross-linking technique in the Fe_3O_4 magnetic carrier technology. The composite magnetic microspheres displayed particles sizes ranging from 230 to 460 μm, with sufficient magnetic field intensity to excite all the dipole moments of magnetic carrier. This TG-chitosan copolymer represents a promising magnetic support to be employed in magnetic carrier technology with good magnetic quality, and as expected, swelling properties changed as aresponse to the pH of the swelling medium, this porperty is useful in potential modulation systems in biomedical fields.

Moreover, synthesis of interpenetrating networks (IPNs), defined as a polymeric network of two or more polymers that form a rigid composite network structure by cross-linking of at least one polymer in the presence of another, has demonstrated improvements in the mechanical strength, loading capacity, and sustained and controlled drug release, and provides space for drug encapsulation in a three-dimensional structure by combination of individual properties of polymers. In this sense, Mali et al. (2017a, b) reported the synthesis of pH-dependent site-specific IPNs of aceclofenac using CMTG and chitosan. In this study, chitosan was used as a base polymer, cross-linked with glutaraldehyde to form a network, and CMTG was the second polymer entangled in cross-linked chains of chitosan. The intercalation of CMTG in the chitosan backbone was carried out due to formation of a polyelectrolyte complex between free $-NH_3^+$ of chitosan and $-COO^-$ of CMTG. The CMTG-chitosan IPNs exhibited enhanced aceclofenac entrapment efficiency as the cross-linker content increased, whereas CMTG concentration displayed a significant effect due to the formation of a thick surface that diminished the loss of the core material by means of the high viscosity of CMTG. Drug delivery in the IPNs showed their pH dependence on swelling properties, suggesting that these composites are suitable for oral site-specific delivery of drugs in order to avoid exposure of drugs to an erratic gastric environment and drug release in the intestine.

14.4.2 Tamarind gum in polymer complexation

The uses of tamarind gum have been explored as a microsphere drug carrier, as other natural polymers; this polysaccharide is cheap, biodegradable, and safe for pharmaceutical formulations. Microsphere systems of naturally derived polymers have been prepared by several techniques, including coacervation phase separation (Farooq et al., 2014).

In aqueous solutions, a great diversity of interactions occurs when proteins and polysaccharides interact. These interactions are defined by and depend strongly on environment conditions, such as pH, temperature, and hydrocolloids ratio (Espinosa-Andrews et al., 2013). Depending upon the composition of the formulations, drastic changes in the structural properties of the protein-polysaccharide may occur. When the aqueous solution of the proteins and the polysaccharides are mixed together, there is a possibility of formation of a liquid water in-water emulsion, complex coacervates (where both the polymers appear in a single concentrated phase), and soluble complexes due to the formation of self-organized structures, where the formation of any arrangement is governed by the thermodynamic compatibility among the proteins and the polysaccharides (Shaw et al., 2017). In this sense, complex coacervates can be classified as soluble or insoluble according to their electrostatic repulsive or attractive forces (González-Martínez et al., 2017; Kaushik et al., 2015).

Complex coacervation between tamarind gum and whey protein isolate (WPI) was first reported by González-Martínez et al. (2017), where the formation of the complex coacervation was attained at a pH value where both hydrocolloids reached their electrical equivalence and an insoluble complex was obtained that was completely neutral. This study states that a maximally electrostatic interaction between the TG polysaccharide and the WPI protein was achieved at a mass ratio of 0.3:1.0 respectively, at pH 3.68. The resultant coacervate phase that was separated was spray-dried and characterized, exhibiting higher crystallinity than the TG polysaccharide. In addition, the complex coacervate exhibited better thermal stability against denaturation of the biopolymers.

Shaw et al. (2015, 2017) reported the use of gelatin-tamarind gum (gelatin-TG) and gelatin-carboxymethyl tamarind gum (gelatin-CMTG) phase-separated hydrogels, formed due to inter- and intra-polymeric interactions, in order to develop vehicles for controlled release of drugs like ciprofloxacin (fluoroquinolone antibiotic), as well as a film supporter for proliferation of human keratinocytes in tissue engineering areas. This polymer complexation led to better mechanical properties, controlled release of drug under triggering stimulus like pH changes, and the improvement of cell proliferation. In addition, both complexed biopolymer hydrogels showed good mucoadhesive properties, due to the presence of the free carboxylic groups in TG and CMTG, and pH-sensitive swelling and drug release behavior (Shaw et al., 2015, 2017).

14.4.3 Tamarind gum in Bigel and Emulgel systems

Gel-based formulations have been explored and used as controlled delivery systems, and can be classified on hydrogels, which consist of a colloidal network of a hydrophilic polymer that traps water molecules (McKee et al., 2014), or oleogels, which may consist of either amphiphilic or hydrophobic crystals (sorbitan monostearate, sorbitan monopalmitate, stearic acid, and stearyl alcohol) that form a network that immobilizes oils (Patel et al., 2014). Formulations where hydrogels and oleogels are mixed to form a new type of gelled structure are regarded as bigels. Since hydrogels are polar and oleogels are apolar, bigels may be regarded as emulsions having both internal and external immobilized phases (Kodela et al., 2017). In this sense, the immobilization of the external phase prevents the motion of the internal phase, and hence the occurrence of coagulation of the internal phase is avoided; as the internal phase is also immobilized, the leaching of the internal phase is minimized (Satapathy et al., 2015). Moreover, bigels exhibit inherent thermodynamic stability compared to the emulsions, even though both bigels and emulsions are biphasic formulations. It is noteworthy that if the external phase of the bigels is externally cross-linked, it will result in the formation of a permanent bigel (Paul et al., 2018).

Paul et al. (2018) have described the suitability of hydrogel-in-oleogel and oleogel-in-hydrogel bigels, by using an oleogel prepared with stearic acid and rice bran oil, and a hydrogel based on tamarind gum with a hydroethanolic solution, for drug delivery systems.

Bigel systems exhibited significant differences in the structural arrangement, while oleogel showed the presence of a hyperbranched fibrous structure formed by the stearic acid and tamarind gum hydrogel that indicated the presence of numerous small droplet-like structures. In the bigels, both distinct phases were observed: a near-globular phase present in a continuous phase and a branched fibrous structure, attributed to the network formed by the stearic acid molecules and irregularly shaped black globular bodies assigned to the presence of hydrophilic tamarind gum gel. Clear differences between the microarchitecture conformation of the oleogel, hydrogel-in-oleogel type of bigel, an oleogel-in-hydrogel type of bigel, and a hydrogel were assessed. Among the physicochemical properties for bigel systems, a reduction in the electrical impedance was observed and was dependent on the hydrogel proportion used in the bigel. In terms of their use for carrying and controlled release of moxifloxacin as a model drug, the formulations exhibited diffusion-mediated drug release, and improved significantly in a composition-dependent manner as the tamarind gum hydrogel proportion increased, maintaining the antimicrobial activity in the gelled matrix (Paul et al., 2018).

On the other hand, emulgels or emulsion gels consist in semisolid multiphase systems with both emulsion and gel properties, in which the liquid phase is immobilized in a structured/gel phase (Farjami and Madadlou, 2019). Gelation of an emulsion can be achieved by two methods: gelation of the continuous phase and aggregation of the emulsion droplets, and it can be formed based on the use of proteins, carbohydrates, or a mixture of these macromolecules (Nasirpour-Tabrizi et al., 2020). In this regard, Rawooth et al. (2020) developed tamarind gum and rice bran oil (RBO)-based emulgels for carrying and delivering ciprofloxacin as a model drug. Their findings showed that these systems were of biphasic nature, with the presence of two types of globular structures associated to the aqueous phase, apparently trapped in a continuous hydrophobic matrix from xyloglucan moieties and those corresponding to the RBO droplets. Moreover, the RBO content in emulgel display an effect on the reduction of hydrogen bonding between the components; despite the in vitro diffusion of ciprofloxacin being decreased with the RBO increase, the corneal permeation was improved with the increase in the RBO content. Therefore, emulgel systems had excellent potential for obtaining controlled and sustained delivery systems for therapies in ocular drug delivery.

14.5 Tamarind gum in industrial applications

14.5.1 Food applications

Tamarind gum has interesting physicochemical properties that give it the possibility of being applied as a food additive. Some recent publications on food applications of tamarind gum are as follows. In baking, tamarind gum was used to optimize the batter characteristics for good performance during leavening and to obtain a suitable final texture of gluten-free rice bread. Tamarind gum was applied at 1% and 2% of concentration, and the optimized formula was obtained using 1% of tamarind gum, 100 g water, 5 min of mixing time, and 60 min of fermentation time (Hong and Kweon, 2020). In another study, the addition of 0.2%, 0.4%, and 0.8% of tamarind gum to gluten-free cakes was probed and the sample with 0.4%

tamarind gum had the better results since increased the dietary fiber, improved the sensory characteristics, and extended the shelf life of the cakes compared to cakes made with 100% rice flour and wheat flour (Wu et al., 2020).

In food packaging, tamarind gum has been used in the production of edible films. The xyloglucan extracted from tamarind seeds was used for elaborating films for packaging cut-up "Sunrise Solo" papaya, which showed good physical characteristics. The film with better results was prepared using 4.5% of xyloglucans and 1.5% of glycerol. To conclude this, the study involved the analysis of the moisture, tensile strength, elongation at break, and mass loss when it was applied (Santos et al., 2019). In another study, tamarind xyloglucan films were added to sesame seed oil (0–20 wt% based on xyloglucans) for preparing emulsion films by different methods. The emulsion films were analyzed by droplet size, permeability, and tensile properties; in addition, these could *have* antimicrobial and/or antioxidant characteristics that may be useful when applied in foods (Rodrigues et al., 2018).

On the other hand, tamarind gum was used as a wall material in microencapsulation of sesame seed oil using ratios of 1:1 and 1:2, respectively. Both types of microcapsules were compared, and the 1:1 tamarind gum-sesame seed oil ratio had a smaller droplet size, higher thermal stability, higher encapsulation efficiency (91%), and higher oil oxidation stability after 6 weeks than the 1:2 ratio microcapsules. Thus, tamarind gum can be used as a wall material for protection of edible oils against oxidation, increasing their shelf life (Alpizar-Reyes et al., 2020).

Tamarind gum was mixed with starches (potato, rice, mung bean, and lotus root), as binary mixtures, in order to improve their physicochemical, textural, and rheological characteristics that allow them to withstand freezing thawing and gelatinization processes, and to delay their retrogradation. In this work, tamarind gum significantly increased the viscosity of mung bean starch gels and improved the texture properties and mouthfeel of lotus root starch gels, which is useful in baked foods (Liu and Xu, 2019). On the other hand, tapioca starch was mixed with tamarind xyloglucan in different ratios (10:0, 9:1, 8:2, 7:3, and 6:4, respectively) using a total concentration of 5% of the mixtures. The addition of tamarind xyloglucan to tapioca starch increased the viscosity and improved the thermal stability during the freeze-thaw process of gelatinized mixtures regarding tapioca starch (Pongsawatmanit et al., 2006).

14.5.2 Pharmaceutical applications

Tamarind gum has many applications in drug formulation, and an important issue in this field is the drug release. Some recent publications of pharmaceutical applications of tamarind gum are as follows.

Tamarind gum (crude and modified) was used as a binder in the tablets formulation of diclofenac sodium by freeze drying, with the aim of reducing oral dissolution times and of improving the drug release. Chemically modified tamarind gum tablets had better results since they enhanced the dissolution rate of the drug and achieved the complete release of the drug (Huanbutta et al., 2019). In another study, CMTG and crude tamarind gum were used to formulate Thai cordial tablets. Tamarind gum was carboxymethylated at different substitution degrees. The tablets were evaluated by their swelling and erosion behavior and were fractured for knowing their breakdown times. CMTG tablets had higher hardness and faster disintegration than crude tamarind gum tablets (Huanbutta and Sittikijyothin, 2017).

Tamarind gum has been used alone or mixed with other biopolymers, to allow controlled oral release of drugs, and some examples are shown below. Tamarind gum was blended with gellan gum through Ca^{2+}-ion cross-linked ionically gelation technique for forming beads of metformin HCl for oral drug delivery. In this work, an optimization was performed where the factors were gellan gum to tamarind gum ratio and $CaCl_2$ concentration, and the response variables were the drug encapsulation efficiency and the cumulative drug release after 10 h. The optimized beads showed good mucoadhesivity and hypoglycemic activity in alloxan-induced diabetic rats due to proper metformin release (Nayak et al., 2014). In a similar optimization study, tamarind gum was blended with alginate by ionotropic gelation with $CaCl_2$ to form composite beads of diclofenac sodium. FTIR and NMR analyses were carried out in order to evaluate biopolymer-drug compatibility. The swelling, degradation, and drug release of the composite beads were influenced by pH changes. Finally, these composite beads allowed the controlled release of the drug for a prolonged time (Nayak and Pal, 2011).

In dental applications, tamarind gum (1%, w/w) was used to formulate oral disintegration tablets of tea powders for oral care. The tablets were analyzed regarding the swelling degree, hardness, friability, disintegration time, adhesiveness, and antimicrobial activity. Tamarind gum tablets of tea powders had a low disintegration time, high mucoadhesivity, and antimicrobial activity against *S. mutans* (Kiniwa et al., 2019). Likewise, tamarind seed gum was used to formulate a thermoreversible gel with lidocaine hydrochloride in order to be applied as local anesthesia into periodontal pocket. The mucoadhesive property of this gel allowed the retention in the site of application and the immediate action of the drug with a release time up to 2 h, which allows a dental procedure to be carried out without pain. This gel is natural, of low cost, and biodegradable, and is an alternative to injected anesthesia (Pandit et al., 2016). In addition, tamarind gum was used to prepare buccal patches of metronidazole. The patches formulation was optimized using as factors: tamarind gum, epichlorohydrin (cross-linker), and propylene glycol (plasticizer), and as response variables: ex vivo drug permeation, mucoadhesiveness strength, folding endurance, and buccal residence time. It was determined that tamarind gum can be used in buccal patches due to its mucoadhesiveness and its drug release mechanism by controlled dissolution (Jana et al., 2010).

The drugs have other routes of application than oral—for example, tamarind gum tablets were used for colon delivery of propranolol HCl to treat blood pressure. These tablets were analyzed by in vitro release studies at the following conditions: 0.1 N of HCl for 1.5 h, then pH 6.8 phosphate buffer for 2 h and pH 7.4 phosphate buffer until complete drug release, and the release profiles were fitted to different pharmacokinetic mathematical models. The prolonged drug release time of tamarind gum tablets, compared to other biopolymers used in this study, had good compression characteristics; tamarind gum is also a cost-effective material (Newton et al., 2015). On the other hand, tamarind gum was used in the formulation of FITC-dextrans microparticles by spray-drying to be applied as a drug in the nasal cavity, which favored its transport to the brain. The microparticles were analyzed by size, morphology, and mucoadhesiveness using laser diffraction, scanning electron microscopy, and a texture analysis, respectively. Through this work, it was found that 10 μm-sized tamarind gum microparticles achieved better deposition in the nasal cavity than smaller particles (Yarragudi et al., 2017).

In ocular applications, tamarind gum and rice bran oil (0%, 5%, 10%, 15%, and 20%) were used to formulate emulgels for ocular delivery of the antibiotic ciprofloxacin HCl. The effect of different rice bran oil concentrations was evaluated in the application of these emulgels. In vitro release study of the drug showed lower release when the emulgels contained rice bran oil; in contrast, ex vivo corneal permeation showed higher drug release as rice bran oil concentration increased in the emulgel (Rawooth et al., 2020). Tamarind xyloglucan nanoaggregates were loaded with tropicamide for ophthalmic delivery. In this study, an optimal formula was found using 0.45% (w/v) of tamarind xyloglucan and 0.55% (w/v) of poloxamer-407 which had higher corneal permeation of the drug than a commercial aqueous formulation. These nanoaggregates had high mucoadhesiveness due to the tamarind gum, and did not irritate the eye (Dilbaghi et al., 2013). Likewise, carboxymethylated tamarind gum nanoparticles were loaded with tropicamide via ionotropic gelation for ocular delivery. An optimization study revealed that carboxymethylated tamarind gum and $CaCl_2$ concentrations had a synergistic effect on particle size and encapsulation efficiency; in addition, these nanoparticles showed ex vivo corneal permeation due to their mucoadhesiveness (Kaur et al., 2012a, b). In another work, tamarind gum and hyaluronic acid were mixed in different ratios, for use as potential excipients of eye drops and to detect a possible synergistic effect with respect to the polymers separately. The mixtures were analyzed using NMR to evaluate the interpolymeric interactions. Tamarind gum-hyaluronic acid (3:2) mixtures formed stable supramolecular aggregates that retained water, had high mucoadhesiveness and low viscosity, stabilized the tear film, and achieved the maximum residence time of the drug in the precorneal area of rabbit eyes (Uccello-Barretta et al., 2010).

Another route of drug application is topical. Tamarind gum was used to produce nanofiber patches using polyvinyl alcohol and fabricated by electrohydrodynamic atomization, and then the patches were loaded with clindamycin (1%–3%) to be applied as wound-dressing materials. These nanofiber patches had skin adherence, were translucent, and had ventilation properties to be applied topically. The diameter of nanofibers was affected by the voltage applied during their formation. The nanofiber patches were analyzed by scanning electronic microscopy, differential scanning calorimetry, and X-ray diffraction, and the antimicrobial activity of clindamycin patches was probed on *S. aureus* (Sangnim et al., 2018). In a similar study, carboxymethylated tamarind gum was added to polyvinyl alcohol to synthesize composite films of ciprofloxacin with mechanical, thermal, and antibiotic properties for skin diseases. Through FTIR spectroscopy, the presence of hydrogen bonding was revealed between the components of the films. These ciprofloxacin films had antimicrobial activity against *E. coli*; on the other hand, these films achieved cell proliferation using human keratinocytes so these can be applied in skin tissue engineering (Yadav et al., 2017). Tamarind gum was also mixed with glycerin and propylene glycol in order to prepare clindamycin transdermal patches which allowed a controlled release of the drug. The patches were evaluated by tensile strength, drug release, and antimicrobial activity against *S. aureus*. The incorporation of different glycerin-propylene glycol ratios to tamarind gum patches affected the properties of the transdermal patches, the 4:6 ratio being the one that had the best results in terms of drug release and antimicrobial activity (Sureewan et al., 2014).

Drug hydrogels are also applied in a topical way. Tamarind gum was mixed with gelatin gum to form a hydrogel that was added with three different types of carbon nanotubes and filled with salicylic acid. Through field emission SEM, these hydrogels were observed as agglomerates where the carbon nanotubes were confined within a dispersed phase of tamarind gum. The carbon nanotubes had interactions with the hydrogel which had large crystallite size, and a mechanical study revealed that these hydrogels had better resistance to the breakdown than the control. The drug was released by the diffusion method and these hydrogels were cytocompatible with human keratinocytes; thus, they can be applied in wound healing and tissue engineering (Maharana et al., 2017).

Carboxymethylated tamarind gum was chemically cross-linked with gelatin to obtain a biocomposite hydrogel that efficiently retained aceclofenac, a drug with an antiinflammatory effect. The drug was compatible with the interpenetrating network of the biocomposite hydrogel. The rate of drug release mainly depended on the degree of cross-linking, the tamarind gum concentration, and the pH of the medium. Finally, the biocomposite hydrogel loaded with aceclofenac had a prolonged antiinflammatory effect on rats (Jana et al., 2016). In a similar study, CTG was cross-linked with citric acid to form ester cross-links between them in order to prepare hydrogel films loaded with a model drug (moxifloxacin hydrochloride). The formation of ester cross-links was analyzed using ATR-FTIR, solid-state ^{13}C NMR study, and differential scanning calorimetry. These hydrogel films achieved high drug retention, had a controlled release of drug, and can be applied topically (Kaur et al., 2010; Mali et al., 2017b).

Tamarind gum was used to prepare ion gels using both synthetic ionic liquids (1-butyl-3-methylimidazolium chloride and 1-butyl-3-methylimidazolium bromide) and bio-based ionic liquids (choline acrylate, choline caproate, and choline caprylate) by heating/cooling processes. These ion gels had viscoelastic behavior with thixotropic nature, and managed to adhere to human finger muscles and skin; thus, ion gels can have applications as sensors and actuators, and even gels with bio-based ionic liquids could have biomedical applications (Sharma et al., 2014).

On the other hand, tamarind xyloglucan has been used to enhance skin regeneration. Xyloglucans were extracted using cold water and a copper complex precipitation, and these were applied to human skin keratinocytes and fibroblast in vitro. After being analyzed, it was determined that tamarind xyloglucans promote the reepithelization and remodeling of the skin through cell proliferation and migration (Nie and Deters, 2013).

14.5.3 Other fields of tamarind gum applications

Tamarind gum, crude and CMTG, was used as a wall material in citronella oil microencapsulation in order to evaluate the oil release rate. Three different formulations of microcapsules were prepared varying the gum-oil ratio (1.25, 1.14, and 0.87). The microcapsules were characterized by SEM, encapsulation efficiency, and the oil release rate was evaluated. CMTG microcapsules allowed slower oil release than those of crude tamarind gum. Citronella oil is used in perfumery and is a natural repellent of insects, so tamarind gum microcapsules guarantee the release of the active compound for its proper functioning (Khounvilay et al., 2019).

In agricultural matter, CMTG was mixed with sodium-acrylate in order to create superabsorbent hydrogels for conditioning soils. The hydrogels' characterization was performed by FTIR spectroscopy, thermal analysis, SEM, and swelling studies. The soils mixed with these superabsorbent hydrogels (0.1%–0.3%) augmented the moisture absorption up to 35%, the porosity up to 7%, and the water retention capacity by planting chickpea seeds, compared to untreated soil. The superabsorbent hydrogels work by conditioning soils because they are excellent water retainers and nutrient carriers; they are also degradable (Khushbu and Kumar, 2019).

In environmental matter, tamarind gum was used to create an amphiphilic graft copolymer with methyl methacrylate by the method of atom transfer radical polymerization (ATRP) using the mixture CuBr/bpy as a catalyst, with the aim to remove toxic dyes. Copolymer characterization was performed by FTIR spectroscopy, ^{1}H NMR spectral analysis, gel permeation chromatography, TGA, DLS, field emission-SEM, and energy dispersive X-ray spectroscopy. Furthermore, adsorption and desorption studies were carried out. The copolymer had excellent sorption capacity of methylene blue and Congo red dyes, and the selective adsorption of the dyes was highly influenced by pH variations which had an effect on the electrostatic and H-bonding interactions between copolymer and dyes (Pal et al., 2012).

References

Ahmad, S., Ahmad, M., Manzoor, K., Purwar, R., Ikram, S., 2019. A review on latest innovations in natural gums based hydrogels: preparations & applications. Int. J. Biol. Macromol. 136, 870–890. https://doi.org/10.1016/j.ijbiomac.2019.06.113.

Ahmed, E.M., 2015. Hydrogel: preparation, characterization, and applications: A review. J. Adv. Res. 6 (2), 105–121. https://doi.org/10.1016/j.jare.2013.07.006.

Ali, A., Ahmed, S., 2018. Recent advances in edible polymer based hydrogels as a sustainable alternative to conventional polymers. J. Agric. Food Chem. 66 (27), 6940–6967. https://doi.org/10.1021/acs.jafc.8b01052.

Alpizar-Reyes, E., Carrillo-Navas, H., Gallardo-Rivera, R., Varela-Guerrero, V., Alvarez-Ramirez, J., Pérez-Alonso, C., 2017a. Functional properties and physicochemical characteristics of tamarind (Tamarindus indica L.) seed mucilage powder as a novel hydrocolloid. J. Food Eng. 209, 68–75. https://doi.org/10.1016/j.jfoodeng.2017.04.021.

Alpizar-Reyes, E., Carrillo-Navas, H., Romero-Romero, R., Varela-Guerrero, V., Alvarez-Ramírez, J., Pérez-Alonso, C., 2017b. Thermodynamic sorption properties and glass transition temperature of tamarind seed mucilage (Tamarindus indica L.). Food Bioprod. Process. 101, 166–176. https://doi.org/10.1016/j.fbp.2016.11.006.

Alpizar-Reyes, E., Román-Guerrero, A., Gallardo-Rivera, R., Varela-Guerrero, V., Cruz-Olivares, J., Pérez-Alonso, C., 2018. Rheological properties of tamarind (Tamarindus indica L.) seed mucilage obtained by spray-drying as a novel source of hydrocolloid. Int. J. Biol. Macromol. 107, 817–824. https://doi.org/10.1016/j.ijbiomac.2017.09.048.

Alpizar-Reyes, E., Varela-Guerrero, V., Cruz-Olivares, J., Carrillo-Navas, H., Alvarez-Ramirez, J., Pérez-Alonso, C., 2020. Microencapsulation of sesame seed oil by tamarind seed mucilage. Int. J. Biol. Macromol. 145, 207–215. https://doi.org/10.1016/j.ijbiomac.2019.12.162.

Bansal, J., Kumar, N., Malviya, R., Sharma, P.K., 2013. Extraction and evaluation of tamarind seed polysaccharide as pharmaceutical in situ gel forming system. Am. Eurasian J. Sci. Res. 9 (1), 1–5.

Bergstrm, E.M., Salmén, L., Kochumalayil, J., Berglund, L., 2012. Plasticized xyloglucan for improved toughness-thermal and mechanical behaviour. Carbohydr. Polym. 87 (4), 2532–2537. https://doi.org/10.1016/j.carbpol.2011.11.024.

Bhattacharya, S., Bal, S., Mukherjee, R.K., Bhattacharya, S., 1993. Some physical and engineering properties of tamarind (Tamarindus indica) seed. J. Food Eng. 18 (1), 77–89. https://doi.org/10.1016/0260-8774(93)90076-V.

Bhattacharya, S., Bal, S., Mukherjee, R.K., Bhattacharya, S., 1994. Functional and nutritional properties of tamarind (Tamarindus indica) kernel protein. Food Chem. 49 (1), 1–9. https://doi.org/10.1016/0308-8146(94)90224-0.

Chandra Mohan, C., Harini, K., Vajiha Aafrin, B., Lalitha Priya, U., Maria Jenita, P., Babuskin, S., Karthikeyan, S., Sudarshan, K., Renuka, V., Sukumar, M., 2018. Extraction and characterization of polysaccharides from tamarind seeds, rice mill residue, okra waste and sugarcane bagasse for its bio-thermoplastic properties. Carbohydr. Polym. 186, 394–401. https://doi.org/10.1016/j.carbpol.2018.01.057.

Chawananorasest, K., Saengtongdee, P., Kaemchantuek, P., 2016. Extraction and characterization of tamarind (tamarind indica L.) seed polysaccharides (TSP) from three difference sources. Molecules 21 (6). https://doi.org/10.3390/molecules21060775.

Choudhary, B., Paul, S.R., Nayak, S.K., Singh, V.K., Anis, A., Pal, K., 2018. Understanding the effect of functionalized carbon nanotubes on the properties of tamarind gum hydrogels. Polym. Bull. 75 (11), 4929–4945. https://doi.org/10.1007/s00289-018-2300-7.

Crispín-Isidro, G., Hernández-Rodríguez, L., Ramírez-Santiago, C., Sandoval-Castilla, O., Lobato-Calleros, C., Vernon-Carter, E.J., 2019. Influence of purification on physicochemical and emulsifying properties of tamarind (Tamarindus indica L.) seed gum. Food Hydrocoll. 93, 402–412. https://doi.org/10.1016/j.foodhyd.2019.02.046.
De, S.K., Aluru, N.R., Johnson, B., Crone, W.C., Beebe, D.J., Moore, J., 2002. Equilibrium swelling and kinetics of pH-responsive hydrogels: models, experiments, and simulations. J. Microelectromech. Syst., 544–555. https://doi.org/10.1109/JMEMS.2002.803281.
Del Real, A., Wallander, D., Maciel, A., Cedillo, G., Loza, H., 2015. Graft copolymerization of ethyl acrylate onto tamarind kernel powder, and evaluation of its biodegradability. Carbohydr. Polym. 117, 11–18. https://doi.org/10.1016/j.carbpol.2014.09.044.
Dickinson, E., 1994. Protein-stabilized emulsions. J. Food Eng. 22 (1–4), 59–74. https://doi.org/10.1016/0260-8774(94)90025-6.
Dilbaghi, N., Kaur, H., Ahuja, M., Kumar, S., 2013. Evaluation of tropicamide-loaded tamarind seed xyloglucan nanoaggregates for ophthalmic delivery. Carbohydr. Polym. 94 (1), 286–291. https://doi.org/10.1016/j.carbpol.2013.01.054.
Espinosa-Andrews, H., Enríquez-Ramírez, K.E., García-Márquez, E., Ramírez-Santiago, C., Lobato-Calleros, C., Vernon-Carter, J., 2013. Interrelationship between the zeta potential and viscoelastic properties in coacervates complexes. Carbohydr. Polym. 95 (1), 161–166. https://doi.org/10.1016/j.carbpol.2013.02.053.
Farjami, T., Madadlou, A., 2019. An overview on preparation of emulsion-filled gels and emulsion particulate gels. Trends Food Sci. Technol. 86, 85–94. https://doi.org/10.1016/j.tifs.2019.02.043.
Farooq, U., Malviya, R., Sharma, P.K., 2014. Advancement in microsphere preparation using natural polymers and recent patents. Recent Pat. Drug Deliv. Formul. 8 (2), 111–125. https://doi.org/10.2174/1872211308666140218110520.
Fathi, M., Mohebbi, M., Koocheki, A., 2016. Introducing Prunus cerasus gum exudates: chemical structure, molecular weight, and rheological properties. Food Hydrocoll. 61, 946–955. https://doi.org/10.1016/j.foodhyd.2016.07.004.
Ferrero, C., 2017. Hydrocolloids in wheat breadmaking: a concise review. Food Hydrocoll. 68, 15–22. https://doi.org/10.1016/j.foodhyd.2016.11.044.
Fry, S.C., 1989. The structure and functions of xyloglucan. J. Exp. Bot. 40 (1), 1–11. https://doi.org/10.1093/jxb/40.1.1.
Ghose, T.P., Krishna, S., 1942. Tamarind seed, a valuable source of commercial pectin. J. Indian Chem. Soc., Ind. Educ. 5, 114–120.
Ghosh, S., Sen, G., Jha, U., Pal, S., 2010. Novel biodegradable polymeric flocculant based on polyacrylamide-grafted tamarind kernel polysaccharide. Bioresour. Technol. 101 (24), 9638–9644. https://doi.org/10.1016/j.biortech.2010.07.058.
Gidley, M.J., Lillford, P.J., Rowlands, D.W., Lang, P., Dentini, M., Crescenzi, V., Edwards, M., Fanutti, C., Grant Reid, J.S., 1991. Structure and solution properties of tamarind-seed polysaccharide. Carbohydr. Res. 214 (2), 299–314. https://doi.org/10.1016/0008-6215(91)80037-N.
González-Martínez, D.A., Carrillo-Navas, H., Barrera-Díaz, C.E., Martínez-Vargas, S.L., Alvarez-Ramírez, J., Pérez-Alonso, C., 2017. Characterization of a novel complex coacervate based on whey protein isolate-tamarind seed mucilage. Food Hydrocoll. 72, 115–126. https://doi.org/10.1016/j.foodhyd.2017.05.037.
Goyal, P., Kumar, V., Sharma, P., 2007. Carboxymethylation of tamarind kernel powder. Carbohydr. Polym. 69 (2), 251–255. https://doi.org/10.1016/j.carbpol.2006.10.001.
Gupta, V., Puri, R., Gupta, S., Jain, S., Rao, G.K., 2010. Tamarind kernel gum: an upcoming natural polysaccharide. System. Rev. Pharm. 1 (1), 50–54. https://doi.org/10.4103/0975-8453.59512.
Hamdani, A.M., Wani, I.A., Bhat, N.A., 2019. Sources, structure, properties and health benefits of plant gums: a review. Int. J. Biol. Macromol. 135, 46–61. https://doi.org/10.1016/j.ijbiomac.2019.05.103.
Hong, Y.E., Kweon, M., 2020. Optimization of the formula and processing factors for gluten-free rice bread with tamarind gum. Foods 9 (2). https://doi.org/10.3390/foods9020145.
Huanbutta, K., Sittikijyothin, W., 2017. Development and characterization of seed gums from Tamarindus indica and Cassia fistula as disintegrating agent for fast disintegrating Thai cordial tablet. Asian J. Pharm. Sci 12 (4), 370–377. https://doi.org/10.1016/j.ajps.2017.02.004.
Huanbutta, K., Yunsir, A., Sriamornsak, P., Sangnim, T., 2019. Development and in vitro/in vivo evaluation of tamarind seed gum-based oral disintegrating tablets after fabrication by freeze drying. J. Drug Deliv. Sci. Technol. 54. https://doi.org/10.1016/j.jddst.2019.101298.

Jana, S., Banerjee, A., Sen, K.K., Maiti, S., 2016. Gelatin-carboxymethyl tamarind gum biocomposites: in vitro characterization & anti-inflammatory pharmacodynamics. Mater. Sci. Eng. C 69, 478–485. https://doi.org/10.1016/j.msec.2016.07.008.

Jana, S., Lakshman, D., Sen, K.K., Basu, S.K., 2010. Development and evaluation of epichlorohydrin cross-linked mucoadhesive patches of tamarind seed polysaccharide for buccal application. Int. J. Pharm. Sci. Drug Res. 2, 193–198.

Kaur, G., Jain, S., Tiwary, A.K., 2010. Chitosan-carboxymethyl tamarind kernel powder interpolymer complexation: investigations for colon drug delivery. Sci. Pharm. 78 (1), 57–78. https://doi.org/10.3797/scipharm.0908-10.

Kaur, H., Ahuja, M., Kumar, S., Dilbaghi, N., 2012a. Carboxymethyl tamarind kernel polysaccharide nanoparticles for ophthalmic drug delivery. Int. J. Biol. Macromol. 50 (3), 833–839. https://doi.org/10.1016/j.ijbiomac.2011.11.017.

Kaur, H., Yadav, S., Ahuja, M., Dilbaghi, N., 2012b. Synthesis, characterization and evaluation of thiolated tamarind seed polysaccharide as a mucoadhesive polymer. Carbohydr. Polym. 90 (4), 1543–1549. https://doi.org/10.1016/j.carbpol.2012.07.028.

Kaushik, P., Dowling, K., Barrow, C.J., Adhikari, B., 2015. Complex coacervation between flaxseed protein isolate and flaxseed gum. Food Res. Int. 72, 91–97. https://doi.org/10.1016/j.foodres.2015.03.046.

Khalil, M.I., Hashem, A., Hebeish, A., 1990. Carboxymethylation of maize starch. Starch Stärke, 60–63. https://doi.org/10.1002/star.19900420209.

Khounvilay, K., Estevinho, B.N., Sittikijyothin, W., 2019. Citronella oil microencapsulated in carboxymethylated tamarind gum and its controlled release. Eng. J. 23 (5), 217–227. https://doi.org/10.4186/ej.2019.23.5.217.

Khounvilay, K., Sittikijyothin, W., 2012. Rheological behaviour of tamarind seed gum in aqueous solutions. Food Hydrocoll. 26 (2), 334–338. https://doi.org/10.1016/j.foodhyd.2011.03.019.

Khushbu, W.S.G., Kumar, A., 2019. Synthesis and assessment of carboxymethyl tamarind kernel gum based novel superabsorbent hydrogels for agricultural applications. Polymer 182. https://doi.org/10.1016/j.polymer.2019.121823.

Kim, B.S., Takemasa, M., Nishinari, K., 2006. Synergistic interaction of xyloglucan and xanthan investigated by rheology, differential scanning calorimetry, and NMR. Biomacromolecules 7 (4), 1223–1230. https://doi.org/10.1021/bm050734+.

Kiniwa, R., Miyake, M., Kimura, S.I., Itai, S., Kondo, H., Iwao, Y., 2019. Development of muco-adhesive orally disintegrating tablets containing tamarind gum-coated tea powders for oral care. Int. J. Pharm.: X 1. https://doi.org/10.1016/j.ijpx.2019.100012.

Kodela, S.P., Pandey, P.M., Nayak, S.K., Uvanesh, K., Anis, A., Pal, K., 2017. Novel agar–stearyl alcohol oleogel-based bigels as structured delivery vehicles. Int. J. Polym. Mater. Polym. Biomater. 66 (13), 669–678. https://doi.org/10.1080/00914037.2016.1252362.

Kulkarni, A.D., Joshi, A.A., Patil, C.L., Amale, P.D., Patel, H.M., Surana, S.J., Belgamwar, V.S., Chaudhari, K.S., Pardeshi, C.V., 2017. Xyloglucan: a functional biomacromolecule for drug delivery applications. Int. J. Biol. Macromol. 104, 799–812. https://doi.org/10.1016/j.ijbiomac.2017.06.088.

Kumar, C.S., Bhattacharya, S., 2008. Tamarind seed: properties, processing and utilization. Crit. Rev. Food Sci. Nutr. 48 (1), 1–20. https://doi.org/10.1080/10408390600948600.

Kumar, S.A., Singh, B.R., Manish, K., Vikas, F., Prakash, J.C., 2018. Applications of tamarind seeds polysaccharide-based copolymers in controlled drug delivery: an overview. Asian J. Pharm. Pharm., 23–30. https://doi.org/10.31024/ajpp.2018.4.1.5.

Limsangouan, N., Milasing, N., Thongngam, M., Khuwijitjaru, P., Jittanit, W., 2019. Physical and chemical properties, antioxidant capacity, and total phenolic content of xyloglucan component in tamarind (Tamarindus indica) seed extracted using subcritical water. J. Food Process. Preserv. 43 (10). https://doi.org/10.1111/jfpp.14146.

Liu, J., Xu, B., 2019. A comparative study on texture, gelatinisation, retrogradation and potential food application of binary gels made from selected starches and edible gums. Food Chem. 296, 100–108. https://doi.org/10.1016/j.foodchem.2019.05.193.

Madgulkar, A.R., Bhalekar, M.R., Asgaonkar, K.D., Dikpati, A.A., 2016. Synthesis and characterization of a novel mucoadhesive derivative of xyloglucan. Carbohydr. Polym. 135, 356–362. https://doi.org/10.1016/j.carbpol.2015.08.045.

Maharana, V., Gaur, D., Nayak, S.K., Singh, V.K., Chakraborty, S., Banerjee, I., Ray, S.S., Anis, A., Pal, K., 2017. Reinforcing the inner phase of the filled hydrogels with CNTs alters drug release properties and human keratinocyte morphology: A study on the gelatin- tamarind gum filled hydrogels. J. Mech. Behav. Biomed. Mater. 75, 538–548. https://doi.org/10.1016/j.jmbbm.2017.08.026.

Mahinroosta, M., Jomeh Farsangi, Z., Allahverdi, A., Shakoori, Z., 2018. Hydrogels as intelligent materials: A brief review of synthesis, properties and applications. Mater. Today Chem. 8, 42–55. https://doi.org/10.1016/j.mtchem.2018.02.004.

Mali, K.K., Dhawale, S.C., Dias, R.J., 2017a. Synthesis and characterization of hydrogel films of carboxymethyl tamarind gum using citric acid. Int. J. Biol. Macromol. 105, 463–470. https://doi.org/10.1016/j.ijbiomac.2017.07.058.

Mali, K.K., Dhawale, S.C., Dias, R.J., 2019. Extraction, characterization and functionalization of tamarind gum. Res. J. Pharm. Technol. 12 (4), 1745–1752. https://doi.org/10.5958/0974-360X.2019.00292.0.

Mali, K.K., Dhawale, S.C., Dias, R.J., Havaldar, V.D., Kavitake, P.R., 2017b. Interpenetrating networks of carboxymethyl tamarind gum and chitosan for sustained delivery of aceclofenac. Marmara Pharm. J. 21 (4), 771–782. https://doi.org/10.12991/mpj.2017.20.

Manchanda, R., Arora, S.C., Manchanda, R., 2014. Tamarind seed polysaccharide and its modifications-versatile pharmaceutical excipients—a review. Int. J. Pharm. Tech. Res. 6 (2), 412–420. http://sphinxsai.com/2014/PTVOL6/PT=02(412-420)AJ14.pdf.

McKee, J.R., Hietala, S., Seitsonen, J., Laine, J., Kontturi, E., Ikkala, O., 2014. Thermoresponsive nanocellulose hydrogels with tunable mechanical properties. ACS Macro Lett. 3 (3), 266–270. https://doi.org/10.1021/mz400596g.

Meenakshi, Ahuja, M., 2015. Metronidazole loaded carboxymethyl tamarind kernel polysaccharide-polyvinyl alcohol cryogels: preparation and characterization. Int. J. Biol. Macromol. 72, 931–938. https://doi.org/10.1016/j.ijbiomac.2014.09.040.

Mishra, A., Malhotra, A.V., 2009. Tamarind xyloglucan: a polysaccharide with versatile application potential. J. Mater. Chem. 19 (45), 8528–8536. https://doi.org/10.1039/b911150f.

Munir, H., Shahid, M., Anjum, F., Mudgil, D., 2016. Structural, thermal and rheological characterization of modified Dalbergia sissoo gum—a medicinal gum. Int. J. Biol. Macromol. 84, 236–245. https://doi.org/10.1016/j.ijbiomac.2015.12.001.

Muzaffar, K., Kumar, P., 2015. Parameter optimization for spray drying of tamarind pulp using response surface methodology. Powder Technol. 279, 179–184. https://doi.org/10.1016/j.powtec.2015.04.010.

Nandi, G., Changder, A., Ghosh, L.K., 2019. Graft-copolymer of polyacrylamide-tamarind seed gum: synthesis, characterization and evaluation of flocculating potential in peroral paracetamol suspension. Carbohydr. Polym. 215, 213–225. https://doi.org/10.1016/j.carbpol.2019.03.088.

Naod, G., Tsige, G.-M., 2012. Comparative physico-chemical characterization of the mucilages of two cactus pears (Opuntia spp.) obtained from Mekelle, Northern Ethiopia. J. Biomater. Nanobiotechnol., 79–86. https://doi.org/10.4236/jbnb.2012.31010.

Nasirpour-Tabrizi, P., Azadmard-Damirchi, S., Hesari, J., Khakbaz Heshmati, M., Savage, G.P., 2020. Rheological and physicochemical properties of novel low-fat emulgels containing flaxseed oil as a rich source of ω-3 fatty acids. LWT 133. https://doi.org/10.1016/j.lwt.2020.110107.

Nayak, A.K., Pal, D., 2011. Development of pH-sensitive tamarind seed polysaccharide-alginate composite beads for controlled diclofenac sodium delivery using response surface methodology. Int. J. Biol. Macromol. 49 (4), 784–793. https://doi.org/10.1016/j.ijbiomac.2011.07.013.

Nayak, A.K., Pal, D., 2018. Functionalization of tamarind gum for drug delivery. In: Functional Biopolymers. Springer Series on Polymer and Composite Materials. Springer, Cham. https://doi.org/10.1007/978-3-319-66417-0_2.

Nayak, A.K., Pal, D., Santra, K., 2014. Tamarind seed polysaccharide-gellan mucoadhesive beads for controlled release of metformin HCl. Carbohydr. Polym. 103 (1), 154–163. https://doi.org/10.1016/j.carbpol.2013.12.031.

Newton, A.M.J., Indana, V.L., Kumar, J., 2015. Chronotherapeutic drug delivery of tamarind gum, chitosan and okra gum controlled release colon targeted directly compressed propranolol HCl matrix tablets and in-vitro evaluation. Int. J. Biol. Macromol. 79, 290–299. https://doi.org/10.1016/j.ijbiomac.2015.03.031.

Nie, W., Deters, A.M., 2013. Tamarind seed xyloglucans promote proliferation and migration of human skin cells through internalization via stimulation of proliferative signal transduction pathways. Dermatol. Res. Pract., 1–14. https://doi.org/10.1155/2013/359756.

Nishinari, K., Takemasa, M., Yamatoya, K., Shirakawa, M., 2009. Xyloglucan. In: Handbook of Hydrocolloids, second ed. Elsevier Inc, pp. 535–566, https://doi.org/10.1533/9781845695873.535.

Olusola, A., Toluwalope, G., Olutayo, O., 2014. Carboxymethylation of Anacardium occidentale L. exudate gum: synthesis and characterization. Scholars Acad. J. Pharm. 3 (2), 213–216.

Orozco, S.M., 2001. El cultivo del tamarindo (Tamarindus indica L.), en el trópico seco de México. In: INIFAP-SAGARPA. Campo Experimental Tecomán. Colima. México, p. 7.

Pal, S., Ghorai, S., Das, C., Samrat, S., Ghosh, A., Panda, A.B., 2012. Carboxymethyl tamarind-g-poly(acrylamide)/silica: a high performance hybrid nanocomposite for adsorption of methylene blue dye. Ind. Eng. Chem. Res. 51 (48), 15546–15556. https://doi.org/10.1021/ie301134a.

Pandit, A., Pol, V., Kulkarni, V., 2016. Xyloglucan based in situ gel of lidocaine HCl for the treatment of periodontosis. J. Pharm., 1–9. https://doi.org/10.1155/2016/3054321.

Pardeshi, C.V., Kulkarni, A.D., Belgamwar, V.S., Surana, S.J., 2018. Xyloglucan for drug delivery applications. In: Fundamental Biomaterials: Polymers. Elsevier Inc, pp. 143–169, https://doi.org/10.1016/B978-0-08-102194-1.00007-4.

Parrotta, J.A., 1990. Tamarindus Indica L., Tamarind: Leguminosae (Caesalpinioideae), Legume Family. US Forest Service, southern Forest Experiment Station, Institute of Tropical….

Patel, A.R., Rajarethinem, P.S., Grędowska, A., Turhan, O., Lesaffer, A., De Vos, W.H., Van De Walle, D., Dewettinck, K., 2014. Edible applications of shellac oleogels: spreads, chocolate paste and cakes. Food Funct. 5 (4), 645–652. https://doi.org/10.1039/c4fo00034j.

Paul, S.R., Qureshi, D., Yogalakshmi, Y., Nayak, S.K., Singh, V.K., Syed, I., Sarkar, P., Pal, K., 2018. Development of bigels based on stearic acid–rice bran oil oleogels and tamarind gum hydrogels for controlled delivery applications. J. Surfactant Deterg. 21 (1), 17–29. https://doi.org/10.1002/jsde.12022.

Phani Kumar, G.K., Battu, G., Lova Raju, K.N.S., 2011. Isolation and evaluation of tamarind seed polysaccharide being used as a polymer in pharmaceutical dosage forms. Res. J. Pharm., Biol. Chem. Sci. 2 (2), 274–290. http://www.rjpbcs.com/pdf/2011_2(2)/35.pdf.

Pongsawatmanit, R., Temsiripong, T., Ikeda, S., Nishinari, K., 2006. Influence of tamarind seed xyloglucan on rheological properties and thermal stability of tapioca starch. J. Food Eng. 77 (1), 41–50. https://doi.org/10.1016/j.jfoodeng.2005.06.017.

Prajapati, V.D., Jani, G.K., Moradiya, N.G., Randeria, N.P., Nagar, B.J., 2013. Locust bean gum: a versatile biopolymer. Carbohydr. Polym. 94 (2), 814–821. https://doi.org/10.1016/j.carbpol.2013.01.086.

Premalatha, M., Mathavan, T., Selvasekarapandian, S., Selvalakshmi, S., Monisha, S., 2017. Incorporation of NH4Br in tamarind seed polysaccharide biopolymer and its potential use in electrochemical energy storage devices. Org. Electron. 50, 418–425. https://doi.org/10.1016/j.orgel.2017.08.017.

Rani, P., Sen, G., Mishra, S., Jha, U., 2012. Microwave assisted synthesis of polyacrylamide grafted gum ghatti and its application as flocculant. Carbohydr. Polym. 89 (1), 275–281. https://doi.org/10.1016/j.carbpol.2012.03.009.

Rao, S., Mathew, M., 2012. Handbook of Herbs and Spices. In: Woodhead Publishing Series in Food Science, Technology and Nutrition, second ed. vol. 2. Elsevier BV, pp. 512–533.

Rao, P.S., Srivastava, H.C., Whistler, R.L., 1973. Chapter XVII—Tamarind. Academic Press, pp. 369–411, https://doi.org/10.1016/B978-0-12-746252-3.50022-8.

Rawooth, M., Qureshi, D., Hoque, M., Prasad, M.P.J.G., Mohanty, B., Alam, M.A., Anis, A., Sarkar, P., Pal, K., 2020. Synthesis and characterization of novel tamarind gum and rice bran oil-based emulgels for the ocular delivery of antibiotics. Int. J. Biol. Macromol. 164, 1608–1620. https://doi.org/10.1016/j.ijbiomac.2020.07.231.

Rezaei, A., Nasirpour, A., Tavanai, H., 2016. Fractionation and some physicochemical properties of almond gum (Amygdalus communis L.) exudates. Food Hydrocoll. 60, 461–469. https://doi.org/10.1016/j.foodhyd.2016.04.027.

Rodrigues, D.C., Cunha, A.P., Silva, L.M.A., Rodrigues, T.H.S., Gallão, M.I., Azeredo, H.M.C., 2018. Emulsion films from tamarind kernel xyloglucan and sesame seed oil by different emulsification techniques. Food Hydrocoll. 77, 270–276. https://doi.org/10.1016/j.foodhyd.2017.10.003.

Sangnim, T., Limmatvapirat, S., Nunthanid, J., Sriamornsak, P., Sittikijyothin, W., Wannachaiyasit, S., Huanbutta, K., 2018. Design and characterization of clindamycin-loaded nanofiber patches composed of polyvinyl alcohol and tamarind seed gum and fabricated by electrohydrodynamic atomization. Asian J. Pharm. Sci. 13 (5), 450–458.

Santos, N.L., Braga, R.C., Bastos, M.S.R., Cunha, P.L.R., Mendes, F.R.S., Galvão, A.M.M.T., Bezerra, G.S., Passos, A.-A.C., 2019. Preparation and characterization of xyloglucan films extracted from Tamarindus indica seeds for packaging cut-up 'sunrise solo' papaya. Int. J. Biol. Macromol. 132, 1163–1175. https://doi.org/10.1016/j.ijbiomac.2019.04.044.

Satapathy, S., Singh, V.K., Sagiri, S.S., Agarwal, T., Banerjee, I., Bhattacharya, M.K., Kumar, N., Pal, K., 2015. Development and characterization of gelatin-based hydrogels, emulsion hydrogels, and bigels: a comparative study. J. Appl. Polym. Sci. 132 (8). https://doi.org/10.1002/app.41502.

Sen, G., Pat, S., 2009. Polyacrylamide grafted carboxymethyl tamarind (CMT-g-PAM): development and application of a novel polymeric flocculant. Macromol. Symp. 277 (1), 100–111. https://doi.org/10.1002/masy.200950313.

Shailaja, T., Latha, K., Sasibhushan, P., Alkabab, A.M., Uhumwangho, M.U., 2012. A novel bioadhesive polymer: grafting of tamarind seed polysaccharide and evaluation of its use in buccal delivery of metoprolol succinate. Pharm. Lett. 4 (2), 487–508. http://scholarsresearchlibrary.com/dpl-vol4-iss2/DPL-2012-4-2-487-508.pdf.

Shankaracharya, N.B., 1998. Tamarind—chemistry, technology and uses—a critical appraisal. J. Food Sci. Technol. 35 (3), 193–208.

Shao, H., Zhang, H., Tian, Y., Song, Z., Lai, P.F.H., Ai, L., 2019. Composition and rheological properties of polysaccharide extracted from tamarind (Tamarindus indica L.) seed. Molecules 24 (7). https://doi.org/10.3390/molecules24071218.

Sharma, M., Mondal, D., Mukesh, C., Prasad, K., 2014. Preparation of tamarind gum based soft ion gels having thixotropic properties. Carbohydr. Polym. 102 (1), 467–471. https://doi.org/10.1016/j.carbpol.2013.11.063.

Sharma, S., Bhardwaj, R., 1997. Tamarind—a suitable fruit crop for dry arid regions. Proc. Natl. Symp. Tamarindus Indica, 4–6.

Shaw, G.S., Biswal, D., Anupriya, B., Banerjee, I., Pramanik, K., Anis, A., Pal, K., 2017. Preparation, characterization and assessment of the novel gelatin–tamarind gum/carboxymethyl tamarind gum-based phase-separated films for skin tissue engineering applications. Polym. Plast. Technol. Eng. 56 (2), 141–152. https://doi.org/10.1080/03602559.2016.1185621.

Shaw, G.S., Uvanesh, K., Gautham, S.N., Singh, V., Pramanik, K., Banerjee, I., Kumar, N., Pal, K., 2015. Development and characterization of gelatin-tamarind gum/carboxymethyl tamarind gum based phase-separated hydrogels: a comparative study. Des. Monomers Polym. 18 (5), 434–450. https://doi.org/10.1080/15685551.2015.1041075.

Shigenobu, Y., Yoshiaki, Y., Hiroshi, U., Kanji, K., Mayumi, S., Kazuhiko, Y., 1999. Gelation of enzymatically degraded xyloglucan extracted from tamarind seed. Sen'i Gakkaishi, 528–532. https://doi.org/10.2115/fiber.55.11_528.

Singh, V., Tripathi, D.N., Malviya, T., Sanghi, R., 2009. Persulfate/ascorbic acid initiated synthesis of poly(acrylonitrile)-grafted tamarind seed gum: a potential commercial gum. J. Appl. Polym. Sci. 111 (1), 539–544. https://doi.org/10.1002/app.29114.

Sureewan, D., Parin, B., Pongsakorn, P., Sittikun, L., Panida, A., Porntip, C., Tanasait, N., 2014. Development and evaluation of tamarind seed xyloglucan for transdermal patch of clindamycin. Adv. Mat. Res., 21–24. https://doi.org/10.4028/www.scientific.net/amr.1060.21.

Tsuda, T., Watanabe, M., Ohshima, K., Yamamoto, A., Kawakishi, S., Osawa, T., 1994. Antioxidative components isolated from the seed of tamarind (Tamarindus indica L.). J. Agric. Food Chem. 42 (12), 2671–2674. https://doi.org/10.1021/jf00048a004.

Uccello-Barretta, G., Nazzi, S., Zambito, Y., Di Colo, G., Balzano, F., Sansò, M., 2010. Synergistic interaction between TS-polysaccharide and hyaluronic acid: implications in the formulation of eye drops. Int. J. Pharm. 395 (1–2), 122–131. https://doi.org/10.1016/j.ijpharm.2010.05.031.

Viveros-García, J.C., Figueroa-Rodríguez, K.A., Gallardo-López, F., García-Pérez, E., Ruíz-Rosado, O., Hernández-Rosas, F., 2012. Systems management and marketing of tamarind (Tamarindus indica L.) in three municipalities of Veracruz. Rev. Mexicana Cienc. Agric. 3 (6), 1217–1230.

White, E.V., Rao, P.S., 1953. Constitution of the polysaccharide from tamarind seed. J. Am. Chem. Soc. 75 (11), 2617–2619. https://doi.org/10.1021/ja01107a018.

Wu, S.-C., Shyu, Y.-S., Tseng, Y.-W., Sung, W.-C., 2020. The effect of tamarind seed gum on the qualities of gluten-free cakes. Processes 8 (3), 318.

Xie, F., Zhang, H., Xia, Y., Ai, L., 2020. Effects of tamarind seed polysaccharide on gelatinization, rheological, and structural properties of corn starch with different amylose/amylopectin ratios. Food Hydrocoll. 105. https://doi.org/10.1016/j.foodhyd.2020.105854.

Yadav, I., Rathnam, V.S.S., Yogalakshmi, Y., Chakraborty, S., Banerjee, I., Anis, A., Pal, K., 2017. Synthesis and characterization of polyvinyl alcohol-carboxymethyl tamarind gum based composite films. Carbohydr. Polym. 165, 159–168. https://doi.org/10.1016/j.carbpol.2017.02.026.

Yamatoya, K., Shirakawa, M., 2003. Xyloglucan: structure, rheological properties, biological functions and enzymatic modification. Curr. Trends Polymer Sci. 8, 27–72.

Yamatoya, K., Tabuchi, A., Suzuki, Y., Yamada, H., 2020. Tamarind seed polysaccharide: unique profile of properties and applications. In: Biopolymer-Based Formulations: Biomedical and Food Applications. Elsevier, pp. 445–461, https://doi.org/10.1016/B978-0-12-816897-4.00020-5.

Yarragudi, S.B., Richter, R., Lee, H., Walker, G.F., Clarkson, A.N., Kumar, H., Rizwan, S.B., 2017. Formulation of olfactory-targeted microparticles with tamarind seed polysaccharide to improve nose-to-brain transport of drugs. Carbohydr. Polym. 163, 216–226. https://doi.org/10.1016/j.carbpol.2017.01.044.

Yuguchi, Y., Urakawa, H., Kajiwara, K., Shirakawa, Yamatoya, K., 2001. Gelation of xyloglucan polysaccharide extracted from tamarind seed. In: Proceedings of the Second International Workshop on Green Polymers. Indonesian Polymer Association, pp. 253–263.

Zhang, J., Xu, S., Zhang, S., Du, Z., 2008. Preparation and characterization of tamarind gum/sodium alginate composite gel beads. Iran. Polym. J. (English Ed.) 17 (12), 899–906. http://journal.ippi.ac.ir/manuscripts/IPJ-2008-12-3700.pdf.

Zhang, J., Zhang, S., Wang, Y., Zeng, J., Zhao, X., 2007. Composite magnetic microspheres of tamarind gum and chitosan: preparation and characterization. J. Macromol. Sci., Part A: Pure Appl. Chem. 44 (4), 433–437. https://doi.org/10.1080/10601320601188356.

CHAPTER

15

Tree gum-based nanostructures and their biomedical applications

K.P. Akshay Kumar[a,b], *Rohith K. Ramakrishnan*[c], *Miroslav Černík*[c], *and Vinod V.T. Padil*[c]

[a]Future Energy and Innovation Laboratory, Central European Institute of Technology, Brno University of Technology, Brno, Czech Republic [b]Department of Applied Chemistry, Cochin University of Science & Technology, Kochi, Kerala, India [c]Institute for Nanomaterials, Advanced Technologies and Innovation (CXI), Technical University of Liberec (TUL), Liberec, Czech Republic

15.1 Introduction

Gums are polysaccharides emanating from various natural sources, and their diverse applications in various fields (environmental, biosensor, catalysis, energy storage, antimicrobial, and biomedical) have been extensively documented (Padil et al., 2018). Tree gums are carbohydrate polymers, and their functionalities, physicochemical properties, and structural-rheological attributes make them attractive and useful for potential food and nonfood applications, as with other synthetic or naturally derived polymers (Gum Arabic, 2018; Kennedy et al., 2011). Commercially accessible tree gum hydrocolloids have been surveyed for the food, pharmaceutical, biomedical, and industrial sectors (Xu et al., 2017; Venkateshaiah et al., 2020a, b).

Tree gums consist of the combination of mono- or disaccharides, e.g., L-arabinose, D-galactose, L-rhamnose, and D-glucuronic acid together with some metal cations including potassium, sodium, magnesium, etc. (Mohammadinejad et al., 2020). The arrangement of the molecules differentiates the physicochemical properties of the gum. Physically, natural gums are amorphous, soluble in water to make a viscous solution or jelly form, and partially soluble in solvents. Chemically tree gums are hydrocolloids mainly with a large number of functional groups (–OH and –COOH) on the backbone of the molecules with glycosidic linkages. These are usually helical in shape and less packed than the linear polysaccharides, making them easily soluble in water. The hydrophilic nature of tree gums holds water molecules within

Micro- and Nanoengineered Gum-Based Biomaterials for Drug Delivery and Biomedical Applications
https://doi.org/10.1016/B978-0-323-90986-0.00008-X

their interfaces via hydrogen bonding and molecular conformation. The gums are limited in many applications due to high water solubility and low mechanical and thermal properties. These limitations are overcome by the conjugate preparation with other synthetic /natural polymers, fillers, and nanoparticles.

Novel design and fabrication approaches pertaining to tree gum-based nanoparticles, electrospun fibers, and 3D nano-architectures, including bio-printed structures, appeal to polymer researchers, food scientists, druggists, and material engineers associated with the New Horizons in Environmental Sciences incorporating a wide variety of fabrication methods, high aspect ratio of materials, mechanical characteristic, and environmentally friendly process, to produce high-performance nanostructures for potential applications. Natural polysaccharides have immense potential as green and sustainable resources for future research. In this chapter, we explore the sustainable biomedical applications of tree gums such as tragacanth gum (TG), gum arabic (GA), gum kondagogu (KG), guar gum (GG), and gum karaya (GK). This chapter aims to underscore the fabrication and key applications of tree-based nanostructures, inclusively, comprising emerging topics of research interest in numerous biomedical fields (antimicrobial, drug delivery, tissue engineering, wound healing, cancer therapy, nano-medicine, etc.) (Fig. 15.1). It also includes a commentary on their prospects in this burgeoning field. In addition, the imminent prospects in the area of environmental, biomedical, and energy appliances for the fabrication of tree gum-based nanostructures via green method can be envisaged. This will be based on the following attributes of these materials, namely stability, fiber diameter, aspect ratio, sustainability, etc.—factors that are expected to play crucial roles in many important areas of research such as chemistry, physics, biology, medicine, environmental science, pharmacy, biotechnology, and nanotechnology. Furthermore, it is illustrated that wide-ranging, exciting research opportunities exist for exploring this topic on electrospinning hydrocolloids encompassing value-added materials via

FIG. 15.1 Schematic representation of tree gum-based nanostructures and their potential biomedical applications.

a sustainable, recyclable, and greener strategy. Future perspectives and challenges will also be provided.

Furthermore, the fabrications and biomedical applications of tree gum-based nanostructures like nanoparticles, nanofibers, and nanogels are discussed. The advantages and effectiveness of these nanostructures on various applications such as targeted drug delivery, tissue engineering, wound dressing, biosensors, and 3D printing are studied by interpreting their cytotoxicity, in vivo and in vitro, and biodegradability properties.

15.2 Tree gum exudates—Structure and properties

GA (gum arabic), GK (gum karaya), KG (gum kondagogu), TG (tragacanth gum), and GG (guar gum) are the major commercial gums from tree exudates. The structural, physicochemical, rheological, food additives, pharmaceutical, and biomedical applications of these gums have been extensively reviewed (Padil et al., 2016a, 2018). The basic structures of the tree gums are presented in Fig. 15.2.

Primarily there are two prominent sources of GA (gum arabic): *Acacia senegal* and *Acacia seyal*. GA, an arabinogalactan-based gum, has many branches, having a core consisting of β-(1–3) galactose, side chains consisting of (1 → 3)-linked β-D-galactopyranosyl units united by (1 → 6)-linkages residues, and other saccharide molecules are L-arabinose, L-rhamnose, and uronic acid (D-glucuronic acid and 4-*O*-methyl-D-glucuronic acid) (Isobe et al., 2020; Williams and Phillips, 2021). The structure of GA is presented in Fig. 15.2. It is easily soluble in water and can form 50% (w/w%) gum solution without any gel formation.

TG (tragacanth gum; *Astragalus gummifer*), a water-soluble or aqueous dispersible tree exudate gum, belongs to the family Fabaceae and consists of heterogeneous structures with high acidic content (Nejatian et al., 2020). The major gum constitutes two fractions as tragacanthin (a water-soluble portion) and bassorin (gel-forming part), respectively. Tragacanthin is a highly branched structure of α-(1–4) galactose residues with arabinose. Bassorin comprises mainly poly(α-D)galacturonic acid, which provides –COOH groups in the structure of TG, and the side-chain contains xylose, fucose, and galactose (Yousuf et al., 2021; Nayak et al., 2020) (Fig. 15.2).

GK (*Sterculia urens*) is a highly branched gum that belongs to the Sterculiaceae family, with a main chain comprising of α-L-Rhap (α-L-rhamnose), and α-D-GalpA (α-D-galacturonic acid). The branched-chain with β-D-Galp (β-D-galactose) and GlcpA (β-D-glucuronic acid) provides the gum with a complex structural assignment (Yousuf et al., 2021) (Fig. 15.2). Special features of GK such as nontoxicity, biocompatibility, biodegradability, gelling properties, high viscosity, solubility in an aqueous medium, possible chemical modification, and ability to construct nanoparticles and nanofibers make it attractive for many industrial, pharmaceutical, and biomedical applications (Nayak et al., 2020).

KG (*Cochlospermum gossypium*) is a tree exudate gum that belongs to the family *Bixaceae* and is categorized under rhamnogalacturonan types. The prevalent saccharides present in the structure of GK are composed of neutral sugars (arabinose, rhamnose, glucose, galactose, and mannose) and acidic sugars (glucuronic acid and galacturonic acid) (Padil et al., 2020) (Fig. 15.2).

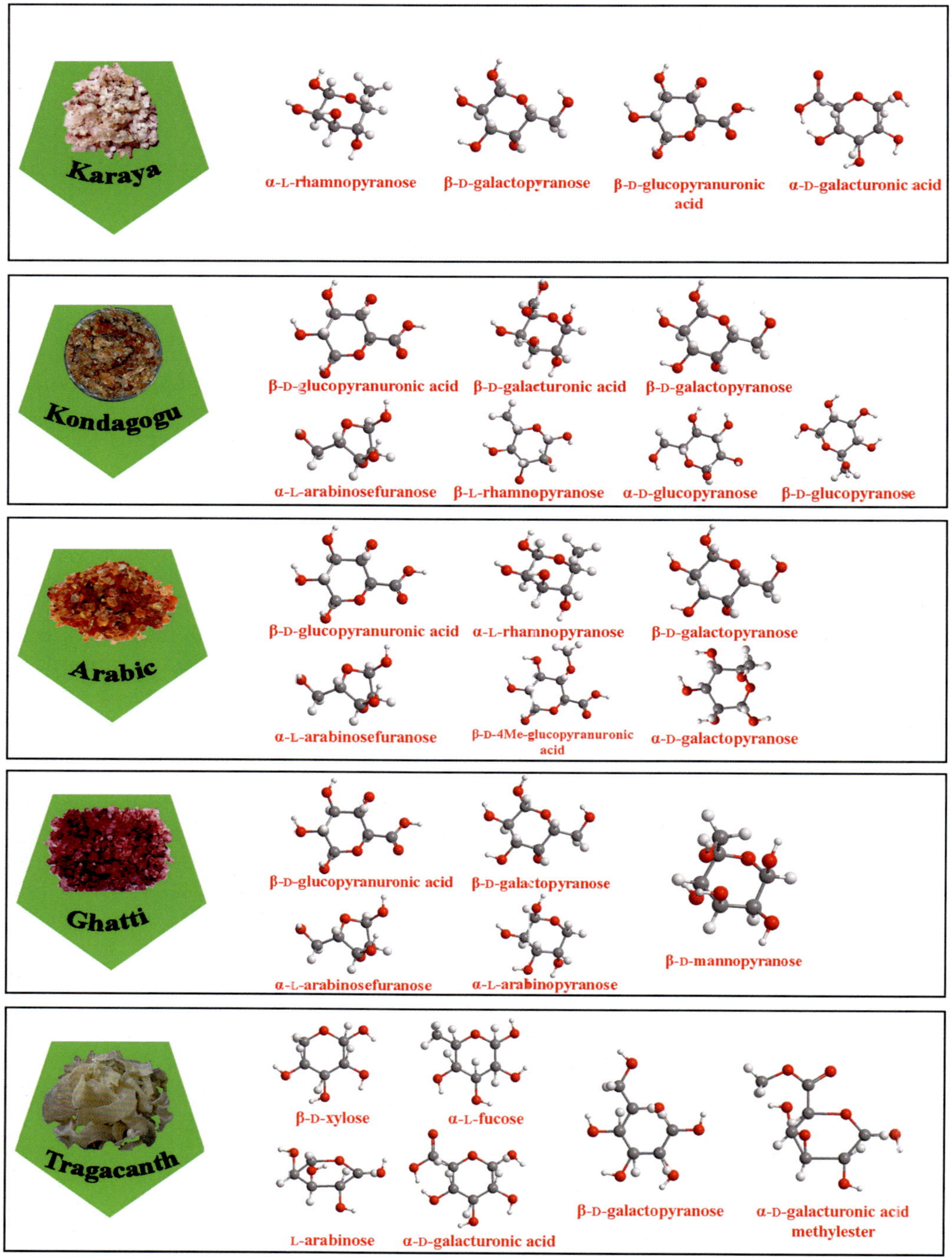

FIG. 15.2 Commercial tree gum polysaccharides and their structural features. *Padil, V.V.T., Wacławek, S., Černík, M., Varma, R.S., 2018. Tree gum-based renewable materials: sustainable applications in nanotechnology, biomedical and environmental fields. Biotechnol. Adv. 36 (7), 1984–2016, ISSN 0734-9750, https://doi.org/10.1016/j.biotechadv.2018.08.008.*

Guar (*Cyamopsis tetragonoloba* L.) is a member of the legume plant family and is extensively used in the food, textiles, and therapeutic industries. Guar belongs to the galactomannan type of gums and its foremost structural features are (1 → 4)-linked-β-D-mannopyranosyl units with single α-D-galactopyranosyl units linked by (1 → 6) linkages (Lee and Yoo, 2021; Jo et al., 2018) (Fig. 15.2). High viscosity, nongelling property, acting as a binder and creamy in foodstuffs, and insoluble dietary fiber make GG a suitable candidate in food and biomedical applications.

15.3 Nanoarchitectures based on tree gums

Nanotechnology is the utilization of a systematic way of transforming knowledge into nanoscale materials from their bulk and controlling the structure-dependent properties and applications in numerous ways. Nanotechnology has been successfully applied to natural polymers and their composites in various fields, including packaging, cosmetics, pharmaceutical, energy, and environmental applications (Sharma et al., 2020; Popescu et al., 2019; Bie et al., 2017; Singh et al., 2020a). These properties are always dependent on the size, shape, and morphology of the materials. Therefore, different types of nanostructures (nanoparticles (NPs), nanofibers (NFs), hydrogel (HG), and nanogel (NG)) based on bio-based polymers have been introduced to improve current systems as well as for future developments in "green" and sustainable applications.

15.3.1 Nanoparticles (NPs)

Various methods of preparation of NPs are illustrated using toxic and nontoxic reducing agents. High toxic reducing agents or organic solvents such as hydrazine, *N,N*-dimethylformamide, sodium borohydride, etc., are not only environmentally friendly but also highly toxic to both humans and the environment. The green synthesis of metallic nanoparticles, using natural gums, has grown into an imperative area in green chemistry by avoiding these stabilizing/reducing hazardous chemicals. These green synthesized metallic nanoparticles have precise magnetic, optoelectronic, and structural, mechanical, and thermal characteristics, which are essential for various applications in the biomedical field.

Metal (platinum, silver, gold, palladium, copper, and zinc) and metallic oxide (Fe_3O_4, CuO, ZnO, MnO_2, etc.) NPs are prepared in green technology by using hydroxyl and carboxyl functional groups of tree gums as a reducing agent for metallic salt and capping agent for nanoparticle stabilization (Dhar et al., 2011; Geetha et al., 2016; Černík and Thekkae Padil, 2013; Pandey and Mishra, 2014; Reddy et al., 2015; Roque et al., 2009; Roy et al., 2019). Green synthesis of variously sized (1–100 nm) and shaped nanoparticles (triangular, spherical, and nanorods) are a cost-effective, simpler, and eco-friendly approach to fabrication (Abdullah et al., 2015; Balasubramanian et al., 2016; Huang et al., 2007).

Green synthesized NPs of Ag, Au, CuO, and ZnO showed antimicrobial properties, to inhibit colonization of pathogenic bacteria and fungi or by helping the repair mechanism of cells involved in wound healing. The fragmentation of bacterial cells by the interaction with NPs could be the reason for antibacterial properties. The oppositely charged particles such as NPs (positively charged) and gram-negative bacterial cells (negatively charged) interrelate with each other via electrostatic interaction. They can accumulate and penetrate the bacterial cell membrane and destroy the cell walls or membranes and interior cell structure. Antiviral properties of NPs are used for the advance of biosensors for diagnosis, discovery, and diversity of the virus. Various forms of particulate matter such as metal ions, gaseous molecules, amino acids, and toxic chemicals could easily be distinguished by employing the specific characteristics of NPs such as optical, electronic, and chemical properties.

In conventional chemotherapy applications, therapeutic agents are being distributed in a large area instead of a specific spot, which causes damage to healthy tissues. Poor bioavailability, no specificity, rapid clearance, adverse side effects, etc., are the drawbacks of the conventional method. Green synthesized NPs have proved a suitable nanocarrier for targeted drug delivery applications. They might also help to ensure a slow and sustained release of the drug to the specific site in the required dosage without affecting healthy cells.

15.3.2 Nanofibers (NFs)

Nanofibers are commonly fabricated by the electrospinning method. Modified forms like side-by-side and coaxial electrospinning methods are used for composite nanofiber synthesis. The extraordinary performance of bio-based nanofibers including porosity, large surface area, functional fibers, mechanical attributes, and thermal stability benefit their superior applications (Padil et al., 2020).

The diameter and surface smoothness depend on the electric charge, gum concentration, and viscosity of the gum composites. During nanofiber fabrication, some active components and metallic nanoparticles are incorporated in action films, bioremediation membranes, and biomedical scaffolds (Makvandi et al., 2020; Delfi et al., 2020; Padil et al., 2019). The bioactive molecules or drugs are even encapsulated with tree gum-based electrospun fibers before or after electrospinning and could be significantly utilized in the development of novel drug delivery systems (Zare et al., 2019). TEM (transmission electron microscopy) analysis, SEM (scanning electron microscopy) analysis, TGA (thermogravimetric analysis), DSC (differential scanning calorimetry) analysis, and FTIR (Fourier transform infrared) analysis are usually used to characterize the nanofibers.

15.3.3 Nanogels (NGs) and hydrogels (HGs)

Hydrogels and nanogels are significant molecular entities and their distinctive properties, characteristics, and applications are of immense importance in the biomedical field. The uniqueness of the hydrogels relies on their biocompatibility, biodegradability, softness, super absorbency, hydrophilicity, viscoelasticity, equilibrium swelling, and responses to various stimuli such as pH, temperature, magnetic field, electric field, solvent composition, and ionic strength, etc. (Rizwan et al., 2017). The hydrogel can be natural, synthetic, or a combination of both, and is formed via cross-linking. Many hydrogels or nanogels can be used as drug carriers or enhance the stability of the encapsulated drug or bioactive molecules for controlled targeting. Natural polymers consisting of chitosan, dextran, pectin, and gums constitute hydrogels and possess weak mechanical properties. Mostly to enhance the stability, and mechanical properties, hydrogels are often grafted or cross-linked or bonded with synthetic polymers or other organic molecules (Mohammadinejad et al., 2019).

Nanogels are nanometer-ranged swellable three-dimensional network hydrogels in water that can encapsulate a large amount of drug and are also able to penetrate small capillary vessels with absorbed biological fluids. Nanogels, due to their outstanding properties, including porosity, hydrophilicity, large surface area, softness, and stimuli-responsivity to pH, temperature, and ionic force, can be used for targeted site drug delivery and bioimaging (Raemdonck et al., 2009; Ilka et al., 2018). Natural hydrogels have some compositional similarity with the native extracellular matrix and can be used as supporting materials for tissue engineering scaffolds (Tong et al., 2020; Pamfil and Vasile, 2018).

Nanoprecipitation is a method used to prepare the drug-loaded nanohydrogels with the highest swelling index in water (Giulbudagian et al., 2014). Nanosized hydrogel particles show both hydrogel and nanoparticle properties simultaneously (Sharma et al., 2016). Nanoprecipitation ensures the maximum therapeutic efficiency with minimum side effects by reducing the premature release during the circulation and ensuring the successive release at the targeting site.

Different types of nanogel synthesis are: (1) polymerization of monomers; (2) physical self-assembly; (3) cross-linking of polymers; and (4) template-assisted nanofabrication (Neamtu et al., 2017). Most of the natural polysaccharide-based nanogels are prepared by physical self-assembly of interacting polymer chains (Fujioka-Kobayashi et al., 2012). The cross-linking agents' medium influences the swelling behavior (Ryu et al., 2010). Nanogels are cross-linked with poly(ethylene glycol)diamine (Huang et al., 1998), pyridyldisulfide (Chacko et al., 2012), diglycidylether, aldehyde (Sadat Hosseini et al., 2016), etc.

Nanogels are suitable for carrying hydrophilic and hydrophobic drugs, nanoparticles, biologically active biomacromolecules, etc. (Kousalová and Etrych, 2018). Biopolymer-based nanogels are biocompatible and biodegradable under physiological conditions. Polydispersity and solubility in aqueous solutions of biopolymer nanogel systems respond to a wide range of pH, ionic strength, and temperature with optimized tissue penetration and bloodstream properties, which enhances their biological applications (Oh et al., 2009).

The major advantages of nanogel for systematic drug delivery depend on their tunable size, surface characteristics (hydrophilic or hydrophobic) which could enhance their quick discharge, clearance, excess drug loading volume, releasing mechanism of the drug into the selected target point, and their capability to infiltrate the small capillary vessels (Kennedy et al., 2011).

15.4 Tree gum-based NPs for biomedical applications

The special features such as distinctive structural, functional, and physicochemical characterizations of bio-based materials would create a novel nanomaterials for potential appliances in various branches of biomedical areas. Tree gum carbohydrate polymers are ecological and multipurpose sustainable ingredients incorporating assorted functional and structural properties to build up nano-assemblies. Many varieties of metal NPs such as Au, Ag, Cu, Pt, Pd, Se, Fe, and metal oxide (CuO, Fe_3O_4, ZnO) have been fabricated using

GA as a reducing and stabilizing agent (Padil et al., 2018). The transformation of gum-based materials can be delivered into the system via NPs, nanogels, granules, or tablet forms. The tree gum-based nanostructures play many essential branches under the biomedical domain, such as drug delivery, tissue engineering, 3D bioprinting, biosensors, and wound healing.

GA has been employed in synthesis and stabilization of many metal/metal oxide NPs due to its excellent properties such as nontoxicity, economical, availability, surface charges capable of protecting NPs in solution, and biodegradable and biocompatible nature (Atgié et al., 2019; Nayak et al., 2012). The production of Ag, Au, and Ag-Au bimetallic NPs employing GA and gamma irradiation (15.0 kGy) resulted in NPs of high purity, tunable size and morphology, which were extremely stable and crystalline (El-Batal et al., 2020). These NPs were highly effective toward both the bacterial (*B. subtilis* and *E. coli*) and the antifungal pathogens (*Candida tropicalis* and *Candida albicans*), separated from patients with diabetic foot infections.

Another study highlighted the potential antibacterial and antibiofilm efficiencies of ZnO NPs (particle size = 180 nm) fabricated using GA via microwave irradiation (Pauzi et al., 2020a). The minimum inhibitory concentration (MIC) and minimum bactericidal concentration (MBC) of ZnO-GA composite were estimated for *Staphylococcus aureus* (31.2 and 62.5 g/mL) and *Escherichia coli* (62.5 and 125 g/mL), respectively. Furthermore, antibiofilm formation of *E. coli* and *S. aureus* preferred a high concentration (>500 g/mL) of GA-ZnO NPs. The major mechanism of GA-ZnO nanoparticles was determined to be the electrostatic interaction between the ZnO NPs and bacterial cell wall via the membrane disruption. This work emphasized the development of nontoxic and ecological ZnO NPs for biomedical application.

Delivery of antioxidants via novel nanoparticle formulation is a challenge either to prevent drug degradation and their side effects or improve the bioavailability. In this context, GA has been introduced to fabricate magnetic nanoparticles (MNPs) to encapsulate *Dunaliella salina* extract (DE), a natural antioxidant and anticancer bioactive material, and for successful oral delivery (Zamani et al., 2019). The benefits such as safety, high dispersibility, and colloidal stability, and the compatibility of GA favored the stability of MNPs and improved their potency to deliver the bioactive material efficiently. The as-synthesized GA-MNPs and DE/GA-MNPs had a particle size of 10–300 nm in various solvents such as water, ethylene glycol, and phosphate buffer saline (pH ~ 7.4). The drug loading capacity into GA-MNPs by using the logarithmic and stationary phases was found to be 79.8 and 186.1/100 mg MNPs, and major interactions between drug and GA/NMPs were demonstrated to be hydrogen bonding, van der Waals forces, as well as hydrophobic interface. The further release profile of the drug from the GA/NMPs suggests that the release of drug extract is much higher under neutral pH (7.2) than acidic pH (4.5) conditions. Therefore, GA/MNPs conserve bioactive compounds under acidic pH conditions to reduce unwanted gastric side effects, although they are capable of releasing them in the neutral pH environment of the intestine. The cytotoxicity study of drug and drug-GA/NMPs logarthmic progression and the stationary growth phases were demonstrated. The results showed that the drug at the stationary phase exhibited higher toxicity than mid-logarithmic on both MCF-7 and HeLa cancerous cell lines. These results confirmed that drug-loaded GA-NMPs are a suitable medium for a safe potential route for oral drug delivery of antioxidants.

Similarly, a hybrid nano-system based on MNPs and GA was produced via the coprecipitation method, as well as solution treatment, and exploited for hyperthermia

(Horst et al., 2017). The particle sizes of MNPs and magnetic core were estimated to be 70–80 nm and 20 nm, respectively. The other properties such as stability of the GA-NMPs, particle size, super-paramagnetism, surface charge, and suitability for hyperthermia application were established.

The detailed antibacterial action of Ag NPs and their influences in an assortment of tests such as micro-broth dilution analysis, antibiofilm formation, growth rate, disruption of cytoplasm, and membrane permeabilization were evaluated (Kora and Sashidhar, 2018). The resultant KG-Ag NPs (spherical, and of average particle size $= 4.5 \pm 3.1$ nm) had shown substantial antibiofilm activity against *E. coli*, *S. aureus*, and *P. aeruginosa*, respectively, which can have consequences in the treatment of drug-resistant bacterial infections caused by biofilms. Additionally, the biocompatibility of KG-Ag NPs with the HeLa cell line was evaluated and they were found to be apt in the biomedical field.

Fluorescent labeled KG-Au NPs with conjugated folic acid has been developed for dual procedures such as targeted drug delivery and cellular imaging purposes (Kumar et al., 2018). KG-Au NPs were coupled to folic acid and fluorescein isothiocyanate (FITC) to produce a targeted and fluorescently labeled Au NPs foliate (particle size = 37 nm; zeta potential $= -23.7$ mV). Various microscopic and spectroscopic techniques were used for characterizing the fluorescently labeled Au NPs foliate. Furthermore, the cell viability, targeting, and fluorescent capability of the produced composite (fluorescently labeled Au NPs foliate) were assessed in diverse cancer cells to establish their targeting capacity. The authors highlighted that the produced Au NPs and fluorescent labeled KG-Au NPs with conjugated folic acid were biocompatible in A549 lung adenocarcinoma and MCF-7 breast cancer cell lines, and could be used in imaging of cancerous cells as well as drug delivery.

15.5 Tree gum composite nanofibers for biomedical applications

Nanofibers based on tree gums are receiving more attention due to their excellent material properties such as great water vapor permeability, a pronounced surface to volume ratio, extraordinary porosity, brilliant pore-interconnectivity, water absorption rate, biocompatibility, hydrophilicity, etc., and are suitable materials for wound dressing (Wei et al., 2016). The majority of tree gums could be subjected to electrospinning to produce nanofibers, which has been investigated in recent times (Padil et al., 2020).

The combination of GA and Ag NPs with other biocompatible and biodegradable synthetic polymers (PVA and PCL) was electrospun into nanofibers (average diameter = 100–600 nm range) for the application of wound dressing (Eghbalifam et al., 2020). In these composite nanofibers, Ag could act as an antimicrobial agent, and PVA and PCL function as additives to enhance the electrospinning process as well as improve the mechanical attributes of the ensured fibers. The developed fiber mat showed porosity (22%–49%), water absorption (119%–540%), and water vapor permeability (1998–2322 g m^{-2} Day^{-1}), respectively. Furthermore, the fiber mat explored the nanofibers' potential antibacterial activity toward common wound infectious bacteria such as *S. aureus*, *E. coli*, and *P. aeruginosa*, and fungal strain (*C. albicans*), and it was demonstrated demonstrated that the developed nanoparticle composite fibers would be the impending solution and superior to the prevailing method for wound healing treatment.

The importance of GA as a viscosity enhancer in the electrospinning of a mixture of gelatin/chitosan/PVA in an acetic acid/water solvent system was established (Tsai et al., 2015). The major impacts of this research highlighted the utilization of greener solvents, higher amount of natural polymeric content (both gelatin and chitosan) incorporation into the final nanofiber composites, and avoidance of toxic and expensive solvents (e.g., trifluoroacetic acid or hexafluoroisopropanol) for the production of gelatin-chitosan fibers. The produced nanofibers with a higher amount of natural polymeric content have extremely advantageous benefits for tissue engineering (Sahay et al., 2012).

An electrospun drug delivery fiber mat scaffold comprised of TG, zein, PLA, and tetracycline hydrochloride (TCH) was fabricated for an antibacterial wound dressing (Ghorbani et al., 2020). The fiber mat diameter (253 ± 15.3 to 547 ± 56.4 nm) with consistent pores had influenced the blending ratio of zein/TG in the electrospinning mixture. The introduction of PLA into the scaffold mat enhanced the overall mechanical properties and biodegradability. Additionally, the cytocompatibility, cell adhesion study (against NIH-3T3 fibroblast cell), and antibacterial action against gram-positive and gram-negative bacteria (*S. aureus*, and *Pseudo aeruginosa*) exposed the developed nanofibrous mat as an efficient delivery route of TCH in wound healing application.

In another study, a combination of TG, PVA, graphene oxide (GO), and TCH nanofiber with fiber diameter (~100 nm) was constructed via electrospinning for a transdermal drug delivery system (Abdoli et al., 2020). GO, due to its solubility in an aqueous medium, large surface functionality and surface area, and biocompatibility, made the resultant fiber scaffold suitable for the biomedical area. Furthermore, the mechanical attributes, and antibacterial, cytotoxicity (high cell viability of nanofiber against the human normal cells from umbilical vein endothelial cells), and biocompatibility studies suggest that this nanocomposite fiber could be a promising candidate for controlled drug delivery.

The coelectrospinning technique was used to generate composite electrospun fibers with fiber diameters (131.6 ± 27.5 nm) based on TG, PCL, and PVA (Zarekhalili et al., 2017). The resultant nanofibers were tested for antibacterial activity against *E. coli* (gram-negative bacteria) and *S. aureus* (gram-positive bacteria). It was revealed that *E. coli* was more susceptible to nanofibers due to the action of sugar residues (such as L-arabinose and L-fucose) present in TG (Ranjbar-Mohammadi et al., 2013). Furthermore, the nanofibers exposed cell proliferation and registered nontoxicity toward the MTT assay and NIH3T3 cells.

For the well-organized nanofibers in the field of nerve tissue regeneration, an aligned and random nanofiber scaffold was constructed on blending with TG and PLLA (poly(L-lactic acid)) with a PLLA/TG ratio of 75:25. The organized nanofibers had outstanding physicochemical and mechanical properties.

A combination of TG, PCL, and curcumin blended nanofibers were fabricated via electrospinning by employing acetic acid/water (90%, v/v) solvent medium at room temperature (Mohammadi et al., 2016; Ranjbar-Mohammadi and Bahrami, 2016). Systematic release studies of curcumin wound healing potential and antibacterial effectiveness of the produced nanofibers were carried out. The TG/PCL/curcumin electrospun fibers were very effective in healing diabetic wounds created in rat models.

An electrospun scaffold of TG/PCL (3:1; 7% (TG), 10%–20% (PCL)) with excellent antibacterial attributes, mechanical strength, porosity, hydrophilicity, and degradation behavior, was contrived (Ranjbar-Mohammadi and Bahrami, 2015). The developed nanofibers had an average diameter of 156 ± 25 nm and the cytotoxicity and antibacterial potencies of these TG/PCL fibers were tested, and their excellent skin and would healing scaffold properties were demonstrated.

In another study, TG was combined with PVA at different TG/PVA ratios (such as TG/PVA: 0:100 to 100:0) and the blended ratio of TG/PVA was optimized (40:60) for the smooth, beadless, uniform diameters of nanofibers production via electrospinning (Ranjbar-Mohammadi et al., 2013). Further, the antibacterial efficacy of TG/PVA was found to be excellent toward gram-negative bacteria (*P. aeruginosa*). The cell proliferation and biological compatibility of the nanofibers were also shown to be impressive.

Electrospinning types of both blending and coaxial were introduced to construct conjugates composites of TG, PLGA, and TCH (as a hydrophilic model drug) (Ranjbar-Mohammadi et al., 2016). Three variants of electrospun fibers such as pristine PLGA, PLGA/TG core-shell, and PLGA/TG-TCH core-shell were fabricated using N-2-hydroxyethyl piperazine (HEP) as a solvent. The higher amount of TG in the blended mixtures resulted in nanowires with smaller fiber diameters. The stability of nanofiber, drug-releasing behavior, cell culture, and proliferation kinetics of nanowires were systematically premeditated. The PLGA/TG-TCH core-shell nanofibers were found to be extremely proficient in the treatment of periodontal disease.

Electrospinning of GK blended with PVA in a greener environment has been carried out. The blend ratio suitable for producing nanofibers with excellent uniformity, shape, and size was maintained by mixing PVA with GK using various weight combinations (Padil and Černík, 2015; Padil et al., 2015c, 2016b). Greener technologies such as heat and plasma treatments were employed to improve the nanofiber performance to endorse cross-linking, water-resistance, water contact angle, BET surface area, and the porosity of the fibers, and all proved to be successful. Furthermore, plasma-treated electrospun fibers displayed potential antibacterial influence against gram-negative and gram-positive bacterial colonies.

KG fibers were produced via electrospinning of the blend mixtures of KG with PVA. Furthermore, KG electrospun fibers (both untreated and plasma-treated) were also applied for antimicrobial efficiency tests, and plasma-treated fibers exhibited enhanced antibacterial potency against *E. coli* and *S. aureus* (Padil et al., 2015a, b).

15.6 Tree gum-based nanostructures for drug delivery

The development of bio-based polymers and their encapsulation efficiency to entrap chemotherapeutic agents for the treatment of cancer such as breast cancer, liver cancer, and pancreatic cancer is one of the prime research areas in bionanotechnology. In this context, GA and sodium alginate (SA)-based nanoparticles were synthesized and encapsulated with curcumin (CUR) to enhance their bioavailability and antioxidant drug release, and were prepared by the ionotropic gelation technique (Hassani et al., 2020). GA-SA/CUR nanoparticles with sizes 10 ± 0.3 nm and 190 ± 0.1 nm were characterized by ATR-FTIR (attenuated total reflectance–Fourier transform infrared), XRD (X-ray diffractometry), DSC (differential scanning calorimetry), DLS (dynamic light scattering-size distribution), TEM (transmission electron microscopy), etc. The nanocomposite exhibited potential antioxidant and anticancer properties against human liver cancer cells (HepG2 cell line), which has been demonstrated by MTT assay. The influence of disaccharide moiety (galactose) from GA might be responsible for the enhanced toxicity of the GA-SA/CUR nanoparticle against the HepG2 cell line due to being highly selective for asialoglycoprotein receptor (ASGPR) on human hepatocytes; in addition, other glycoproteins present in GA further contributed to the action as both an antimicrobial and anticancer tool (Nasir et al., 2010; Sarika et al., 2015).

Chemically modified GA with a gelation cross-linked hydrogel scaffold was synthesized for biomedical applications (Sarika et al., 2014). The interaction between gum Arabic aldehyde with chitosan via Schiff base formation is demonstrated (aldehyde groups GA with amino groups of gelatin) (Fig.15.3). The formed scaffold showed high functionality, biocompatibility, cytotoxicity, and cell adhesion studies, which can be used for cancer diagnosis and drug screening.

Due to its special structural properties and biodegradability, GK has been used as a drug carrier in many pharmaceutical delivery formulations such as tablets, micro/nanoparticles,

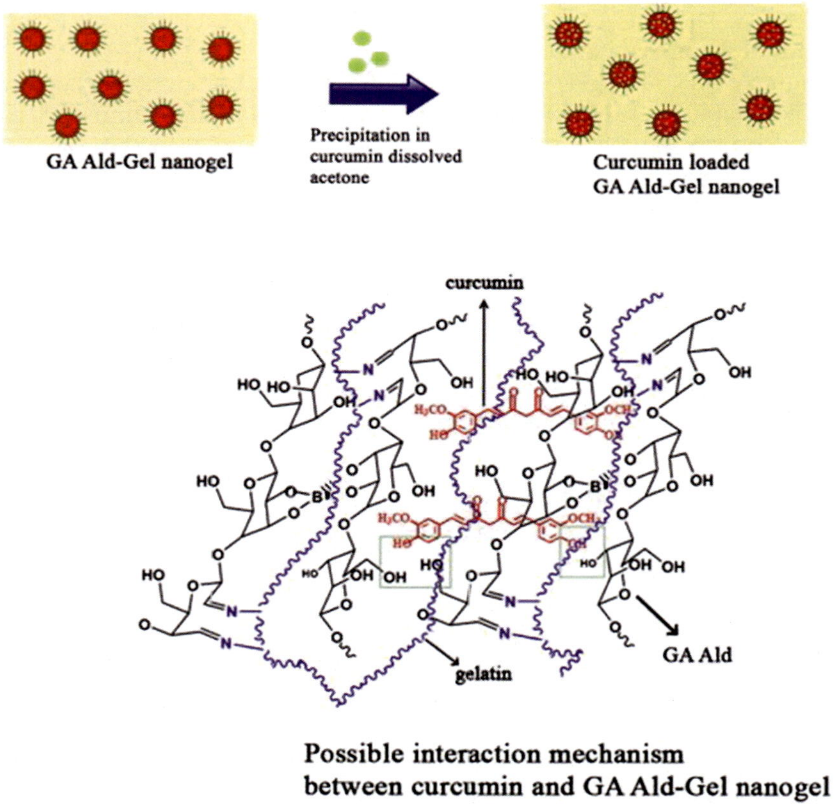

FIG. 15.3 The interaction between curcumin encapsulated into GA Ald (GA-aldehyde) gel nanogel matrix and its mechanism. *Reprinted with permission from Sarika, P.R., Nirmala, R.J., 2016. Curcumin loaded gum arabic aldehyde-gelatin nanogels for breast cancer therapy. Mater. Sci. Eng. C 65, 331–337. https://doi.org/10.1016/j.msec.2016.04.044.*

beads, nanofibers, and hydrogel (Pamfil and Vasile, 2018; Vellora Thekkae Padil et al., 2015; Bera et al., 2015; Liu et al., 2014; Singh and Sharma, 2011; Zare et al., 2019).

GK hydrogel has been synthesized using acrylic acid via a free radical mechanism, in which N,N^{I}-methylene bisacrylamide (MBA) was used as a cross-linking agent (Bashir et al., 2018). The synthesized hydrogels were characterized using FTIR, XRD, DSC, FESEM, and mechanical and swelling capacity. Further exploration of the hydrogel for the loading and release study of the hydrophobic drug (quercetin) has been established. This is an important milestone to encapsulate and deliver a hydrophobic drug using a GK-based hydrogel. The non-Fickian diffusion mechanism was confirmed by the drug release kinetics from the hydrogel. The research emphasized that such natural gum-based hydrogen could be a very promising route for oral drug delivery of hydrophobic drugs.

Many natural polymers are being used for drug carriers and delivery for cancer treatments due to their exceptional properties such as biocompatibility, much lower toxicity, low material cost, availability, renewability, antioxidant nature, low immunogenicity, biodegradability, variety of functionality, and satisfactory stability in both in vitro and in vivo environments (Soleimani et al., 2021). Although natural polymers have certain disadvantages, such as low mechanical strength, complicated structures, and high solubility in an aqueous environment, the majority of the drawbacks could be dealt with via chemical modification or drafting with other molecules of interest and the polymers could be pertinent to stimuli-responsive drug delivery systems (Eghbalifam et al., 2020; Samadian et al., 2020). The major boost for using natural gums as drug carriers is the elimination of intricate problems such as drug-adverse events and detrimental immune responses. Furthermore, many tree gums can be functionalized in the amendment of NPs, NFs, and hydrogels to keep the nanoparticles or drug molecules highly stable in their environment, thus improving their performance in drug delivery. The numerous functionality available in tree gums polysaccharides such as $-CONH_2$, $-NH_2$, $-COOH$, $-OH$, $-SO_3H$, and $-CH_3CO$ are suitable for chemical modification or polymerization into desired materials for pH-responsive, sustained drug release systems (Jana et al., 2017).

2D Janus composite (dual-stimuli-responsive) nanostructures based on GA, chitosan, and poly(ε-caprolactone)-*block*-poly(ethylene glycol) with Au NPs were fabricated via deposition-surface functionalization-chemical exfoliation processes (Kim and Lu, 2016). The produced nanostructures had shown a potential pH-responsive stimulus. The potential of 2D Janus composite nanostructures with poly (ε-caprolactone)-*block*-poly (ethylene glycol) grafted chitosan was further explored for the release studies of doxorubicin (Dox) induced by photothermal stability under NIR-laser irradiation. The fabricated nanocomposite could be used for dual purposes such as cancer phototherapy as well as for two-photon-induced photothermal cancer imaging.

Magnetic (Fe_3O_4) NPs were synthesized using GA as a reducing and stabilizing agent and used for cancer theranostic agents (Zhang et al., 2009). The interaction between magnetic NPs with various functional groups of GA ($-NH_2$ and $-COO$) was demonstrated. The GA-magnetic composite with rhodamine B could easily be coopted into 9 L glioma cells. The research highlighted that the resultant magnetic composite could be used as a magnetic resonance imaging-visible drug carrier in accomplishing both magnetic tumor targeting and intracellular drug delivery. A similarly pH-responsive doxorubicin (Dox) loaded GA-magnetic (Fe_3O_4) nanocomposite was synthesized. The developed nanocomposite had good stability at physiological pH (7.4) and delivery of an anticancer drug (Dox) at acidic pH (Banerjee and Chen, 2008).

TG-based stimuli responses (pH- and temperature-dependent) drug delivery composed of Fe_3O_4 NPs functionalized with (3-amino propyl) triethoxy silane (APTES) and chitosan (CS) for curcumin was established (Shafiee et al., 2019). The nanocomposite (Fe_3O_4@APTES/TG/CS) has been characterized by various spectroscopic and microscopic techniques. A Taguchi method has been applied for the standardization and drug entrapment into the nanocomposite. Furthermore, the nanocomposite was studied in two different pH (7.4 and 3.4) buffered solutions at 37 and 40 °C to explore the kinetics and in vitro drug release profile of the drugs, respectively. The results showed that more than 60% of curcumin was delivered at pH 3.4 for the period of 2 h at 40 °C by the Fe_3O_4@APTES/TG/CS nanocomposite.

Magnetic (Fe_3O_4) nanostructured hydrogel composed of TG, maleic anhydride, and polyacrylic acid via free radical polymerization was fabricated. The developed pH-responsive hydrogel was established as a chemo/hyperthermic route to treat the tumor (Sayadnia et al., 2021). The hydrogel complex was incorporated into an anticancer drug (Dox) and its loading efficiency and encapsulation effectiveness were confirmed. The other aspects such as drug release study, the effect of pH and temperature on the targeting of the cancerous cell, and the chemistry of the functional components in the hydrogel system were studied for the effective development of this nano-assembly for cancer therapy.

GG-based nanostructures and composite hydrogels have been applied to various stimuli-response drug delivery formulations such as pH, thermo, enzyme, and multistimuli drug delivery systems (Jana et al., 2017). Recently, an electro-responsive transdermal drug delivery system based on polyacrylamide-grafted-gum ghatti copolymer was synthesized via free radical polymerization, and its controlled release of quetiapine fumarate, an antipsychotic drug, was investigated (Birajdar et al., 2019).

15.7 Hydrogels/nanogels based on tree gums for biomedical applications

Natural polymer-based hydrogels and nanogels are in high demand as they are beneficial for developing a greener strategy for the biomedical area due to their biodegradability, abundance, and biocompatibility.

Nanogel based on GA-aldehyde with gelatin and its encapsulation and controlled release of curcumin toward breast cancer cells was demonstrated (Sarika and Nirmala, 2016). The nanogel formation by inverse miniemulsion technique, encapsulation efficiency of curcumin into nanogel, and the release profiles and cytotoxicity of the curcumin-loaded nanogels in MCF7 cells—human breast cancer cell line were systematically characterized and established. The bioavailability, water solubility, and stability of curcumin have improved the overall performance of the nanogel (GA-aldehyde with gelatin). This system could use for the safe drug delivery application for the treatment of cancer cells.

The pH- and thermal-responsive GG-based Ag NPs composite hydrogel was synthesized (Dai et al., 2016). During the Ag NPs synthesis, sodium borohydride was involved as a reducing and cross-linking agent in the GG-hydrogel structure. The research highlighted that the prepared hydrogel had a pH-responsive sol-gel transition in acidic and basic conditions, respectively, in addition to their thermal responsiveness (hydrogel to 70 °C yielded the sol state, even though consequent cooling to 20 °C reinstated the hydrogel). The resultant GG-Ag NPs-based hydrogel has shown the properties such as self-healing, facile synthesis protocol, and injectable properties, and can be employed for various biomedical fields.

A GG-based anticancer drug loaded (5-fluorouracil (5-FU)) into lysine-β-cyclodextrin (L-β-CD) and the composite material was successfully demonstrated as a pH-responsive drug carrier (Praphakar et al., 2018). The biocompatibility and hydrophobic/hydrophilicity of the composite and interaction between the functionality among the host-guest interaction of the components in the composite and drug molecules played major roles in the potential rate-controlled drug delivery. A similar approach was carried out in the formulation of 5-FU loaded GG/poly(acrylamide-*co*-acrylamidoglycolic acid) (PGAGA) and demonstrated as a pH responsive anticancer drug, which showed low swelling ratio at low pH value (Reddy et al., 2018).

The development of nanogel-based natural gums incorporated anticancer drugs and their delivery is an important emerging area in the field of "smart" drug delivery. Despite its biodegradability and biocompatibility, polysaccharide-based nanogel possesses an assortment of functional groups, high specific surface area, the balance of hydrophobicity/hydrophobicity, and tunable sizes to carry bioactive molecules or drugs for biomedical applications including drug delivery schemes (Debele et al., 2016; Das and Pal, 2015; Kwon and Matsuda, 2005; Dheer et al., 2017; Peters et al., 2018).

A nanogel consists of cisplatin, an anticancer drug, TG, and lecithin via the nanoemulsion method for controlled delivery of the drug (Verma et al., 2020). The shape and particle size of the prepared nanogel were found to be polygonal and 50–60 nm, respectively. Additionally, the effects of the drug release were controlled in different pH (2.0–7.4), and the functionality such as carboxyl in the TG structure, the time for the maximum amount of drug release, and their circulation time (24 h) in the body have been demonstrated. The resultant nanogel-based tree gum would be a potential system for anticancer treatment.

A sustainable and controlled drug delivery mechanism without any adverse side effects based on TG, poly(2-acrylamide-2-methylpropane sulfonic acid) (PAMPSA), and poly(vinyl alcohol) (PVA) with an anticancer drug (methotrexate) hydrogel was established (Singh et al., 2020b). The nanocomposite hydrogel was characterized by XRD, FTIR, and SEM analysis, and drug-releasing characteristic properties were also analyzed by blood compatibility, mucoadhesion, and mechanical strength. The biocompatibility and biodegradability of the resultant hydrogel composite offer potential therapeutic applications.

A superadsorbent biosensor nanohydrogel was structured by grafting acrylic acid (AA) on TG. In this radical polymerization reaction, *N,N′*-methylene bisacrylamide (MBA) was used as a cross-linker and potassium persulfate was used as the redox initiator. The resultant nanohydrogel with TG nanoparticles, CdTe quantum dots (QDs), and glucose oxidase (GOx) complex for the detection of glucose in diabetic patients was assembled (Qasemi and Ghaemy, 2020).

Subsequently, TG-based nanohydrogel with itaconic acid for the controlled release of ampicillin was formulated. This nanohydrogel was synthesized by free radical polymerization reaction under microwave radiation using TG, itaconic acid, MBA (cross-linker), and potassium persulfate (initiator) (Pathania et al., 2018). The major factors for the synthesis, such as reaction time, temperature, amount of solvent, pH, monomer concentration, and cross-linker, were optimized to acquire maximum nanogel yield. The results demonstrated that pH is an important factor for ampicillin drug release and an acidic pH (2.2) was observed to be the maximum amount of the drug (93.8%) released. The drug release followed a Korsmeyer-Peppas kinetic model. The generated nanohydrogel system can be applied for the effective delivery of several varieties of antibiotics.

GK, carbopol, and GO-based composite hydrogel was fabricated by radiation-induced cross-linking (Singh and Singh, 2018). The prepared composite hydrogel was used for the release study of gemcitabine, an anticancer drug. The influence of GO on the formation of the hydrogel, and its gel strength, swelling capacity, drug loading, and release study, were investigated. Further various drug release and aspects such as non-Fickian diffusion mechanism and kinetics of drug discharge were also analyzed by the Korsmeyer-Peppas model for drug release in GO-based hydrogel. Furthermore, GK/carbopol/GO-based hydrogel was investigated for its mechanical, blood compatibility, antioxidant activity, mucoadhesion, and gel strength, which would be useful for biomedical applications, especially for site-specific drug delivery of anticancer drugs.

Natural biopolymers such as KG polysaccharide and a protein (gelatin) were blended and freeze-dried to form a 3D KG-gelatin sponge (Rathore et al., 2016). The properties of the resultant sponges were identified by SEM, FTIR, thermal study, and swelling capacity, and their in vitro drug loading/release mechanism and antibacterial efficiency were assessed. The KG-gelatin sponge exhibited 95% of drug release (ciprofloxacin) at the end of 48 h and antibacterial efficacy against *S. aureus* and *E. coli*. Furthermore, the sponge's biocompatibility was explored by testing cell adhesion and proliferation using NIH 3T3 fibroblast and human keratinocytes cell lines. The overall result highlighted that the developed sponge had potential as a wound dressing scaffold.

The progress of tree gums in expanding application fields such as pharmaceutical, biomedical, tissue engineering, and food and other nonfood applications have been considered in this chapter. However, Table 15.1 also describes the expanding progress of tree gum nanostructures (NPs, NFs, and nanohydrogels and nanocomposites) in biomedical

TABLE 15.1 Tree gum nanostructures and their preparation, properties, and potential biomedical applications.

Tree gums or conjugated polymers	NPs/composites/membranes/ hydrogel/nanogel based on tree gum structures	Method of preparation/applications	References
GA (gum arabic)	Ag NPs (silver nanoparticles)	– GA-based semi-IPN hydrogel with Ag NPs via in situ reduction method – Antibacterial, biodegradable superabsorbent hydrogel for biomedical application	Gils et al. (2010)
GA	Fe_3O_4 NPs; 20 nm; nonspherical morphology	– Wet chemical process – Isolation/detection of *S. aureus* – Application in clinical and diagnosis	Chockalingam et al. (2010)
GA (GA/citrus fruits flavonoids Hesperidin (HDN)	Ag NPs; particle size of GA-AgNPs = 107 nm; and GA-AgNPs-HDN = 182 nm	– Green synthesis – Amoebicidal and bactericidal properties – Noncytotoxicity	Anwar et al. (2019)

TABLE 15.1 Tree gum nanostructures and their preparation, properties, and potential biomedical applications—cont'd

Tree gums or conjugated polymers	NPs/composites/membranes/hydrogel/nanogel based on tree gum structures	Method of preparation/applications	References
GA (GA-chitosan-quercetin NPs)	Particle size of the NPs = 267.3 ± 43.6 nm	– Ionic gelation – Encapsulation of bioactive compounds and their bioavailability	Kim et al. (2019)
GA	ZnO nanofluids; 200–350 nm	– Precipitating method assisted with microwave heating – Antibacterial activities (*Staphylococcus aureus* and *Escherichia coli*)	Pauzi et al. (2020b)
GA (GA-SA hydrogel)	ZnO NPs; 70–100 nm	– Hydrothermal process and covalent cross-linking – Low toxicity, biocompatible, antibacterial against *Pseudomonas aeruginosa* and *Bacillus cereus* and wound-healing effect	Chopra et al. (2015) and Raguvaran et al. (2017)
GA (GA-PVA hydrogel)	Ag NPs; 10–40 nm	– In situ synthesis of (AgNPs) and gamma radiation-induced cross-linking – Antibacterial action against *E. coli*	Juby et al. (2012)
GA (GA/chitosan-gelatin-PVA) nanofibers	Fiber diameter = 100–200 nm	– Blend electrospinning – Cytocompatibility for tissue engineering applications	Tsai et al. (2015)
GK	Interpenetrating biopolymer-based network of GK and CM locust bean gum	– Williamson synthesis – Controlled release of carvedilol phosphate for the treatment of hypertension	Laha et al. (2019)
TG	ZnO; particle size = 62 nm; star-like morphology	– In situ synthesis – Potential antibacterials (tested for *S. aureus*, *E. coli*, and *Candida albicans*)	Ghayempour and Montazer (2017)
TG (TG/chitosan complex)	Nanocarbon dots; particle size = 70–90 nm	– Process – Much lower cytotoxicity and biocompatibility – Uses in bio-sensing and labeling – Applications in medical diagnostics and bio-imaging	Moradi et al. (2018)
TG	ZnO; urchin-like; 55–80 nm	– Ultrasonic irradiation – Exhibited antibacterial (*S. aureus* and *E. coli*) and antifungal (*C. albicans*) activity	Ghayempour et al. (2016)

Continued

TABLE 15.1 Tree gum nanostructures and their preparation, properties, and potential biomedical applications—cont'd

Tree gums or conjugated polymers	NPs/composites/membranes/ hydrogel/nanogel based on tree gum structures	Method of preparation/applications	References
TG (TG, poly (Ɛ-caprolactone) (PCL) and poly (vinyl alcohol) (PVA)) nanofibers	168 ± 29 nm (fiber diameter); Tensile strength (2.7 MPa); Young's modulus (56 MPa)	– Electrospinning – Scaffold for diabetic wound healing	Ranjbar Mohammadi et al. (2020)
TG (polyvinylpyrrolidone (PVP)/TG) fibers	Uniform, cylindrical, and smooth fibers with diameters 465 ± 185 to 1171 ± 342 nm in range	– Electrospinning – Biomedical applications	Martín-Alfonso et al. (2019)
TG (TG-chitosan (Ch)-/ SiO_2) and TG-Ch/SiO_2/ Ag NPs nanocomposites	SiO_2 NPs (22 nm particle size); Ag NPs (27 nm)	– Preparation of nanocomposite films by solution casting and solvent evaporation – Antibacterial composites	Mallakpour and Abbasi (2020)
TG (TG-lecithin/Au NPs mucoadhesive hybrid NPs with Amphotericin B)	358.3 ± 1.7 nm (particle size)	– Solvent diffusion and stirring methods – Nano-carriers-based drug delivery of enhancing AmpB oral bioavailability	Jabri et al. (2018)
TG (curcumin) (PCL/TG) nanofibers	Fiber diameter = 164 ± 34 to 191 ± 24 nm	– Electrospinning technique – Biodegradable, mechanically robust nanofibers for biomedical application	Ranjbar-Mohammadi and Bahrami (2016)
KG (carboxyl methylated)	Au NPs; particle size = 11 ± 2 nm	– Stirring at 65 °C – Antibacterial action	Seku et al. (2019)
KG (DDSA-KG/PVA) nanofibers	Nanofiber diameter = ~200 nm	– Electrospinning – Antibacterial nanofibers	Padil et al. (2015a)
GG (PVA/GG nanofibers)	Fe_3O_4; particle size = 10.6 ± 4.4 nm; nanofiber (PVA/GG/Fe_3O_4) diameter = 150 nm	– Electrospinning – Biodegradable wound dressing material	Lubambo et al. (2015)
GG (GG/curcumin) tablet	–	– Wet granulation – Colon cancer treatment using guar gum, and enhanced bioavailability of curcumin	Elias et al. (2010)
GG	GG-folic acid/methotrexate NPs; average size = 325 nm	– Emulsion cross-linking method – Drug delivery to the GI tract for targeting colon cancer	Sharma et al. (2013)

applications (prominence on pharmaceutical drug delivery, cancer treatment, biosensor, imaging, tissue engineering, wound healing, etc.).

15.8 Conclusions

Several commercial tree gum-based nanostructures have been developed recently with various properties and their biomedical applications justify the importance of this chapter. The modified natural gums, and their nanoparticles, nanofibers, and nanohydrogels, provide a safe and stable medium for encapsulation of drugs and their systematic drug release that has engrossed excellent effectiveness. Generally, tree gum-based nanostructures have superb properties such as nontoxicity, good carriers for drug or bioactive molecules, bio-friendliness, cost-effectiveness, biocompatibility, and biodegradability. To achieve a green and sustainable future, the development of nanostructured materials based on natural or biodegradable polymers, specifically tree gum-based nanostructures, would be an ideal candidate for many potential applications in the biomedical field.

Acknowledgments

The authors gratefully acknowledge the financial support from the Ministry of Education, Youth and Sports in the Czech Republic under the "Inter Excellence Action Programme and Project Hybrid Materials for Hierarchical Structures" (HyHi, Reg. no. CZ.02.1.01/0.0/0.0/16_019/ 0000843), and Research Infrastructure NanoEnviCz, under Project no. LM2018124, supported by the Ministry of Education, Youth and Sports of the Czech Republic and the European Union—European Structural and Investment Funds in the framework of the Operational Programme Research, Development, and Education. This work was also supported by the project "Tree Gum Polymers and Their Modified Bioplastics for Food Packaging Application" granted by the Bavarian-Czech-Academic-Agency (BTHA) (registration numbers LTAB19007 and BTHA-JC-2019-26) and the Ministry of Education, Youth and Sports in the Czech Republic under the "Inter Excellence—Action program" within the framework of the project "Bio-based Porous 2D Membranes and 3D Sponges Based on Functionalized Tree Gum Polysaccharides and their Environmental Application" (registration number LTAUSA19091)—TUL internal no. 18309/136.

References

Abdoli, M., Sadrjavadi, K., Arkan, E., Zangeneh, M.M., Moradi, S., Zangeneh, A., Shahlaei, M., Khaledian, S., 2020. Polyvinyl alcohol/gum tragacanth/graphene oxide composite nanofiber for antibiotic delivery. J. Drug Deliv. Sci. Technol. 60, 102044.

Abdullah, M.F., Ghosh, S.K., Basu, S., Mukherjee, A., 2015. Cationic guar gum orchestrated environmental synthesis for silver nano-bio-composite films. Carbohydr. Polym. 134, 30–37.

Anwar, A., Masri, A., Rao, K., Rajendran, K., Khan, N.A., Shah, M.R., Siddiqui, R., 2019. Antimicrobial activities of green synthesized gums-stabilized nanoparticles loaded with flavonoids. Sci. Rep. 9 (1), 3122.

Atgié, M., Garrigues, J.C., Chennevière, A., Masbernat, O., Roger, K., 2019. Gum arabic in solution: composition and multi-scale structures. Food Hydrocoll. 91, 319–330.

Balasubramanian, S., Bezawada, S.R., Raghavachari, D., 2016. Green, selective, seedless and one-pot synthesis of triangular Au nanoplates of controlled size using bael gum and mechanistic study. ACS Sustain. Chem. Eng. 4 (7), 3830–3839.

Banerjee, S.S., Chen, D.-H., 2008. Multifunctional PH-sensitive magnetic nanoparticles for simultaneous imaging, sensing and targeted intracellular anticancer drug delivery. Nanotechnology 19 (50), 505104.

Bashir, S., Teo, Y.Y., Ramesh, S., Ramesh, K., 2018. Synthesis and characterization of karaya gum-g-poly (acrylic acid) hydrogels and in vitro release of hydrophobic quercetin. Polymer 147, 108–120.

Bera, H., Boddupalli, S., Nayak, A.K., 2015. Mucoadhesive-floating zinc-pectinate–sterculia gum interpenetrating polymer network beads encapsulating ziprasidone HCl. Carbohydr. Polym. 131, 108–118.
Bie, Y., Yang, J., Nuli, Y., Wang, J., 2017. Natural karaya gum as an excellent binder for silicon-based anodes in high-performance lithium-ion batteries. J. Mater. Chem. A 5 (5), 1919–1924.
Birajdar, R.P., Patil, S.B., Alange, V.V., Kulkarni, R.V., 2019. Electro-responsive polyacrylamide-grafted-gum ghatti copolymer for transdermal drug delivery application. J. Macromol. Sci. A 56 (4), 306–315.
Černík, M., Thekkae Padil, V.V., 2013. Green synthesis of copper oxide nanoparticles using gum karaya as a biotemplate and their antibacterial application. Int. J. Nanomedicine 8, 889.
Chacko, R.T., Ventura, J., Zhuang, J., Thayumanavan, S., 2012. Polymer nanogels: a versatile nanoscopic drug delivery platform. Adv. Drug Deliv. Rev. 64 (9), 836–851.
Chockalingam, A., Babu, H., Chittor, R., Tiwari, J., 2010. Gum arabic modified Fe3O4 nanoparticles cross linked with collagen for isolation of bacteria. J. Nanobiotechnol. 8 (1), 30.
Chopra, M., Bernela, M., Kaur, P., Manuja, A., Kumar, B., Thakur, R., 2015. Alginate/gum acacia bipolymeric nanohydrogels—promising carrier for zinc oxide nanoparticles. Int. J. Biol. Macromol. 72, 827–833.
Dai, L., Nadeau, B., An, X., Cheng, D., Long, Z., Ni, Y., 2016. Silver nanoparticles-containing dual-function hydrogels based on a guar gum-sodium borohydride system. Sci. Rep. 6 (1), 36497.
Das, D., Pal, S., 2015. Modified biopolymer-dextrin based crosslinked hydrogels: application in controlled drug delivery. RSC Adv. 5 (32), 25014–25050.
Debele, T.A., Mekuria, S.L., Tsai, H.-C., 2016. Polysaccharide based nanogels in the drug delivery system: application as the carrier of pharmaceutical agents. Mater. Sci. Eng. C 68, 964–981.
Delfi, M., Ghomi, M., Zarrabi, A., Mohammadinejad, R., Taraghdari, Z.B., Ashrafizadeh, M., Zare, E.N., Agarwal, T., Padil, V.V.T., Mokhtari, B., Rossi, F., Perale, G., Sillanpaa, M., Borzacchiello, A., Kumar Maiti, T., Makvandi, P., 2020. Functionalization of polymers and nanomaterials for biomedical applications: antimicrobial platforms and drug carriers. Prosthesis 2 (2), 117–139.
Dhar, S., Mali, V., Bodhankar, S., Shiras, A., Prasad, B.L.V., Pokharkar, V., 2011. Biocompatible gellan gum-reduced gold nanoparticles: cellular uptake and subacute oral toxicity studies. J. Appl. Toxicol. 31 (5), 411–420.
Dheer, D., Arora, D., Jaglan, S., Rawal, R.K., Shankar, R., 2017. Polysaccharides based nanomaterials for targeted anti-cancer drug delivery. J. Drug Target. 25 (1), 1–16.
Eghbalifam, N., Shojaosadati, S.A., Hashemi-Najafabadi, S., Khorasani, A.C., 2020. Synthesis and characterization of antimicrobial wound dressing material based on silver nanoparticles loaded gum arabic nanofibers. Int. J. Biol. Macromol. 155, 119–130.
El-Batal, A.I., Abd Elkodous, M., El-Sayyad, G.S., Al-Hazmi, N.E., Gobara, M., Baraka, A., 2020. Gum arabic polymer-stabilized and gamma rays-assisted synthesis of bimetallic silver-gold nanoparticles: powerful antimicrobial and antibiofilm activities against pathogenic microbes isolated from diabetic foot patients. Int. J. Biol. Macromol. 165, 169–186.
Elias, E.J., Anil, S., Ahmad, S., Daud, A., 2010. Colon targeted curcumin delivery using guar gum. Nat. Prod. Commun. 5 (6), 1934578X1000500.
Fujioka-Kobayashi, M., Ota, M.S., Shimoda, A., Nakahama, K., Akiyoshi, K., Miyamoto, Y., Iseki, S., 2012. Cholesteryl group- and acryloyl group-bearing pullulan nanogel to deliver BMP2 and FGF18 for bone tissue engineering. Biomaterials 33 (30), 7613–7620.
Geetha, A., Sakthivel, R., Mallika, J., Kannusamy, R., Rajendran, R., 2016. Green synthesis of antibacterial zinc oxide nanoparticles using biopolymer Azadirachta indica gum. Orient. J. Chem. 32 (2), 955–963.
Ghayempour, S., Montazer, M., 2017. Ultrasound irradiation based in-situ synthesis of star-like tragacanth gum/zinc oxide nanoparticles on cotton fabric. Ultrason. Sonochem. 34, 458–465.
Ghayempour, S., Montazer, M., Rad, M.M., 2016. Tragacanth gum biopolymer as reducing and stabilizing agent in biosonosynthesis of urchin-like ZnO nanorod arrays: a low cytotoxic photocatalyst with antibacterial and antifungal properties. Carbohydr. Polym. 136, 232–241.
Ghorbani, M., Mahmoodzadeh, F., Yavari Maroufi, L., Nezhad-Mokhtari, P., 2020. Electrospun tetracycline hydrochloride loaded zein/gum tragacanth/poly lactic acid nanofibers for biomedical application. Int. J. Biol. Macromol. 165, 1312–1322.
Gils, P.S., Ray, D., Sahoo, P.K., 2010. Designing of silver nanoparticles in gum arabic based semi-IPN hydrogel. Int. J. Biol. Macromol. 46 (2), 237–244.
Giulbudagian, M., Asadian-Birjand, M., Steinhilber, D., Achazi, K., Molina, M., Calderón, M., 2014. Fabrication of thermoresponsive nanogels by thermo-nanoprecipitation and in situ encapsulation of bioactives. Polym. Chem. 5 (24), 6909–6913.

Anon., 2018. Gum Arabic. Elsevier.
Hassani, A., Mahmood, S., Enezei, H.H., Hussain, S.A., Hamad, H.A., Aldoghachi, A.F., Hagar, A., Doolaanea, A.A., Ibrahim, W.N., 2020. Formulation, characterization and biological activity screening of sodium alginate-gum arabic nanoparticles loaded with curcumin. Molecules 25 (9), 2244.
Horst, M.F., Coral, D.F., Fernández van Raap, M.B., Alvarez, M., Lassalle, V., 2017. Hybrid nanomaterials based on gum arabic and magnetite for hyperthermia treatments. Mater. Sci. Eng. C 74, 443–450.
Huang, H., Remsen, E.E., Wooley, K.L., 1998. Amphiphilic core-shell nanospheres obtained by intramicellar shell crosslinking of polymer micelles with poly(ethylene oxide) linkers. Chem. Commun. 13, 1415–1416.
Huang, J., Li, Q., Sun, D., Lu, Y., Su, Y., Yang, X., Wang, H., Wang, Y., Shao, W., He, N., Hong, J., Chen, C., 2007. Biosynthesis of silver and gold nanoparticles by novel sun dried Cinnamomum camphora leaf. Nanotechnology 18 (10), 105104.
Ilka, R., Mohseni, M., Kianirad, M., Naseripour, M., Ashtari, K., Mehravi, B., 2018. Nanogel-based natural polymers as smart carriers for the controlled delivery of timolol maleate through the cornea for glaucoma. Int. J. Biol. Macromol. 109, 955–962.
Isobe, N., Sagawa, N., Ono, Y., Fujisawa, S., Kimura, S., Kinoshita, K., Miuchi, T., Iwata, T., Isogai, A., Nishino, M., Deguchi, S., 2020. Primary structure of gum arabic and its dynamics at oil/water interface. Carbohydr. Polym. 249, 116843.
Jabri, T., Imran, M., Shafiullah, R.K., Ali, I., Arfan, M., Shah, M.R., 2018. Fabrication of lecithin-gum tragacanth mucoadhesive hybrid Nano-carrier system for in-vivo performance of amphotericin B. Carbohydr. Polym. 194, 89–96.
Jana, S., Maiti, S., Jana, S., 2017. Stimuli-responsive guar gum composites for colon-specific drug delivery. In: Biopolymer-Based Composites. Elsevier, pp. 61–79.
Jo, W., Bak, J.H., Yoo, B., 2018. Rheological characterizations of concentrated binary gum mixtures with xanthan gum and galactomannans. Int. J. Biol. Macromol. 114, 263–269.
Juby, K.A., Dwivedi, C., Kumar, M., Kota, S., Misra, H.S., Bajaj, P.N., 2012. Silver nanoparticle-loaded PVA/gum acacia hydrogel: synthesis, characterization and antibacterial study. Carbohydr. Polym. 89 (3), 906–913.
Kennedy, J.F., Phillips, G.O., Williams, P.A. (Eds.), 2011. Gum Arabic. Royal Society of Chemistry, Cambridge.
Kim, J.-H., Lu, T.-M., 2016. Bio-inspired janus composite nanoscrolls for on-demand tumour targeting. RSC Adv. 6 (21), 17179–17187.
Kim, E.S., Kim, D.Y., Lee, J.-S., Lee, H.G., 2019. Mucoadhesive chitosan–gum arabic nanoparticles enhance the absorption and antioxidant activity of quercetin in the intestinal cellular environment. J. Agric. Food Chem. 67 (31), 8609–8616.
Kora, A.J., Sashidhar, R.B., 2018. Biogenic silver nanoparticles synthesized with Rhamnogalacturonan gum: antibacterial activity, cytotoxicity and its mode of action. Arab. J. Chem. 11 (3), 313–323.
Kousalová, J., Etrych, T., 2018. Polymeric nanogels as drug delivery systems. Physiol. Res. 67, s305–s317.
Kumar, S.S.D., Mahesh, A., Antoniraj, M.G., Rathore, H.S., Houreld, N.N., Kandasamy, R., 2018. Cellular imaging and folate receptor targeting delivery of gum kondagogu capped gold nanoparticles in cancer cells. Int. J. Biol. Macromol. 109, 220–230.
Kwon, I.K., Matsuda, T., 2005. Co-electrospun nanofiber fabrics of poly(L-lactide-co-ε-caprolactone) with type I collagen or heparin. Biomacromolecules 6 (4), 2096–2105.
Laha, B., Goswami, R., Maiti, S., Sen, K.K., 2019. Smart karaya-locust bean gum hydrogel particles for the treatment of hypertension: optimization by factorial design and pre-clinical evaluation. Carbohydr. Polym. 210, 274–288.
Lee, H., Yoo, B., 2021. Agglomeration of galactomannan gum powders: physical, rheological, and structural characterizations. Carbohydr. Polym. 256, 117599.
Liu, G., Song, Y., Wang, J., Zhuang, H., Ma, L., Li, C., Liu, Y., Zhang, J., 2014. Effects of nanoclay type on the physical and antimicrobial properties of PVOH-based nanocomposite films. LWT Food Sci. Technol. 57 (2), 562–568.
Lubambo, A.F., Ono, L., Drago, V., Mattoso, N., Varalda, J., Sierakowski, M.R., Sakakibara, C.N., Freitas, R.A., Saul, C.-K., 2015. Tuning Fe3O4 nanoparticle dispersion through PH in PVA/guar gum/electrospun membranes. Carbohydr. Polym. 134, 775–783.
Makvandi, P., Ghomi, M., Padil, V.V.T., Shalchy, F., Ashrafizadeh, M., Askarinejad, S., Pourreza, N., Zarrabi, A., Nazarzadeh Zare, E., Kooti, M., Mokhtari, B., Borzacchiello, A., Tay, F.R., 2020. Biofabricated nanostructures and their composites in regenerative medicine. ACS Appl. Nano Mater. 3 (7), 6210–6238.
Mallakpour, S., Abbasi, M., 2020. Hydroxyapatite mineralization on chitosan-tragacanth gum/silica@silver nanocomposites and their antibacterial activity evaluation. Int. J. Biol. Macromol. 151, 909–923.
Martín-Alfonso, J.E., Číková, E., Omastová, M., 2019. Development and characterization of composite fibers based on tragacanth gum and polyvinylpyrrolidone. Compos. Part B 169, 79–87.

Mohammadi, M.R., Rabbani, S., Bahrami, S.H., Joghataei, M.T., Moayer, F., 2016. Antibacterial performance and in vivo diabetic wound healing of curcumin loaded gum tragacanth/poly(ε-caprolactone) electrospun nanofibers. Mater. Sci. Eng. C 69, 1183–1191.

Mohammadinejad, R., Maleki, H., Larrañeta, E., Fajardo, A.R., Nik, A.B., Shavandi, A., Sheikhi, A., Ghorbanpour, M., Farokhi, M., Govindh, P., Cabane, E., Azizi, S., Aref, A.R., Mozafari, M., Mehrali, M., Thomas, S., Mano, J.F., Mishra, Y.K., Thakur, V.K., 2019. Status and future scope of plant-based green hydrogels in biomedical engineering. Appl. Mater. Today 16, 213–246.

Mohammadinejad, R., Kumar, A., Ranjbar-Mohammadi, M., Ashrafizadeh, M., Han, S.S., Khang, G., Roveimiab, Z., 2020. Recent advances in natural gum-based biomaterials for tissue engineering and regenerative medicine: a review. Polymers 12 (1), 176.

Moradi, S., Sadrjavadi, K., Frahdian, N., Hossainzadeh, L., Shahlaei, M., 2018. Easy synthesis, characterization and cell cytotoxicity of green nano carbon dots using hydrothermal carbonization of gum tragacanth and chitosan biopolymers for bioimaging. J. Mol. Liq. 259, 284–290.

Nasir, O., Wang, K., Föller, M., Bhandaru, M., Sandulache, D., Artunc, F., Ackermann, T.F., Ebrahim, A., Palmada, M., Klingel, K., Saeed, A.M., Lang, F., 2010. Downregulation of angiogenin transcript levels and inhibition of colonic carcinoma by gum arabic (Acacia Senegal). Nutr. Cancer 62 (6), 802–810.

Nayak, A.K., Das, B., Maji, R., 2012. Calcium alginate/gum arabic beads containing glibenclamide: development and in vitro characterization. Int. J. Biol. Macromol. 51 (5), 1070–1078.

Nayak, A.K., Hasnain, M.S., Pal, K., Banerjee, I., Pal, D., 2020. Gum-based hydrogels in drug delivery. In: Biopolymer-Based Formulations. Elsevier, pp. 605–645.

Neamtu, I., Rusu, A.G., Diaconu, A., Nita, L.E., Chiriac, A.P., 2017. Basic concepts and recent advances in nanogels as carriers for medical applications. Drug Deliv. 24 (1), 539–557.

Nejatian, M., Abbasi, S., Azarikia, F., 2020. Gum tragacanth: structure, characteristics and applications in foods. Int. J. Biol. Macromol. 160, 846–860.

Oh, J.K., Lee, D.I., Park, J.M., 2009. Biopolymer-based microgels/nanogels for drug delivery applications. Prog. Polym. Sci. 34 (12), 1261–1282.

Padil, V.V.T., Černík, M., 2015. Poly (vinyl alcohol)/gum karaya electrospun plasma treated membrane for the removal of nanoparticles (Au, Ag, Pt, CuO and Fe3O4) from aqueous solutions. J. Hazard. Mater. 287, 102–110.

Padil, V.V.T., Nguyen, N.H.A., Rozek, Z., Ševců, A., Černík, M., 2015a. Synthesis, fabrication and antibacterial properties of a plasma modified electrospun membrane consisting of gum kondagogu, dodecenyl succinic anhydride and poly (vinyl alcohol). Surf. Coat. Technol. 271, 32–38.

Padil, V.V.T., Stuchlík, M., Černík, M., 2015b. Plasma modified nanofibres based on gum kondagogu and their use for collection of nanoparticulate silver, gold and platinum. Carbohydr. Polym. 121, 468–476.

Padil, V.V.T., Nguyen, N.H.A., Ševců, A., Černík, M., 2015c. Fabrication, characterization, and antibacterial properties of electrospun membrane composed of gum karaya, polyvinyl alcohol, and silver nanoparticles. J. Nanomater. 2015, 1–10.

Padil, V.V.T., Wacławek, S., Černík, M., 2016a. Green synthesis: nanoparticles and nanofibres based on tree gums for environmental applications. Ecol. Chem. Eng. S 23 (4), 533–557.

Padil, V.V.T., Senan, C., Wacławek, S., Černík, M., 2016b. Electrospun fibers based on arabic, karaya and kondagogu gums. Int. J. Biol. Macromol. 91, 299–309.

Padil, V.V.T., Wacławek, S., Černík, M., Varma, R.S., 2018. Tree gum-based renewable materials: sustainable applications in nanotechnology, biomedical and environmental fields. Biotechnol. Adv. 36 (7), 1984–2016.

Padil, V.V.T., Senan, C., Černík, M., 2019. "Green" polymeric electrospun fibers based on tree-gum hydrocolloids. In: Materials for Biomedical Engineering. Elsevier, pp. 127–172.

Padil, V.V., Cheong, J.Y., AkshayKumar, K.P., Makvandi, P., Zare, E.N., Torres-Mendieta, R., Wacławek, S., Černík, M., Kim, I.-D., Varma, R.S., 2020. Electrospun fibers based on carbohydrate gum polymers and their multifaceted applications. Carbohydr. Polym. 247, 116705.

Pamfil, D., Vasile, C., 2018. Nanogels of natural polymers. In: Thakur, V., Thakur, M., Voicu, S. (Eds.), Polymer Gels. Gels Horizons: From Science to Smart Materials. Springer, Singapore, pp. 71–110. https://doi.org/10.1007/978-981-10-6080-9_4.

Pandey, S., Mishra, S.B., 2014. Catalytic reduction of P-nitrophenol by using platinum nanoparticles stabilised by guar gum. Carbohydr. Polym. 113, 525–531.

Pathania, D., Verma, C., Negi, P., Tyagi, I., Asif, M., Kumar, N.S., Al-Ghurabi, E.H., Agarwal, S., Gupta, V.K., 2018. Novel nanohydrogel based on itaconic acid grafted tragacanth gum for controlled release of ampicillin. Carbohydr. Polym. 196, 262–271.

Pauzi, N., Zain, N.M., Kutty, R.V., Ramli, H., 2020a. Antibacterial and antibiofilm properties of ZnO nanoparticles synthesis using gum arabic as a potential new generation antibacterial agent. Mater. Today Proc. 41, 1–8.

Pauzi, N., Zain, N.M., Yusof, N.A.A., 2020b. Gum arabic as natural stabilizing agent in green synthesis of ZnO nanofluids for antibacterial application. J. Environ. Chem. Eng. 8 (3), 103331.

Peters, J.T., Hutchinson, S.S., Lizana, N., Verma, I., Peppas, N.A., 2018. Synthesis and characterization of poly(N-isopropyl methacrylamide) core/shell nanogels for controlled release of chemotherapeutics. Chem. Eng. J. 340, 58–65.

Popescu, M.-C., Dogaru, B.-I., Sun, D., Stoleru, E., Simionescu, B.C., 2019. Structural and sorption properties of bio-nanocomposite films based on κ-carrageenan and cellulose nanocrystals. Int. J. Biol. Macromol. 135, 462–471.

Praphakar, R.A., Jeyaraj, M., Mehnath, S., Higuchi, A., Ponnamma, D., Sadasivuni, K.K., Rajan, M., 2018. A PH-sensitive guar gum-grafted-lysine-β-cyclodextrin drug carrier for the controlled release of 5-flourouracil into cancer cells. J. Mater. Chem. B 6 (10), 1519–1530.

Qasemi, S., Ghaemy, M., 2020. Novel superabsorbent biosensor nanohydrogel based on gum tragacanth polysaccharide for optical detection of glucose. Int. J. Biol. Macromol. 151, 901–908.

Raemdonck, K., Demeester, J., De Smedt, S., 2009. Advanced nanogel engineering for drug delivery. Soft Matter 5 (4), 707–715.

Raguvaran, R., Manuja, B.K., Chopra, M., Thakur, R., Anand, T., Kalia, A., Manuja, A., 2017. Sodium alginate and gum acacia hydrogels of ZnO nanoparticles show wound healing effect on fibroblast cells. Int. J. Biol. Macromol. 96, 185–191.

Ranjbar Mohammadi, M., Kargozar, S., Bahrami, S.H., Rabbani, S., 2020. An excellent nanofibrous matrix based on gum tragacanth-poly (Ɛ-caprolactone)-poly (vinyl alcohol) for application in diabetic wound healing. Polym. Degrad. Stab. 174, 109105.

Ranjbar-Mohammadi, M., Bahrami, S.H., 2015. Development of nanofibrous scaffolds containing gum tragacanth/poly (ε-caprolactone) for application as skin scaffolds. Mater. Sci. Eng. C 48, 71–79.

Ranjbar-Mohammadi, M., Bahrami, S.H., 2016. Electrospun curcumin loaded poly(ε-caprolactone)/gum tragacanth nanofibers for biomedical application. Int. J. Biol. Macromol. 84, 448–456.

Ranjbar-Mohammadi, M., Bahrami, S.H., Joghataei, M.T., 2013. Fabrication of novel nanofiber scaffolds from gum tragacanth/poly(vinyl alcohol) for wound dressing application: in vitro evaluation and antibacterial properties. Mater. Sci. Eng. C 33 (8), 4935–4943.

Ranjbar-Mohammadi, M., Zamani, M., Prabhakaran, M.P., Bahrami, S.H., Ramakrishna, S., 2016. Electrospinning of PLGA/gum tragacanth nanofibers containing tetracycline hydrochloride for periodontal regeneration. Mater. Sci. Eng. C 58, 521–531.

Rathore, H.S., Sarubala, M., Ramanathan, G., Singaravelu, S., Raja, M.D., Gupta, S., Sivagnanam, U.T., 2016. Fabrication of biomimetic porous novel sponge from gum kondagogu for wound dressing. Mater. Lett. 177, 108–111.

Reddy, G.B., Madhusudhan, A., Ramakrishna, D., Ayodhya, D., Venkatesham, M., Veerabhadram, G., 2015. Green chemistry approach for the synthesis of gold nanoparticles with gum kondagogu: characterization, catalytic and antibacterial activity. J. Nanostruct. Chem. 5 (2), 185–193.

Reddy, G.V., Reddy, N.S., Nagaraja, K., Rao, K.S.V.K., 2018. Synthesis of PH responsive hydrogel matrices from guar gum and poly(acrylamide-co-acrylamidoglycolic acid) for anti-cancer drug delivery. J. Appl. Pharm. Sci. 8 (8), 84–91.

Rizwan, M., Yahya, R., Hassan, A., Yar, M., Azzahari, A., Selvanathan, V., Sonsudin, F., Abouloula, C., 2017. PH sensitive hydrogels in drug delivery: brief history, properties, swelling, and release mechanism, material selection and applications. Polymers 9 (12), 137.

Roque, A.C.A., Bicho, A., Batalha, I.L., Cardoso, A.S., Hussain, A., 2009. Biocompatible and bioactive gum arabic coated iron oxide magnetic nanoparticles. J. Biotechnol. 144 (4), 313–320.

Roy, S., Shankar, S., Rhim, J.-W., 2019. Melanin-mediated synthesis of silver nanoparticle and its use for the preparation of carrageenan-based antibacterial films. Food Hydrocoll. 88, 237–246.

Ryu, J.H., Chacko, R.T., Jiwpanich, S., Bickerton, S., Babu, R.P., Thayumanavan, S., 2010. Self-cross-linked polymer nanogels: a versatile nanoscopic drug delivery platform. J. Am. Chem. Soc. 132 (48), 17227–17235.

Sadat Hosseini, M., Hemmati, K., Ghaemy, M., 2016. Synthesis of nanohydrogels based on tragacanth gum biopolymer and investigation of swelling and drug delivery. Int. J. Biol. Macromol. 82, 806–815.

Sahay, R., Kumar, P.S., Sridhar, R., Sundaramurthy, J., Venugopal, J., Mhaisalkar, S.G., Ramakrishna, S., 2012. Electrospun composite nanofibers and their multifaceted applications. J. Mater. Chem. 22 (26), 12953–12971.

Samadian, H., Maleki, H., Allahyari, Z., Jaymand, M., 2020. Natural polymers-based light-induced hydrogels: promising biomaterials for biomedical applications. Coord. Chem. Rev. 420, 213432.

Sarika, P.R., Nirmala, R.J., 2016. Curcumin loaded gum arabic aldehyde-gelatin nanogels for breast cancer therapy. Mater. Sci. Eng. C 65, 331–337.

Sarika, P.R., Cinthya, K., Jayakrishnan, A., Anilkumar, P.R., James, N.R., 2014. Modified gum arabic cross-linked gelatin scaffold for biomedical applications. Mater. Sci. Eng. C 43, 272–279.

Sarika, P.R., James, N.R., Kumar, P.R.A., Raj, D.K., Kumary, T.V., 2015. Gum arabic-curcumin conjugate micelles with enhanced loading for curcumin delivery to hepatocarcinoma cells. Carbohydr. Polym. 134, 167–174.

Sayadnia, S., Arkan, E., Jahanban-Esfahlan, R., Sayadnia, S., Jaymand, M., 2021. Tragacanth gum-based <scp>pH</Scp> -responsive magnetic hydrogels for "smart" chemo/hyperthermia therapy of solid tumors. Polym. Adv. Technol. 32 (1), 262–271.

Seku, K., Gangapuram, B.R., Pejjai, B., Hussain, M., Hussaini, S.S., Golla, N., Kadimpati, K.K., 2019. Eco-friendly synthesis of gold nanoparticles using carboxymethylated gum cochlospermum gossypium (CMGK) and their catalytic and antibacterial applications. Chem. Pap. 73 (7), 1695–1704.

Shafiee, S., Ahangar, H.A., Saffar, A., 2019. Taguchi method optimization for synthesis of Fe3O4@chitosan/tragacanth gum nanocomposite as a drug delivery system. Carbohydr. Polym. 222, 114982.

Sharma, M., Malik, R., Verma, A., Dwivedi, P., Banoth, G.S., Pandey, N., Sarkar, J., Mishra, P.R., Dwivedi, A.K., 2013. Folic acid conjugated guar gum nanoparticles for targeting methotrexate to colon cancer. J. Biomed. Nanotechnol. 9 (1), 96–106.

Sharma, A., Garg, T., Aman, A., Panchal, K., Sharma, R., Kumar, S., Markandeywar, T., 2016. Nanogel—an advanced drug delivery tool: current and future. Artif. Cells Nanomed. Biotechnol. 44 (1), 165–177.

Sharma, R., Jafari, S.M., Sharma, S., 2020. Antimicrobial bio-nanocomposites and their potential applications in food packaging. Food Control 112, 107086.

Singh, B., Sharma, N., 2011. Design of sterculia gum based double potential antidiarrheal drug delivery system. Colloids Surf. B: Biointerfaces 82 (2), 325–332.

Singh, B., Singh, B., 2018. Modification of sterculia gum polysaccharide via network formation by radiation induced crosslinking polymerization for biomedical applications. Int. J. Biol. Macromol. 116, 91–99.

Singh, J., Kumar, S., Dhaliwal, A.S., 2020a. Controlled release of amoxicillin and antioxidant potential of gold nanoparticles-xanthan gum/poly (acrylic acid) biodegradable nanocomposite. J. Drug Deliv. Sci. Technol. 55, 101384.

Singh, B., Sharma, K., Rajneesh, D.S., 2020b. Dietary fiber tragacanth gum based hydrogels for use in drug delivery applications. Bioact. Carbohydr. Diet. Fibre 21, 100208.

Soleimani, K., Derakhshankhah, H., Jaymand, M., Samadian, H., 2021. Stimuli-responsive natural gums-based drug delivery systems for cancer treatment. Carbohydr. Polym. 254, 117422.

Tong, X., Pan, W., Su, T., Zhang, M., Dong, W., Qi, X., 2020. Recent advances in natural polymer-based drug delivery systems. React. Funct. Polym. 148, 104501.

Tsai, R.Y., Kuo, T.Y., Hung, S.C., Lin, C.M., Hsien, T.Y., Wang, D.M., Hsieh, H.J., 2015. Use of gum arabic to improve the fabrication of chitosan-gelatin-based nanofibers for tissue engineering. Carbohydr. Polym. 115, 525–532.

Venkateshaiah, A., Cheong, J.Y., Shin, S.-H., Akshaykumar, K.P., Yun, T.G., Bae, J., Wacławek, S., Černík, M., Agarwal, S., Greiner, A., Padil, V.V.T., Kim, I.-D., Varma, R.S., 2020a. Recycling non-food-grade tree gum wastes into nanoporous carbon for sustainable energy harvesting. Green Chem. 22 (4), 1198–1208.

Venkateshaiah, A., Cheong, J.Y., Habel, C., Wacławek, S., Lederer, T., Černík, M., Kim, I.-D., Padil, V.V.T., Agarwal, S., 2020b. Tree gum–graphene oxide nanocomposite films as gas barriers. ACS Appl. Nano Mater. 3 (1), 633–640.

Verma, C., Negi, P., Pathania, D., Anjum, S., Gupta, B., 2020. Novel tragacanth gum-entrapped lecithin nanogels for anticancer drug delivery. Int. J. Polym. Mater. Polym. Biomater. 69 (9), 604–609.

Wei, Q., Xu, F., Xu, X., Geng, X., Ye, L., Zhang, A., Feng, Z., 2016. The multifunctional wound dressing with core–shell structured fibers prepared by coaxial electrospinning. Front. Mater. Sci. 10 (2), 113–121.

Williams, P.A., Phillips, G.O., 2021. Gum arabic. In: Handbook of Hydrocolloids. Elsevier, pp. 627–652.

Xu, L., Sitinamaluwa, H., Li, H., Qiu, J., Wang, Y., Yan, C., Li, H., Yuan, S., Zhang, S., 2017. Low cost and green preparation process for α-Fe2O3@gum arabic electrode for high performance sodium ion batteries. J. Mater. Chem. A 5 (5), 2102–2109.

Yousuf, B., Wu, S., Gao, Y., 2021. Characteristics of karaya gum based films: amelioration by inclusion of schisandra chinensis oil and its oleogel in the film formulation. Food Chem. 345, 128859.

Zamani, H., Rastegari, B., Varamini, M., 2019. Antioxidant and anti-cancer activity of dunaliella salina extract and oral drug delivery potential via nano-based formulations of gum arabic coated magnetite nanoparticles. J. Drug Deliv. Sci. Technol. 54, 101278.

Zare, E.N., Makvandi, P., Borzacchiello, A., Tay, F.R., Ashtari, B., Thekkae, V.P.V., 2019. Antimicrobial gum bio-based nanocomposites and their industrial and biomedical applications. Chem. Commun. 55, 14871–14885.

Zarekhalili, Z., Bahrami, S.H., Ranjbar-Mohammadi, M., Milan, P.B., 2017. Fabrication and characterization of PVA/gum tragacanth/PCL hybrid nanofibrous scaffolds for skin substitutes. Int. J. Biol. Macromol. 94, 679–690.

Zhang, L., Yu, F., Cole, A.J., Chertok, B., David, A.E., Wang, J., Yang, V.C., 2009. Gum arabic-coated magnetic nanoparticles for potential application in simultaneous magnetic targeting and tumor imaging. AAPS J. 11 (4), 693.

CHAPTER

16

Application of micro- and nanoengineering tragacanth and its water-soluble derivative in drug delivery and tissue engineering

Azam Chahardoli[a], Nasim Jamshidi[b], Aliasghar Varvani[c], Yalda Shokoohinia[d,e], and Ali Fattahi[e]

[a]Department of Biology, Faculty of Science, Razi University, Kermanshah, Iran [b]Students Research Committee, Kermanshah University of Medical Sciences, Kermanshah, Iran [c]Students Research Committee, Faculty of Pharmacy, Kermanshah University of Medical Sciences, Kermanshah, Iran [d]Ric Scalzo Institute for Botanical Research, Southwest College of Naturopathic Medicine, Tempe, AZ, United States [e]Pharmaceutical Sciences Research Center, Health Institute, Kermanshah University of Medical Sciences, Kermanshah, Iran

16.1 Introduction

Micro- and nanotechnologies are defined as technologies designed, engineered, and performed at micrometer or nanometer scales (Xu et al., 2020). Recently, micro- and nanotechnology in biological and biomedical applications have attracted significant attention, which may be due to extraordinary progress in the research and development of these technological devices and tools (particularly in the field of imaging, cells-probe, and biomolecules). These technologies are a great advancement in cellular and molecular biology and biomedical sciences, and help to further increase and develop multidisciplinary research in chemistry, physics, engineering, biology, and medicine, which have increased demand in the field of nanotechnology (Lim et al., 2010).

Micro- and Nanoengineered Gum-Based Biomaterials for Drug Delivery and Biomedical Applications
https://doi.org/10.1016/B978-0-323-90986-0.00005-4

A variety of materials have been synthesized as micro- and nanoparticles with different composition, shape, size, and surface chemistry, and a wide range of biomedical applications such as drug delivery, imaging, tissue engineering, diagnosis and, basic research (Ali et al., 2020; Suri et al., 2013). Micro- and nanoparticle-based drug delivery systems have been prominent and essential in research, clinical medicine, and experimental pharmaceutics (Kohane, 2007). Although these two size classes are a portion of a continuum, various physicochemical, anatomical, and physiological factors lead to their different properties, and hence multiple applications (Kohane, 2007).

Microparticles are most commonly applied in the delivery of drugs, antiinflammatories, antibiotics, chemotherapeutics, proteins, vitamins, and vaccines (Ahadian et al., 2020; Suri et al., 2013). Nano-scale drug-delivery systems are promising and versatile systems such as nanoparticles, nanotubes, dendrimers, nanocapsules, and nano-gels that can be applied to deliver different drugs include small-molecule drugs and bio-macromolecules (proteins, peptides, plasmid DNA and synthetic oligodeoxynucleotides) (Goldberg et al., 2007).

Advantages of nano-scale drug-delivery systems include delivery of drugs to the target sites with reduced dosage frequency in a spatial/temporal controlled mode to reduce the side effects, increase the efficiency and safety of drugs, improve bioavailability, enhance drug stability, and extend a drug's effect in the target tissue (Zahin et al., 2020). Furthermore, nanoparticles with specific properties such as low toxicity, appropriate characteristics, contrasting agent features, targeted delivery potential, and exact regulator over behavior through the external stimuli such as magnetic have been applied for improving engineered tissues and incapacitating problems in tissue engineering and regenerative medicine (Fathi-Achachelouei et al., 2019).

The purpose of tissue engineering is to develop biological alternatives to repair, replace, retain, or improve tissue and organ functions with the help of engineering materials sciences and medical biology (Hasan et al., 2018). In fact, the approach of micro- and nanotechnologies in response to the biomedical demands of tissue engineering is the production of nanoparticles, nanofibers, and scaffolds in nanometric properties with the ability to mimic native tissues, which is achieved through the optimization of biomaterials and their geometry structures (Limongi et al., 2017). Additionally, nano-scale systems have been used to provide different functions in tissue engineering, including improvement of mechanical, electrical, and biological properties, biosensing and molecular detection, assist the growth of different types of tissues, patterning of cells and also gene delivery, viral transduction, and DNA transfection (Hasan et al., 2018).

Recently, polymeric-based biodegradable micro- and nanoparticles have been used as potential devices or platforms for the delivery of drugs and tissue engineering due to their capability for target specificity, the capacity to deliver genes, proteins, and peptides, the ability to deliver poorly soluble drugs and their sustained release, performing as carriers for nucleic acids (in gene therapy), good biocompatibility, low cytotoxicity, greater permeation and retention effect, and prevention from enzymatic degradation of bioactive agents and maintenance of their bioactivity (Fathi-Achachelouei et al., 2019; Ali et al., 2020; Suri et al., 2013). Both synthetic and natural polymers have attracted massive attention and provided a revolution in drug delivery and tissue engineering over the past several decades. However, natural polymers' unique characteristics, e.g., biocompatibility, biodegradation, and abundance, prioritized them over synthetic polymers (Guo et al., 1998; Sheorain et al., 2019). In contrast, the synthesis of polymers is labor-intensive and causes toxicity and environmental pollution (Bai et al., 2018; Nazarzadeh Zare et al., 2019; Peng et al., 2018).

It should also be noted that the chemical and biochemical amendments of natural polymers are more straightforward than synthetic polymers (Gupta et al., 2018; Maeki et al., 2018). Polysaccharides are a complex group of polymers consisting of several monosaccharides conjugated by a glycosidic bond to form branched or unbranched chains (Pierre et al., 2017; Prasad et al., 2018; Tang et al., 2017). Among the hydrophilic polysaccharides, natural gums are one of the most abundant raw materials derived from inexpensive, nontoxic, easily accessible, and environmentally friendly renewable sources (Izydorczyk et al., 2005; Nazarzadeh Zare et al., 2019). Generally, natural gums can be classified based on their obtained sources, and including marine, microbial, and plant origins as marine gums (e.g., carrageenan gum), microbial gums (e.g., gellan gum), plant exudate gums (e.g., arabic, tragacanth, ghatti, and karaya gums) or seed gums (e.g., guar and bean gums). They can also be categorized based on their structure as linear chains (e.g., bean gum) and branched chains (e.g., tragacanth, guar, arabic, and karaya gums) and their surface charges as anionic (e.g., tragacanth, karaya, xanthan, arabic, gellan, and carrageenan gums), cationic (e.g., modified guar gum), and nonionic (e.g., arabinans, bean, and tamarind gums) (Mohammadinejad et al., 2020).

Some of the famous gums, e.g., tragacanth (white to dark brown), arabic gum, salai gum, and karaya gum (gray to dark) are obtained through the detachment of plant cellulose under a process called gummosis (Mishra and Malhotra, 2009; Thombare et al., 2016). The gum has two different physical forms, including ribbons and flakes, which are much more attractive than the shell (López-Franco et al., 2009; Robbins, 1988).

Tragacanth gum (TG) was discovered by Theophrastus several centuries before Christ (López-Franco et al., 2009). The name of tragacanth comes from the Greek word *tragos* meaning goat and *akantha* meaning horn, i.e., taken from the appearance of the exuded gum, which tends to compose ribbons akin to a goat's horn (Ahmadi Gavlighi, 2013). TG as an anionic polysaccharide is derived from the stems and branches of 15 different species of Astragalus genus such as *Astragalus gummifer*, *A. microcephalus*, *A. gossypinus*, *A. microcephalus*, *A. adscendens*, *A. brachycalyx*, *A. tragacantha*, and *A. kurdicus* (López-Franco et al., 2009; Mohammadinejad et al., 2020).

The best areas for TG production are the mountainous and arid regions of Iran and the Anatolia region in Turkey, which also produced less in Afghanistan and Syria (Anderson, 1989). As a result, it should be noted that geographical and seasonal variations affect the chemical composition of gum and cause significant changes in the methoxy content, sugar content, as well as ratio of soluble and insoluble gum compounds (Nazarzadeh Zare et al., 2019). The major part of the TG structure is composed of branched acidic heteropolysaccharides, including D-galacturonic acid (Nazarzadeh Zare et al., 2019; Mohamadnia et al., 2008). It is noteworthy that TG is a nonallergenic, noncarcinogenic, nonmutagenic, and nonteratogenic hydrophilic polysaccharide (Ghayempour et al., 2015; Mohamed et al., 2018).

TG has been used as a medicine for thousands of years. In accordance with the Scientific Committee for Food of the European Community and the FDA Code of Federal Regulations, TG with *E*-number E413 can be used as a food additive (in between doses of 0.2 and 1.3 wt%) (Ghayempour et al., 2015). In addition, it has been broadly used in food processing as polysaccharide food coatings (Jafari et al., 2018), as a thickening factor to improve viscosity, and as an emulsifier for easy emulsification of foods due to its good stability against heat and acidity (Mostafavi et al., 2016; Tonyali et al., 2018). In pharmaceutical formulations, TG as an emulsifying agent and suspending factor has been used widely to form creams, emulsions, and tablets (Farzi et al., 2013; Samavati et al., 2014; Samavati and D-Jomeh, 2013; Whistler and BeMiller, 2012). In addition, in medicinal fields, TG as different forms of microgels, hydrogels, nanoparticles, nanohydrogels, composites, nanocomposites, and nanofibers has been applied in cosmetics, wound healing, wound dressing, drug delivery, tissue engineering, and bioimaging applications, among others (Cikrikci et al., 2018; Fayazzadeh et al., 2014; López-Cebral et al., 2017; Maia et al., 2018; Mohammadi et al., 2016; Moradi et al., 2018; Otadi and Mobayen, 2011; Pathania et al., 2018; Ranjbar-Mohammadi et al., 2016; Ranjbar-Mohammadi and Bahrami, 2015; Singh and Sharma, 2014; Zarekhalili et al., 2017). Moreover, this polymer is used as a topical agent to relieve rheumatoid arthritis, toothache, and neck pain, or in an oral drug to treat stomach pain in some countries (Bagheri et al., 2015). TG has also been applied for the synthesis of numerous metal and metal oxide nanoparticles as both a reductant and stabilizer, such as ZnO, Ag, NPs Cu-Fe_2O_4, Ni-Cu-Mg ferrite, Ni-Cu-Zn ferrite Al_2O_3, and carbon (Atrak et al., 2018; Darroudi et al., 2013; Fardood et al., 2017; Indana et al., 2016; Moradi et al., 2018; Ramazani et al., 2017; Taghavi et al., 2017). In environmental fields, TG as an adsorbent in different forms has been developed for the removal of heavy metals such as Pb(II) Ag(I), Cr(VI) Co(II), Zn(II), Cd(II), and Cr(III) and cationic and anionic dyes such as methylene blue, Congo red, crystal violet, and methyl orange (Masoumi and Ghaemy, 2014; Rahimdokht et al., 2019; Sahraei et al., 2017; Sahraei and Ghaemy, 2017).

In this chapter, we review the recent biomedical applications of TG in the fields of drug delivery (based on dosages forms, microparticles, nanoparticles, hydrogels, and fibers), tissue engineering, and wound healing individually or in combination with other various biodegradable polymers.

16.2 Composition and chemical structure of TG

TG is a very branched and heterogeneous hydrophilic carbohydrate (Ahmadi Gavlighi, 2013). The molecular weight of TG is 840 kDa, which is determined by Svedberg's manner, and the molecular weight of the soluble fraction (from Iranian species of TG (*A. microcephalus*)) is about 3000 kDa, which is determined using size exclusion chromatography (Emam-Djomeh et al., 2019). TG is a complex of acidic polysaccharides bounded with low amounts of protein (<4%), starch, and cellulosic material, as well as possessing several cations such as calcium, magnesium, and potassium (Ahmadi Gavlighi, 2013).

TG comprises two main parts: bassorin (~60%), which is the insoluble section of gum (Fig. 16.1) that swells in water and forms gel; and tragacanthin (arabinogalactan), which is water-soluble (Fig. 16.2). The water-soluble tragacanthin is neutral and makes up about 30%–40% of the gum's weight (Phillips and Williams, 2009). Acidic hydrolysis of tragacanthin indicated that it is composed of L-fucose (L-Fuc), D-galactose (D-Gal), D-glucose (D-Glc) L-(L-Ara), D-xylose (D-Xyl), and D-mannose (D-Man), in a 3:5:5:52:29:6 M ratio (Nazarzadeh Zare et al., 2019).

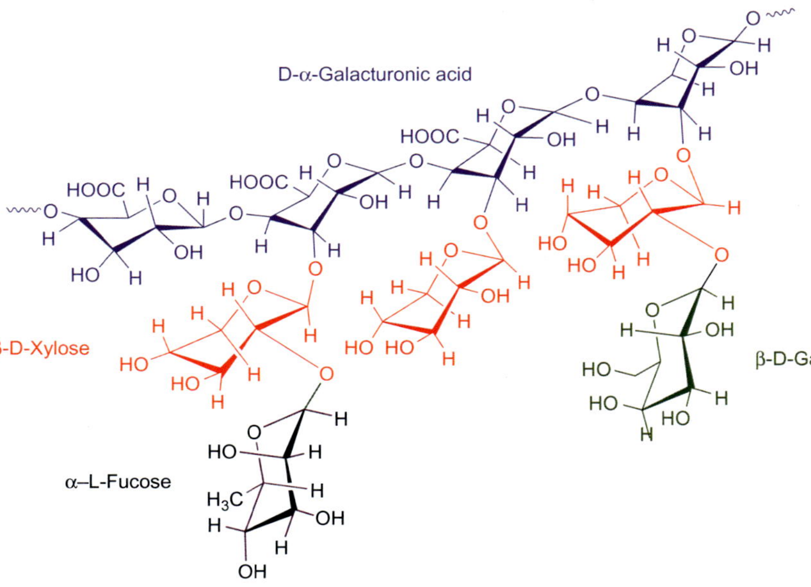

FIG. 16.1 The structure of tragacanthic acid or bassorin.

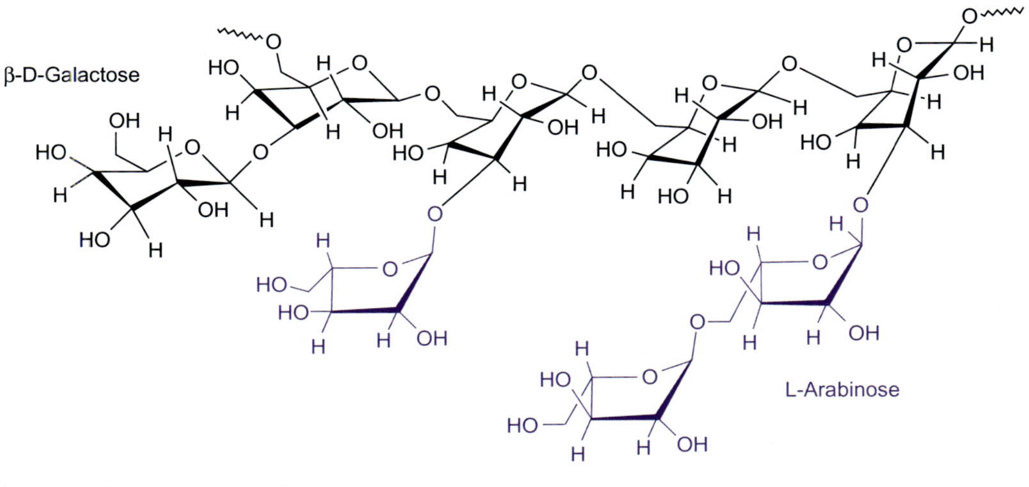

FIG. 16.2 The structure of arabinogalactan or tragacanthin.

Bassorin is included D-galacturonic acid, D-xylose, D-galactose, L-fucose, and a tiny amount of L-rhamnose (Koshani and Aminlari, 2017), where the main portion of D-galacturonic acid is methyl ester (Gralén and Kärrholm, 1950).

It should be noted that bassorin has a lower viscosity than gum tragacanth but has a higher viscosity than tragacanthic acid. The rheological properties of bassorin and tragacanthin were entirely different; while a 1% bassorin solution at 25 °C has a high viscosity gel-like structure, tragacanthin behaves like a semidilute to concentrated random polymers (López-Franco et al., 2009; Mohammadifar et al., 2006). Arabinogalactan ortragacanthin has high molecular weight and includes L-rhamnose (L-Rha), L-Fuc, L-Ara, D-Xyl, D-Man, D-Gal, and D-Glc in a 1:1:68:2:5:22:1 M ratio, and is soluble in a mixture of ethanol-water (7:3) (Balaghi et al., 2010; Jalali et al., 2018; Kiani et al., 2012; Montazer et al., 2016; Tischer et al., 2002). The arabinogalactan fraction is basically included core units of galactopyranose, arabinopyranose, and 2-*O*-substituted β-arabinofuranose with side-chain sequences of α-L-Ara *f*-(1 → 2)-α-Ara *f*-(1 → 4)-L-Ara*p* and α-L-Ara *f*-(1 → 2)-α-Ara *f*-(1 → 5)-L-Ara*fp* (Emam-Djomeh et al., 2019).

Arabinogalactan has been used as foam adhesive, thickeners, bulking agents, film-formers, emulsifying agents, additives, and therapeutic agents. Numerous studies have confirmed that it is safe and has low toxicity. Arabinogalactan is commonly found in larchwood in Europe and can also be extracted from heartwood with high efficiency (Stephen, 1983).

Tragacanthic acid is a main component of the complex basorine, composed of D-galacturonic acid methyl ester, D-xylose, and L-fucose (Monjezi et al., 2019; Phillips and Williams, 2009). The main residues of sugar in the external chains are single β-D-xylopyranose and 2-*O*-D-galactopyranosyl-D-xylopyranose, as well as disaccharide units such as 2-*O*-L-glucopyranosyl (Emam-Djomeh et al., 2019) . It is found that in most astragalus species, the tragacanthin and bassorin possess methyl groups, which indicate methoxylated galacturonic acid, while the insoluble bassorin portion has fewer methoxyl substitutions compared to the soluble tragacanthin part (Ahmadi Gavlighi et al., 2013).

16.3 Properties

TG solution is acidic in the pH range of 5–6, and the maximum initial viscosity is at pH 8, while its maximum stability is close to pH 5. In fact, TG is relatively stable in a wide range of pH from down to very acidic conditions (at around pH 2) compared to other gums (Ahmadi Gavlighi, 2013). The TG quality depends on its viscosity, which affects its suspending, emulsifying, or stabilizing capacity. Hence, the viscosity values rely heavily on the type of sugars and counter-ions, molecular weight, concentration, temperature, and shear rate (Emam-Djomeh et al., 2019).

The effects of shear rate and temperature from 5 to 55 °C on the viscosity of TG and its fractions have been studied, which established the lack of a conformational transition (Mohammadifar et al., 2006). In comparison with different industrial gums, TG has the highest activation energy of thermal decomposition (Emam-Djomeh et al., 2019). Previous literature based on assessing energies of activation indicated that the viscosity of bassorin was less sensitive to temperature than tragacanthin (Silva et al., 2017). However, by increasing temperature, the energy needed for the flowing of molecules and subsequently the viscosity of TG solution reduce, and heating for a long time leads to degradation of the gum and reduces its viscosity forever (Emam-Djomeh et al., 2019).

Furthermore, TG has emulsifying and stabilizing properties in acidic oil-in-water emulsions and displays two functions by having the ability to facilitate emulsification and providing stabilization of the emulsion after its formation (Wang, 2000). In addition, TG has well-defined surface activity features and rapidly reduces the surface tension of water at low concentration (<0.25%) and facilitates emulsification (Balaghi et al., 2010). Based on the reported literature, TG obtained from different species of astragalus possesses different chemical compositions and physicochemical properties; these features affect its rheological properties, which include the viscosity of tragacanth solutions and its stabilizing effects in emulsions (Balaghi et al., 2011).

It has been observed that the terminal deoxyhexoxyl groups (i.e., fucose) or the methoxyl groups of the homogalacturonan in the TG structure can affect emulsion stabilization (Ahmadi Gavlighi, 2013). The emulsification capacity of TG is because of its residual surface activity, improved emulsion viscosity, and zeta potential (−21 mV) (Rezvani et al., 2012), which may be derived from negatively charged carboxylic groups of galacturonic acid (Azarikia and Abbasi, 2010). The stabilizing capability of TG is due to the steric repulsion force, and by changing the pH, its action can be controlled. In addition, the polyelectrolyte nature of TG is due to the existence of a uronic acid, which has carbonyl and carboxylic acid (Emam-Djomeh et al., 2019).

16.4 Characterization

The physicochemical properties of TG's hetero-polysaccharides have been characterized by various techniques, including gas chromatography, gel permeation chromatography, 1D (13C-^{1}H NMR) and 2D NMR (HMBC, HMQC, COSY, and NOESY) spectroscopy, high-pressure anion exchange chromatography, steady/dynamic rheology, and capillary viscometry (Emam-Djomeh et al., 2019).

Furthermore, different techniques such as FT-IR, XRD, and TGA-DTA have been used to identify and confirm the structure of the TG in the scientific literature (Monjezi et al., 2018). Based on the FT-IR spectra reported in the literature, the absorption bands at 3415, 1750, and 1640 cm^{-1} can be assigned to the stretching frequencies of the hydroxyl (—OH) and carbonyl (—C=O) groups, respectively. In addition, the two peaks shown in 2937 and 2867 cm^{-1} are related to the symmetric and asymmetric stretching frequencies of the methylene group (CH_2). Stretching absorptions related to C—O bonds of the polyol, pyranose rings, and ether are displayed at 1244 and 628 cm^{-1} (Akkol et al., 2011; Sadeghi et al., 2014).

In the XRD pattern of tragacanth gum, a wide diffraction peak appears in $2\Theta = 19$ degree and a sharp band in $2\Theta = 28$ degree, confirming the microcrystalline structure of the gum due to its intramolecular hydrogen bonding (Moghaddam et al., 2019). The three main stages of weight loss can be seen in the TGA curve of the gum. The first weight loss that occurs in the area below 150 °C is related to the evaporation of water that is physically absorbed by the surface. In addition, the two-weight reductions that were seen at 200–300 °C and 400 °C are attributed to the removal of surface-bound organic groups that comprise about 90% of the gum weight (Nazarzadeh Zare et al., 2019).

16.5 Chemical modification of TG

Polymers' surface modification is performed for several important reasons, including the following (Bashir et al., 2016):

(1) To place a specific functional group on the surface for a particular purpose (Chourasia and Jain, 2004).

(2) To make polymers more resistant to moisture or heat. Modification of polymers reduces their degradation and makes them more resistant to heat and moisture (Konwarh et al., 2013).
(3) To increase the solubility of polymers in water and organic solvents (Chen, 2008).
(4) To increase the flexibility, transparency, and compatibility of polymers (Bashir et al., 2016).
(5) To reduce the toxicity of polymers (Bashir et al., 2016).

In 2006, Kaffashi et al. used TG and TG/collagen composite to transfer pentoxifylline drugs. According to the results, the TG/collagen composite strength is greater than that of the TG, and the composite is more effective for achieving the structure required for proper drug delivery (Kaffashi et al. (2006)). For modification with carboxyl derivatives, the carboxyl or –COONa group is attached to the substrate surface. This increases the hydratability and solubility of gum. For example, cashew gum with carboxymethyl groups has a negative charge on its surface that can form a physical complex with cationic polymers like chitosan (Bashir et al., 2016). Gums with the hydroxyethyl groups can be composed of natural gums through nucleophilic substitution. These compounds have much better colloidal properties, greater solubility, better thermal stability in solution, excellent dispersion in water, and greater compatibility with cationic, anionic, and nonionic surfactants than raw gum (Bashir et al., 2016).

16.6 Biomedical applications

16.6.1 Drug delivery

Drug delivery refers to formulations, technologies, and systems approaches for safely delivering or transporting a drug or pharmaceutical compound in the body to reach its favorable therapeutic effect (Badakhshanian et al., 2016). Recently, enormous developments have taken place in the field of delivery systems for the treatment of various diseases through the delivery of therapeutic agents or natural-based active compounds to their target tissues (Patra et al., 2018). Polysaccharides such as tragacanth have a tendency to be internalized and degraded quickly, thus assisting a moderate intracellular release. The presence of hydrophilic groups (hydroxyl, carboxyl, and amino groups) in their structure improves the bio-adhesion with epithelia and mucous membranes, which is a beneficial approach for improving the bioavailability of drugs in drug delivery systems (Naji-Tabasi et al., 2017).

In the following sections, we refer to the TG-based different drug delivery systems and study them in detail. In addition, Table 16.1 summarizes further reports about TG-based drug delivery systems with different properties.

TG-based traditional dosage form systems for drug delivery (tablets, capsules, and gels)

Among all drug delivery systems, the oral-based drug delivery is the most favored and suitable option as this providing a maximum active surface area for the administration of different drugs (Ratnaparkhi et al., 2013). The interest in these dosage forms is due to the knowledge of toxicity and inefficiency of drugs when administered in the form of tablets and

TABLE 16.1 TG-based drug delivery systems.

TG-based carrier	Carrier type	Loaded drug	Size (nm)	Drug entrapment efficiency	Drug release efficiency	Release time (h)	References
TG-chitosan	Nanoformulation	Thymol	150–200		94.74% at pH 7	24	Sheorain et al. (2019)
Lectin-TG-AuNPs	Nanoparticle	Amphotericin B	358.3 ± 1.78	78.91 ± 2.44%	–	–	Jabri et al. (2018)
TG-AuNPs/AgNPs	Nanoparticle	Naringin/ hesperidin	221 182.8	72%/73.66%	–	–	Anwar et al. (2019)
TG-AuNPs	Nanoparticle	Naringin	85	76.39 ± 2.61%	–	–	Rao et al. (2017)
TG-g-poly(vinylimidazole)/ $Fe_3O_4@SiO_2$	Nanocomposite	Quercetin	16–54	75%	40% at pH 2.0	8	Hemmati et al. (2016)
TG-graphene oxide	Nanocomposite	Rivastigmine	–	91%	97% at pH 7.4	6.6	Rahmani et al. (2018)
TG/collagen	Composite	Pentoxifylline		–	at pH 7.4	8	Kaffashi et al. (2006)
TG-quetiapine fumarate	Tablet	Quetiapine fumarate	–	–	>85% at pH 1.2 and pH 7.4	15	Salamanca et al. (2018)
TG-glutaraldehyde-poly(vinyl alcohol)	Nanohydrogels	Indomethacin	120–140	70–85%	50–80% at pH 9	24	Hosseini et al. (2016)
TG-g-poly(itaconic acid)	Nanohydrogel	Ampicillin	40–100	–	80%–93% at pH 2.2	24	Pathania et al. (2018)
TG-polyvinyl alcohol	Cryo-gels Xero–gels	Silymarin			74% 50% at pH 7.4	8	Niknia and Kadkhodaee (2017)
TG-sodium alginate	Hydrogels	Insulin	–	–	70% at pH 6.8	6	Cikrikci et al. (2018)
TG-glucono-δ-lactone	Microparticle hydrogels	Insulin	601 ± 19 at pH 4.6 811 ± 20 at pH 4.3	89% at pH 4.3–4.6			Nur and Vasiljevic (2018)
TG-g-poly(acrylic acid)	Hydrogel	Amoxicillin	–	–	at pH 7.4 > pH 2.2	24	Singh and Sharma (2014)

Continued

TABLE 16.1 TG-based drug delivery systems—cont'd

TG-based carrier	Carrier type	Loaded drug	Size (nm)	Drug entrapment efficiency	Drug release efficiency	Release time (h)	References
TG-cl-poly(lactic acid-*co*-itaconic acid)	Hydrogel	Amoxicillin	>100	73%	96% at pH 2.4	6	Gupta et al. (2018)
Galacturonic acid/arabinogalactan-poly(vinyl pyrrolidone)-*co*-poly (2-acrylamido-2-methylpropane sulfonic acid	Hydrogel	Amoxicillin	–	–	at pH 2.2 > pH 7.4	5	Singh and Sharma (2017)
Whey protein-TG	Composite hydrogels	Black carrot extract	–	–	29.39% at pH 7.0	24	Ozel et al. (2017)
TG/poly(methyl methacrylate-alt-maleic anhydride)-*g*-polycaprolactone	Hydrogel	Quercetin	–	58%	40%–80% at pH 7	7	Hemmati et al. (2015)
TG-*g*-poly(dimethylaminoethyl methacrylate)–poly (ε-caprolactone)–poly(ethylene glycol)	Hydrogel	Quercetin	–	86.71	at pH 2.2	30	Hemmati and Ghaemy (2016)
TG-*g*-poly(acrylic acid)	Hydrogel	Pantoprazole	–	–	at pH 9.2	30	Saruchi et al. (2014a, b)
Tragacanthin-tyramine	Hydrogel	Bovine serum albumin and insulin	120–160 μm	–	65% 90%	288 288	Dehghan-Niri et al. (2015)
TG-cl-poly(acrylic acid-*co*-acrylamide)	Hydrogel	Losartan	–	79%	at pH 9.2 > pH 7 > pH 2	34	Saruchi et al. (2014b)
TG-*co*-sodium alginate-cl-poly (vinyl alcohol)	Hydrogel	Moxifloxacin	–	–	at pH 2.2	24	Singh et al. (2016)
TG-cl-poly(vinyl pyrrolidone-*co*-acrylic acid)	Hydrogel	Ciprofloxacin	–	69.80%	at pH 7.4 > pH 2.2	24	Singh and Sharma (2017)

TG/poly(acrylic acid)	Hydrogel	Amphotericin B	–	–	at pH 7.0	24	Mohamed et al. (2018)
TG-acacia gum-polyvinyl alcohol-*co*-polyvinyl pyrrolidone	Hydrogel	Gentamicin	–	–	At distilled water and simulated wound fluid	24	Singh (2018)
TG-polyvinyl alcohol-*co*-polyvinyl pyrrolidone	Hydrogel	Gentamicin Lidocaine	–	63.98% 40.96%	At pH 2.2 buffer, simulated wound fluid, phosphate buffer	6	Singh et al. (2017b)
Tyramine-methoxyl-TG	Hydrogel	Bovine serum albumin	–	–	88.5% at pH 7.4	240	Tavakol et al. (2016)
TG/poly(ε-caprolactone)	Nanofiber	Curcumin	191 ± 24	3%	65% at pH 7.4	480	Ranjbar-Mohammadi and Bahrami (2016)
TG/poly(ε-caprolactone)	Nanofibrous scaffold	Curcumin	–	3%	65% at pH 7.4	480	Mohammadi et al. (2016)
Zein-TG	Nanofiber	Saffron extract	95–271	3.57%–9.52%	43.88% at pH 6.8	6	Dehcheshmeh and Fathi (2019)
TG-	Nanofiber	Peppermint oil	Thickness of 58 nm and length of μm	18.3%	92.38%	18	Ghayempour and Montazer (2019)
Poly lactic glycolic acid-TG	Nanofiber	Tetracycline hydrochloride	180 ± 24	5%	68.10% at pH of 7.4	1800 (75 days)	Ranjbar-Mohammadi et al. (2016b)

capsules as a conventional oral route; these conventional dosage forms create a wide range of oscillation in drug concentration in the tissues and bloodstream with poor efficiency and unwanted toxicity (Ratnaparkhi et al., 2013). The limitations of conventional oral dosage forms can be poor patient adaptation and an enhanced risk of losing the drug dose with a short half-life, which leads to repeated administration and unavoidable fluctuations of drug concentration, in turn resulting in under-medication or over-medication in narrow therapeutic index drug and obtaining a typical peak-valley plasma concentration time profile, which leads to impossible attainment of the steady-state condition (Ratnaparkhi et al., 2013).

A drug delivery system using TG along with $CaCl_2$, gelatin and 1,8 octandiamine curing agent and TG/collagen matrices as tablet form has been investigated to improve the gel behavior and drug delivery of this system for pentoxifylline release (Kaffashi et al., 2006). The obtained results in this study indicated that the gel strength of TG was increased by forming a porous composite and slower drug release at the body temperature for this composite containing 1:2 TG/collagen (Kaffashi et al., 2006). The tablet formulated with TG in comparison with those prepared tablets with gum from *Bombax buonopozense* and acacia gum was applied for paracetamol release study (Ngwuluka et al., 2012). Based on the *in vitro* drug release studies, the tablets with TG and acacia gum indicated more than 95% and 85% release of paracetamol after 45 min, respectively, while the tablets with bombax gum indicated 52.5% of drug release after 60 min. Thus, the results of this study showed that the applied concentration of 3.5% for binders which were appropriate for TG and acacia gum was not suitable for bombax gum in order to formulate an immediate release dosage form (Ngwuluka et al., 2012).

Sustained release matrix tablets of metoprolol tartrate have been prepared using various ratios of TG and xanthan gum using the direct compression technique, and the potential of these natural gums has been examined for the fabrication of sustained release tablets (Rasul et al., 2010). Obtained results in this study indicated that increasing the content of TG and xanthan gum in the formulation led to slow release of drug, and decreasing the polymer content resulted in increased release of metoprolol tartrate, but drug release from the tablets with TG was faster than the matrices containing xanthan gum, which has been attributed to the low matrix-forming ability of TG (Rasul et al., 2010). Based on the authors' reports, the formulation containing 10% of xanthan gum and 30% of TG released more than 80% of the drug in 6 h, while other formulations including 20% of both TG and xanthan gum and formulations with 30% of xanthan gum and 10% of TG (the most similar to that of the marketed reference) released 80% of the drug in nearly 8 and 10 h, respectively, at a slower rate due to a thick gel structure in the formulation, which delays the release of drug from matrix tablet (Rasul et al., 2010).

TG along with hydroxypropylmethylcellulose (HPMC), and acacia as a carrier, has been applied in the forms of one- and three-layer matrix tablets to sustain the release of verapamil hydrochloride (Siahi et al., 2005). TG indicated satisfactory prolonged release either alone or in combination with the other polymers. In addition, the authors in this study stated that the position of the polymers in the three-layer tablets can have a distinct effect on the drug release (Siahi et al., 2005). In addition, the effect of HPMC in combination with TG, guar gum, and xanthan gum has been evaluated on the release of diclofenac sodium matrix tablet (Iqbal et al., 2010). Reported results indicated that tablets containing HPMC (80 mg) along with both TG and guar gum (40/40 mg) and along with both TG and xanthan gum (40/40 mg) released 29.84% and 22.48% of diclofenac sodium after 2 h, and 94.36% and 82.45% after 12 h, respectively (Iqbal et al., 2010).

TG, as a surfactant, has been used for the preparation of microcapsules containing alkannin capsules (Assimopoulou and Papageorgiou, 2004). It has been indicated that the use of TG as a surfactant for this order led to the formation of porous spheres with a different texture and with larger mean particle size, higher efficiency, and increased loading compared with other surfactants (sodium dodecyl sulfate), which has been attributed to its less potent surfactant activity, which reduces the loss of the active substance to the aqueous phase (Assimopoulou and Papageorgiou, 2004). Nevertheless, the authors did not suggest using TG as a surfactant for the preparation of microcapsules due to it providing a wide range of particle sizes and a quick release rate (Assimopoulou and Papageorgiou, 2004). Nicardipine hydrochloride has been formulated as sustained release floating capsules in combination with different polymers such as TG, HPMC 4000, and methylcellulose, which has shown an average release of 24.2% after 3 h and then 93.9% at the end of 6 h (Moursy et al., 2003).

TG-polyvinyl alcohol-based cryogels and xerogels have been prepared and designed for the delivery study of silymarin (Niknia and Kadkhodaee, 2017). The results of prepared gels indicated that xerogels had a densely packed agglomerated microstructure with a continuous network of cracks, higher mechanical strength, bulk density, and silymarin retention compared to cryogels with highly porous structure. In addition, it has been demonstrated that increasing the content of TG adversely impacts both structural features and physicomechanical properties of the dried gels, while improving the mucoadhesive property and the release profile of silymarin (Table 16.1) (Niknia and Kadkhodaee, 2017).

TG-based microparticles for drug delivery

De-esterified tragacanth microspheres in combination with Eudragit S-100 coated capsules have been fabricated using a microemulsion method for delivery of 5-fluorouracil (5-FU) to the colon (Ahmadi et al., 2018). These microspheres were spherical in shape with sizes ranging from 3.55 ± 0.22 to 7.0 ± 0.478 μm. The maximum loading efficiency in this formulation has been reported as 44.1% at a drug-to-polymer ratio of 1:5 and cross-linker ($CaCl_2$) concentration of 0.7%. The reducing drug-to-polymer ratio (from 1:1 to 1:5) has been introduced as an important parameter in increased loading efficiency. These loaded microspheres have shown 70% release of 5-FU at colon pH level (pH 7.4) and less than 5% release of 5-FU at the stomach and small intestine pH levels (pH 1.5) as well as about 25% of the drug released in the presence of pectinase enzyme. The slower release of 5-FU at the stomach pH than that at the colon pH has been attributed to a lower swell capacity of de-esterified tragacanth hydrogel at an acidic pH due to carboxylic acid deionization. Therefore, the presence of pectinase enzyme, the cross-linker concentration, and environmental pH affect the release rate. Furthermore, in this study, the IC_{50} of drug-loaded microspheres (80 μg/mL) against HT29 cells was higher than that of free drug (50 μg/mL), which its reason was attributed to the profile of release (Ahmadi et al., 2018).

In another work, Nur and Vasiljevic (2018) synthesized microparticles from a mixture of two polyelectrolytes based on TG and used it purposefully to transmit oral insulin (Nur and Vasiljevic, 2018). In this study, insulin particles were inserted into the tragacanth hydrogel. Nur also examined the factors affecting particle sizes, such as PH and polyelectrolyte concentration, and for this purpose, utilized the zeta potential and SEM technique. The interaction of insulin and tragacanth was evaluated at different concentrations (0.1%, 0.5%, or 1%, w/w) and several different pH (3.7, 4.3, 4.6, and 6) by differential scanning calorimetry (DSC) and ATR Fourier transform infrared (ATR-FTIR) techniques. It is noteworthy that the best particle size (>800 nm) was obtained at pH 4.6 at 0.5% tragacanth concentration. According to the acid gelation analysis, insulin was enclosed in the tragacanth hydrogel. As the DSC thermogram showed, the interaction between the polyelectrolyte and the protein occurred in the PH range of 4.3–4.6 and the insulin-tragacanth thermogram is shifted to the unmodified tragacanth, which confirmed the insulin loading in tragacanth. FT-IR spectra demonstrated the protein connecting on the tragacanth surface. In addition, it was exhibited that loading efficiency increased with a decrease of pH from 6 to 3.7 and it reached the maximum particularly at pH 4.3 or 4.6 (Table 16.1). Loading efficiency increased at pH 4.3 from 65% with 0.1% (w/w) of TG to 89% at concentration 0.5% (w/w) of TG. The results showed that insulin affinity is highest for carboxylic groups of TG at pH 4.3 or 4.6, as shown by the study of loading efficiency. Thus, the best results were obtained with a TG concentration of 0.5% (w/w) at pH 4.3 with a loading efficiency of 89% (Nur and Vasiljevic, 2018).

Using TG and casein (CAS) as the wall material, microcapsules were fabricated containing carotene by the complex coacervation method to achieve a nontoxic micro-particulate system with long residual action and improved stability for controlled release of carotene (Fig. 16.3) (Jain et al., 2016). The coacervation process occurs due to electrostatic interaction between protein (Pr) and polysaccharides (Ps) chain, so the Pr:Ps ratio plays a key role in this process. Actually, Pr:Ps has an optimum value that could be measured by turbidity measurement. pH in which coacervation occurs can determine the efficacy of the encapsulation process by affecting the charge densities of the biopolymers. In this study, the best pH for coacervation has been reported as being between 4.3 and 6.5. These carotene-loaded microcapsules with particle size 159.71 ± 2.16 μm indicated coacervates yield of 82.51%, entrapment efficiency of

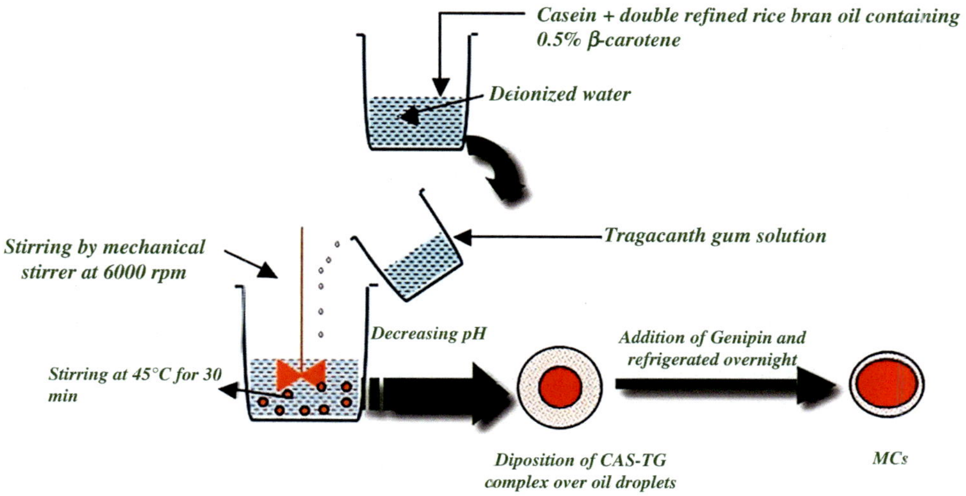

FIG. 16.3 Synthesis of β-carotene-loaded microcapsules by the complex coacervation method. *Reprinted from Jain, A., Thakur, D., Ghoshal, G., Katare, O.P., Shivhare, U.S., 2016. Characterization of microcapsulated β-carotene formed by complex coacervation using casein and gum tragacanth. Int. J. Biol. Macromol. 87, 101–113. https://doi.org/10.1016/j.ijbiomac.2016.01.117 with permission from the publisher.*

79.36%, and biphasic release pattern with the initial burst release (29.25%) followed by slower and sustained release nature after 1 h (Jain et al., 2016).

The Pr:Ps ratio plays an important role in the releasing profile of MCs. Therefore, as the hydrophobicity of the biopolymers increases, the release is delayed as the goal of any controlled release formulations. According to the authors, the initial burst release may be due to the quick release of drug absorbed on the surface or the existence of drug just below the microcapsules (MCs) layer, which may occur due to loss of wall integrity during the production or drying of particles. Furthermore, the slower and sustained release is attributed to the mixed mechanism, including penetration, solubilization, and diffusion of β-carotene oil through tragacanth/casein matrix of microcapsules and also the presence of the tragacanth/casein layer on the droplet of oil containing β-carotene. Accordingly, it has been demonstrated that the long residual action and stability of encapsulated β-carotene improved after it was fabricated in the form of microcapsules (Jain et al., 2016).

Different microgels have been synthesized using different weight ratios of an azide-functionalized TG as a cross-linker, and poly(methyl methacrylate-*alt*-maleic anhydride)-g-polycaprolactone as a synthetic copolymer; after characterizing by different techniques, their properties such as swelling behavior, gel percent and thermal, as well as loading and *in vitro* release of quercetin at pH 7, were assessed (Hemmati et al., 2015). Temperature, gel content, immersion time, and pH were introduced as effective factors on the swelling behavior of these microgels. It has been indicated that the drug loading capacity of these amphiphilic conetwork microgels increased with reducing the cross-linker content and total release has been achieved at a range of 40–80% after 7 h at pH 7; this may be due to the network structure of microgels (Hemmati et al., 2015).

Asghari-Varzaneh et al. (2017) microencapsulated iron salt ($FeSO_4 \cdot 7H_2O$) into TG hydrogel through the solvent vaporization technique (Asghari-Varzaneh et al., 2017). Three important parameters including the amount of TG, the value of ferrous sulfate, and the ratio of alcohol to the mixture were checked to gain the highest encapsulation performance. In addition, the maximum encapsulation efficiency was attained in the presence of 22% of TG, 5% of ferrous sulfate, and an 11:1 ratio of alcohol to the mixture. The microstructure of iron microcapsules was confirmed by SEM. According to the SEM photographs and XRD pattern, microcapsules showed two crystalline and amorphous structures. In addition, the presence of Fe in the TG microcapsules was confirmed with FTIR spectra. Based on the particle size distribution diagram, the average particle size of the synthesized microcapsules is 55.31 μm. Based on the evaluation of iron release in simulated gastric fluid (pH 2), the complete release happened in the stomach during 3 h, due to the complete dissolving of TG in this condition, which is required for its absorption in the duodenum. Thus, the fast release of iron may be due to the water-soluble fraction of TG. In addition, fabricated microcapsules gradually released the iron in the water (80% during 24 h); this slow and gradual release of iron may be due to the presence of a water-insoluble fraction of TG as gel forms (Asghari-Varzaneh et al., 2017).

TG-based nanoparticles for drug delivery

Nanoparticle-based delivery systems are promising candidates or advanced systems for controlled drug delivery systems (Patra et al., 2018; Dang and Guan, 2020). These nano-based drug delivery systems can increase the effectiveness of the drug by helping to reduce toxicity and regulate the biodistribution of the drugs, increasing drug half-life, ensuring sustained or controlled release of drug, and enhancing the solubility of some hydrophobic drugs (Dang and Guan, 2020).

In order to enhance bactericidal potentials of Naringin, tragacanth stabilized/reduced gold nanoparticles (AuNPs) were loaded with Naringin (NRG) as antimicrobial agents against brain-eating amoebae such as *Naegleria fowleri* and *Acanthamoeba castellanii* and multidrug-resistant bacteria, including neuropathogenic *Escherichia coli* K1 and methicillin-resistant *Staphylococcus aureus*-MRSA (Fig. 16.4) (Anwar et al., 2019). These NRG-loaded NPs has been shown to significantly decrease the viability of *N. fowleri* and *A. castellanii* at 50 μg/mL and inhibit the encystation of *A. castellanii* by 85% at 100 μg/mL without significant excystation and bactericidal effects at 50 μg/mL. NRG-loaded NPs have indicated a low cytotoxicity effect (23%) against HeLa at a concentration of 100 μg/mL (Anwar et al., 2019). In another study, the antibacterial effects of NRG loaded TG-AuNPs have been shown against different bacterial strains such as *B. subtilis*, *M. luteus*, *E. coli*, and *P. aeruginosa* using Tetrazolium microplate assay (Rao et al., 2017). The higher NGR loading efficiency (76.39 ± 2.61%) in this study has been attributed to the attendance of TG in nanoformulation and the interaction of its —C—O groups with TG-AuNPs. IC50 values of TG-AuNPs-NR loaded with Naringin have been reported as 288.23 ± 3.68, 253.93 ± 3.47, 390.54 ± 3.91 and

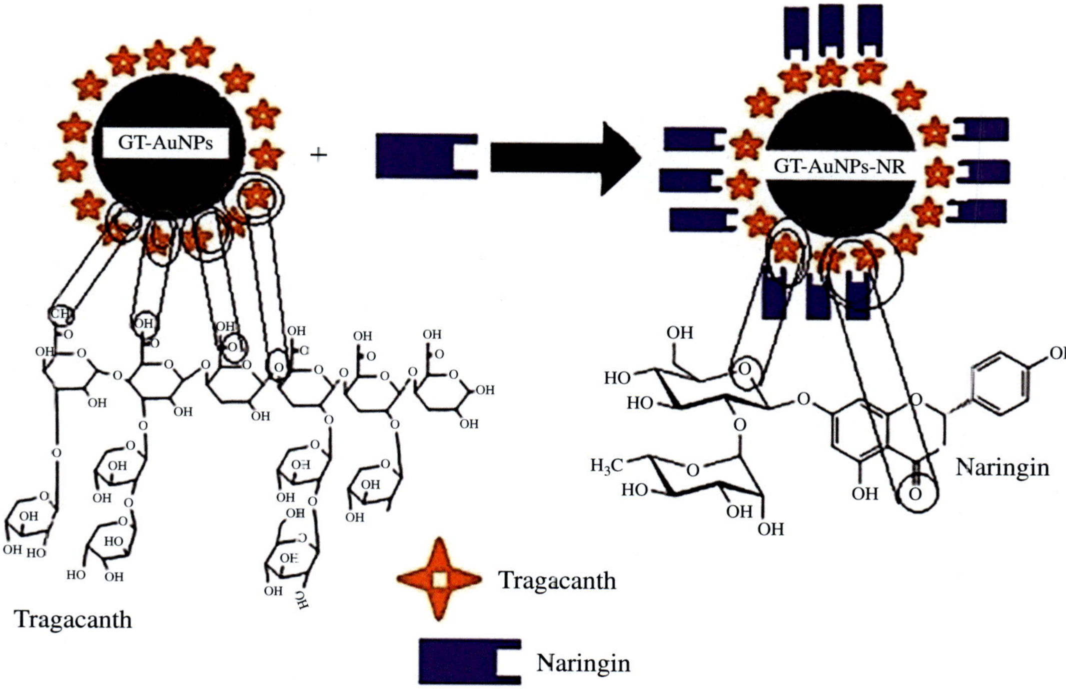

FIG. 16.4 TG-stabilized green-gold nanoparticles as cargos for Naringin loading. *Reprinted from Rao, K., Imran, M., Jabri, T., Ali, I., Perveen, S., Shafiullah, Ahmed, S., Shah, M.R., 2017. Gum tragacanth stabilized green gold nanoparticles as cargos for Naringin loading: a morphological investigation through AFM. Carbohydr. Polym. 174, 243–252. https://doi.org/10.1016/j.carbpol.2017.06.071 with permission from the publisher.*

320.73 ± 3.54 μg/mL against *B. subtilis*, *M. luteus*, *E. coli*, and *P. aeruginosa*, respectively, compared to Naringin with higher IC50 values; these enhanced bactericidal effects of loaded NPs have been attributed to their small size and increased internalization through AuNPs cargos (Rao et al., 2017).

In addition, lecithin-TG mucoadhesive hybrid AuNPs as a nano-carrier system has been designed to increase the oral bioavailability of Amphotericin B in an *in vivo* study in rabbits (Jabri et al., 2018). The higher entrapment efficiency of both NPs, including lecithin-Amphotericin B (86.54 ± 1.32%) and TG-lecithin-amphotericin B (78.91 ± 2.44%), has been attributed to enhanced lipophilicity of Amphotericin B that leads to maximum solubilization in the lipophilic part of the lecithin. Based on obtained results in this study, the maximum absorption of the drug in the TG-lecithin-Amphotericin B formulation by maximum plasma concentration of 271.47 ± 12.54 may result from its improved mucoadhesive nature due to the presence of TG, which is responsible for enhancing the interaction of TG-lecithin-Amphotericin B formulation with gastrointestinal mucosa (Jabri et al., 2018).

Given the ability of tragacanth as a natural biopolymer to transport physiologically important peptides and proteins, the Nur Research Group in 2016, used TG as an oral peptide for protein transfer (Nur et al., 2016). In addition, physical and mucoadhesive features of tragacanth are required for protein/peptides delivery. In this research, mucoadhesivity of tragacanth was compared with PVP, chitosan, and alginate. Differential scanning calorimetry (DSC) was used to scrutinize the thermal behavior of tragacanth (Nur et al., 2016).

In another study by Fattahi et al. (2013), the oligochitosan/water soluble tragacanth (OCH–WST) nanoparticles with spherical morphology and size of 132.5 ± 6.77 nm were synthesized as gene carriers and their transfection efficiency on HepG2 and Hela cell lines was evaluated (Fattahi et al., 2013). Based on the authors' report, prepared NPs had no significant cytotoxicity on Hela and HepG2 cell lines, but NPs–DNA complexes (nanoplexes) indicated higher transfection efficiency in both cell lines than chitosan polyplexes, due to their weaker complexation. In addition, they showed that transfection of nanoplexes in HepG2 cells increased more than that in Hela cells, while it decreased in the presence of galactose in HepG2 cells, which indicated receptor-mediated endocytosis of nanoplexes (Fattahi et al., 2013).

TG-based fibers/nanofiber systems for drug delivery

Recently, fibrous technologies including membranes, filters, patches, and films and their engineered platforms have attracted a lot of attention (Wang et al., 2018). Among the various nanomaterials with increased potential applications, nanofibers are the one of the highly regarded and outstanding nanomaterials due to their very remarkable features, such as high surface area-to-volume ratio and high porosity (Lim, 2017). Nanofibers can be attained by different techniques, including pulling of nonpolymer, assembling from individual CNT molecules, depositing materials on linear templates, and air-blast atomization of mesophase pitch (Heydary et al., 2015).

The electrospinning technique has obtained considerable attention as a simple, popular, versatile, and inexpensive manner to produce soft polymer-based fibers and nanocomposites (Dehcheshmeh and Fathi, 2019). Electrospinning includes a syringe pump, nozzle, collector, and high voltage supply (Wu et al., 2017). Production of nanofibers using the electrospinning technique depends on different parameters such as process conditions, solution criteria, and environmental factors, which directly affect the fibers' morphology (Dehcheshmeh and Fathi,

2019). Produced nanofibers have been widely considered for potential applications in various fields, such as drug delivery, tissue engineering, military protective clothing, photocatalytic degradation, lithium-ion batteries, supercapacitor electrodes, thermal energy storage, solar cells, and removing heavy metals from wastewater (Wang et al., 2017; Xu et al., 2017).

Tragacanth nanofibers have been synthesized by using a sonochemical/microemulsion method for loading peppermint oil and investigating its release profile (Fig. 16.5) (Ghayempour and Montazer, 2019). These nanofibers with a relatively smooth surface, one-dimensional shape, and 58 nm thicknesses have shown low cytotoxicity against human fibroblast cells and good antibacterial activities against *S. aureus* and *E. coli*. Peppermint oil with 18.3% has been loaded in tragacanth nanofibers. In this study, suggested sonochemical microemulsion has been introduced as one of the effective factors on the loading percentage of the nanofibers because of the production of a microemulsion with peppermint oil trapped in the center of tragacanth as the outer layer. Based on the obtained results, 92.38% of peppermint oil has been released from nanofibers after 18 h based on the non-Fickian model, in which the short-term release of peppermint oil has been attributed to the hard mechanical conditions by 18 h stirring in the phosphate buffer solution that finally leads to the release of all peppermint oil (Ghayempour and Montazer, 2019).

In another study, the loading capacity and release potential of saffron extract in the synthesized core-shell nanofiber with sizes of 95–271 nm using TG and zein by the coaxial electrospinning technique have been investigated (Dehcheshmeh and Fathi, 2019). It has been reported that by increasing the saffron percentage, loading efficiency was significantly enhanced from 68% to 89.77%. Saffron extract has been released at a rate of 21.66%, 27.75%, 43.88%, and 16.12% in saliva, hot water, gastric, and intestinal media, respectively, under different mechanisms of Kopcha, Peppas-Sahlin, Peppas-Sahlin, and Ritger-Peppas models, respectively (Dehcheshmeh and Fathi, 2019).

Nanofibrous scaffolds containing TG, poly(vinylalcohol) (PVA) and poly(ε-caprolactone) (PCL) have been synthesized through the two nozzles electrospinning process. These hybrid nanofibers had a diameter of 132 ± 27 nm and have shown 95.19% antibacterial activity against *S. aureus* bacteria without cytotoxic effect on NIH3T3 fibroblast cells (Zarekhalili et al., 2017).

Other nanofibrous scaffolds containing TG and PCL as skin scaffolds have been fabricated and their mechanical properties, biocompatibility hydrophilicity, and fibroblast cell proliferation have been evaluated for wound dressing application (Ranjbar-Mohammadi and Bahrami, 2015). Based on the obtained results, produced scaffolds with the composition of 3:1.5 of TG/PCL had increased hydrophilicity and increased the human fibroblast and NIH 3T3 fibroblast cells adhesion and proliferation in these nanofibers (Ranjbar-Mohammadi and Bahrami, 2015).

TG-based hydrogel/nanogel systems for drug delivery

Hydrogels or nano-hydrogels are fabricated based on polysaccharides—for example, chitosan, guar gum, and sodium alginate—for controlled drug release (Pathania et al., 2018). Furthermore, in recent years, TG has been applied as an efficient polysaccharide for the synthesis of hydrogels and nanohydrogels for the controlled release of different drugs. In this section, we will refer to some studies in this field, and others were summarized in Table 16.1.

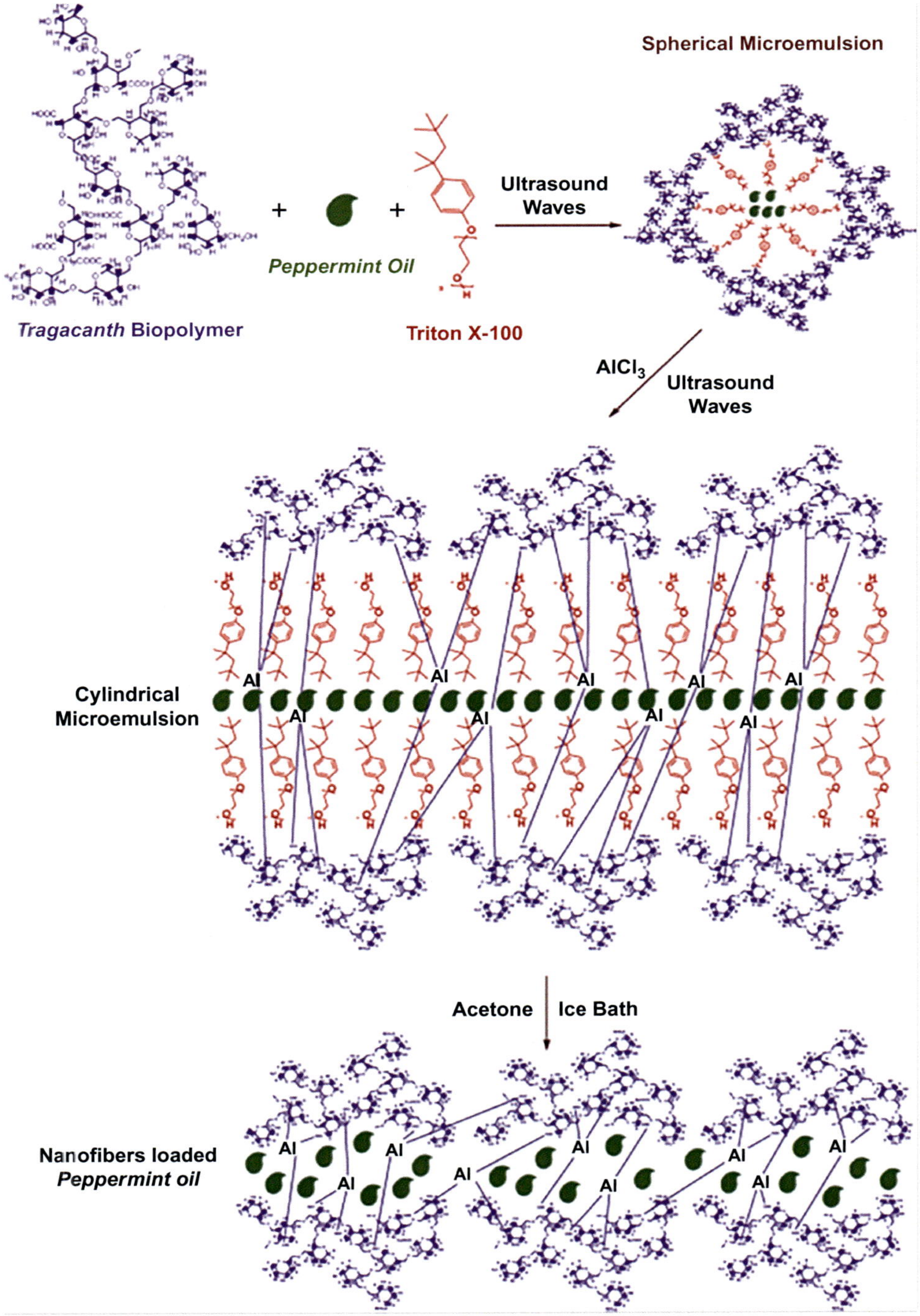

FIG. 16.5 The suggested mechanism of the loading peppermint oil into tragacanth nanofibers. *Reprinted from Ghayempour, S., Montazer, M., 2019. A novel controlled release system based on tragacanth nanofibers loaded peppermint oil. Carbohydr. Polym. 205, 589–595. https://doi.org/10.1016/j.carbpol.2018.10.078 with permission from the publisher.*

Synthesis of hydrogel film using TG, sodium alginate, and polyvinyl alcohol has been performed by a γ-radiation method for evaluation of their drug delivery of moxifloxacin and their application as wound healing and dressing agents (Singh et al., 2016). In addition, biological properties of this hydrogel film, including antioxidant activity, microbial permeability, thrombogenic ability, hemolysis, water vapor transmission rate, O_2 permeability, and bio-adhesion and mechanical properties, have been investigated in this study. This biocompatible hydrogel film without microbial permeability property has shown free radical scavenging activity of 39%, hemolysis of 0.83%, thrombogenicity of 82.43% with a water vapor permeability of 197.39 $g/m^2/day$, and oxygen permeability of 6.433 mg/L. Based on the release study of moxifloxacin, it has been reported that the release of this antibiotic happened through a non-Fickian mechanism and fitted in the Hixson-Crowell model (Singh et al., 2016).

Natural TG was applied for preparation of hydrogel beads (Fig. 16.6) using graphene oxide (GO) nanosheets, $CaCO_3$ as pores generator, and ionic cross-linking agents such as Ca^{+2} and Ba^{+2} for delivery of Rivastigmine (RIV) (Rahmani et al., 2018). The maximum loading for RIV (3 mg) was calculated as 65% for TG without GO and 91% for TG with 5 mg GO. It was shown that the entrapment efficiency of these hydrogel beads improves by incorporating GO nanosheets. This behavior was attributed to higher swelling of TG with 5 mg GO beads, which created more interaction between the functional groups of the hydrogel beads and RIV molecules (Rahmani et al., 2018). The drug release from these hydrogel beads under simulated gastric has been reported at <45% at pH 1.2 after 400 min and in the intestinal condition has been calculated at <97% at pH 7.4 after 240 min. The lower drug release at pH 1.2 has been attributed to the lower swelling degrees of hydrogel beads at this pH, whereas at high pH values (7.4) with higher swelling degrees, greater drug release has been attributed to the

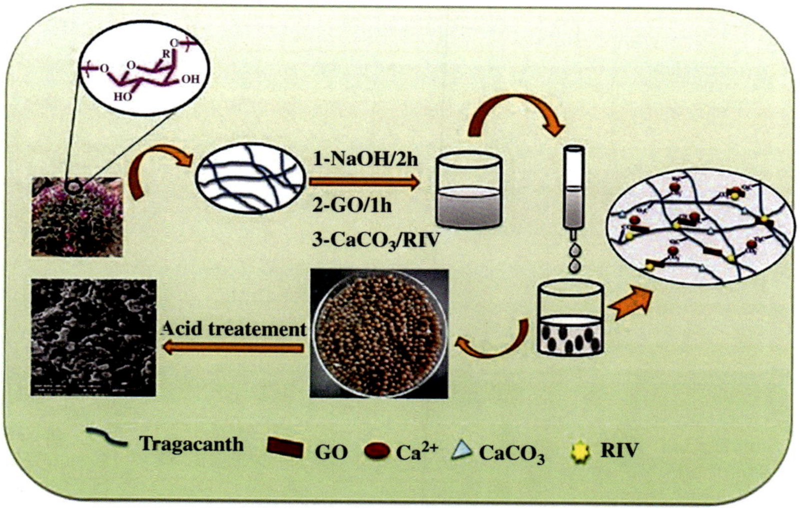

FIG. 16.6 Schematic for synthesis and preparation of TG-based hydrogel beads. *Reprinted from Rahmani, Z., Sahraei, R., Ghaemy, M., 2018. Preparation of spherical porous hydrogel beads based on ion-crosslinked gum tragacanth and graphene oxide: study of drug delivery behavior. Carbohydr. Polym. 194, 34–42. https://doi.org/10.1016/j.carbpol.2018.04.022 with permission from the publisher.*

ionization of carboxylic acid groups of GO and TG at high pH values. Based on the cytotoxicity results of prepared hydrogel beads on human fibroblast cells using MTT assay, the authors revealed a 98% of cell survival rate in <125 μg/mL beads concentration and when more GO was incorporated, the cell survival rate decreased. Therefore, in this study, incorporating GO in the hydrogel was introduced as an important parameter for increasing swelling capacity, entrapment efficiency, and ensuring a controlled release of the entrapped drug (Rahmani et al., 2018).

In another study by Singh and co-worker, TG-based mucoadhesive hydrogels have been fabricated and designed for the delivery of ciprofloxacin (Singh and Sharma, 2017). The drug release process has been best fitted in the Korsmeyer-Peppas model and performed through a non-Fickian diffusion mechanism. Based on this mechanism, the maximum drug release occurred at pH 7.4 (255.10 mg/L) compared to pH 2.2 (117.51 mg/L). The one of the factors affecting the mechanical strength and structure of the polymer network of this hydrogel has been introduced the pH of the swelling medium. Swelling of the hydrogels has been reported as a more effective factor in releasing the drug from this ciprofloxacin loaded hydrogel. Furthermore, this hydrogel with a thrombose percentage of 47.26% and hemolytic index of 0.75% has been recognized as a blood-compatible agent (Singh and Sharma, 2017).

TG hydrogel has been produced using a green synthesized silver nanocomposite by *Terminalia chebula* leaf extract using acrylamide as a monomer via a redox polymerization protocol (Rao et al., 2018). TEM analysis indicated spherical shape and the average size of 5 nm of Ag-NPs formed in TG hydrogels, and SEM images indicated the smooth surface of these Ag-NPs-loaded TG hydrogels. These formulated hydrogels have shown good antibacterial activity against *B. subtilis* and *E. coli*. The authors in this study suggested the application of produced hydrogels as a wound dressing (Rao et al., 2018).

TG as a cross-linker along with Fe_3O_4/SiO_2 nanoparticles as a magnetic core, N-vinyl imidazole (VI) as a functional monomer, and quercetin as a template has been applied for the synthesis of superparamagnetic molecularly imprinted polymer nanogel with a core-shell structure using a sol-gel method in order to achieve quercetin recognition and controlled release (Fig. 16.7) (Hemmati et al., 2016). Based on the authors' reported results, imprinted magnetic nanogel with high binding affinity and specific binding selectivity adsorbs the quercetin quickly with a maximum capacity of 175.43 mg/g and adsorption equilibrium time of 2 h, which was found from the Langmuir model. This nanogel released about 30% of the total loaded quercetin in the first 8 h at pH 7, while the released ratio of quercetin in the first 8 h increased up to 40% at pH 2. This pH-sensitive behavior of imprinted magnetic nanogel has been interpreted by the authors as follows: at low pH, the quercetin and PVI molecules were very protonated and thus interaction between two molecules became restricted because of the electronic repulsion of protonated groups. On the other hand, at high pH, the protonated degree of each atom of oxygen and nitrogen of two related molecules was slowly reduced, which led to the weakened electrostatic interaction between them and thus increased quercetin release (Hemmati et al., 2016).

In addition, nanohydrogels have been synthesized using TG in the presence of glycerol diglycidyl ether (GDE) and functionalized multiwalled carbon nanotubes (CNTs) (Fig. 16.8) for evaluation of load and release amount of indomethacin (Badakhshanian et al., 2016). The swelling behavior of these nanohydrogels depended on the pH, gel content and contact time, and also on the presence of CNTs. The authors' results indicated that

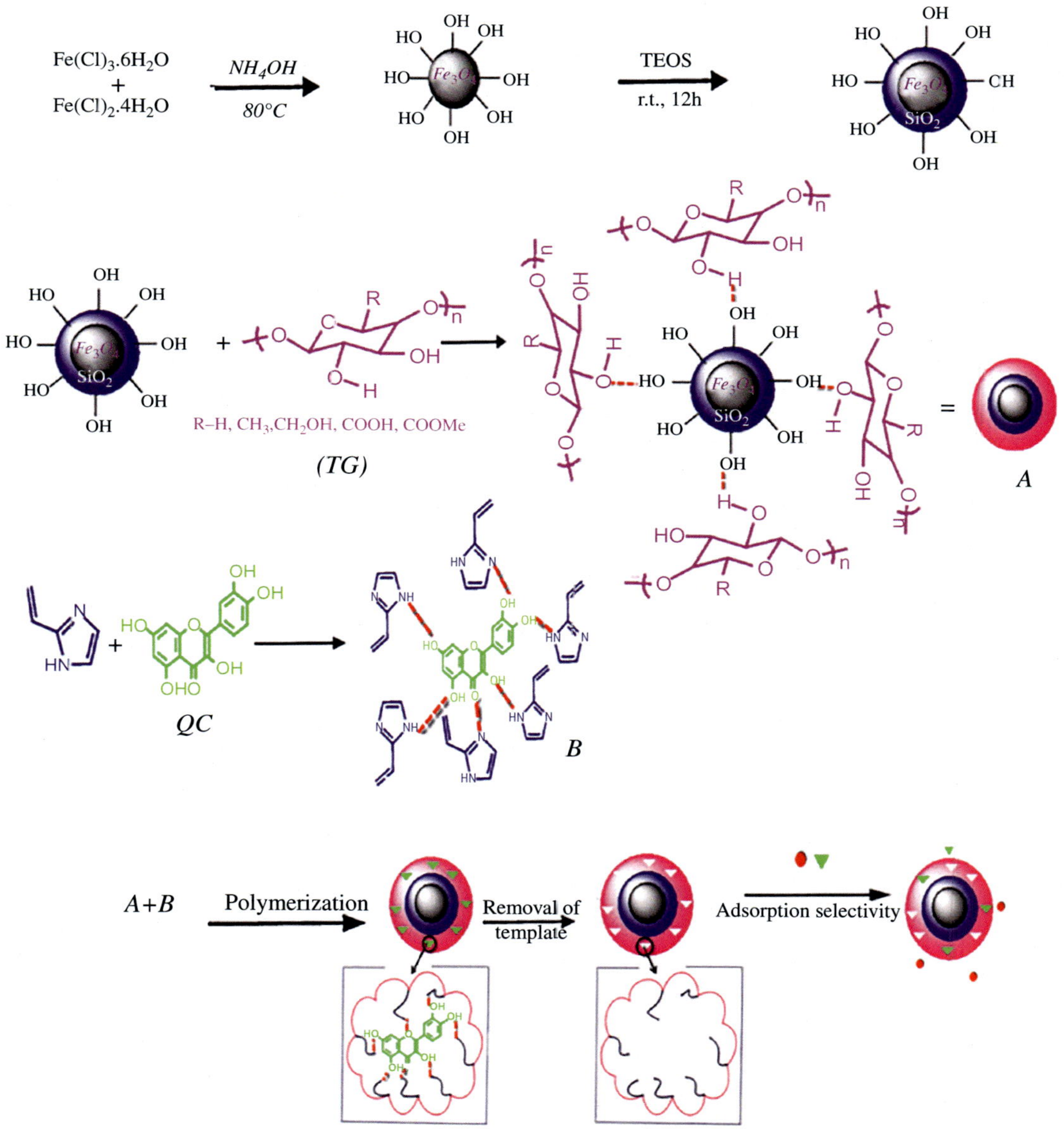

FIG. 16.7 A schematic representation of the preparation procedure for quercetin (QC) imprinted $Fe_3O_4@SiO_2$-based polymer in the five steps: (1) synthesis of Fe_3O_4 and its surface modification with TEOS; (2) fixation of $Fe_3O_4@SiO_2$ on TG by the formation of hydrogen bonding (A); (3) binding VI to template QC through hydrogen bonding (B); (4) inter-penetrating polymerization of VI in the presence of $Fe_3O_4@SiO_2$/TG; and (5) removal of the template. *Reprinted from Hemmati, K., Masoumi, A., Ghaemy, M., 2016. Tragacanth gum-based nanogel as a superparamagnetic molecularly imprinted polymer for quercetin recognition and controlled release. Carbohydr. Polym. 136, 630–640. https://doi.org/10.1016/j.carbpol.2015.09.006 with permission from the publisher.*

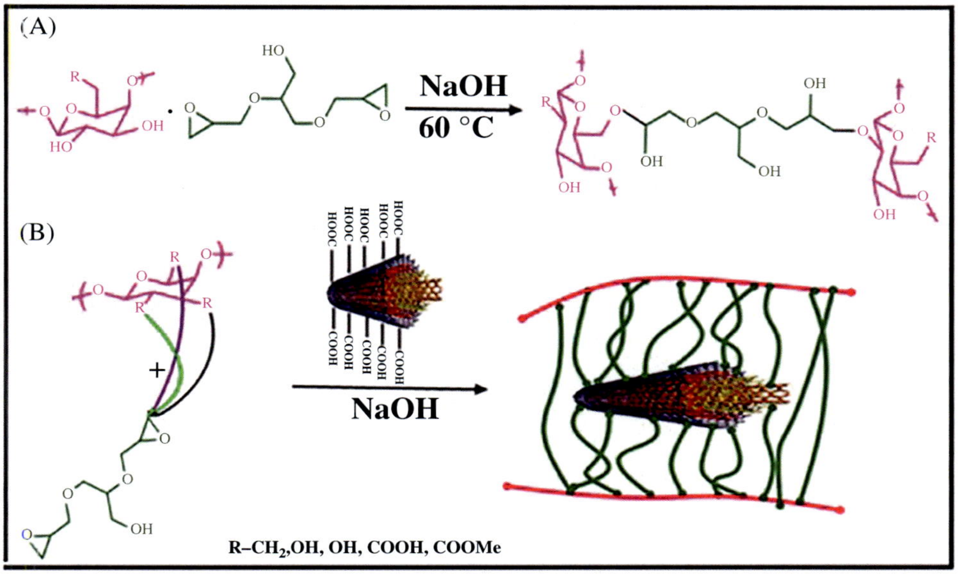

FIG. 16.8 Illustrations of synthesis of TG-GDE (A) and synthetic procedure for preparation of TG-GDE-CNT (B). *Reprinted from Badakhshanian, E., Hemmati, K., Ghaemy, M., 2016. Enhancement of mechanical properties of nanohydrogels based on natural gum with functionalized multiwall carbon nanotube: study of swelling and drug release. Polymer 90, 282–289. https://doi.org/10.1016/j.polymer.2016.03.028 with permission from the publisher.*

nanohydrogels containing CNTs had higher mechanical and swelling properties. The maximum loading for 0.06 mg of indomethacin has been calculated as 65% in the nanohydrogel containing TG (1 g) and GDE (1.5 g) and 94% in the nanohydrogel containing TG (1 g), GDE (1.5 g), and CNTs (0.01 g) after 90 min immersion. According to the release study, a 30-min delay has been reported in the release of indomethacin from nanohydrogel containing TG-GDE at pH of 2, 7.4, and 9, and from nanohydrogel containing TG-GDE-CNT at pH of 2 and 7.4. The maximum release for these two nanohydrogels has been reported as 88% and 98%, respectively, at pH 9 after 3 h; the higher sensitivity of TG-GDE-CNT to pH 9 has been attributed to the ionization process of carboxylic acid groups in CNT (Badakhshanian et al., 2016).

The release of ampicillin has been controlled using synthesized nanohydrogel with a size range of 40–100 nm from grafted TG with itaconic acid in the presence of potassium persulfate as an initiator and N,N1-methylene-bis-acrylamide (MBA) as a cross-linker by microwave radiations (Pathania et al., 2018). Based on the release study under different pH values of 2.2, 5.4, and 9.4, and in the distilled water, the authors concluded that pH 2.2 is the optimal pH for maximum release of ampicillin. This ampicillin-loaded nanohydrogel has been indicated to be more effective antibacterial activity against *E. coli* with an inhibition zone of 19.3 mm compared to the ampicillin drug alone with an inhibition zone of 15 mm (Pathania et al., 2018).

Furthermore, the loading efficiency of Indomethacin (IND) as a model drug in the pH-responsive nanohydrogels formulation has been investigated by Hosseini et al. (2016).

The authors in this study produced pH-responsive nanohydrogels based on TG using different cross-linkers such as PVA, glutaraldehyde (GA), glyceroldiglycidylether (GDE), and 3-aminopropyltriethoxysilane (APTES) modifier for drug delivery target (Hosseini et al., 2016). The swelling behavior of these nanohydrogels was evaluated in different conditions, including reaction time of 2–24 h, temperatures of 27, 37, and 60 °C, and pH of 2.2, 7.4, and 9. These hydrogels with heterogeneous structure showed a rough surface that was more porous. The swelling ratio of the prepared nanohydrogels improved upon increasing contact time, temperature, and pH. This can be attributed to the increase of cross-link molecular weight, mesh size values of the polymer network, and slow hydrolysis of ester bonds of TG side chains, which prolonged the contact times and opening of the created pores by ionic repulsion of the constituted ions. Loading of IND as a model drug showed dependence on the network structure of nanohydrogels. The loading efficiency of IND increased with loading time dependence on the higher swellability of nanohydrogels and was in the range of 70%–85%. The total IND release was calculated at 50%–80% at pH 9 after 24 h; this was dependent on the network structure of nanohydrogels and the swelling behavior of hydrogels at various pH values (Hosseini et al., 2016).

TG-based grafted systems for drug delivery

Biodegradable polymers such as gum polysaccharides possess weak mechanical strength; therefore, to solve this problem and enhance their mechanical strength, grafting of these polymers was performed with different monomers. Graft copolymers and their different modifications have been used in different fields (Gupta et al., 2018). A hydrogel has been synthesized by graft copolymerization of itaconic acid (IA) and lactic acid (LA) onto TG using a microwave-assisted method, and their antioxidant and antibacterial activity, as well as drug delivery potential of amoxicillin, were evaluated (Gupta et al., 2018). The prepared hydrogel showed the maximum swelling percentage of 311.61% (after 6 h at room temperature), mild free radical scavenge activity of 43.85% (at a concentration of 640 μg/mL), highest amoxicillin loading of 73% (after 24 h in double distilled water), maximum drug release of about 96% (after 6 h at pH 2.2), and good antibacterial activity against *S. aureus* (Gupta et al., 2018).

Hemmati and Ghaemy (2016), using combined methods of ring-opening polymerization, atomic transfer radical polymerization, and click reaction, synthesized a graft copolymer (pH/temperature-responsive) from TG and amphiphilic alkyne terminated terpolymers consist of methylated poly (ethylene glycol), polycaprolactone, and poly (dimethylaminoethyl methacrylate) (Hemmati and Ghaemy, 2016). In an aqueous solution, the graft copolymers self-assembled into single micelles, and by changes of pH, they assembled into micellar aggregates. In this study, the quercetin release from these micelles was dependent on pH and increased noticeably at the acidic conditions with a pH of 2.2. Rapid release occurred after 5 h and was then sustained, and slow release occurred with a release rate of about 90% at pH 2.2 after 30 h (Hemmati and Ghaemy, 2016).

TG-grafted poly(acrylic acid) hydrogel has been fabricated by graft copolymerization of TG with acrylic acid using glutaraldehyde as a cross-linker and ascorbic acid-potassium persulfate as an initiator. The prepared hydrogel was applied for the controlled colon-specific release of pantoprazole sodium as an antiulcerative drug. Based on the reported results, higher drug release happened at pH 9.2 based on the non-Fickian diffusion mechanism followed by the pH 7.0- and 2.0-based Case II diffusion mechanism. This occurrence may be due to the ionization of the hydroxyl and carboxylic groups in the hydrogel. In another study by Saruchi and coworkers, an interpenetrating polymer network was prepared, including TG, poly(acrylic acid), and poly(acrylamide) for *in vitro* controlled release of losartan potassium as an antihypertensive drug at various pH conditions and a temperature of 37 °C. Drug release from this prepared matrix occurred based on the non-Fickian mechanism at pH 2.0, 7.0, and 9.2 (Saruchi et al., 2014a, 2014b).

16.6.2 Wound healing

The wound healing potential of TG has been studied in different forms such as hydrogels, creams, and composites (Nazarzadeh Zare et al., 2019). The moxifloxacin-loaded hydrogel wound dressings were prepared by Singh et al. using TG, PVA, and sodium alginate by the dual cross-linking method (Singh et al., 2017a). At first, the interactions between TG and alginate fiber happened by calcium ions; subsequently, covalent cross-linking was performed in the presence of PVA by gamma irradiation. In this study, the sterile, biocompatible, and mucoadhesive nature of dressings was exhibited. Release of moxifloxacin happened based on the Fickian diffusion mechanism and the release profile fitted in to Korsmeyer-Peppas model of drug release. Furthermore, it indicated that drug release was greater in simulated wound fluid than distilled water (Singh et al., 2017a).

The authors of this research in their previous study, using PVA, TG, and sodium alginate, synthesized wound dressing hydrogel films by the radiation method (Singh et al., 2016). They indicated the noticeable features of the prepared hydrogel wound dressings, which included good wound fluid uptake, blood compatibility, nonhemolytic and nonthrombogenic nature, impermeability to microorganisms, ameliorated gas permeability, and controlled release of antibiotic moxifloxacin for 24 h (without burst release) (Singh et al., 2016).

A dressing bandage has been prepared based on tragacanth mucilage and experimented with as an *in vivo* model. The complete wound closure period in the treated animals with saturated bandages of 6% TG were 12 and 13 days compared to untreated and Vaseline gauze-treated animals with a wound closure period of 24 and 22 days, respectively (Moghbel and Naji, 2008). In addition, the wound healing effects of the prepared creams from TG (6 and 9 wt %) have been examined on the full-thickness wound in rabbits. The results demonstrated the better performance of cream containing 6% TG in the treated animals compared to the other groups and the untreated controls (Moghbel et al., 2005).

Singh's research team employed TG, PVP, and PVA to prepare sterile TG-PVA-PVP hydrogel dressings for the care of wound, infection, and injury pain together (Singh et al., 2017b). Singh et al. loaded antibiotic drug (gentamicin) and analgesic drug (lidocaine) on the surface of TG-PVA-PVP hydrogels (Table 16.1). The successful synthesis of these hydrogel polymers was affirmed by AFM, DSC, XRD, FTIR, TGA, cryo-SEM, ^{13}C NMR, and turgescence studies. The authors also considered the mechanism of medicine release, kinetic models of medicine deliverance, network parameters, and some other specifications of polymers such as microbial permeation, antioxidant abilities, and oxygen permeability. Their results displayed injury fluid absorption and gradual medicine release capacity of hydrogel films. Further, antioxidant and antimicrobial nature and the synergic consequence of mucoadhesive of hydrogel dressings make them suitable candidates for wound healing (Singh et al., 2017b). Singh (2018) in another study prepared the mucoadhesive gentamicin-loaded-TG-acacia gum (AG)-PVA-PVP hydrogel dressings through the radiation-induced cross-linking manner for use in wound healing (Table 16.1). These provided dressings absorbed simulated wound fluid (4.62 ± 0.31 g/g gel) and indicated antioxidant and antibacterial activity due to the existence of the gums and antibiotic drug; thus they were suggested for wound application (Singh, 2018).

In addition, curcumin-loaded PCL/TG nanofibers have been synthesized by electrospinning in order to evaluate controlled release of curcumin (Ranjbar-Mohammadi and Bahrami, 2016). PCL/TG nanofibers containing 3% curcumin exhibited sustained curcumin release (65%) up to a period of 20 days, improved mechanical performance, increased cell proliferation, and attachment onto the scaffold's surface compared to TG/PCL nanofibers (Ranjbar-Mohammadi and Bahrami, 2016). In another study, Ranjbar-Mohammadi et al. (2016) fabricated curcumin-loaded nanofibers containing TG, PVA, and PCL for investigation of their healing of the full-thickness wound in diabetic rats (Fig. 16.9). The pathological study in treated animals with these nanofibrous scaffolds after

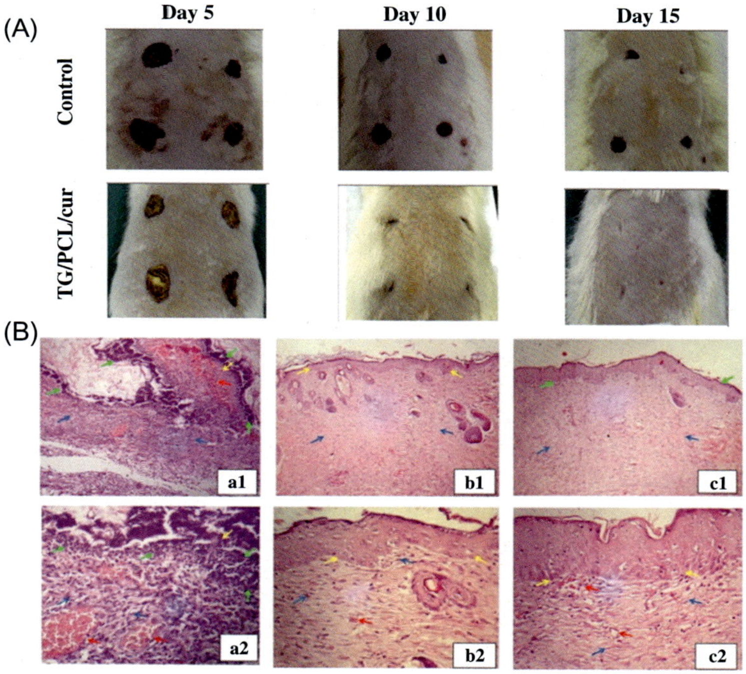

FIG. 16.9 (A) Wound closure after 5, 10, and 15 days in diabetic rats treated with curcumin-loaded poly(ε-caprolactone)/tragacanth gum (TG/PCL/cur) nanofibers compared to control samples and (B) histological evaluation of wounds treated by PCL/TG/Cur nanofibers after 15 days. (a) Hematoxylin and eosin staining for skin wound samples of control (open wound), (b) PCL/TG/Cur. and (c) PCL/TG/Cur/cell. Granulation tissue, epithelial regeneration, angiogenesis, and collagen fibers were indicated by *blue*, *yellow*, *red*, and *green arrows*, respectively (Magnification ×100). *Reprinted from Ranjbar-Mohammadi, M., Rabbani, S., Bahrami, S.H., Joghataei, M.T., Moayer, F., 2016. Antibacterial performance and in vivo diabetic wound healing of curcumin loaded gum tragacanth/poly(ε-caprolactone) electrospun nanofibers. Mater. Sci. Eng. C 69, 1183–1191. https://doi.org/10.1016/j.msec.2016.08.032 with permission from the publisher.*

15 days demonstrated fast wound closure with well-formed granulation tissue controlled by collagen deposition, fibroblast proliferation, the formation of sweat glands and hair follicles, and complete early regenerated epithelial layer (Fig. 16.9B) (Ranjbar-Mohammadi et al., 2016). Furthermore, the same authors synthesized TG/PVA nanofibers and indicated their good antimicrobial properties and biocompatibility as well as attachment and proliferation of human fibroblast lines (AGO) on these nanofiber scaffolds. The authors, by referring to these features, demonstrated that prepared nanofibers are active in wound dressings, and the protection of the wound area from its surroundings can prevent infection and dehydration and accelerate the healing route by providing an optimal environment for healing, eliminating any additional wound exudates, and allowing continuous tissue regeneration (Ranjbar-Mohammadi et al., 2013). In the study of Fayazzadeh et al. (2014), the topical application of TG was investigated for evaluation of its skin wound healing effects on adult male rats. The authors indicated that the rate of wound closure on day 7 in the TG-treated group was higher than in the control groups and it completely closed on day 10 of the experiment (Fayazzadeh et al., 2014).

16.6.3 Tissue engineering

Natural compounds are an attractive choice for tissue engineering because of their desirable features. They are inexpensive and easily accessible. They induce less inflammation and allergic reactions, and are more biocompatible and biodegradable than synthetic compounds (Mohammadinejad et al., 2020). TG has some desirable properties such as water solubility, biocompatibility, and biodegradability with no allergic and carcinogenic effects, so it is one of the most widely used natural compounds for tissue engineering (Alvandimanesh et al., 2017; Mohammadinejad et al., 2020).

Based on the importance of natural hydrogels in bone tissue engineering, Haeri et al. (2016) synthesized the TG-based hydrogel and compared its osteoinductive ability with collagen hydrogel and tissue culture plate (TCPS) by culturing adipose-derived mesenchymal stem cells (AT-MSCs) on these hydrogels and investigating their biocompatibility, alkaline phosphatase (ALP) activity, osteo-related genes, and calcium content (Haeri et al., 2016). The results reported by the authors have shown the highest ALP activity and mineralization of TG hydrogel and the highest proliferation of AT-MSCs, expression of Runx2, osteocalcin, and osteonectin in the cultured cells on this hydrogel compared to the collagen hydrogel and TCPS. Thus, the study suggested the application of this hydrogel as an appropriate scaffold for the proliferation and osteogenic differentiation of stem cells in the orthopedic field (Haeri et al., 2016).

Bone implant is a method for treating skeletal fissures or replacing lost bone; to be a perfect scaffold, it must be prepared from biological materials that mimic the structure and features of natural bone. Nevertheless, the fabrication of living tissue constructs with function, mechanics, and architecture similar to natural bone is a major challenge in the regeneration and treatment of bone tissue in dentistry and orthopedics (Anita Lett et al., 2019). Therefore, Anita Lett et al. (2019) using the TG (as a natural binder) produced pure and porous hydroxyapatite (HAP) scaffolds for bone replacements by a polymeric replication method (Fig. 16.10). The bioactivity of this prepared scaffold was evaluated by immersing the scaffold in

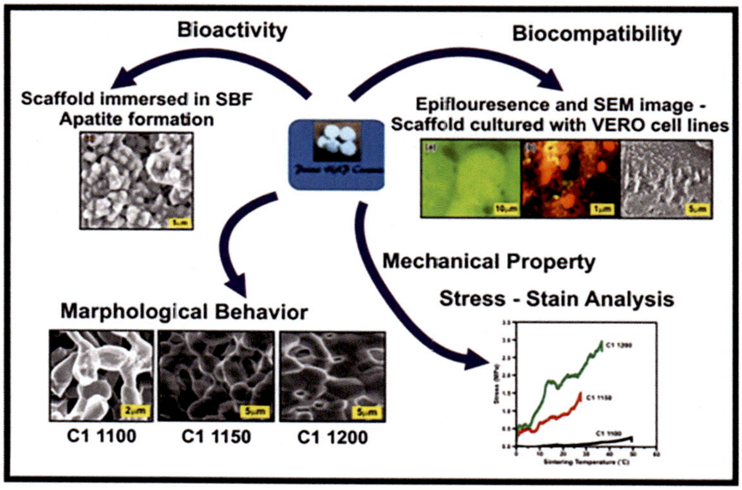

FIG. 16.10 Schematic diagram of fabricated HAP scaffolds. *Reprinted from Anita Lett, J., Sundareswari, M., Ravichandran, K., Latha, B., Sagadevan, S., 2019. Fabrication and characterization of porous scaffolds for bone replacements using gum tragacanth. Mater. Sci. Eng. C 96, 487–495. https://doi.org/10.1016/j.msec.2018.11.082 with permission from the publisher.*

simulated body fluid (SBF) for 7 days and its biocompatibility was established by seeding of Vero cells on the scaffold. Based on the obtained results, it was shown that the porous and pure HAP scaffolds with 15% (wt) TG concentration possessed highly interconnecting pores with maximum compressive yield strength of 2.945 MPa. Given that it was shown that these scaffolds support Vero cell attachment after 24 h, the authors suggested that the 3D structure of HAP scaffolds can provide a platform for cell attachment, proliferation, and differentiation. According to the authors of this study, the fabricated HAP scaffolds are compatible with natural bone and should be further considered for use in bone implants (Anita Lett et al., 2019).

In a study by Paknejad et al. (2012), TG has been used as a carrier for lactoferrin (a nonheme iron-binding glycoprotein is a fraction of whey protein and extracted from milk) in combination with anorganic bovine bone to evaluate ossification in rabbits calvaria with experimentally induced bony defects (Paknejad et al., 2012). The results demonstrated that the application of lactoferrin with an organic bovine bone cannot improve and accelerate bone regeneration in the treated groups (Paknejad et al., 2012).

Furthermore, aligned and random nanofibrous scaffolds containing poly(L-lactic acid) (PLA) and TG have been prepared in various ratios of 100:0, 75:25, and 50:50, respectively, by the electrospinning method for application in the regeneration of peripheral nerve damage (Ranjbar-Mohammadi et al., 2016a); ; . Based on the authors' results, nanofibers at a ratio of 50:50 (PLA/TG) demonstrated smooth and uniform morphology with diameters of 226 ± 73 nm and scaffolds containing PLA/TG at a ratio of 75:25 indicated the most balanced properties, which have been used for *in vitro* culture of nerve cells (PC12) to evaluate their potential in nerve regeneration. In addition, results have indicated higher proliferation, better cellular phenotype, more cell differentiation, and improved neurite outgrowth, on electrospun aligned nanofibers (PLA/TG 75:25) compared to random nanofibers (Ranjbar-Mohammadi et al., 2016a).

Tetracycline hydrochloride (TCH)-loaded blend and core-shell nanofibers with smooth and bead-less morphology were produced from TG and polylactic glycolic acid (PLGA) by the blend electrospinning and coaxial electrospinning method for periodontal regeneration in controlled drug delivery systems (Fig. 16.11) (Ranjbar-Mohammadi et al., 2016b). Drug release results revealed that the TCH release rate from both nanofibrous membranes can be effectively controlled. Drug release in the incorporated TCH into core-shell nanofibers was more prolonged (75 days) with only 19% of burst release of TCH within the first 2 h. Thus, the authors recommended that these core-shell nanofibers with prolonged drug release and good mechanical properties, antibacterial activity, and biocompatibility can be promising candidates as scaffolds for the treatment of periodontal diseases (Ranjbar-Mohammadi et al., 2016b).

Tragacanth for cell encapsulation

Encapsulation is defined as a process for entrapping active substances by means of wall material and mainly for protecting them against physicochemical damages and improving stability. Natural compounds are mostly sensitive to environmental conditions and could be easily damaged; therefore, encapsulation could improve their bioavailability and stability (Nedovic et al., 2011; Đorđević et al., 2014). In recent years, TG has been used as a biopolymer to form micro- and nanocapsules with different content and in different manners called the encapsulation process. Micro- or nanocapsules produced in this manner could involve some natural compounds or some kinds of drugs, or even some essential nutrients.

For example, tragacanth nanocapsules containing chamomile extract have been produced through the ultrasound irradiation-assisted water/oil/water (W/O/W) microemulsion technique on cotton fabric through a UV curing approach (Ghayempour and Montazer, 2017). In this study, the encapsulation process of the aqueous chamomile extract into a water-soluble polymer was performed by trapping of the extract in an oil-based material by a surfactant, and then triton X-100 was added to this prepared microemulsion including chamomile extract and almond oil (as the core material) (Fig. 16.12). Triton X-100 makes micelles with a hydrophilic center and hydrophobic external surface, which the chamomile extract traps in the center and almond oil surrounds the micelles. After that, TG was added to the mixture, which covers the chamomile extract and almond oil as double-layer micelles (with a hydrophobic center and hydrophilic surface). This microemulsion was then transferred on the cotton fabric (a cellulosic-based polymer) by the presence of aluminum chloride ($AlCl_3$) (Fig. 16.12). The encapsulation in this method occurred due to cross-linking between the carboxylic groups of TG and Al ions (Ghayempour and Montazer, 2017).

The spherical shape and size (60–80 nm) of these nanocapsules were shown by FESEM analysis. The fabricated NE&TC tissue displayed good persistence with sensible release behavior. The encapsulation efficiency of the chamomile extract into TG has been reported as

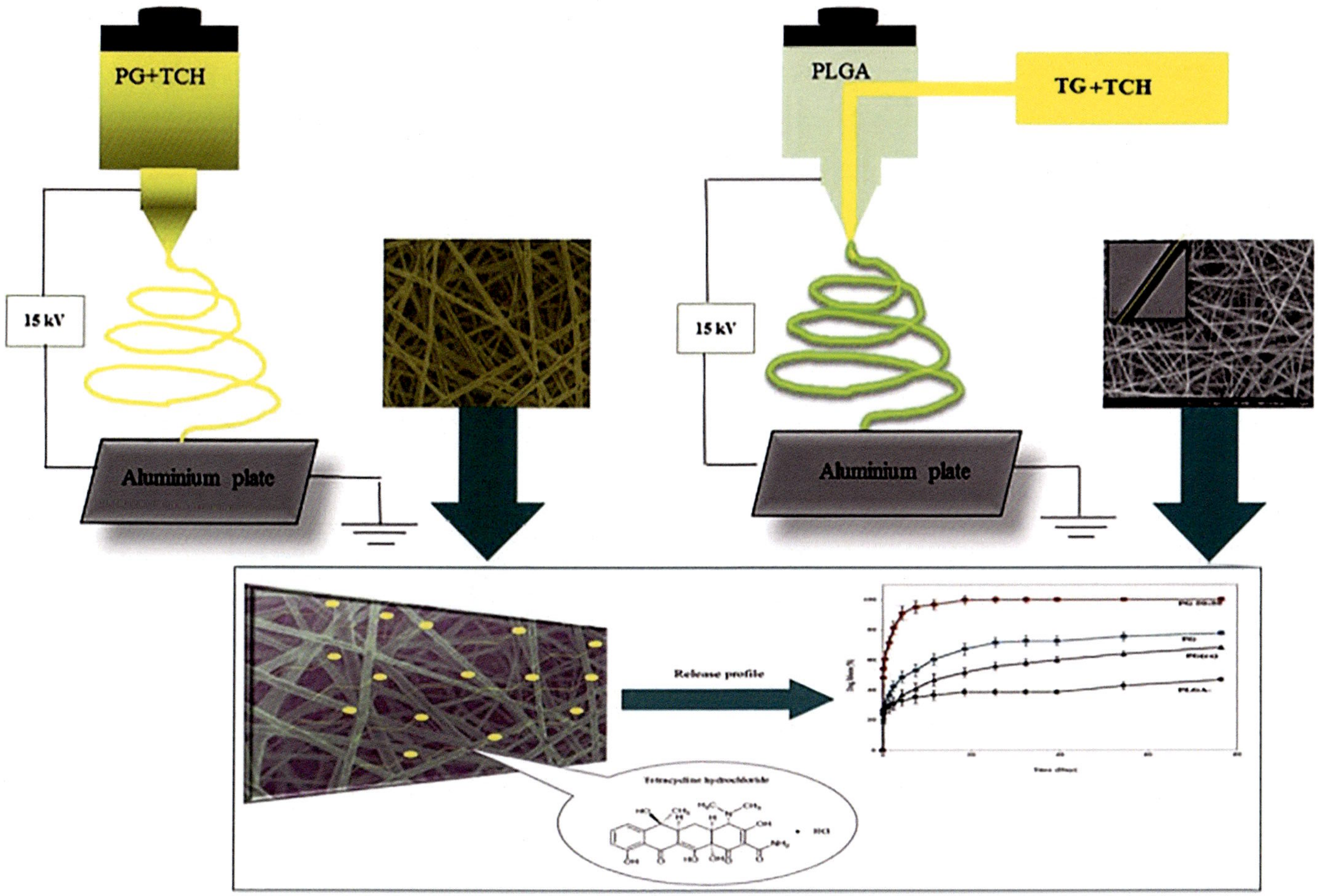

FIG. 16.11 Schematic representation of the synthesis of nanofibers and their release properties. *Reprinted from Ranjbar-Mohammadi, M., Zamani, M., Prabhakaran, M.P., Bahrami, S.H., Ramakrishna, S., 2016. Electrospinning of PLGA/gum tragacanth nanofibers containing tetracycline hydrochloride for periodontal regeneration. Mater. Sci. Eng. C 58, 521–531. https://doi.org/10.1016/j.msec.2015.08.066 with permission from the publisher.*

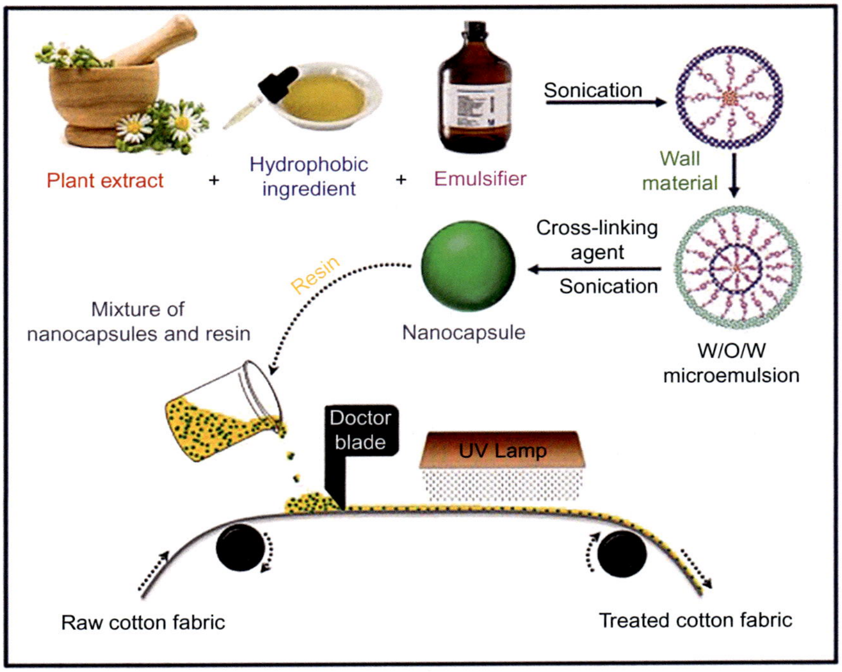

FIG. 16.12 The suggested mechanism for simultaneous nano-encapsulation and treatment of chamomile extract on the cotton fabric through the microemulsion method. *Reprinted from Ghayempour, S., Montazer, M., 2017. Tragacanth nanocapsules containing Chamomile extract prepared through sono-assisted W/O/W microemulsion and UV cured on cotton fabric. Carbohydr. Polym. 170, 234–240. https://doi.org/10.1016/j.carbpol.2017.04.088 with permission from the publisher.*

93%, which depended on the contents of the wall material and emulsifier to prevent the aggregation of particles and to reduce the surface tension. The treated cotton fabric with nanocapsules has indicated a good release behavior of 96 h, bactericidal activity of 91% and 89% against *S. aureus* and *E. coli*, and a 94% fungicidal effect against *C. albicans*; it also demonstrated high stability against washing and rubbing tests (Ghayempour and Montazer, 2017).

Nanocapsules containing peppermint oil as an antibacterial, antifungal agent has been produced using triton X-100 as an emulsifier and $AlCl_3$ or $CaCl_2$ as cross-linking agents (Ghayempour et al., 2015). It has been reported by the authors that the size of the nanocapsules depends on the concentration of the wall material (TG), emulsifier and cross-linking agents, but the duration of the encapsulation process does not actually affect the size of the nanocapsules. Antibacterial and antifungal effects were investigated using the shake flask method on *E. coli*, *S. aureus*, and *C. albicans*. Results show that after 12 h, the full effects of antibacterial and antifungal activity of nanocapsules were reached. However, because of the outer membrane in gram-negative bacteria, *E. coli* was less affected than gram-positive bacteria (Ghayempour et al., 2015).

Encapsulation of *Aloe vera* extract is another example of encapsulating natural products. Nanocapsules containing *A. vera* extract have been produced using sonochemical-based microemulsion with aim of wound healing (Ghayempour et al., 2016b). *A. vera*-loaded microcapsules (MCs) have been formed using *A. vera* extract, almond oil, and triton X-100, initially, to form a microemulsion and after adding TG, triton X-100, and $AlCl_3$ (as a cross-linking agent), nanocapsules were formed (Fig. 16.13). Antimicrobial activity against *E. coli*, *S. aureus*, and *C. albicans* was investigated using the dynamic shake flask method. The results show the reduction of 84%, 91%, and 80% growth of *E. coli*, *S. aureus*, and *C. albicans*, respectively, by affecting the inner membrane and organs of the microorganism. The cytotoxicity of the formulation was tested using morphological changes on fibroblast cells of normal primary human skin. Optical microscopic images show the fusiform appearance of fibroblast cells after 24 h with cell viability of 98%, and no cytotoxicity was observed (Ghayempour et al., 2016a).

Probiotics could also be encapsulated using TG as a biopolymer during the extrusion method (Foroutan et al., 2017). These encapsulations have been performed using three types of polymers mixture, including ATC (alginate-tragacanth-chitosan), AGC (alginate-gum arabic-chitosan), and APC (alginate-pectin-chitosan). In this study, the size of the prepared MCs was between 1 and 1.5 mm. Based on the results obtained, it has been reported that the larger MCs ensure better protection against environmental damages, although the MCs' size also should not be larger than 1 mm because of decreasing uniformity. Moreover, the encapsulation process related to the ATC has been indicated to have higher efficacy than the others due to the presence of chitosan, which covers the negatively charge alginate capsules, which in turn results in physical and chemical stability. The rationality of the probiotic encapsulation is to enhance the viability of the probiotics in low PH environments like a

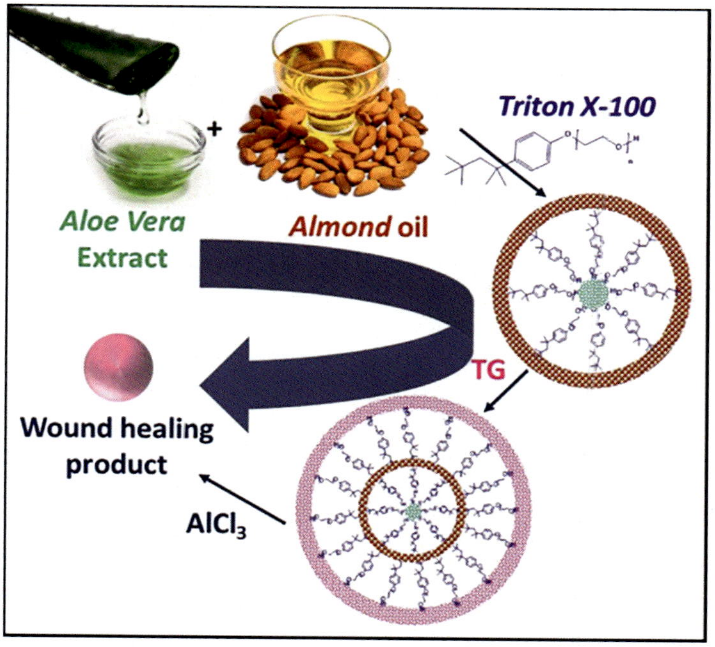

FIG. 16.13 A schematic of nanocapsules containing *Aloe vera* fabricated by the sonochemical microemulsion method to prepare a wound healing product. *Reprinted from Ghayempour, S., Montazer, M., Mahmoudi Rad, M., 2016. Encapsulation of* Aloe vera *extract into natural tragacanth gum as a novel green wound healing product. Int. J. Biol. Macromol. 93, 344–349. https://doi.org/10.1016/j.ijbiomac.2016.08.076 with permission from the publisher.*

gastric fluid. Therefore, the encapsulation with different natural polymers used in this study enhanced the viability with no difference between each other. On the other hand, all three polymers mentioned above are similar to each other in enhancing the viability of probiotics (Foroutan et al., 2017).

Carotenoids, as lipid-soluble nutrients, which are responsible for the red, yellow, or orange color of some fruits or vegetables, can be encapsulated in order to enhance the stability and storage of these fruits or vegetables. The coacervation process is the method of choice to encapsulate β-carotene, which occurs between TG as a polysaccharides chain with a negative charge and casein (CAS) as the protein chain with positive patches on it (Jain et al., 2016).

In another study, ferrous sulfate, as an essential nutrient, was encapsulated in the TG using the solvent evaporation method (Asghari-Varzaneh et al., 2017). In this study, the effects of three parameters including ferrous sulfate content (w/v%), TG content (w/v%), and alcohol to mixture ratio (w/w%) on the encapsulation efficacy were also characterized. These parameters affect the encapsulation efficacy both individually and in association with each other. Based on the authors' results, the optimized condition for encapsulation efficacy includes ferrous sulfate content of 22%, TG content of 5%, and alcohol to the mixture ratio of 11:1 (w/w%) (Asghari-Varzaneh et al., 2017).

Structural changes in TG could optimize the physicochemical properties of micro- and nanocapsules. De-esterified tragacanth (DET) is an analog of TG that contains de-acetylated tragacanthic acid as a major fraction and arabinogalactan as a minor fraction. De-acetylated tragacanthic acid consists of galacturonic acid as the backbone with fucose and xylose as side branches (Alvandimanesh et al., 2017). Because of some advantages like cell adhesion, DET could be a novel polymer for cell encapsulation. Cell therapy was introduced as a novel treatment of chronic disease. However, it has some limitations, such as the inflammatory response of the immune system and stability. To overcome these problems, encapsulating therapeutic cells into microcapsules is applied. Microcapsules (MCs) produced by a coflow extrusion method and βTC3 cell line (a mouse pancreatic β cell line) were employed for cell encapsulating because of their ability in insulin secretion, which could be investigated effectively during the encapsulation process. Acceptable cell encapsulation depends on reducing the shocks along the process and optimizing the mass transfer by changing the size and sphericality of MCs (Alvandimanesh et al., 2017). According to the Taguchi method, the optimal condition depends on decreasing the size and increasing the sphericality of the MCs. On the other hand, air pressure is the most effective parameter in controlling the size, and the concentration of DET and calcium chloride solution is the most effective parameter in controlling sphericality. The results reported by the authors showed that the optimum MCs' size and sphericality were 214.58 μm and 60.75%, respectively (Alvandimanesh et al., 2017).

Cell adhesion, differentiation, and immobilization in tragacanth scaffolds

TG also plays an important role in tissue engineering and cell differentiation. In one study, TG was used with calcium alginate (CA) to encapsulate MG-63 as a bone cell for bone tissue engineering (Kulanthaivel et al., 2017). Briefly, the soluble part of TG was extracted using the aqueous extraction method and CA-TG (calcium alginate-gum tragacanth) beads were prepared using the ionic gelation method. The SEM results attained have confirmed that the spherical shape of the bead changes to an irregular shape with increasing TG concentration, wrinkles of bead surface alters with increasing the alginate concentration, and the size of the beads increases with increasing TG concentration (Kulanthaivel et al., 2017). The authors, using FT-IR analysis, revealed that no significant change occurs in the molecular structure of the TG and CA. Furthermore, they indicated that with increasing TG concentration, the size of beads and their capacity for cell loading increased, i.e., more cell encapsulation occurred with increasing TG concentration. The authors employed the dye exclusion assay and flowcytometry, and concluded that the biocompatibility of the beads increased due to increasing TG concentration and also increased in cell counts. They reported that to maintain the cell growth and proliferation in beads, cells should adhere to the live cells in live tissues, and this property indicates the efficacy of encapsulation. Thus, their results showed that increasing TG concentration leads to an increase in the ratio of the adhered cells to live cells, and finally in the TG-CA beads, osteogenic and angiogenic activity increases compared to in the CA beads (Kulanthaivel et al., 2017).

Cell immobilization is a method for the restriction of cells in a certain media in order to protect them against catalytic processes (Xumeng et al., 2016). TG is used as a natural polymer that is potentially useful for cell immobilization as media, but its low viscosity makes it susceptible to mechanical damage. In this study, ionomer formation with trivalent cations is used for increasing the viscosity of TG (Otady et al., 2005). In addition, in this study, the whole-cell penicillin G acylase system is served as a model for evaluating the immobilization system. TG as a matrix was applied for the immobilization of *E. coli* with penicillin G acylase activity to catalyze the conversion of benzyl penicillin to 6-aminopenicillanic acid. According to the results reported by the authors, TG is a suitable polymer for cell immobilization systems because of its biocompatibility, mechanical and PH stability, and ability to enhance hydrolysis rate (Otady et al., 2005).

Furthermore, TG is an appropriate matrix for enzyme immobilization and cell growth (Nazarzadeh Zare et al., 2019). In a study by Otadi and Mobayen (2011), the immobilization of glucose oxidase was performed on TG gel capsules. Based on their results, the optimum conditions for effective encapsulation of glucose oxidase were TG (1%, w/v), $FeCl_3$ (10%, w/v), $AlCl_3$ (1%, w/v), and gelation time of 2 h at 50 °C. In addition, the reported kinetics parameters (Km) for both free and immobilized glucose oxidase were calculated as 9.00 and 12.375 mmol, respectively (Otadi and Mobayen, 2011).

References

Ahadian, S., Finbloom, J.A., Mofidfar, M., Diltemiz, S.E., Nasrollahi, F., Davoodi, E., Hosseini, V., Mylonaki, I., Sangabathuni, S., Montazerian, H., Fetah, K., Nasiri, R., Dokmeci, M.R., Stevens, M.M., Desai, T.A., Khademhosseini, A., 2020. Micro and nanoscale technologies in oral drug delivery. Adv. Drug Deliv. Rev. 157, 37–62. https://doi.org/10.1016/j.addr.2020.07.012.

Ahmadi Gavlighi, H., 2013. Tragacanth Gum: Structural Composition, Natural Functionality and Enxymatic Conversion as Source of Potential Prebiotic Activity. Technical University of Denmark.

Ahmadi Gavlighi, H., Meyer, A.S., Zaidel, D.N.A., Mohammadifar, M.A., Mikkelsen, J.D., 2013. Stabilization of emulsions by gum tragacanth (Astragalus spp.) correlates to the galacturonic acid content and methoxylation degree of the gum. Food Hydrocoll. 31 (1), 5–14. https://doi.org/10.1016/j.foodhyd.2012.09.004.

Ahmadi, E., Sadrjavadi, K., Mohammadi, G., Fattahi, A., 2018. De-esterified tragacanth microspheres loaded into eudragit s-100 coated capsules for colon-targeted delivery. Iran. J. Pharm. Res. 17 (2), 470–479. http://ijpr.sbmu.ac.ir/article_2235_b1e89f4c164e02d68e9525a38d41b23d.pdf.

Akkol, E.K., Koca, U., Pesin, I., Yilmazer, D., 2011. Evaluation of the wound healing potential of achillea biebersteinii Afan. (Asteraceae) by in vivo excision and incision models. Evid. Based Complement. Alternat. Med. 2011. https://doi.org/10.1093/ecam/nep039.

Ali, Q., Malik, S., Malik, A., Hafeez, N.M., Salman, S., 2020. Role of modern technologies in tissue engineering. Arch. Neurosci. 7 (1), e90394. https://doi.org/10.5812/ans.90394.

Alvandimanesh, A., Sadrjavadi, K., Akbari, M., Fattahi, A., 2017. Optimization of de-esterified tragacanth microcapsules by computational fluid dynamic and the Taguchi design with purpose of the cell encapsulation. Int. J. Biol. Macromol. 105, 17–26. https://doi.org/10.1016/j.ijbiomac.2017.06.059.

Anderson, D.M., 1989. Evidence for the safety of gum tragacanth (*Asiatic astragalus* spp.) and modern criteria for the evaluation of food additives. Food Addit. Contam. 6 (1), 1–12. https://doi.org/10.1080/02652038909373733.

Anita Lett, J., Sundareswari, M., Ravichandran, K., Latha, B., Sagadevan, S., 2019. Fabrication and characterization of porous scaffolds for bone replacements using gum tragacanth. Mater. Sci. Eng. C 96, 487–495. https://doi.org/10.1016/j.msec.2018.11.082.

Anwar, A., Masri, A., Rao, K., Rajendran, K., Khan, N.A., Shah, M.R., Siddiqui, R., 2019. Antimicrobial activities of green synthesized gums-stabilized nanoparticles loaded with flavonoids. Sci. Rep. 9 (1). https://doi.org/10.1038/s41598-019-39528-0.

Asghari-Varzaneh, E., Shahedi, M., Shekarchizadeh, H., 2017. Iron microencapsulation in gum tragacanth using solvent evaporation method. Int. J. Biol. Macromol. 103, 640–647. https://doi.org/10.1016/j.ijbiomac.2017.05.047.

Assimopoulou, A.N., Papageorgiou, V.P., 2004. Preparation and release studies of alkannin-containing microcapsules. J. Microencapsul. 21 (2), 161–173. https://doi.org/10.1080/02652040310001637839.

Atrak, K., Ramazani, A., Taghavi Fardood, S., 2018. Green synthesis of amorphous and gamma aluminum oxide nanoparticles by tragacanth gel and comparison of their photocatalytic activity for the degradation of organic dyes. J. Mater. Sci. Mater. Electron. 29 (10), 8347–8353. https://doi.org/10.1007/s10854-018-8845-2.

Azarikia, F., Abbasi, S., 2010. On the stabilization mechanism of Doogh (Iranian yoghurt drink) by gum tragacanth. Food Hydrocoll. 24 (4), 358–363. https://doi.org/10.1016/j.foodhyd.2009.11.001.

Badakhshanian, E., Hemmati, K., Ghaemy, M., 2016. Enhancement of mechanical properties of nanohydrogels based on natural gum with functionalized multiwall carbon nanotube: study of swelling and drug release. Polymer 90, 282–289. https://doi.org/10.1016/j.polymer.2016.03.028.

Bagheri, S.M., Keyhani, L., Heydari, M., Dashti-R, M.H., 2015. Antinociceptive activity of Astragalus gummifer gum (gum tragacanth) through the adrenergic system: a in vivo study in mice. Journal of Ayurveda and Integrative Medicine 6 (1), 19–23. https://doi.org/10.4103/0975-9476.146543.

Bai, H., Li, Z., Zhang, S., Wang, W., Dong, W., 2018. Interpenetrating polymer networks in polyvinyl alcohol/cellulose nanocrystals hydrogels to develop absorbent materials. Carbohydr. Polym. 200, 468–476. https://doi.org/10.1016/j.carbpol.2018.08.041.

Balaghi, S., Mohammadifar, M.A., Zargaraan, A., 2010. Physicochemical and rheological characterization of gum tragacanth exudates from six species of iranian astragalus. Food Biophys. 5 (1), 59–71. https://doi.org/10.1007/s11483-009-9144-5.

Balaghi, S., Mohammadifar, M.A., Zargaraan, A., Gavlighi, H.A., Mohammadi, M., 2011. Compositional analysis and rheological characterization of gum tragacanth exudates from six species of Iranian Astragalus. Food Hydrocoll. 25 (7), 1775–1784. https://doi.org/10.1016/j.foodhyd.2011.04.003.

Bashir, A., Warsi, M.H., Sharma, P.K., 2016. An overview of natural gums as pharmaceutical excipient: their chemical modification. World J. Pharm. Pharm. Sci. 5, 2025–2039.

Chen, Y.L., 2008. Preparation and Characterization of Water-Soluble Chitosan Gel for Skin Hydration. Universiti Sains Malaysia.

Chourasia, M.K., Jain, S.K., 2004. Polysaccharides for colon targeted drug delivery. Drug Deliv. 11 (2), 129–148. https://doi.org/10.1080/10717540490280778.

Cikrikci, S., Mert, B., Oztop, M.H., 2018. Development of pH sensitive alginate/gum tragacanth based hydrogels for oral insulin delivery. J. Agric. Food Chem. 66 (44), 11784–11796. https://doi.org/10.1021/acs.jafc.8b02525.

Dang, Y., Guan, J., 2020. Nanoparticle-based drug delivery systems for cancer therapy. Smart Mater. Med. 1, 10–19. https://doi.org/10.1016/j.smaim.2020.04.001.

Darroudi, M., Sabouri, Z., Kazemi Oskuee, R., Khorsand Zak, A., Kargar, H., Hamid, M.H.N.A., 2013. Sol-gel synthesis, characterization, and neurotoxicity effect of zinc oxide nanoparticles using gum tragacanth. Ceram. Int. 39 (8), 9195–9199. https://doi.org/10.1016/j.ceramint.2013.05.021.

Dehcheshmeh, M.A., Fathi, M., 2019. Production of core-shell nanofibers from zein and tragacanth for encapsulation of saffron extract. Int. J. Biol. Macromol. 122, 272–279. https://doi.org/10.1016/j.ijbiomac.2018.10.176.

Dehghan-Niri, M., Tavakol, M., Vasheghani-Farahani, E., Ganji, F., 2015. Drug release from enzyme-mediated in situ-forming hydrogel based on gum tragacanth–tyramine conjugate. J. Biomater. Appl. 29 (10), 1343–1350. https://doi.org/10.1177/0885328214568468.

Đorđević, V., Balanč, B., Belščak-Cvitanović, A., Lević, S., Trifković, K., Kalušević, A., Kostić, I., Komes, D., Bugarski, B., Nedović, V., 2014. Trends in encapsulation technologies for delivery of food bioactive compounds. Food Eng. Rev. 7 (4), 452–490. https://doi.org/10.1007/s12393-014-9106-7.

Emam-Djomeh, Z., Fathi, M., Askari, G., 2019. Gum tragacanth (*Astragalus gummifer* Labillardiere). In: Emerging Natural Hydrocolloids: Rheology and Functions. John Wiley & Sons, Inc, USA, pp. 299–326.

Fardood, S.T., Atrak, K., Ramazani, A., 2017. Green synthesis using tragacanth gum and characterization of Ni–Cu–Zn ferrite nanoparticles as a magnetically separable photocatalyst for organic dyes degradation from aqueous solution under visible light. J. Mater. Sci. Mater. Electron. 28 (14), 10739–10746. https://doi.org/10.1007/s10854-017-6850-5.

Farzi, M., Emam-Djomeh, Z., Mohammadifar, M.A., 2013. A comparative study on the emulsifying properties of various species of gum tragacanth. Int. J. Biol. Macromol. 57, 76–82. https://doi.org/10.1016/j.ijbiomac.2013.03.008.

Fathi-Achachelouei, M., Knopf-Marques, H., da Silva, C.E.R., Barthès, J., Bat, E., Tezcaner, A., Vrana, N.E., 2019. Use of nanoparticles in tissue engineering and regenerative medicine. Front. Bioeng. Biotechnol. 7, 113.

Fattahi, A., Sadrjavadi, K., Golozar, M.A., Varshosaz, J., Fathi, M.H., Mirmohammad-Sadeghi, H., 2013. Preparation and characterization of oligochitosan-tragacanth nanoparticles as a novel gene carrier. Carbohydr. Polym. 97 (2), 277–283. https://doi.org/10.1016/j.carbpol.2013.04.098.

Fayazzadeh, E., Rahimpour, S., Ahmadi, S.M., Farzampour, S., Anvari, M.S., Boroumand, M.A., Ahmadi, S.H., 2014. Acceleration of skin wound healing with Tragacanth (Astragalus) preparation: an experimental pilot study in rats. Acta Med. Iran. 52 (1), 3–8. http://acta.tums.ac.ir/index.php/acta/article/download/5703/4544.

Foroutan, N.S., Tabandeh, F., Khodabandeh, M., Mojgani, N., Maghsoudi, A., Moradi, M., 2017. Isolation and identification of an indigenous probiotic Lactobacillus strain: its encapsulation with natural branched polysaccharids to improve bacterial viability. Appl. Food Biotechnol. 4 (3), 133–142. https://doi.org/10.22037/afb.v4i3.16471.

Ghayempour, S., Montazer, M., 2017. Tragacanth nanocapsules containing chamomile extract prepared through sono-assisted W/O/W microemulsion and UV cured on cotton fabric. Carbohydr. Polym. 170, 234–240. https://doi.org/10.1016/j.carbpol.2017.04.088.

Ghayempour, S., Montazer, M., 2019. A novel controlled release system based on Tragacanth nanofibers loaded peppermint oil. Carbohydr. Polym. 205, 589–595. https://doi.org/10.1016/j.carbpol.2018.10.078.

Ghayempour, S., Montazer, M., Mahmoudi Rad, M., 2015. Tragacanth gum as a natural polymeric wall for producing antimicrobial nanocapsules loaded with plant extract. Int. J. Biol. Macromol. 81, 514–520. https://doi.org/10.1016/j.ijbiomac.2015.08.041.

Ghayempour, S., Montazer, M., Mahmoudi Rad, M., 2016a. Encapsulation of Aloe vera extract into natural Tragacanth gum as a novel green wound healing product. Int. J. Biol. Macromol. 93, 344–349. https://doi.org/10.1016/j.ijbiomac.2016.08.076.

Ghayempour, S., Montazer, M., Mahmoudi Rad, M., 2016b. Simultaneous encapsulation and stabilization of Aloe vera extract on cotton fabric for wound dressing application. RSC Adv. 6 (113), 111895–111902. https://doi.org/10.1039/c6ra22485g.

Goldberg, M., Langer, R., Jia, X., 2007. Nanostructured materials for applications in drug delivery and tissue engineering. J. Biomater. Sci. Polym. Ed. 18 (3), 241–268. https://doi.org/10.1163/156856207779996931.

Gralén, N., Kärrholm, M., 1950. The physicochemical properties of solutions of gum tragacanth. J. Colloid Sci. 5 (1), 21–36. https://doi.org/10.1016/0095-8522(50)90003-1.

Guo, J.H., Skinner, G.W., Harcum, W.W., Barnum, P.E., 1998. Pharmaceutical applications of naturally occurring water-soluble polymers. Pharm. Sci. Technol. Today 1 (6), 254–261. https://doi.org/10.1016/S1461-5347(98)00072-8.

Gupta, V.K., Sood, S., Agarwal, S., Saini, A.K., Pathania, D., 2018. Antioxidant activity and controlled drug delivery potential of tragacanth gum-cl-poly (lactic acid-co-itaconic acid) hydrogel. Int. J. Biol. Macromol. 107, 2534–2543. https://doi.org/10.1016/j.ijbiomac.2017.10.138.

Haeri, S.M.J., Sadeghi, Y., Salehi, M., Farahani, R.M., Mohsen, N., 2016. Osteogenic differentiation of human adipose-derived mesenchymal stem cells on gum tragacanth hydrogel. Biologicals 44 (3), 123–128. https://doi.org/10.1016/j.biologicals.2016.03.004.

Hasan, A., Morshed, M., Memic, A., Hassan, S., Webster, T.J., Marei, H.E.S., 2018. Nanoparticles in tissue engineering: applications, challenges and prospects. Int. J. Nanomedicine 13, 5637–5655. https://doi.org/10.2147/IJN.S153758.

Hemmati, K., Ghaemy, M., 2016. Synthesis of new thermo/pH sensitive drug delivery systems based on tragacanth gum polysaccharide. Int. J. Biol. Macromol. 87, 415–425. https://doi.org/10.1016/j.ijbiomac.2016.03.005.

Hemmati, K., Masoumi, A., Ghaemy, M., 2015. PH responsive tragacanth gum and poly(methyl methacrylate-co-maleic anhydride)-g-poly(caprolactone) conetwork microgel for in vitro quercetin release. Polymer 59, 49–56. https://doi.org/10.1016/j.polymer.2014.12.050.

Hemmati, K., Masoumi, A., Ghaemy, M., 2016. Tragacanth gum-based nanogel as a superparamagnetic molecularly imprinted polymer for quercetin recognition and controlled release. Carbohydr. Polym. 136, 630–640. https://doi.org/10.1016/j.carbpol.2015.09.006.

Heydary, H.A., Karamian, E., Poorazizi, E., Heydaripour, J., Khandan, A., 2015. Electrospun of polymer/bioceramic nanocomposite as a new soft tissue for biomedical applications. J. Asian Ceramic Soc. 3 (4), 417–425. https://doi.org/10.1016/j.jascer.2015.09.003.

Hosseini, M.S., Hemmati, K., Ghaemy, M., 2016. Synthesis of nanohydrogels based on tragacanth gum biopolymer and investigation of swelling and drug delivery. Int. J. Biol. Macromol. 82, 806–815. https://doi.org/10.1016/j.ijbiomac.2015.09.067.

Indana, M.K., Gangapuram, B.R., Dadigala, R., Bandi, R., Guttena, V., 2016. A novel green synthesis and characterization of silver nanoparticles using gum tragacanth and evaluation of their potential catalytic reduction activities with methylene blue and Congo red dyes. J. Anal. Sci. Technol. 7. https://doi.org/10.1186/s40543-016-0098-1.

Iqbal, Z., Khan, R., Nasir, F., Khan, J.A., Ahmad, L., Khan, A., Shah, Y., Dayo, A., 2010. Preparation and in-vitro evaluation of sustained release matrix diclofenac sodium tablets using HPMC KM 100 and gums. Arch. Pharm. Pract. 1 (2), 9–17.

Izydorczyk, M., Cui, S.W., Wang, Q., 2005. Polysaccharide gums: structures, functional properties, and applications. In: Food Carbohydrates: Chemistry, Physical Properties, and Applications. 293. Routledge.

Jabri, T., Imran, M., Shafiullah, Rao, K., Ali, I., Arfan, M., Shah, M.R., 2018. Fabrication of lecithin-gum tragacanth muco-adhesive hybrid nano-carrier system for in-vivo performance of amphotericin B. Carbohydr. Polym. 194, 89–96. https://doi.org/10.1016/j.carbpol.2018.04.013.

Jafari, S., Hojjati, M., Noshad, M., 2018. Influence of soluble soybean polysaccharide and tragacanth gum based edible coating to improve the quality of fresh-cut apple slices. J. Food Process. Preserv. 42 (6). https://doi.org/10.1111/jfpp.13638.

Jain, A., Thakur, D., Ghoshal, G., Katare, O.P., Shivhare, U.S., 2016. Characterization of microcapsulated β-carotene formed by complex coacervation using casein and gum tragacanth. Int. J. Biol. Macromol. 87, 101–113. https://doi.org/10.1016/j.ijbiomac.2016.01.117.

Jalali, S., Montazer, M., Malek, R.M.A., 2018. A novel semi-bionanofibers through introducing Tragacanth gum into PET attaining rapid wetting and degradation. Fibers Polym. 19 (10), 2088–2096. https://doi.org/10.1007/s12221-018-8276-y.

Kaffashi, B., Zandieh, A., Khadiv-Parsi, P., 2006. Drug release study of systems containing the Tragacanth and collagen composite: release characterization and viscoelastic measurements. Macromol. Symp., 120–129. https://doi.org/10.1002/masy.200690088.

Kiani, A., Shahbazi, M., Asempour, H., 2012. Hydrogel membranes based on gum tragacanth with tunable structure and properties. I. Preparation method using Taguchi experimental design. J. Appl. Polym. Sci. 124 (1), 99–108. https://doi.org/10.1002/app.35038.

Kohane, D.S., 2007. Microparticles and nanoparticles for drug delivery. Biotechnol. Bioeng. 96 (2), 203–209. https://doi.org/10.1002/bit.21301.

Konwarh, R., Karak, N., Misra, M., 2013. Electrospun cellulose acetate nanofibers: the present status and gamut of biotechnological applications. Biotechnol. Adv. 31 (4), 421–437. https://doi.org/10.1016/j.biotechadv.2013.01.002.

Koshani, R., Aminlari, M., 2017. Physicochemical and functional properties of ultrasonic-treated tragacanth hydrogels cross-linked to lysozyme. Int. J. Biol. Macromol. 103, 948–956. https://doi.org/10.1016/j.ijbiomac.2017.05.124.

Kulanthaivel, S., Rathnam, S.V., Agarwal, T., Pradhan, S., Pal, K., Giri, S., Maiti, T.K., Banerjee, I., 2017. Gum tragacanth-alginate beads as proangiogenic-osteogenic cell encapsulation systems for bone tissue engineering. J. Mater. Chem. B 5 (22), 4177–4189. https://doi.org/10.1039/c7tb00390k.

Lim, C.T., Han, J., Guck, J., Espinosa, H., 2010. Micro and nanotechnology for biological and biomedical applications. Med. Biol. Eng. Comput. 48 (10), 941–943. https://doi.org/10.1007/s11517-010-0677-z.

Lim, C.T., 2017. Nanofiber technology: current status and emerging developments. Prog. Polym. Sci. 70, 1–17.

Limongi, T., Tirinato, L., Pagliari, F., Giugni, A., Allione, M., Perozziello, G., Candeloro, P., Di Fabrizio, E., 2017. Fabrication and applications of Micro/nanostructured devices for tissue engineering. Nano-Micro Lett. 9 (1). https://doi.org/10.1007/s40820-016-0103-7.

López-Cebral, R., Civantos, A., Ramos, V., Seijo, B., López-Lacomba, J.L., Sanz-Casado, J.V., Sanchez, A., 2017. Gellan gum based physical hydrogels incorporating highly valuable endogen molecules and associating BMP-2 as bone formation platforms. Carbohydr. Polym. 167, 345–355. https://doi.org/10.1016/j.carbpol.2017.03.049.

López-Franco, Y., Higuera-Ciapara, I., Goycoolea, F.M., Wang, W., 2009. Other exudates: tragancanth, karaya, mesquite gum and larchwood arabinogalactans. In: Handbook of Hydrocolloids, second ed. Elsevier Inc, pp. 495–534, https://doi.org/10.1533/9781845695873.495.

Maeki, M., Kimura, N., Sato, Y., Harashima, H., Tokeshi, M., 2018. Advances in microfluidics for lipid nanoparticles and extracellular vesicles and applications in drug delivery systems. Adv. Drug Deliv. Rev. 128, 84–100. https://doi.org/10.1016/j.addr.2018.03.008.

Maia, F.R., Musson, D.S., Naot, D., Da Silva, L.P., Bastos, A.R., Costa, J.B., Oliveira, J.M., Correlo, V.M., Reis, R.L., Cornish, J., 2018. Differentiation of osteoclast precursors on gellan gum-based spongy-like hydrogels for bone tissue engineering. Biomed. Mater. 13 (3). https://doi.org/10.1088/1748-605X/aaaf29.

Masoumi, A., Ghaemy, M., 2014. Removal of metal ions from water using nanohydrogel tragacanth gum-g-polyamidoxime: isotherm and kinetic study. Carbohydr. Polym. 108 (1), 206–215. https://doi.org/10.1016/j.carbpol.2014.02.083.

Mishra, A., Malhotra, A.V., 2009. Tamarind xyloglucan: a polysaccharide with versatile application potential. J. Mater. Chem. 19 (45), 8528–8536. https://doi.org/10.1039/b911150f.

Moghaddam, R.H., Dadfarnia, S., Shabani, A.M.H., Tavakol, M., 2019. Synthesis of composite hydrogel of glutamic acid, gum tragacanth, and anionic polyacrylamide by electron beam irradiation for uranium (VI) removal from aqueous samples: equilibrium, kinetics, and thermodynamic studies. Carbohydr. Polym. 206, 352–361. https://doi.org/10.1016/j.carbpol.2018.10.030.

Moghbel, A., Hemmati, A.A., Agheli, H., Rashidi, I., Amraee, K., 2005. The effect of tragacanth mucilage on the healing of full-thickness wound in rabbit. Arch. Iran. Med. 8 (4), 257–262.

Moghbel, A.A.H., Naji, M., 2008. Design and formulation of tragacanth dressing bandage for burn healing. Jundishapur Sci. Med. J. 7 (57), 274–283.

Mohamadnia, Z., Zohuriaan-Mehr, M.J., Kabiri, K., Razavi-Nouri, M., 2008. Tragacanth gum-graft-polyacrylonitrile: synthesis, characterization and hydrolysis. J. Polym. Res. 15 (3), 173–180. https://doi.org/10.1007/s10965-007-9156-0.

Mohamed, H.A., Radwan, R.R., Raafat, A.I., Ali, A.E.H., 2018. Antifungal activity of oral (Tragacanth/acrylic acid) amphotericin B carrier for systemic candidiasis: in vitro and in vivo study. Drug Deliv. Transl. Res. 8 (1), 191–203. https://doi.org/10.1007/s13346-017-0452-x.

Mohammadifar, M.A., Musavi, S.M., Kiumarsi, A., Williams, P.A., 2006. Solution properties of targacanthin (water-soluble part of gum tragacanth exudate from *Astragalus gossypinus*). Int. J. Biol. Macromol. 38 (1), 31–39. https://doi.org/10.1016/j.ijbiomac.2005.12.015.

Mohammadinejad, R., Kumar, A., Ranjbar-Mohammadi, M., Ashrafizadeh, M., Han, S.S., Khang, G., Roveimiab, Z., 2020. Recent advances in natural gum-based biomaterials for tissue engineering and regenerative medicine: a review. Polymers 12 (1). https://doi.org/10.3390/polym12010176.

Monjezi, J., Jamaledin, R., Ghaemy, M., Makvandi, P., 2019. Antimicrobial modified-tragacanth gum/acrylic acid hydrogels for the controlled release of quercetin. J. Appl. Chem. 13 (1), 57–71.

Monjezi, J., Jamaledin, R., Ghaemy, M., Moeini, A., Makvandi, P., 2018. A performance comparison of graft copolymer hydrogels based on functionalized-tragacanth gum/polyacrylic acid and polyacrylamide as antibacterial and antifungal drug release vehicles. Am. J. Nanotechnol. Nanomed. Res. 1 (1), 010–015.

Montazer, M., Keshvari, A., Kahali, P., 2016. Tragacanth gum/nano silver hydrogel on cotton fabric: in-situ synthesis and antibacterial properties. Carbohydr. Polym. 154, 257–266. https://doi.org/10.1016/j.carbpol.2016.06.084.

Moradi, S., Sadrjavadi, K., Farhadian, N., Hosseinzadeh, L., Shahlaei, M., 2018. Easy synthesis, characterization and cell cytotoxicity of green nano carbon dots using hydrothermal carbonization of gum Tragacanth and chitosan biopolymers for bioimaging. J. Mol. Liq. 259, 284–290. https://doi.org/10.1016/j.molliq.2018.03.054.

Mostafavi, F.S., Kadkhodaee, R., Emadzadeh, B., Koocheki, A., 2016. Preparation and characterization of tragacanth-locust bean gum edible blend films. Carbohydr. Polym. 139, 20–27. https://doi.org/10.1016/j.carbpol.2015.11.069.

Moursy, N.M., Afifi, N.N., Ghorab, D.M., El-Saharty, Y., 2003. Formulation and evaluation of sustained release floating capsules of nicardipine hydrochloride. Pharmazie 58 (1), 38–43.

Naji-Tabasi, S., Razavi, S.M.A., Mehditabar, H., 2017. Fabrication of basil seed gum nanoparticles as a novel oral delivery system of glutathione. Carbohydr. Polym. 157, 1703–1713. https://doi.org/10.1016/j.carbpol.2016.11.052.

Nazarzadeh Zare, E., Makvandi, P., Tay, F.R., 2019. Recent progress in the industrial and biomedical applications of tragacanth gum: a review. Carbohydr. Polym. 212, 450–467. https://doi.org/10.1016/j.carbpol.2019.02.076.

Nedovic, V., Kalusevic, A., Manojlovic, V., Levic, S., Bugarski, B., 2011. An overview of encapsulation technologies for food applications. Procedia Food Sci., 1806–1815. https://doi.org/10.1016/j.profoo.2011.09.265.

Ngwuluka, N.C., Kyari, J., Taplong, J., Uwaezuoke, O.J., 2012. Application and characterization of gum from bombax buonopozense calyxes as an excipient in tablet formulation. Pharmaceutics 4 (3), 354–365. https://doi.org/10.3390/pharmaceutics4030354.

Niknia, N., Kadkhodaee, R., 2017. Gum tragacanth-polyvinyl alcohol cryogel and xerogel blends for oral delivery of silymarin: structural characterization and mucoadhesive property. Carbohydr. Polym. 177, 315–323. https://doi.org/10.1016/j.carbpol.2017.08.110.

Nur, M., Vasiljevic, T., 2018. Insulin inclusion into a tragacanth hydrogel: an oral delivery system for insulin. Materials (Basel) 11 (1). https://doi.org/10.3390/ma11010079.

Nur, M., Ramchandran, L., Vasiljevic, T., 2016. Tragacanth as an oral peptide and protein delivery carrier: characterization and mucoadhesion. Carbohydr. Polym. 143, 223–230. https://doi.org/10.1016/j.carbpol.2016.01.074.

Otadi, M., Mobayen, S., 2011. The survey of kinetic behavior of immobilized glucose oxidase on gum tragacanth carrier. World Appl. Sci. J. 14, 15–19.

Otady, M., Vaziri, A., Seifkordi, A.A., Kheirolomoom, A., 2005. Gum tragacanth gels as a new supporting matrix for immobilization of whole-cell. Iran. J. Chem. Chem. Eng. 24 (4), 1–7.

Ozel, B., Cikrikci, S., Aydin, O., Oztop, M.H., 2017. Polysaccharide blended whey protein isolate-(WPI) hydrogels: a physicochemical and controlled release study. Food Hydrocoll. 71, 35–46. https://doi.org/10.1016/j.foodhyd.2017.04.031.

Paknejad, M., Rokn, A.R., Yaraghi, A.A.S., Elhami, F., Kharazifard, M.J., Moslemi, N., 2012. Histologic and histomorphometric evaluation of the effect of lactoferrin combined with anorganic bovine bone on healing of experimentally induced bony defects on rabbit calvaria. Dent. Res. J. (Isfahan) 9, S75–S80.

Pathania, D., Verma, C., Negi, P., Tyagi, I., Asif, M., Kumar, N.S., Al-Ghurabi, E.H., Agarwal, S., Gupta, V.K., 2018. Novel nanohydrogel based on itaconic acid grafted tragacanth gum for controlled release of ampicillin. Carbohydr. Polym. 196, 262–271. https://doi.org/10.1016/j.carbpol.2018.05.040.

Patra, J.K., Das, G., Fraceto, L.F., Campos, E.V.R., del Pilar Rodriguez-Torres, M., Acosta-Torres, L.S., Diaz-Torres, L.A., Grillo, R., Swamy, M.K., Sharma, S., 2018. Nano based drug delivery systems: recent developments and future prospects. J. Nanobiotechnol. 16 (1), 1–33.

Peng, H., Ning, X., Wei, G., Wang, S., Dai, G., Ju, A., 2018. The preparations of novel cellulose/phenylboronic acid composite intelligent bio-hydrogel and its glucose, pH-responsive behaviors. Carbohydr. Polym. 195, 349–355. https://doi.org/10.1016/j.carbpol.2018.04.119.

Phillips, G.O., Williams, P.A., 2009. Handbook of hydrocolloids. Elsevier.

Pierre, G., Punta, C., Delattre, C., Melone, L., Dubessay, P., Fiorati, A., Pastori, N., Galante, Y.M., Michaud, P., 2017. TEMPO-mediated oxidation of polysaccharides: an ongoing story. Carbohydr. Polym. 165, 71–85. https://doi.org/10.1016/j.carbpol.2017.02.028.

Prasad, K., Mondal, D., Sharma, M., Freire, M.G., Mukesh, C., Bhatt, J., 2018. Stimuli responsive ion gels based on polysaccharides and other polymers prepared using ionic liquids and deep eutectic solvents. Carbohydr. Polym. 180, 328–336. https://doi.org/10.1016/j.carbpol.2017.10.020.

Rahimdokht, M., Pajootan, E., Ranjbar-Mohammadi, M., 2019. Titania/gum tragacanth nanohydrogel for methylene blue dye removal from textile wastewater using response surface methodology. Polym. Int. 68 (1), 134–140. https://doi.org/10.1002/pi.5706.

Rahmani, Z., Sahraei, R., Ghaemy, M., 2018. Preparation of spherical porous hydrogel beads based on ion-crosslinked gum tragacanth and graphene oxide: study of drug delivery behavior. Carbohydr. Polym. 194, 34–42. https://doi.org/10.1016/j.carbpol.2018.04.022.

Ramazani, A., Fardood, S.T., Hosseinzadeh, Z., Sadri, F., Joo, S.W., 2017. Green synthesis of magnetic copper ferrite nanoparticles using tragacanth gum as a biotemplate and their catalytic activity for the oxidation of alcohols. Iran. J. Catal. 7 (3), 181–185. http://ijc.iaush.ac.ir/article_48951_bddc2a241e85539f819691e6f893d847.pdf.

Ranjbar-Mohammadi, M., Bahrami, S.H., 2015. Development of nanofibrous scaffolds containing gum tragacanth/poly (ε-caprolactone) for application as skin scaffolds. Mater. Sci. Eng. C 48, 71–79. https://doi.org/10.1016/j.msec.2014.10.020.

Ranjbar-Mohammadi, M., Bahrami, S.H., 2016. Electrospun curcumin loaded poly(ε-caprolactone)/gum tragacanth nanofibers for biomedical application. Int. J. Biol. Macromol. 84, 448–456. https://doi.org/10.1016/j.ijbiomac.2015.12.024.

Ranjbar-Mohammadi, M., Bahrami, S.H., Joghataei, M.T., 2013. Fabrication of novel nanofiber scaffolds from gum tragacanth/poly(vinyl alcohol) for wound dressing application: in vitro evaluation and antibacterial properties. Mater. Sci. Eng. C 33 (8), 4935–4943. https://doi.org/10.1016/j.msec.2013.08.016.

Ranjbar-Mohammadi, M., Prabhakaran, M.P., Bahrami, S.H., Ramakrishna, S., 2016a. Gum tragacanth/poly(l-lactic acid) nanofibrous scaffolds for application in regeneration of peripheral nerve damage. Carbohydr. Polym. 140, 104–112. https://doi.org/10.1016/j.carbpol.2015.12.012.

Ranjbar-Mohammadi, M., Rabbani, S., Bahrami, S.H., Joghataei, M.T., Moayer, F., 2016. Antibacterial performance and in vivo diabetic wound healing of curcumin loaded gum tragacanth/poly(ε-caprolactone) electrospun nanofibers. Mater. Sci. Eng. C 69, 1183–1191. https://doi.org/10.1016/j.msec.2016.08.032.

Ranjbar-Mohammadi, M., Zamani, M., Prabhakaran, M.P., Bahrami, S.H., Ramakrishna, S., 2016b. Electrospinning of PLGA/gum tragacanth nanofibers containing tetracycline hydrochloride for periodontal regeneration. Mater. Sci. Eng. C 58, 521–531. https://doi.org/10.1016/j.msec.2015.08.066.

Rao, K., Imran, M., Jabri, T., Ali, I., Perveen, S., Shafiullah, Ahmed, S., Shah, M.R., 2017. Gum tragacanth stabilized green gold nanoparticles as cargos for Naringin loading: a morphological investigation through AFM. Carbohydr. Polym. 174, 243–252. https://doi.org/10.1016/j.carbpol.2017.06.071.

Rao, K.M., Kumar, A., Krishna Rao, K.S.V., Haider, A., Han, S.S., 2018. Biodegradable Tragacanth gum based silver nanocomposite hydrogels and their antibacterial evaluation. J. Polym. Environ. 26 (2), 778–788. https://doi.org/10.1007/s10924-017-0989-2.

Rasul, A., Iqbal, M., Murtaza, G., Waqas, M.K., Hanif, M., Khan, S.A., Bhatti, N.S., 2010. Design, development and in-vitro evaluation of metoprolol tartrate tablets containing xanthan-tragacanth. Acta Pol. Pharm. Drug Res. 67 (5), 517–522. http://www.ptfarm.pl/pub/File/acta_pol_2010/5_2010/517-522.pdf.

Ratnaparkhi, M.P., Gupta, P., Jyoti, 2013. Sustained release oral drug delivery system-an overview. Int. J. Pharma Res. Rev. 2 (3), 11–21.

Rezvani, E., Schleining, G., Taherian, A.R., 2012. Assessment of physical and mechanical properties of orange oil-in-water beverage emulsions using response surface methodology. LWT-Food Sci. Technol. 48 (1), 82–88. https://doi.org/10.1016/j.lwt.2012.02.025.

Robbins, S.R.J., 1988. A review of recent trends in selected markets for water-soluble gums (ODNRI Bulletin).

Sadeghi, S., Rad, F.A., Moghaddam, A.Z., 2014. A highly selective sorbent for removal of Cr(VI) from aqueous solutions based on Fe_3O_4/poly(methyl methacrylate) grafted Tragacanth gum nanocomposite: optimization by experimental design. Mater. Sci. Eng. C 45, 136–145. https://doi.org/10.1016/j.msec.2014.08.063.

Sahraei, R., Ghaemy, M., 2017. Synthesis of modified gum tragacanth/graphene oxide composite hydrogel for heavy metal ions removal and preparation of silver nanocomposite for antibacterial activity. Carbohydr. Polym. 157, 823–833. https://doi.org/10.1016/j.carbpol.2016.10.059.

Sahraei, R., Sekhavat Pour, Z., Ghaemy, M., 2017. Novel magnetic bio-sorbent hydrogel beads based on modified gum tragacanth/graphene oxide: removal of heavy metals and dyes from water. J. Clean. Prod. 142, 2973–2984. https://doi.org/10.1016/j.jclepro.2016.10.170.

Salamanca, C.H., Yarce, C.J., Moreno, R.A., Prieto, V., Recalde, J., 2018. Natural gum-type biopolymers as potential modified nonpolar drug release systems. Carbohydr. Polym. 189, 31–38. https://doi.org/10.1016/j.carbpol.2018.02.011.

Samavati, V., D-Jomeh, Z.E., 2013. Multivariate-parameter optimization of aroma compound release from carbohydrate-oil-protein model emulsions. Carbohydr. Polym. 98 (2), 1667–1676. https://doi.org/10.1016/j.carbpol.2013.07.074.

Samavati, V., Emam-Djomeh, Z., Mehdinia, A., 2014. Thermodynamic and kinetic study of volatile compounds in biopolymer based dispersions. Carbohydr. Polym. 99, 556–562. https://doi.org/10.1016/j.carbpol.2013.08.059.

Saruchi, Kaith, B.S., Jindal, R., Kapur, G., 2014a. Synthesis of gum tragacanth and acrylic acid based hydrogel: its evaluation for controlled release of antiulcerative drug pantoprazole sodium. J. Chin. Adv. Mater. Soc. 2 (2), 110–117.

Saruchi, Kaith, B.S., Jindal, R., Kumar, V., Bhatti, M.S., 2014b. Optimal response surface design of gum tragacanth-based poly[(acrylic acid)-co-acrylamide] IPN hydrogel for the controlled release of the antihypertensive drug losartan potassium. RSC Adv. 4 (75), 39822–39829. https://doi.org/10.1039/c4ra02803a.

Sheorain, J., Mehra, M., Thakur, R., Grewal, S., Kumari, S., 2019. In vitro anti-inflammatory and antioxidant potential of thymol loaded bipolymeric (tragacanth gum/chitosan) nanocarrier. Int. J. Biol. Macromol. 125, 1069–1074. https://doi.org/10.1016/j.ijbiomac.2018.12.095.

Siahi, M.R., Barzegar-Jalali, M., Monajjemzadeh, F., Ghaffari, F., Azarmi, S., 2005. Design and evaluation of 1-and 3-layer matrices of verapamil hydrochloride for sustaining its release. AAPS PharmSciTech 6, 626–632.

Silva, C., Torres, M.D., Chenlo, F., Moreira, R., 2017. Rheology of aqueous mixtures of tragacanth and guar gums: effects of temperature and polymer ratio. Food Hydrocoll. 69, 293–300. https://doi.org/10.1016/j.foodhyd.2017.02.018.

Singh, B., 2018. Gamma radiation synthesis and characterization of gentamicin loaded polysaccharide gum based hydrogel wound dressings. J. Drug Delivery Sci. Technol. 47, 200–208. https://doi.org/10.1016/j.jddst.2018.07.014.

Singh, B., Sharma, V., 2014. Influence of polymer network parameters of tragacanth gum-based pH responsive hydrogels on drug delivery. Carbohydr. Polym. 101 (1), 928–940. https://doi.org/10.1016/j.carbpol.2013.10.022.

Singh, B., Sharma, V., 2017. Crosslinking of poly(vinylpyrrolidone)/acrylic acid with tragacanth gum for hydrogels formation for use in drug delivery applications. Carbohydr. Polym. 157, 185–195. https://doi.org/10.1016/j.carbpol.2016.09.086.

Singh, B., Varshney, L., Francis, S., Rajneesh, 2016. Designing tragacanth gum based sterile hydrogel by radiation method for use in drug delivery and wound dressing applications. Int. J. Biol. Macromol. 88, 586–602. https://doi.org/10.1016/j.ijbiomac.2016.03.051.

Singh, B., Varshney, L., Francis, S., Rajneesh, 2017a. Designing sterile biocompatible moxifloxacin loaded trgacanth-PVA-alginate wound dressing by radiation crosslinking method. Wound Med. 17, 11–17. https://doi.org/10.1016/j.wndm.2017.01.001.

Singh, B., Varshney, L., Francis, S., Rajneesh, 2017b. Synthesis and characterization of tragacanth gum based hydrogels by radiation method for use in wound dressing application. Radiat. Phys. Chem. 135, 94–105. https://doi.org/10.1016/j.radphyschem.2017.01.044.

Stephen, A., 1983. Other plant polysaccharides. In: The Polysaccharides. Elsevier, pp. 97–193.

Suri, S., Ruan, G., Winter, J., Schmidt, C.E., 2013. Microparticles and nanoparticles. In: Biomaterials Science: An Introduction to Materials, third ed. Elsevier Inc, pp. 360–388, https://doi.org/10.1016/B978-0-08-087780-8.00034-6.

Taghavi, F.S., Ramazani, A., Golfar, Z., Woo, J.O.O.S., 2017. Green synthesis of α-Fe_2O_3 (hematite) nanoparticles using Tragacanth gel. J. Appl. Chem. Res. 11, 19–27.

Tang, W., Shen, M., Xie, J., Liu, D., Du, M., Lin, L., Gao, H., Hamaker, B.R., Xie, M., 2017. Physicochemical characterization, antioxidant activity of polysaccharides from Mesona chinensis Benth and their protective effect on injured NCTC-1469 cells induced by H_2O_2. Carbohydr. Polym. 175, 538–546. https://doi.org/10.1016/j.carbpol.2017.08.018.

Tavakol, M., Vasheghani-Farahani, E., Mohammadifar, M.A., Soleimani, M., Hashemi-Najafabadi, S., 2016. Synthesis and characterization of an in situ forming hydrogel using tyramine conjugated high methoxyl gum tragacanth. J. Biomater. Appl. 30 (7), 1016–1025. https://doi.org/10.1177/0885328215608983 Article information.

Thombare, N., Jha, U., Mishra, S., Siddiqui, M.Z., 2016. Guar gum as a promising starting material for diverse applications: a review. Int. J. Biol. Macromol. 88, 361–372. https://doi.org/10.1016/j.ijbiomac.2016.04.001.

Tischer, C.A., Iacomini, M., Gorin, P.A.J., 2002. Structure of the arabinogalactan from gum tragacanth (*Astralagus gummifer*). Carbohydr. Res. 337 (18), 1647–1655. https://doi.org/10.1016/S0008-6215(02)00023-X.

Tonyali, B., Cikrikci, S., Oztop, M.H., 2018. Physicochemical and microstructural characterization of gum tragacanth added whey protein based films. Food Res. Int. 105, 1–9. https://doi.org/10.1016/j.foodres.2017.10.071.

Wang, W., 2000. Tragacanth and karaya. In: Handbook of Hydrocolloids. Elsevier, pp. 231–246.

Wang, Q., Yu, D.G., Zhang, L.L., Liu, X.K., Deng, Y.C., Zhao, M., 2017. Electrospun hypromellose-based hydrophilic composites for rapid dissolution of poorly water-soluble drug. Carbohydr. Polym. 174, 617–625. https://doi.org/10.1016/j.carbpol.2017.06.075.

Wang, B., Ahmad, Z., Huang, J., Li, J.S., Chang, M.W., 2018. Development of random and ordered composite fiber hybrid technologies for controlled release functions. Chem. Eng. J. 343, 379–389. https://doi.org/10.1016/j.cej.2018.03.021.

Whistler, R.L., BeMiller, J.N., 2012. Industrial Gums: Polysaccharides and Their Derivatives, third ed. Elsevier Inc., pp. 1–642, https://doi.org/10.1016/C2009-0-03188-2.

Wu, S., Wang, B., Ahmad, Z., Huang, J., Chang, M.W., Li, J.S., 2017. Surface modified electrospun porous magnetic hollow fibers using secondary downstream collection solvent contouring. Mater. Lett. 204, 73–76. https://doi.org/10.1016/j.matlet.2017.06.015.

Xu, Y., Li, J.J., Yu, D.G., Williams, G.R., Yang, J.H., Wang, X., 2017. Influence of the drug distribution in electrospun gliadin fibers on drug-release behavior. Eur. J. Pharm. Sci. 106, 422–430. https://doi.org/10.1016/j.ejps.2017.06.017.

Xu, H., Gao, M., Tang, X., Zhang, W., Luo, D., Chen, M., 2020. Micro/nano technology for next-generation diagnostics. Small Methods 4, 1900506.

Xumeng, G., Liangcheng, Y., Jianfeng, X., 2016. Cell Immobilization: Fundamentals, Technologies, and Applications. Wiley, pp. 205–235, https://doi.org/10.1002/9783527807833.ch7.

Zahin, N., Anwar, R., Tewari, D., Kabir, M.T., Sajid, A., Mathew, B., Uddin, M.S., Aleya, L., Abdel-Daim, M.M., 2020. Nanoparticles and its biomedical applications in health and diseases: special focus on drug delivery. Environ. Sci. Pollut. Res. 27 (16), 19151–19168.

Zarekhalili, Z., Bahrami, S.H., Ranjbar-Mohammadi, M., Milan, P.B., 2017. Fabrication and characterization of PVA/gum tragacanth/PCL hybrid nanofibrous scaffolds for skin substitutes. Int. J. Biol. Macromol. 94, 679–690. https://doi.org/10.1016/j.ijbiomac.2016.10.042.

CHAPTER

17

Development of Persian gum-based micro- and nanocarriers for nutraceutical and drug delivery applications

Rassoul Kadkhodaee[a] *and Nassim Raoufi*[b]

[a]Department of Food Nanotechnology, Research Institute of Food Science and Technology, Mashhad, Iran, [b]Department of Food Science and Biotechnology, Zhejiang Gongshang University, Hangzhou, P.R. China

17.1 Introduction

Plant gums and related polysaccharides are a diverse group of hydrocolloids that are known for their thickening and gel forming ability. Their unique inherent and functional properties have made them suitable for design and manufacture of drug delivery vehicles and systems of biological applications. Owing to their remarkable advantages including being naturally abundant, low-cost, biodegradable, renewable, nontoxic, and easily modifiable by enzymatic, physical, and chemical approaches, they are preferred over synthetic polymers. In addition, they are biologically active materials with the ability to mimic the natural extracellular matrix microenvironment and thus can readily be accepted by the human body. Despite synthetic polymers, plant hydrocolloids maintain the physical health without any side effect on human organs. The building blocks of plant gums are different sugar units such as glucose, galactose, mannose, xylose, arabinose, and uronic acids that are connected together by *O*-glycosidic linkages forming polysaccharide chains. Therefore, they may undergo digestion in the gastrointestinal tract with body friendly degradation byproducts or may exit the body without being digested (Bhatia, 2016; Gopinath et al., 2018).

Plant gums occur abundantly in nature. They are viscous exudates secreted from the trunk, stems, and branches of trees and shrubs of a vast number of plant genera, and are usually

Micro- and Nanoengineered Gum-Based Biomaterials for Drug Delivery and Biomedical Applications
https://doi.org/10.1016/B978-0-323-90986-0.00003-0

found in the form of opaque or glossy mass of different shapes and colors. These exudates have a pathologic function protecting the plant against environmental stimulation or damage by microorganisms, insects, or mechanical injuries (Mirhosseini and Amid, 2012). The exudate gums of the Rosaceae family, especially the *prunus* genus, have been extensively studied. This family includes a large number of stone fruit species which all are of economic importance and classified under different subgenera, mainly *Amygdalus* (peaches and almonds), *Prunophora* (plums and apricots), and *Cerasus* (cherries) (Biswajit et al., 2011). Among the exudates produced by the trees and shrubs from the *Prunus* genus, those obtained from almond trees have recently received great attention by the food and pharmaceutical industries due to their unique functional properties. Two types of almond tree exudates are more popular than the others: one is known as "sweet almond gum" or "almond gum" (AG), obtained from *Prunus dulcis*, and the other is known as "bitter almond gum" or "Persian gum" (PG), produced by *Amygdalus scoparia* Spach. PG is also referred to as Zedo, Farsi, Shirazi, and wild or mountain almond gum. However, it is sometimes mistermed as Angum gum; the word "Angum" itself is a general expression in Persian language denoting the sticky exudates from trees (Fadavi et al., 2017). The *P. dulcis* subspecies is widely found in Asia, the Mediterranean region, and California (Kester et al., 1991), while *A. scoparia* Spach mainly grows in arid and semiarid lands (dry calcareous soils and rocky mountains) (Mozaffarian, 1996) of North Africa, the Middle East, and South Asia (Fadavi et al., 2014; Sabeti, 1975). In Iran, it is mostly spread in the Irano-Turanian, Zagros, and Gulf of Oman (northern parts) regions. Iran is the main producer of PG, exporting large amounts of this gum annually for medical and industrial applications (Azarikia and Abbasi, 2016).

Chemically, both types of almond tree exudates are complex branched heteropolysaccharides consisting of multiple sugar units linked together. They possess similar functionalities and have been introduced as natural polymer excipients for drug delivery applications (Farooq et al., 2014). PG is a highly branched polysaccharide with a slight solubility in water (<7%) (Raoufi et al., 2016). The solubility of AG has been also reported to be 10% (Rezaei et al., 2016b). Both gums are slightly acidic in nature and exist as a salt-mixture of calcium, magnesium, potassium, sodium, zinc, and iron (Fadavi et al., 2014; Mahfoudhi et al., 2012). The acidic nature of gums suggests that pH adjustment is likely necessary for oral and buccal drug delivery systems. Unlike PG, which has mainly been considered for food applications, AG has gained a potential interest as an excipient for pharmaceutical purposes. However, intact PG, due to its interesting physiochemical and functional features such as wide pH stability, thickening, stabilizing, emulsifying, and suspending capability, individually or in combination with other biopolymers, may become a promising future excipient for the development of a wide range of pharmaceutical formulations.

17.2 Medicinal applications of *Amygdalus scoparia*

All parts of *A. scoparia* tree (bark, branches, leaves, and fruits) have been used as herbal remedies for asthma, dropsy, hand and foot pains, hair loss, osteomalacia, hemorrhoids, wounds, and skin allergies since ancient times (Maleki and Akhani, 2018). The kernels contain amygdalin ($C_{20}H_{27}NO_{11}$) and various fatty acid compositions depending on the growing area

(Heydari and Hosseini, 2017). The ratio of unsaturated to saturated fatty acids in *A. scoparia* kernel oil (>82% unsaturated) is much higher than that of olive oil (Farhoosh and Tavakoli, 2008). It has positive effects on the control of triglyceride in patients with dyslipidemia, and can be used as a supplement or dietary intervention for hypertriglyceridemia (Zibaeenezhad et al., 2017). It has also been reported that either alcoholic or aqueous extract of the kernel has *in vitro* antimicrobial, antioxidant, and antitumor activities (Gomaa, 2013). The oil has also traditionally been used for treating cardiovascular diseases, rheumatism, cancer, kidney stone, asthma, spasm, headaches, and bronchitis (Gharaghani and Eshghi, 2015; Hashemnia et al., 2015). Oral administration of *A. scoparia* kernel extracts (200 mg/kg body weight) for 15 days led to antihyperglycemic effects in streptozotocin-induced diabetic mice, reducing their blood glucose level to the normal range through regeneration of pancreatic B cells. This was ascribed to the action of flavonoids and the increase of insulin response (Hashemnia et al., 2015). The bark and leaves of *A. scoparia* are used for healing cough (Gharaghani and Eshghi, 2015), and its shoot extract is employed for antidiabetic (Esfandiari et al., 2017), antioxidant, and antifungal applications (Hashemi and Raeisi, 2018). The root is also utilized for treatment of burn wound infections due to having phenols and flavonoids as antibacterial agents (Roointan et al., 2020). PG has traditionally been used for curing bladder and kidney stones, relief of toothache (Zargari, 1997) and swollen joints, and as an antipyretics and appetite stimulant (Seyfi et al., 2019).

17.3 Chemical composition, structure, and properties

The dry exudates from *A. scoparia* come in a wide range of colors including white, yellow, yellow-orange, orange-red, red, amber, and brown (Abbasi, 2017; Ghasemzadeh and Modiri, 2020; Golkar et al., 2017) with different shapes and sizes depending on the growing area and the chemical composition. They mainly consist of 87.17%–98.4% carbohydrate (Fadavi et al., 2014; Golkar et al., 2018), along with 0.2%–1.33% protein, 0.03%–0.2% fat, and 0.04%–3.63% ash (Fadavi et al., 2014; Ghasemzadeh and Modiri, 2020; Hadian et al., 2020; Sepeidnameh et al., 2018; Seyfi et al., 2019). It has been found that increasing the darkness leads to a reduction in pH, ash, carbohydrate, and protein content together with a drop in molecular weight and thermal stability (Abbasi and Rahimi, 2015; Fadavi et al., 2014). The carbohydrate fraction is known to be mostly an arabinogalactan with an arabinose to galactose ratio of 2:1 as well as minor quantities of xylose, rhamnose, and mannose (Abbasi, 2017; Fadavi et al., 2014). It also contains 9.8% uronic acids (D-galacturonic acid and D-glucuronic acid) (Seyfi et al., 2019), which are responsible for its negative charge density. The backbone of PG is composed of α-L-arabinofuranosyl (α-L-Araf) and β-D-galactopyranosyl (β-D-Galp) units that are connected together via (1 → 3) linkages with other monomers attached to the main skeleton as side chains through (1 → 6) linked -D-Glap and (1 → 3) linked -L-Araf residues (Hadian et al., 2020; Molaei and Jahanbin, 2018). The total elemental composition of PG is 38.5% carbon, 35.3% oxygen, 5.8% hydrogen, and 0.3% nitrogen (Seyfi et al., 2019). A considerable amount of water (5.06%–12.2%, w/w) is usually associated with PG (Fadavi et al., 2014; Raoufi et al., 2020), indicating the hygroscopic nature of the gum due to a large number of hydroxyl groups in its structure. Monovalent ions of K^+ and Na^+ and divalent ions of Mg^{2+}, Zn^{2+},

Fe^{2+}, and Ca^{2+} have been found in PG (Abbasi, 2017; Fadavi et al., 2014). These metal ions are probably attached to carboxyl groups in the form of carboxylate (Seyfi et al., 2019). The pH of aqueous dispersion of PG naturally falls in the range of 4.3–5.6, representing some buffering capacity (Abbasi and Rahimi, 2015; Chahibakhsh et al., 2019; Ghasempour et al., 2012; Golkar et al., 2017). Its ζ-potential has been reported to be −30.55 mV at pH 6.0–7.0 (Azarikia and Abbasi, 2016; Raoufi et al., 2016). PG is stable within a wide range of pH from 2 to 11 and at temperatures below 90 °C (Dabestani and Yeganehzad, 2019).

From a rheological point of view, the aqueous dispersion of PG exhibits non-Newtonian behavior with a pattern similar to that of unlinked polymers (Fadavi et al., 2014). Like other anionic polysaccharides, its apparent viscosity is salt sensitive owing to the charge screening effect of counter ions (Hosseini et al., 2017). At high concentrations, PG forms a gel. This could be due to Ca^{2+}-mediated interchain bridging and hydrogen bonding. The Ca^{2+} content of light gum nodules is more than that of the darker ones (Fadavi et al., 2014), and thus they may give gel at lower concentrations. However, at concentrations below 11%–12% (w/w), the gel is highly weak and sticky, and easily undergos syneresis by centrifugation (Rahimi and Abbasi, 2014).

Solubility studies showed that PG is composed of two main fractions: one is water-soluble (<7%) and the other is water-insoluble (>93%). These fractions can be separated by centrifugation. The sediment represents 70%–75% of the total insoluble fraction (Hadian et al., 2016). It is, however, extremely heterogeneous and cannot completely be separated by centrifugal force even at very high speeds (Raoufi et al., 2019). Filtration seems to be an efficient supplementary method to remove the insoluble residues, although some insoluble particles may pass through the filter depending on its pore size. Applying both centrifugation and filtration showed that >93% of PG is composed of a water-insoluble fraction (Raoufi et al., 2016). Investigations on the structure of both fractions revealed that the soluble part contains polysaccharide entities with molecular weight less than 5.11×10^6 g/mol (Raoufi et al., 2016). The insoluble fraction consists of much larger molecules that are associated together through intra- and intermolecular cross linkages, leading to the formation of compacted and stiff particles (Raoufi et al., 2020). FTIR analysis unveiled the presence of $-OH$, $-CH_2$, $-CH_3$, $-COO^-$, C—O, C=O, and –CON groups in PG structure (Dabestani et al., 2018). These functional groups are responsible for noncovalent bonding and generation of insoluble polymer associations. There is a small protein component covalently linked to the PG backbone in both soluble and insoluble fractions (Raoufi et al., 2020), which could be attributed to the weak buffering capacity of the gum. Like PG, AG is also an anionic polysaccharide with a native pH of 5.25 at 1% (w/w) concentration in water (Farooq et al., 2014). It contains 82.26%–92.36% carbohydrate (46.82% arabinose, 35.49% galactose, 10.9% xylitol, and 5.97% uronic acids), 1.41%–2.45% protein, 0.23%–0.85% fat, 2.3%–3.86% ash, and 12.23% moisture content (Bashir and Haripriya, 2016; Mahfoudhi et al., 2012). Both soluble and insoluble fractions of AG have been demonstrated to be strong electron donors and thus can act as antioxidant by terminating radical chain reactions (Bouaziz et al., 2017a). They also have antimicrobial effect on *Enterobacter* spp., *Escherichia coli*, *Staphylococcus aureus*, *Pseudomonas aeruginosa*, *Salmonella typhimurium*, *Listeria monocytogenes*, *Micrococcus luteus*, and *Bacillus subtilis* together with an antihypertensive property. The functional properties of the insoluble fraction were shown to be stronger than those of the soluble one (Bouaziz et al., 2015, 2017a, 2017b). Owing to the similar structure of AG and PG, they may have comparable functional properties. Therefore, where appropriate, AG will also be dealt with in this chapter.

17.4 Improving solubility and functionality of PG

As mentioned earlier, a large fraction of PG (>90%) is water-insoluble due to inter- and intrachain linkages, which lead to the formation of an inflexible structure with a high molecular weight (Raoufi et al., 2019; Rezaei et al., 2016b). These molecular associations can be disintegrated using mechanical and biological methods. As a result, the size and molecular weight are decreased and more functional groups become available which favor the interaction with water and thus solubility. It has been shown that sonication of PG dispersion at high temperatures (~60 °C) and long times (>10 min) results in random scission of glycosidic linkages at the midpoint of the backbone, breakage of side chains, and cleavage of C–O–C pyranose rings. These changes give rise to the disentanglement of polysaccharide chains and increased homogeneity. The depolymerized chains have a worm-like coil conformation, more available carboxyl groups, high molecular density and solubility (>90%) with a molecular weight of 0.76×10^6 g/mol (Raoufi et al., 2019), a zeta potential of −30.0 mV, and a pK_a value of 2.1 (Raoufi et al., 2020).

Like PG, AG is also an anionic polysaccharide with a native pH of 5.25 at 1% (w/w) concentration in water (Farooq et al., 2014). It contains 82.26%–92.36% carbohydrate (46.82% arabinose, 35.49% galactose, 10.9% xylitol, and 5.97% uronic acids), 1.41%–2.45% protein, 0.23%–0.85% fat, 2.3%–3.86% ash, and 12.23% moisture content (Bashir and Haripriya, 2016; Mahfoudhi et al., 2012). Both soluble and insoluble fractions of AG have been demonstrated to be strong electron donors and thus can act as antioxidant by terminating radical chain reactions (Bouaziz et al., 2017a). They also have antimicrobial effect on *Enterobacter* spp., *Escherichia coli*, *Staphylococcus aureus*, *Pseudomonas aeruginosa*, *Salmonella typhimurium*, *Listeria monocytogenes*, *Micrococcus luteus*, and *Bacillus subtilis* together with an antihypertensive property. The functional properties of the insoluble fraction were shown to be stronger than those of the soluble one (Bouaziz et al., 2017b). Owing to the similar structure of AG and PG, they may have comparable functional properties. Therefore, where appropriate, AG will also be dealt with in this chapter.

Gamma irradiation is another method for solubilization of PG. Unlike sonication, which results in progressive viscosity reduction, gamma-ray intensities up to 4 kGy increase both apparent and intrinsic viscosities (Teimouri et al., 2016), which could be related to the difference in their mechanism of action.

Enzymatic hydrolysis of PG by *Penicillium occitanis* (Pol6) fungus has been reported to result in the formation of two highly branched oligosaccharide fractions mainly composed of arabinose and galactose at a molar ratio of 2:1 with traces of xylose, rhamnose, glucose, and mannose (Bouaziz et al., 2014). Although this study was intended to produce oligosaccharide derivatives with wound healing properties, it clearly shows that enzymatic digestion is a feasible approach for solubilization of the gum.

In an attempt to chemically modify PG and increase its solubility, Samari-Khalaj and Abbasi (2017) partially substituted the gum with acrylamide. They succeeded to attain a 64% rise in solubility when 6% (w/w) insoluble fraction of PG was heated with 0.08 mol acrylamide at 60 °C for 3 h. The intrinsic viscosity and molecular weight were both significantly decreased. Stoichiometric calculations indicated that there was 4.95 μg acrylamide per every gram of the gum. This is much less than the tolerable daily intake level and thus can be safely used for food and pharmaceutical applications.

17.5 Drug delivery systems

The disclosure of molecular basis of different diseases has increased the demand for developing new drugs and designing specific methods to address such disorders. However, formulating a delivery system that can release the drug at a specific site in the body over a reasonable period of time with no serious disruption in the functionality of the organ is a challenging issue in the pharmaceutical industry. Hydrocolloids, apart from their medicinal properties, play a central role as functional excipients in formulation of drug delivery systems for controlled and targeted release of active pharmaceutical ingredients. Their ability to decrease the interfacial tension at oil/water interface, form hydrogel network and film, enhance viscosity, and control the flow of liquids, suspensions, and emulsions and mask unpleasant tastes makes them a prime choice as carriers of drugs, binders, and coaters of tablets. Hydrocolloids are able to modify the bio-accessibility of drugs and enhance their bioavailability when formulated as hydrogels or micro/nanoparticles. They can also be used for biodistribution purposes in the form of dense nanoparticles or administration of hydrophobic drugs owing to their micelle forming property. In addition, they can be employed for gene therapy applications in order to transport genetic materials into cells and development of stimuli-responsive drug delivery systems (Gandhi et al., 2012).

In the following sections, PG-based micro/nano-sized systems for delivery of drugs and nutraceuticals will be dealt with in detail.

17.5.1 Emulsions

Oil-in-water (O/W) emulsions are the most prevalent form of delivery systems for lipophilic drugs. Emulsion-based vehicles are usually composed of water as the continuous phase, along with a hydrophilic surfactant shell, and a hydrophobic oil core containing solubilized active ingredient (McClements, 2011). Emulsions are thermodynamically unstable systems and undergo breakdown through different mechanisms (flocculation, coalescence, Ostwald ripening and phase separation) by lapse of time depending on the nature of stabilizer, size and charge density of droplets, viscosity, pH and ionic strength of the continuous phase, etc. (McClements, 2005). Some biopolymers can adsorb at the oil/water interface due to their amphiphilic properties or act as thickening agents, modulating the interaction between droplets. Both mechanisms are crucial to the stability of an emulsion system.

PG, due to the presence of a protein moiety covalently linked to its backbone, contains both hydrophilic and hydrophobic groups and thus can adsorb at the oil/water interface and stabilize the emulsion (Fig. 17.1). It also contributes to the stability by increasing the viscosity of the continuous phase (Mehrnia et al., 2017). Additionally, the insoluble fraction of PG can easily adsorb onto the interface as solid particles leading to the formation of a type of emulsion known as Pickering emulsion, which is more stable than the conventional ones. It has been reported that the emulsifying capacity of PG is lower than that of gum Arabic and whey protein on account of having lower protein content (Jafari et al., 2012; Mehrnia et al., 2017). This may result in larger droplets size in PG-stabilized emulsions. However, owing to their higher viscosity, they would show greater stability over a prolonged storage time than those stabilized by gum Arabic and whey protein (Jafari et al., 2012, 2013; Mehrnia et al., 2017).

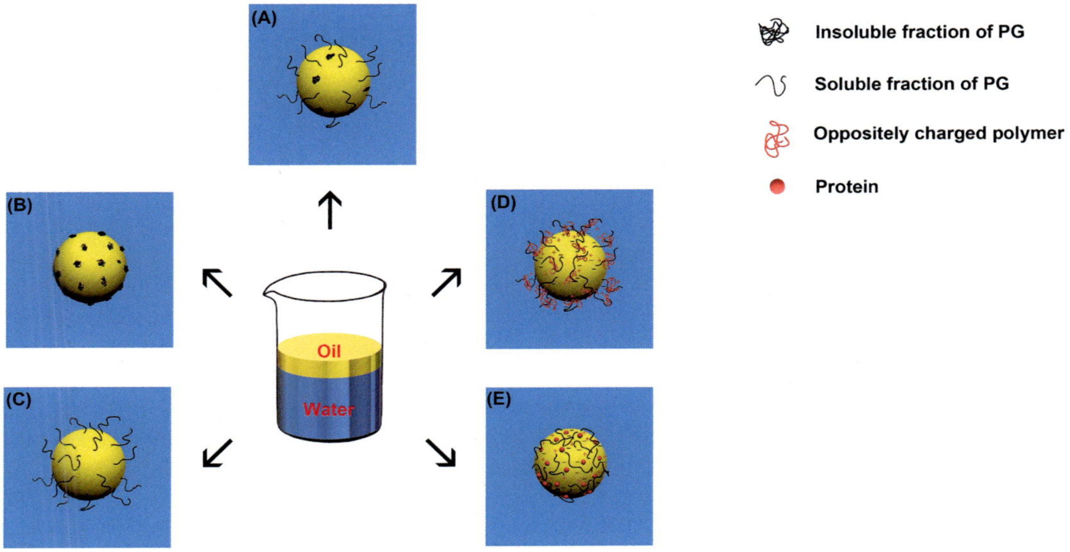

FIG. 17.1 Schematic representation of oil-in-water emulsion systems stabilized by (A) whole PG, (B) insoluble fraction of PG (Pickering emulsion), (C) soluble fraction of PG, (D) soluble fraction of PG + oppositely charged polymer, and (E) protein + soluble fraction of PG.

Comparing the surface activity of AG with gum Arabic revealed that their kinetics of adsorption at O/W interface are different. The dynamic of interfacial adsorption and reorganization of AG was found to be remarkably slower than that of gum Arabic. As a result, at low concentrations, AG-stabilized emulsions exhibited lower stability than those stabilized by gum Arabic, while at concentrations greater than 5%, AG showed a superior stabilizing effect (Mahfoudhi et al., 2014). This may be attributed to the insoluble fraction of AG, which acts like a weak gel, preventing droplets mobility. Moreover, by virtue of their Pickering effect, the solid particles of the insoluble fraction adsorb onto the oil/water interface, inhibiting droplet-droplet interactions through steric and electrostatic repulsion.

In an oil-water mixture containing PG, the polysaccharide molecules may reduce the interfacial tension by adsorbing onto the interface through their protein moieties. Once adsorbed, the large polysaccharide chains with negatively charged carboxylic groups protrude from the surface outward preventing the droplets to approach one another via steric and electrostatic repulsive forces, thereby enhancing the stability of the emulsion. Owing to the trace amount of protein in PG structure, the concentration of polysaccharide plays a major role in the long-term stability of the emulsion as the nonadsorbed molecules enhance the viscosity of the continuous phase which decreases the mobility of droplets and hence the probability of droplet-droplet interactions. It has been reported that no phase separation would occur in PG-stabilized emulsions at concentrations above 3.0% (Jafari et al., 2012). In addition, as pointed out before, the insoluble fraction of PG is believed to have a Pickering stabilizing effect. The solid polysaccharide particles may therefore adsorb onto the interface, leading to the formation of an integrated layer around the oil droplets and thus enhance the total stability of emulsion. PG-stabilized emulsions have been shown to be stable in the pH range of 3–7 (Golkar et al., 2018). Moreover, these emulsions are stable at ionic strength of less than 200 mM NaCl at pH 4.5 and temperature up to 80 °C. They are also completely stable in the presence of $CaCl_2$ (<300 mM). It is interesting to note that PG results in less interfacial tension compared to gum Arabic at 150–300 mM $CaCl_2$ (Golkar et al., 2018). Therefore, in the gastric and intestinal domains where the concentrations of NaCl and $CaCl_2$ are less than 50 and 6 mM, respectively (Minekus et al., 2014), PG-stabilized emulsions are expected to remain unchanged.

Although the stability of emulsions increases by rising PG concentration, this polysaccharide is generally considered as a weak emulsifying agent due to its limited protein content. The emulsifying capacity of PG can be enhanced by conjugating it with proteins. Golkar et al. (2015) reported that the β-lactoglobulin-PG conjugates prepared through the Maillard reaction had higher emulsifying stability index than gum Arabic. The highest Amadori compounds were observed after 4 days of incubation followed by the formation of browning melanoidins from day 6 onwards. The emulsions made with the conjugates obtained in different days revealed that increasing the incubation time would improve the stability and decrease the creaming index. Although the emulsifying activity of conjugates was expected to be independent of the incubation time, the highest activity was observed on the second day. The authors related this to the generation of insoluble high molecular weight products with lower translational as well as probably rotational diffusivity towards the oil/water interface as the Maillard reaction progressed. In an another study, PG was chemically modified by octenyl succinic anhydride (OSA) substitution and used for emulsification of sunflower oil (Mohammadi et al., 2016). OSA has been approved by the Food and Drug Administration (FDA) and is employed to convert the hydrophilic hydrocolloids into amphiphilic molecules by substituting a number of hydrophobic groups on their hydrophilic backbone. In addition to chemical methods, PG may also be modified by physical approaches. Ultrasonic depolymerization has been shown to enhance the emulsifying property of PG. Sonication of PG was reported to increase its emulsifying activity index (amount of oil emulsified per unit mass of emulsifier) by six times. The emulsions remained physically stable for up to 4 weeks (Bhosale and Singhal, 2006).

The stabilizing property of PG has also been investigated for double emulsions. In an attempt to make a water-in-oil-in-water nano-emulsion, the inner and outer phases were stabilized by polyglycerol polyricinoleate (PGPR) and PG, respectively (Mehrnia et al., 2017). The droplets size and viscosity of this system were found to be higher than those stabilized by the same concentration of whey protein or gum Arabic, but the release of the core bioactive compound in the stomach and duodenum was much lower. It was claimed that PG could control the release of core material with no need for adjusting the osmotic pressure between inner and outer phases. The reason is not fully understood, but it may be due to the difference in the stabilizing mechanism exerted by PG due to its Pickering effect.

When a combination of PG with other biopolymers or small molecule surfactants is used, it is their interactions and the net charge density of the interface that determine the stability of the emulsion. For an emulsion stabilized by a nonionic emulsifier, the presence of low concentrations of PG may decrease the stability by inducing depletion flocculation, which gives rise to size evolution of oil droplets and phase separation of the colloidal system. At high concentrations, on the other hand, PG increases the viscosity and thus enhances the stability through reducing droplet-droplet interactions. For an emulsion stabilized by charged biopolymers or surfactants, the type of interactions (repulsive or attractive) may influence the stability. If attractive interactions take place, the droplets will gain more stability due to steric and electrostatic forces. In the case of repulsive interactions, the stability may change with PG concentration and the degree of interface charge screening. Electrostatic bonding of PG to whey protein isolate (WPI) through the layer-by-layer technique indicated that the resulting emulsion showed the highest stability at a pH near the isoelectric point of protein (Hadian et al., 2020; Sepeidnameh et al., 2018). Above or below this pH, the emulsion underwent instability due to depletion flocculation and interdroplet bridging, respectively. The stability also decreased at high ionic strength (>25 mM $CaCl_2$ or >75 mM NaCl) owing to a charge screening effect. On the other hand, the Schiff base of proteins and PG has been found to have a high emulsifying stability index at neutral pH (Golkar et al., 2017). In a multilayer microemulsion stabilized by WPI and the soluble fraction of PG, increasing the ratio of PG improved the physical stability in the pH range of 4–6. Moreover, PG enhanced the chemical stability of oil droplets against oxidation (Sepeidnameh et al., 2018).

As mentioned earlier, sonication can increase the solubility of PG and thus its emulsifying property (Raoufi et al., 2019). Raeisi et al. (2019) reported that ultrasonic emulsification of fish oil and garlic essential oil in water containing Tween 80, PG and chitosan led to a multilayer nano-emulsion. They found that increasing the ratio of PG decreased the size of droplets to 23.19 nm and reduced the polydispersity index. This was related to the ability of PG to increase the viscosity and develop a three-dimensional network in which the droplets are entrapped. The electrostatic interaction of PG with chitosan molecules was also postulated to contribute to the reinforcement of the interface. Similarly, Hashtjin and Abbasi (2015) observed that ultrasonic emulsification of oil in water containing Tween 80 and soluble fractions of PG and gum tragacanth at a ratio of 3:1 resulted in nano-emulsions with droplets of 13 nm in size. Although the enhanced stability of emulsions observed in these studies was not attributed to the effect of sonication on PG, it seems that partial sonolysis of the gum and its conformational changes could improve its surface activity (Raoufi et al., 2019).

17.5.2 Capsules and coacervates

Encapsulation is a useful technique for delivery of drugs and bioactive compounds. The main purpose of encapsulation is to protect the core material and promote its bioavailability and controlled release. Biopolymers can be used for encapsulation of bioactive ingredients, individually or in combination with other biopolymers and small molecules. As for PG, the intact gum alone may not be a proper choice for encapsulation purposes by virtue of being mostly in the form of inflexible cross-linked chains. Instead, the whole PG or more preferably its soluble fraction is usually used along with other biopolymers for the entrapment and encapsulation of drugs and nutraceuticals (Fig. 17.2A). When mixed with oppositely charged biopolymers, the carboxyl groups of PG electrostatically interact with the cationic groups of other biopolymers to form interpolymeric complexes (Raoufi et al., 2016). Considering the stiffness of PG backbone this complexation is favored provided its original conformation is maintained and the entropy loss is minimized (Turgeon and Laneuville, 2009). These structures can be utilized for stabilizing colloidal systems (Abbasi and Mohammadi, 2013; Azarikia and Abbasi, 2016; Behbahani and Abbasi, 2017; Dabestani and Yeganehzad, 2019; Hadian et al., 2016; Khanniri et al., 2017; Raoufi et al., 2016; Teimouri et al., 2017), but they

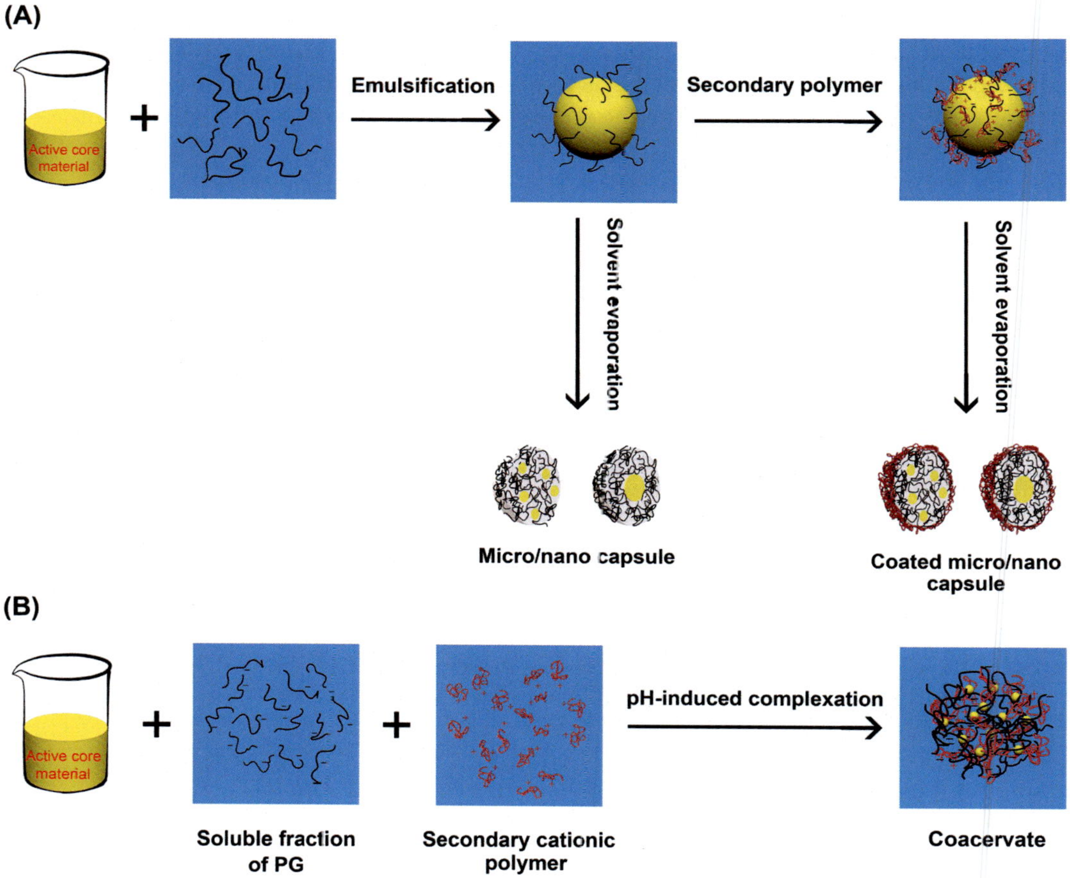

FIG. 17.2 Schematic representation of PG-based micro/nano-encapsulation (A) and coacervation (B).

are not flexible enough to surround bioactive components and form microcapsules. Moayyedi et al. (2018) investigated the survivability of *Lactobacillus rhamnosus* ATCC 7469 cells microencapsulated in WPI, WPI + inulin, and WPI + inulin + PG matrices produced by different methods (spray drying, freeze drying and electrospraying). They found that in the presence of PG cell damages were significantly decreased. PG also protected microbial cells from the harsh conditions of freeze drying by forming a viscous layer around them and retained their survivability during storage. It was reported that the low glass transition temperature of PG increased the stickiness of microcapsules and thus their aggregation, which would protect the cells against environmental conditions. Bassijeh et al. (2020) encapsulated astaxanthin in a double layered matrix composed of WPI and PG at pH 4, where the protein is positively charged and can electrostatically bind to carboxyl groups of the gum. Their results indicated that increasing the ratio of PG decreased the size of microcapsules as well as their moisture content and enhanced the encapsulation efficiency, solubility, water dispersibility, and chemical stability of astaxanthin.

The encapsulation efficiency of sodium selenite in a mixed matrix of PG (1%–5%) and gum Arabic (25%–29%), fabricated by evaporation technique, as well as the size of capsules were demonstrated to be adversely affected by PG concentration (Jalalizand and Goli, 2021). This can be related to the insolubility and inflexibility of PG chains, which adversely affect its film-forming ability. However, some efforts have been made to improve this property by using cross-linkers. Taban et al. (2020) used citric acid and transglutaminase enzyme to cross-link PG chains together or to gelatin. Stable nanocapsules were obtained through enzyme- or citric acid-mediated cross-links between PG or PG-gelatin molecules. However, the size of PG-gelatin nanocapsules produced through enzymatic cross-linking was larger which was attributed to their more open structure due to the less number of linkages. Citric acid-mediated cross-linking of PG also led to the highest encapsulation efficiency compared to enzymatic cross-linking of PG-gelatin.

PG on account of being mostly insoluble in water is not able to form coacervates with proteins, and thus the coacervation yield may be extremely low. Coacervates are a special form of capsules made by electrostatic interactions between oppositely charged polymers, mainly proteins and polysaccharides, giving rise to liquid-liquid phase separation in the aqueous system. To fabricate this type of molecular association, polysaccharide must have special characteristics including specific molecular weight, random coil conformation, and chain flexibility (Turgeon and Laneuville, 2009). As explained above, sonication has been proved to be an efficient technique for chain breakage of polysaccharide backbone, leading to the formation of highly soluble residues with low molecular weight and very flexible structure. The products of PG sonolysis have been shown to form coacervates effectively with oppositely charged patches on WPI molecules (Fig. 17.2B). Mixing sonicated PG with WPI results in two types of coacervates depending on the pH; one in the pH range of 4.4–4.85, which is associated with the formation of interpolymer complexes with β-lactoglobulin, and the other in the pH range of 3.27–4.0, corresponding to intrapolymer association or coacervation between PG short chain species and α-lactalbumin (Raoufi et al., 2020). Emamverdian et al. (2020) studied the coacervation of gelatin with the soluble fraction PG over a wide pH range at different blending ratios. They found that the optimum coacervation occurred at gelatin-PG mixing ratio of 1:1 and pH 4 with a ζ-potential of −6 mV. The coacervation yield was calculated to be 58%. It was claimed that the resulting coacervate could potentially be used for the delivery of bioactive compounds. Similarly, Mousazadeh et al. (2018) reported that the highest coacervation yield for chickpea protein isolate and soluble fraction of PG was obtained at pH 3 and mixing ratio of 1:1. Differential scanning calorimetry showed that binding PG to the protein increased its denaturation temperature as well as its initial conformation due to electrostatic and hydrophobic interactions as confirmed by isothermal titration calorimetry.

17.5.3 Micro- and Nano-particles/Fibers

Gum-based micro- and nano-particles are promising vehicles for carrying and delivering bioactive compounds. Polysaccharide particles can be fabricated using top-down and bottom-up methods (Fig. 17.3A). The former involves mechanical size reduction of bulk material by crushing, milling, high shearing, etc., while the latter begins at molecular scale and builds up to desired size by physical, chemical, and biological means. From a morphological point of view, the shape of particles varies from sphere, through to irregular, to fiber, rod, and plate depending on the type of polymer, procedure, and processing conditions.

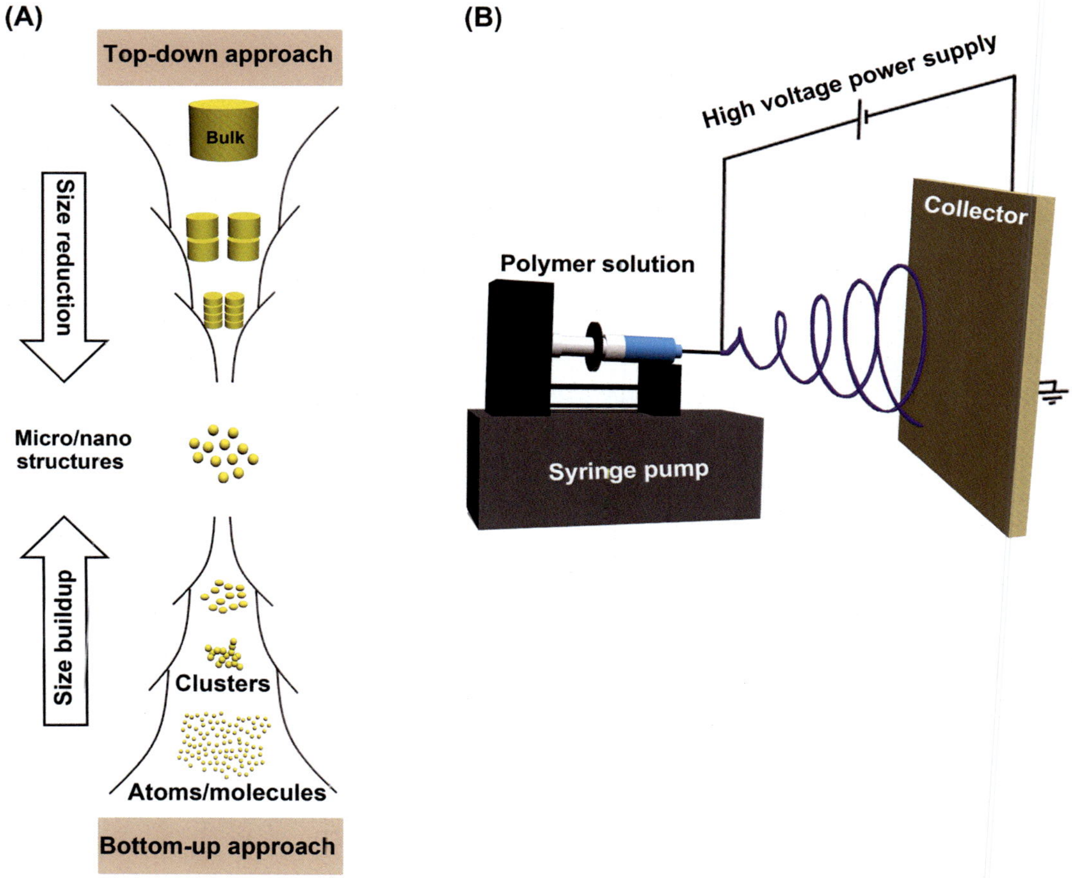

FIG. 17.3 Schematic representation for fabrication of micro/nano-particles (A) and electrospinning process (B).

Electrospinning is a simple and straightforward method for producing submicron- and nanofibers. The basic setup for electrospinning process involves a high voltage power supply, a reservoir tipped with a needle containing polymer solution, and a conductive collector (Fig. 17.3B). Upon applying an electric field between the needle and collector, high charges accumulate on the surface of the pendant droplet formed at the tip of a needle. If the electrostatic repulsion among surface charges exceeds the surface tension of the polymer solution, the droplet is deformed into a Taylor cone, from which a charged jet is ejected and stretched into fibers of nanoscale diameters due to whipping instability which are eventually deposited on the collector. Plant gums like PG and AG that are mostly insoluble cannot be electrospun due to the lack of enough chain entanglement. It has been shown that more than 2.5 entanglements per polymer chain are required to obtain stable uniform fibers (Shenoy et al., 2005). On the other hand, it has also been demonstrated that the polymer relaxation time and thus the Deborah number (ratio of polymer relaxation time to timescale of the process) should be larger than a critical value to overcome the capillary stresses and achieve fiber formation (Gadkari, 2017). Rezaei et al. (2016b) reported that they could electrospin a mixture of AG/polyvinyl alcohol (PVA) into fibers with a diameter depending on the ratio of the two polymers. Increasing the ratio of AG/PVA from 70:30 to 80:20 decreased the thickness of fibers from 116 to 77 nm. The electrospinnability of the mixture was attributed to the formation of etheric bonds between AG and PVA. This fibrous structure could potentially be used for encapsulation and stabilization of sensitive bioactive compounds. In continuation to this study, curcumin and curcumin-β-cyclodextrin inclusion complex were encapsulated into AG/PVA nanofibers and their release behavior was investigated under simulated oral and gastrointestinal conditions. It was found that the water absorption and swelling of the gum was the main driving force for the release of curcumin from nanofibers. In simulated saliva environment, 57% and 75% curcumin and curcumin-β-cyclodextrin inclusion complex were released after 180 min, respectively. In simulated gastrointestinal media, a burst release was observed during the first 120 min in gastric conditions followed by a sustained release in the intestinal domain, leading to 70% and 85% release for curcumin and curcumin-β-cyclodextrin inclusion complex after 480 min, respectively (Rezaei and Nasirpour, 2019).

Samrot et al. (2018) fabricated nanoparticles of Indian almond (*Terminalia catappa* L.) gum by a two-step process: it was first carboxymethylated and then cross-linked using trisodium trimetaphosphate at alkaline pH and high temperature. The nanoparticles were loaded by curcumin using ultrasonication technique with an encapsulation efficiency of 82%. It was found that curcumin showed a burst release at very low pH (2.2) followed by a gradual increase at later times, while at higher pH values (≥4.7) it was released steadily during the incubation period. The minimum inhibitory concentration of loaded curcumin for *Pseudomonas aeruginosa* was reported to be 4 μg/mL. The IC_{50} was estimated to be 75 μg/mL against the Hep G2 cell line, which was similar to that of unloaded curcumin.

Jaison et al. (2020) encapsulated doxorubicin and iron oxide in AG matrix through antisolvent precipitation method by isopropyl alcohol with an entrapment efficiency of 88.29%. The particles had an average size of 181.5 nm and a zeta potential of −28.8 mV. It was shown that the nanoparticles were pH-responsive, releasing their core material under acidic conditions (pH 5.5) due to hydrolysis of steric bonds and protonation of carboxyl groups resulting in reduced electrostatic attraction between AG and doxorubicin. *In vitro* cellular uptake and cytotoxicity studies revealed the delivery of nanoparticles into cytosols and maximum inhibition of HeLa cell lines.

17.5.4 Composite films

Cast film of pure PG is poorly transparent, highly brittle and crispy, fairly soluble in water (55.7%), with a glass transition of 121.6 °C, no melting point, and mechanical resistance (Razmjoo et al., 2021). Scanning electron microscopy (SEM) showed that the film surface is smooth and homogeneous (Ghiasi et al., 2020). The swelling property of the film was found to be less than gum tragacanth and more than gelatin films (Khodaei et al., 2019).

In order to improve the functional properties of PG film, it has been blended with other polymers and composite films of various physical, mechanical, and thermal features have been developed (Fig. 17.4). Casting a blend of PG with PVA, starch, or sodium caseinate resulted in composites with low moisture content and solubility in water owing to the strong hydrogen bonding between the two polymers. These interchain cross-links usually enhance the coherence and integrity of film matrix and reduce its water vapor permeability.

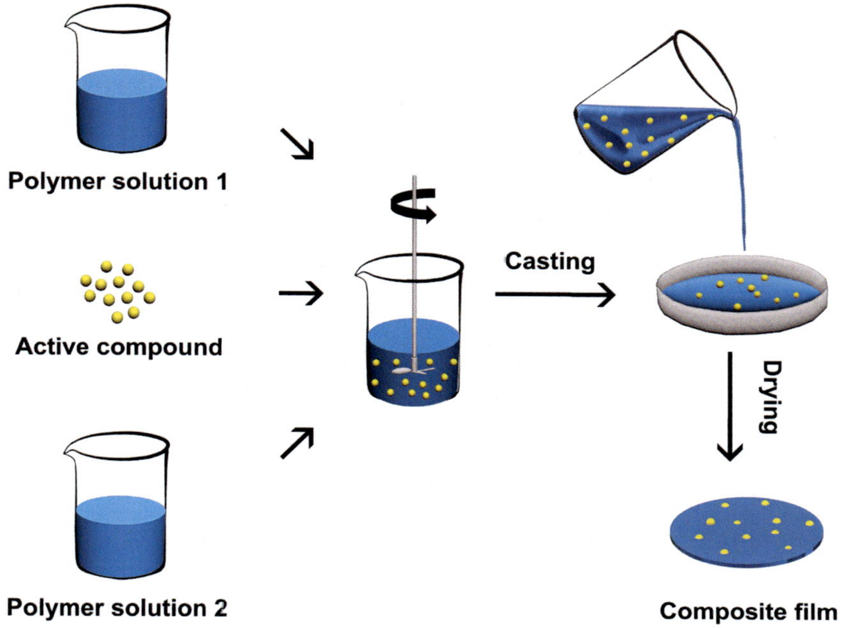

FIG. 17.4 Schematic diagram for casting PG-based micro/nano-composite film.

Incorporation of PG also changes the thermal property of polymer films in a way that increases the glass transition temperature (T_g) and reduces the melting point. The rise of T_g is attributed to the increased number of hydrogen bonds between PG and the other polymer which diminishes the mobility of molecules. In addition, this decreases the degree of crystallinity, making the film structure more amorphous. It has been shown that the mechanical parameters including elongation at break, tensile strength, and Young's modulus of PG/PVA film was significantly improved compared to the pure PG film due to the strong interfacial adhesion between the two polymers (Razmjoo et al., 2021). Addition of a certain amount of glycerol can also enhance the mechanical strength of PG film (Pak et al., 2020). On the other hand, the composite films of PG with starch, gelatin, and gum tragacanth retained higher moisture content which led to weaker mechanical properties and higher stretchability. The surface of the films was smooth, with reduced number of cracks and gaps (Khodaei et al., 2019; Pirouzifard et al., 2019; Razmjoo et al., 2021). PG-based composite films are usually thicker due to low water loss on drying. This affects their solubility in the mouth and during digestion (Khanzadi et al., 2015).

As for PG-PVA composite film, it was observed that the most interactions between the functional groups of the two polymers occurred at low percentage of PG due to its rigid and inflexible structure. Increasing the ratio of PG in the blend reduced the number of interactions. Therefore, the excess number of available functional groups resulted in higher solubility, moisture content, and water vapor permeability, and decreased the glass transition temperature of the composite film. Furthermore, the mechanical properties of the blend film progressively weakened by increasing the ratio of PG (Razmjoo et al., 2021).

17.5.5 Tablets and pellets

Many plant hydrocolloids have been used as binders in tablet and pellet formulations. They impart the structural strength required during processing, handling, and packaging of tablets. Owing to their swelling property, they act as disintegrants whereby they increase the exposed surface area of medicine and improve the dissolution property of a drug formulation.

PG is a good swelling agent adsorbing 11.07–14.26 times of its weight in water (Rahimi and Abbasi, 2014). As a wall material, it therefore has the potential to retard the release of bioactive materials and thus could potentially be used in formulation of colon-targeted delivery carriers. However, it has not yet been studied for pharmaceutical applications. Considering the structural similarity of PG with the exudate gums from other subgenera of *Amygdalus*, particularly AG, which has widely been studied for drug delivery purposes, the same applications can be postulated for PG.

AG has been utilized in tablet and pellet formulations as a swelling agent, disintegrant, binding agent, and pelletizer. Like PG, AG is found in irregular shapes and different colors with semicrystalline structure (Bashir and Haripriya, 2016; Rezaei et al., 2016a). Scanning electron microscopic observations revealed that the gum surface may be rough or smooth (Anushiravani et al., 2020; Farooq et al., 2014), indicating the critical effect of soil and environment on the exudate characteristics. The rough particles have been shown to cause extended-release dosage form owing to the entrapment of drug inside their pores (Farooq et al., 2014). AG has been used as a binder in diclofenac sodium tablets with no chemical interactions with the drug, nominating it as a safe excipient for pharmaceutical applications (Rahim et al., 2014; Sarojini et al., 2010). It has a high swelling property and thus can retard the release of active ingredient making it useful for delivery of colon-targeted drugs (Kadiyam and Muzib, 2015). Swelling increases the distance that the drug has to traverse to get released. Moreover, the gelatinous viscous layer formed around the tablet due to the swelling retards the diffusion rate of the drug (Ramesh et al., 2017, 2019). Apart from being a release retardant, AG also acts as a spheronizing agent in the process of pelletization. Incorporation of AG into the formulation of furosemide pellets showed that increasing its ratio enhanced the plasticity and cohesiveness of the extruded mass leading to more spherical and free-flowing pellets with smooth and regular surfaces with no need for the addition of microcrystalline cellulose, i.e., the most conventional spheronizing agent (Ramesh et al., 2020).

Shellac is a natural nontoxic resinous compound secreted by lac insects (*Kerria lacca*). It is used in the food and pharmaceutical industries as a coating material, especially as an enteric coating to prevent the disintegration and dissolution of drug formulations in the gastric environment (Pearnchob et al., 2003). Shellac, however, because of its insolubility at low pH, aggregates in the stomach, which may affect the release and bioavailability of the drug. On the other hand, the stability of the active ingredient in the harsh acidic conditions of stomach is crucial for formulating a delivery system targeting the colon. It has been reported that coating quercetin-loaded shellac nanoparticles by AG and Tween 80 using the antisolvent precipitation technique imparted considerable gastric stability to the particles with no sign of aggregation. The particles also revealed high antioxidant activity and bioavailability (Sedaghat Doost et al., 2018, 2019). Taking into account these results, it is presumed that incorporation of PG and AG into pellets and tablets may protect acid-sensitive drugs in the gastric environment.

The good film-forming and swelling properties of PG make it a proper potential candidate to be used for making multiparticulate dosage formulations, individually or in combination with other polysaccharides. Multiparticulate dosage formulations, unlike conventional immediate release dosage forms, are able to maintain stable plasma level of drug over an extended period of time. They not only decrease the daily dosing regimen and thus missed medication doses, but also are appropriate for patients suffering from swallowing difficulty. Multiparticulate dosage formulations are composed of several polymer-coated discrete micro- or nano-particles that are compressed into tablets to meet sustained release of the drug. The particles are in fact micro/nanocapsules acting as metric reservoirs. Upon swallowing, these formulations are disintegrated rapidly into micro/nanoparticles, whose small size leads to their even distribution over a large surface area of the gastrointestinal tract, which greatly enhances the bioavailability of the active pharmaceutical ingredient and prevents local irritation and toxicity due to the burst release of the drug (Al-Hashimi et al., 2018).

Peng et al. used ethyl cellulose along with Poloxamer 127 to coat Huperzine A, an acetylcholinesterase inhibitor proved to relieve the symptoms of Alzheimer's disease, into microparticles by an ultra-fine particle processing system and compressed them into tablet form. The results showed that the release profile of the drug from tablet was similar to that from microparticles, indicating the fast disintegration of the tablet in dissolution medium. The multiparticulate tablet exhibited sustained release with more profound relative bioavailability (125.18%) compared to the commercial tablet (Peng et al., 2019). Patel and Amin developed a solid multiparticulate dosage form for colon-targeted delivery of mebeverine hydrochloride based on guar gum microspheres. *In vivo* studies in rabbit gastrointestinal tract showed that the tablet coated by Eudragit S100 remained stable in stomach and small intestine conditions, and reached the colon after 5 h where it disintegrated leading to the release of the drug-loaded microspheres (Patel and Amin, 2011).

17.6 Other biological applications of PG

PG has been shown to reduce triglyceride levels in hyperlipidemic patients. This has been attributed to the flavonoids and phenolic compounds including apigenin and kaempferol that naturally occur in PG. They can induce peroxisome proliferators-activated receptors, which are involved in the expression of lipoprotein lipase and apo C-II genes, hepatic uptake, and esterification of free fatty acids and mitochondrial oxidation of free fatty acid. PG is also a prebiotic polysaccharide, being able to increase the production of glucagon-like peptide 1, and thus can reduce insulin resistance in type-2 diabetics. Moreover, as a dietary fiber and satiety-promoting carbohydrate, PG can be used to aid body weight loss. Sonicated PG, owing to the large number of hydroxyl and carboxyl groups that are available in its structure, can be used for gel or hydrogel formation, alone or along with other biopolymers. These three-dimensional networks have the potential to be employed for solubilization and delivery of hydrophobic drugs. Moreover, previous studies showed that natural oligosaccharides or the enzymatic or chemical hydrolysis products of exudate gums have prebiotic effects.

Additionally, the whole gum or its hydrolyzates have been used in the pharmaceutical formulations including tablets as mucoadhesive components, antitumor drugs, or epithelial wound healing. Bouaziz et al. investigated the wound healing capacity of enzymatically hydrolyzed AG, alone and in combination with a commercial cream (CICAFLORA), on rats. They observed accelerated healing in the presence of AG oligosaccharides, which was related to the stimulation of new blood vessels creation (neovascularization) and collagen formation (Bouaziz et al., 2014).

The application of AG as a tissue-equivalent phantom material for photoacoustic imaging and diagnostic radiation purposes has also been proved (Ababneh et al., 2016). Additionally, it has an antiulcerogenic effect on the gastrointestinal epithelium that helps regulate asthma in patients with refractory asthma (Anushiravani et al., 2020).

17.7 Concluding remarks and future trends

PG, due to its unique functional properties, is a promising natural biopolymer for fabrication of a wide range of drug and nutraceutical delivery systems. It is a food-grade plant exudate and a traditional herb that has been used since ancient times. This naturally abundant polysaccharide is low-cost, biocompatible, nontoxic, and inert with no adverse impact on the appearance and flavor of food and pharmaceutical products as well as the functionality of body organs. The findings of recent studies clearly show that PG as a phytochemical—either the whole gum or its oligosaccharide hydrolyzates—has antimicrobial, antioxidant, wound healing, and prebiotic effects, and thus can find various pharmaceutical applications. It can also be used as a base for the development of drug delivery systems, from simple hydrogels, through film composites, to complex micro/nano colloidal systems including emulsions, particles, and coacervates. The swelling and plasticizing properties of PG make it a good excipient, binder, release retardant, and spheronizing agent for different drug formulations as well as targeted delivery purposes. This polysaccharide has an appropriate mucoadhesive property and hence can enhance the bioavailability and bioaccessibility of active compounds and increase their uptake. In spite of the efforts made so far, the medicinal and functional features of this plant exudate have not fully been understood. Therefore, further studies are needed to be carried out in order to explore the potential of PG as a vehicle for drug delivery purposes.

References

Ababneh, B., Tajuddin, A.A., Hashim, R., Shuaib, I.L., 2016. Investigation of mass attenuation coefficient of almond gum bonded Rhizophora spp. particleboard as equivalent human tissue using XRF technique in the 16.6–25.3 keV photon energy. Australas. Phys. Eng. Sci. Med. 39 (4), 871–876.

Abbasi, S., 2017. Challenges towards characterization and applications of a novel hydrocolloid: Persian gum. Curr. Opin. Colloid Interface Sci. 28, 37–45.

Abbasi, S., Mohammadi, S., 2013. Stabilization of milk-orange juice mixture using Persian gum: efficiency and mechanism. Food Biosci. 2, 53–60.

Abbasi, S., Rahimi, S., 2015. Persian gum. In: Mishra, M. (Ed.), Encyclopedia of Biomedical Polymers and Polymeric Biomaterials. CRC Press, pp. 5919–5928.

Al-Hashimi, N., Begg, N., Alany, R.G., Hassanin, H., Elshaer, A., 2018. Oral modified release multiple-unit particulate systems: compressed pellets, microparticles and nanoparticles. Pharmaceutics 10 (4), 176.
Anushiravani, M., Azad, F.J., Taghipour, A., Mirsadraee, M., Afshari, J.T., Salari, R., Najafzadeh, M.J., Khoshkhui, M., Engel, R.M., Farshchi, M.K., 2020. The effect of Plantago major seed and almond gum on refractory asthma: a proof-of-concept study. J. Herb. Med. 19, 100297.
Azarikia, F., Abbasi, S., 2016. Mechanism of soluble complex formation of milk proteins with native gums (tragacanth and Persian gum). Food Hydrocoll. 59, 35–44.
Bashir, M., Haripriya, S., 2016. Assessment of physical and structural characteristics of almond gum. Int. J. Biol. Macromol. 93, 476–482.
Bassijeh, A., Ansari, S., Hosseini, S.M.H., 2020. Astaxanthin encapsulation in multilayer emulsions stabilized by complex coacervates of whey protein isolate and Persian gum and its use as a natural colorant in a model beverage. Food Res. Int. 137, 109689.
Behbahani, M.S., Abbasi, S., 2017. Stabilization of flixweed seeds (Descurainia sophia L.) drink: Persian refreshing drink. Food Biosci. 18, 22–27.
Bhatia, S., 2016. Natural Polymer Drug Delivery Systems: Nanoparticles, Plants, and Algae. Springer International Publishing, pp. 1–225.
Bhosale, R., Singhal, R., 2006. Process optimization for the synthesis of octenyl succinyl derivative of waxy corn and amaranth starches. Carbohydr. Polym. 66 (4), 521–527.
Biswajit, D., Ahmed, N., Singh, P., 2011. Prunus diversity-early and present development: a review. Int. J. Biodivers. Conserv. 3, 721–734.
Bouaziz, F., Ben Romdhane, M., Boisset Helbert, C., Buon, L., Bhiri, F., Bardaa, S., Driss, D., Koubaa, M., Fakhfakh, A., Sahnoun, Z., Kallel, F., Zghal, N., Ellouz Chaabouni, S., 2014. Healing efficiency of oligosaccharides generated from almond gum (Prunus amygdalus) on dermal wounds of adult rats. J. Tissue Viability 23 (3), 98–108.
Bouaziz, F., Koubaa, M., Chaabene, M., Barba, F.J., Ghorbel, R.E., Chaabouni, S.E., 2017a. High throughput screening for bioactive volatile compounds and polyphenols from almond (Prunus amygdalus) gum: assessment of their antioxidant and antibacterial activities. J. Food Process. Preserv. 41 (4), e12997.
Bouaziz, F., Koubaa, M., Ellouz Ghorbel, R., Ellouz Chaabouni, S., 2017b. Biological properties of water-soluble polysaccharides and hemicelluloses from almond gum. Int. J. Biol. Macromol. 95, 667–674.
Bouaziz, F., Koubaa, M., Helbert, C.B., Kallel, F., Driss, D., Kacem, I., Ghorbel, R., Chaabouni, S.E., 2015. Purification, structural data and biological properties of polysaccharide from *Prunus amygdalus* gum. Int. J. Food Sci. Technol. 50 (3), 578–584.
Chahibakhsh, N., Hosseini, E., lslam, M. S., & Rahbar, A. R., 2019. Bitter almond gum reduces body mass index, serum triglyceride, hyperinsulinemia and insulin resistance in overweight subjects with hyperlipidemia. J. Funct. Foods 55, 343–351.
Dabestani, M., Yeganehzad, S., 2019. Effect of Persian gum and xanthan gum on foaming properties and stability of pasteurized fresh egg white foam. Food Hydrocoll. 87, 550–560.
Dabestani, M., Kadkhodaee, R., Phillips, G.O., Abbasi, S., 2018. Persian gum: a comprehensive review on its physicochemical and functional properties. Food Hydrocoll. 78, 92–99.
Emamverdian, P., Moghaddas Kia, E., Ghanbarzadeh, B., Ghasempour, Z., 2020. Characterization and optimization of complex coacervation between soluble fraction of Persian gum and gelatin. Colloids Surf. A Physicochem. Eng. Asp. 607, 125436.
Esfandiari, A., Roustaei, M., Haghighifarjam, E., 2017. The effects of Amygdalus scoparia extract on ultrastructure changes of photoreceptor layer in diabetic rats. Comp. Clin. Pathol. 26 (5), 1027–1032.
Fadavi, G., Mohammadifar, M.A., Zargarran, A., Mortazavian, A.M., Komeili, R., 2014. Composition and physicochemical properties of zedo gum exudates from Amygdalus scoparia. Carbohydr. Polym. 101 (1), 1074–1080.
Fadavi, G., Ghiasi, M., Zargarran, A., Mohammadifar, M.A., 2017. Some physicochemical and rheological properties of zedo (Farsi) gum exudates from Amygdalus scoparia. Nutr. Food Sci. Res. 4, 33–40.
Farhoosh, R., Tavakoli, J., 2008. Physicochemical properties of kernel oil from Amygdalus scoparia growing wild in Iran. J. Food Lipids 15 (4), 433–443.
Farooq, U., Sharma, P.K., Malviya, R., 2014. Extraction and characterization of almond (Prunus dulcis) gum as pharmaceutical excipient. Am. Eurasian J. Agric. Environ. Sci. 14 (3), 269–274.
Gadkari, S., 2017. Influence of polymer relaxation time on the electrospinning process: numerical investigation. Polymers 9 (10), 501.

Gandhi, K.J., Deshmane, S.V., Biyani, K.R., 2012. Polymers in pharmaceutical drug delivery system: a review. Int. J. Pharm. Sci. Rev. Res. 14 (2), 57–66.

Gharaghani, A., Eshghi, S., 2015. Prunus scoparia, a potentially multi-purpose wild almond species in Iran. Acta Hortic. 1074, 67–72.

Ghasempour, Z., Alizadeh, M., Bari, M.R., 2012. Optimisation of probiotic yoghurt production containing zedo gum. Int. J. Dairy Technol. 65 (1), 118–125.

Ghasemzadeh, H., Modiri, F., 2020. Application of novel Persian gum hydrocolloid in soil stabilization. Carbohydr. Polym. 246, 116639.

Ghiasi, F., Golmakani, M.-T., Eskandari, M.H., Hosseini, S.M.H., 2020. A new approach in the hydrophobic modification of polysaccharide-based edible films using structured oil nanoparticles. Ind. Crop Prod. 154, 112679.

Golkar, A., Nasirpour, A., Keramat, J., Desobry, S., 2015. Emulsifying properties of Angum gum (Amygdalus scoparia Spach) conjugated to β-lactoglobulin through Maillard-type reaction. Int. J. Food Prop. 18 (9), 2042–2055.

Golkar, A., Nasirpour, A., Keramat, J., 2017. Improving the emulsifying properties of β-lactoglobulin–wild almond gum (Amygdalus scoparia Spach) exudate complexes by heat. J. Sci. Food Agric. 97 (1), 341–349.

Golkar, A., Taghavi, S.M., Dehnavi, F.A., 2018. The emulsifying properties of Persian gum (Amygdalus scoparia Spach) as compared with gum arabic. Int. J. Food Prop. 21 (1), 416–436.

Gomaa, E.Z., 2013. In vitro antioxidant, antimicrobial, and antitumor activities of bitter almond and sweet apricot (Prunus armeniaca L.) kernels. Food Sci. Biotechnol. 22 (2), 455–463.

Gopinath, V., Saravanan, S., Al-Maleki, A.R., Ramesh, M., Vadivelu, J., 2018. A review of natural polysaccharides for drug delivery applications: special focus on cellulose, starch and glycogen. Biomed. Pharmacother. 107, 96–108.

Hadian, M., Hosseini, S.M.H., Farahnaky, A., Mesbahi, G.R., Yousefi, G.H., Saboury, A.A., 2016. Isothermal titration calorimetric and spectroscopic studies of β-lactoglobulin-water-soluble fraction of Persian gum interaction in aqueous solution. Food Hydrocoll. 55, 108–118.

Hadian, M., Labbafi, M., Hosseini, S.M.H., Safari, M., Vries, R., 2020. A deeper insight into the characteristics of double layer oil-in-water emulsions stabilized by Persian gum and whey protein isolate. J. Dispers. Sci. Technol. https://doi.org/10.1080/01932691.2020.1816178.

Hashemi, S.M.B., Raeisi, S., 2018. Evaluation of antifungal and antioxidant properties of edible coating based on apricot (Prunus armeniaca) gum containing Satureja intermedia extract in fresh wild almond (Amygdalus scoparia) kernels. J. Food Meas. Charact. 12 (1), 362–369.

Hashemnia, M., Nikousefat, Z., Yazdani-Rostam, M., 2015. Antidiabetic effect of Pistacia atlantica and Amygdalus scoparia in streptozotocin-induced diabetic mice. Comp. Clin. Pathol. 24 (6), 1301–1306.

Hashtjin, A.M., Abbasi, S., 2015. Nano-emulsification of orange peel essential oil using sonication and native gums. Food Hydrocoll. 44, 40–48.

Heydari, R., Hosseini, M., 2017. Determination of the fatty acid composition of Amygdalus scoparia kernels from Iran using gas chromatography-mass spectrometry. Chem. Nat. Compd. 53 (3), 538–539.

Hosseini, E., Mozafari, H.R., Hojjatoleslamy, M., Rousta, E., 2017. Influence of temperature, pH and salts on rheological properties of bitter almond gum. Food Sci. Technol. 37 (3), 437–443.

Jafari, S.M., Beheshti, P., Assadpoor, E., 2012. Rheological behavior and stability of d-limonene emulsions made by a novel hydrocolloid (Angum gum) compared with Arabic gum. J. Food Eng. 109 (1), 1–8.

Jafari, S.M., Beheshti, P., Assadpour, E., 2013. Emulsification properties of a novel hydrocolloid (Angum gum) for d-limonene droplets compared with Arabic gum. Int. J. Biol. Macromol. 61, 182–188.

Jaison, D., Chandrasekaran, G., Mothilal, M., 2020. pH-sensitive natural almond gum hydrocolloid based magnetic nanocomposites for theragnostic applications. Int. J. Biol. Macromol. 154, 256–266.

Jalalizand, F., Goli, M., 2021. Optimization of microencapsulation of selenium with gum Arabian/Persian mixtures by solvent evaporation method using response surface methodology (RSM): soybean oil fortification and oxidation indices. J. Food Meas. Char. 15, 495–507.

Kadiyam, R., Muzib, Y.I., 2015. Colon specific drug delivery of tramadol HCl for chronotherapeutics of arthritis. Int. J. Pharm. Investig. 5 (1), 43–49.

Kester, D.E., Gradziel, T.M., Grasselly, C., 1991. Almonds (Prunus). Acta Hortic. 290, 701–760.

Khanniri, E., Sohrabvandi, S., Masomeh Arab, S., Shadnoush, M., Mortazavian, A.M., 2017. Effects of stabilizer mixture on physical stability of non-fat Doogh, an Iranian traditional drink. Koomesh 19 (1), 144–153.

Khanzadi, M., Jafari, S.M., Mirzaei, H., Chegini, F.K., Maghsoudlou, Y., Dehnad, D., 2015. Physical and mechanical properties in biodegradable films of whey protein concentrate-pullulan by application of beeswax. Carbohydr. Polym. 118, 24–29.

Khodaei, D., Oltrogge, K., Hamidi-Esfahani, Z., 2019. Preparation and characterization of blended edible films manufactured using gelatin, tragacanth gum and Persian gum. LWT Food Sci. Technol. 117, 108617.

Mahfoudhi, N., Chouaibi, M., Donsì, F., Ferrari, G., Hamdi, S., 2012. Chemical composition and functional properties of gum exudates from the trunk of the almond tree (Prunus dulcis). Food Sci. Technol. Int. 18 (3), 241–250.

Mahfoudhi, N., Sessa, M., Chouaibi, M., Ferrari, G., Donsì, F., Hamdi, S., 2014. Assessment of emulsifying ability of almond gum in comparison with gum arabic using response surface methodology. Food Hydrocoll. 37, 49–59.

Maleki, T., Akhani, H., 2018. Ethnobotanical and ethnomedicinal studies in Baluchi tribes: a case study in Mt. Taftan, southeastern Iran. J. Ethnopharmacol. 217, 163–177.

McClements, D.J., 2005. Food Emulsions: Principles, Practice, and Techniques. CRC Press.

McClements, D.J., 2011. Edible nanoemulsions: fabrication, properties, and functional performance. Soft Matter 7 (6), 2297–2316.

Mehrnia, M.A., Jafari, S.M., Makhmal-Zadeh, B.S., Maghsoudlou, Y., 2017. Rheological and release properties of double nano-emulsions containing crocin prepared with Angum gum, Arabic gum and whey protein. Food Hydrocoll. 66, 259–267.

Minekus, M., Alminger, M., Alvito, P., Ballance, S., Bohn, T., Bourlieu, C., Carrière, F., Boutrou, R., Corredig, M., Dupont, D., Dufour, C., Egger, L., Golding, M., Karakaya, S., Kirkhus, B., Le Feunteun, S., Lesmes, U., Maclerzanka, A., MacKie, A., Brodkorb, A., 2014. A standardised static in vitro digestion method suitable for food-an international consensus. Food Funct. 5 (6), 1113–1124.

Mirhosseini, H., Amid, B.T., 2012. A review study on chemical composition and molecular structure of newly plant gum exudates and seed gums. Food Res. Int. 46 (1), 387–398.

Moayyedi, M., Eskandari, M.H., Rad, A.H.E., Ziaee, E., Khodaparast, M.H.H., Golmakani, M.T., 2018. Effect of drying methods (electrospraying, freeze drying and spray drying) on survival and viability of microencapsulated lactobacillus rhamnosus ATCC 7469. J. Funct. Foods 40, 391–399.

Mohammadi, S., Abbasi, S., Scanlon, M.G., 2016. Development of emulsifying property in Persian gum using octenyl succinic anhydride (OSA). Int. J. Biol. Macromol. 89, 396–405.

Molaei, H., Jahanbin, K., 2018. Structural features of a new water-soluble polysaccharide from the gum exudates of Amygdalus scoparia Spach (Zedo gum). Carbohydr. Polym. 182, 98–105.

Mousazadeh, M., Mousavi, M., Askari, G., Kiani, H., Adt, I., Gharsallaoui, A., 2018. Thermodynamic and physiochemical insights into chickpea protein-Persian gum interactions and environmental effects. Int. J. Biol. Macromol. 119, 1052–1058.

Mozaffarian, V., 1996. A Dictionary of Iranian Plant Names. Latin-English–Persian, Farhang Mo'aser.

Pak, E.S., Ghaghelestani, S.N., Najafi, M.A., 2020. Preparation and characterization of a new edible film based on Persian gum with glycerol plasticizer. J. Food Sci. Technol. 57 (9), 3284–3294.

Patel, M.M., Amin, A.F., 2011. Process, optimization and characterization of mebeverine hydrochloride loaded guar gum microspheres for irritable bowel syndrome. Carbohydr. Polym. 86, 536–545.

Pearnchob, N., Siepmann, J., Bodmeier, R., 2003. Pharmaceutical applications of shellac: moisture-protective and taste-masking coatings and extended-release matrix tablets. Drug Dev. Ind. Pharm. 29 (8), 925–938.

Peng, T., Shi, Y., Zhu, C., Feng, D., Ma, X., Yang, P., Bai, X., Pan, X., Wu, C., Tan, W., Wu, C., 2019. Huperzine A loaded multiparticulate disintegrating tablet: drug release mechanism of ethyl cellulose microparticles and pharmacokinetic study. Powder Technol. 355, 649–656.

Pirouzifard, M., Yorghanlu, R.A., Pirsa, S., 2019. Production of active film based on potato starch containing Zedo gum and essential oil of Salvia officinalis and study of physical, mechanical, and antioxidant properties. J. Thermoplast. Compos. Mater. 33 (7), 915–937.

Raeisi, S., Ojagh, S.M., Quek, S.Y., Pourashouri, P., Salaün, F., 2019. Nano-encapsulation of fish oil and garlic essential oil by a novel composition of wall material: Persian gum-chitosan. LWT Food Sci. Technol. 116, 10849.

Rahim, H., Khan, M.A., Badshah, A., Chishti, K.A., Khan, S., Junaid, M., 2014. Evaluation of prunus domestica gum as a novel tablet binder. Braz. J. Pharm. Sci. 50 (1), 195–202.

Rahimi, S., Abbasi, S., 2014. Characterization of some physicochemical and gelling properties of Persian gum. Innov. Food Technol. 1, 13–27.

Ramesh, K.V., Mohanalayam, S., Sarheed, O., Islam, Q., Krishna, G.G., 2017. Design and evaluation of single and multi-unit sustained release dosage forms of captopril. Asian J. Pharm. 11 (2), 118–128.

Ramesh, K.V., Achamma, M., Yadav, H.K., Elmarsafawy, T.S., Islam, Q., 2019. Inclusion complexation in sulfobutyl ether beta cyclodextrin and dispersion in gelucire for sustained release of nifedipine employing almond gum. J. Drug Deliv. Therap. 9 (6), 70–78.

Ramesh, K.V., Yadav, H.K., Usman, S., Elmarsafawy, T.S., 2020. Studies on the influence of formulation and processing factors on the drug release from multiparticulate systems. Asian J. Pharm. 14 (1), 17–25.

Raoufi, N., Fang, Y., Kadkhodaee, R., Phillips, G.O., Najafi, M.N., 2016. Changes in turbidity, zeta potential and precipitation yield induced by Persian gum-whey protein isolate interactions during acidification. J. Food Process. Preserv. 41 (3), e12975.

Raoufi, N., Kadkhodaee, R., Fang, Y., Phillips, G.O., 2019. Ultrasonic degradation of Persian gum and gum tragacanth: effect on chain conformation and molecular properties. Ultrason. Sonochem. 52, 311–317.

Raoufi, N., Kadkhodaee, R., Fang, Y., Phillips, G.O., 2020. pH-induced structural transitions in whey protein isolate and ultrasonically solubilized Persian gum mixture. Ultrason. Sonochem. 68, 105190.

Razmjoo, F., Sadeghi, E., Rouhi, M., Mohammadi, R., Noroozi, R., Safajoo, S., 2021. Polyvinyl alcohol—zedo gum edible film: physical, mechanical and thermal properties. J. Appl. Polym. Sci. 138 (8), 49875.

Rezaei, A., Nasirpour, A., 2019. Evaluation of release kinetics and mechanisms of curcumin and curcumin-β-cyclodextrin inclusion complex incorporated in electrospun almond gum/PVA nanofibers in simulated saliva and simulated gastrointestinal conditions. BioNanoScience 9 (2), 438–445.

Rezaei, A., Nasirpour, A., Tavanai, H., 2016a. Fractionation and some physicochemical properties of almond gum (Amygdalus communis L.) exudates. Food Hydrocoll. 60, 461–469.

Rezaei, A., Tavanai, H., Nasirpour, A., 2016b. Fabrication of electrospun almond gum/PVA nanofibers as a thermostable delivery system for vanillin. Int. J. Biol. Macromol. 91, 536–543.

Roointan, A., Kamali-Kakhki, R., Fathalipour, M., Hashemi, Z., Zarshenas, M.M., Soleimani, M., Mirjani, R., 2020. Antibacterial activity of prunus scoparia root methanol extract against most common burn wound pathogens. Iran. J. Med. Sci. 45 (6), 444–450.

Sabeti, H., 1975. Forests. Trees and Shrubs of Iran, Ministry of Agriculture and Natural Resources.

Samari-Khalaj, M., Abbasi, S., 2017. Solubilisation of Persian gum: chemical modification using acrylamide. Int. J. Biol. Macromol. 101, 187–195.

Samrot, A.V., Suvedhaa, B., Sahithya, C.S., Madankumar, A., 2018. Purification and utilization of gum from Terminalia catappa L. for synthesis of curcumin loaded nanoparticle and its in vitro bioactivity studies. J. Clust. Sci. 29 (6), 989–1002.

Sarojini, S., Kunam, D.S., Manavalan, R., Jayanthi, B., 2010. Effect of natural almond gum as a binder in the formulation of diclofenac sodium tablets. Int. J. Pharm. Sci. Res. 1 (3), 55–60.

Sedaghat Doost, A., Muhammad, D.R.A., Stevens, C.V., Dewettinck, K., Van der Meeren, P., 2018. Fabrication and characterization of quercetin loaded almond gum-shellac nanoparticles prepared by antisolvent precipitation. Food Hydrocoll. 83, 190–201.

Sedaghat Doost, A., Kassozi, V., Grootaert, C., Claeys, M., Dewettinck, K., Van Camp, J., Van der Meeren, P., 2019. Self-assembly, functionality, and in-vitro properties of quercetin loaded nanoparticles based on shellac-almond gum biological macromolecules. Int. J. Biol. Macromol. 129, 1024–1033.

Sepeidnameh, M., Hosseini, S.M.H., Niakosari, M., Mesbahi, G.R., Yousefi, G.H., Golmakani, M.T., Nejadmansouri, M., 2018. Physicochemical properties of fish oil in water multilayer emulsions prepared by a mixture of whey protein isolate and water-soluble fraction of Farsi gum. Int. J. Biol. Macromol. 118, 1639–1647.

Seyfi, R., Kasaai, M.R., Chaichi, M.J., 2019. Isolation and structural characterization of a polysaccharide derived from a local gum: Zedo (Amygdalus scoparia Spach). Food Hydrocoll. 87, 915–924.

Shenoy, S.L., Bates, W.D., Frisch, H.L., Wnek, G.E., 2005. Role of chain entanglements on fiber formation during electrospinning of polymer solutions: good solvent, non-specific polymer-polymer interaction limit. Polymer 46 (10), 3372–3384.

Taban, A., Saharkhiz, M.J., Khorram, M., 2020. Formulation and assessment of nano-encapsulated bioherbicides based on biopolymers and essential oil. Ind. Crop Prod. 149, 112348.

Teimouri, S., Abbasi, S., Sheikh, N., 2016. Effects of gamma irradiation on some physicochemical and rheological properties of Persian gum and gum tragacanth. Food Hydrocoll. 59, 9–16.

Teimouri, S., Abbasi, S., Scanlon, M.G., 2017. Stabilisation mechanism of various inulins and hydrocolloids: milk-sour cherry juice mixture. Int. J. Dairy Technol. 71 (1), 208–215.

Turgeon, S.L., Laneuville, S.I., 2009. Protein + polysaccharide coacervates and complexes. From scientific background to their application as functional ingredients in food products. In: Kasapis, S., Norton, I.T., Ubbink, J.B. (Eds.), Modern Biopolymer Science: Bridging the Divide Between Fundamental Treatise and Industrial Application. Academic Press, pp. 327–363.

Zargari, A., 1997. Medicinal Plants, sixth ed. Vol. 3 University of Tehran Press.

Zibaeenezhad, M.J., Shahamat, M., Mosavat, S.H., Attar, A., Bahramali, E., 2017. Effect of Amygdalus scoparia kernel oil consumption on lipid profile of the patients with dyslipidemia: a randomized, open label controlled clinical trial. Oncotarget 8 (45), 79636–79641.

CHAPTER

18

Guar gum-based hydrogel and hydrogel nanocomposites for biomedical applications

Chinmoy Baruah and Jayanta K. Sarmah

Department of Chemistry, School of Basic Sciences, The Assam Kaziranga University, Jorhat, Assam, India

18.1 Introduction

Science and technology have gone a long way hand in hand to develop tailor-made materials according to the needs of humankind. Materials intended for human use need critical observation during their various stages of processing and manufacturing. Material building and processing are crucial steps for an analytical chemist before transportation as they include the applicability of the material in the practical field. Major attention has been paid in the development of polymer-based hydrogel composites comprising of injectable and self-healing properties for various biomedical applications (Taylor and In Het Panhuis, 2016; Xu et al., 2016). Carbohydrate-based polymers are the preferred choice in this regard owing to their biodegradability, biocompatibility, and excellent rheological properties (Sharma et al., 2018; Ahmad et al., 2019). Hydrogel is classified as a 3D network polymer gel that is capable of retaining large amounts of water which thereby sustains it for a long time, and is hydrophilic in nature (Warren et al., 2017; Ahmed et al., 2013) The fundamental definition for new material to behave as a hydrogel is that water present inside the material must be 10% by weight, and these hydrogels possess a highly flexible nature and elastic properties (Dannert and Dias, 2019). The presence of $–NH_2$, –COOH, –OH, $–CONH_2$, and $–SO_3H$ in these groups makes the hydrogel more hydrophilic. The hydrogel undergoes volume change transition or sol-gel transition. These kinds of conformational changes are reversible and cause hydrogels to revert to their original state on expiry of such actions. The critical observation comes into play when we have to take a closer look at how the hydrogels respond to

Micro- and Nanoengineered Gum-Based Biomaterials for Drug Delivery and Biomedical Applications
https://doi.org/10.1016/B978-0-323-90986-0.00002-9

external stimuli, and this observation depends practically on monomer, charge density, and degree of cross-linking (Ye et al., 2020; Koetting et al., 2015; Patel and Amiji, 1996). Hydrogels are categorized as hydrogels prepared from biopolymers and polyelectrolytes (Takashi et al., 2007; Yang et al., 2002; Maolin et al., 2000; Hacker and Mikos, 2011). Hydrogels are classified based on their physical, chemical, and biochemical responses. Physical gels are capable of undergoing liquid to gel transition in the presence of the external stimuli (Echeverria et al., 2015). It was reported that chemical gels use covalent bonding, which helps them in providing mechanical strength and resistance sustaining ability in comparison to weaker materials. In the case of the bio-based hydrogel systems, enzymes are found to be responsible for the gelation process. A range of natural and synthetic polymers are being used for the synthesis of hydrogels. Natural polymers, e.g., chitosan, alginates, guar gum, etc., have been extensively used in the synthesis of hydrogels for biomedical applications (Sharma and Tiwari, 2020). Nontoxicity, biocompatibility, water retention capacity, easy availability, and low cost are some of the important criteria in the selection of a polymer or polymers for a successful hydrogel synthesis. Guar-gum (Gg) is a natural polymer and there has been a recent trend in the use of guar gum and guar gum-based hydrogels and hydrogel composites in biomedical applications (Maiti et al., 2021; Shafiq et al., 2020).

18.2 Chemistry of hydrogels

18.2.1 Covalent linkage hydrogel

Hydrogels based on dynamic covalent bonding have been able to attract attention among researchers and led to the rapid development in the field of polymer science and life science disciplines (Ye et al., 2020). The dynamic covalent bonded hydrogels have excellent self-healing ability and responsiveness to external stimuli. As such, these materials are also regarded as smart materials with great promise to be applicable in biomedical areas (Fig. 18.1). However, many challenges remain to be addressed in the future.

18.2.2 Noncovalent linkage hydrogel

In recent times, noncovalently linkaged hydrogels have been gaining rapid attention. Noncovalent interactions in the hydrogels can be brought in by using nontoxic cross-linking agents (Parhi, 2017). The physical cross-linking agents in many cases may perform dual role as the cross-linker and also as a reductant in metal incorporated hydrogel composites. The proper choice of cross-linker may also bring in stimuli-responsive and self-healing capacity in the hydrogels in addition to the noncovalent linkages in the resultant hydrogels (Cui et al., 2020). Hydrogels with noncovalent linkages are preferred over covalently linkaged hydrogels due to better control on the hydrogel's property. The reversible sol-gel property is attributable to the noncovalent linkage in the hydrogels (Huang et al., 2021). Unlike the covalently linkaged hydrogels, the strength of the noncovalently linkaged hydrogels can be enhanced by changing the pH of the media. Incorporation of dynamic and reversible cross-linking not only accounts for the high strength in the hydrogel but also brings in extraordinary properties suitable for biomedical applications (Fig. 18.2).

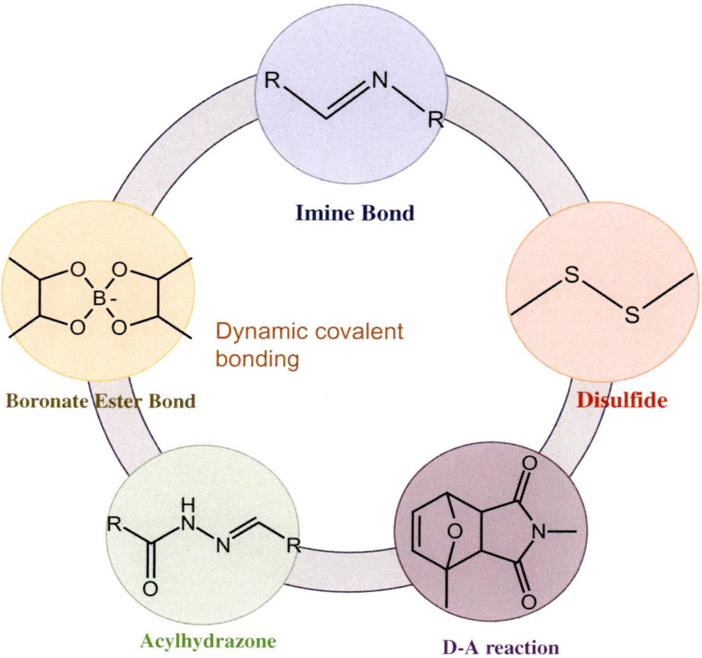

FIG. 18.1 The various dynamic covalent linkages in hydrogels. *From Ye, J., Fu, S., Zhou, S., Li, M., Li, K., Sun, W., Zhai, Y., 2020. Advances in hydrogels based on dynamic covalent bonding and prospects for its biomedical application. Eur. Polym. J. 139 (5), 110024. https://doi.org/10.1016/j.eurpolymj.2020.110024.*

18.3 Synthetic routes of hydrogel

18.3.1 Solution casting

The principle which is defined to describe the solution casting method is based on Stokes Law (Das and Swain, 2018). Here, the polymer and prepolymer solution are mixed and a resultant solution is obtained. The polymer (matrix phase) gets quickly dissolved in the solution phase and the resultant solution is then casted onto a glass plate to obtain a thin film of the resultant polymer. Metal nanoparticles incorporated hydrogels are prepared in a similar method. The nanoparticles may be introduced in the hydrogel in situ or it may be synthesized in a separate solution. In the latter case, both the solutions, i.e., the polymer solution and the nanoparticle solution, are mixed using a magnetic stirrer followed by sonication to obtain a homogeneous solution. The resultant solution containing the nanoparticles-embedded polymer is then casted onto a glass plate to obtain the hydrogel film. Solution casting is a method which is also used to prepare membranes by casting. The basic component of the system is a polymer, and the solvent, but a variety of additives of relative composition can be needed (Li et al., 2020). The selection of the polymer plays a primary role that should always be kept in mind. In addition, the polymer must be soluble in the selected solvent of appropriate concentration. Uniform dispersion of graphene oxide on to a polymer matrix can be achieved using the solution casting method. Many advantages are associated with the solution casting

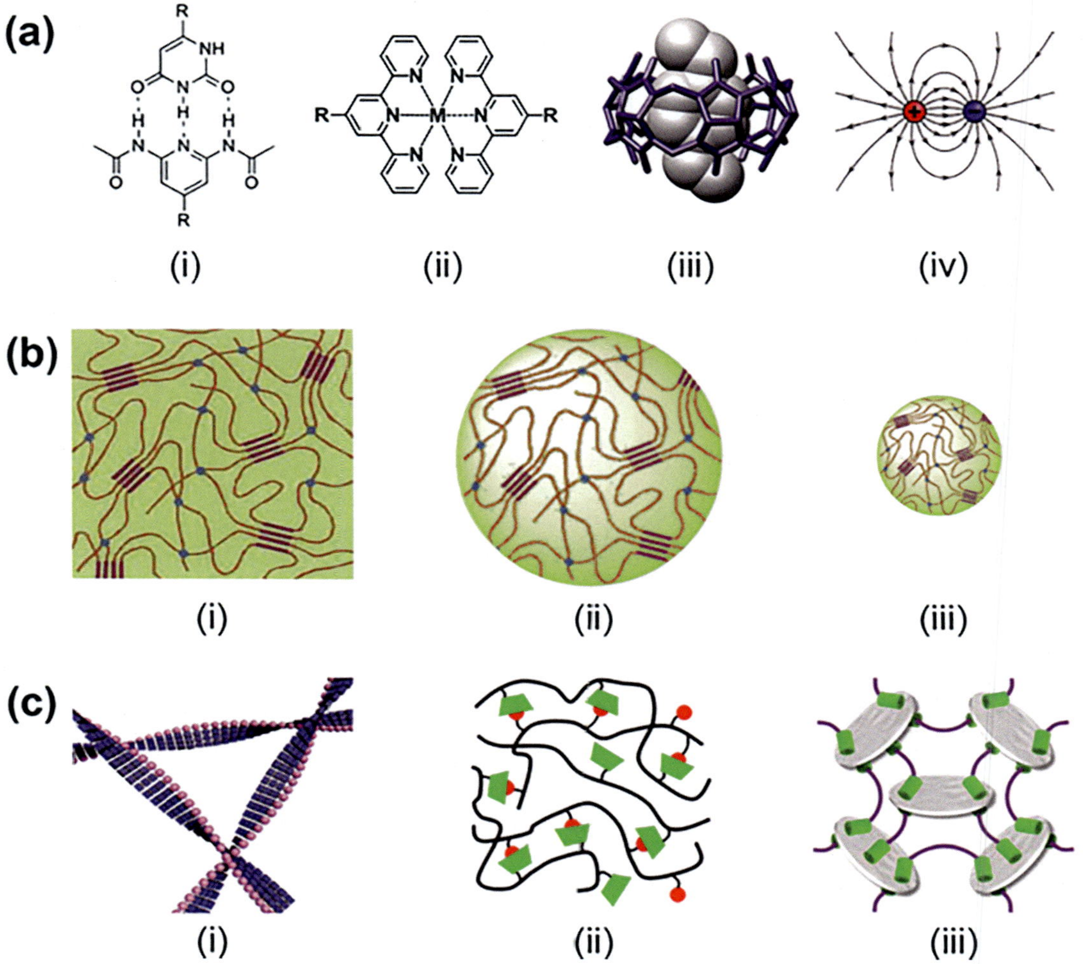

FIG. 18.2 Showing various noncovalent linkages in (A) hydrogels (i) hydrogen bond, (ii) M ◇ L linkage, (iii) host-guest recognition, and (iv) electrostatic interaction. (B) Hydrogels by their size: (i) macro hydrogel, (ii) micro hydrogel, and (iii) nano hydrogel. (C) Hydrogels by their building blocks: (i) molecular hydrogel, (ii) supramolecular polymer-based hydrogel, and (iii) supramolecular hybrid hydrogel. *From Dong, R., Pang, Y., Su, Y., Zhu, X., 2015. Supramolecular hydrogels: synthesis, properties and their biomedical applications. Biomater. Sci. 3, 937–954. https://doi.org/10.1039/c4bm00448e.*

methods, viz., it can be processed at room temperature, reinforcing agents may be introduced with ease, and control over the film thickness is enabled (Gurunathan et al., 2015).

18.3.2 Free radical polymerization

Free radical polymerization is employed for the development of polymer-based nanocomposite hydrogels and comes under the category of chain-growth polymerization.

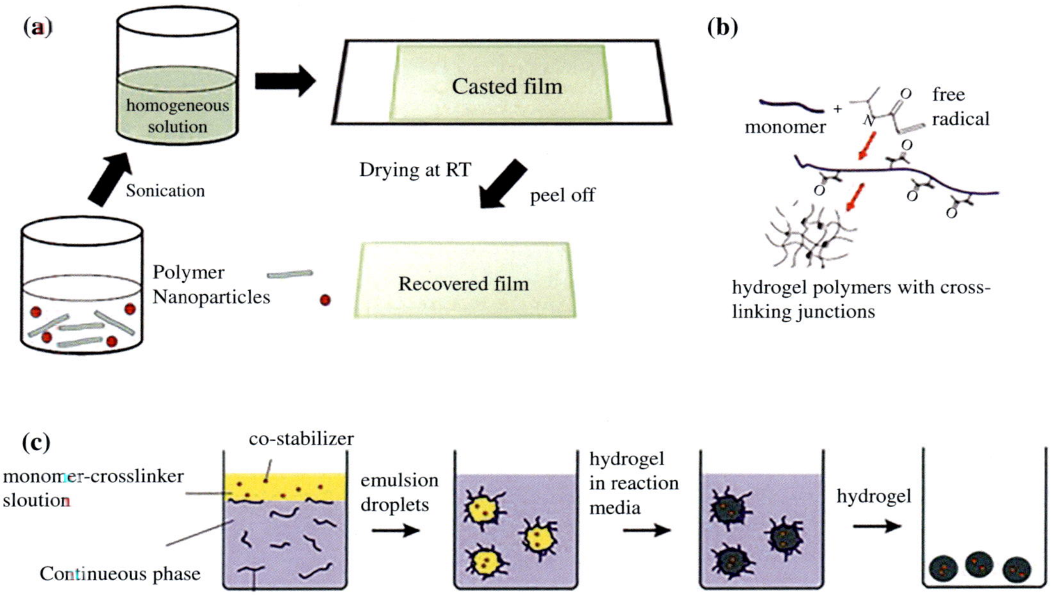

FIG. 18.3 (A) Solution casting, (B) free radical polymerization, and (C) emulsion polymerization.

This method involves free radicals generated by initiators, which then react with the monomer and lead to the growth of the polymeric chain (Fig. 18.3B). It is one of the useful methods for the synthesis of hydrogels and hydrogel composites owing to its ease of preparation and control over reaction condition (Das et al., 2018a, b; Shantha and Harding, 2002; Lopes et al., 2019). The first step of this process is called initiation, which leads to the formation of the chain through the generation of various active centers. The initiation of the free radical polymerization is generally easy on unsaturated double bonded compounds and in compounds with carbon oxygen bonds, viz., aldehydes and ketones. During the process of polymerization, a polymer spends most of its time propagating. The final step of the free radical polymerization process is the termination of the growing chain and realization of the final product (Lopes et al., 2019). Control of the concentration of the initiator different chain length of the polymer can be obtained. For example, long chain polymers can be obtained at low initiator concentrations (Su, 2013).

18.3.3 Emulsion polymerization

In this method the monomer, water as the dispersing media or continuous media, an initiator, and an emulsifier are made to react to effect polymerization. It is one of the important polymerization methods adopted at both laboratory and industrial levels. Emulsion polymerization is considered as an important method to form poly-vinyl acetate, poly-vinyl chloride, poly-vinyl alcohol, etc. from their corresponding monomers. In the process of emulsification, stable oil droplets (o/w emulsion) are being formed in the presence of an emulsifier. Based on the type of emulsion, both hydrophilic and hydrophobic drug molecules may be incorporated in the hydrogel during the course of reaction. The inverse emulsion polymerization method is also used for hydrogel synthesis (Mohsin and Attia, 2015).

The stability of the emulsion has been a concern in this method of preparation of hydrogel (Ahmed, 2015). Angellier et al. (2004) have successfully explained that the stability of the emulsion droplets could be improved using blends of emulsifiers. Various researchers also took the help of response surface methodology (RSM) for optimizing the synthesis conditions of hydrogels by emulsion polymerization. Statistical tools like box–behnken design (BBD) is being used in this process. The synthesis of PVA-AA-Tm hydrogel was optimized by Zheng et al. (2011). The inverse emulsion polymerization technique is used for the synthesis of interpenetrating polymer network (IPN) hydrogels and is receiving much attention from researchers in hydrogel development for various applications. The current trend is on the manufacture of biocompatible-based hydrogels because of their high natural tissue content as compared to other biomaterials with high water content, and also they have low interfacial tension with water and biological fluids (Peppas et al., 2000).

Emulsion polymerization is accompanied by anionic polymerization further initiated by covalent bases such as –OH groups. It is processed in an aqueous medium and terminates with the addition of cations responsible for neutralizing the chains (Fig. 18.3). The supplied monomer in an aqueous solution further acidified by hydrochloric acid at pH from 1 to 3.5, containing both surfactant and stabilizer.

18.4 Polymers for hydrogel systems

18.4.1 Semisynthetic polymer

Polymers that are derived from naturally occurring polymers by chemical modifications are called semisynthetic polymers. Rayon, cellulose nitrate, and cellulose acetate are few examples to be noted. Cellulose is a naturally occurring polymer which upon acetylation with acetic anhydride in the presence of sulfuric acid forms a cellulose diacetate polymer. Semisynthetic polymers are used for making various consumer products in the cosmetics industry. The function of such polymers may include stabilizers, thickeners, etc. The polymers are also used widely in biomedical applications. However, due to the modification, in most cases, their antibacterial activities are found to be compromised as compared to their natural counterparts. The basic idea for modification of the natural polymer in drug delivery applications is to improve the behavior of the polymeric carrier in different pH regions (Liechty et al., 2010).

18.4.2 Synthetic polymer

There are a plenty issues with the biocompatibility of the synthetic polymers in biomedical applications. After extensive research on their biodegradability and with modifications, such polymers are widely used by researchers and in industries for various applications. PLA, PLGA, PVA, PEG, etc., are the most commonly used synthetic polymers. The synthetic polymers are preferred over the natural counterparts in regards to their large-scale production and stability during long-term storage (Sung and Kim, 2020).

18.4.3 Natural polymer

Natural polymers offer many advantages over synthetic and semisynthetic polymers in material development. These are highly biodegradable, biocompatible, nontoxic, easily available, with inherent antibacterial properties in most cases which makes them ideal for biomedical applications. Polysaccharide-based polymers in particular are gaining much attention in hydrogel synthesis and applications (Álvarez et al., 2019). Chitosan, alginates, gelatine, carrageenan, guar gum, starch, etc. are some examples. Various research groups across the globe have been working intensively on the development of polymer-based nanosystems using polymers of natural and synthetic origin as well. In addition, blends of two or more polymers, either natural or a combination of both natural and synthetic, are being used and their activities are being studied and compared for various biomedical applications (Vasile et al., 2020).

18.5 Chemistry of guar gum

Guar gum is a carbohydrate-based polymer, which is obtained from *Cyamopsis tetragonoloba* seeds and it belongs to the Leguminosae family. Guar gum is chemically a high molecular weight natural polysaccharide, which consist of galactan and mannan units and linked through glycosidic linkages (Fig. 18.4). Structurally, guar gum is composed of a linear chain of β-D-mannonopyranosyl units linked (1 → 4) with α-D-galactopyranosyl units as a side chain (Das et al., 2018a, b; Dhas et al., 2014).

Guar gum and their derivatives have some excellent properties like the film-forming ability, biodegradability, pH stability, being environmental friendly, wound healing capability, thickening, emulsifying, high viscosity, and binding and gelling properties (Cui et al., 1994; Dhas et al., 2014; Dodi et al., 2016). Because of its natural, nontoxic, cost-effective, and mucoadhesive nature, it has been widely used in the intestinal drug delivery via the oral route (Gerber et al., 2013), antigen delivery through oral administration, and as a potential oral immunization technique against tuberculosis (Kaur et al., 2015). Moreover, excellent swelling behavior of guar gum at lower pH protects the antigen in the harsh gastric environment. Guar gum has a great affinity toward mannan receptors due to the presence of mannose (sugar)

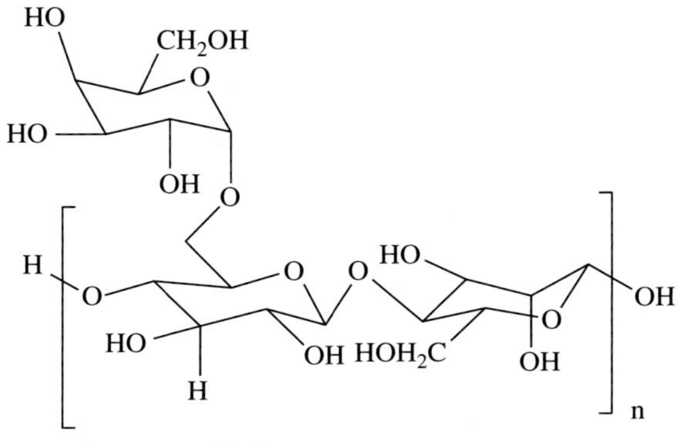

FIG. 18.4 Chemical structure of guar gum (Gg).

unit. Mannan receptors has a carbohydrate recognition domain, which shows a specificity toward mannose unit and acts as endocytic receptors, which can bind mannosylated structures and be able to mediate the cellular immune response in an in vitro manner (Fig. 18.4) (Gerber et al., 2013). The surface exposed hydroxyl groups are made to react with other substances to get new derivatives of Gg with desirable characteristics and functionalities (Boddohi and Kipper, 2010; George et al., 2019; Giri et al., 2015; Singha et al., 2017; Rao et al., 2008).

Guar gum and its derivatives have some excellent properties like the film-forming ability, biodegradability, pH stability, environmental friendly, wound healing capability, thickening, emulsifying, high viscosity, binding, and gelling properties (Cui et al., 1994; Dhas et al., 2014; Dodi et al., 2016). Because of its natural, nontoxic, cost effective and mucoadhesive nature, it has been widely used in the intestinal drug delivery via oral route (Gerber et al., 2013), antigen delivery through oral administration and as potential oral immunization technique against tuberculosis (Kaur et al., 2015). Moreover, excellent swelling behavior of guar gum at lower pH protects the antigen in harsh gastric environment. Guar gum has a great affinity toward mannan receptors due to the presence of mannose (sugar) unit (Adikwu, 2009; Collnot et al., 2012). Mannan receptors has a carbohydrate recognition domain, which shows a specificity toward mannose unit and acts as endocytic receptors, which can bind mannosylated structures and be able to mediate the cellular immune response in a in vitro manner (Gerber et al., 2013). Due to the presence of surface exposed hydroxyl groups on guar gum, one can modify guar gum to new derivatives of desirable characteristics and functionalities (Boddohi and Kipper, 2010; George et al., 2019; Giri et al., 2015; Singha et al., 2017; Rao et al., 2008).

18.6 Applications of gg-based hydrogels

18.6.1 Gg-hydrogels in drug delivery system (DDS)

Guar gum and Gg-modified hydrogels have very diverse applications in different practical fields. This is possible only because of their rich heritage in the structural existence and their way of becoming compatible under different usages. High water content of Gg owing to extensive hydrophilicity due to the presence of a large number of hydroxyl groups on its backbone results in excellent flexibility in the hydrogel. The water retention behavior of Gg may be further improved according to intended applications by modifications of its surface or by blending with other polymers. Thus hydrogels, due to the presence of Gg, are ideal for the applications in targeted delivery of potent drug molecules especially to the colon and mammary gland in cancer treatments. The beauty of using Gg is that potent drugs could be delivered at the target sites at specific rates in some well-defined time periods (Gamal-Eldeen et al., 2006; Krotkiewski, 1984; Kono et al., 2014). Guar gum-based hydrogels thus are found to be superior in terms of delivery of anticancer and anti-HIV drugs to conventional formulations (Day et al., 2020; Sarmah et al., 2011). The performance of the DDS can be enhanced by bringing in porosity in the hydrogel or by reinforcing with suitable reinforcing agents. High porosity results in highly permeable hydrogels which are permeable toward different drugs that are easily transportable in proper conditions. The physical integrity of the hydrogel is an issue in long-term storage applications. Double or in some cases triple network hydrogel systems are useful in bringing stability in the long-term storage applications of the drug-loaded hydrogels. Guar gum-modified PVA hydrogel (Deka et al., 2020), silver nanoparticles-reinforced Gg hydrogels (Dai et al., 2017), and silver nanoparticles and curcumin-loaded Gg hydrogels (Talodthaisong et al., 2020) have been successfully synthesized and optimized for enhanced mechanical strength and for stability in long-term storage applications.

Stimuli-responsive Gg-hydrogels are also highly attractive in drug delivery applications due to their ability to respond to the change in pH of the media as well as sol-gel transition when subjected to alternate heating and cooling cycles (Klouda, 2015; Koetting et al., 2015). The pH- and thermo-responsive properties of the Gg hydrogels can be brought about taking a cross-linking approach using suitable cross-linker as well as without a cross-linking agent. Guar gum-borax cross-linked hydrogels are one of the best ways of producing hydrogels which can bring both pH- and thermo-responsive properties. However, the ratio of borax to Gg is very important in hydrogel formation and one should also be careful not to exceed the borax concentration by much. Such approaches can also induce a rapid self-healing ability in the hydrogels.

pH change acts as a major environmental parameter for drug delivery systems in different specific pathological body sites. The human body is prone to variation in pH along the gastrointestinal tract and in some certain tissues and subcellular compartments. Guar gum modified poly-acrylic acid (PAA), poly-methamethylacrylate (PMAA), and poly-L-glutamic acid hydrogels are synthesized and characterized for drug delivery applications. A schematic representation indicating applications of guar gum is represented in Fig. 18.5.

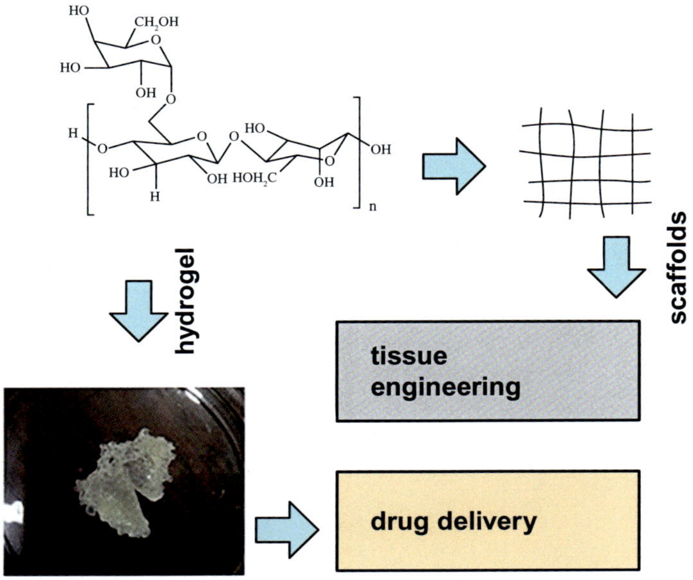

FIG. 18.5 Showing the various applications of guar gum hydrogel.

18.6.2 Gg-hydrogels in wound dressing applications

In situ gelling as well as formation of metal nanoparticles in presence of suitable reagents which can perform the dual function as the reducing agent for the metal precursor and also as the cross-linker is an exciting approach for the formation of stimuli-responsive Gg and Gg-modified hydrogels. Embedding silver nanoparticles in the Gg hydrogel in situ or ex situ can bring in antimicrobial property in the hydrogels (Fig. 18.6). The antimicrobial property alone cannot cause the hydrogel to be considered as a wound dressing material. It should have a high degree of swelling and also be able to withstand the daily strains occurring in a human's body parts. Guar gum has excellent film forming property and its mechanical properties can be tuned by reinforcing with metal nanoparticles. Thus incorporating silver nanoparticles to Gg hydrogel not only brings improved mechanical properties but also excellent antimicrobial properties necessary for wound healing. In addition, due to the presence of a large number of hydroxyl groups on its surface, Gg hydrogel is able to retain lots of water in it, which equally implies that the Gg hydrogel would be able to absorb a large quantity of wound extrudate. The incorporation of silver nanoparticles in the guar gum film can improve the elongation at break and as a result of it the Gg-Ag film would be highly stretchable. These outstanding properties of the Gg make it the chosen polymer in wound dressing applications.

18.6.3 Guar gum-based bio-sensors

Biosensors are composed of both physical and chemical ones. Biosensors act as devices capable of sensing and reporting of a biophysical property of a system or they may be classified as devices for procuring analytical data from biochemical ones. All biosensors have a detector type recognition biological tool responsible for analyzing biological information. These sensors cover a wide range of applications like home diagnostics, environmental monitoring, and point of care testing. Some of the known biological recognition parts are bio-elements consisting of different forms like enzymes, antibodies, living cells, and tissues. These biosensors can respond specifically to the analyte of importance. Guar gum is used in biosensors in

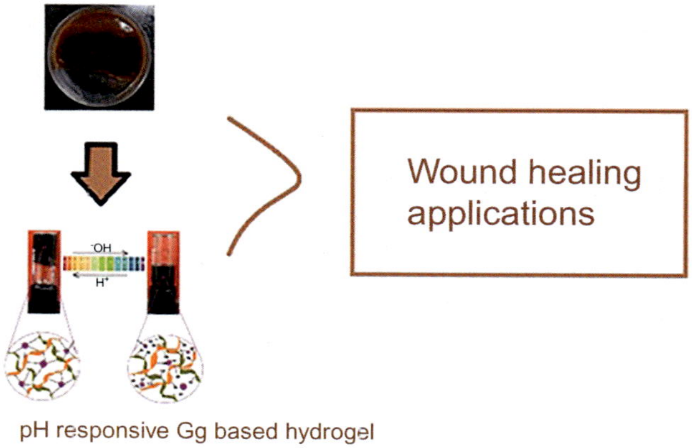

FIG. 18.6 Gg-based hydrogel with pH-responsive property for wound healing applications.

the form of hydrogel that is used to apply a coating over the sensing material. The coated sensing material in turn prevents any undesirable reactions with the biomolecules. Gg hydrogels also play an active role in immobilizing the active sensing materials for optimum output in bio-sensing applications.

18.7 Guar gum-based hydrogels

18.7.1 Guar gum-polyacrylic acid hydrogels

Guar gum (Gg) and its modified derivatives, such as carboxymethyl guar gum (CGg), have been largely used in biomedical applications. Modification of Gg brings enhancement in properties needed for drug delivery, wound dressing applications. Addition of a suitable cross-linking agent or materials like polyacrylic acid (PAA) makes the resultant GG hydrogel stimuli-sensitive, pH-responsive, and temperature-responsive. Such attributes make them smart materials. One such smart material based on cationic Gg and PAA was reported by Hunag and his coworkers (Huang et al., 2007b). The resultant smart hydrogel exhibited excellent swelling ability and controlled drug release at pH 7.4. The same group of researchers also developed a cationic guar gum-PAA hydrogel membrane with different compositions of Gg and PAA (Huang et al., 2007a). The resultant hydrogel exhibited better thermal and mechanical properties. With the increase in PAA content the tensile strength of the hydrogel membrane increased up to 41.1 MPa.

The synthesis of the hydrogel started with the preparation of the homogeneous solution of acrylic acid (AA) monomer and carboxy guar gum (CGg) samples in various weight ratios of CGg and AA. The polymerization reaction was initiated by adding DMPAP to a CGg-AA mixture and the resultant solution was irradiated under UV radiation in an inert atmosphere to get the CGg-PAA hydrogel.

A novel rapid self-healing, stretchable guar gum-PAA hydrogels as wearable strain sensors for the detection of human motion has been synthesized by Deng and his co-workers (Deng et al., 2021). The novel Gg-PAA stretchable hydrogel was synthesized by free radical polymerization using KPS as an initiator. Graphene oxide was also added as one of the major components in the hydrogel which provided the improved strength and stretchability. The presence of guar gum contributed to the high order stretchability performance of the hydrogel.

There are a plenty of H-bond generated between the poly PAA groups and the oxygen containing functional groups of GO. This structure provides lot of strengths to the hydrogel under stress. The self-healing ability in the GG-PAA hydrogel was endowed by the addition of Janus NS which contains polypyrrole and PDMAEMA in its structure that provides multiple bonding for the self-healing process. The schematic representation of the self-healing performance of the novel hydrogel is illustrated in Fig. 18.7.

Highly conducting and antibacterial IPN hydrogel comprising of Gg-PAA-PAN has been successfully reported by Sharma and his coworkers (Kaith et al., 2015). The work describes copolymerization of grafted Gg with AA by using APS as an initiator and hexamine as a cross-linker through a free radical polymerization technique in order to obtain Gg-cl-PAA hydrogel, which was then further transformed to Gg-cl-PAA-*IPN*-PAN, an interpenetrating

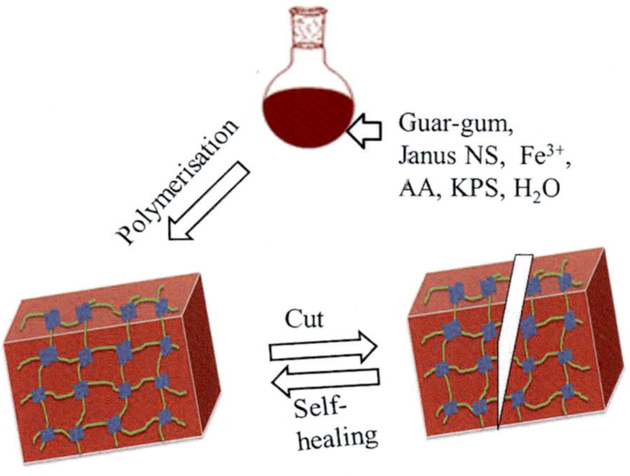

FIG. 18.7 Self-healing attributes of the novel guar-gum-PAA hydrogel.

hydrogel. Gg is highly susceptible to microbial degradation and exhibits an uncontrolled rate of hydration. However, the IPN hydrogel with PAA and PAN, exhibited excellent water retention as well as microbial resistance in both gram-positive and gram-negative bacteria.

The interpenetrating network of Gg-PAA-PAN was made by the cross-linking of PAN under acidic condition and neutral conditions. Preoptimized APS, hexamine, and a thermal initiator were added at the polymerization step which results in a change in solution color from brown to slightly green. The thermal property of the hydrogel was also significantly improved compared to the pristine Gg (Fig. 18.8).

18.7.2 Guar gum-based metal nanocomposite

In the last 2 decades, there has been a widespread focus on polymer/inorganic metal nanocomposites. Introduction of metal nanoparticles into the Gg matrix brings a lot of improvement in the mechanical properties, swelling and release behavior, and antimicrobial properties. These attributes make metal containing Gg nanocomposite a versatile material for diverse applications. Over the years, silver-containing pharma products have been developed and the demand for many more products is ever increasing. Simultaneously various research groups across the globe are working on development of novel AgNPs-based materials. The very important reason for considering Ag as a model metal in metal-based nanocomposites is due to its applicability in photovoltaics, biological, and chemical products. There has been an enhanced growth in the development of Gg-based silver nanocomposites in recent times. Guar gum provides a matrix medium which is able to provide a uniform distribution of the AgNPs under controlled operative conditions and human skills. The Gg matrix in a way immobilizes the AgNPs and thereby restricts their aggregation over time.

Single and double network nanocomposites

Metal nanocomposites of guar gum and guar gum-modified double network matrices have been successfully reported (Deka et al., 2020; Ren et al., 2018). Although there is plenty of

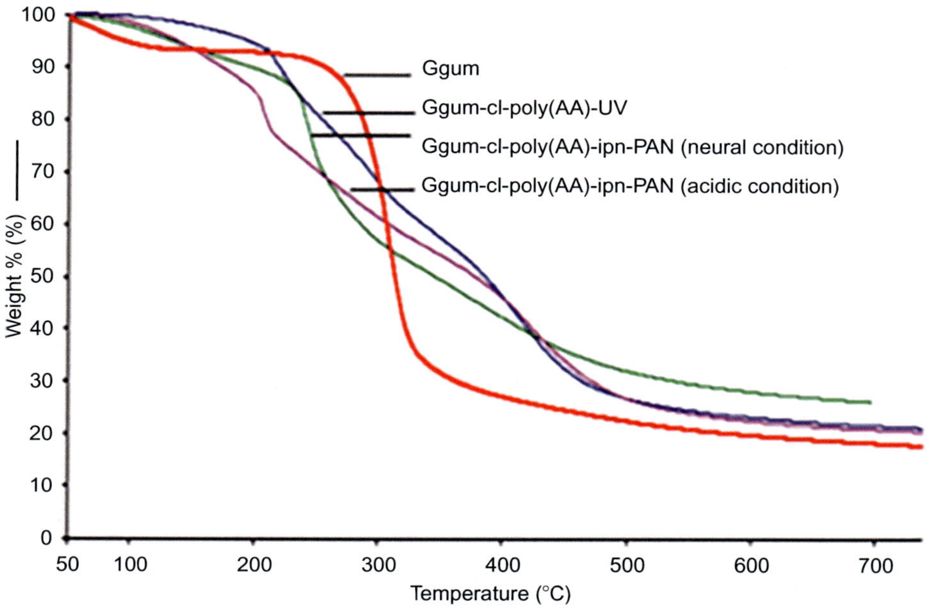

FIG. 18.8 Showing the thermogravimetric curves of the various hydrogels and guar gum. *From Kaith, B.S., Sharma, R., Kalia, S., Bhattid, M.S., 2015. Response surface methodology and optimized synthesis of guar gum-based hydrogels with enhanced swelling capacity. RSC Adv. 4, 40339–40344. https://doi.org/10.1039/c4ra05300a.*

information on the synthesis of AgNP-based hydrogels in the open literature, these hydrogels face degradation of the polymer and silver function during long-term storage. There is great demand for hydrogels that can provide both efficacy and stability of silver nanoparticles over time. Double network hydrogel comprising of two polymers, Gg and PVA, can provide improved storage and loss modulus, and mechanical and swelling properties. Recently there has been a report on preparation of a novel PVA-Gg double network hydrogel composite containing AgNPs (Deka et al., 2020). The Gg-PVA-Ag nanocomposite was synthesized in one pot within 3 min. Sodium borohydride was used as the reducing agent for Ag^+ to Ag^o and at the same time as a cross-linker in the hydrogel formation. The hydrogel exhibited sol-gel transformation at different pH. The pH-responsive behavior of the modified Gg nanocomposite was attributed to the borate-boric acid transformation of the hydrogel composite at high to low pH. Guar gum-AgNPs hydrogel was synthesized in the similar way by Le Dai and his coworkers (Dai et al., 2017). The hydrogel composite exhibited an excellent self-healing property owing to the dynamic borate-boric acid transition at different pH. Recently, Palem and his coworkers synthesized a graft polymer of Gg and polyacrylamidoglycolic acid (AA) and then they synthesized the nanocomposites of Gg-AA-AgNPs (Palem et al., 2019). The researchers used sodium borohydride for performing the dual function as a cross-linker and reducing agent for $AgNO_3$. The hydrogel nanocomposite could release 80% of the loaded anticancer drug 5-flurouracil (5-FU) over a period of 18 h. The sustained release behavior was attributed to the lower aggregation of the AgNPs in the grafted guar gum and also to the relaxation of the polymeric chains at high pH. Borax cross-linked guar gum hydrogel containing silver nanoparticles and curcumin were successfully synthesized and characterized (Talodthaisong et al., 2020). The resultant hydrogel nanocomposite exhibited excellent self-healing, injectable, and antibacterial properties. The nanocomposite was able to create a zone of inhibition in both gram-positive and gram-negative bacteria (Fig. 18.9).

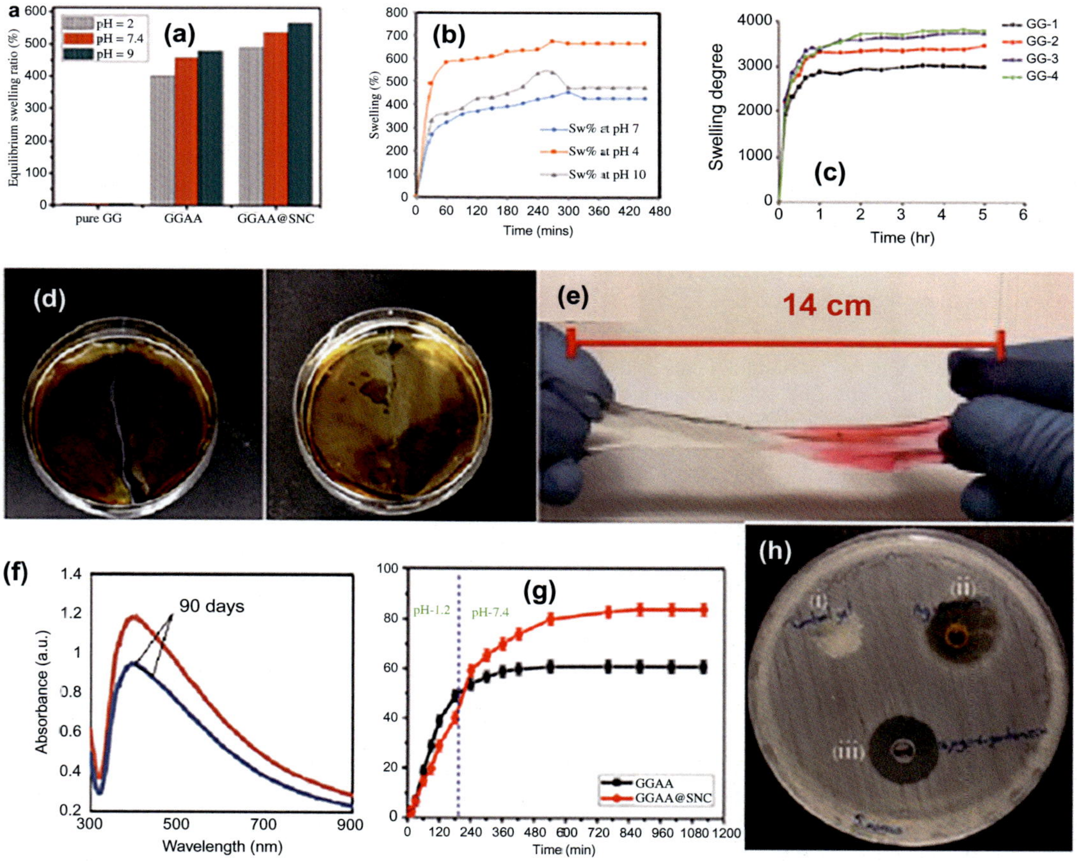

FIG. 18.9 Showing the (A–C) swelling, (D and E) self-healing and stretchability, (F) long-term storage stability, (G) drug release applications, and (H) antimicrobial activities against bacterial strains by guar gum hydrogels and guar gum hydrogel nanocomposite. *Reprinted from Dai, L., Zhang, L., Wang, B., Yang, B, Khan, I., Khan, A., Ni, Y., 2017. Multifunctional self-assembling hydrogel from guar gum. Chem. Eng. J. 330 (15), 1044–1051. https://doi.org/10.1016/j.cej.2017.08.041; C. Talodthaisong, W. Boonta, S. Thammawithan, R. Patramanon, N. Kamonsutthipaijit, J.A. Hutchison, S. Kulchat, 2020. Composite guar gum-silver nanoparticle hydrogels as self-healing, injectable and antibacterial biomaterials. Mater. Today Commun. 24, 100992. https://doi.org/10.1016/j.mtcomm.2020.100992; Deka, R., Sarmah, S., Patar, P., Gogoi, P., Sarmah, J.K., 2020. Highly stable silver nanoparticles containing guar gum modified dual network hydrogel for catalytic and biomedical applications. Carbohydr. Polym. 248, 116786.*

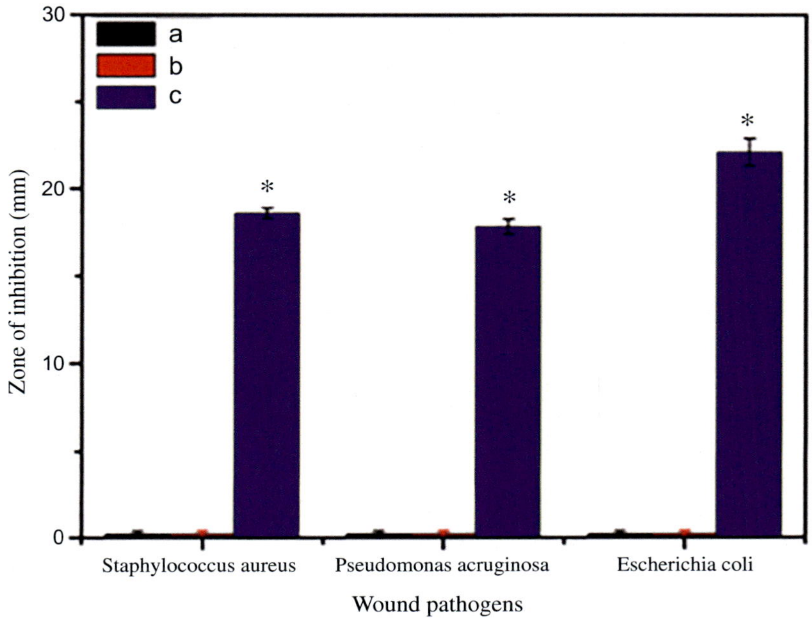

FIG. 18.10 Showing the antimicrobial performance of the (A) pristine Gg, (B) Gg grafted poly acrylamidoglycolic acid, and (C) Gg grafted poly acrylamidoglycolic acid-AgNPs on various bacterial strains. *From Palem, R.R., Rao, K.M., Kang, T.J., 2019. Self-healable and dual-functional guar gum-grafted-polyacrylamidoglycolic acid-based hydrogels with nano-silver for wound dressings. Carbohydr. Polym. 223 (1), 115074. https://doi.org/10.1016/j.carbpol.2019.115074.*

Dual network hydrogel consisting of guar gum and alginate containing silica nanoparticles was successfully synthesized and exhibited excellent self-healing abilities (Rao et al., 2019). The mechanical properties of the hydrogel compared to those of guar gum were significantly improved. Glutaraldehyde was utilized for cross-linking in the hydrogel. The hydrogel was able to recover its original shape after application of stress. G′ and G″ was also able to recover its initial value after completion of multiple cycles. Hydrogels of guar gum grafted poly acrylamidoglycolic acid containing silver nanoparticles exhibited excellent wound healing abilities (Palem et al., 2019). The hydrogel also exhibited responses to change in its structure to different pH and temperatures. The hydrogel was highly stretchable injectable through a syringe and could regain its original shape from two cut pieces of it (Fig. 18.10). The Gg grafted composite exhibited improved stress-strain behavior, high swelling ratio, and bacterial resistance.

18.7.3 Self-assembled Gg-hydrogel

Self-assembled Gg-based hydrogel was synthesized in less than 10 min by Zhang et al. (2017). Hydrogel was formed by introducing the carbonyl groups in Gg molecule via an oxidation reaction in presence of $NaIO_4$. This facilitates the carbonyl groups of Gg for reaction

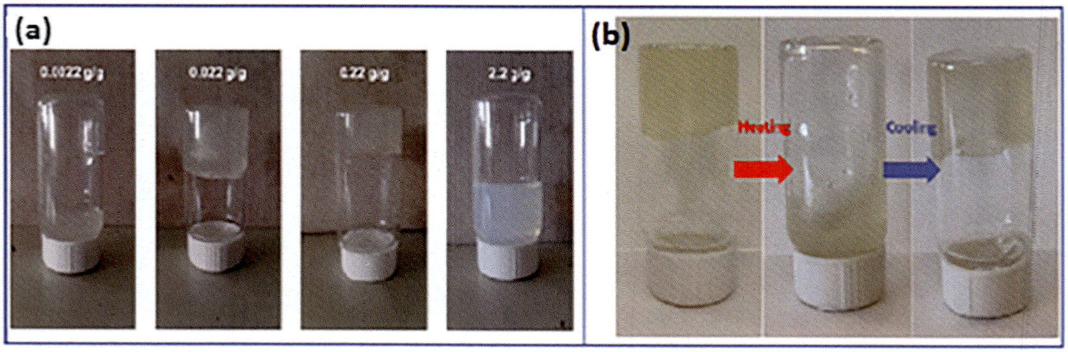

FIG. 18.11 Showing the (A) effect of various ratios of $NaIO_4$/Gg on the formation of self-assembled Gg hydrogel and (B) sol-gel behaviors of the Gg hydrogel during alternate heating and cooling cycles. *From Dai, L., Zhang, L., Wang, B., Yang, B, Khan, I., Khan, A., Ni, Y., 2017. Multifunctional self-assembling hydrogel from guar gum. Chem. Eng. J. 330 (15), 1044–1051. https://doi.org/10.1016/j.cej.2017.08.041.*

with vicinal groups forming cyclic acetals leading to gelation (Llanos and Sefton, 1991). However, gelation is dependent on the wt% of Gg. In general, Gg with 1.0 wt% and above is considered sufficient for formation of hydrogel. In the formation of self-assembled Gg-based hydrogel using $NaIO_4$, the concentration of $NaIO_4$ also influences the gelation process significantly. Thus, one can consider an optimized ratio of $NaIO_4$ to Gg for fast and efficient formation of self-assembled Gg hydrogel. Cross-linked Gg hydrogel was also reported by Fantozzi and coworkers in the presence of polyethene glycol diglycidyl ether (PEGDGE) as the cross-linking agent (Fantozzi et al., 2010). However, the reaction time to produce this hydrogel was 15 h, which is very high compared to that for the formation of Gg in presence of $NaIO_4$. The self-assembled hydrogels did exhibit self-healing and thermo-reversibility properties. Two cut pieces of the Gg hydrogel could regain their original structure into one within a standing time of 30 min and the integrity of the hydrogel structure could be maintained even on shaking the gel. The Gg hydrogel on subjection to alternative heating and cooling cycles could produce sol-gel transformation. These outstanding properties of the Gg-based hydrogel under study show promise in drug delivery and tissue engineering applications (Jacob et al., 2015; Appel et al., 2015). Guar gum in general possesses reversible sol-gel transitions. However, Gg in addition to poly(N-isopropyl acrylamide) could significantly enhance the thermo-responsive properties of the resultant Gg/poly(N-isopropyl acrylamide) hydrogels (Li et al., 2008). Moreover, the addition of Gg introduced extensive hydrophilicity in the hydrogel (Fig. 18.11). The swelling ratio in deionized water was approximately 3000% after 8 h. Such attributes of Gg make it a very attractive polymer for designing various nanocomposites for biomedical applications.

18.8 Conclusions

Fundamental knowledge about guar gum, important methods for preparation and characterization of guar gum hydrogels and hydrogel nanocomposites, and their applications in drug delivery, wound dressing, self-healing devices and other related biomedical applications have been set out in this chapter. The importance of guar gum as a polymer and its inherent properties in deciding the fate of the composites have also been explored. Improvements in the mechanical, swelling, and in long-term storage applications were observed in the guar gum-based hydrogels and composites. The release of the potent anticancer drug molecules was efficiently sustained from the guar gum systems. Guar gum hydrogels and hydrogel composites as self-healing, injectable, and multiple-stimuli-responsive materials have emerged and show promise for future smart materials in biomedical applications.

Conflict of interest

The authors declare no conflict of interest.

References

Ahmad, S., Ahmad, M., Manzoor, K., Purwar, R., Ikram, S., 2019. A review on latest innovations in natural gums based hydrogels, preparations & applications. Int. J. Biol. Macromol.

Ahmed, E.M., 2015. Hydrogel: preparation, characterization, and applications. J. Adv. Res. 6, 105–121.

Ahmed, M.E., Aggor, S.F., Awad, M.A., El-Aref, T.A., 2013. An innovative method for preparation of nanometal hydroxide superabsorbent hydrogel. Carbohydr. Polym. 91, 693–698.

Álvarez, L.P., Rubio, L.R., Lizundia, E., Vilas-Vilela, J.L., 2019. Polysaccharide-Based Superabsorbents: Synthesis, Properties, and Applications. Springer International Publishing, pp. 1393–1431.

Angellier, H., Choisnard, L., Molina-Boisseau, S., Ozil, P., Dufresne, A., 2004. Optimization of the preparation of aqueous suspensions of waxy maize starch nanocrystals using a response surface methodology. Biomacromolecules 5, 1545–1551.

Appel, E.A., Tibbitt, M.W., Webber, M.J., Mattix, B.A., Veiseh, O., Langer, R., 2015. Self-assembled hydrogels utilizing polymer–nanoparticle interactions. Nature 6.

Boddohi, S., Kipper, M.J., 2010. Engineering nanoassemblies of polysaccharides. Adv. Mater. 22, 2998–3016.

Cui, W., Mazza, G., Biliaderis, C.G., 1994. Chemical structure, molecular size distributions, and rheological properties of flaxseed gum. J. Agric. Food Chem. 42, 1891–1895.

Cui, W., Pi, M.-H., Li, Y.-S., Shi, L.-Y., Ran, R., 2020. Multimechanism physical cross-linking results in tough and self-healing hydrogels for various applications. ACS Appl. Polym. Mater. 2 (8), 3378–3389.

Dai, L., Long, Z., Chen, J., An, X., Cheng, D., Khan, A., Ni, Y., 2017. Robust guar gum/cellulose nanofibrils multilayer films with good barrier properties. Appl. Mater. Interfaces 330 (15), 1044–1051. https://doi.org/10.1016/j.cej.2017.08.041.

Dannert, C., Stokke, B.T., Dias, R.S., 2019. Nanoparticle-hydrogel composites: from molecular interactions to macroscopic behavior. Polymers 11 (2), 275.

Das, A., Abdullah, M.F., Kundu, S., Mukherjee, A., 2018a. Synthesis of guar gum propionate nanoparticles for antimicrobial applications. Mater. Today Proc. 5, 9683–9689.

Das, D., Pham, T.T.H., Noh, I., 2018b. Characterizations of hyaluronate-based terpolymeric hydrogel synthesized via free radical polymerization mechanism for biomedical applications. Colloids Surf. B Biointerfaces.

Das, R., Swain, S.K., 2018. Polymer-based nanocomposite for sensor devices. Energy Environ. Appl., 205–218.

Day, C.M., Hickey, S.M., Song, Y., Plush, S.E., Garg, S., 2020. Novel tamoxifen nanoformulations for improving breast cancer treatment: old wine in new bottles. Molecules 25 (5), 1182.

Deka, R., Sarmah, S., Patar, P., Gogoi, P., Sarmah, J.K., 2020. Highly stable silver nanoparticles containing guar gum modified dual network hydrogel for catalytic and biomedical applications. Carbohydr. Polym. 248, 116786.

Deng, Z., Lin, B., Wng, W., Bai, L., Chen, H., Yang, L., Yang, H., Wei, D., 2021. Stretchable, rapid self-healing guar gum-poly(acrylic acid) hydrogels as wearable strain sensors for human motion detection based on Janus graphene oxide. Int. J. Biol. Macromol. 191, 627–636.

Dhas, T.S., Kumar, V.G., Karthick, V., Govindaraju, K., Narayana, T.S., 2014. Biosynthesis of gold nanoparticles using Sargassumswartzii and its cytotoxicity effect on HeLa cells. Spectrochim. Acta A Mol. Biomol. Spectrosc. 133, 102–106.
Dodi, G., Pala, A., Barbu, E., Peptanariu, D., Hritcu, D., Popa, M.I., Tamba, B.I., 2016. Carboxymethyl guar gum nanoparticles for drug delivery applications: preparation and preliminary in-vitro investigations. Mater. Sci. Eng. C 63, 628–636.
Echeverria, C., Fernandes, S.N., Godinho, M.H., Borges, J.P., Soares, P.I.P., 2015. Stimulus-responsive hydrogels: theory, modern advances, and applications. Gels 93, 1–49.
Fantozzi, F., Arturoni, E., Barbucci, R., 2010. The effects of the electric fields on hydrogels to achieve antitumoral drug release. Bioelectrochemistry 78, 191–195.
Gamal-Eldeen, A.M., Amer, H., Helmy, W.A., 2006. Cancer chemopreventive and anti-inflammatory activities of chemically modified guar gum. Chem. Biol. Interact. 161 (3), 229–240.
George, A., Shah, P.A., Shrivastav, P.S., 2019. Guar gum: versatile natural polymer for drug delivery applications. Eur. Polym. J. 112, 722–735.
Gerber, A., Bundschuh, M., Klingelhofer, D., Groneberg, D.A., 2013. Gold nanoparticles: recent aspects for human toxicology. J. Occup. Med. Toxicol. 8 (1), 32.
Giri, A., Bhunia, T., Goswami, L., Panda, A.B., Bandyopadhyay, A., 2015. Fabrication of acrylic acid grafted guar gum multiwalled carbon nanotube hydrophobic membranes for transdermal drug delivery. RSC Adv. 5 (41), 736–41744.
Gurunathan, T., Mohanty, S., Nayak, S.K., 2015. A review of the recent developments in biocomposites based on natural fibres and their application perspectives. Compos. A Appl. Sci. Manuf. 77, 1–25.
Hacker, M.C., Mikos, A.G., 2011. Synthetic Polymers, Principles of Regenerative Medicine, second ed., pp. 587–622.
Huang, Y., Lu, J., Xiao, C., 2007a. Thermal and mechanical properties of cationic guar gum/poly(acrylic acid) hydrogel membranes. Polym. Degrad. Stab. 92, 1072–1081.
Huang, X., Wang, X., Shi, C., Liu, Y., Wei, Y., 2021. Research on synthesis and self-healing properties of interpenetrating network hydrogels based on reversible covalent and reversible non-covalent bonds. J. Polym. Res. 28 (1).
Huang, Y., Yu, H., Xiao, C., 2007b. pH-sensitive cationic guar gum/poly (acrylic acid) polyelectrolyte hydrogels: swelling and in vitro drug release. Carbohydr. Polym. 69, 774–783.
Jacob, R.S., Ghosh, D., Singh, P.K., Basu, S.K., Jha, N.N., Das, S., Sukul, P.K., Patil, S., Sathaye, S., Kumar, A., Chowdhury, A., Malik, S., Sen, S., Maji, S.K., 2015. Self healing hydrogels composed of amyloid nano fibrils for cell culture and stem cell differentiation. Biomaterials 54, 97–105.
Kaith, B.S., Sharma, R., Kalia, S., Bhattid, M.S., 2015. Response surface methodology and optimized synthesis of guar gum-based hydrogels with enhanced swelling capacity. RSC Adv. 4, 40339–40344.
Kaur, M., Malik, B., Garg, T., Rath, G., Goyal, A., 2015. Development and characterization of guar gum nanoparticles for oral immunization against tuberculosis. Drug Deliv. 22 (3), 328–334.
Klouda, L., 2015. Thermoresponsive hydrogels in biomedical applications: a seven-year update. Eur. J. Pharm. Biopharm. 97, 338–349.
Koetting, M.C., Peters, J.T., Steichen, S.D., Peppas, N.A., 2015. Stimulus-responsive hydrogels: theory, modern advances, and applications. Mater. Sci. Eng. R. Rep. 93, 1–49.
Kono, H., Otaka, F., Ozaki, M., 2014. Preparation and characterization of guar gum hydrogels as carrier materials for controlled protein drug delivery. Carbohydr. Polym. 111, 830–840.
Krotkiewski, M., 1984. Effect of guar gum on body-weight, hunger ratings and metabolism in obese subjects. Br. J. Nutr. 52 (1), 97–105.
Li, X., Wu, W., Liu, W., 2008. Synthesis and properties of thermo-responsive guar gum/poly(N-isopropylacrylamide) interpenetrating polymer network hydrogels. Carbohydr. Polym. 71, 394–402.
Li, B., Wang, J., Gui, Q., Yang, H., 2020. Drug-loaded chitosan film prepared via facile solution casting and air-drying of plain water-based chitosan solution for ocular drug delivery. Bioact. Mater. 5 (3), 577–583.
Liechty, W.B., Kryscio, D.R., Slaughter, B.V., Peppas, N.A., 2010. Polymers for drug delivery systems. Annu Rev Chem Biomol Eng. 1, 149–173.
Llanos, G.R., Sefton, M.V., 1991. Immobilization of poly(ethylene glycol) onto a poly(vinyl alcohol) hydrogel, synthesis and characterization. Macromolecules 24, 6065–6072

Lopes, J., Fonseca, R., Viana, T., Fernandes, C., Morouco, P., Moura, C., Biscaia, S., 2019. Characterization of biocompatible poly(ethylene glycol)-dimethaacrylate hydogels for tissue engineering. Appl. Mech. Mater. 890, 290–300.

Maiti, S., Khillar, P.S., Mishra, D., Nambiraj, N.A., Jaiswal, A.K., 2021. Physical and self-crosslinking mechanism and characterization of chitosan-gelatin-oxidized guar gum hydrogel. Polym. Test.

Maolin, Z., Jun, L., Min, Y., Hongfei, H., 2000. The swelling behavior of radiation prepared semi-interpenetrating polymer networks composed of polyNIPAAm and hydrophilic polymers. Radiat. Phys. Chem. 58, 397–400.

Mohsin, M.A., Attia, N.F., 2015. Inverse emulsion polymerization for the synthesis of high molecular weight polyacrylamide and its application as sand stabilizer. Int. J. Polym. Sci.

Palem, R.R., Kummarab, M.R., Kanga, T., 2019. Self-healable and dual-functional guar gum-grafted-polyacrylamidoglycolic acid-based hydrogels with nano-silver for wound dressings. Carbohydr. Polym. 223, 115074.

Parhi, R., 2017. Cross-linked hydrogel for pharmaceutical applications: a review. Adv. Pharm. Bull. 7 (4), 515–530.

Patel, V.R., Amiji, M.M., 1996. Preparation and characterization of freeze-dried chitosan-poly(ethylene oxide) hydrogels for site-specific antibiotic delivery in the stomach. Pharm. Res. 13 (4), 588–593.

Peppas, N.A., Bures, P., Leobandung, W., Ichikawa, H., 2000. Hydrogels in pharmaceutical formulations. Eur. J. Pharm. Biopharm. 50, 27.

Rao, K.S.V.K., Chung, I., Ha, C.S., 2008. Synthesis and characterization of poly(acrylamidoglycolic acid) grafted onto chitosan and its polyelectrolyte complexes with hydroxyapatite. React. Funct. Polym. 68, 943–953.

Rao, Z., Liu, S., Wu, R., Wang, G., Sun, Z., Bai, L., Wang, W., Chen, H., Yang, H., Wei, D., Niu, Y., 2019. Fabrication of dual network self-healing alginate/guar gum hydrogels based on polydopamine-type microcapsules from mesoporous silica nanoparticles. Int. J. Biol. Macromol. 129, 916–926.

Ren, X., Yang, Q., Yang, D., Liang, Y., Dong, J., Ren, Y., Lu, X., Xue, L., Li, L., Xu, L., 2018. High-strength double network hydrogels as potential materials for artificial 3D scaffold of cell migration in vitro. Colloid. Surface A 549, 50–57.

Sarmah, J.K., Mahanta, R., Bhattacharjee, S.K., Mahanta, R., Biswas, A., 2011. Controlled release of tamoxifen citrate encapsulated in cross-linked guar gum nanoparticles. Int. J. Biol. Macromol. 49 (3), 390–396.

Shafiq, S., Khan, S.M., Ibrahim, S.M., Abubshait, S.A., Nazir, A., Abbas, M., Iqbal, M., 2020. Novel chitosan/guar gum/PVA hydrogel: preparation, characterization and antimicrobial activity evaluation. Int. J. Biol. Macromol. 164, 499–509.

Shantha, K.L., Harding, D.R.K., 2002. Synthesis and evaluation of sucrose containing polymeric hydrogels for oral drug delivery. J. Appl. Polym. Sci. 84, 2597–2604.

Sharma, S., Tiwari, S., 2020. A review on biomacromolecular hydrogel classification and its applications. Int. J. Biol. Macromol. 162, 737–747.

Sharma, G., Sharma, S., Kumar, A., Al-Muhtaseb, A.H., Ayman, M.N., Ghfar, A., Tessema, G., Florian, M., Stadler, J., 2018. Guar gum and its composites as potential materials for diverse applications. Carbohydr. Polym. 199, 534–545.

Singha, N.R., Mahapatra, M., Karmakar, M., Dutta, A., Mondal, H., Chattopadhyay, P.K., 2017. Synthesis of guar gum-g-(acrylic acidcoacrylamide-co-3-acrylamido propanoic acid) IPN via in situ attachment of acrylamidopropanoic acid for analyzing super adsorption mechanism of pb(II)/cd(II)/cu(II)/MB/MV. Polym. Chem. 8, 6750–6777.

Su, W.F., 2013. Principles of Polymer Design and Synthesis. vol. 82 Springer-Verlag Berlin Heidelberg, pp. 137–183.

Sung, Y.K., Kim, S.W., 2020. Recent advances in polymeric drug delivery systems. Biomater. Res. 24 (12).

Takashi, L., Hatsumi, T., Makoto, M., Takashi, I., Takehiko, G., Shuji, S., 2007. Synthesis of porous poly(N-isopropylacrylamide) gel beads by sedimentation polymerization and their morphology. J. Appl. Polym. Sci. 104 (2), 842.

Talodthaisong, C., Boonta, W., Thammawithan, S., Patramanon, R., Kamonsutthipaijit, N., Hutchison, J.A., Kulchat, S., 2020. Composite guar gum-silver nanoparticle hydrogels as self-healing, injectable and antibacterial biomaterials. Mater. Today Commun.

Taylor, D.L., In Het Panhuis, M., 2016. Self-healing hydrogels. Adv. Mater. 28 (41), 9060–9093.

Vasile, C., Pamfil, D., Stoleru, E., Baican, M., 2020. New developments in medical applications of hybrid hydrogels containing natural polymers. Molecules 25 (7), 1539.

Warren, D.S., Sam Sutherland, P.H., Jacqueline, Y.K., Geoffrey, R.W., Sean, M.M., 2017. The preparation and simple analysis of a clay nanoparticle composite hydrogel. J. Chem. Educ., 04–20.

Xu, D., Huang, J., Zhao, D., Ding, B., Zhang, L., Cai, J., 2016. High-flexibility, high-toughness double-cross-linked chitin hydrogels by sequential chemical and physical cross-linkings. Adv. Mater. 28 (28), 5844–5849.
Yang, L., Chu, J.S., Fix, J.A., 2002. Colon-specific drug delivery: new approaches and in vitro/in vivo evaluation. Int. J. Pharm. 235, 1–15.
Ye, J., Fu, S., Zhou, S., Li, M., Li, K., Sun, W., Zhai, Y., 2020. Advances in hydrogels based on dynamic covalent bonding and prospects for its biomedical application. Eur. Polym. J. 139, 110024.
Zhang, M., Song, L., Jiang, H., Li, S., Shao, Y., Yang, J., Li, J., 2017. Biomass based hydrogel as an adsorbent for the fast removal of heavy metal ions from aqueous solutions. J. Mater. Chem. A 5, 3434–3446.
Zheng, Y., Liu, Y., Wang, A., 2011. Fast removal of ammonium ion using a hydrogel optimized with response surface methodology. Chem. Eng. J. 171, 1201–1208.

Index

Note: Page numbers followed by *f* indicate figures and *t* indicate tables.